The Credit Scoring Toolkit

The Credit Scoring Toolkit

Theory and Practice for Retail Credit Risk Management and Decision Automation

Raymond Anderson

OXFORD
UNIVERSITY PRESS

OXFORD
UNIVERSITY PRESS

Great Clarendon Street, Oxford OX2 6DP

Oxford University Press is a department of the University of Oxford.
It furthers the University's objective of excellence in research, scholarship, and education by publishing worldwide in

Oxford New York

Auckland Cape Town Dar es Salaam Hong Kong Karachi
Kuala Lumpur Madrid Melbourne Mexico City Nairobi
New Delhi Shanghai Taipei Toronto

With offices in

Argentina Austria Brazil Chile Czech Republic France Greece
Guatemala Hungary Italy Japan Poland Portugal Singapore
South Korea Switzerland Thailand Turkey Ukraine Vietnam

Oxford is a registered trade mark of Oxford University Press
in the UK and in certain other countries

Published in the United States
by Oxford University Press Inc., New York

First published 2007

British Library Cataloguing in Publication Data
Data available

Library of Congress Cataloging in Publication Data
Data available

Typeset by Newgen Imaging Systems (P) Ltd, Chennai, India
Printed in Great Britain
on acid-free paper by
CPI Group (UK) Ltd., Croydon, CR0 4YY

ISBN 978–0–19–922640–5

10 9 8

To
Nofunisela Mirriam Nyikiza

Housekeeper and Friend,
for your many years of service

Contents

List of figures

List of tables

List of equations

Preface

When people ask me, 'What do you do for a living?' they often note my hesitation, and add, 'It is not a trick question!', as I pause to phrase my answer. Most people are in the enviable position of being able to say doctor, plumber, secretary, bus driver, or rat catcher, and it is usually fairly clear what they do.

My first response is usually 'banker' even though I have not been through the normal banker's training routine, but I think it is a better description than statistician, computer programmer, business analyst, or any other term that comes to mind. When children ask me, they may well stop after that simple answer; but with adults, I often see a smile crossing their faces, as the oft-repeated collective-noun joke crosses their minds—assuming they do not automatically start lambasting me about their latest problems with their robber band of bankers.

Thereafter, assuming they are not already bored enough to change the conversation to the stock market (during a bull run), global warming, local politics, or the latest news on skirmishes in the East, they might actually ask, 'What is it that you do . . . exactly?' It took me a long time to come up with an adequate response, 'When the bank turns down your loan request, and you blame the computer, blame me! I am the guy telling the computer what to do.' Well, I cannot really take full credit for that. The task is a little bigger, with a lot more people involved.

The Scorecard Builder's Prayer (ver. 3.01)

O scoring, who art in regression,
Guessing be thy name.
Thy assumptions come,
Thy will be done in future as it was in the past,
Give us this day our expected bad rates,
And forgive us our lousy model weights,
As we forgive those who supply us with poor data.
Lead us not into write-offs,
And deliver us from the auditors.
For thine is the #NAME, the #DIV/0,
and the #VALUE!
Forever and ever, Amen.

Background and literature

Credit scoring is a discipline that has developed, and been widely adopted, since the early 1960s. Today, these models are the grease that supports decision-making in countless businesses around the world, yet the amount of literature available about the field is limited. As can be seen from the list below, prior to 2000, there were very few books on the topic; but since then, the list has been growing at the rate of about one per year. Even so, as at 2005, there were still less than 15 (some are not mentioned), and each varied in terms of the focus area, how up-to-date it was, the background of the authors, target audience, and whether the book was still in print. Of the available books, the following comments can be made:

1992—Lewis, E. 'An Introduction to Credit Scoring'. Thirty years of experience was summarised into the first, and one of the more readable, texts dedicated to credit scoring, which is still widely used as a reference work.

1995—Hoyland, C. 'Data-Driven Decisions for Consumer Lending'. Much like the above, except it focuses more on the practical application of the scores. It is also useful, in that it is well illustrated, with examples and tables.

1998—Mays, E., ed. 'Credit Risk Modelling: Design and Application'. Collection of articles by various well-respected authors, on different aspects of the topic.

2000 2nd edn 2003—McNab, H., and Wynn, A. 'Principles and Practices of Consumer Credit Risk Management'. A summary of practices within the consumer credit industry, largely credit cards, that was originally developed as a set of course notes. The primary focus is business practices, with credit scoring secondary.

2001—Mays, E., ed. 'Handbook of Credit Scoring'. Similar to the 1998 book, with repeats of several of the articles.

2002—Thomas, L., Edelman, D., and Crook, J. 'Credit Scoring and its Applications'. Very comprehensive, but highly academic and inaccessible to the layman. Prior knowledge of statistical and mathematical notation is assumed.

2003—Thomas, L., Edelman, D., and Crook, J., eds. 'Readings in Credit Scoring'. A collection of topical papers from various sources, especially the Credit Scoring and Control conferences in Edinburgh.

2004—Mays, E., ed. 'Credit Scoring for Risk Managers: The Handbook for Lenders'. A collection of articles, mostly by Ms Mays, covering various aspects of credit scoring. While accessible, it does not hang together as a coherent whole.

2005—Siddiqi, N. 'Credit Risk Scorecards: Developing and Implementing Intelligent Credit Scoring'. Focuses on the development of in-house scoring capabilities, covering a broad range of development and implementation issues. Very few references.

There is also information available on the Internet, but it unfortunately resides in countless scattered articles, and it is very difficult to get a full picture. Much of it is also couched in specialist jargon that is difficult to interpret. One also has to endure umbongi fatigue[1] from reading countless marketing blurbs, while trying to find something meaningful.

Purpose of this book

This book's purpose is to provide an overview of credit scoring and automation of credit decision processes. When I first started writing it in 2003, only four of the above books were available to me, and there seemed to be a distinct gap in the market, for some not-quite-so-light bedside reading, for people outside the field who needed some knowledge of the topic. I had taken up writing as a hobby four years earlier, including travel and history articles, personal anecdotes, and open-mike poetry, and when someone suggested the possibility of writing a textbook, I was curious about whether my skills could be used to this end. These were enhanced by an underlying desire simultaneously to inform, entertain, influence, and somehow capitalise on my own anal retentiveness, which has consistently fostered attention to detail in which nobody else is interested (that's supposed to be a joke!).

The intended audience for the book was initially second or third year university students, studying towards business degrees, who needed an overview of predictive statistics and their application in the consumer-credit industry. At the same time, it was hoped that propeller-heads . . . uhhh, I mean the statisticians and mathematicians that develop the scorecards . . . would get some value from learning about how their scoring models fit within the business. By the way, no offence is meant here, as the author falls within this category. Propeller-heads rule!

One of the titles considered for an early version was 'The Working Man's Guide to Credit Scoring', but this was trashed because of potential political incorrectness, and because, over time, the content became more sophisticated. The audience broadened to include academics, managers, directors, and even regulators and law-makers, who are increasingly required to understand credit processes, and the statistical models used to support them. From an initial focus on consumer credit, it also grew to cover aspects of micro-finance, and lending to businesses ranging from small and medium enterprises, to middle-market companies. From an initial focus on first-world English-speaking countries, it grew to include non-English-speaking areas and developing countries, albeit most of the focus is still upon the United States and the United Kingdom, because of the large amount of available information. This non-geocentric approach had the advantage of allowing broad principles to be derived, instead of focusing upon country-specific circumstances. For a time, 'The Credit Scoring MBA' was considered as a title, to emphasise the breadth of topics covered; but unfortunately, this was discarded due to potential confusion with a recognised Masters of Business Administration degree.

Finally, the title morphed to 'The Credit Scoring Toolkit: Theory and Practice for Retail Credit Risk Management and Decision Automation'. The book's primary goal is to inform readers regarding the concepts and language used in credit scoring and associated disciplines, so that they can both understand the concepts, and communicate with people with many years of experience in their subject areas.

[1] 'Umbongi' is the Zulu word for praise singer.

Writing this book was a learning experience for me not only with regards to the subject matter, but also in terms of writing an academic textbook—especially given that its original purpose was to act as course notes. Much time and effort has gone into both finding relevant information, and quoting the sources. If the same information was found in three or more places, no specific reference is made, but any books and web-based articles of an academic nature are still cited in the bibliography. In spite of the academic bent, wherever possible, attempts were made to use a conversational writing style, explain specialist jargon, and explain the heavy maths and stats to the best of my abilities. This was not always feasible though. In my defence, a comprehensive glossary cum dictionary has been provided, which should assist both English and non-English speakers.

Acknowledgments

There are times when you might think you have done it all yourself, but if you think about it, you probably have countless people to thank, many of whom have since moved on to new positions. The first person to thank is *Harry Greene*, who suggested that the book be written, but may have regretted it, after seeing how much of my life it consumed. I would also like to thank *Neville Robertson*, who invited me into the credit field in 1996, which totally expanded my horizons. Thanks must also go to *Suzanne Parry*, whose challenge provided the impetus for me to take up writing as a hobby, and to *Etienne Cilliers*, *Paul Middleton*, and *David Hodnett*, who provided me with the latitude to carry on with the book as it grew. People working within the credit scoring and analytical areas contributed directly to the book, and put up with me during the time it was being written, by indulging discussions on trivial topics. These were *Denis Dell*, *Lizelle Bezuidenhout*, *Suben Moodley*, *Hanlie Roux*, *Richard Crawley-Boevey*, *Derek Doody*, *Charlotte Crowther*, *Garth Zietsman*, and *Derrick Nolan*. Others that provided input were *Steve Barker*, *Mike Waiting*, *Ninian Gordon*, *Andre Tredoux*, *Pierre Kloppers*, *Dave Brimblecombe*, *Henrietta van Greuning*, and *Brian Hutchinson*.

A person who must also be acknowledged is *Jes Freemantle*, a scorecard development consultant from whom much was learnt about the process. Without interaction with him, many of the more technical concepts encountered while researching this book would have been much more difficult to decipher. Likewise, thanks must also go to *Helen McNab*, who consulted on the setting of strategies and use of scoring within the business, and whose book was the bible while writing parts of this work. I would also like to thank *Mark Schreiner*, who is known as the scoring guru of micro-finance for the third world, which is a totally different world to mainstream consumer credit. Conversations with him challenged many of the assumptions I had made in the past, and added new dimensions.

Finally, I would like to thank all of the individuals that I came to know through the *Edinburgh Credit Scoring and Control* conferences, who have provided support and/or input over the period. These include, amongst others, *Lyn Thomas*, *David Edelman*, *Jonathan Crook*, *Ross Gayler*, and *Graham Platts*.

Several people have taken the time to proofread and provide input on various aspects of this book. I apologise if any names are omitted, but would particularly like to acknowledge: *Garth Zietsman*, *Denis Dell*, *Henrietta van Greuning*, and *Daniel Morais* (Standard Bank, RSA), *Hanlie Roux* (Nedcor, RSA), *Riaan de Jong* and *Tania de la Rey* (North West University, RSA), *Mark Schreiner* (micro-finance scoring consultant, USA), *David Edelman* (Caledonia Consulting, UK), *Jes Freemantle* (Stratus Credit Consultancy, UK), *Lyn Thomas* (University of Southampton, UK), *Ross Gayler* (Baycorp Advantage, Australia), *Graham Platts* (Experian-Scorex, Monaco), *Anthony Sumner* (Experian, UK), *Corne Strydom* (SAS Monocle, RSA), and *Barbara Hulme* (practising psychologist, significant other, and a huge distraction in my life).

Acknowledgements

Outline

Many textbooks start off with an outline that tells the reader what is going to be covered, but in many cases, the reader will only appreciate it after the whole textbook is read. That does not mean it is not useful—the reader can always refer to it, and/or the table of contents, when trying to track down a topic. It also acts as a useful tool for the author to check whether the sections have been put in a logical order, which ultimately also helps the reader. I am guilty as charged; these summaries have helped me to organise the sections, and hopefully they will also provide you with a quick overview of the book, and a quick reference if you lose your bookmark.

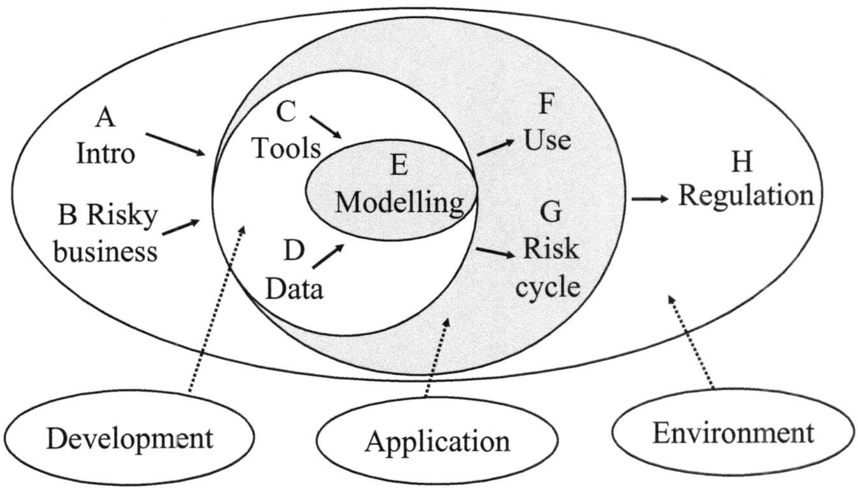

Figure 1. Module flow.

Module A Setting the scene

> [The use of credit scoring technologies] has expanded well beyond their original purpose of assessing credit risk. Today they are used for assessing the risk-adjusted profitability of account relationships, for establishing the initial and ongoing credit limits available to borrowers, and for assisting in a range of activities in loan servicing, including fraud detection, delinquency intervention, and loss mitigation. These diverse applications have played a major role in promoting the efficiency and expanding the scope of our credit delivery systems and allowing lenders to broaden the populations they are willing and able to serve profitably.
>
> *Alan Greenspan*, U.S. Federal Reserve Chairman, in an October 2002 speech to the American Bankers Association.[1]

Our modern world depends upon credit. Entire economies are driven by people's ability to 'buy-now, pay-later'. Indeed, two hundred years ago it was a privilege to borrow money, but in today's industrialised societies it is considered a right. Providing credit is a risky business though, as borrowers differ in their ability and willingness to pay. At the extreme, lenders may lose the full amount, and perhaps even get sucked in for more. In other instances, they may lose only a part, or just incur extra costs to get the money back. It is a gamble, and lenders are always looking for means of improving their odds.

Over the last fifty years, automation has extended beyond the back-office functions of accounting and billing, and moved into the domain of decision-making. Its influence has been greatest on credit provision, where the much improved risk assessments have empowered lenders to lend where they once feared to tread, and improved processes have aided accessibility for the general public. For people applying for credit, it is often a black box though—they know what goes in and comes out, but not what happens inside. If you apply for a loan, you are told either 'yes' or 'no'. If 'yes', you are told the amount you can borrow and the repayment terms. If 'no', you either slink away with your tail between your legs, or go to the lender next door and try again. And if the latter, or if you do not like the repayment terms, it is often difficult to get an adequate explanation of 'Why?' Much of the problem arises because of a poor understanding of what goes on behind the scenes, even by lenders' own employees. This textbook covers the topic, and the first module has three chapters:

(1) **Credit scoring and the business**—Covers what credit scoring is, where it fits within the business and the economy, and how it has affected us.

---

[1] http://www.federalreserve.gov, quoted in Mays (2004:4).

(2) **History of credit**—A micro-history of the provision of credit, credit scoring, credit bureaux, and credit rating agencies.
(3) **Mechanics of credit scoring**—An overview of how credit scoring works, especially as regards scorecard development.

Chapter 1 starts the module, using a *FAQs framework* to address several questions: (i) What is credit scoring?—which treats the parts ('credit' and 'scoring') before considering the whole ('credit scoring'), and delves into the economic rationale, including concepts such as asymmetric information, adverse selection, moral hazard, and information rents; (ii) Where is it used?—a brief look at the processes, data sources (customer, internal, and external), credit risk management cycle (CRMC), and behavioural propensities (risk, response, revenue, and retention); (iii) Why is it used?—especially the quality, speed, and consistency of decisions, and how these have affected lenders and consumers; and (iv) How has it influenced the credit industry?—its broader impact, with particular attention paid to data, risk assessment, decision rules, process automation, and regulation.

Wherever possible in this textbook, historical background has been provided to put concepts in context. *Chapter 2* is dedicated to *history*, including: (i) credit provision—from the first documented use of credit in ancient Babylon, through to the evolution of credit cards and risk-based pricing; (ii) credit scoring—from the time it was first proposed in 1941, through the establishment of Fair Isaac in 1958, to the evolution of bureau scores, and scoring's use in securitisation; (iii) credit bureaux—from the origins of Dun and Bradstreet in the 1840s, through to the more recent evolution of Experian, Equifax, and Transunion; and (iv) credit rating agencies—including Moody's Investor Services from 1909, as well as Standard and Poor's and Fitch IBCA.

Credit scoring is a technical area, and *Chapter 3* touches on the *mechanics*: (i) scorecards—form and presentation, development, how good are the predictions?, how does scorecard bias arise?, and what can be done about it?; (ii) measures used—whether as part of the business process, assessment of scorecard performance, or the default probability and loss severity measures used in finance functions; (iii) development process—covering project preparation, data preparation, modelling, finalisation, decision-making and strategy, and security; and (iv) changes that can affect the scorecards—including the economy, marketing, operations, and societal attitudes towards debt.

These provide the 25¢ tour of credit scoring, after which the reader should have a broad overview of the topic. It may be enough by itself, or just set the scene for the rest of the book.

Module B Risky business

> While microprocessors used in workstations are doubling their capacity practically every year, demands posed by the user population grow much faster.
>
> *Dimitris Chorafas* (1990)

When the term 'credit scoring' is uttered, different people think of different things: *customers*, the credit application form and the ensuing call to the credit bureau, and possibly the last time they were refused credit; *statisticians*, the predictive-modelling tools used to derive the risk rankings; *lenders*, the cut-off and limit strategies used to improve their bottom line; and for *IT staff*, the systems required to calculate the scores, apply the strategies, and deliver the final decisions.

This section focuses on the business aspects—the strategic justifications for why, when, where, and how it should be used. In some cases, the topics are grouped together just because they seem to fit together, yet they are quite distinct:

(4) **Theory of risk**—Frameworks for considering risks to the broader organisation, where credit risk is only one of them.
(5) **Decision science**—Credit scoring allows case-by-case risk management, but use of scientific methodologies allows even greater value to be extracted.
(6) **Assessing enterprise risk**—A look at lending to businesses of any size, including traditional frameworks, and recent developments.

Risk is a part of any endeavour, but over the past few decades, it has become a specialist function within organisations. *Chapter 4* looks at broader *risk frameworks*: (i) the risk lexicon—highlighting risk linkages, the playing field (company proposition, physical resources, and market, economic, social, and political factors), and risk types (primarily business, credit, market, and operational, but also others falling under business environment, business dealings, extraterritorial, personal, and intelligence) and (ii) data and models—looks at data types (which can vary by source, time, inputs, indicators, and view), and model types (statistical, expert, hybrid, and pure judgement). Some risk types are easier to model than others, and frameworks are presented showing how the type of credit risk model used is typically a function of structure and technology, and the volume of deals and profit per deal.

In order to reduce risk, businesses strive for greater control, which can be aided by having proper policies, procedures, structure, and infrastructure. Businesses have made increasing use of scientific methods to provide greater structure. *Chapter 5* looks at *decision science*, including: (i) adaptive control—where processes are adjusted to maintain consistent output and (ii) experimentation and analysis—including champion/challenger, optimisation, simulation, and strategy inference. A framework is presented, illustrating that the strategy chosen should be determined by an event's probability and potential impact.

Credit scoring originated in the consumer credit arena, but is increasingly replacing (or supporting) traditional enterprise risk assessments. *Chapter 6* covers lending to *business enterprises*: (i) basic credit risk assessment—covering the traditional 5 Cs, data sources (securities prices, financial statements, payment history, environmental assessment, and human input), and risk assessment tools (agency grades, business report scores, and public/private firm, hazard, and exposure models); (ii) SME lending—and forces driving lenders from relationship to transactional lending; (iii) financial ratio scoring—covering pioneers, predictive ratios, rating agencies, and internal grades; (iv) credit rating agencies—their letter grades, derivation, and issues (small numbers, population drift, downward rating drift, business cycle sensitivity, and risk heterogeneity within the grades); (v) modelling with forward looking data—covering straightforward historical analysis, structural approaches (Wilcox's gambler's ruin, Black and Schole's options-theoretic), and the reduced-form approach (proposed by Jarrow and Turnbull, which is based primarily on the credit spreads of bonds' market prices).

Module C Stats and maths

> To chop a tree quickly, spend twice the time sharpening the axe.
>
> *Chinese proverb*

The concept of data mining evolved during the 1990s, as classical statistics, artificial intelligence (AI), and machine learning techniques were harnessed to search data for non-obvious patterns, and knowledge. It is similar to conventional mining in that: (i) vast volumes have to be processed just to yield a few gems and (ii) it requires its own picks and shovels, assayers' scales, and people who know how to use them. Credit scoring might have started thirty years earlier, but is nonetheless considered part of the same arsenal (under 'classical statistics'). Computing power was limited in the early days though, and the use of predictive statistics to drive production processes—especially selection processes—brought with it new challenges. As a result, some practices are specific to this environment, and may provide a competitive advantage. Even so, businesses' interest today lies less in statistical tricks, and more in making better use of data, and getting maximum value out of the resulting scores.

Nonetheless, credit scoring cannot be discussed without covering the statistical techniques used. Such concepts are normally covered when discussing the Scorecard Development Process (Module E), but here they are instead treated as basic building blocks, primarily because many of them are used at different stages in the process, and thereafter. These include

(7) **Predictive statistics**—Methods for providing estimates of unknown values, whether future events or outcomes, that are difficult to determine (high cost or destructive).
(8) **Measures of separation and accuracy**—Calculations used to provide indications of the power and stability of both predictors and predictions, and the accuracy of predictions.
(9) **Odds and ends**—A collection of topics, including descriptive modelling techniques, forecasting tools, some statistical concepts, and basic scorecard development reports.
(10) **Minds and machines**—A look at the required people (scorecard developers, project team, steering committee) and software (scorecard development, decision engines).

As indicated, credit scoring has been built upon *predictive statistics*. *Chapter 7* starts by describing some of the statistical notation, and moves on to: (i) on overview of the techniques—including modelling and data considerations when using them; (ii) parametric techniques—linear regression, linear probability modelling (LPM), discriminant analysis (DA), and logistic regression; and (iii) non-parametric techniques—recursive partitioning algorithms (RPAs, used to derive decision trees), neural networks (NNs), genetic algorithms, K-nearest neighbours, and linear programming; (iv) critical assumptions—covering treatment of missing data, statistical assumptions for parametric techniques (relating to variables and model residuals) and how violations can be addressed; and (v) a comparison of results—which provides no clear winners, although logistic regression leads the fray based purely on popularity.

Besides just developing the models, the results have to be measured. *Chapter 8* looks at *measures of separation/divergence* used to assess both power and drift, including: (i) the misclassification matrix and a graphical representation; (ii) the Kullback divergence measure, including the weight of evidence upon which it is built, information value, and stability index; (iii) the Kolmogorov–Smirnov statistic and associated graph; (iv) correlation coefficients and equivalents—covering Pearson's product-moment, Spearman's rank-order, the Lorenz curve, Gini coefficient, and receiver operating characteristic; and (v) Pearson's chi-square—which measures the difference between frequency distributions. Further, section (vi) deals with measures of accuracy—starting with probability theory (and Bernoulli trials), before covering the binomial test (and its normal approximation), Hosmer–Lemeshow statistic, and log-likelihood measure.

Chapter 9 covers *odds and ends* that do not fit neatly elsewhere, including: (i) descriptive modelling techniques used for variable reduction—cluster analysis (for records) and factor analysis (for variables); (ii) forecasting tools—including transition matrices/Markov chains and survival analysis; (iii) an explanation of some statistical concepts—such as correlations, interactions, monotonicity, and normalisation; and (iv) basic scorecard development reports—characteristic analysis, score distribution, and the new business strategy table.

Finally, there are issues relating to the *minds and machines* used to develop credit-scoring models. *Chapter 10* covers: (i) people and projects—scorecard developers, external vendors/consultancies, internal resources, project team, and steering committee and (ii) software—for scorecard development (which may be user-friendly, but have limited transparency and flexibility), and applying the models and making decisions within the business (decision engines).

A polysyllabic overview

It might also help to briefly describe some of the high-level terms used in this domain. As can be seen from the above, it is impossible to keep the discussion monosyllabic, but most of the words only just rival 'television' in terms of the number of syllables.

Predictive/descriptive/forecasting—Defines the model's purpose. *Predictive*—develop models that provide an estimate of a target variable (regression techniques, RPAs, NNs). *Descriptive*—find patterns that describe the data, whether the records (cluster analysis) or the variables (factor analysis). *Forecasting*—tools used for prediction at an aggregated level, including movements between states (Markov chains/transition matrices) and mortality rates (survival analysis).

Parametric/non-parametric—Defines whether the modelling technique or test makes assumptions about the data. *Parametric*—makes assumptions, such as a normal distribution, linearity, homoscedasticity, and independence (linear regression, logistic regression,

DA). *Non-parametric*—makes no assumptions, and it is used where the parametric equivalent cannot be used (RPAs, AI).

Statistical, operations research, AI—Defines the discipline where the technique originated. *Statistical*—linear regression, logistic regression, and RPAs. *Operations research*—linear programming, and other methods used for resource allocation and logistics. *AI*—newer approaches, such as NNs, genetic algorithms, K-nearest neighbour, and machine learning.

Algorithmic/heuristic—Defines the development procedure. *Algorithmic*—defined by a formula, or set of steps (regression techniques, RPAs). It also applies to the use of strict policy rules in any part of the business process. *Heuristic*—based upon empirical data analysis, but uses trial and error, to come up with a result that has no explicit rationalisation (NNs, genetic algorithms). The term also applies where expert judgement is used to set rules of thumb or flexible guidelines.

Deterministic/probabilistic—Defines the level of certainty in the relationship. *Deterministic*—outcomes can be exactly determined using a formula/algorithm, which is more often the case in hard sciences such as physics. *Probabilistic*—definite outcomes cannot be determined, but probabilities can be derived (associated with stochastic processes and 'fuzzy' logic).

Labels such as these are used in different environments—finance, engineering, science, psychology, and so on, and the techniques that are appropriate in each will vary according to the problem. Credit scores are developed using predictive models, which are usually parametric, statistical, algorithmic, probabiListic, regression models, used to represent a stochastic process with a binary good/bad outcome. Non-parametric and heuristic AI techniques may also be used, but are not as widely accepted.

An apology must be made here! It is one thing to digest a single multi-syllable word, but quite something else to handle so many in quick succession. Hopefully though, these explanations will allow the reader a better understanding of the following chapters, and other literature on the topic.

Module D Data!

> Without data the modern commercial opportunities would be very limited. Data and information (and they are different from each other) are fundamental to the success of any business today and they increasingly provide a commercial competitive edge.
>
> *McNab and Wynn* (2003:17)

Decisions can only be as good as the information upon which they are based, which is why spies, industrial espionage, and private investigators exist—along with other under-handed ways of trying to get the upper hand. Unfortunately, there is often a laxity about the data that is gathered, which may be insufficient, of poor quality, or difficult to interpret. Poor intelligence has been the undoing of countries and companies, generals, CEOs, and others in highly competitive situations. Information is crucial!

Most literature on credit scoring focuses on the statistical methods used, and pays scant attention to data. Indeed, the starting point—and often more difficult task—is to ensure that relevant and reliable information is available for both scorecard development and business processes where the scorecards are applied. This does not mean that the role statistical methods play is less relevant; only that the cumulative organisational investment in obtaining and managing credit intelligence is greater. Indeed, data problems can result not only in financial losses, but also lost sleep and lost sanity.

Advances in technology since 1960 have significantly increased the quantity and quality of available credit intelligence, especially in terms of (i) the number of data sources; (ii) the amount of relevant information provided; and (iii) the ease with which it can be acquired, analysed, and summarised. There has also been a credit explosion in both developed and developing economies, especially for people who did not previously qualify. This chapter covers data in some detail, under the following headings:

(11) **Data considerations**—Factors that must be in place before a scorecard can be built, and issues relating to the characteristics used as predictors.
(12) **Data sources**—Discusses the types of information obtained from the customer, internal systems, and the credit bureaux.
(13) **Scoring structure**—Looks at scorecard customisation and hosting, data integration, and matching data from various sources.
(14) **Information sharing**—Describes the types of credit registries, the reason for their existence, how they operate, and what motivates or inhibits lender participation.
(15) **Data preparation**—The first stage of the scorecard development process, covering assembly, the good/bad definition, sample windows, and sample selection.

Chapter 11 looks at *data considerations*—starting with (i) transparency—a prerequisite for credit scoring; (ii) data quantity—depth and breadth (including minimum requirements), issues around homogeneity/heterogeneity, and accessibility; (iii) data quality—relevant, accurate, complete, current, and consistent; (iv) data design—data types (both statistical and practical classifications, as well as manipulation and special cases like missing data and division by zero), and form design issues for both categorical and numeric characteristics.

Nothing would be possible without *data sources*, which are the focus of *Chapter 12*: (i) customer supplied—including the application form and supporting documentation; (ii) internal systems—which provide both performance and predictors; (iii) credit bureau data—enquiries/searches, publicly available information, and shared performance; (iv) fraud warnings—known frauds, third-party data, and special information sharing arrangements; (v) bureau scores—which summarise available bureau data into propensity measures, especially risk; (vi) geographic indicators—including geographic aggregates and lifestyle codes; and (vii) other miscellaneous sources.

There are a lot of data issues that do not fit neatly elsewhere, which are covered in *Chapter 13* under *scoring structure*: (i) customisation—looks at generic and bespoke scorecards, and factors influencing which is most appropriate; (ii) hosting—whether to execute the scorecard on internal or external systems; (iii) data integration—which may be independent, discrete, or consolidated; (iv) credit risk scoring—looks at customisation and integration for each stage in the CRMC; and (v) matching—covers issues on how records from various sources are linked.

Perhaps the greatest advances in credit risk assessment have come from lenders' cooperation. *Chapter 14* provides a broad view of *information sharing*: (i) credit registries—public versus private registries (including which operate where, and why), and positive versus negative data; and (ii) do I or don't I?—principles of reciprocity governing such arrangements, and motivators/inhibitors to participation.

Finally, *Chapter 15* looks at *data preparation*, the first real stage of the scorecard development process: (i) data acquisition—for application data, bureau data, own current and past dealings, performance data, and initial data assembly; (ii) the good/bad definition—which is split into selection statuses and performance statuses, and also covers definition setting (consensus, prescribed, or empirical), and what a good/bad definition should be; (iii) observation and outcome windows—considerations when setting sampling windows, including maturity, censoring, and decay, especially for application and behavioural scoring; and (iv) sample design—covers sample types (training, holdout, recent, etc.), minimum and maximum sample sizes, and stratified random samples.

Module E Scorecard development

> If one has sufficient data and wishes to make a scoring model, the following objective helps: aim to make a model that has equal power but is simpler or more transparent than its alternatives. That is, instead of focussing on increasing power, which often leads to overfitting, focus on simplifying a well-known model's structure or data inputs. This is a much more promising way to add value, playing on the fact that most models are overfit.
>
> *Falkenstein* (2002:185)

One wonders if Mr. Falkenstein was aware that he effectively restated the centuries-old philosophy known as Ockham's Razor, or the 'principle of parsimony', according to which 'of two alternative explanations for the same phenomena [sic] the more complicated is likely to have something wrong with it, and therefore, other things being equal, the more simple is likely to be correct'.

William Ockham was a fourteenth-century philosopher, whose arguments caused Aristotelian nominalism to triumph over Platonic realism, and who is associated with his own brand of nominalism. He was also known for contesting the power of the papacy, outside of religious affairs.—*Collins English Dictionary*, 21st Century Edition.

This was qualified by Albert Einstein, who commented, 'Everything should be as simple as possible, but not simpler'. These quotations are not just of passing interest, but highlight the need for structure and simplicity, no matter what the endeavour. While this text might not seem to be sticking to that principle, it is at least making an effort.

Falkenstein et al. (2002:20) make reference to studies in economics, which have shown that 'naïve' models consistently outperform more sophisticated alternatives, where '"naïve" does not mean uninformed or arbitrary, but parsimonious and informed by theory'.

Philosophy aside, by this stage we are like a medical intern reporting for the first emergency room rotation—all the right training and equipment, but little practical experience beyond television ER dramas. In the ideal world, one should be able to jump right in and assist, but may freeze when the first real-life trauma case arrives. Likewise, when developing models, a set of data and a statistical technique are not enough. One needs to know what to do with them, otherwise the results will be similar to the emergency room scenario above.

Let us recap briefly. Module A sets the scene, covering economic theory and history. Module B views credit risk within the broader risk framework, issues with risk quantification, and assessment of business enterprises. Module C looks at statistical theory and scorecard development tools, which should assist when the more practical aspects of the scorecard development process are considered. And Module D covers data, including the data assembly process, required to provide the predictor and target variables (which is often the hard part; sample design and construction can take weeks, or even months). This section moves on to scorecard development, both: (i) milestones, where contact with the business is required and (ii) process, some of which requires no business input.

Milestones

Unfortunately, scorecard developers and project teams will never have a full view of the business. Just as a ship's engineer relies upon information from the bridge, scorecard developers rely upon management for insights about the business's past, and its proposed future. Questions have to be asked whenever inconsistencies arise, and assumptions must be documented as part of the development. For this reason, the entire scorecard development process must be as interactive as possible. Key milestones that should require presentations to, and possibly approval from, company decision-makers, are

Start-up—Initial meetings to determine responsibilities, project scope, possible data sources, and problems that may be encountered,

Data assembly—Data sources and sample sizes, where appropriate,

Good/bad definition—Not just good, bad, and indeterminate, but also any accounts that are supposed to be excluded from the development,

Scorecard splits—Determine whether or not any groups need to be treated separately. Past scorecard splits, and input from the business, provide the best starting point,

Final scorecards—The results of the development, including point scores associated with the different attributes for each scorecard, and any validation that has been done,

Strategies—Decision to be applied in each scenario, where scores are part of the scenarios. These may be simple cut-offs, but are often more complex.

The final deliverable is not just the scorecard and strategies, but also documentation covering various aspects of the scorecard development process, including data sampling, scorecard splits, characteristic analyses, statistical methods, scorecard validation, and the specifications necessary for implementation into the delivery system—whether by hard coding (possibly including the program code), or just modifying parameters.

Ultimately, the decision-makers will be most interested in scorecard implementation and strategies, and the latter may change over time. At any point post-implementation, the scorecard developer and project team may be brought back in, to ensure that the scorecards are working to design, and to keep management apprised of scorecard effectiveness.

Scorecard Development Process

The development process involves more than just these milestones. This module assumes that data assembly is finished, and covers all aspects required to develop a scorecard, whether presented to business or not. Much of it is conceptually difficult, but a skilled scorecard developer can work through it quite quickly. Unfortunately, there are a number of different ways in which scorecards can be developed, with a variety of factors influencing the choices. The primary influences are (i) the amount of available data; (ii) the implementation platform; and (iii) available skills. It is impossible to cover all of the different possibilities, and many scorecard developers will contest—perhaps rightly—what is being written here. Fortunately, this text is not aiming for the lofty heights of a scientific treatise, but instead hopes to provide the reader with some insight into the choices that are available.

The scorecard development process is illustrated in Figure E.1., which splits it into a full and simple process, the latter being a recurring and time consuming sub-process. This module gives each stage individual treatment:

(16) **Transformation**—Analyse available data and turn it into something useful, which traditionally involves (i) fine classing; (ii) coarse classing; and (iii) conversion.
(17) **Characteristic selection**—Choose candidates for consideration, which are predictive, logical, stable and available, compliant, customer related, and uncorrelated.
(18) **Segmentation**—Determine whether different scorecards are required, and how many. The split may be driven by market, customer, data, process, or model-fit factors.
(19) **Reject inference**—For an application scoring development, or any model used to drive a selection process, performance of rejected accounts should be inferred.
(20) **Calibration**—Use of banding or scaling to ensure score results have the same meaning across scorecards, and to provide default probabilities.
(21) **Validation and delivery**—Test for overfitting and potential model instability using holdout and recent samples, and prepare the scorecards for presentation to business.
(22) **Development management issues**—Scheduling and streamlining of scorecard developments.

The first part of the development process is to put data into a usable form. *Chapter 16* covers *transformation*: (i) methodologies—both univariate and bivariate, especially the latter's dummy variable and weight of evidence approaches; (ii) classing—the characteristic analysis report, and binning of both categorical and numeric characteristics; (iii) use of statistical measures—including the chi-square statistic, Gini coefficient, and information value; (iv) pooling algorithms—adjacent, non-adjacent, and monotone adjacent; and (v) some practical examples—court judgements, industry, and occupation.

The number of variables at the start can be significant, but can be reduced prior to starting the development. *Chapter 17* focuses on *characteristic selection*: (i) considerations for inclusion—including significance, correlation, available and stable, logical, compliant, and customer-related; (ii) measures of significance—again the chi-square statistic, Gini coefficient, and information value; (iii) data reduction methods—factor analysis, correlation assessment,

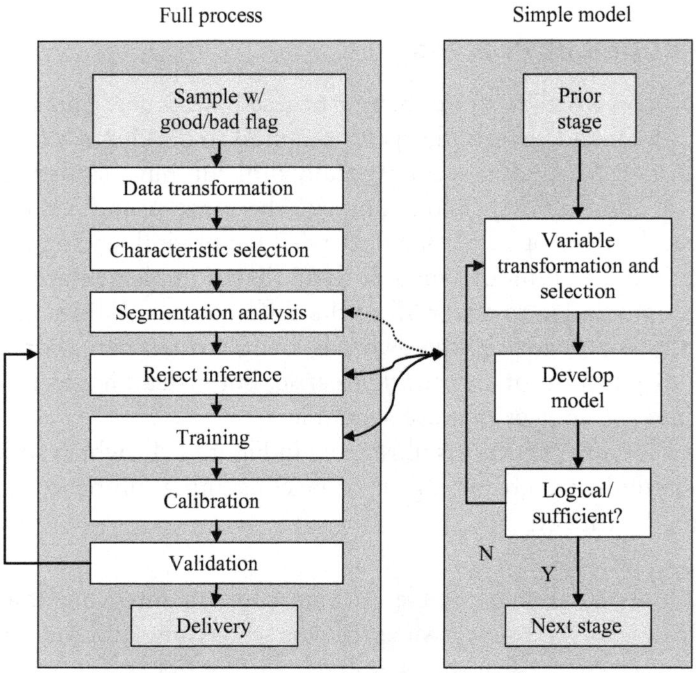

Figure E.1. Scorecard development process

or treat during training; and (iv) variable feed—covers stepping (forward, backward, and stepwise) and staging (independent and dependent).

Companies are used to splitting their customer base for marketing, and the same applies for credit. *Chapter 18* looks at the *segmentation*: (i) drivers—including marketing, customer, data, process, and model-fit factors; (ii) identifying interactions—whether through manual review or use of an RPA; and (iii) addressing interactions—use of scorecard splits and identifying which is best.

With any selection process, there will be discarded cases that might have yielded decent results had they been kept. In credit scoring, *reject inference* is used to guess what rejects' performance would have been, had they been accepted. *Chapter 19* covers the topic, including: (i) why reject inference?—the logic behind it, intermediate model types (known good/bad and accept/reject), and the potential benefits (or lack thereof); (ii) population flows—a tool for assessing changes to the frequency distribution; (iii) performance manipulation tools—including reweighting, reclassification, and parcelling; (iv) special categories—policy rejects, not-taken-ups, indeterminates, and limit increases; and (v) reject inference methodologies—random supplementation, augmentation, extrapolation, cohort performance, and bivariate inference.

There is no specific section covering *training*, as most of the concepts are covered elsewhere. Thus, *Chapter 20* moves on to *calibration*: (i) banding into groups—including use of the Calinski–Harabasz statistic, benchmarking, and marginal risk boundaries; (ii) linear shift and scaling—minor changes to ensure scores from different scorecards have the same meaning, conversion into numbers that can be better used by business (and some of the features required), and a possible method of achieving it using linear programming.

Checks and balances are required not only immediately after the development, but also ongoing thereafter. *Chapter 21* covers validation, which, for the most part, uses Basel II frameworks: quantitative (conceptual soundness) and qualitative (predictive power, explanatory accuracy, stability) factors; expected loss parameters (PD, EAD, LGD, and M); and process components (data, estimation, application, and mapping). The chapter itself focuses primarily on (i) actions—review of developmental evidence (including scorecard presentation), ongoing validation, and backtesting (including analysis of score shifts); and (ii) disparate impact—which looks more specifically at American anti-discrimination requirements.

Finally, a couple of scorecard development *management issues* are covered in *Chapter 22*: (i) *scheduling*, emphasis must be put on getting value for effort spent; and (ii) *streamlining*, piggybacking on what has been done before, to speed development.

Module F Implementation and use

> There are three important systems and programming issues that relate to credit-scoring projects (1) scorecard installation, (2) connectivity with credit information, and (3) once the scoring process has been completed, scorecard tracking.
>
> *Wiklund* (2004)

It is now assumed that there are one or more scorecards, and this module moves on to their implementation and use within the business. Much of it follows the outline provided by Wiklund (2004), but other issues are also covered. It is split into four sections.

(23) **Implementation**—Issues for greenfield developments when scoring is first used, and immediate issues relating to data, resources, and migration for brownfields.
(24) **Overrides, referrals, and controls**—Checks and balances, used to ensure that the scores are used appropriately, and effectively.
(25) **Monitoring**—Reports used to track what is happening within the business, for both front-end and back-end reporting.
(26) **Finance**—Tools used to estimate and provide for losses, and others that allow lenders to focus upon profitability, including the use of risk-based pricing.

Chapter 23 looks at scorecard *implementation*, both: (i) decision automation—high-level considerations (especially for greenfield developments), relating to the level of automation, responsibility, employee communications, and customer education (including decline reasons and the appeals process) and (ii) implementation and testing—including data, resources, and migration issues, and testing 'actual versus expected' for scorecard and strategy parameters, and for operational drift.

Credit scoring is not perfect, and issues may arise because of rare but severe events, data evolution, and known scorecard weaknesses—especially when there is information not captured by the system. *Chapter 24* looks at *overrides, referrals, and controls*: (i) policy rules—and instances where rules should be used instead of scores; (ii) overrides—subjective intervention, both high- and low-score; (iii) referrals—verification (documentation/security procedures, fraud suspicion triggers, account conditions); and (iv) controls—including the playing field (risks that may arise, and tools that can be used to protect against them), and scorecard/strategy and override controls.

A control that receives separate treatment is *monitoring*, covered in *Chapter 25*: (i) portfolio analysis—including delinquency distribution and transition matrix reports; (ii) performance tracking—scorecard performance, vintage/cohort analysis (and new-account, life-cycle, and portfolio effects), and score misalignment reports; (iii) drift reporting—including population stability, score shifts, and characteristic analysis by booking rates; (iv) selection process—decision process (track applications through the process), score distribution by system or final decision, policy rules (and how they have affected the decision), and manual overrides (by reason code and by their influence on the final decision).

Finally, *Chapter 26* covers reports required by the *finance* function: (i) loss provisioning—the distinction between general and specific provisions, and types of approaches; (ii) direct estimation—using the net-flow method or transition matrices; (iii) component approaches—which split the problem into loss probability and loss severity; (iv) scoring for profit—including profit drivers, profit-based cut-offs, and profit modelling approaches; and (v) risk-based pricing—mechanics and implementation, behavioural changes, strategic issues, and how it affects customers (especially higher-risk borrowers' increased use of home loans to finance consumption).

For those familiar with scoring, it might seem as though strategy setting has been overlooked. The basics are covered in Module A (Section 3.2.1, Process and Strategy) and Module B (Chapter 5, Decision Science), while this module covers some of the more sophisticated approaches (Section 26.4, Scoring for Profit; and Section 26.5, Risk-Based Pricing).

Module G Credit risk management cycle

> When written in Chinese the word crisis is composed to two characters. One represents danger, and the other represents opportunity.
>
> *John F. Kennedy* (1917–1963)

Companies evolve, so it is no surprise that the terms 'organism' and 'organisation' have the same root. The concepts differ, only in that one is the product of nature, and the other the product of men. Organisms have to worry about nourishment, reproduction, and predators, while organisations must compete for resources, attract customers, and control a myriad of risks. This applies to all 'for profit' entities (and others), including banks, finance houses, credit card issuers, retailers, and other consumer-credit providers. These companies are unique though, in that there are well-defined stages, collectively referred to as the credit risk management cycle (CRMC), where risks peculiar to the industry are managed. Essentially, this is an account management cycle, from the day it is a glimmer in the lender's eye, until it passes through to its grave. In the 1960s, scoring was associated with just one part of this cycle (new-business application processing), but it is now being applied throughout.

The CRMC is not to be confused with other concepts related to the economic cycle: (i) 'credit cycle', the expansion and contraction of credit and (ii) the 'credit risk cycle', changes in overall credit quality.

Before these stages are considered, a brief review of basic marketing is in order. Textbooks put forward basic frameworks, like the marketing mix, or 5 'P's, which can be used to define any market offering:

Product—The good or service being offered.
Package—Product presentation, including packaging materials and labelling.
Price—Positioning, in terms of luxury, mass, or somewhere in between.
Promotion—Communications, to prompt the product's purchase by the market.
Place—Distribution channel(s), used to deliver the product.

The framework is general, and applies primarily to consumer goods, such as toothpaste, automobiles, perfumes, or fashion denims. The goods on offer are picked off the shelf, and paid for at the checkout counter, no questions asked. If the buyer instead wants to 'buy now, pay later', there are other risks, other processes, other costs, and other questions that must be asked. These may vary, depending upon whether lending is the company's primary business (bank, finance house, or card issuer), or a secondary activity used to support sales (retailer, motor

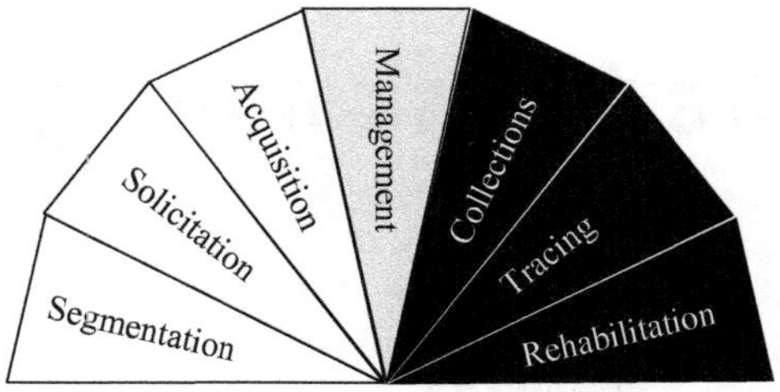

Figure G.1. Credit risk management cycle.

dealer/manufacturer, utility or service provider). In either case, providing credit adds another dimension—a cycle that also has promotion and distribution aspects, but extends further into an ongoing service relationship dedicated to the money, as opposed to what is being purchased.

For retail credit, McNab and Wynn (2003)[1] split the CRMC into five stages: marketing, application processing, account management, collections, and recoveries. Marketing and recoveries can be further split to create seven operations, as shown in Figure G.1.

Segmentation—Identifies customers to be targeted, their needs, and appropriate products.
Solicitation—Designs and executes marketing campaigns, used to invite potential customers to do business.
Acquisition—New-business processing, which obtains and processes applications, delivers the goods if they are accepted, and handles communications and queries if not.
Management—Functions required during normal account operations, especially limit management, but also handling repayments, billing, queries, billing, and others.
Collections—Focuses on early-stage delinquencies, and on maintaining the customer relationship.
Tracing—Attempts to find and contact absconders, who move without providing a change of address or other contact details.
Rehabilitation—Deals with late-stage delinquencies, to get the money back (or as much as possible), which may lead to legal action and/or loss of the customer relationship.

In corporate credit, risk transfer is treated as a separate stage. In retail credit, it is done at portfolio level, and could be done as part of account acquisition, management, collections, or recoveries, whether through insurance, securitisation, hedging, or outright sale of assets.

Credit risk management function

All of the decisions made during the CRMC have an impact upon risk, and many lenders will have a specialist area that works with various business units to manage it. This '*credit risk*

[1] This section is borrows heavily from McNab and Wynn (2000 and 2003), and from associations with Helen McNab and Scoreplus Ltd.

management' area would perform functions like (i) working with *marketing* on setting eligibility criteria for new products (whether for through-the-door or pre-approved customers), identification of prospective customers, new product pricing, package eligibility, and so on; (ii) setting *new business strategies*, and policies for application processing—including cut-offs and limits, pricing, and repayment terms at different levels of risk; (iii) setting *account-management strategies*, for limit increases and authorisations; and (iv) setting *collections policies and strategies*. It may also provide a *decision support* function and 'decision tools'—the models and software required to calculate scores and apply strategies and policies, as well as monitor what happens, and make changes as required.

Other business functions

While not directly related to credit risk management, there are several other areas with which the credit function must interact:

Compliance/legal—Ensures that no laws, statutes, or regulations are broken. This is particularly important in areas of illegal discrimination, data protection, and 'know your customer' legislation.

IT/systems—Ensures the smooth operation of mainframe and networked computers, and communications used to perform functions across the business. At one time, the risk management and other functions were highly dependent upon them, but this changed as computers became cheaper and smaller—minis, PCs, notebooks, etc.

Management information—Required to manage and understand customer behaviour, and to report information from across the organisation to the company executive, and others. This may be a part of the IT function, but most companies have split it off separately.

Accounting, finance, planning, and audit—Other functions within the company that are responsible for accounting, understanding profitability, setting high-level strategies, and ensuring that the results are understood.

The matrix

The CRMC is widely referred to within the retail lending industry. Indeed, it applies to almost any credit product or market, and is used as a conceptual framework when positioning discussions about problem areas within the business, especially when combined with the key process components: (i) *data*—for analysis, modelling, and reporting; (ii) *systems*—for gathering data and delivering products and decisions; (iii) *models*—for representing risk, revenue, retention, and response; (iv) *strategy*—for rule-sets that leverage upon data, by using models and policies to drive decision-making; (v) *analytics*—for manual review of summary statistics to turn data into knowledge; and (vi) *reporting*—for monitoring results to ensure that all runs

Table G.1. CRMC versus process components—discussion matrix

Function	Component					
	Data	Systems	Models	Strategy	Analytics	Reporting
Marketing						
Application processing						
Account management					✓	
Collections						
Recoveries						

according to plan. This is by no means the full set; there are entire departments whose names could be listed across the top. Two others that have a direct interest in risk are

Fraud—Prevents fraud when it can; identifies fraud when it happens; and brings in the law enforcement agencies when necessary.
Risk management—Considers all risks, where credit risk is only one of them. Ultimately, business targets must be met, while keeping risks at acceptable levels.

These are then presented in a matrix, such as that in Table G.1., which indicates an analytics issue in account management.

This module

The above section provided a broad overview of many functions that must be performed by any credit provider. The module itself is split out into five sections, each of which gives certain aspects of the CRMC individual treatment, in particular,

(27) **Marketing**—Advertising media, quality versus quantity, pre-screening, and data used.
(28) **Application processing**—Operations of selection processes: gather, sort, and action.
(29) **Account management**—Takers, askers, givers, repeaters, and leavers.
(30) **Collections and recoveries**—Default reasons and recovery processes, triggers and strategies.
(31) **Fraud**—Trends, types, and tools.

All of these are becoming increasingly dependent upon statistically derived models, and decision automation, to drive their business processes. Fraud is really an operational risk, which does not really belong in this group, but must be considered across the CRMC.

Marketing is the tout responsible for identifying and attracting prospective customers, which is covered in *Chapter 27*: (i) advertising media—which can be defined as broad-based or personal, or as print, tele-, cyber, or person-to-person, with a focus on maximising the 'bang per buck'; (ii) quantity versus quality—a conflict that arises between marketing as credit, and which affects processes' ability to cope; (iii) pre-screening—which involves list scrubbing and use of other metrics to target customers (the 4 Rs); and (iv) data—including types of data, and its assembly into a data mart.

Application processing is the gatekeeper for through-the-door customers, covered in *Chapter 28*. It is treated using headings that would apply to any selection process: (i) gather—acquisition and preparation of completed forms; (ii) sort—obtain the necessary information, use it to provide an assessment, and then make a decision; (iii) action—communicate the decision and carry out the required actions, and exploit opportunities for up-sells, down-sells, cross-sells, approval in principle, and credit insurance.

Chapter 29 moves on to *account management*, the bartender who ensures existing customers' needs are served. While it includes a range of functions, including billing and payment processing, here it relates primarily to limit management: (i) types of limits—agreed, shadow, and target limits (along with brief mention of debt counselling services relating to cash-flow triage); (ii) over-limit management—to deal with those who take without asking, including pay/no pay decisions for cheque accounts and authorisations for credit cards, and the informed customer effect (customers facing equally bad choices will choose that which is best understood); and (iv) more limit and other functions—including limit increase requests, limit increase campaigns, limit reviews, cross-sales, and win-back campaigns.

Collections and recoveries are the heavies, who deal with problematic customers and guard the back door. Collections play the good cop, who tries to put the customer on the right track. In contrast, recoveries play the bad cop, whose only interest is in getting the money back. *Chapter 30* is split into (i) overview—delinquency reasons, underlying processes, core system requirements, and agencies; (ii) triggers and strategies—where triggers include excesses, missed payments, and dishonours, and strategies can vary by message tone, content, delivery, timing, and extent; (iii) scoring—special issues relating to definitions, time frames, and usage.

Finally, *Chapter 31* looks at *fraud*, the town detective who deals with cheating customers. This area has always been challenging, and modern technology is making it even more so. After highlighting fraud trends, the chapter moves on to (i) fraud types—split by product, relationship (first-, second-, or third-party), process (application, transaction), timing (short or long term), misrepresentation (embellishment, identify theft, fabrication), acquisition (lost or stolen, not received, skimming), usage (counterfeit, not present, altered), and technology (ATM, Internet); (ii) detection tools—negative files, shared databases, rule-based verification, scoring, and pattern detection; (iii) prevention strategies—for the application process, transaction media, and account management; and (iv) scoring—its usage for both application and transaction fraud. Of particular note, is that fraudsters' *modus operandi* are quick to counter lenders' moves, and to seek and exploit new opportunities. In recent years, this has been best evidenced by the growth of card-not-present fraud, especially for Internet transactions.

Module H Regulatory environment

The years since the 1960s have been characterised by increasing regulation of financial institutions. This has impacted on credit scoring, either promoting it or controlling its use. This module looks at the various types of legislation, and their impact. Rather than covering the legal environment in one particular country though, the goal is to provide a framework, or frameworks, within which the regulations can be analysed.

The module is split out into six sections, a conceptual overview followed by five separate sections each covering a regulatory pillar that directly affects the provision of consumer credit, and the use of credit scoring:

(32) **Regulatory concepts**—Best practice, good governance, business ethics, social responsibility, and the compliance hierarchy of statutes, legal precedents, industry codes, policies and procedures, and unwritten codes.
(33) **Anti-discrimination**—Covers what information may be used in a lending decision, and prohibits the use of fields that are discriminatory (race, religion, etc.), or any information relating to parties other than the prospective borrower.
(34) **Fair lending**—Ensures that lenders take adequate steps, to ensure that borrowers can afford the loan repayments, and that the terms are fair in the circumstances.
(35) **Data privacy**—Governs the sharing of data between lenders, what may be kept on credit bureau, what must be divulged to customers, and so on.
(36) **Capital adequacy**—Focuses primarily on the New Basel Accord for banks, which allows the use of their own internal ratings to calculate reserve requirements.
(37) **Know your customer**—Increased personal identification requirements, primarily to prevent money laundering and criminal activities, but also terrorist activities.
(38) **National differences**—An overview of some of the laws in force within various English-speaking countries, in particular the United States, United Kingdom, Australia, Canada, and South Africa.

The module's starting point is some *basic concepts*, presented in *Chapter 32*: (i) best practice—ways of doing things that have a proven record of success; (ii) corporate good governance—limit the executive's power, and ensure transparency; (iii) business ethics and social responsibility—act according to what is right or wrong, respect other stakeholders, and give back to those being served; (v) the compliance hierarchy—including statutes, legal precedents, industry codes, policies and procedures, and unwritten codes.

Lenders depend upon data for their risk assessment, which presents a power imbalance that can be abused, and *data privacy* issues. These are covered in *Chapter 33*, which covers: (i) background—including a historical overview covering the Tournier case of 1924, OECD data privacy guidelines, the Council of Europe convention, and the EU data protection directive; (ii) data privacy principles—relating to the manner of collection, reasonable data, data quality, use of data, disclosure to third parties, subjects' rights, and data security.

Credit scoring is used to discriminate between potentially good and bad business, which can lead to claims of unfair discrimination. *Chapter 34* covers *anti-discrimination* legislation, including (i) what does it mean?—which provides different views on what is acceptable, from 'credit scoring is unfair' through to 'most characteristics can be used, as long as they form part of a holistic assessment, and there are no reasonable alternatives' and (ii) problematic characteristics—where treatment varies, including age, gender, marital status, government assistance, and unlisted phone numbers.

In general, credit scoring promotes *fair lending*, but it can be abused. *Chapter 35* illustrates the distinction between: (i) predatory lending—which victimises borrowers for the personal gain of the lender; (ii) irresponsible lending—practices that involve questionable ethics and fail to consider the effect of debt on borrowers; and (iii) responsible lending—acting in the best interest of borrowers, including conducting due diligence, aiming for financial inclusion, ensuring transparency, and educating customers. Responsible lending is obviously the ideal, and while most countries have legislation to guard against predatory lending, the treatment of irresponsible lending is mixed.

On the good governance front, credit scoring has also become a cornerstone of determining banks' *capital adequacy* requirements for retail credit risk. *Chapter 36* provides a historical overview, and then covers: (i) Basel I—implemented in 1988, which set simple requirements for sovereign, bank, residential property, and other lending; (ii) the new Basel accord—which allows a similar standardised approach, but most affected banks are instead opting for the foundation and advanced 'internal ratings based' approaches; and (iii) the risk-weighted asset calculation—based on Merton's model, it uses internal ratings and within-portfolio correlations to derive a capital requirement, that is then adjusted for factors such as size, maturity, future margin income, and double default.

Something for which credit scoring is neither a regulatory target nor a tool is *Know Your Customer* legislation, which is intended to guard against criminal activities. This is the focus of *Chapter 37*, which defines racketeering, organised crime, and money laundering, before covering due diligence requirements—with respect to customer acceptance policy, customer identification, treatment of high-risk accounts, and day-to-day identification of abnormal transactions.

Although most English-speaking countries are surprisingly similar in this domain, there are some notable differences. *Chapter 38* briefly discusses the situations in (i) the United States—the Fair Credit Reporting Act (1970) and Equal Credit Opportunity Act (1974); (ii) Canada—privacy (especially Quebec's Bill 68) and human rights legislation; (iii) the United Kingdom—the Consumer Credit Act (1974) and Data Privacy Act (1984/1988); (iv) Australia—Privacy Act (1988) and Privacy Amendment Act (2000); and (v) South Africa—National Credit Act (2006). Other pieces of legislation in each country are also mentioned. The major credit bureau and personal identifiers used in each are also discussed.

Finally . . . .

Enjoy the book.

This book is based on an extensive literature review, and is naturally influenced by the author's interpretations and own experience. While every effort has been made to ensure accuracy, please do not hesitate to submit any queries, suggestions or corrections relating to the text to the author or publisher.

Author's contact details:
Raymond Anderson
To contact author: TCSAcademy@yahoogroups.com
To receive updates: TheCreditScoringToolkit-subscribe@yahoogroups.com
Web Address: http://finance.groups.yahoo.com/group/TheCreditScoringToolkit/

Publisher's contact details
Oxford University Press
To contact publisher: Science.books.uk@oup.com

Module A

Setting the scene

1 Credit scoring and the business

It is utterly implausible that a mathematical formula should make the future known to us, and those who think it can would once have believed in witchcraft.

Jacob Bernoulli, in Ars Conjectandi (1713)

Books, websites, pamphlets, etc. often have a section containing frequently asked questions (FAQs). This should be restated as 'frequently answered questions'. Countless websites use this format, whether provided by lenders, credit bureaux, or government bodies. The same format is used for much of what follows, as it is very effective. This first chapter focuses on the theory of credit scoring, and attempts to answer questions like:

What is credit scoring?—What is credit? What is scoring?
Where is it used?—Where is the data obtained from? What aspects of customer behaviour are assessed? How is it applied over the risk management cycle?
Why is it used?—How has it affected lenders? How has it affected consumers? What did it replace?
How has it affected credit provision?—What has brought about the growth in credit? Where does credit scoring fit?

1.1 What is credit scoring?

Common sense is the most widely spread human characteristic, which is why each of us has so little.

Unattributed

In order to answer the question, 'What is credit scoring?' let us first break it into two components, 'credit' and 'scoring'.

What is credit?

In the current context 'credit' simply means, 'buy now, pay later', whether the purchase is for short-term consumption, durable goods and other assets that provide users with valuable services, or productive enterprises. The word 'credit' comes from the old Latin word 'credo', which means, 'trust in', or 'rely on'. If you lend something to somebody, then you have to have trust in him or her to honour the obligation.

Many people today view access to credit as a right, but it comes with its own obligations. Borrowers must pay the price of (i) creating the impression of trust; (ii) repaying according to the agreed terms; and (iii) paying a risk premium for the possibility they might not repay. This gives rise to concepts like: *creditworthiness*—borrowers' willingness and ability to repay; and *credit risk*—the potential financial impact of any real or perceived change in borrowers' creditworthiness.

According to Thomas et al. (2002), creditworthiness is often mistakenly viewed like a personal attribute like height, weight, or eye colour, as something that can be directly measured. Indeed, application, behavioural, and bureau risk scores are sometimes mistakenly believed to be creditworthiness measures, but almost all of them ignore the loss severity and profitability elements. A person of higher than average risk may nonetheless be creditworthy at the right price. You may not be creditworthy to one lender, but just adjust the risk/return dynamic—increase the interest rate, lower the amount, shorten the term, etc.—and presto!

As regards credit risk, and any other situation involving trust (which includes all economic activities involving contracts or liabilities), the contracting parties must be aware of the possibility that things may not be as they seem. Where trust is low, lenders will increase their charges to cover the risks. Trust can however, be enhanced through collateral, other security, or more information. In ages past, credit was often only extended against collateral, but the cost of realising its value is high. The modern information age allows lenders to enhance trust, by using data about borrowers' financial and other circumstances, whether at time of application or ongoing thereafter.

Just as credit is a human construct, and a commodity with no physical form other than documentary evidence, so too is the information upon which it is based. The two are not only similar, but heavily intertwined, to the extent that lenders' activities extend far beyond lending money, to managing and transacting in the information needed to manage the risks. This leads to concepts like *information goods* and *information economies*, and a variety of others used in economics and law:

Information rents—Extra benefits that can be gained from 'signals' not available to competitors, if strategies are used to take advantage of the discrepancies (economics).

Asymmetric information—Differences in information available to different game players, especially those that provide competitive advantage (game theory/economics).

Adverse selection—Poor choices that result from information asymmetries, especially where these are consciously exploited by other parties (economics/insurance).

Moral hazard—Risk of parties to a contract changing their behaviour once a contract is in place (law/economics).

In sales situations, it is the buyers that are most at risk of *adverse selection*, because sellers are more familiar with the item being sold. In contrast, in credit and insurance markets, the sellers are most at risk, because their customers have better knowledge of their own personal circumstances. Likewise, for *moral hazard*, just as people may engage in riskier behaviour when they know that they are insured, so too may borrowers become less financially responsible, once they have funds in hand.

The concept of adverse selection was first proposed by Akerlof (1970), in his article, 'The Market for "Lemons": Quality Uncertainty and the Market Mechanism.' 'Lemons' refers to potentially faulty goods being sold, such as used cars, but the concept also applies to customers in credit and insurance markets.

All of these informational aspects are particularly pertinent for banks, which have huge investments in information, and rely upon it for a competitive advantage. The better the information, the greater the rent that can be achieved. This will, of course, also be dependent upon: (i) methods used to interrogate the data; and (ii) actions taken based upon the knowledge that is gained. Indeed, the situation can be compared to other conflict situations where intelligence gathering is critical, like military intelligence and business intelligence. In like fashion, 'credit intelligence' can be used to describe the gathering, interpretation, and delivery of information to support credit decisions, where the information relates to borrowers, either individually or collectively. Many lenders refer to it as 'customer insights', if only because it is more politically palatable, but this plays down the lender/borrower conflict, and focuses on the positive-sum game.

What is scoring?

Scoring refers to the use of a numerical tool to rank order cases (people, companies, fruit, countries) according to some real or perceived quality (performance, desirability, saleability, risk) in order to discriminate between them, and ensure objective and consistent decisions (select, discard, export, sell). Available data is integrated into a single value that implies some quality, usually related to desirability or suitability. Scores are usually presented as numbers that represent a single quality, while grades may be presented as letters (A, B, C, etc.) or labels (export quality, investment grade) to represent one or more qualities.

Scoring has become ubiquitous in processes where predictions are needed, which can only be stated as probabilities (stochastic), and not cut and dried certainties (deterministic). If it were possible to develop scorecards that provide perfect predictions of the future, such technology would be better off applied at the racetrack (see Table 1.1). Prescience is an unachievable ideal, and lenders have to do the best they can with the available information.

Instead, predictive scoring models are used to assess the relative likelihood of a future event, based upon past experience. Most scoring models are derived using historical data, but in the absence of data, judgmental models may be used. When computers are used automatically to combine scores and strategies to make decisions, it provides a form of artificial intelligence (AI), which substantially reduces the cost of decision-making.

Predictive models could be developed to aid gambling on the gold price, horse races, and other punters' favourites. These endeavours are affected by a multitude of variables though, many of them impossible to capture, and whose relative importance may change rapidly and unpredictably. The resulting model might be statistically valid, but it would be expensive to maintain, and its predictive power probably insufficient to justify a wager. Care must be taken to ensure that scoring is a suitable alternative.

Table 1.1. A day at the races

Racetrack	Scoretrack	In the stables
Jockey	People	Analysts and decision-makers
Feedstock	Data	Application, past dealings, credit bureaux
Horse	Infrastructure	Computers, scorecards, strategies
Odds	Odds	Historical bad rates, good/bad odds
Training	Learning	Knowing and developing tools
Racetrack	Market	Prospective, new, existing
Other horses	Competitors	Credit market
Finish line	Measurement	Reject rates, default rates
Winnings	Winnings	Future profits, customer satisfaction, market share

What is credit scoring?

With that explanation of the parts, an attempt can now be made to address the initial question, 'What is credit scoring?' Simply stated, it is the use of statistical models to transform relevant data into numerical measures that guide credit decisions. It is the industrialisation of trust; a logical further development of the subjective credit ratings first provided by nineteenth century credit bureaux, that has been driven by a need for objective, fast, and consistent decisions, and made possible by advances in technology. There are limits though, due to its data-dependent and backward-looking nature. Credit scores dominate in automated high-volume low-value environments, while credit ratings still imply some degree of subjective input, especially for larger loans to companies, governments, and others.

Credit scoring was first used in the 1960s, to determine whether people applying for credit would repay the debt, honour the obligation, and—in general—act in a manner deemed acceptable by the treasury's gatekeeper. At that time, it was associated exclusively with 'accept/reject' decisions generated by the new-business application process (application scoring), and many people still use the term in that limited sense. In the twenty-first century, however, the label is used more broadly to describe any use of statistical models to extend and manage credit generally. This includes the measurement of risk, response, revenue, and retention (the 4 Rs), whether for marketing, new-business processing, account management, collections and recoveries, or elsewhere (the credit risk management cycle, or CRMC).

Although most commonly associated with the risk-assessment models, it is difficult to divorce 'credit scoring' from other aspects of the decision-making process:

Data—Credit related information about the customer, whether obtained directly from the customer, internal systems, or the credit bureau.

Risk assessment—Not only credit scoring models, but also policy rules and judgmental input, used to assess each case.

Decision rules—Strategies used to guide accept/reject, pricing, pay/no pay, collections, and other decisions.

While credit scoring plays an integral role, its role has become so ubiquitous that people are starting to take it for granted. Instead, the focus has shifted onto expanding the data sources used as inputs into the scores, and enhancing how the scores are used to drive the business.

What did credit scoring replace?

In traditional lending, underwriters make judgmental assessments of prospective borrowers according to the 5 Cs: *character* (of the applicant), *capacity* (to borrow), *capital* (as backup), *collateral* (as security), and *conditions* (external factors). These assessments are based upon underwriters' own experience, and what they have learnt from their mentors, taking into consideration not only historical information, but also a forward-looking view of the borrowers' prospects. The key is obtaining information through customer relationships, which makes it very difficult for customers to take their business elsewhere.

The use of credit scoring has caused a shift away from *relationship lending* to *transactional lending*. The former is appropriate in communities where lender and borrower had personal knowledge of each other, but is inefficient in an era of high customer mobility and extended branch networks. At the same time, there has been a shift away from *secured lending*, based upon collateral and guarantees, to *unsecured lending*, that relies upon information, and repayment out of the next month's income. Data has replaced experience, causing underwriters and human judgment to play less of a role. The 5 Cs still apply, but credit scores can now capture much of them by extracting maximum value out of available information. Scoring lacks only the forward-looking element, but in retail credit it is questionable whether underwriters can accurately read the signs. Ultimately, credit scores should have a high correlation with underwriters' assessments—and a cost advantage for most consumer and small-business lending.

This does not mean human judgment and collateral have disappeared; they are still alive and well—just leading less stressed lifestyles. Relationship lending is still used by smaller lenders, who believe that the relationship provides them with a competitive advantage. Transactional lending is favoured by larger lenders dealing in high-volume low-value products, because of the (potential) economies of scale. Underwriters are still used, but mostly where potential profits are high, the scorecards are not sufficient, and/or when a customer disputes the system's decision. And finally, collateral is still used where the loan size is so great that (i) the customer's ability to repay it out of income is questionable; and (ii) the hassles of managing the collateral, and realising its value, can be cost-justified.

1.2 Where is credit scoring used?

Credit scores provide the greatest value, when they are used to guide decisions that affect the customer. In decision processes, lenders define different scenarios using scores and policy, and then the action to be taken in each case—like accept/reject, maximum loan value or repayment, interest rate, loan term, etc. Alternatively, underwriters may consider scores as one of several inputs into a credit decision. The cost benefits of decision automation are placing incredible pressure upon organisations to limit the use of underwriters, to cases where their specialist knowledge is absolutely essential, especially where there is significant information that cannot be captured within the scoring process, and potential profits are high.

Credit scores go under a variety of different names, depending upon where and how they are used. The labels usually refer to: (i) the information source; (ii) the task being performed; or (iii) what is being measured. The most common labels used are:

Application score—Used for new business origination, and combines data from the customer, past dealings, and the credit bureaux.
Behavioural score—Used for account management (limit setting, over-limit management, authorisations), and usually focuses upon the behaviour of an individual account.
Collections score—Used as part of the collections process, usually to drive predictive diallers in outbound call centres, and incorporates behavioural, collections, and bureau data.
Customer score—Combines behaviour on many accounts, and is used for both account management and cross-sales to existing customers.
Bureau score—A score provided by the credit bureau, usually a delinquency or bankruptcy predictor that summarises the data held by them.[1]

There are a couple of major items that should be noted here: (i) all of these scores cover aspects of *customer behaviour*; and (ii) as the cost of *bureau information* reduces, it is being increasingly used in (or with) the other scores. Many lenders will combine their own scores and bureau scores using a decision matrix, while others will integrate the bureau information directly into their own scores.

Credit scoring was first proposed to American *finance houses* and *credit card issuers*, for application processing during the late 1950s and early 1960s. It was a hard sell to the finance houses, because most found it difficult to believe that statistical models could do a better job than experienced underwriters. The situation was much different for credit cards though, as the product was new, and there was little experience. Card issuers gained the dual benefit, of reduced losses and speed of decision-making, in a rapidly growing market. Over time, there was phenomenal growth in where credit scoring was applied, in terms of the range of credit products, processes, and organisation types. Markets where credit scoring is used today include, but are not limited to:

Unsecured—Credit cards, personal loans, overdrafts.
Secured—Home loan mortgages, motor vehicle finance.
Store credit—Clothing, furniture, mail order.
Service provision—Phone contracts, municipal accounts, short-term insurance.
Enterprise lending—Working-capital loans, trade credit.

The greatest benefits and fastest advancements were achieved in unsecured lending and store credit, where there is a heavy reliance upon information.

[1] According to Mays (2004), FICO scores assess the probability that any of the facilities will become 90 days delinquent within the next two years.

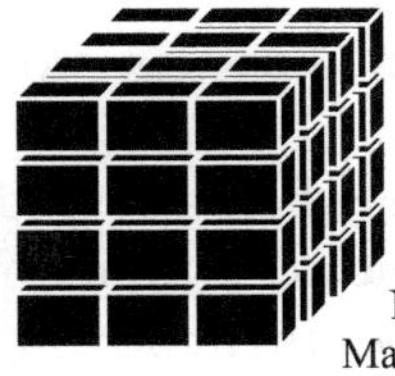

Figure 1.1. Scoring aspects.

All of these markets have certain common features, as shown in Figure 1.1. Application, behavioural, collections, customer, and bureau scores can all be slotted somewhere within this framework. First, each will use the same generic *data sources*—customer, internal, and external. Second, there are several aspects of *customer behaviour* that can be modelled—risk, response, retention, and revenue. And third, there are various stages in the CRMC where scoring can be used: marketing, new business processing, account management, and collections.

Score triggers

A major defining feature in credit scoring is what triggers the score calculation in practice: which varies with the CRMC stage:

Request—Customer application for a specific product, or increased facilities.

Time—Regular recalculation, such as monthly.

Entry—Calculated on first entry into a specific stage of the CRMC, and the resultant score is retained for future use.

Transaction—Customer already has the product, and scores are calculated each time the customer transacts.

Trigger	Usual area
Request	Application
Time	Behavioural
Entry	Recovery
Transaction	Fraud
Event warning	Behavioural/fraud
Campaign	Response

Event warning—Something occurs that indicates a not insignificant change in risk, which prompts the lender to recalculate the score. This might occur because of: (i) lack of activity; (ii) risky activity; or (iii) an external event, perhaps advised by the credit bureau.

Campaign—Scores used to target specific customers, especially in marketing and collections, as part of specific campaigns. Scorecard life is usually short, perhaps allowing one or two uses, and lenders will develop them on an ad hoc basis.

A distinction that should be made here is between selection and reaction processes. Lenders use selection processes when choosing cases for inclusion within a group, whether the groups are defined by accept/reject or the terms to be offered. In general, these are instances where the *lender has control*. Selection processes have what are referred to either as 'selection' and 'outcome' mechanisms (Feelders 2000), or 'input' and 'output' mechanisms (Bugera et al. 2002). Included under this heading are new business processing, marketing campaigns, and some collections campaigns.

In contrast, reaction processes are those where the lender has less control, and is instead *responding to circumstances*. This might include: (i) limiting the potential damage caused by an adverse event, such as missed payments; or (ii) improving operational efficiencies, by setting higher shadow limits. This applies to most account-management and collections activities that are the result of customer actions or inaction, once an account is open.

1.2.1 Data sources

As indicated, credit scoring is highly dependent upon data. It is obtained from a number of different sources, and then assessed prior to making a decision. For retail credit, the sources include, but are not limited to:

Customer—Application forms, financial statements, asset details.
Internal data—Past dealings, other account behaviour.
Bureau data—Credit bureaux (private and public), other lenders, court records.

One would expect the most important source of information to be the *customer*, but over time, lenders have become increasingly sophisticated at accessing data from other sources. Technology now allows a lender readily to access its own data on *past and current dealings*, but there will always be a segment, with which it has had little or no experience. As a result, *credit bureaux* play a major role by: (i) facilitating the gathering of information from public sources, especially court records; and (ii) allowing lenders to share information on customers' performance.

Scorecard types

The reliability of credit scores will vary, depending upon what data is used to develop them. The best results are obtained with bespoke scorecards, which have been specifically tailored for a lender, product, and process. Where this is not possible, lenders can use generic scorecards, which have been developed for general use across companies and/or processes. The latter term is usually applied to bureau scores, or application scores developed using data from elsewhere. Lenders can also combine forces to use *pooled data* for a new generic, especially if they are operating in the same market, and none of them has sufficient data for a bespoke solution.

Failing all else, a lender could also try to: (i) tap into underwriter's experience to develop an *expert model*; (ii) use a *prior-stage* score, such as using a behavioural score in the collections process; or (iii) use a scorecard developed for a *related market or product*. The latter may be appropriate where the product is new, or data has not been collected in an appropriate form, yet the transactions will be processed through the same system as where the scorecard currently resides. In such instances, scores should not be used to drive fully automated decisions.

Instead, underwriters should either have significant latitude to override the system, or alternatively use the score or preset strategy for guidance only.

1.2.2 Credit risk management cycle

Different triggers may be used at different stages, which are covered only briefly here because they are covered in detail in Module G (Credit Risk Management Cycle). The cycle includes:

Marketing solicitation—Pre-screening for offer setting and mail shots.
New business processing—Accept/reject, pricing, and cross-sales.
Account management—Limit setting, pay/no pay, customer retention.
Collections and recoveries—Customer rehabilitation, tracing, legal.

Marketing solicitation

One of lenders' most expensive tasks is to get new business. A combination of risk and response scoring is used, to avoid the cost of making offers to prospects that are either clearly rejects, or will not respond. The amount of information that can be used in marketing tends to be much more limited than for other functions, in particular due to privacy and competition issues.

New business processing

Lenders will not accept all through-the-door customers. Instead, any prospective customers will be put through a selection process. Application scores are used to rank the creditworthiness of people applying for credit, which uses information from the application form, past dealings, credit bureaux, and elsewhere. Decisions on whether or not to accept the applicant may be based solely upon risk, but may also consider aspects that relate to profitability.

Account management

Once the business is on board, lenders have to manage it. Behavioural scores are almost exactly the same as application scores, except they are associated mostly with account management of existing customers. Scores are refreshed at regular intervals, usually monthly, and used for limit setting, over-limit management, pay/no pay and authorisations decisions, collections, cross-sales, etc. The greatest benefit is gained from lenders' own account performance data, but increasingly, lenders are also trying to incorporate customer and bureau information. There are however, issues relating to cost and frequency of update.

Collections and recoveries

The final stages in the CRMC are collections and recoveries. These areas have special needs due to the urgency of the task, especially as relates to information requirements—like recent customer contacts, promises to pay, etc. Collections scores are used primarily as tools to drive predictive diallers, which are used to manage outbound call centres, while recoveries scores are used both for predictive diallers, and for valuing portfolios of defaulted accounts.

1.2.3 Behavioural propensities

Credit scoring is usually associated with the use of statistical techniques to assess the risk of non-payment, but it goes further than that. Indeed, sometimes credit scoring is classed as a form of propensity scoring, where customers' propensity to behave in certain ways is measured. Propensity means 'inclination' or 'tendency', and although it is commonly associated with marketing and response scoring, it really covers the entire spectrum of human behaviour (Table 1.2). It is split into the 4 Rs of credit scoring:

Risk—Will the customer do something to put us at risk of financial loss?
Response—Will the customer respond to an offer?
Retention—Will the customer stay, or move on?
Revenue—How much income is expected?

Risk

Scoring's greatest benefit has been in the realm of risk assessment. 'Risk' is being used here in the traditional sense of investing, relating to whether an investment will be diminished or destroyed. It covers not only loss probability, but also loss severity. There are three basic types of risk scoring being used by businesses: credit, fraud, and insurance scoring.

Credit risk is the primary area where scoring is used, and is logically what the term 'credit scoring' is usually associated with. Credit risk scores are used primarily to predict delinquencies, and include most application, behavioural, customer, collections, and bureau scores. They are often the only scores used to make decisions, but value can also be gained by combining them with response, retention, and revenue scores; or alternatively, by deriving probability-of-default (PD), exposure-at-default (EAD), and loss-given-default (LGD) estimates (see Section 3.2.2).

Table 1.2. Aspects of customer behaviour

Risk	Credit	Will he pay?
	Fraud	Will he cheat?
	Insurance	Will he claim?
Response	Response	Will he call?
	Cross-sell	Will he buy others?
Retention	Churn	Will he use me and leave?
	Attrition	Will he leave?
Revenue	Utilisation	Will he use it?
	Profit	Will it be worth it?

Fraud and credit risk may seem closely related, but fraud risk is viewed as an operational risk, and is treated totally separately. The impact of fraud on the financial services sector has been huge, costing millions each year. Application fraud scorecards are difficult to develop, because of the low numbers of known frauds, which is exacerbated because it is difficult to differentiate between 'unable to repay', and 'no intention of repaying.' There are also fraud-scoring systems that can run on a daily basis, looking for inconsistencies at transaction level.

Finally, insurance risk is the risk that an applicant will claim on an insurance policy. This falls outside of the field of credit, but is closely related, because it often relies on credit information, in particular that obtained from the credit bureaux. The use of credit data in insurance underwriting is contentious, yet extremely strong correlations have been shown between credit data and short-term insurance claims (household, personal, and motor vehicle). The most likely explanations are that individuals who keep their credit affairs in order are: (i) likely to take better care of their possessions; and (ii) less likely to give up their no claim bonus for a small claim.

Response

It costs a lot of money to attract new customers, especially where lenders do targeted marketing campaigns using mail shots or other channels. Response scoring is used to limit mailings to those people who are most likely to result in a profitable relationship for the company. This was one of the first applications of scoring; Sears used it in the early 1950s to decide who it would send its mail-order catalogues to, and it is still widely used today. Lenders also try to grow their businesses using cross-sales, and use scores to assess which other products would be best suited for a customer, based upon demographic characteristics, existing account holdings, demographic details, and other information.

Retention

We also wish to know whether customers will keep their business with us, as the cost of account acquisition can be high. Churn scoring is used at the time of application to assess whether or not the newly acquired customer will stay long enough for the account to be profitable, especially where special offers are made. Customers may avail themselves of the offer, but not hang around afterwards—leaving the lender with costs, but no revenue. Lenders may also use attrition scoring to predict inactivity or closure of existing accounts, and then design strategies to keep them active and open.

Revenue

The final area of interest is revenue. Lenders wish to focus upon customers who will be profitable, and may use some simple modelling to assess whether the potential revenue will be sufficient for the lender to make a profit. This can be done by modelling profit or revenue directly, or by using the level of utilisation (balance, activity) as a surrogate. Profit scoring should be the ultimate goal, but it is affected by the number of decisions made along the way, in various parts of the business—limit increases, collections, marketing, etc. It is also difficult to implement, because of problems apportioning costs at the account level.

1.3 Why is credit scoring used?

What information consumes is rather obvious: it consumes the attention of its recipients. Hence, a wealth of information creates a poverty of attention, and a need to allocate that attention efficiently among the overabundance of information sources that might consume it.

Herbert Alexander Simon, cognitive psychologist (1916–2001), in Simon (1971:40–41)

In 1990, D.N. Chorafas suggested that in the following decade, returns from information technology investments would be driven by: (i) AI and networking, as opposed to data processing and stand-alone computing; and (ii) keeping ahead of competition, instead of traditional ROI measures. These trends were already well established at the time, and have continued into the twenty-first century. Credit scoring and decision automation fall into the AI camp. In the early days, there were significant competitive advantages for the early adopters, but today they are the standard in volume-driven retail credit markets. Even so, their adoption in new areas still causes significant tremors in the way lending is done.

Much has been written on the impact of decision automation on retail lending (consumer and small business), yet there are very few good summaries. Stanton (1999) provided one of the better ones, and noted the following:

Shift to high-tech—A structural change in the market saw a shift of volumes, values, and profits from traditional (relationship) to high-tech (transactional) lenders. Those who were quickest and best at updating their systems forced a shift of higher-risk applicants to lenders less able, increasing the latter's chances of adverse selection.

Organisational instability—Companies became unsettled as the new processes were implemented. There was a move away from bricks and mortar to travelling salespeople with laptops, from over-the-counter service to Internet banking, and from clerks shuffling snail-mail to PCs and email. As new ways of doing things evolve, the old ways become obsolete—people, organisational structures, and even whole companies.

Changed skills requirements—The investment is not once off, but ongoing and requires the development of a completely different set of skills, than required previously by lenders. The labourers that once shovelled the coal are replaced by technicians that watch the gauges and turn the valves.

Credit market growth—Credit scoring and decision automation have significantly lowered the cost of extending credit, improved lenders' capacity to service smaller loans, and increased service levels generally. Infrastructure investments are huge, and there has been much industry consolidation to gain economies of scale. This includes credit bureaux, as smaller operators were either swallowed by bigger fish, or beached (Furletti 2002).

Table 1.3. Impact on business areas

	Operations	Finance	Strategy	Human resources	Customer
For	Fast	Cost savings	Controllable	Improved human capital allocation	Lower costs
	Consistent	Reduced bad debts			Improved credit access
	Objective and defendable	Reduced security	Adaptable		More choices and channels
	Comprehensive				Mobility between lenders
	Greater reach				
Against	Complexity	Capital intensive	Backward-looking	Skills sensitive	Impersonal
	Data dependent	Lost information rents		Staff acceptance	Demands communication

It is not a foregone conclusion that every company or division will wish to automate its decision processes. Many lenders are still comfortably engaged in relationship lending, and believe that they have a competitive advantage in their niche. Unfortunately, lenders' ability to grow in these niches is limited, and any that wish to grow beyond a certain level, must consider automating their processes. The following section looks at the advantages and disadvantages of credit scoring and decision automation across the different business areas, as well as for the customer. Table 1.3 splits it out under five headings.

Operations—Speed, consistency, objectivity, and comprehensiveness of the decisions, allowing greater geographical reach, but at the cost of complexity.
Finance—Reduced bad debts, less collateral management, and manpower cost savings, but at the expense of a significant capital investment, and lost information rents.
Strategy—Improved control over strategies, allows monitoring at detailed level, adaptable to changing circumstances, but with the proviso that the models are backward-looking.
Human resources—Allows more productive staff allocation, but at the cost of change management, problems gaining staff acceptance, and scarcity of the new skills required.
Customer—Improved access to credit, lower cost, and greater mobility between lenders, but at the expense of losing a personal relationship with the bank.

1.3.1 How has it affected lenders?

The development of computers brought the industrial revolution to the financial services industry, as accounting and billing systems were automated. The advent of credit scoring then brought this industrialisation into the realm of decision-making. The greatest benefits have been achieved in application processing, but it is also being used in marketing, account management, collections and recoveries, and fraud. The benefits are greatest in high-volume low-value environments, and include:

Accuracy—Adverse selection is reduced, because of improved decision-making, and even where humans can provide better assessments, the difference is usually small.
Speed—Near instantaneous responses can be provided, for requests that used to take weeks.
Consistency—Service delivery can be standardised across vast branch networks, allowing greater control.
Objectivity—Decisions can be defended, where there is the possibility of unfair discrimination.
Responsiveness—Strategies can be updated quickly, in fast changing environments.
Intelligence—Improved analytics that allow lenders to tell what is happening in the business.
Reach—Loans can be made from a distance with little customer contact, whether through branch networks, or electronic channels.
Flexibility—Ability to forecast, price for risk, value portfolios, and even trade debt.
Lower costs—Operating costs are reduced when volumes are high.
Reduced collateral—Unsecured lending can be done, where consumers can borrow against their future income stream, without the need for collateral.

The term 'discrimination' usually has the negative connotation that results from personal prejudices and inappropriate assumptions. This is better embodied in the expression 'unfair discrimination'. Fair discrimination can result from full and unbiased assessments, especially those based upon the empirical analysis made possible by credit scoring.

The list of benefits is extensive, but this does not mean there are not problems. Some of the issues include:

Complexity—The systems are complex, and errors made in scorecards, strategy, or infrastructure can either result in large losses, or be difficult and expensive to rectify.
Change management—There are huge changes to the way business is done, which requires significant communication with both staff and customers.
Capital intensive—Large capital investments may be required to install the required infrastructure.
Backward-looking—It makes the assumption that the future will be like the past, and results in a situation analogous to 'driving a car by looking through the rear view mirror'.[2]
Skills sensitive—Specialist skills are required that are difficult to develop and retain.
Increased competition—Lenders may lose their information rents, especially where positive performance data is shared via the credit bureaux, as barriers to entry are removed.

The rest of this section looks at some of the above points in more detail, in particular some of the shortfalls of judgmental decisions, along with the objectivity and reach of automated processes, and the possibilities that are created for mass customisation.

---

[2] Dennis Ash, quoted in Burns and Ody (2004).

Shortfalls of judgmental decisions

When credit scoring was first proposed, many experienced underwriters scoffed at the idea that it could be as effective as human judgment. Even so, human decisions have been shown to have faults, as noted by Falkenstein et al. (2000) who quotes a number of different studies:

(i) People tend to overestimate their own knowledge (Alpert and Raiffa 1982).
(ii) Their confidence increases with the importance of the task, and they recall their successes more readily than their failures (Barber and Odean 1999).
(iii) The feedback for judgmental decisions is usually anecdotal, and not well structured (Nisbet et al. 1982), because the focus is on providing explanations, without a single default objective.
(iv) While good at identifying important factors, they are not able to integrate them optimally (Meehl 1954).
(v) When evaluating a set of 60 financial statements, of which half had defaulted, 43 loan officers (16 and 27 from small and large banks respectively) got 74 per cent correct, but it was still not as good as just using a simple 'liabilities to assets' ratio (Libby 1975).

Falkenstein (2002:182) also highlights a law of diminishing returns, in that 'after a certain amount of attention, the quality of the credit rating actually decreases'. The comment was made in the context of corporate credit, where it is greatest for large exposures—especially where company politics plays a role, and the views of many individuals are accommodated. The same concept also applies to retail credit, where too much effort is spent trying to analyse limited details.

Objective/defendable

Credit scoring's empirical nature has the advantage of minimising human bias. This is particularly important in a world where such accusations can cause huge public relations damage, or even lawsuits. Objective decision-making is something that would naturally be expected of people responsible for making loan decisions, but they will unfortunately never be immune to the use of stereotypes and generalisations; this is a fundamental part of the human condition.

The use of generalisations, assumptions, and associations is not something that is peculiar to humans. Animals use association to assess danger, the possibility of food, or restricted movement. In a 1995 experiment, Watanabe, Sakamoto, and Wakita showed that pigeons could distinguish between paintings by Monet and Picasso. Different paintings by the two artists were put side by side, with birdseed at the base of Picasso's. When new paintings were put out with no birdseed, the birds would automatically gravitate to the Picassos.

Almost immediately from birth, people start developing assumptions that guide their lives—especially when they encounter the same patterns of event and outcome, situation and result. These assumptions are not generated by personal experiences only—they also stem from what is learnt from personal communications (friends, family, workplace), the print media (books, newspapers, magazines, Internet), and the airwaves (radio and television). Although they truly are assumptions, when ill-formed, they are more often referred to as prejudices or biases.

In the age of fair-credit legislation and 'I'll sue' litigation, this poses problems. Thus, if decisions are driven by scores and policy rules, then only the overrides are left subject to human foibles. Even so, it is still possible for automated decisions to be biased. Credit scoring is a powerful tool, but also an imperfect tool, and there are limitations on what can be achieved. Problems may arise with any number of factors, including but not limited to:

Data—Problems occur with obtaining the data, or it changes significantly.
Scorecards—The model is biased, and provides substandard results.
Strategies—The strategies being employed are inappropriate for current conditions.
Implementation—The model has not been implemented correctly.
Use—The model is being used for purposes for which it was not intended.

Reach—new markets/products

The 'grass is always greener' proverb also applies to credit. Once a company has reached the grass (sic) ceiling in terms of market share, it starts looking for new territories to explore—up-market, down-market, or in new regions. There are two primary issues to consider when looking to new markets: (i) whether the organisation's *experience* is appropriate; and (ii) whether it has the appropriate *tools*. What worked for the rich may not work for the poor; what worked in Johannesburg may not work in Johore. Past experience may not translate well into new areas, and making inroads can be expensive, to the point of ruin. This applies to both relationship and transactional lending. Conservative strategies must be applied, unless the company wishes to be aggressive in order to gain market share. Perhaps the only advantage of relationship lending in this space, is that underwriters might be quicker to gain some insight that may not be readily apparent; or it may be possible to hire some who already have experience with the market.

Wiklund (2004) provides a comparison of the differences underwriting credit cards and motor vehicle finance. Underwriters would take into consideration applicant stability for both, but for the former would consider past performance on revolving credit, and for the latter past performance on instalment finance and whether or not the debt burden is acceptable.

In contrast, with credit scores, the change of territory violates the base assumption that the future will be like the past. The validity of any existing models will be in question in the new and unfamiliar territory. It may be possible to apply existing scorecards with a higher cut-off, but often the scorecards may not be at all valid. All is not lost though. Assuming that the lender can piggyback on existing processes to gather and store data, and that there are sufficient volumes, it should be possible to develop new models within two to three years.

According to Rhyne (2001), a Chilean consumer-finance company used a scorecard built on data from salaried people in Chile to score self-employed people in Bolivia. The mistake caused the company's bankruptcy, after losses of several million dollars.

Mass customisation

While credit scoring might seem like a 'mass production' one-size-fits-all approach, with little regard for individual needs, this need not be the case. As technology has improved, it has become increasingly possible to tailor products to customer needs. Stan Davis first presented the seemingly contradictory concept of 'mass customisation' in his visionary book 'Future Perfect', at a time (1987) when technology could not yet support it. The basic premise is that tailored products (shoes, dresses, credit cards) can be provided to customers at the same price as their off-the-shelf equivalents. According to Allen et al. (2003:24),

> Transactional lending's apparent lack of personal touch could also be overcome. Pine et al. (1995) point out the use of hard data could lead to "mass customization". Such customisation, however, relies on managing customer needs as opposed to managing products. Products could be priced individually just as Dell computers are individually created to suit individual needs, but are still able to turn the company a profit. Furthermore, relationship databases could be established. This would combine the informational benefits of relationship lending, with the cost efficiencies of transactional lending.

Thus, when a person sees a message coming up on an ATM screen offering a given product, it will be done based upon the assessment of a need for that particular customer.

1.3.2 How has it affected Joe and Jane Public?

Just as Henry Ford was able to produce automobiles that could be purchased by a mass market—including his own production-line workers—improvements to credit processes have expanded credit markets. At one stage the use of debt was viewed as something to be avoided, but today it is accepted that access to credit can be a major force for empowering people, allowing them access to goods and services that either make their lives easier, aid them in some productive endeavour, or both. As a result, there have been increasing demands to improve access to credit in underserved markets; both at home, and in developing and third-world countries. The combination of credit scoring, and huge investments in data and delivery systems, have drastically improved the lives of the general public:

Improved access to credit—Affordable credit is now being offered, to people that previously did not have access.

Lower rates—Improved risk assessment and lower processing costs have led to cheaper interest rates, and even if expensive, the rates are more likely to be appropriate.

More choice—There is a much greater range of options available to borrowers, with respect to products, delivery channels, and lenders.

Improved mobility—Data sharing via the credit bureau has given borrowers greater mobility between lenders, especially where people are geographically mobile.

Convenience—The impersonal nature often makes asking for loans easier, especially for those who would be cowed by having to enter the banking halls, and explain their situation to the bank manager and other bank staff.

According to Thomas (2000), the average adult in the USA is being scored once a week, either on new or existing accounts. This is not to say that everybody is happy. There are also a lot of concerns.

Impersonal—Many people dislike the idea of faceless machines having such power over their lives, especially when they believe they are truly creditworthy, and their personal circumstances are being ignored.

Disputes—Many lenders and information providers have inadequate mechanisms for borrowers to contest: (i) data held about them; and (ii) the decisions that are made.

Blacklisting—Once borrowers have an adverse credit record, it may take some time before the slate is cleared, which will either increase their cost of credit, or block access.

Privacy—There are public concerns about the extent of data available, and how it is used. In many environments, the data may only be used for the purpose for which it was collected.

Causation—Credit scoring models focus on correlations, and do not try to identify the underlying causes. Even so, both the industry and regulators have accepted that it is sufficient to rely upon correlations within available data, albeit with certain caveats.

Credit scoring was initially greeted with scepticism. Lenders doubted it could replace the judgment of experienced underwriters; customers disliked it because it stripped all emotion from the credit granting process; and regulators distrusted it because of potential impacts on personal privacy, or access to credit for the disadvantaged. Even so, today most of these issues have been addressed, and credit scoring has become accepted as an empirical, objective, and valid means of assessing credit risk.

Impersonal—advantages and disadvantages

A scene used in several silent movies was of a woman tied to railway tracks, in the path of an oncoming train. On at least one occasion, the villain was a nefarious handlebar-moustachioed banker, who hoped to marry the widow to get title to the family farm, but his advances were spurned. She was, of course, saved by a man in a white hat, but perhaps it would have been better had this banking relationship not been so . . . personal. Of course, even in those olden days the banker/client relationship was (almost) never so dramatic. Banks would accept deposits from local townsfolk and farmers, and lend out the money. If anybody in the community needed to borrow, they would approach the banker, who would assess the offered collateral, repayment capability, standing in the community, character, etc. He was up there with the town's mayor, sheriff, and church minister, as an upright member of the community, and was often more feared.

While the faceless nature of decision automation is usually seen as a negative, it also has its benefits. Less time and emotion are required for customers to enter the hallowed halls of the bank, face employees cap in hand, explain personal circumstances that are being kept even from close family, and assemble and submit documentation required to support the request. As a result, lenders are further improving borrowers' access to credit, by simplifying the application process, especially where the amount being extended is small. The convenience is greatest with facilities offered via ATMs and the Internet, as there is nobody to witness the request, and the only evidence is computer entries with the lender and the credit bureaux. As the size of the loan increases though—or in the absence of a credit history—stricter standards are applied, and the documentation requirements increase.

1.4 How has credit scoring affected credit provision?

According to Barron and Staten (2003:11), 'broader access to credit markets is widely recognized as the consequence of four simultaneous and interdependent factors', which relate to access to customer data, better and cheaper data processing, the use of statistical techniques for risk assessment, and changes to interest rate ceilings that make risk-based pricing more feasible. These should be restated along two dimensions: (i) *decision components*—data, risk assessment, decision-making, and delivery; (ii) *change areas*—practices, technology, and regulation. These are illustrated in Table 1.4, where the cells provide key factors or examples.

As can be seen, credit scoring is a practice that has been adopted for automating credit risk assessment. What are not so evident, however, are the interconnections. It is difficult to ascribe the accelerating growth of credit to any one factor. It can, however, probably be safely said that—low interest rates and a benign economy aside—the forces driving credit growth have been those relating to *data* and *automation*, with improved *risk assessment* and an empowering *legal environment* as supporting factors. The following sections do not cover all of the above, but look at:

Data—Increasing amounts of information, resulting from automation, data sharing, and empowering data privacy legislation.

Risk assessment—Use of credit scoring to drive transactional lending, as opposed to the relationship lending of old.

Decision-making—Lenders are no longer limited to the accept/reject decision, but can also use scores to set prices, and value portfolios for securitisation.

Process automation—Evolving technology that has made computers—including processors, data storage, and networks—faster, smaller, and cheaper.

Legislation—Fair-lending legislation and Basel II have promoted the use of credit scoring, while data privacy legislation and practices allow the sharing of data.

Table 1.4. Credit growth drivers

Decision component	Change area		
	Practices	Automation	Regulation
Data	Sharing	Collection	Privacy
Risk assessment	Credit scoring	Calculation	Fair credit
Decision-making	Risk-based pricing Securitisation	Decision agents	Rate ceilings
Delivery	Cross-sales	ATM, Internet	Distance lending

Data

A phenomenon that started during the latter half of the twentieth century was the growing power and sophistication of computers, along with improved capabilities for gathering, processing, storing, analysing, and communicating data. All of this led Varian (1998), to propose a Malthusian view of information.

In 1798, Thomas Malthus anonymously published his work, 'Essay on the Principle of Population', in which he proposed that at the then rapid rates of population growth in Europe, available food and other resources would soon be depleted. Although his prophecy was incorrect, Malthus was the first to highlight the relationship between overpopulation and misery. Population growth was geometric (doubling every 25 years in Europe at that time), while increases in subsistence (food production) were linear. Resources would be depleted unless population was checked by famine, war, pestilence, or birth control. Malthus's theories on the political economy were highly influential, and heavily influenced economic theory and social doctrine in the nineteenth century, as well as England's 1834 Poor Law Amendment Act. Interestingly, Malthus's ideas also found their way into Charles Dickens' portrayal of Ebenezer Scrooge in *A Chirstmas Carol*, where reference is made to 'the surplus population'.

Varian proposed that in modern information economies the growth in data is geometric, while the increase in consumption is linear ('Malthus's law of information'):

> This is ultimately due to the fact that our mental powers and time available to process information is constrained. This has the uncomfortable consequence that the fraction of the information produced that is actually consumed is asymptoting towards zero.

While Varian does have a point regarding individuals' ability to process information, he made the same mistake as Malthus. Malthus failed to consider changes to farming techniques that boosted food production, while Varian disregards tools used to get greater value out of data. This applies particularly to credit scoring, and the computer systems that are used to assemble, assess, decide, and deliver.

Varian (1998) also proposed the equivalent of 'Gresham's Law' for information. Sir Thomas Gresham was an English merchant financier during the Tudor era, who contended that bad money drives out good money—referring specifically to commodity currencies like those using gold or silver, whose value may be debased by changing the alloy or shaving the rims. In like fashion, Varian proposed that bad information can drive out good information, meaning that information that is both low cost and low quality can force out high-quality information—note Microsoft Encarta versus Encyclopaedia Britannica.

Risk assessment

In the early days, credit underwriters reviewed the applicant's financial and other details manually, but credit scoring provided a tool to condense information into a single number. This has been an incredible tool for empowering lenders to make decisions, and has provided them with much greater control over the business. Figure 1.2 is meant to illustrate the development of a scorecard using historical information, both on observed characteristics and subsequent outcomes, and the model's subsequent use as part of a business process.

The primary goal is to provide a tool for ranking accounts according to the relative odds of them being 'good' or 'bad' at the end of the period, where good is the desired outcome, and bad is to be avoided. The major assumption is that *future will be like the past*, or at least sufficient enough for the models to provide value. Unfortunately, credit scoring: (i) is highly *backward-looking* and not able to provide a forward view; (ii) is unable to assess *exogenous data* (not provided as part of the system); and (iii) is not well suited for assessing *rare but severe events*.

As a result, credit scoring is not a law unto itself and cannot be used in isolation. There are two possibilities, based upon the level of automation that can be achieved: (i) *high-volume low-value environments*—policies can be put in place to cover not only risk, but also statutory, strategy, and other issues, and underwriters may still have the final say when system decisions are disputed by customers; and (ii) *other environments*—where amounts at risk or potential profits are high, scores may be provided to underwriters as decision aids.

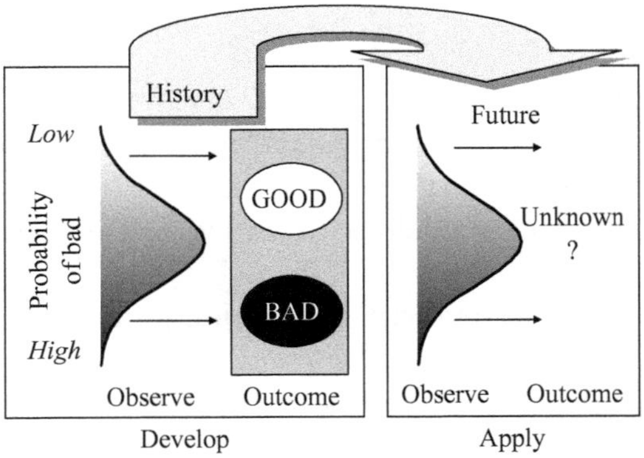

Figure 1.2. Historical data use.

Decision rules

Credit scores provide little value by themselves. They have to be combined with strategies (rule sets) that are used to guide decisions. Initially, the scores focussed almost exclusively on providing an accept/reject decision, and much effort was expended upon choosing an appropriate cut-off. Over time, however, lenders have become more sophisticated, not only in how, but also where the scores are used. Application scores are being used for risk-based pricing. Behavioural scores are used for limit setting, over-limit management, and pay/no pay decisions. Collections scores are used to drive automatic diallers, and decisions on whether to consider the client rehabilitated, or dead (figuratively). Fraud scores are used to identify cases to be referred for further investigation. And propensity scores are used in marketing, to guide who should receive the mail-shot, and who not, or to set the terms offered.

Decisions provided by credit scoring are not always the final word. Lenders' policies and/or *staff can override* them, and even then, *customers can contest*. Care must be taken here, because: (i) policies may undermine the scorecards; (ii) loan officers are usually limited in their ability to assess large quantities of information; and (iii) customers often have a better understanding of their own circumstances than lenders. In any event, disputes and overrides can provide a vital feedback mechanism for the people behind the computers, to find out what is actually happening in the field—and there should be a significant investment in override monitoring.

Something that must be stressed here is that these changes have given lenders greater flexibility, and they are becoming more adept at using the tools in an increasing number of ways, especially: (i) improved account management and collections; (ii) risk-based pricing and securitisation of consumer and small-business loan portfolios. At the same time, changes to interest-rate ceilings have made lending to previously underserved markets feasible (sub-prime, MicroFinance).

Lenders must take care when securitising portfolios. According to Allen et al. (2003), 'Banks are perceived as having superior information concerning the clients to whom they lend. Dahiya et al. (2003) show that the market reacts negatively when a bank sells a loan in its portfolio. The perception is well founded: firms whose loans are sold have a higher probability of bankruptcy than firms that do not. This special relationship is lost as soon as the loans are treated as transactional retail exposures'.

Process automation

Technology has been evolving at a fantastic rate. Everything is becoming faster, smaller, and cheaper. This applies to every decision component, at every stage of the credit risk management process. It also applies to delivery mechanisms, both for marketing (cross-sales, target marketing) and the channels used to deliver decisions and products (networking, Internet, ATMs, distance lending).

Credit scoring is primarily associated with application processing, where it has been the key driver behind decision automation. Many companies' first experience—especially during the 1960s and 1970s—was highly manual, with staff members filling in scorecards, tallying the results, and applying the cut-off set by head office. Today the norm is automation, to the maximum extent allowed by current technology. Relationship lending has given way to transactional lending, especially by ever larger banks, which use distance lending to reach customers with increased job and geographical mobility. This does not mean that the underwriter's role has been done away with: (i) lenders limit automation, where the volumes are low, or the relationship can be used to competitive advantage in a niche market; (ii) customers may dispute system generated decisions; and (iii) there are many instances, where scores are known to be insufficient by themselves.

Regulation

Lenders and consumers have been the main players driving credit growth, but legislators have also been playing an increasing role (see Module G, Regulatory Environment). Changes have been implemented that both facilitate access to credit, and provide guidelines for participants' practices. In general, there has been an increasing movement towards the use of best practice, good governance, business ethics, and social responsibility. There is also a compliance hierarchy, encompassing statutes, legal precedent, industry codes of practice, company policies and procedures, and unwritten codes used by businesses.

The types of legislation that either affect, or are affected by, credit scoring are: (i) *data privacy*, which sets out limitations relating to manner of collection, data relevance, data quality, its use, information disclosure, subjects' rights, and data security; (ii) *anti-discrimination*, to prevent prejudice on the basis of race, colour, language, religion, national origin, gender, etc.; (iii) *fair lending*, to guard against predatory and irresponsible lending, and instead promote responsible lending; (iv) *capital adequacy*, to ensure that banks hold sufficient capital to protect against unforeseen losses; and (v) *know your customer*, to guard against criminal and terrorist activities, which also helps to protect against fraud.

1.5 Summary

This chapter has provided an overview of the broader theoretical aspects of credit scoring, including what motivates its use, where it fits, how it has affected lending, and how the public benefits. The lending of money is something that requires trust and, as in any instance where two parties contract, there is the potential for *adverse selection* (making the wrong choice), and *moral hazard* (change of behaviour once the deal is done). Lenders often rely on collateral, and guarantees, to enhance that trust, but improvements in data and automation have shifted the focus onto information. *Information asymmetries* (differences in information) that give borrowers an advantage over lenders will always exist, but this advantage can be reduced by accessing data from outside sources. Interest charges will also be reduced, as banks then have less scope to earn *information rents* (extra benefit that can be obtained by exploiting information not available to competitors), but this is offset by increased volumes. It also allows borrowers greater flexibility to move between lenders, and geographically.

Credit scoring is a tool used to collapse the wealth of available data into something more manageable. It came about as lenders applied predictive statistics to historical data, and derived models of customers' propensity to behave in certain ways: to repay a loan (*risk*); to respond to an offer (*response*); to move their business elsewhere (*retention*); and to use the product in a fashion that would be profitable for the lender (*revenue*). Its greatest benefit has been for assessing credit risk, especially *loss probabilities*—and more recently *loss severity*. When it was first proposed, its primary use was for application processing, but it has become increasingly used throughout the CRMC.

While credit scoring can provide more accurate decisions, that is only one part of the story. Scorecards often provide answers similar to humans, but lack human foibles, and provide greater speed and consistency. As a result, larger transactional lenders have greater flexibility and reach than they have ever known—especially banks. This does not mean that credit scores are a panacea though. They suffer because of huge data demands, complex systems, skills requirements, and staff and customer acceptance. Many lenders—especially smaller lenders—still opt to avoid the pain, and stick with traditional relationship lending where they rely upon personal contact.

While lenders have seen significant benefits, so too have the customers that they serve, as competition has meant that most of the benefits are being passed on. Customers have improved: (i) access to *affordable credit*; (ii) *choice*, from a range of products; (iii) *mobility*, as it has become easier to move relationships and credit histories between lenders; and (iv) *convenience*, as the task of applying has become easier, and often less embarrassing where the answer is 'No!' Once again though, there are also concerns: (i) it is very *impersonal*; (ii) many lenders do not handle *disputes* well; (iii) negative public perception about *blacklisting*; (iv) *data privacy* issues; and (v) credit scoring cannot identify causes, but only uses *correlations*.

In spite of the concerns, there has been a huge amount of credit market growth. Credit scoring has played a role, as it has allowed lenders to improve their risk assessment processes, and venture into previously underserved markets. At the same time, the amount and quality of available data has improved substantially, especially as a result of data sharing arrangements via the credit bureaux. Increasing automation has reduced administrative costs, not only for data collection, but also risk assessment, decisioning, and delivery. And finally, changes to legislation have empowered lenders, by giving them access to information, and the ability to price for risk.

2 Credit micro-histories

Most people think of history in terms of the Magna Carta, Christopher Columbus, the Second World War, or other events that define human existence. There is also a new and increasingly popular genre called micro-histories, which chronicle seemingly inconsequential events, which may even have been part of a temporary phenomenon. The history of credit scoring is a micro-history, but the phenomenon is far from temporary (Table 2.1). It is treated under four headings:

(i) **History of credit**—Evolution of lending money, from ancient Babylon to today.
(ii) **History of credit scoring**—Use of statistics within credit decision-making, and the origins of Fair Isaac (FI) and Experian-Scorex.
(iii) **History of credit bureaux**—Origins of companies like Equifax, Experian, and TransUnion.
(iv) **History of credit rating agencies**—Origins of Moody's Investors Services (MISs), Standard & Poor's (S&P), and Fitch IBCA.

Table 2.1. History of credit, scoring, reporting, rating

Date	Event
2000 BC	First use of credit in Assyria, Babylon, and Egypt.
1100s	First pawnshops in Europe established by charitable institutions, and by 1350 they were being run as commercial concerns.
1536	Charging of interest deemed acceptable by the Protestant church.
1730	First advertisement for credit placed by Christopher Thornton of Southwark, London who offered furniture that could be paid off weekly.
1780s	First use of cheques in England.
1803	First consumer reports by Mutual Communications Society in London.
1832	First publication of the *American Railroad Journal*.
1841	Mercantile Agency is first American credit reporting agency.
1849	Harrod's established as one of the world's first department stores.
1851	First use of credit ratings for trade creditors by John M. Bradstreet.
1856	Singer Sewing Machines offers consumer credit.
1862	Poor's Publishing publishes *Manual of the Railroads of the United States*.
1869	First American consumer bureau is Retailers Commercial Agency (RCA) in Brooklyn.
1886	Sears established, and launches its catalogue in 1893.

Table 2.1. *Continued*

Date	Event
1906	National Association of Retail Credit Agencies formed in the USA.
1909	John M. Moody publishes first credit rating grades for publicly traded bonds.
1913	Henry Ford uses production lines to produce affordable automobiles.
1927	Establishment of Schufa Holdings AG, first credit bureau in Germany.
1934	First public credit registry (PCR) established in Germany.
1936	R.A. Fisher's use of statistical techniques to discriminate between iris species.
1941	David Durand writes report, suggesting statistics can assist credit decisions.
194?	Henry Wells uses credit scoring at Spiegel Inc.
1950	Diners Club and American Express launch first charge cards.
1950s	Sears uses propensity scorecards for catalogue mailings.
1956	FI consultancy established in California, USA.
1958	First use of application scoring by American Investments.
1960s	Widespread adoption of credit scoring by credit card companies.
1966	Credit Data Corp. becomes first automated credit bureau.
1970	Fair Credit Reporting Act governs credit bureaux.
1974	Equal Credit Opportunity Act causes widespread adoption of credit scoring.
1975	FI implements first behavioural scoring system for Wells Fargo.
1978	Stannic implements first vehicle finance scorecards in South Africa.
1982	CCN offers Credit Account Information Sharing (CAIS), its consumer credit bureau service.
1984	FI develops first bureau scores used for pre-screening.
1987	MDS develops first bureau scores used for bankruptcy prediction.
1995	Mortgage securitisers Freddy Mac and Fannie Mae adopt credit scoring.
2000	Moody's KMV introduces RiskCalc for financial ratio scoring (FRS).
2000s	Basel II implemented by many banks.

2.1 History of credit[1]

Humans are social animals, inclined by nature to association with others of their own species and to community life, perhaps with the exception of a few monks, hermits, and ornery mountain men. A by-product of this has been the further evolution from Homo sapiens to Homo economicus—social animals that barter and trade items that assist them in their ongoing survival, or just make life a bit more enjoyable. Related to this is the economic concept of utility, whereby more value is associated with use of an item today than tomorrow. Although there is no documentation in this regard, it is likely that Homo economicus figured out how to enjoy

[1] Internet References:
http://www.moneycontrol.com/cards/cardsinfo/credith.php
http://www.didyouknow.cd/creditcards.htm
http://www.bangladeshinfo.com/business/credit_card01.php

goods prior to paying for them very early on, as illustrated by this hypothetical pre-history:

Credit in pre-history—a possibility?

Cro-Magnon tribes traded goods in the days before money. The Uburs have needed skins for winter warmth, but have not been able to gather enough nuts and wild grains to trade with the Azars. The Azars think that the results of their sabre-tooth tiger kills are worth more than what the Uburs have to offer, and threaten to trade instead with the Ebo-Agu. To counter this, the Uburs state that they will provide two times as much food for each skin, if only they can provide the food next year. The Azars agree, but take the Uburs's chief's daughter as security.

It has several features that differentiate it from the normal bartering situation: (i) the time period between the delivery and payment; (ii) the amount of the compensation is greater; and (iii) there is collateral pledged to ensure repayment. Sounds like any normal lending transaction, except with no money. Money is not a precondition for the charging of interest, which can be charged on any commodity.

2.1.1 Ancient history

The economic aspect of mankind's nature has driven the development of civilisations. The first systems of writing were developed in Sumeria (3100–2350 BC) following on the establishment of the first cities and development of the wheel. Writing was used to record agreements,[2] laws, commandments,[3] and oral histories.[4] Nobody can say with certainty what provided the greatest impetus, but given that records of commercial transactions provide so much of the early evidence of writing, the role of business activity must have been great. Knowledge of reading and writing was the way to power and wealth. Writing was also the catalyst that allowed the development of states, as previously there was no means of efficient communications to send out commands to the provinces—unless the king wished to travel himself, or rely upon the memory of some horseman.

It is also this period that provides some of the earliest evidence of lending transactions. According to the *Encyclopaedia Britanica*, the first indication of anything resembling a bank in the modern sense is the text of a Babylonian document dating from about 2000 BC:

> The shekels of silver have been borrowed by Mas-Schamach, the son of Adadrimeni, from the Sun-priestess Amat-Schamach, daughter of Warad-Enlil. He will pay the Sun-God's interest. At the time of the harvest he will pay back the sum and the interest upon it.

---

[2] Goody (1977). Jack Goody is an anthropologist who emphasises writing's very practical early uses, such as the recording of sales, loans, property, and so on.

[3] Wells H.G. (1922). *A Short History of the World.*

[4] The 'Epic of Gilgamesh', the adventurous story of a Babylonian king in 2700 BC, is a collection of tales that were passed down orally for hundreds of years before being collected and written down by a storyteller.

The mention of a priestess is strange, but it seems that these early banks were not private initiatives, but an incidental service run by the cult—a wealthy and organised institution that dominated the society. Put into modern terms, this was a piece of commercial paper, and the priestess an accredited agent for the institution.

Although these early bankers might have seen themselves as sitting next to God, their time in this position was extremely limited. For most of history, the life of a moneylender has not been an easy one. Western culture is filled with negative references—from Jesus chasing the merchants and moneylenders out of the temple at Jerusalem, to Shakespeare's portrayal of Shylock in the Merchant of Venice.

Business ethics has been an oxymoron for most of human history. Clark J. (unk.) notes that Aristotle [384–322 BC] made a distinction between household trade (*oikonomikos*) that was essential for societal functioning, and trade for profit (*chrematisike*) that was devoid of virtue. This view only started changing with the rise of the Calvinists and English Puritans, whose views provided the framework for Adam Smith's *Wealth of Nations* in 1776. Smith has become a cult icon for 'greed is good' economists, while his ethical and moral leanings—evidenced in his *The Theory of Moral Sentiments*, published in 1759—are ignored.

The major contributing factors were: (i) the extremely high interest rates that were charged; and (ii) draconian penalties that were imposed for non-payment, because it was associated with theft or fraud. According to Gratzer (2006), under the Twelve Tables of 451 BC it was Roman custom for borrowers to commit both life and property to their creditors; not only of self, but also family. If there were multiple creditors, they had 'the right to dismember the debtor's body'. The *lex poetelia* of 326 BC brought some liberalisation, by abolishing creditors' right to 'ill-treat, kill, or sell the debtor and his family as slaves'. Non-payment was criminalised though, and a defaulter could be kept in 'the creditor's private prison in a kind of debtor's servitude, but he regained his freedom if the debt was settled'. Some 200 years later, *lex julia* (under either Caesar or Augustus), allowed debtors to avoid disgrace through voluntary bankruptcy and cession of assets (*cessio bonorum*), but abuse caused it to be limited to legitimate causes (fire, shipwreck, theft, etc.), while failure to disclose assets would invoke even harsher penalties (prison and debtor's servitude). Thus, Roman legislation provided for: (i) equality of losses between creditors; (ii) separation in law of person and property; and (iii) a distinction between honest and dishonest debtors. With the fall of empire, this legal framework degenerated and merged with the Germanic invaders' common-law traditions, which treated bankrupts as severely as those of early Rome. Their custom even allowed the keeping of wives and children as hostages. The process of liberalisation repeated itself, but this time it took more than a millennium.

Given such harsh penalties for non-payment, it is understandable that lending was condemned by many ethical teachers, including Buddha, Jesus, Mohammed, Plato, Aristotle, and St Thomas Aquinas (see Table 2.2).[5] It was nonetheless practised in many societies that instead put limitations on the rates that could be charged (see Table 2.3). In biblical times, it was done by businessmen who charged for 'changing' money, in order to circumvent religious ethic.

[5] www.macroknow.com/books/philosophy/usury.htm

Table 2.2. Usury—ancient prohibitions

There must be no lending at interest because it will be quite in order for the borrower to refuse absolutely to return both interest and principal. Plato, The Laws.
The trade of the petty usurer is hated with most reason: it makes a profit from currency itself, instead of making it from the process which currency was meant to serve. Aristotle, Politics.
Take thou no usury of him: but fear God . . . Thou shalt not give him thy money upon usury, nor lend him thy victuals for increase. Leviticus 25:36–37.
To take interest for money lent is unjust in itself, because this is to sell what does not exist, and this evidently leads to inequality, which is contrary to justice. St Thomas Aquinas, On Law, Morality, and Politics.
Those that live in usury shall rise up before God like men whom Satan has demented by his touch; for they claim that trading is no different from usury. Koran, The Cow 2:275.

Table 2.3. Usury—ancient limitations

Code of Hammurabi (2130–2188 BC)	33%
Hindoo Law—Damdupat	Capital
Early Roman	Prohibited
Constantine (reign 306–337)	12½%
Justinian (reign 517–565)	4–8%
Charlemagne (in 806)	Prohibited

It was condemned by the early Christian church in the fifth century, made a criminal offence by Charlemagne in the ninth century, and suffered an anti-usury movement that caused it to be banned by Pope Clement V in 1311.

2.1.2 Middle ages to nineteenth century

Most early credit extension was done for the purposes of trade, and the extent to which it was used depended upon the economy of the day. In the 1100s, there were large trade fairs across Europe, and people travelled long distances to purchase spices, materials, weapons, and other goods. Traders moved from fair to fair, and often used credit to buy goods in one place for sale in the next. In Italy, trade agents at the fairs even made the task easier, by recording transaction details—purchases, sales, and repayments. The bill of exchange was developed, probably by Florentine Jews, as a means of transferring funds without the risk or expense of moving gold or other precious items. These were the first commercial credit instruments, which by the thirteenth century were widely used not only for short-term credit transactions, but also foreign exchange. Usury laws were circumvented, because the interest was hidden in the handling fees. There was also an active trade in the bills, which circulated almost like paper money.

The 1100s also saw the establishment of the first pawnshops in Europe, but—as difficult as it is to believe—these were charitable institutions that did not charge interest. Within a few years, the profit potential behind this popular institution became apparent, and some people started charging. By 1350, commercial pawnshops were being established in various European countries. The interest rates were often high, and in a society dominated by the church, this fuelled a controversy regarding the morality of charging interest.

Although banned from 1311, contradictions in the church's arguments and loopholes in legislation were exploited to continue the practice, and over time the demands of economic growth caused a pro-usury movement. By 1516 the idea of an institution charging interest was widely accepted, and by 1536 it was deemed acceptable by the Protestant church. John Calvin deemed interest sinful only if it caused personal harm, which did not include business loans.

Bankruptcy legislation

Mediaeval bankruptcy legislation first emerged in the twelfth-century Italian city states, where it focused on the protection of creditors financing trade; bankrupts were subjected to punishments including incarceration, torture, servitude, and death. According to TBYA (2006) and di Martino (2002), such treatment was the norm for most early European legislation, including England's first official bankruptcy laws in 1542 (during the reign of Henry VIII), and subsequent acts in 1571, 1604, and 1624. It was only in 1705 that English legislation started taking a more lenient stance, recognising insolvents as innocent victims of a malevolent economy, and allowing debts to be discharged. In contrast, Napoleon's commercial code of 1807 demonised bankruptcy even further, and influenced its treatment not only in France, but also Spain, Portugal, and Italy. The United States first implemented/repealed bankruptcy legislation in: (a) 1800/1803, in response to land speculation; (b) 1841/1843, after the financial panics of 1837 and 1839; and (c) 1867/1878, after the upsets of the Civil War. Each of these focused more on rehabilitation, and catered for some forgiveness of debts, with the 1867 law providing some protection for corporations. The Bankruptcy Act of 1898 went further, to provide distressed companies protection from creditors.

According to di Martino (2002), bankruptcy legislation can be rated by its ability to: (i) reduce default risk; and (ii) maximise the value of assets available to creditors in the event of default. Under the Anglo-Saxon model, the discharge option is only available to non-fraudulent bankrupts, and allows them to resume economic activity more quickly. Claims are only against current assets and not future income, and some allowance is made for keeping a portion of assets, which can provide seed capital for new ventures. In contrast, the Napoleonic model only allows for the resumption of entrepreneurial activity after full repayment of debts, and there is greater motivation for defaulting borrowers to mask the risk and hide assets. This, at least theoretically, increases the economic costs of insolvency, reduces the number and quality of honest entrepreneurs, and reduces economic growth.

Mortgages

The pledge of property for the repayment of debt is as old as debt itself, and was already catered for in ancient Roman law. In twelfth-century England, land was gaged to Jewish moneylenders by noblemen raising money for the crusades, and pilgrims to Jerusalem would cede property to the Knights Templar in return for letters of credit to aid their passage. According to Maurer

(2006), the jurist Ranulf de Glanvill, who was killed at Acre in 1190 while serving under King Richard, wrote a treatise called *Tractatus de legibus et consuetudinibus regni Angliae* [treatise on the laws and customs of the realm of England] between 1187 and 1189. In it, he made a distinction between two types of gages, depending upon whether the rents and issues of the land were used to reduce the debt. These later became known as: (i) *vifgage*, or 'live pledge' (*vivum vadium*), where the land plays a role, and is allowed; and (ii) *mortgage*, a 'dead pledge' (*mortuum vadium)*, where the land plays no role, which is both usurious and immoral.

Vifgages were effectively a form of lease, which allowed lending without falling foul of usury laws. Interpretation of the agreement was strict though, allowing for forfeiture of property if repayment was just one day late, yet debtors were still responsible for the debt. Not unsurprisingly, debtors turned to the 'court of equity' (or chancery court), which based decisions on equity instead of law, begging for a grace period to repay debt, and any additional costs and interest ('equity of redemption'). Such petitions were often granted, which made it almost impossible for lenders to dispose of gaged assets. Courts also recognised that land pledges were just collateral, restricted lenders' rights to interest on the loan, and required them to account for lands' income while in their possession. Lenders thus had poor protection in law, and became loath to lend. Mortgages had the advantage of expunging the pledge once the debt was repaid, while providing lenders full title to the land if it was not. Borrowers also did not expose themselves to the harsh penalties inflicted on bankrupts. By the fourteenth century, mortgages had become the norm, but vifgages were still in use as late as the seventeenth century.

The word 'mortgage' initially meant 'death pledge'. It first appeared in old French in 1287, and in Middle English in 1390, when it appeared in John Gower's *Confessio Amantis* (The Lover's Confession), a 33,000 line Middle English poem written at the request of Richard II between 1386 and 1390. There it was used in the sense of a pledge in marriage, '*Forthi scholde every good man knowe; And thenke, hou that in mariage; His trouthe plight lith in morgage; Which if he breke, it is falshode.*' With respect to credit, most scholarly works make reference to the explanation provided by Sir Edward Coke:

> It seemeth that the cause why it is called mortgage is, for that it is doubtful whether the Feoffor [holder of freehold land] will pay at the day limited such summe or not, & if he doth not pay, then the Land which is put in pledge vpon condition for the payment of the money, is taken from him for euer, and so dead to him vpon condition, &c. And if he doth pay the money, then the pledge is dead as to the Tenant [mortgagee], &c.
>
> *English jurist Sir Edward Coke* (1552–1634), *in 'The First Part of the Institutes of the Lawes of England', 1628.*

Simply stated, the property is dead to the mortgager, if the debt is not repaid; and the pledge is dead to the mortgagee, if it is. This is different to the earlier explanation, but just as credible.

Overdrafts and cheques

Prior to the 1800s, only the rich had access to unsecured credit from institutions. The Royal Bank of Scotland is said to have invented the overdraft, when in 1728—one year after its founding—it allowed a merchant named William Hog to withdraw £1,000 more than he had in his account. The bank saw further opportunities, and began offering a 'cash-credit' service to its wealthier customers. The practice soon spread to other banks in Scotland, and then

England. The interest charged by most of them was the maximum 5 per cent allowed by law. Indeed, English law prohibited charging rates higher than 5 per cent until 1832, when significant amounts of capital had to be mobilised to finance the industrial revolution.

Overdraft usage was given a further push by the advent of cheques—a variation on the bill of exchange—being used against the accounts. First used in England during the 1780s, they were slow to become commonly accepted, and did not come into general use until after 1875. Prior to this, many obligations were settled by one part cash, and two parts bills of exchange. During the next quarter century, the use of cheques grew as the number of bank branches grew, and as traders realised the ease with which they could make payments between each other, no matter the location. By the early twentieth century, cheques had supplanted cash for most payments, other than the purchase of property, payment of wages, and minor household expenses (Thomson 1926). The concept did not catch on in the USA at that time, which instead developed finance houses, to meet growing demand for consumer credit (Lewis 1992).

Merchants, department stores, and mail order

While the overdraft could be viewed as a combined effort of the upper-middle class and their bankers, there was very little available for the less fortunate. There were some local shops where people could buy things on the slate or tab, but these were small operations that were doing a local service, perhaps with no interest. There are a few exceptions to this. One of the first consumables offered on credit was clothing. From the 1700s to the early 1900s English tallymen would sell clothes and be repaid weekly, and keep a tally of the repayments on a wooden stick—one side of the stick representing the debt and the other the repayments.

> According to Birch (2002), tally sticks were first used in Norman England, as a means of keeping track of whether local sheriffs had submitted sufficient taxes to the king. Assessments were notched into wooden twigs that were split in two, so each 'had a durable record', making it easy for the Exchequer to reconcile. By the time of Henry II (1154–1189), tallies were being discounted to raise funds, and a market developed. The Exchequer promoted the market, as tallies were cheaper for merchants to transport than other transaction media. Tally sticks continued to be used until 1826, and when those remaining were burnt in 1834, the Houses of Parliament were accidentally burnt to the ground.

The year 1730 saw the first newspaper advertisement for consumer credit placed by furniture retailer Christopher Thornton, in Southwark, London. It looked much like modern 'buy now pay later' adverts placed by furniture stores, reading 'rooms may be furnished with chests of drawers or looking glasses at any price, paying them weekly, as we shall agree'. At the time, interest rates were low, and the companies made their profits from the sale of goods. Sellers often waited for significant periods of time before being repaid. The Wedgwood furniture company in London waited for three years to be repaid by one buyer, but there is not much motivation to repay, when interest rates are 2 per cent.

In England, the United States, and other countries, the industrial revolution fuelled the growth of personal wealth, and a middle class with more money to spend. They could now afford items that were previously out of reach, and many shops were quick to offer credit arrangements for the new sewing machine, potbelly stove, or gramophone. This growth also fuelled changes in the way people shopped and bought. The mid-1800s saw the establishment of the first department stores (including Jenners of Edinburgh in 1838 and Harrod's of London in 1849) and the first documented cases of a psychological disorder called kleptomania—what is today better known as shoplifting. In 1856, the Singer Sewing Machine Company was one of the first to offer consumer credit in the USA (Tamilia 2002), and by the late 1800s, finance houses were growing the market. Another innovation of the late 1800s was mail order. Montgomery Ward (MW) was established in 1872 in Chicago, and by 1904 was mailing three million 4-pound catalogues. Sears was founded in 1886 as a watch retailer, published its first catalogue in 1893, and opened its first store in 1925.

Most lending during this period was personal, relying upon character assessments of the people borrowing. In small communities it might have been possible to investigate a new customer, but where trade and commercial activities were involved it became difficult. The mid-1800s thus saw the rise of credit reporting agencies (see Section 2.3).

Carruthers and Cohen (2002) highlight how small communities can be geographically distributed, as shown by Boyce's (1995) analysis of Britain's nineteenth-century shipping industry. Creditors' reputations influenced their ability to secure business, and credit.

2.1.3 Twentieth century

The late nineteenth century saw the invention of the horseless carriage; a noisy toy, only affordable to the rich, which scared horses and children alike. Most companies saw it as an upmarket product, but by 1913, Henry Ford's production lines were producing automobiles affordable for the man on the street, including his own employees. Even so, it was a major investment that most people had to buy on credit. This provided even more growth for the finance houses, as banks did not think it wise to lend for movable assets that, unlike houses, could disappear into the sunset (Lewis 1992).

There was a time when most consumer goods were made out of metal. During the twentieth century, however, plastics made significant inroads, to replace metal in a variety of consumer

Table 2.4. Genealogies and milestones—credit cards

Date	Event
1914	Western Union introduces embossed metal plate first charge card in the United States.
1920s	Introduction of 'shopper's plates', early version of modern store cards.
1950	Diners Club and American Express launch first charge cards.
1951	Diners Club launches first credit card in New York city.
1960	Bank Americard established, later to become Visa.
1966	Master Charge established, later to become MasterCard.
1966	Barclaycard established in the United Kingdom.

products, especially in motor vehicles during the 1960s. Believe it or not, this also applies to what today is called 'plastic money'. The forerunner of the modern day charge card was an embossed metal plate, first issued in 1914 by Western Union (the American company best known today from old cowboy movies, where it is delivering telegraphs, transferring money, or being targeted by desperados robbing a train). This metal plate was issued only to preferred customers, who paid an annual subscription fee. By the 1920s, some American department stores were also issuing 'shopper's plates', early versions of today's retail store cards. These could only be used in the issuing store, and any purchases had to be repaid at the end of the month.

The growth of credit also brought with it demands for regulation. It was already being done indirectly through banking regulation, but the early 1900s saw the introduction of direct legislation to protect and inform the public regarding credit transactions. In England, the Moneylenders Act of 1927 governed advertising by and licensing of moneylenders. In 1932, Scotland introduced laws to govern hire purchase agreements, and the Hire Purchase Act of 1938 extended this to England. Parts of this legislation were highly prejudicial to finance companies.

Credit cards—1950s–1990s

The end of the Second World War saw the start of a post-war boom in North America and Europe, which brought with it significant urbanisation and movements of people between cities. It also saw a growth in the demand for consumer credit, and significant amounts of excess capital, that made further investments in unsecured credit possible, by both banks and finance houses (Lewis 1992).

The period also saw the advent of a new payment medium. In 1950, Diners Club and American Express both introduced the first charge cards. In 1951, Diners Cub issued its first credit card to 200 customers, who were only able to use it at 27 restaurants in New York. These early cards were all swipe cards, where transaction slips had to be processed. It was only in 1970, that standards for magnetic strips were agreed upon, and all sorts of possibilities opened up for plastic money.

In 1960, the Bank of America issued its Bank Americard, which in 1977 was renamed to Visa. Initially, they only issued cards under their own name, but in 1966 they started licensing them to other banks around the world. Many oil companies also issued cards to promote petroleum sales during the early 1960s. At first, it was possible to send unsolicited cards in the mail, but this practice was outlawed because of card theft, and the problems it caused for individuals (Lewis 1992). In 1966, Barclays Bank in England established Barclaycard. In 1967, four California banks came together to form the Western State Bankcard Association, and offered the Master Charge card, which was renamed MasterCard in 1979.

Most credit card lending was initially done at a fixed interest rate. It was only during the late 1970s, that card issuers started to recognise a need for differential pricing. According to Barron and Staten (2003), annual charges to penalise non-revolvers were started during the 1980s, and reduced the pressure on interest rates charged for credit card debt. Further charges, like late-payment, cash-advance, and over-limit fees, were added during the same decade—which was the start of a 'user pays' trend, to halt cross-subsidisation of different services via the interest charge. Tiered pricing soon followed, first by American Express in 1991, with charge-volume based pricing. Risk-based pricing was introduced as risk assessment capabilities improved. The end result was a massive reduction in credit card interest rates; between 1990 and 1992 the proportion of bank-card issuers charging over 18 per cent dropped from 70 to 44 per cent.

Legislation—1970s onwards

During the 1960s and 1970s, federal policy in the United States encouraged lenders to make credit available to a broader population, especially poorer households. This combined with improved risk assessment capabilities to increase the proportion of households with access to credit from 55 to 74 per cent between 1956 and 1998 (Barron and Staten 2003). The increasing use of credit brought with it public concerns about both how information was obtained, and how it was assessed. The Fair Credit Reporting Act of 1970 (FCRA) set forth rules for American credit bureaux, to ensure data privacy and accuracy.[6] This also limited the scope of the information that could be used to credit related information, including positive information, and stopped 'newspaper clipping' (Furletti 2002). The effect was to speed industry consolidation, as many small operators could not justify the required systems.

The Equal Credit Opportunity Act of 1974 (ECOA) went further, to prohibit unfair discrimination in the granting of credit. Every credit decision is discriminatory in some fashion, but now the decisions had to be entirely objective, and not include any human bias. In effect, this practically banned judgmental credit assessments for consumer credit and guaranteed anybody in the business of developing application risk scorecards a salary for life.

In 1996, the FCRA was amended to protect consumers further. Credit bureaux were made liable for misinformation, and allowed consumers to sue anybody obtaining a credit report 'without a permissible purpose'. Time limits were also set for bureaux to investigate erroneous information; potential creditors were allowed to screen customers before or after making offers; and banks were allowed to share information, as long as they could stop it upon customer request.

Although these pieces of legislation are specific to the United States, other countries have adopted the same or similar principles—sometimes under different names. The FCRA principles may be referred to as 'data protection' or 'data privacy', and ECOA principles as either 'equal opportunity' or 'unfair discrimination'. Even so, many countries still have little or no legislation in this area, in spite of a growing demand for credit, but the situation is evolving quickly. In China, credit bureaux are a recent phenomenon, while Russian law prohibits the sharing of credit information between lenders.

The primary issue in the early years of the twenty-first century is not consumer protection legislation, but the Basel II Accord, which tries to protect the banking system by enforcing good governance. The accord puts a significant focus on credit scoring for risk assessments, which are used to determine banks' minimum capital requirements.

---

[6] A 'national uniformity provision' was included in 1996 to pre-empt any state legislation. The Fair and Accurate Credit Transactions (FACT) Act of 2003 made this provision permanent to ensure that national standards are maintained.

2.2 History of credit scoring

Over the period since 1960, credit scoring has had a dramatic impact upon the way credit decisions are made. Its success has been totally dependent upon the advent of computers, which brought the industrial revolution into lenders' back-offices, with similar effects. The process is as follows:

(1) Functions that can be profitably automated are identified. The precondition is always that volumes and economies of scale will be sufficient to justify the expense.
(2) They are deconstructed and/or re-engineered, which requires that inputs, hardware, and processes can be defined to provide outputs of acceptable quality. This is easiest for processes that are simple and repeated, and hardest for complex tasks where volumes are small. As volumes increase, so too does the level of complexity that can be tackled.
(3) Consumers benefit from lower costs, better quality, and greater consistency. The market grows, because goods are now affordable to a broader populace, and many people can afford multiples.
(4) Goods tailored to personal needs become rarer and more expensive. There are concerns about quality being compromised. Indeed, just as a tailor can make a better suit, a human can make a better decision—but the extra cost may not be warranted.
(5) Workers are displaced, often with great discomfort (including Luddite uprisings).
(6) Increasing sophistication of production and communication allows mass customisation, as products can be modified to suit individual needs.

This has also happened to retail credit decision-making. The high volumes, relative simplicity, and amount of attention credit assessments received over the 120 years from 1840 to 1960, helped define information requirements, and made the task of decision automation that much

Table 2.5. Genealogies and milestones—credit scoring consultancies

Name	Year	Notes
Fair Isaac (FI)		
FI	1956	Founded San Francisco CA, by Bill Fair and Earl Isaac
	1958	First scorecard development, for American Investments
	1984	Develops first bureau score for pre-screening
	1995	First use of scoring by mortgage securitisers
Experian-Scorex		
Management Decision Systems (MDS)	1974	Founded by John Coffman and Gary Chandler
	1982	MDS purchased by CCN
Scorex	1984	Founded in Monaco by Jean-Michel Trousse
MDS	1987	MDS develops first monthly bureau score, for bankruptcy
Experian-Scorex	2003	Created as subsidiary of Experian, after purchase of Scorex

easier. Even so, many credit underwriters were of the opinion that 'credit is an art, not a science', and that a computer could never have the insight required to make a credit decision. They were, however, proved wrong. Credit scoring is here to stay, and has since moved from revolution to evolution (Table 2.5).

2.2.1 Pioneers—1935–1959

The roots of credit scoring lie in a strange place. In 1936, the English statistician Sir Ronald Aylmer Fisher published an article on the use of a technique called 'linear discriminant analysis' to classify different species of irises, which he also later used to classify skulls, using only their physical measurements. Fisher's work focused on the sciences, but provided the basis for the predictive statistics used in a multitude of other disciplines. In 1941, David Durand showed that the same techniques could be used to discriminate between good and bad business. According to Johnson (2004), the study 'examined 7200 reports on good and bad instalment loans made by 27 firms' using data on age, gender, stability (time at address and employment), occupation and industry, and major assets (bank accounts, real estate, life policies).[7]

> Johnson (2004) also notes that 'it is of more than a passing interest that the model allocates 2.72 points if the customer had a bank account, and 2.63 points if she was a woman', adding that those few women who were accepted in those days were exceptionally good credit risks.

Later that same year, the United States found itself involved in the Second World War, and a large portion of the American workforce entered military service. This was a severe loss to mail-order and finance houses in an age when credit underwriters made judgmental decisions.[8] Unlike other wartime industries, the housewives of the day could not easily move in to fill the gap. Instead, experienced credit underwriters wrote down some of their many 'rules of thumb'—effectively developing a paper-based expert system, for use by non-experts.

One company did, however, develop a proper credit scoring system. Henry Wells was an executive at Spiegel Corporation, who recognised that sound statistical techniques could be employed to develop decision models, and spearheaded the development of the first credit scoring system (Lewis 1992). Likewise, in 1946 E.F. Wonderlic—president of Household Finance Corporation—used his knowledge of statistics (gained from training in psychology), to develop a 'Credit Guide Score'. Unfortunately, the score was never really accepted by the organisation, even though he proved that it worked (Johnson 2004).

In the early days, two factors inhibited the adoption of credit scoring: (i) organisational resistance to the use of computers in decision-making, much like cowboys trying to race steam locomotives; and (ii) the statistical calculations and application of scorecards in the workplace were tedious, and difficult to explain. Johnson (2004) highlights that during the post-war decade, several companies developed scorecards through judgmental analysis of charged-off accounts, which at least provided some consistency in the credit granting process—especially

[7] University of New South Wales, Bank Management Lectures, Section 3.5.

[8] Thomas (2000).

in a rapidly expanding economy, where credit skills were scarce. The next company to use statistically derived models in its business was Sears in the 1950s, but this was for response scorecards, used to decide to whom it should send its mail-order catalogues.

Perhaps the best-known pioneers of credit scoring are engineer Bill Fair and mathematician Earl Isaac, who founded their consultancy, Fair Isaac (FI), in San Francisco in 1956. Their initial contract was to create a billing system for Carte Blanche, a credit card offered by Hilton Hotels. It was two years later that they first introduced the concept of credit scoring to 50 credit grantors via a mailshot, but only one—American Investments—replied, and in 1958, they produced their first application-risk scorecards. Most of FI's early efforts were directed at other finance houses, but it was a hard sell, due to entrenched attitudes. Before long, however, many of them started to realise the potential, and adopted credit scoring as a part of the decision process (Lewis 1992).

2.2.2 Age of automation—1960–1979

In 1963, FI started a long-term relationship with department store MW. This success helped them entrench their position in the market, and they moved on to serve many other credit providers within the United States. Like so many other large retailers, MW had credit departments at each store, and the advent of computers presented the possibility of centralising the credit function. With their successes other retailers followed, including R.H. Macy, Gimbel's, Bloomingdale's, and J.C. Penney. MW was amongst the first companies to use behavioural scoring, which allowed them to have what Lewis (1992) described as 'one of, if not the, most efficient credit operation in the world twenty-odd years later'.

> Montgomery Ward was the world's first mail-order house in 1872, and opened its first store in 1926. Its last mail-order catalogue was published in 1985, and the company was closed in 2000, after changing hands several times.
> http://www.chipublib.org/004chicago/ timeline/mtgmryward.html

During the mid-1960s, the oil companies were having problems generally with their credit operations, caused by card theft, fraud, and credit losses. They decided to adopt more conservative approaches, and implemented credit scoring. The three travel and entertainment (T&E) cards—Diners Club, American Express, and Carte Blanche—also implemented scoring during this period. At this point, a lot of credit cards were issued without a fee, which brought significant volumes and competition into the market. Many of the banks to which Master Charge and Bank Americard were licensed were receiving huge volumes, and experiencing huge losses. According to Lewis (1992), it was the losses that were the driving factor behind implementing credit scoring, not the volumes. The scorecards made better decisions, with default rates dropping by as much as 50 per cent once they were implemented. Banks, retailers, and others were quick to appreciate the usefulness of the new tool.

Credit scoring was not an overnight success. Some people did not like the total reliance upon statistical models, which removed any human element from the decision process. Scorecard developers were also often unable to explain, in easily understandable terms, why certain characteristics were favoured over others. Irrespective, credit scoring continued to gain acceptance, and the combination of the FCRA of 1970, and ECOA of 1974 practically guaranteed its widespread adoption—and lifetime employment for anybody involved in credit scoring.

At this point, one of the restrictions was the processing power required. The computers were big, hot, required special dust-free environments, and were stone age by today's standards. The IBM 7090 mainframe computer was leading edge technology in 1963, yet was only capable of handling 25 variables for 600 applicants at one time (Myers and Forgy 1963).

The IBM 7090 was a scientific computer, intended for use in 'large scale scientific and technological applications', such as design and simulation by NASA, NORAD, US military, and commercial avionics. Introduced in 1959, it was the second transistorised computer (replacing vacuum tubes), and the first used for commercial applications. It was withdrawn in 1969.

As their cost and speed improved, lenders were able to justify developments for other products with lower volumes. During the late 1970s and 1980s, scoring was applied to personal loans, overdrafts, motor vehicle finance, and even small-business loans, but much was done manually. It was only in 1972, that the first fully automated implementation of credit scoring was done by FI, for Wells Fargo.

Wells Fargo was also a pioneer in the use of risk-based pricing for small businesses. Allen et al. (2003) quoted Feldman (1997), who stated that 'Wells Fargo charges . . . a range of interest rates from prime plus one to prime plus eight percent based on the business's credit scores.' The same gradations would not be possible, if only human judgment were used.

In 1962, Cyert, Davidson, and Thompson provided the first academic work that started addressing the probability theory around behavioural scoring. This described the problem as a 'Markov chain', where accounts moved from one delinquency status to another, over time. Rather than considering the number of accounts however, they instead focused on the dollar value (Thomas 2000). It was only in 1975 that FI implemented the first proper behavioural-scoring system, also at Wells Fargo.

In 1974/5, John Coffman and Gary Chandler formed a company called MDS (Management Decision Systems), and were the first to develop bureau scores for bankruptcy prediction in 1987. MDS was purchased by CCN in 1982, but is still known for its bankruptcy scores in the United States. In South Africa, a company called Stannic first scored motor vehicle finance applications in 1978, even before General Motors Acceptance Corporation in the United States in 1979.

2.2.3 Age of expansion—1980s onwards

For the most part, credit scoring was an American preserve prior to 1980, while most other countries relied upon traditional relationship lending, and risk-assessment procedures. According to McNab and Wynn (2000:10), the 1980s saw significant changes in the way lending was done in the United Kingdom: (i) banks started marketing products to non-customers;

(ii) there was phenomenal credit card growth; and (iii) there was a shift in focus from large corporate lending, where the goal was loss avoidance, to consumer lending, where the focus was profit maximisation.

The statistical techniques used for most early developments were discriminant analysis (DA) and linear probability modelling (LPM), but increases in computing power and developments in statistical software during the 1980s also allowed scorecard developers to experiment with other statistical techniques. Logistic regression is now the most widely used method, while expert systems and neural networks (NNs), have been tried with varying degrees of success (Thomas 2000). The 1980s also saw the use of scoring beyond the traditional credit and response areas, moving into retention, attrition, collections, insurance, and other types of scoring.

According to Mays (2004), the first 'credit bureau' score, 'PreScore', was developed by FI in 1984/5, for pre-screening of mailing lists, using bureau data. The concept gained broad acceptance after MDS developed bankruptcy-scoring models for all three major bureaux in 1987, and thereafter FI developed competing delinquency scores between 1989 and 1991 (see Table 2.6).

In 1984, Jean-Michel Trousse, who had been the 20th employee of FI in 1974, and founded their European office, formed the Monaco-based Scorex. He left to start the new firm on his 10th anniversary with FI, because his colleagues did not share his vision that the future of credit scoring lay in partnerships with the credit bureaux.

At the outset, Scorex was 40 per cent owned by Grattan, a UK mail-order business, who had recently acquired a Scottish manual credit file that it captured and merged with its own data, to set up Wescot, the UK's third consumer credit reference agency. Grattan then merged with Next, and after financial troubles, both Wescot and 40 per cent of Scorex were sold to Equifax in 1989.

During the late 1980s, some attempts were being made by lenders to develop in-house credit scoring capabilities, and in 1987, seven people left TSB (now part of Lloyds TSB) to form Scorelink—the credit scoring arm of Infolink. Unfortunately, the arrangement fell apart, and the individuals moved on elsewhere, three of them to the fledgling Scorex, and the rest to other areas in the credit scoring industry.

Through into the early 1990s, Scorex serviced a growing client base in Greece, Italy, and France, and expanded the company by opening offices in South Africa, Canada, and Spain. In 1996, Equifax sold its stake in Scorex UK to CCN, with a six-month period where Scorex was

Table 2.6. Early bureau score developments

For/by	MDS	FI
Equifax	Delphi	Beacon, in 1989
TransUnion	Delinquency Alert System	Empirica, in 1990
TRW	Gold Report	FICO, in 1991

servicing both (a 50/50 joint venture was entered into for other countries). Jean-Michel died in a plane crash in 2001, and in 2003 Scorex was taken over by Experian, who merged it with its Decision Support arm, to form Experian-Scorex. In 2004, Experian-Scorex had 850 people in 29 offices, servicing clients in 60 countries.

During this period, there were also massive improvements in computing technology. According to Dahlin (2000), data storage costs reduced by a factor of 7 between 1985 and 1990, and by a further factor of 21 by 1995. This made it increasingly possible for lenders to invest in data warehousing, and data mining. Further, some vendors started providing credit scoring software to lenders, for their own in-house developments.

During the 1990s, credit scoring was introduced into other areas that had long been the preserve of judgmental assessments. Lending to previously underserved low-income areas became feasible, because of the combined effects of improved transparency, technology, and pricing. For home loans, modelling was initially difficult because of a lack of sufficient bad debts, and reluctance by mortgage lenders to accept the new technology. Policy rules were favoured instead, whether based upon own or industry past experiences. According to Makuch (2001:10) 'mentored AI' systems based on underwriter judgment were developed over the period 1990–95, which improved processing speed, but did nothing to improve risk assessment.

According to Stanton (1999), credit scoring was first used by mortgage securitisers Freddie Mac and Fannie Mae in 1995 (the scorecards were developed by FI, for whom it was a huge publicity win), and by 1996 they were asking lenders wanting to sell them loans to include a credit score. Within two years, scoring was being used to assess 40 per cent of all mortgage applications in the United States. This substantially changed the nature of that market. Previously, securitisers focused only on low-risk loans. According to Edelberg (2003), they initially accepted riskier loans with no variation in terms, but as credit scores became accepted, they started varying terms according to risk. From then onwards, risk-based pricing was used increasingly for securitised loans of all types, while its use for non-securitised loans lagged.

Similar changes occurred with credit card lending. According to Scroggins et al. (2004), prior to the 1990s, most card issuers only offered a single product, charged an average rate of 18 per cent, and had an annual fee of \$25–\$30, with only modest fees for late payments. Efforts to gain market share then led lenders to lower or drop annual fees, and replace them with late-payment and over-limit fees. Interest rates were reduced, but only with the adoption of more sophisticated pricing models. Late payers suffered though, as the vast majority of card issuers now hike interest rates, usually after the first missed payment, and some increase rates if the bureau score indicates late payments elsewhere. These trends are reflected in the fees stated as a percentage of total credit card revenue, which were 16.1, 27.9, and 33.4 in 1996, 2000, and 2003 respectively.

According to Thomas (2000), the use of credit scoring for small-business credit evolved, as lenders came to realise there was little difference, *ceteris paribus*, between lending to an individual, and to a one-man business. FI launched its Small Business Scoring Service (SBSS) in 1993,[9] a trade credit service called CreditFYI.com in 1998, and a loan credit service called LoanWise.com in 1999.[10]

---

[9] Asch (2000).

[10] Allen et al. (2003:13).

Credit scoring is not normally associated with risk assessments of larger companies, but in 2000, MIS launched RiskCalc, a credit scoring model, used to provide expected default frequencies for small and middle market companies, based upon their financial statement data. The initial version was limited to the United States and Canada, but over time, models have been developed for the United Kingdom, Korea, Japan, Singapore, the Nordic countries (Denmark, Finland, Norway, and Sweden), South Africa, and others. In 2002, MIS bought KMV, known for its credit analysis techniques, based upon Merton's equity valuation model. Moody's KMV was created as an MIS subsidiary, which focuses upon providing lenders with credit analysis tools for assessing businesses, and has been developing ways of integrating the Moody's and KMV approaches.

2.3 History of credit bureaux

As will be covered in Chapter 12 (Data Sources), lenders have three main data sources: the customer, internal systems, and external agents. This section covers the history of the primary class of external agents—the credit bureaux, which are retail lenders' conduit for credit intelligence from the outside world. While even the earliest lenders used spies to gather intelligence about borrowers' activities, modern credit reporting was born of the industrial revolution, and over the past two centuries, has become a mainstream industry in its own right. Today, four companies dominate the industry; Dun & Bradstreet (D&B) is the major player in business credit reporting, while Equifax, Experian, and TransUnion dominate the consumer market (Table 2.7).

2.3.1 Early to mid-1800s

Contrary to popular belief, formalised information sharing between credit providers was not an American innovation, but originated first in the United Kingdom—both for consumer and commercial credit.

United Kingdom

The Mutual Communication Society of London was formed in 1803, as a collaborative effort between several tailors, who compiled information on people that did not pay their bills, and published a newsletter that was distributed to members. No direct link can be shown, but it is likely that other similar arrangements evolved, and in 1842 the London Association for the Protection of Trade was formed, for a broader group of traders. At its peak, it covered 2000 merchants, covering London's West End 'carriage trade', or wealthy clients who travelled in carriages (Olegario 2002). This was renamed the United Association for the Protection of Trade in 1965, and eventually became known as UAPT-Infolink. It was the major competitor of CCN, and was purchased by Equifax in 1994.

During the early nineteenth century, the consumer credit market was tiny, but there was a lot of activity in both trade credit and financing of business ventures. Much credit was

Table 2.7. Genealogies and milestones—credit bureaux

Name	Year	Notes
Dun & Bradstreet		
Mercantile Agency	1841	Founded, New York NY, by Lewis Tappan.
	1849	Benjamin Douglass takes over, and expands.
John M. Bradstreet Co.	1849	Founded, Cincinnati OH.
	1851	First use of credit rating grades.
R.G. Dun & Co.	1859	Robert G. Dun incorporates Mercantile Agency.
Dun & Bradstreet	1933	Merger orchestrated by Arthur Whiteside.
Experian		
Manchester Guardian Society	1827	Founded, Manchester, UK.
Chilton Corp.	1897	Founded, Dallas TX. Publishes 'Red Book'.
Michigan Merchants	1932	Founded, later to become Credit Data Corp.
TRW	1968	Purchases Credit Data Corp., and changes name to TRW-Credit Data.
TRW	1976	Information Systems and Services (IS&S) division produces first business credit report.
CCN	1980	Founded, when Great Universal Stores (GUS) spins off information services division
	1884	Purchases Manchester Guardian Society
TRW	1989	Purchases Chilton Corp.
Experian	1996	Founded, through TRW divestiture of TRW-CD & IS&S. Purchased by GUS, who merges it with CCN.
Equifax		
London Assn. for the Protection of Trade	1842	Founded, London, UK
RCA	1869	Founded, Brooklyn, NY
RCC	1899	Founded, Atlanta, GA
	1934	Purchases RCA
United Assn. for the Protection of Trade	1965	LAPT renamed
Equifax	1975	RCC renamed to Equifax
	1994	Purchases UAPT-Infolink and Canadian Bonded Credits
TransUnion		
TransUnion	1968	Founded, as holding company for Union Tank Car Company (UTCC)
	1969	Purchases the Credit Bureau of Cook County

extended based upon letters of recommendation, and the only way for lenders to check on potential borrowers was to hire private investigators. Barings Brothers even hired local American agents to investigate their customers across the pond, 'but this was a costly arrangement [limited] to the very largest firms' (Olegario 2002). Co-operative arrangements evolved, one of which was the Manchester Guardian Society in 1827. It collated business information and financial reports at a time when the industrial revolution was gaining force, and the British Empire was growing rapidly. It operated under the same name for 157 years, until it was bought by CCN in 1984.

United States of America

The expansion of the American economy during the mid-1800s brought a credit boom. According to Olegario (2002), some merchants' associations were formed, but these were limited in their geographical coverage, and focused on trade creditors as opposed to consumers.[11] One of the first was a group of 1820s New York wholesalers who hired an investigator, but the arrangement was short-lived. Credit reporting agencies were first formed during the 1830s, and 'were better suited to the peculiar needs of American society, where [customers' business dealings] were frequently dispersed over a wide area'. Most operated on a hub and spoke system, which was well suited to the task. One of these was the Mercantile Agency, a New York based company founded by Lewis Tappan (1788–1873) in 1841, towards the end of a worldwide depression, and during a year when bankruptcy legislation was implemented.

> The depression started in England in 1839, and is the backdrop for many of Charles Dickens's novels. During the early 1840s it caused several failures amongst the fledgling banks in the United States, and caused New South Wales to move quickly from boom in 1840 to bust in 1841, causing a sharp fall-off in strikes by labour unions. By the mid-1840s, the industrial boom and railway expansion was again underway.

According to Sylla (2001), Tappan was a dry goods and silk merchant who had compiled a great deal of information on the creditworthiness of his customers, and decided to change his business to the provision of commercial information. The only major competing credit reporting agency during this era was the John M. Bradstreet Company, which was established in 1849 in Cincinnati, OH, and pioneered the use of credit ratings in 1851.

> Mercantile's handwritten credit reports are part of the R.G. Dun and Company Collection, covering the period 1841 to 1892, held by the Baker Library at Harvard University. Little or no documentation has survived from other credit reporting agencies of the nineteenth century. According to the Harvard University History Department, the collection comprises 2,580 volumes organised by city and region covering the United States, western territories, Canada, and the West Indies. Credit reports were done at least semi-annually and were collated by the New York office. Entries cover 'the duration of the business, net worth, sources of wealth, character and reputation of the owners, their partners, and successors'.

The Mercantile office remained New York bound until Benjamin Douglass, a clerk who also married Tappan's granddaughter, took it over in 1849. Douglass took advantage of transportation and

[11] Olegario (2002) obtained much of the historical information from Hidy's 1939 article, 'Credit Rating before Dun & Bradstreet.'

communication improvements to expand. He hired credit reporters who gained sound business skills, and over a 10-year period set up a network of 2,000 correspondents to provide information about businesses around the United States, that was published in a 'Reference Book'. Many of the correspondents were local attorneys, who in exchange received referrals for collections work. Four of them went on to become American presidents (Lincoln, Grant, Cleveland, and McKinley).

The Mercantile Agency was incorporated as R.G. Dun & Company in 1859, after Douglass turned it over to Robert Graham Dun, Tappan's grandson and his brother-in-law. Competition with Bradstreet was fierce, and in an 1861 advertisement, Dun claimed that their ratings were predictive. The growth of the market indicates that these ratings had significant value in the eyes of their customers, even if more recent analysis shows limited value (Carruthers and Cohen 2002/6). By the 1850s, Dun had 2,000 correspondents throughout the United States and Canada. The number of companies reported on grew from: 1859—20K; 1870—430K; 1880—764K; 1890—1.176M; and 1900—1.285M (Olegario 2002). In like fashion, by the 1870s, the subscriber base had grown to 7,000, and 10 years later it was at 40,000. The reference book was published quarterly from 1873. At the same time, John M. Bradstreet Company also continued to grow.

According to Carruthers and Cohen (2006), Dun provided separate assessments of 'general credit' (soft data, largely from intuitive assessments; high, good, fair, or limited), and 'pecuniary strength' (hard data on net worth; grew from 8 to 15 categories from 1864 to 1883, mostly to accommodate smaller companies). Interestingly, subscribers had to hand reference books back before receiving updates, making it difficult for them to do retrospective evaluations of grades' worth.

2.3.2 Late 1890s onwards—consumer

It took a lot longer for consumer credit reporting agencies to catch on in the United States, the first being *Retailers Commercial Agency* (RCA) of Brooklyn, New York, in 1869. It was many years later though, that they started proliferating. The Credit Clearing House was established in 1888, and was effectively the first successful national wholesalers' association. In 1897, James Chilton formed the *Chilton Corp.* in Dallas, Texas, and collected information from merchants on shoppers' payment habits, in a notebook he called his 'Red Book' (the Chilton Corp. was purchased by TRW in 1989). Later, in 1899, Cator and Guy Woolford in Atlanta, Georgia formed the *Retail Credit Company* (RCC). RCC purchased RCA in 1934, and in 1975, it changed its name to Equifax.

During the first half of the twentieth century, the number of American credit bureaux increased phenomenally, but these almost always focussed upon a specific industry and geographic area. The National Federation of Retail Credit Agencies was thus formed on 24 February 1906, to promote sharing across these barriers, and grew rapidly. Membership was less than 100 bureaux in 1916, but grew to 800 by 1927, and 1600 by 1955 (Staten and Cate 2004).

Most of the credit bureaux in 1919 represented retailers, who at the time provided 80 per cent of consumer credit. This was largely because usury legislation prevented other lenders from competing with retailers, who were disguising their finance charges in the prices charged for goods sold on credit. In 1916, many states had relaxed their usury laws, which brought banks and finance companies into the market—some of whom were offering revolving credit.

Growth in the number of credit bureaux was also aided by American legislation limiting branch banking, as there were few risks of increased competition and even the sharing of positive information between banks was common at this early date (Japelli and Pagano 1999).

Over time, the retailers' share of consumer credit reduced from 1919's 80 per cent, to 67, 40, and 5 per cent in 1929, 1941, and 2000 respectively. In spite of the reductions, retailers were still able to benefit from the 1920s explosion in instalment credit demand for consumer durables. This was accompanied by a surge in the number of credit bureaux, and the onset of the depression did not kill the trend. The Michigan Merchants Co. was formed in 1932, which later became Credit Data Corporation (purchased by TRW in 1969).

This is not to say that all was smooth sailing during the depression. The competition between R.G. Dun and John M. Bradstreet in the trade creditor's market was been fierce, and the depression caused Dun's CEO—Arthur Whiteside—to broker a merger between the two companies, to form *Dun and Bradstreet* (D&B) in 1933.

2.3.3 1960s onwards

The American post-war boom saw further growth of consumer credit reporting (Furletti 2002). Little had changed in the prior 40 or so years though: they were still small community-based companies, or co-operatives, that served a specific type of lender—bank, finance company, or retailer—and they would provide information over the phone on delinquencies and defaults. The agencies would also comb through newspapers for 'notices on arrests, promotions, marriages, and deaths', and include them in people's files.

These practices continued through the 1960s, and by 1969 there were 2,200 credit bureaux in the United States, collecting data from public records and 400,000 creditors, that maintained files on 1.1 million consumers.[12] Advances in computing technology during the 1960s led to the automation of countless labour-intensive back-office functions. This extended to credit bureaux and aided industry consolidation, which brought the number of credit bureaux down to about 200 by 2005, in spite of continuing credit growth. Credit Data Corporation was the first to automate in 1966.

TransUnion was founded in 1968, as the holding company of the *Union Tank Car Company* (UTCC). One of its business areas was business intelligence, and it sensed broader opportunities in credit reporting. In 1969, it diversified by taking over the Credit Bureau of Cook County, which at the time had 3.6 million card files in 400 seven-drawer cupboards.

[12] *The Consumer Reporting Reform Act of 1994*, 103 S. Rpt. 209 103 Cong. 2 Sess. (1993), cited in Cate et al. (2003).

The UTCC is the railway equipment leasing company, which owns the UTLX tankers and hoppers that abound in North America.[13] Its genealogy can be traced back to the Star Tank Line, founded by J.J. Vandergift in 1866, to ship oil from the Pennsylvania oil-fields to Chicago. John D. Rockefeller's Standard Oil bought it in 1873, and made Vandergift Vice President. The name was changed to Union Tank Line in 1878; and in 1891 the Standard Oil Trust was formed, and UTC incorporated, to avoid anti-trust measures. Standard Oil remained its sole customer though, and the Trust was eventually dissolved in 1911. In 1904, UTCC operated 10,000 railway cars, more than any other private operator, and by the 1920s, 30,000. In the 1930s, it started producing its own tank cars, and shipping chemicals. The Marmon group bought it in 1981, and in 2005, it had an 80,000-strong railcar fleet—*Encyclopedia of Chicago, Wikipedia.*

Credit bureaux in the United States were largely unregulated, prior to the passing of the Fair Credit Reporting Act in 1970, which according to Cate et al. (2003) 'was a notably even-handed attempt to balance the need for accessible credit data, with consumers' concerns about privacy'. This set of ground rules aided both consolidation and growth within the industry, and by the late 1970s, TransUnion and Equifax had emerged as leaders. They were later joined by TRW (today's Experian) to form the 'big three'.

TRW (Thompson Ramo Woolridge) first entered the credit reporting industry in 1968, when it purchased Credit Data Corp., and renamed it TRW Credit Data. Its focus was consumer credit, and in 1989 it also purchased Chilton Corp. In 1976, TRW's IS&S division collaborated with the National Association of Credit Managers, to create its first business credit report. The division grew by acquisition into direct marketing/target marketing, and real-estate information and loan services.

The 1968 purchase was a major shift from TRW's core technology businesses. Thompson Products (automobile and aircraft engines) and Ramo Woolridge (scientific research and project management) were both engaged in research and development of ICBM missiles for the US government during the 1950s. The latter spawned Space and Technology Laboratories (STL) in 1957, which remained a wholly owned subsidiary when the two merged into TRW in 1958, but later became a division of the parent. [A History of the Department of Defense Federally Funded Research and Development Centres, Office of Technology Assessment, Congress of the United States, June 1995]. Si Ramo championed the development of a 'cashless' system in the 1960s, and envisaged businesses offering credit and financial information.

The CCN was formed in Nottingham, England in 1980, when GUS split off its information services division—which had been supporting a mail-order operation, in existence since 1900. By 1982, they were already offering a credit bureau service (CAIS—Credit Account

[13] The author grew up in a small Alberta city called Medicine Hat, and worked for the Canadian Pacific Railroad for two years in the mid-1970s. Two of the major local industries made fertiliser from natural gas, and the UTLX hoppers were a common sight.

Information Sharing), and their credit scoring capabilities were obtained with the purchase of MDS in the United States that same year. In 1984, they expanded their bureau operations through the purchase of the Manchester Guardian Society.

In 1996, as a part of a drive to focus on core businesses, TRW divested all of its credit reporting interests into a new company, called Experian. It was immediately purchased by GUS, and merged with CCN under the Experian name, to enhance brand recognition, and stop ongoing confusion with CNN, the satellite television news service. Prior to 1996, CCN was already established in a variety of countries, including a small office in the United States, but this allowed them a much larger footprint, as one of the American 'big three'.

According to Furletti (2002), by the early 1980s technology had already evolved to such an extent that credit bureau were able to provide subscribers with more accurate information electronically, than over the phone. They had transformed themselves from paper-based local associations serving specific industries, to high-tech companies serving the broader economy.

2.3.4 International

At the turn of the twentieth century, credit reporting agencies were also being established in other countries outside the United States and the United Kingdom. For example, in 1901, D&B established an office in Cape Town, South Africa, to provide information on local traders to suppliers in the United States. Over time, other agencies were established around South Africa, some as the initiatives of local chambers of commerce (as in Durban, Kroonstad, and East London),[14] but almost all of these were acquired by D&B during the 1970s and 1980s. D&B divested in 1986, in a management buyout that saw the local operations renamed Information Trust Corporation (ITC). It was sold to M-Net in 1990, and then to TransUnion in 1993. D&B returned and took a minority stake in 1994. The company was renamed TransUnion ITC in 2002, and dropped ITC from the name in 2006.

The late 1920s saw the establishment of the first private credit bureau on the European continent. *Schufa Holding AG* was formed in 1927 by a group of banks and retailers, and is today the largest bureau in Germany. The *Union Professionnelle du Credit* (UPC) was formed in Belgium during the 1930s, the *Consorzio per la tutela del credito* (CTC) in Italy in 1964, and the *Bureau Krediet Registratie* in the Netherlands in 1965.

The depression saw the Deutsche Bundesbank establish the first ever public credit registry (PCR) in 1934, as a response to the systemic risks highlighted by the depression. Today the German *Evidenzzentrale für Millionenkredite* only covers loans larger than €1.5 million. Other public credit registries were established in: 1946 France, *Service Central des Risques*; 1950 Chile, *Archivo Deudas Generales*; 1951 Turkey; 1961 Finland; 1962 Italy, *Centrale dei Rischi*; 1962 Spain, *Central de Información de Riesgos*; 1964 Burundi; 1964 Mexico, *Servicio Nacional de Información de Crédito*; 1966 Jordan; 1967 Belgium; and 1968 Peru, *Central de Riesgo*. The number of countries with credit bureaux has continued to grow, and in 2005 there were over 50 countries with private credit bureaux, over 70 countries with public credit registries, and 20 countries with both.

---

[14] Consumer Affairs Committee (2003).

2.4 History of credit rating agencies

The origin of the credit rating industry dates back almost as far as that of the credit bureaux. Both started as publishing endeavours, aimed at specific markets. Credit reporting was aimed at trade creditors, trying to ensure that other companies were safe to do business with. In contrast, credit rating was aimed at bond investors looking for places to park their funds safely (Table 2.8).

According to Sylla (2001), the Dutch, English, and Americans had been issuing and purchasing (mostly government) bonds for 300, 200, and 100 years respectively, prior to bond rating grades being used for the first time in 1909. He asks the question, 'Why would investors be willing to purchase bonds where there is a distinct lack of information, as existed prior to bond ratings?' This applies especially to companies. As Sylla explains it, investment bankers put their reputations on the line with every issue, and demanded access to information, including seats on the board. They were the 'consummate insiders'. While this would not have been an issue to Europeans of the age, who provided much of the capital for American expansion, it caused resentment amongst emerging American investors, who saw this as an unfair advantage. The end result was mandatory disclosure laws in the 1930s, and formation of the Securities Exchange Commission. In many respects, the advent of bond ratings and greater company transparency can be seen to have weakened the position of investment bankers.

The origins of this process started during the early to mid-nineteenth century, with the growth of American railroads. It took only four years from the 1828 founding of the Baltimore and Ohio Railroad for the industry to have its own dedicated publication, *The American Railroad Journal* (ARJ). In the very early days, the railroads were small and localised in settled areas, and were able to obtain funds through common stock issues, and bank credit. After

Table 2.8. Genealogies and milestones—credit rating agencies

Name	Year	Notes
Standard & Poor's (S&P)		
Poor's Publishing Co.	1862	Founded, by Henry Varnum Poor
S&P	1941	Poor's Publishing and Standard Statistics merge
Moody's Investor Services (MIS)		
John Moody & Co.	1900	Founded, by John Moody, but fails in 1907
John Moody	1909	First use of rating grades for bonds
Moody's Investor Services	1914	Incorporation of MIS
	1962	MIS purchased by D&B
Moody's KMV	2002	Created as MIS subsidiary after merger of Risk Management Services and KMV
Fitch IBCA		
Fitch Publishing Co.	1913	Founded, by John Knowles Fitch
IBCA	1978	Founded
Fitch IBCA	1997	Merger of Fitch Publishing and IBCA

1850 though, the railroads expanded into frontier regions, and the expansion became so great that it required capital from both domestic and European sources.[15]

> Sylla (2001) states that the railroads were perhaps the world's first big businesses, but in doing so he overlooks enterprises like the Dutch East India Company (VOC)—which predated the railroads by over two hundred years—even though he mentions the VOC as being the world's first issuer of common stock.

It was in this environment, that Henry Varnum Poor (1812–1905) took over editorship of the ARJ in 1849, and the journal soon found itself catering for the needs of investors. Poor continued as editor until 1862, and then after the end of the Civil War he and his son started Poor's Publishing Company. Its flagship publication was the *Manual of the Railroads of the United States*, an authoritative annual that contained multi-year financial statements and operating statistics for most of the industry.

John Moody (1868–1958) was a much later entrant into this market, and was working for a Wall Street brokerage house when he founded John Moody & Company in 1900.[16] Its flagship publication was Moody's *Manual of Industrial and Miscellaneous Securities*, which 'provided information and statistics on stocks and bonds of financial institutions, government agencies, manufacturing, mining, utilities, and food companies'. By 1903, it had coast-to-coast circulation, and the monthly Moody's Magazine followed it in 1905. Unfortunately though, the company was not able to survive the stock market crash of 1907, and was sold.

Two years later, John Moody was back with a new idea. Rather than just publishing available information, why not provide an analysis that is summarised in a letter-rating grade. Credit ratings had already been used for over fifty years for trade and loan credit, and by the 1890s, some had evolved into letter-rating grades similar to those used today. Moody's innovation was to apply grades to publicly traded securities. His 1909 *Analyses of Railroad Investments*, an annual publication, was immediately popular with investors. The ratings were expanded to industrials and utilities in 1913, to cities and municipalities in 1914 (when MIS was incorporated), and to state and local governments in 1919. By 1924, Moody's ratings covered almost all of the American bond market. MIS was sold to D&B in 1962, and still operates under that name. In 2002, MIS purchased KMV, and merged it with Moody's Risk Management Services to create Moody's KMV, which specialises in risk management solutions.

Poor's Publishing was not absent from this game. After watching the popularity of bond ratings rise over several years, it started its own bond rating service in 1916. It was not as comprehensive as Moody's, and did not cover state and local bonds until the 1950s. Poor's Publishing merged with Standard Statistics in 1941, to create Standard and Poor's (S&P), which was purchased by McGraw-Hill Publishing in the 1960s.

The other major player in this market is Fitch Ratings.[17] John Knowles Fitch founded the Fitch Publishing Company in 1913 in New York City. It was a publisher of financial statistics,

---

[15] The nineteenth century was a period when the United States, Russia, and Japan were developing economies, and the destination of much European capital.

[16] See the 'About Us' section of Moody's website.

[17] See the 'About Us' section of Fitch Ratings' website.

including the *Fitch Bond Book* and the *Fitch Stock and Bond Manual*. It started doing credit rating in 1924, as an extension of its financial publishing. Fitch Investor Services merged with IBCA (London) in 1997 to become Fitch IBCA.[18] In 2000, the company went on the acquisition trail and bought both Duff and Phelps Credit Rating Company (Chicago) and Thomson Bankwatch (Toronto).[19]

In 1975, the American Securities and Exchange Committee (SEC) deemed the three major agencies—S&P, Moody's, and Fitch—to be 'nationally recognised statistical rating organisations' (NRSRO). Today, the first two are acknowledged as dominating the North American market, while Fitch IBCA is a major player in many other countries.

According to Ong (2002), 'As the art of credit rating evolved into modern times, agency ratings have become inextricably tied to pricing and risk management.' Indeed, their grades have become a cornerstone of risk assessment in the wholesale market, to the extent that many large institutional investors limit themselves to 'investment grade' bond issues, and banks benchmark their own internal risk grades against those provided by the rating agencies. With the advent of NRSROs, regulators are also better able to ensure the health of the entire financial system, as the rating grades have become key to determining banks' capital requirements under Basel II.

2.5 Summary

The use of credit is something that has probably been part of human activities ever since man started trading goods and services, but the first documented use of credit was in Babylon in about 2000 BC. It is something that has been crucial within many economies, to finance trade and commercial endeavours, but has seldom been a glorious activity. Most early religions frowned on the charging of interest, largely because of the extremely high interest rates being charged, and draconian penalties for non-payment. It was only in the 1500s that the protestant church accepted the lending of money for interest.

The real growth of credit accompanied the industrial revolution, not only to finance factories and infrastructure, but also the consumer goods they produced. Initially, retailers selling the goods provided much of the credit, but over time, finance houses and banks entered the market. Most credit was offered through fixed-term loans and overdrafts, but in the 1960s, credit cards and revolving credit started being used.

While the concept of credit scoring had been touted in the 1940s, it was only in the 1960s that it started gaining wide acceptance. FI implemented their first scorecard at American Investments in 1958, and became the champion of the new technology. It was aided by: (i) improvements in computing power, that made scorecard development easier; and (ii) phenomenal growth in the credit card market, that was accompanied by high loss rates on new business. Initially, the focus was to reduce bad debts, but there were huge benefits from process

[18] IBCA was formed in 1978, and was purchased by Fimalac S.A. in 1992. With the Fitch IBCA merger Fimalac become the holding company for the merged firm.

[19] Thomson Bankwatch specialises in rating financial institutions, covering over 1,000 companies in 95 countries.

automation. The first scorecards were developed using DA and LPM, but over time, logistic regression became more feasible, and is today the most popular statistical technique.

During the 1970s, lenders started realising that it could provide value not only for other products, but also other stages of the risk management process. Behavioural scoring first started being used in the early 1980s for account-management functions, and *risk-based pricing* evolved along with the *securitisation* of home loans, in the mid-1990s. The year 2000 saw Moody's launch of RiskCalc, which was the first commercial use of credit scoring to assess company financial statements.

Lending relies on trust, which in turn relies upon knowledge about whomever the money is being lent to, especially as regards their capacity to repay. Credit bureaux provide credit intelligence to retail lenders and trade creditors, while credit rating agencies focus on bond investors and the wholesale market. The first private credit bureaux were founded in the United Kingdom in the early 1800s, and then in the United States in the mid-1800s. Bradstreet and Company was the first to provide credit ratings for trade creditors in 1851. Today, the major private credit bureaux in the world are: D&B, for trade credit; and Equifax, Experian, and TransUnion, for consumer credit. The first public credit registry was established in Germany in 1934, and registries exist in many countries, either to help monitor the financial system, or to provide credit bureau services, where no private bureau exists.

The credit rating agencies had their origins in publishing companies that provided financial reports on bond issuers for investors in the United States and Europe. Today, the three main companies are: S&P, established in 1862 as Poor's Publishing Company, and merged with Standard Statistics in 1941; Moody's Investor Services, established in 1909 after a previous failure; and Fitch IBCA, established as Fitch Publishing Company in 1913, and merged with IBCA in 1997. Moody's was the first true 'bond rating' service, while Poor's and Fitch only started providing ratings in 1916 and 1924 respectively.

Into the future, credit scoring techniques will be applied to further areas, including other types of business and types of prediction. For banks, its use is being promoted by the Basel II Accord, which will further increase their risk assessment capabilities. Lenders are continually pushing the envelope—both upwards and downwards—in terms of the amounts that they will lend, based upon automated decision rules, and will continue to do so, as long as automated data sources improve, and companies learn how to harness their power. Beyond those limits, there will also always be room for credit scoring tools to support judgmental decisions, by underwriters or loan officers.

3 The mechanics of credit scoring

Our first two chapters focused on: (i) the use of credit scoring within the business; and (ii) the history of credit, credit bureaux, credit rating agencies, and credit scoring. Ultimately, credit scoring is used to aid decision-making, but is often more associated with the statistical techniques and processes used to develop the scorecards. This next chapter delves more into the latter, and provides an overview of topics covered in subsequent modules. The primary goal is to familiarise the reader with terminology used in the rest of the book, skipping over the detail where possible.

(i) **What are scorecards?** How are they presented? How are they developed? Can they be biased? How does it arise? What can be done about it?
(ii) **What measures are used** for: (i) process and strategy; (ii) default probability and loss severity; and (iii) scorecard performance?
(iii) **What is the scorecard development process?** Who needs to be involved, and what tasks have to be performed?
(iv) **What can affect the scorecards?** Changes in the environment, including economic, market, operational, target, and unattributed drift.

3.1 What are scorecards?

Most people understand a scorecard as a piece of paper that allows a scorekeeper, spectator, or participant to keep track of competitors' performance in a sporting activity. The final score is then used to determine who won. In credit scoring, it started much the same way, only a 'win' is an accepted loan application.

The difference is how the scorecards are derived, and applied. Credit scoring is the use of predictive models (algorithms), to rank cases by their probability of being 'good' or 'bad' at a future date, based upon lenders' past experiences. The logic is simple: a 'good' customer is one to be welcomed with open arms (low risk); and a 'bad' customer is one that would have been turned away, had the lender known better (high risk). Scoring algorithms can take many forms, but the most common are regression formulae of the form:

Equation 3.1. Regression formula $y = b_0 + b_1 x_1 + b_2 x_2 + \cdots + b_n x_n + e$

This formula has four main components:

x = *independent/predictor variable*, which may be the original variable, a transformed variable, or dummy 0/1 flags indicating whether or not that attribute applies.

b = *regression coefficient* or *parameter estimate*, factor by which that variable will be weighted, which for dummy variables indicates relative importance.
y = *dependent/response/target function* associated with the outcome, which is usually: (i) 0 for bad and 1 for good; (ii) a logistic unit (logit); or (iii) a probability unit (probit).
e = *residual/error term*, which is the portion that cannot be explained, and is ignored when the model is applied in practice.

The regression coefficients are derived to provide a model that best explains the relationship between predictors and the response function. In retail credit, traditional scorecards use classed variables, where the points are only allocated if that condition holds true.

Condition	Score	
If age < 29 then	Deduct	20
If age ≥ 45 then	Add	50
If no home phone then	Deduct	30
If existing customer then	Add	20
And so on		

3.1.1 How are they presented?

The final scorecard can be presented to the layman as a series of statements, as shown above, or in a tabular format as shown in Table 3.1. The table is comprised of characteristics (shown as

Table 3.1. Application scorecard example

Characteristic	Attributes					Points
Years @ address	<3 years 30	3–6 years 36	>6 years 38		Blank 35	38
Years @ employer	<2 years 30	2–8 years 39	9–20 years 43	>20 years 64	Blank	43
Home phone	Given 47				Not Given 30	30
Accom. status	Own 41	Rent 30	Parents 39		Other 36	41
Bankers	Us 49	Them 42			Blank 42	49
Credit card		Bank or travel 75	Retail or garage 43		Blank 43	43
Judgments on bureau	Clear 20	1 −16	2 −30	3 −54		20
Past experience	None 3	New 13	Up-to-date 36	Arrears −1	Write-off Reject	3
					Final score	267

rows) and attributes (shown as columns). An attribute is a non-overlapping range of numbers, or set of values, and the points associated with that attribute are assigned for each case where it is true. The points are then totalled; the higher the total, the lower the risk.

In the example, the attributes for a hypothetical applicant have been highlighted, the points allocated, and the final score calculated. The exact cut-off used for this scorecard is not known, but they were always 200 or less, so it is almost certain that an applicant with a score of 267 would have been accepted.

The scorecard in Table 3.1 is based upon a FICO scorecard that was used by Stannic, a South African motor vehicle finance company, from 1978 through to 1983. Like many of the early scorecards, they were score sheets that underwriters completed and tallied, which left the company open to potential fraud. When a new scorecard was developed, it was programmed into HP 41C calculators, which were distributed to branches.[1]

Traditional scorecards have the advantage of transparency, which is why they have been the preferred format since credit scoring was introduced. Other formats have been used, but with varying success. Some are just as valid, but have their own advantages and disadvantages.

3.1.2 How are they developed?

Different ways of developing credit scoring models have evolved over the years, and not all models take on the traditional form . . . . There are other means of transforming the data, and deriving estimates. Predictive modelling techniques can be classified into two broad camps: (i) *parametric*, which make assumptions about the data; and (ii) *non-parametric*, which do not require any assumptions. Both are given much greater attention in Chapter 7 (Predictive Statistics 101).

Parametric techniques

Traditional scorecards are developed primarily using parametric techniques: discriminant analysis (DA), linear probability modelling (LPM), and logistic regression. While powerful, they require assumptions that do not always hold true. For all intents and purposes, DA and LPM can be treated as one, as LPM is used as a part of DA. LPM is quick and easy, and was the tool of choice for many years. Referring back to the regression formula in Equation 3.1, with LPM the response function is:

Equation 3.2. Linear probability modeling $y \cong G/(G+B)$

where G and B are the counts of goods and bads respectively. LPM has been subjected to a great deal of criticism though, because it is not usually considered suitable for modelling binary outcomes. Even so, it is still commonly used, because most of the criticisms have been

[1] The author's first exposure to credit scoring was to program these calculators for Stannic in 1983, after which many years were spent doing other programming work unrelated to credit scoring. He entered the credit scoring field full-time only in 1996.

addressed by how the technique is applied. In contrast, *logistic regression* is slower, but better suited to modelling dualities. It works by deriving an estimate of the natural log odds:

Equation 3.3. Logistic (logit) regression $y \cong e^{G/B}$

Another form that is used is probit, which assumes a Gaussian as opposed to logistic distribution.

Non-parametric techniques

Because of the assumptions required when using parametric techniques, many people have tried to use non-parametric techniques. Many of these come from the field of machine learning, and include *neural networks* (NNs), *genetic algorithms*, and *K-nearest neighbours*. The greatest criticisms of these techniques relate to: (i) lack of transparency; and (ii) potential overfitting. Most are not associated with traditional scorecards, but can be used to develop them. Other techniques that have been tried are *decision trees* and *linear programming*, but these have not been very successful.

Which is best?

There are many statistical arguments for and against the various techniques, yet no clear winners. Comparisons of their ranking ability (power) have been inconclusive. LPM was the methodology of choice when computers were big and slow, but this changed as computing power increased. Today, logistic regression is the most widely accepted technique, largely because: (i) it is well suited for modelling binary outcomes; and (ii) the scores can be easily converted into, or calibrated onto, probability estimates. It may, however, not be appropriate for all circumstances. LPM has specific advantages under certain conditions, and is still well accepted. In many instances, the choice is determined by skills availability, and appropriateness for a particular task. NNs, in particular, are well suited for fast changing environments, such as fraud detection.

3.1.3 How good are the predictions?

Credit scoring has a huge data dependence. The goal is to rank order cases according to the probability of a future event, and perhaps even provide estimates of that probability. To do this lenders need: (i) *transparency*, information that can be used for an assessment; (ii) *structure*, it is available in a form that is readily analysable; (iii) *data quantity*, sufficient cases, especially bads (default, close, etc.), to develop a model; and (iv) *data quality*, relevant, accurate, complete, current, and consistent. Any problems can compromise the reliability of the final model.

As indicated elsewhere, it is not possible to develop a predictive model that provides certainty. There will always be idiosyncratic and exogenous factors that cannot be captured. Even so, as data improves, so too will the flat maximum—an unquantifiable level of predictive power that cannot be exceeded (see Figure 3.1). For any particular set of data though, there will be several different models, or sets of coefficients, that come close (first illustrated by Wilks (1938), and Dawes and Corrigan (1974)).

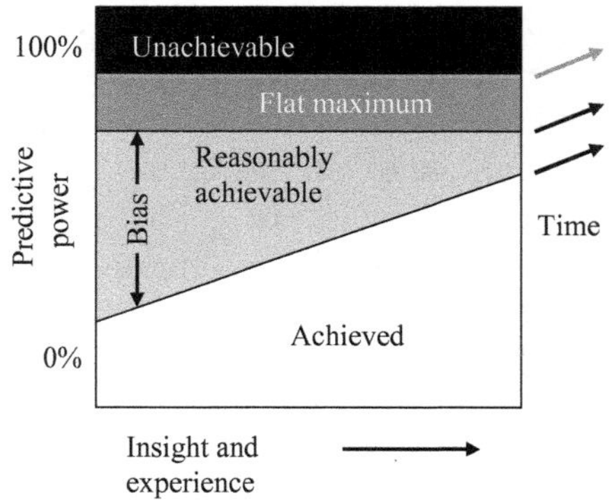

Figure 3.1. Bias and flat maximum.

According to Falkenstein et al. (2000), Wilks examined a situation where there were two positively correlated characteristics. Assuming X and Y are independent and normally distributed, and $Z_1 = X + 2Y$ and $Z_2 = 2X + Y$, then the correlation between Z_1 and Z_2 is 0.8. The coefficient weightings provide surprisingly similar results, highlighting that it is more important to find the right data to use as X and Y, than to determine the weights.

By implication, lenders have to invest heavily in data, before getting into analysis and decision-making. It also helps to explain why the results provided by different modelling techniques are so similar. Indeed, in many cases, scorecard developers will develop several different models, and choose the one best suited to business needs.

For any one development, the goal is to get as close to the flat maximum as possible, hopefully exceeding that which is reasonably achievable using current technology. The standard is what is reasonably achievable, and an assessment will be considered biased, to the extent that the standard is not achieved. The term 'bias' refers to a tendency or inclination, but usually connotes irrationality—especially when it applies to unjustifiable personal beliefs, where no attempt is made to correct them. In predictive modelling, the term is used similarly, and in both the personal and modelling cases:

(a) The biases can arise, amongst others, from limited or inaccurate information (data), poor reasoning, use in inappropriate circumstances, lack of experience with a significant subgroup, inappropriate sampling, or inability to adjust to a changed environment.
(b) Faults may cause relevant data to be overlooked/ignored, or be given more or less weight than it should.
(c) The fault is often blamed upon the person (beliefs) or model (points, structure), without looking deeper at the experiences that created the biases.
(d) The problem may be so great, that the person/model is not qualified to do the assessment.

This same line of thinking applies to any type of predictive assessment, whether done using human judgment, or a statistically-derived model. If people with greater insight and experience can provide better assessments, should the same not apply to models developed with greater care and attention, or using better data? And just as human assessments can be subjected to the 'reasonable-man test', should scoring models not also be subjected to a reasonable-model test, relative to results provided by an old or competing model (as opposed to naïve-model test, that compares against a constant value, or an extrapolated trend-line)? In rare instances, the competing model will be developed specifically for the test, possibly including new data, a simpler set of assumptions, and/or another methodology.

3.1.4 How does scorecard bias arise?

Mays (2004) provides three possible sources of bias in scorecard developments: (i) omitted variables; (ii) errors in the predicted variables; and (iii) sample selection. These headings are also used here, but expanded upon. The list is by no means exhaustive, as bias can arise from any number of sources in data selection, model development, changing environments, and so on.

Data quality

Predictive models are heavily reliant upon the data used for their development, and if the data is substandard, it affects the quality of the final result. The major sources of problems are missing data, misrepresentation, and miscapture (see Section 11.3):

Missing data—No value is provided for one or more characteristics in a record, either because the applicant did not complete the field, or the infrastructure did not accommodate it at that time. With some modelling techniques, the entire record may have to be discarded.

Misrepresentation—Incorrect information, provided with the express purpose of deceiving the lender. An obvious example is income, which many applicants overstate, because it is commonly used to set limits. Such key fields are often subject to greater verification.

Miscapture—Where data is captured manually, there may be finger trouble that causes inaccuracies. This can be addressed by having quality checks on data capture, or by changing emphasis onto data sources that are more reliable.

Omitted characteristics

There will often be characteristics, both known and unknown, which are unavailable for use, either in the scorecard development or implementation. Instead, lenders will make the best use of available characteristics, which may not be optimal. There are several reasons for characteristics being omitted:

Compliance—The data may not be used for legal, regulatory, or ethical reasons.
Poor quality—The data quality may be so bad that it cannot be used.

Lack of infrastructure—The infrastructure is not in place to obtain the information, whether it is a communications link (credit bureau, other product area), or tables to interpret it (conversion of addresses into geo-codes).

Ignorance—Being unaware that the information can provide value. This can be addressed during initial system design, and ongoing investigations into possible new characteristics.

Sample selection

Sampling is done to ensure that a representative group of cases is used for the scorecard development. They are usually chosen using stratified-random sampling . . . separate sampling of several predefined groups. There may, however, still be problems, where data is either improperly included or excluded.

Improper inclusions—Records should only be included if they are representative of cases to which the model will be applied in future, especially where it is used in decision-making. For application scorecards exclusions might include: VIP accounts; products or markets that have been discontinued; and cases that will be policy declines in future, such as where minimum-income thresholds have increased.

Improper exclusions—There are two types of cases that will be missing from the development, rejects and non-entrants:

(i) **Rejects**—This is the most commonly mentioned bias, whenever credit scoring is discussed. With selection processes, lenders only have true performance for cases that were accepted. In order to represent non-selected cases adequately, reject inference is required to provide an educated guess of what the performance would have been otherwise (see Chapter 19, Reject Inference).

(ii) **Non-entrants**—Groups that have not applied in the past, as would be the case where there are conscious changes to the markets being targeted. This is almost impossible to address within a development, but lenders may be able to shift emphasis onto bureau scores.

Transformation

In most scorecard developments, the characteristics will be transformed to allow the model to capture best their relationship with the target variable. In many instances, this step may be omitted, or be done improperly.

No transformation—When assessing numeric characteristics, regression formulae of the form $y = a + bx$ assume that there is a linear relationship between x (numeric) and y (risk measure), but this is seldom the case. Examples are asset and revenue growth rates; the highest risks usually lie at both ends of the spectrum, either negative or very high growth.

Improper transformation—There are a number of methods that can be used to transform characteristics into a more suitable form: dummy variable, weights of evidence, logarithmic, exponential, standardisation using z-statistic, polynomial expansion, and others. The first two rely on binning, while the others require mathematical calculations.

Misapplication

Very often, companies will borrow scorecards that have been developed in other areas, with the hope that they will provide value. Problems can arise from:

Inappropriate use—The model is not appropriate in the current environment, whether because of differences in the underlying population, the infrastructure used to process the data, or even the end goal of the model.

Shock events—There may be instances where some significant event causes defaults, or lack thereof, such as natural disasters, corporate closures, and price collapses or surges (agricultural, property,[2] commodity).

3.1.5 What can be done about it?

There are means of addressing these issues, whether tailoring how the scorecards are used, using different types of models, or accessing new data sources. Some of the first questions that should be asked are:

Will the scores drive or guide decisions?—Demands are less when cases are being scored for guidance.

Why should internal ratings be used when external ratings are available?—Lenders can avoid significant development costs by using generics, but will lose the information rents possible from exploiting their own data on existing customers.

Are other forms of models possible?—Scorecards can also be developed using expert input, whether based on raw data, or scores already available elsewhere (generics, other products).

In retail credit, lenders strive to automate as many decisions as possible. The motivation for human input arises only where the models are known to be weak, the value at risk is large, and/or the potential profit is high. Instances where judgmental assessments dominate are in

---

[2] Mays (2004) highlights how increasing property prices will have a disproportionately beneficial impact upon home loans with high loan-to-value ratios.

sovereign, corporate, and project-finance lending, where the borrowers' financial situation is complex, the information is not standard and/or difficult to interpret, and volumes are extremely low. If there is a scoring model, it will only be used for guidance, and the underwriter will assess other information that has not been incorporated in the score. The situation is similar at the other end of the spectrum in underserved markets, where there is little or no economic activity to aid transparency. In these and other instances, lenders—and even governments—are trying to put together the infrastructure and legal frameworks necessary to aid the provision of credit.

The internal versus external rating issue is primarily the bespoke versus generic debate: Most internal ratings are provided by bespoke models developed specifically for a lender, while external ratings are generics based upon the experience of many lenders, that are also available to competitors. For retail credit, the latter may be provided by credit bureaux, or co-operatives. Generics are often used by lenders that: (i) are small, and cannot provide sufficient data for a bespoke development; (ii) are looking to enter new markets, where they have no experience; or (iii) lack the technological sophistication to develop and implement a bespoke system.

Finally, not all models require data. Lenders can try to develop expert models, based upon the experience of their underwriters. The use of expert models is tenuous, as it is widely accepted that, by far, the best results are achieved using statistical models. Experts are known to be good at identifying the relevant factors, but are not very good at determining the appropriate weights to be assigned to each. Even so, expert models are often used as interim measures, until sufficient data has amassed for a proper scorecard development. Often, these will take the form of hybrids, which use both statistically-derived models and expert input.

3.2 What measures are used?

The qualitative approach emphasises explanation, narrative, and anecdotes, as opposed to the quantitative approach on prediction, models, and statistics. This would all be a matter of personal preference, except that statistics dominate anecdotes because the bottom line is a statistic—a portfolio with lower average credit losses, other things being equal, makes more money regardless of how many compelling anecdotes exist.

Eric Falkenstein (2002).

Credit scoring is a highly statistical and mathematical discipline, which demands its own measures. These are treated here under three headings:

(i) **Process and strategy**—Used by individuals, who are using the scores to drive strategies, and monitor the results.
(ii) **Scorecard performance**—Used to assess the model's power and stability (see Chapter 8, Measures of Separation/Divergence).
(iii) **Default probability and loss severity**—Used for risk-based pricing, and various finance functions (see Chapter 26, Finance).

3.2.1 Process and strategy

Credit scores are developed as tools for managing the business, which broadly speaking has two aspects:

Selection—How many cases enter the system, and the immediate result?
Outcome—What happens to those cases subsequently?

The selection aspect only applies to selection processes, such as application scoring for new-business origination. In these instances, volumes, reject rates, accept rates, and take-up rates have to be measured. They have a major bearing on the profitability and growth (or shrinkage) of the portfolio, and will be a function of the lenders' marketing strategies, business model, competition, and customers' circumstances.

Once in the system, the focus shifts to the outcome, or subsequent account performance. In credit scoring, just as in gambling, the term 'odds' is used; the casino comes to the workplace—but instead of monetary 'winnings to wager' odds, it is used in the context of good/bad odds, bad rates, default probabilities, or probability of good (P(Good)). These are all a function of the bad definition, and will vary according to the product, process, and company. The usual interpretation of 'bad' is, 'If I knew then what I know now, I would not have done the business!' and vice versa for good. Lewis (1992) indicated that for the bulk of risk developments he was involved with, the good/bad odds fell into the 10 to 20 times range, but ranged from 1/1 to 125/1. In high-risk markets, the odds are usually lower, but are offset by higher profits on good accounts.

An example of these calculations is provided in Table 3.2. The good/bad odds and P(Good) calculations include only goods and bads, as they are the only accounts included in the modelling process (see Section 15.2, Good/Bad Definition). In the example, there are 60,000 goods and 3,000 bads, giving a good/bad odds rate of 20 times, and a P(Good) of 0.952 (or 95.2%). The bad rate calculation also includes the 9,000 indeterminates, providing a total of 72,000 accounts, in which instance the 3,000 bads provide a bad rate of 4.2%.

Table 3.2. Odds and bad rate calcs

Description	All accounts	Good/bad odds	Bad rate
Good	60,000	60,000	60,000
Indeterminate	9,000		9,000
Bad	3,000	3,000	3,000
Exclude	900		
Total	72,900		72,000
Good/bad odds		20.0	
P(Good)		0.952	
Bad rate			4.2%

Reject inference

In selection processes, both 'accepts' and 'rejects' must be gauged. Unfortunately though, there is usually no performance data for rejects. If accept performance only is used, there will be a *sample selection bias*, because a significant subgroup to which the scorecard will be applied in practice, has been ignored. Reject inference attempts to address this potential bias, resulting in two sets of performance measures: *known performance*, for accepted applicants where performance is readily available; and *inferred performance*, informed guesses, provided by the reject-inference process.

If the existing selection process is providing any value, then known performance will always be better than any inferred performance, usually by a factor of two or more. Scorecard developers have some discretion in setting the multiple, and it is considered good practice to be a bit harsh on past rejects, in order to reduce the size of the 'swap set'—the set of cases that might receive a different decision from the new model. Care should, however, always be taken—especially where there are large numbers of rejects—as reject inference is fallible, and the inferred performance might distort the results. Indeed, many scorecard developers and other commentators dispute the value that can be added by reject inference. Over time though, lenders are learning how to use *cohort performance*—meaning outcome-performance data on other loans held by the customer—to enhance the estimates.

The combined set of known and inferred performance ('all') is then used for the scorecard development. An illustration is provided in Table 3.3, where of the 15,000 through-the-door applicants 80% were accepted, with an outcome all good/bad odds performance of 3 to 1 (which for interest, would be an extremely high-risk portfolio). For the 3,000 rejects, there is an inferred odds rate of 0.5 to 1—six times worse than the known performance group. When combined, the population of 15,000 has an odds ratio of 2 to 1, which provides 5,000 bads for the development.

Strategy

All of the above examples are at portfolio level. In practice however, these measures are applied to different segments, especially those defined by score, and especially for setting

Table 3.3. Inferred performance for rejects

	Known	Inferred	All
Good	9,000	1,000	10,000
Bad	3,000	2,000	5,000
Total	12,000	3,000	15,000
Odds	3.0	0.5	2.0
P (Good)	0.750	0.333	0.667
Bad rate	25%	67%	33%
% Row	80%	20%	100%

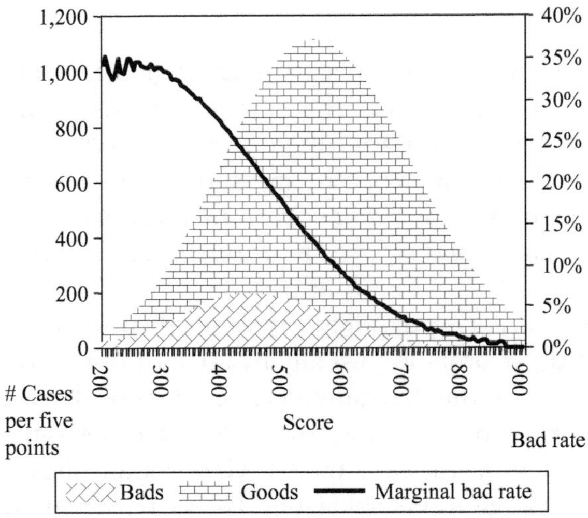

Figure 3.2. Bad rate by score.

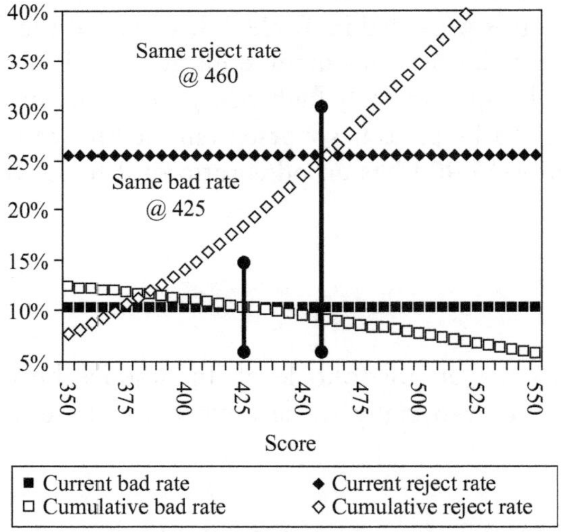

Figure 3.3. Cut-off strategies.

strategy. Figure 3.2 shows the good/bad distributions and marginal bad rates (inclusive of inferred performance) for a hypothetical scorecard. When comparing different models applied to the same set of data: (i) the greater the distance between the two distributions, the better; and (ii) the steeper the bad rate graph's slope, the better. This is not sufficient by itself, as the scores provide the greatest value when used in strategies, the choice of which will depend upon the lender's goals. Figure 3.3 shows what the cut-offs would be under two traditional (hypothetical) scenarios:

Same reject rate—Used if the lender wishes to reduce bad debts. A score cut-off of 465 will match the historical reject rate of 25.5 per cent, and reduce the bad rate from 10.3 to 9.1 per cent (an 11.6 per cent reduction).

Same bad rate—Used if the lender wishes to gain market share and grow the business. A score cut-off of 425 will match the historical bad rate of 10.2 per cent, and reduce the reject rate from 25.5 to 18.4 per cent (a 9.5 per cent improvement).

Such approaches are simple, and are favoured when application scoring is first implemented. Lenders can achieve core objectives, without massive changes to the structure of the business. There are also a lot of choices in between these two points, and when circumstances are very favourable—especially where lenders have invested heavily in their back-end processes—lenders may risk even lower cut-offs.

Ideally, lenders should try to maximise profit. Lewis (1992) was the first to highlight the obvious approach of setting the cut-off to the lowest score with a contribution greater than or equal to zero, which implies accepting any account that provides a profit. As a highly simplified example, if each good account results in profit of $1 and each bad in a loss of $19, then the optimal cut-off is where the marginal good/bad odds are 19 to 1. The task then becomes one of coming up with reliable profit and loss figures at account level, which presents its own challenges.

These are the Model T versions of strategy-setting using credit scores. Traditional approaches assume that the same offer is made to each customer, and that risk is the only factor considered in the decisions. Over time, lenders have become more comfortable with credit scoring, and have learnt how to: (i) take potential profitability into consideration; (ii) incorporate other aspects of customer behaviour (response, retention, and revenue); (iii) use it to adjust loan terms, especially for risk-based pricing; (iv) use it at other stages of the risk management process (marketing, account management, recoveries); (v) apply scientific approaches to better achieve business goals (champion/challenger, experimentation, optimisation, simulation); and (vi) use it for other purposes, such as forecasting and portfolio valuation.

3.2.2 Scorecard performance

Credit scoring provides an extremely valuable tool for measuring risk, but at the same time, the results need to be measured. The particular aspects of interest are power and accuracy, both of which are subject to drift. Power refers to a score's ranking ability, or the extent to which it discriminates between good and bad. It is the primary attribute that lenders require of scorecards; the greater the power, the greater the value they can provide in business processes. In contrast, accuracy refers to how closely the odds or bad rate estimates approximates what happens in practice. Because it is so dependent upon economic, operational, marketing, and other exogenous factors that cannot be captured by credit scores, it is secondary to power; and can really only be achieved through calibration, based on newer data, long-term averages, (or 'central tendencies'), supplementary economic modelling, or even judgmental overlays. It is of primary interest in finance functions, especially where the scores are being used for pricing, forecasting, capital reserving, or other calculations. And finally, drift is the extent to which things have changed over time, which has implications for power, accuracy, and the overall effectiveness of the scorecards within the business. These changes are illustrated in the accompanying figures: Figure 3.4 shows possible changes in the account distribution, while Figure 3.5 shows changes in the model's power and accuracy.

Power loss	Accuracy loss	Account distribution
No	No	Changes in 'all' are accompanied by proportional changes in good and bad, across the entire range.
No	Yes	A constant change in the good/bad odds along the full range of possible scores.
Yes	No	Slope of the score to odds curve reduces, without a change in the overall good/bad odds.
Yes	Yes	Both the slope and the overall good/bad odds change.

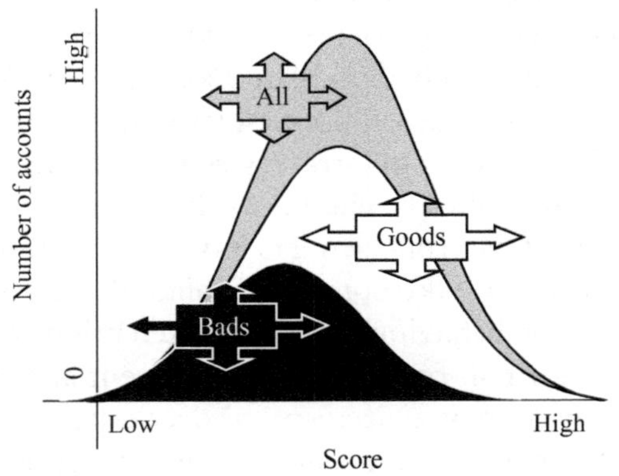

Figure 3.4. Score distributions.

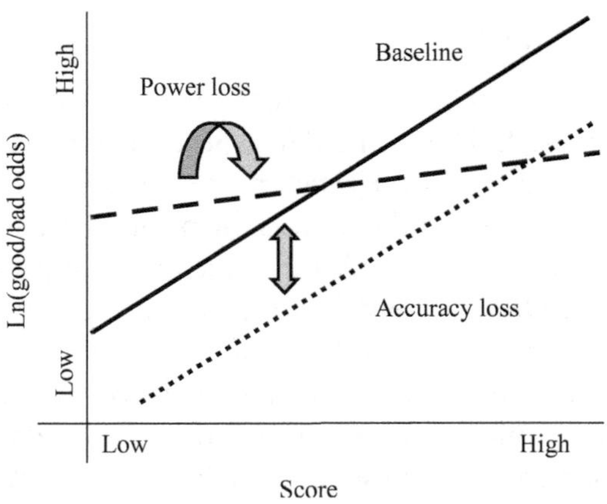

Figure 3.5. Power and accuracy loss.

While a loss of accuracy can be corrected by recalibrating the scorecard or modifying strategies (cut-off, limit, etc.), the only way to correct for a loss of power is by modifying or redeveloping the scorecards. In any event, there must be strict procedures in place, to determine when drift moves beyond acceptable boundaries. There are only a few measures used to assess model accuracy. Most lenders will focus on relative changes to the outcome measures at the portfolio level (like changes in the overall bad rate), but it is possible to use *binomial probabilities*, the *Hosmer–Lemeshow statistic*, or the *log-likelihood measure*. The latter splits the error into its power and accuracy components, using naïve models as reference points.

In contrast, there are a lot of tools available to measure power and drift, the generic terms for which are measures of separation, measures of divergence, or power-divergence measures. When used to measure scorecard power, they are gauges of the graph's slope in Figure 3.5. The most commonly used measures are *rank-order correlation coefficients* that provide values between +1 and –1, where +1 means it is always right, –1 means it is always wrong, and 0 means there is no relationship. Such measures include the *Gini coefficient* (also called Somer's D) and *Spearman rank-order correlation* coefficient, while the *Receiver Operating Characteristic* is similar. Other measures include: the *Kolgomorov–Smirnov statistic*, which provides the maximum difference between the cumulative percentage of goods and bads, across the range of possible scores; the *chi-square Statistic*, which measures the difference between observed and expected values, where the expected values for each score range assume the average odds; and the *Information Value* (Kullback divergence measure), which measures the difference between two distributions. Most of these same measures can also be used to measure drift in the score distribution, in particular the *chi-square statistic*, *Kolgomorov–Smirnov Statistic*, and *Stability Index* (Kullback divergence measure, applied to changes in a distribution over time).

3.2.3 Default probability and loss severity

Credit scoring's primary strength is its ability to rank risk. Increasingly, however, lenders have to estimate expected losses (ELs)—and even profits—whether for risk-based pricing or portfolio valuation. The EL is the amount that the lender expects to lose, based upon available data. It is made up of two parts: *probability-of-default* (PD), which is the risk of non-payment according to some definition; and *loss severity*, the extent of the loss in the event of default, which is affected by the *exposure-at-default* (EAD), *loss-given-default* (LGD), and *maturity* of the loan (M).

Equation 3.4. Expected loss $\$EL = PD\% \times \$EAD \times LGD\% \times f(M)$

Probability-of-default (PD%)—An obligor (borrower) risk rating, which is related to individual economic and environmental circumstances.

Exposure-at-default ($EAD)—A monetary value related to the outstanding balance, agreed loan limit, the lender's shadow/target limits, and loan product characteristics.

Loss-given-default (LGD%)—Proportion of the EAD that the lender expects to lose if default occurs, which is heavily influenced by collateral and other security.

Maturity (f (M))—An adjustment that is a function of the remaining loan term or repayment schedule, which applies in the wholesale market for maturities of longer than one year.

Care must be taken here, as there will always be a positive correlation between default probability and loss severity, which is not captured in many models. This is best illustrated by considering an economic downturn, when both increase: (i) as asset values reduce, counterparties are more likely to walk away from them, which results in an increase in both LGD and PD;[3] (ii) LGDs increase, because the time frames required to collect, if at all, become longer; (iii) because the number of defaults are higher, the LGD values will be dominated by those occurring during downturns, which will result in conservative results for capital allocation and pricing calculations; and (iv) EADs may be higher, because lenders are more likely to: (a) take greater advantage of any credit lines currently available; (b) request increases; and/or (c) abuse the facilities. On this last point, there are also contrary tendencies, because lenders relax and tighten their lending policies according to the perceived risk, both for individual borrowers, and during the cycle. This leads to higher EADs during upturns, and for companies perceived as low risk.[4]

Some further comments can be made with respect to some of the individual elements. First, according to Miu and Ozdemir (2005:30) it is common practice to split EAD into drawn EAD_d and undrawn EAD_u components, and the 'forward-looking dollar amount' is calculated as $EAD = d \times EAD_d + u \times EAD_u$. The components are calculated for defaulters using aggregated drawn and undrawn values at time of default and one year prior:

$$EAD_d = \min (d_T, d_{T-1})/d_{T-1}, \text{ and}$$

$$EAD_u = \min (1, \max (0, d_T - d_{T-1})/u_{T-1}).$$

Note that these formulae assume defaulters have—on average—been managed at or within their limits.

Second, there are two primary approaches for LGD estimation: (i) *workout*, which discounts post-default cash flows; and (ii) *market*, which uses the market value of a security at time of default. The latter is infeasible for retail portfolios. Third, the final term, $f(M)$, is used to recognise the higher risk of *longer maturities*, and applies mostly to corporate, inter-bank, and sovereign lending. It is usually dropped, because in most cases: (i) its impact is negligible; (ii) lenders annualise the PD, EAD, and LGD values; and (iii) at least part of it may be already reflected in the EAD and LGD values. For loans with known repayment schedules, M is calculated as the weighted-average time-to-maturity, using the scheduled cash flows. If M cannot be derived, then the termination date of the agreement should be used, as a conservative estimate. M is not used directly in the formula, but is instead used to calculate an adjustment, the '$f(M)$' (function of maturity) shown in Equation 3.4, which is usually only slightly greater than 100 per cent.

---

[3] This can be especially difficult for real estate markets, where asset correlations are high. Property owners are prone to jump ship simultaneously, especially when their wage or rental incomes fail to cover loan repayments.

[4] See Miu and Ozdemi (2005:32), who also made points (ii) and (iii) (p. 28, fn 25).

Fourth, losses can be split into: (i) the loss of principal; (ii) collections and recovery costs (workout and legal); and (iii) the cost of funds (Schuermann 2004). Discounting the post-default cash flows usually captures the latter.

According to Miu and Ozdemir (2005), if the post-default cash flow volatility has already been captured elsewhere, then a risk-free discount rate should be used. Otherwise, there should be a risk premium.

EAD and LGD could theoretically be modelled using statistical methods, but the low numbers of defaults may make it infeasible. LGD is especially problematic due to problems obtaining data on the amount and timing of post-default cash flows, unless appropriate systems are in place. Irrespective, any bank hoping to use the advanced approach under Basel II needs to come up with these estimates, whether using own or pooled data.

And finally, in general, the possible post-default outcomes are *cure* (rehabilitation), *restructure* (renegotiation), and *liquidation*, and if the latter, funds may be recovered by realising the collateral's value, calling upon guarantees, any residual remaining after all senior debt has been settled, or other sources.[5] Studies have shown that the LGD tends to be a function of: type of debt (bank loan, bond, store credit); contractual terms (seniority, collateral); market segment (higher in sectors with greater assets); and economic conditions (better when times are good). The LGD will also vary depending upon the lender's bargaining power, experience in managing distressed borrowers, and ability to realise collateral's value.

Finance calculations

Credit scoring was originally used for accept/reject decisions in fixed-offer scenarios, but is increasingly being used in more innovative ways. It has become the basis for the expected-loss calculation, which is used for risk-based decisioning and 'value at risk' (VaR) models. Risk-based decisioning includes: *risk-based pricing*, where prices and loan terms are adjusted according to the level of risk, which is especially common where the resulting portfolios will be securitised; and *risk-based processing*, where other actions are adjusted, such as the level of documentation, or number of security checks when processing applications.

In contrast, VaR models do not affect decisions on a deal-by-deal level, but instead focus on the portfolio. They are used to provide estimates of worst-case losses that can arise from market fluctuations, assuming a given time frame and confidence interval . . . the greater the EL and loss volatility, the greater the possible unexpected loss. At the extreme, the loss may be catastrophic, resulting from events that might occur once in a millennium. VaR models have become the basis for determining banks' capital requirements, and have been adopted as part of the Basel II regulatory framework. The formula in Equation 3.4 is still quite simplistic, as it

---

[5] The concepts relating to post-default outcomes and treatment presented in these two paragraphs were influenced by presentations by, and discussions with, Christian Endter and Evren Üçok of Mercer Oliver Wyman, during early 2006.

does not recognise the potential variation that can occur in the underlying values.[6] Regulators will thus increase capital requirements, to ensure that there is sufficient capital to handle unexpected losses (see Chapter 36, Capital Adequacy).

Bad versus default definition

When dealing with corporate bonds, the definition of good and bad is clear-cut—either the obligor defaults, or does not. When dealing with loan accounts however, the situation is different—and there are usually differences between the default definition and the good/bad definition used for a scorecard development. The scorecard good/bad definition focuses upon providing the *best possible risk ranking*; whereas the default definition is used for *finance calculations* (but could be used as a good/bad definition).

Why is this? The primary goal of credit scoring models is to provide tools that aid case-by-case decision-making, and any extra benefits that come from aiding the finance function are secondary, unless the two are so intertwined they are indistinguishable. When deciding upon the scorecard definition, it must be ensured that: (i) the scores discriminate between cases the lender wants, and does not want, taking into consideration those that may not be clear cut; and (ii) there are sufficient bads to develop a model. Even so, the good/bad and not-default/default statuses should be very highly correlated, to the extent that if the good/bad scores cannot be used directly, it should still be possible to map them onto default probabilities.

While there may be some flexibility around the good/bad definition, the default definition will be set either by company policy or regulation, and may vary by type of organisation. Perhaps the best illustration is the Basel II default definition, which classifies accounts as being in default, if at any time during the previous year:

(i) The customer was 90 days-past-due (for cheque accounts, 90 continuous days in excess of agreed limit) on any material obligation to the bank.
(ii) Other factors made it clear that there was a high probability of loss, such as when a specific loss provision was raised, financial difficulties caused the borrower to request a (distressed) loan restructuring, the account was passed on to a third-party recoveries agency, the obligor was put under bankruptcy protection, or the lender filed for the obligor's bankruptcy.
(iii) A loss was incurred, either because any or all of the debt was written- off, or sold for less than the outstanding balance.

Definitions versus estimates

Something that should be highlighted is the difference between a current-status and worst-ever definition, and a point-in-time versus through-the-cycle estimate. These concepts are sometimes confused. Definitions are the basis for the target variables used in scorecard

[6] Some people maintain that risks where accurate probabilities can be determined are no longer risks but a 'cost of doing business,' and that it is only the random exogenous or idiosyncratic risks that are threats.

developments and reporting. With a *current-status definition*, a case tests positive only if the condition holds true at the end of the period, whereas with a *worst-ever definition*, it tests positive if the condition holds true at any point over the period. Basel II requires that banks use a worst-ever definition (covering a one-year period), while scorecards may be developed using either current- or worst-ever, and may have elements of both.

In contrast, estimates are probabilities derived using those definitions. A *point-in-time* (PIT) estimate refers to immediate probabilities, typically one-year, that will fluctuate up and down over the course of an economic cycle. In contrast, a *through-the-cycle* (TTC) estimate is one that approximates a stressed bottom-of-the-cycle scenario, with a horizon of five years or more. According to Aguais (2005), no risk estimate will ever be purely PIT or TTC, but will always be a combination of the two. For example, default estimates based upon account performance or the value of traded securities tend towards PIT, while rating agency grades tend towards TTC. All of them will vary over an economic cycle, some more than others. Companies' own internal grades are made up of a combination of 'subjective assessments, statistical models, market information, and agency ratings' that have a mixture of different time horizons. While it may be ideal to provide separate PIT and TTC estimates for each obligor, this is beyond the capabilities of today's banks, and has not been required by Basel II.

Aguais et al. (2003) highlight that the two terms are relatively new additions to the credit lexicon, and were first used with respect to companies' credit ratings. 'Through-the-cycle' was first used in Moody's and Standard & Poor's (S&P) literature—in 1995 and 1996 respectively—to highlight how they take economic fluctuations into account, and 'point-in-time' was coined in a 1998 article by two Federal Reserve researchers, William Treacy and Mark Carey, to illustrate the difference between the approaches used by the rating agencies and many banks.

Rather than focusing upon PIT or TTC, lenders should use whatever provides the best estimates over the longer term. These can then be transformed into PIT or TTC estimates at portfolio level, depending upon what they will be used for. If the goal is to determine capital reserving requirements, a TTC estimate will be used to provide estimates that *ceteris paribus* keep reserve requirements stable over the economic cycle. In contrast, if the goal is account management, a PIT estimate will be used to keep the decision-making consistent, as the level of risk changes over the cycle. Furthermore, if used for pricing, the term used for the estimate should approximate the deal term.

3.3 What is the scorecard development process?

Up until now, the focus has been on what lenders are trying to achieve. This section shifts the focus onto the 'how'. Credit scoring is commonly understood as the use of statistical models in credit decision processes. The scorecard development process is an adaptation of what might be found documented in many textbooks on statistics (Figure 3.6). Care must always be taken in this process, because each step is dependent upon the validity of what was done upstream. Some of the stages covered over the following pages are:

(i) **Project preparation**—Goal definition, feasibility study, and player identification.
(ii) **Data preparation**—Data scope, good/bad definition, sample windows, sample size, generated characteristics, and matching.
(iii) **Scorecard modelling**—Transformation, characteristic selection, reject inference, segmentation, and training.
(iv) **Finalisation**—Validation, calibration, strategy setting, loading, testing, and monitoring.
(v) **Decision-making and strategy**—Level of automation, change management, overrides, policy rules, referrals, and strategy enhancement.
(vi) **Security**—Documentation, confidentiality, and change control.

3.3.1 Project preparation

Prior to any real scorecard development work being done, there are a lot of decisions to be made, which require a lot of research. These initial project preparation stages can be broken up into: goal definition, feasibility study, and player identification.

Goal definition

Besides determining market, product, and process is to be served, lenders also have to define objectives. What is the goal? Key objectives might be to improve credit decisions, reduce the cost of decision-making, and/or have consistency across the branch network. This applies to every type of scoring, and may relate to any number of factors. Lenders have to consider:

Economy—What are the immediate business goals? Lenders may wish to reduce their losses; or, alternatively, grow the book while keeping losses under control.

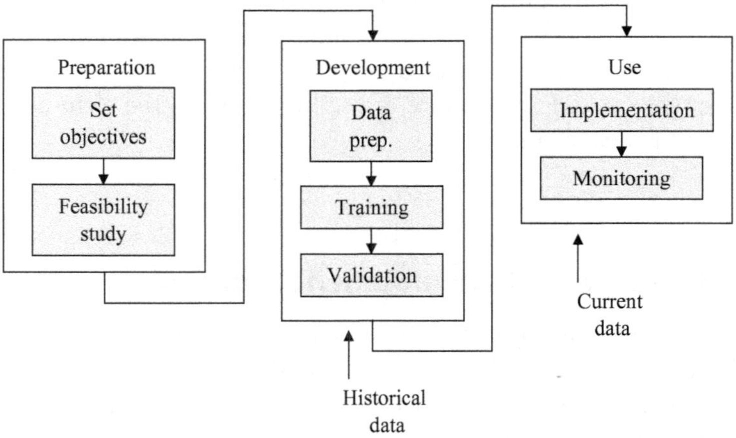

Figure 3.6. Development process.

Customers—What are the customer requirements? Customers may want personalised service, but will sacrifice it for lower-cost and higher-speed decisions.
Competitors—What are other lenders doing? The decision to use scoring may be proactive or reactive, depending upon current practice within a specific market.
Legislation—What does the law require? Objective models may be required to comply with existing or expected laws, especially to ensure fair and consistent decision-making.

Feasibility study

Lenders may have their goals, but these are not always achievable. A cost/benefit analysis should be done, but this is often a holy grail, as benefits are much more difficult to quantify than costs. Many of the benefits are subjective and/or long-term, relating to an overall change in the way of doing business—especially where lenders are reacting to competitive pressures. As a result, many of these analyses are based on whether or not the company can afford the costs. Beyond these purely financial aspects, and assuming that the project has organisational commitment, there may be other substantial constraints:

Data—Will there be sufficient data available to develop a model?
Resources—Will there be money and people available?
Technology—Is the technology available to support us?

Statistical models are dependent upon data, and problems with data may compromise the models. Factors that must be considered are the data sources, whether there is enough data, and how to get it into the appropriate form. Resources relate primarily to people. Are there scorecard developers, business analysts, IT staff, and others available to assemble the data, build the scorecard, and then implement it? There are trade-offs between the use of internal staff and external consultancies, which can have a significant impact upon costs. And finally, technology is evolving at an extremely high rate, and is expensive. Even so, its cost has been reducing, which has lessened the economies of scale needed to justify decision automation. This applies not only to processing power and data storage, but also to networking and communications costs.

Player identification

The two main groups are the steering committee and the project team. The steering committee is responsible for driving the project, and the two major players are the sponsor and the champion—both of whom are part of the company executive, and may or may not be the same person. Cheques are signed by the sponsor, while the champion has to ensure buy-in from all parties. The champion must have sufficient clout to get co-operation and resources; otherwise, the project can die a premature death. This person is usually the same person who motivated the feasibility study. Other people on the steering committee will include those from: (i) the targeted area, where the scorecards will be implemented; (ii) downstream areas, where the changes may have an effect; and (iii) other business functions, including strategy/marketing, sales/distribution, compliance/legal, finance, and information technology.

While the steering committee may drive the project, they will have little direct involvement. Once feasibility has been determined, the next step is to identify the individuals who will be responsible for different aspects of scorecard development and implementation. This is the project team, which is comprised of a:

Project manager—Reports to the champion, and must advise of any resource requirements or shortfalls.
Scorecard developer—Develops the actual scorecard.
Internal analysts—Assist in assembling and understanding data.
Functional experts—Assist in understanding the business and affected areas, and will be key when deciding upon the strategies to be employed.
Technical resources—Responsible for final implementation of the scorecard.

When lenders are doing bespoke scorecard developments for internal use, most of these tasks are performed by internal staff. For years though, the *project management and scorecard development* tasks were outsourced to scorecard vendors, including Fair Isaac (FI), Experian, Equifax, and others. As credit scoring became more widespread though, lenders found that in-house developments could be much cheaper, and even yield better results. Even so, they may still call upon external assistance to provide developmental support, and to obtain, assemble, and interpret external data.

Besides the people physically involved in the scorecard development, it is also necessary to include *functional experts*, who are business specialists that provide input into past and planned changes that could affect the model results. These same individuals will also provide input into the strategies and policies that will be employed along with scorecard. Finally, the *technical resources* are systems or IT staff, who will be responsible for building the system (greenfield), or implementing the scorecards (brownfield). This would include individuals with programming, networking, systems design, and other capabilities.

3.3.2 Data preparation

When doing university-level courses in statistics, much emphasis is put upon the statistical and mathematical aspects. It is only once trainee statisticians start working in the field, that they truly realise the importance of data preparation and sample design. Extensive coverage is given to the topic in Chapter 15 (Data Preparation), so this section just touches briefly on:

Project scope—Which cases are to be included?
Good/bad definition—What is to be predicted?
Sample windows—What will be the observation and outcome periods?
Sample size—How many cases will be included?
Generated characteristics—Are special characteristics required?
Matching—How is the data to be brought together?

Some of these issues have already been touched on briefly. The first decision relates to the project scope, which defines which cases should be included. Ultimately, the data should only include records that are representative of cases, where the score will influence the decision-making. Exclusions would include, amongst others, statutory declines, and those that are the responsibility of other business units.

Predictive models are developed using historical data, to explain the relationship between data observed at one point in time (independent/predictor), and a later outcome (independent/target/ response). The good/bad definition provides the target variable for what the model is trying to predict, and crystallises lenders' views on what is desired and undesired behaviour.

> The use of dualities to represent the universe is called Manicheanism, which divides everything into binary pairs that are either/or or polarised. It forms the basis of most organised religions today by separating everything into good and evil, starting with Zarathustra (628–551 BC). Other combinations are black/white, mind/body, night/day, etc. This type of thinking was good enough for early man to understand his world (Arsham 2002), but is used here as the cornerstone for creating a continuum.

The definition may be: (i) *prescribed*, by accounting or an external agency; (ii) *subjective*, based upon judgmental inputs from the lenders' own experience; or (iii) *empirical*, the result of data analysis. A 'bad' definition that is too strict may limit the number of bads available for a development, but one that is too lenient may weaken the scorecard. For credit risk, the definitions are dominated by missed payments or repeated limit violations. The definition will also include classifications of one or more of 'reject', 'indeterminate', 'inactive' or 'not-taken-up', and 'exclude'.

For many developments, the data will not be available from one place, but is instead obtained from many data sources, such as the application processing system, account-management system(s), collections systems, customer-information files, credit bureaux, etc. In order to create the customer records required for the scorecard development, data must be linked using some matching key, whether internal (customer number), or national (personal identifier), or, failing all else, the name, address, and birth date. Where systems have been automated, this data may be readily available, but if not, it is up to the project team to ensure appropriate matching.

A major constraint for developing statistical data-driven models is the amount of available data, which will impact upon the sample size. The minimum requirements commonly quoted are 1,500 goods and 1,500 bads, with a further 1,500 rejects for selection processes. These are not large numbers, but bads are often rare and can be a severely limiting constraint. Where multiple scorecards are used for the same portfolio, these constraints will apply to each. It is possible to use fewer cases, but that increases issues related to potential overfitting.

The sample window defines the observation and outcome periods used for the development. Factors that must be considered are: (i) *maturity*, enough time must go by for customers to have the opportunity to show their true colours; (ii) *censoring*, if the time period is too short, valuable information may be lost; and (iii) *decay*, if the time period is too long, the observations may no longer be representative of the current business.

While the data provided could be sufficient, there may be cause for generated characteristics, such as combination, ratio, and 'time elapsed'. Combination characteristics, such as

'Marital Status and Number of Dependents', are used to address interactions, where the relative importance of one characteristic varies depending upon the value of another. Ratio characteristics are used to normalise data for size, such as 'Repayment to Gross Income'; and 'time elapsed' characteristics may be used to work out the customer age, account age, or time since last delinquent behaviour. Utmost care must be taken to ensure that generated characteristics can be implemented in practice.

And finally, data preparation must provide both training and validation samples. The *training sample* is the data used to derive the point allocations, while *validation samples* are used to check the data and final model. Both might include: (i) a *historical sample*, with performance that is used to test whether the final model will work in practice; and (ii) a *recent sample* with no performance, which is used to ensure that the data has not changed substantially since the observation window. The historical validation sample (*out-of-sample*) can take two forms: *hold-out*, from the same time period; and *out-of-time*, from a different time period. Out-of-time samples are preferable, but if infeasible, hold-out samples are usually sufficient. Indeed, there may even be problems obtaining sufficient data for a hold-out sample, in which case, techniques like *bootstrapping* and *jackknifing* can be used, both of which involve continuous resampling of the training sample.

3.3.3 Scorecard modelling

Once data has been assembled, the development of a predictive model can be started. The first step is to choose a modelling technique. According to Siddiqi (2006), the issues that should be considered include: *data quality*, as missing data could force the use of decision trees; *type of target variable*, as linear regression is best suited for continuous outcomes, and logistic or probit regression for binary; *sample size*, as decision trees require more data; *implementation platform*, because the final model has to be implemented in a business system; *model transparency*, which may be required both by regulation and the business, and often forces the use of traditional scorecards; and *monitoring capabilities*, as the lender has to track performance over time. Most lenders will have greater familiarity with certain techniques, and may favour them in spite of potential problems.

Thereafter, there are a number of different stages, which are covered in greater detail in Module E (Scorecard Development Process):

Transformation—Conversion of data into a form that can be used!
Characteristic selection—Which characteristics can provide value?
Reject inference—How would the rejects have performed if accepted?
Segmentation—Do certain subgroups require their own separate scorecards?
Training—What weight should be allocated to each variable?

The first step is to transform the data into a usable form. Even though there is plenty of data, it is often inappropriate for use within the model. There are a number of different transformation techniques available, but most retail (consumer and small-business credit) credit scoring systems have been developed to handle traditional scorecards. As a result, the most common transformation technique is to: (i) create *fine classes* for analysis; (ii) group these further into *coarse classes* of similar risk; and (iii) either convert the coarse classes into *dummy variables*; or calculate a new characteristic containing a *relative risk measure* (like the weight of evidence) for each.

Another task often performed is characteristic selection, which limits the number of characteristics initially considered in the model development. Some scorecard developers will focus on finding characteristics that are correlated with the target variable but not with each other, in order to minimise multicollinearity—especially where sample sizes are small. This can be aided by using factor analysis to group the characteristics, and possibly even use these factors in the scorecard development. Others will use common sense to select the characteristics, and ignore the multicollinearity, instead relying upon the sample size, and ensuring that the point allocations make logical sense, to keep the standard error in check.

For selection processes, there is no historical performance available for rejects, and reject inference is used to make educated guesses. In the early days, no reject inference was done, but over the past few years, lenders have become more sophisticated, and many new terms have entered the credit scoring lexicon. Distinctions should be made between: (i) *performance manipulation* techniques, including reweighting, reclassification (rule- or score-based), and parcelling (polarised, random, or fuzzy); and (ii) *reject inference* techniques, including random supplementation, augmentation, extrapolation, cohort performance, and bivariate two-step; and (iii) *model types*, including known good/bad, accept/reject, and all good/bad. Today, the most sophisticated approaches involve: (i) stratified-fuzzy parcelling; (ii) extrapolation; (iii) known good/bad and accept/reject models in a two-step approach; and (iv) use of cohort performance. Special care must be taken where the number of rejects is very large, as the inferred performance may severely distort the results.

When developing credit scoring models, the cases included must be similar enough to be treated together, but different enough for models to distinguish between them. Different scorecards may be required for a single portfolio, and the segmentation, or scorecard splits, may be affected by five types of factors:[7]

(i) **Marketing**, the lender wishes to apply different strategies going forward, and requires greater confidence in one area (usually much higher risk) than another.
(ii) **Customer**, instances where certain characteristics do not apply, often related to a lack of credit related data.
(iii) **Data**, differences relating to what data is available, or when and how it becomes available, especially for different channels or application forms.
(iv) **Process**, where the cases receive different treatment, whether because of operational, technological, legal, or other factors.
(v) **Model-fit**, all of the above, and others, where the relative importance of the predictors varies between groups.

[7] The starting point for this was a framework provided by Thomas et al. (2001), who suggested strategic (marketing), operational (customer), and interactional (model-fit).

Care must be taken in all of these cases, as there must be sufficient data to develop each scorecard, and bads are often in short supply.

Once the segmentation has been decided upon, model training can begin. This is the glory aspect of scorecard development, where a parametric technique (like logistic regression or DA), or non-parametric technique (like a NN) is applied. For a traditional scorecard, it is where the points (a combination of variable transformation and regression coefficients) are allocated. It is an iterative process, as the scorecard developer may have to generate many models, and/or make many cosmetic changes.

The key factor is to ensure that the points correspond to the relative risk of each group, and to *avoid overfitting*, especially where the predictors are correlated and sample sizes are small. Scorecard developers will guard against: (i) gaps where no points are allocated; (ii) wrong-sign problems, where the points are the opposite of what is expected; (iii) point allocations that decrease where an increase is expected, and vice versa; and (iv) *t*-statistics, or other measures, indicating that variables' relationships with the response function are insignificant. There will also be issues relating to: (i) *controlling* for certain factors, like company strategies; and (ii) *staging*, determining the order in which characteristics will be considered for possible inclusion.

3.3.4 Finalisation

After training has been finished, the next step is to finalise the model, and get it into production. This is covered in both Module E (Scorecard Development Process) and Module F (Implementation and Use), including:

Validation—Will the scorecard work in practice?
Calibration—Can the scores be used to provide estimates?
Strategy setting—How are the scores to be used?
Loading—Physical implementation of the scorecards, in whatever form!
Testing—Is the system working according to design?
Monitoring—Are the scorecards providing the expected results?

After the scorecard has been developed, the next step is validation, to ensure that the model will work on the intended population. Checks will be made to test: (i) the loss of predictive power, when applied to a validation sample; and (ii) score drift, when applied to a recent sample. Lenders may also *benchmark* the results against other models, especially those developed by external agencies (credit bureau or rating agency). Where an existing model is to be replaced, the size and composition of the swap-set should be considered. In all cases, and throughout the development, documentation must be kept of what assumptions were made.

Calibration is used to: (i) ensure that the scores provided by different scorecards have the same meaning; and possibly (ii) determine or refine the probability estimates to be associated with each score. The easiest way is to create grades by banding score ranges, but many lenders want flexibility at score level. This can be done by doing score transformations, or alternatively by mapping the scores onto probabilities.

Scores provide little value without associated strategies. In its simplest form, strategy setting may involve a simple one-dimensional cut-off for an accept/reject decision using an application-risk score. In more complex forms, there may be: (i) multiple cut-offs for different risk grades; or (ii) combination with other factors, including bureau scores or response, retention, and/or revenue dimensions. The strategy may be chosen to cause minimal upset to the existing business process; or alternatively, to make best advantage of the new or updated tool.

Once the scorecard has been completed, the next stage is loading it into the system where it is to be applied. In modern environments, this involves setting up the scorecard details within a parameterised system, along with any changes to strategies that will accompany the new scorecards. In other environments however, it could involve creating, and possibly distributing: (i) paper-based score sheets; (ii) electronic calculators or spreadsheets; or (iii) computer program code, whether on PC, network, or mainframe.

Once the scorecard has been loaded, especially where the calculations will be done electronically, the next step is testing (also referred to as verification). This ensures that the system is working according to design; as opposed to other validations, that determine whether the design is correct. All stages of the decision process should be tested, including data, scores, and strategy. Initial loading and testing is best done in a separate test environment, but this is not always possible.

Validation and testing are performed not only at implementation, but also as part of ongoing monitoring thereafter. This includes: (i) *drift reporting*, to measure how much the data and score distributions have changed; (ii) *back-testing*, to ensure that the scorecards have the expected predictive power and accuracy, once performance data is available; (iii) *decision process*, to measure the scores' impact on the business; (iv) *adherence*, to ensure that scores and policies are being applied as intended; and (v) *portfolio analysis*, to measure how well the business is doing generally.

3.3.5 Decision-making and strategy

Once scorecards have been developed, some decisions have to be taken as to how they will be implemented and used within the business. These aspects are covered in more detail in Module F (Implementation and Use) and Chapter 5 (Decision Science), and include:

Level of automation—How much human input will be required?
Change management—Have customers and staff been told?
Overrides—Is it possible to change the decision?
Policy rules—What rules will be applied along with the scores?
Referrals—What happens when manual input is required?
Strategy enhancement—Can the strategies be improved?

When considering credit scoring, lenders have to decide upon the desired level of automation. There is a trade-off between the fixed cost of automating, and the variable cost per assessment thereafter. The most low-tech approach is where staff members calculate scores using a score sheet, and then apply a strategy according to a cut-off or strategy table. Volume lenders, however, will always try to automate to the maximum extent feasible, including data acquisition, score calculation, strategy determination, and decision delivery. There will, however, be instances where lower levels of automation are appropriate.

When organisations undertake significant change, they usually have to do change management to improve its acceptance. Staff should be kept informed of what impact the changes will have upon them, their work, and their continued business dealings. In credit scoring, the amount of change management required will depend upon whether scoring is being implemented for the first time, or minor changes are being made to an existing system.

Scorecards are risk assessment tools, but are not sufficient by themselves. Overrides may occur either through company policy or human judgment. There are four types of decision: (i) the *pre-score decision*, which may reject or redirect a case, prior to a score being calculated; (ii) the *score decision*, determined purely using the score; (iii) the *system decision*, which includes any automated policies; and (iv) the *final decision*, which is subject to a judgmental overlay (also called a manual override). *Low-score overrides* are most likely to result from customer disputes and/or where the underwriter has local knowledge, while *high-score overrides* are most common where there are fraud or extreme-event warnings. Where manual overrides are allowed, extra effort must be put into monitoring the extent of and reasons for overrides.

Policy rules may be: (i) *product rules*, which determine eligibility; (ii) *credit rules*, to cover factors not adequately incorporated in the scorecards; and (iii) *fraud-prevention rules*, which may require more or less verification. Continuous monitoring is needed to ensure that remaining policy rules are still serving the desired function, and new policies may be required to address identified scorecard faults. Referrals are cases that require manual input due to a combination of policy and/or score, which may be required for the purposes of *validation* (fraud checks), *manual review* (calling for extra information), or *adjustment of normal terms and conditions*.

Finally, lenders' strategies need not be cast in stone. Strategy enhancement can be done using decision science tools, such as: (i) *champion/challenger*, experimentation by applying new strategies to small groups; (ii) *simulation*, use of data analysis to simulate the results of changed strategies; (iii) *optimisation*, search for optimal solutions amongst a variety of different strategies; and (iv) *strategy inference*, to better incorporate customer responses to the changed offers/actions.

3.3.6 Security

Finally, lenders will have to undertake certain actions to ensure the security of the system. This includes protecting the organisation against itself, insiders, and outsiders:

Documentation—How was the model derived?
Confidentiality—Who can they tell about it?
Change control—How can changes be made?

Scorecard developments rely not only upon data, but also upon the myriad of assumptions and decisions made along the way. Lenders need some means of reviewing these, which requires documentation covering most stages of the scorecard development process, especially: (i) project scope and objectives; (ii) sample design, (iii) scorecard modelling; and (iv) strategies.

Credit scoring requires a huge investment in infrastructure, and has become critical in driving many credit processes. Staff members, contractors, and consultants should be subject to confidentiality agreements to protect proprietary information from industrial espionage. The information may be of value not only to competitors, but also fraudsters. Authority levels should be put in place to restrict access to scorecards, strategies, systems, and associated documentation. Further, highly sensitive documentation should be kept under lock and key.

Such systems are not static, and there will be times when changes are required to various aspects of the process. This also requires a change control procedure that sets out: (i) *authority levels*, for both authorising and loading the required changes; and (ii) *controls*, to ensure that they are appropriately tested. Where changes are required to correct errors, the authority levels may be low. In contrast, any changes to scorecards or strategies should require authorisation from senior management.

3.4 What can affect the scorecards?

In times of change, learners inherit the Earth, while the learned find themselves beautifully equipped to deal with a world that no longer exists.

Eric Hoffer

The quotation 'Nothing is permanent but change' is attributed to Heraclitus, a fifth century BCE Greek philosopher. His world was one where most people experienced change either as passive observers or unwilling participants. For most of human history change has been unwelcome, as it only caused uncertainty and insecurity. The ancient Chinese curse, 'May you live in interesting times,' stems from an era of significant upheaval when life was uncertain (which applies to much of China's history). In spite of these ancient roots, it was only during the mid-twentieth century that 'the only constant is change' maxim gained widespread currency, as technology increasingly affected people's lives. Today, much of society has become accustomed to change, and some even welcome the constant state of future shock. Even so, in credit scoring, change violates the base assumption that 'the future will be like the past'. It will never be totally true, but usually suffices. Eventually however, lenders have to update their world-view to reflect the current environment—market, economy, company infrastructure, information sources, etc.

Any variation from the base assumption is called 'drift'. It can be rapid, like when dating; or gradual, like in a marriage. In both cases, the same actions may provide different results depending on the circumstances, and adjustments have to be made. The advantage in business though, is that the extent of the changes can be more readily ascertained, as long as there is effective monitoring and feedback. Lenders then have to determine how to adapt their tools and strategies for the current environment.

The types of drift most commonly referred to are: *population drift*, changes to the customer base where the economy and market are the major drivers; and *score drift*, changes to the scores and their distribution that can arise from population or infrastructure drift. Failure to

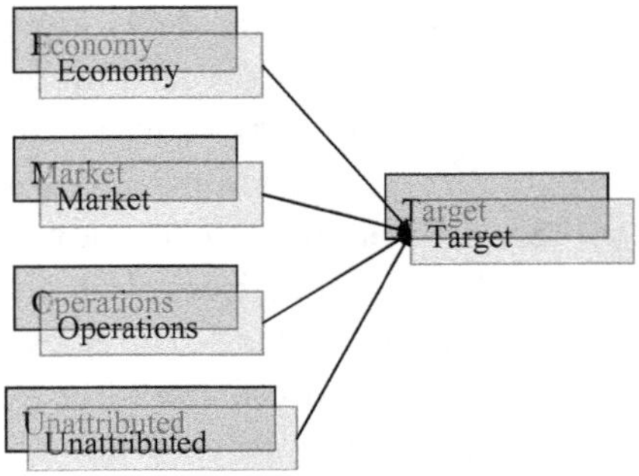

Figure 3.7. Environmental drift.

recognise it can result in major strategic risks, as it implies that the tools used to drive decisions may have unseen faults (Figure 3.7). Here, drift is treated according its possible origins:

Economy—Upturns and downturns, with changes in employment, interest rates, inflation, etc.
Market—Changes to customer demographics, including lenders' conscious moves up, down, or across markets, as well as changes to product features that affect client behaviour.[8]
Operations—Changes to forms, processes, systems, or calculations, used at any point, whether for new or existing business.
Target—People's handling of debt changes over time, as do their responses to various stimuli. Score-driven strategies may also influence the outcomes they are meant to predict.
Unattributed—Changes that cannot be attributed to anything, other than scorecards' age.

3.4.1 Economic drift

Changes to the economy are one of the biggest motivators behind updating scorecards, especially where interest, unemployment, and/or GDP growth rates are affected. Drift may also be directly influenced by the general availability of credit within the economy, which can vary dramatically over time. Between 1988 and 2001, consumer credit in the United Kingdom grew from 50 per cent of national income to 70 per cent (Bridges and Disney 2001). Such massive changes make variations in the economic cycle even greater than would occur naturally, sometimes with strange consequences. During the mid-1990s, competition in the American market caused lenders chasing market share to lower their qualifying criteria, thus spurring both credit growth and delinquencies in a growth economy (Zandi 1998). The expansion ended in 2001, which impacted upon the reliability of any scorecards built prior to that (Wiklund 2004).

One must then ask the question, 'Can scorecards built at one point in the economic cycle be used at other points?' The answer is usually a qualified 'yes'. In general, scorecards are relatively robust, and need not be discarded because of minor changes in the economy. The next

[8] A special case is portfolio drift, which results from movements of existing accounts between portfolios. It has little impact on the business, other than to shift responsibility for the account.

question is, 'Which scorecards are more robust, those built during good times or bad times?' The general belief is that scorecards built during a recession fare better, because the number of bad accounts available for the scorecard development is greater, and the resulting scorecard should be valid for a greater number of possible economic scenarios. Unfortunately, this view is difficult to prove, and nothing can be found in the literature to support it. In either case, economic changes tend to have an effect on a population's overall risk, but scorecards should still continue to rank risk correctly, albeit slightly less so (Lewis 1992, Schreiner 2002a).

The exceptions

There have however, been instances where this did not hold true. According to Hoyland (1995), the end of the Second World War saw a 40-year period where the professional and middle classes in the United Kingdom and the United States were immune to economic downturns, but this all changed with the recession in the late 1980s. In the United Kingdom, accountants and architects with exemplary credit histories were shattered by job losses and tumbling housing prices. This shift caused significant changes in the credit market, thus making any scorecards built during the good-times era less reliable. This is not the norm though, as most recessions have the greatest impact on semi-skilled and blue-collar workers, especially those on temporary contracts, as employers cut back on costs.

Coincidentally, Crook et al. (1992) did an analysis on a UK consumer-lending product, to compare scorecards developed using applications from a good year and bad year, 1988 and 1989, respectively. Both scorecards were developed using a one-year outcome period. Each was then applied to the other dataset, keeping the reject rate the same. The end result was that 25 per cent of those rejected by each scorecard were accepted by the other (i.e. 3.7 per cent of a total 16.5 per cent rejects)—the changes are significant (see Table 3.4). These results may be exaggerated because of the particular circumstances, but provide an indication of a worst-case scenario. Also, in this particular instance, it would be unwise to use the bad year scorecard in practice, as it was an abnormal period.

Treatment of economic variables

Several authors have suggested the inclusion of economic variables, such as aggregated unemployment claims and court judgments, to reduce the economic sensitivity of credit scoring models (Mays 2004). Indeed, unemployment claims have featured in a regression model for the prediction of bankruptcy (Thomas 2000). Not everybody is a convert though. Zandi

Table 3.4. Economic changes

	1989 Apps	
1988 Apps	Accepts (%)	Rejects (%)
Accepts	79.9	3.7
Rejects	3.7	12.8

(1998) recommends having a separate scorecard, comprised only of economic variables, to create a credit confidence index for each region, which could then be combined with normal credit scores to adjust cut-offs up and down. The scores for each region would be updated whenever new economic data becomes available. How well this works when there are significant changes in the broader economy, is another question.

While this expression is a rip-off of the 'business confidence index' concept, there is validity. Indeed, if some form of business confidence index already exists, it might be possible to use it, either as a predictor or to adjust cut-offs.

Wiklund (2004) suggests taking economic changes and regional differences into consideration, as part of the sample design. If the sample comes from a part of the economic cycle that is markedly different than current conditions, it might be possible to augment it with data from a comparable period. If changes in the bad rates have been observed, then the model's predictions may be adjusted to reflect recent trends. Finally, there may be different performances and trends across regions that must at least be understood. The lender may opt to model the differences, or neutralise them. Care must be taken, as the source may not be local economies and cultures, but product features, service delivery, and/or operational efficiencies in each region. Products like 'bankcards and auto finance loans' are becoming more generic though, making this less of an issue. Wiklund also highlights that much research is currently being done into how to 'make scores more robust across economic environments', but as of yet no best practice has evolved.

While most lenders adjust strategies to recognise changes in the broader macroeconomy, some will model cross-sectional data for regional economies—like unemployment, gross domestic product, and house price data—where such data is available. This demands that the lender: (i) ensures that the data is updated regularly; and (ii) guards against larger national shifts, possibly by indexing the values relative to the national average.

3.4.2 Market drift

A company's customers could be compared to members of a club. Many will join and be very active for a couple of years, make a new circle of friends, and eventually move on. The club's culture—quiet, rowdy, nerdy, sporty—will vary as different personalities pass through, and leave their mark. In like fashion, companies experience market drift as the make-up of the client base changes, primarily resulting from the interaction between marketing strategies and competitor actions. In the most extreme case, a merger or acquisition may present a lender with a multitude of new customers, in market segments where they have little experience, and existing scorecards and strategies may be useless, or require significant tweaking.

The greatest catalyst driving market drift is marketing strategies, which vary over time with respect to: *product*—features may increase or decrease the appeal of home loans, revolving credit, credit cards, etc.; *pricing*—interest rate and repayment terms, service and penalty fees; *promotion*—advertising media and target markets; *place*—ease and convenience (branch network, ATMs); and *distribution*—speed of delivery, and service levels. Changes to these strategies are not always well thought out, and often have unintended consequences. All aspects of the credit risk management cycle (CRMC) need to undergo periodic review, to ensure that they are aligned with the current customer base. Factors that should be considered are:

Affordability—Are customers able to afford their debt levels in the current environment? Initial affordability assessments may prove incorrect, especially where there are adverse economic shocks. Special repayment arrangements may have to be made.

Access to credit—Are they receiving offers from elsewhere? Better customers are more fickle, and likely to move their business elsewhere for a better offer. Lenders may have to revisit their pricing, and improve customer retention strategies.

Price sensitivity—Are they willing to pay a premium? Risky customers are less price sensitive, and often more profitable than those traditionally more sought after. This comes at the price of requiring better management, systems, and controls to ensure repayment.

Financial sophistication—Can they manage their financial affairs? The individuals may not have borrowed before, and not be aware of the responsibilities or consequences in the event of non-payment.

Community or parental support—Are there others that can assist in need? For example, parents and related parties may provide guarantees (student loans, enterprise credit), or it may be possible to rely upon peer pressure from the community (micro-finance).

Repayment culture—Do they honour commitments? People's attitudes towards debt vary by income, geography, upbringing, peer group, and other factors.

Repayment mechanism—Are the repayments being secured in an appropriate fashion? Debit orders are the most common, but sometimes lenders want greater control over their portion of the next month's income (payday lending, home-collected credit, micro-lending).

Contactability—Can the customers be contacted in need? It may not be possible to contact people to advise of missed payments, especially if there is no home phone, and little access to a work phone. This poses a challenge for the collections area.

These issues need to be considered throughout the CRMC, whether marketing, application processing, account management, collections and recoveries, or elsewhere. There are usually multiple solutions that need to be evaluated, and perhaps used together.

In post-Apartheid South Africa, retailers started selling to the emerging black market on the basis of 'six months to pay', meaning six one-monthly instalments, but instead customers made a single payment at the end of six months. This confounded the retailers' traditional risk assessments, which normally treat somebody as a bad debt after three months. In like fashion, in Kenya and elsewhere, rural customers are often past due, but regularise the account on their trip to town every three to six months.

3.4.3 Operational drift

As the world changes, so too does the organisation and its infrastructure, including computer systems, data sources, data being supplied, internal procedures, or other factors. These changes can have unexpected consequences at any point in the risk management cycle. For example, with respect to application scoring: (i) changes in procedures used to monitor data capture will affect data quality; (ii) a change of application form wording or layout may change the way in which customers interpret a question; and (iii) a scored field may have been inadvertently deleted, or the way in which it is calculated may change. Such factors affect scorecards' risk ranking capabilities. The impact may be significant, but more often the drift is unnoticeable, or so small that it is difficult to associate with a specific factor. Even so, over time, small changes will lead to a situation where the scorecards have to be redeveloped.

In many respects, this is the most insidious and dangerous type of drift, as there is usually no real reason that the change should have occurred. It might arise because of an error or misunderstanding, where one division makes a change, not realising what impact it will have elsewhere. Communication is critical, as is proper impact analysis when assessing proposed changes to computer systems. Operational drift is worst when it is the result of an implementation error. Many of the variables used in scorecards involve calculations of some form, and it is hoped that those used for scorecard development and operational implementation are the same. If not, there may be substantial differences between expected (per scorecard design) and actual (system-calculated) scores, in which case the scorecard will either be unusable, or have a substantially shortened shelf life.

New data

A special case of operational drift is new and improved data, which has provided the bulk of improvements in predictive power since the 1960s. New-business processing used to put heavy reliance upon a few application form details, and perhaps a phone call to the credit bureau. Today, there is a wealth of information—especially behavioural data—available electronically, which has made many of the application-form details redundant:

Internal data—Performance on other products, and information from other stages in the risk management process.

External data—Better presentation of data by credit bureaux, including bureau scores, improved matching routines, new characteristics including geocodes and aggregates, and an increased acceptance of data sharing.

These are covered in much greater detail in Chapter 12 (Data Sources). Lenders are under pressure to obtain maximum benefit from new data sources, as and when they become available, and try to incorporate them into their scorecards as soon as feasibly possible. It is easier said than done though, as time is needed for the observation data to stabilise, and for accounts to mature. Where a specific item is known to be highly predictive, like 'three or more months delinquent on another product', it can be introduced as a policy rule in the existing system.

3.4.4 Target drift

People are not Pavlovian creatures that always respond in the same way to a ringing bell. Reactions vary, not only according to the situation, but also over time. Culture changes, people's attitudes towards debt and obligations change, and people's reactions to stimuli—such as the 7 PM phone call (just after supper and before their favourite prime-time TV show) reminding them of the late payment on last Christmas's credit card bill—change. For the purposes of this text, this is referred to as target drift, meaning change in people's behaviour given the same set of circumstances. Indeed, in the United States, it was shown that over the period from 1995 to 1997, retail borrowers' payment behaviour deteriorated because of the 'falling social, information, and legal costs of default' (Gross and Souleles 2002).[9]

Target drift can also be magnified by lenders' actions. Predictive models are used to drive strategies that affect borrowers' behaviour, and hence the results the models are supposed to predict. This 'strategy effect' is akin to a circular reference in a spreadsheet, where a calculation somehow feeds back into itself, whether after one or many links, causing different results to pop up each time it is recalculated.

This is analogous to the 'Heisenberg uncertainty principle' (physics/quantum mechanics) and 'Hawthorne effect' (industrial psychology). Both refer to instances where the fact that something is being watched affects the outcome, but the former relates to subatomic particles, and the latter to workers' performance.

Its extent depends on the stage of the risk cycle. In *new business scoring* it is relatively subdued, but exists in that, over time, staff, customers, and intermediaries get a feel for who will be accepted and who not. Staff may discourage customers whom they do not think qualify, based on a couple of over-the-counter questions, or discard applications that they believe have little chance of being accepted. Likewise, customers of similar backgrounds may share their experiences, causing more or less of the same to apply. The model can thus influence the through-the-door population, albeit such changes will be slow to occur.

In contrast, for *account management and collections* the effect is greater. Scores form a core component of campaigns, that: (i) increase limits, for low-risk accounts; and (ii) redirect resources, to reduce delinquencies amongst high-risk accounts. The latter applies especially to collections and recoveries. As problematic accounts are targeted they will improve, but then problems arise in areas from which resources were pirated. In like fashion, if a new scorecard is developed three months later: (i) the campaigned accounts would then show as lower risks; (ii) resources would again be shifted; and (iii) their risks may return to close to their original levels. This may sound like a fire-fighting approach, but is critical in instances where there are limited resources for a rapidly changing environment. It requires consistent and effective monitoring, and a risk-ranking infrastructure that can be quickly updated for changing circumstances.

[9] Cited in Allen et al. (2003).

3.4.5 Unattributed drift

Scorecards have something in common with people and the products they make . . . Even if they are maintained in ideal conditions, they can still get old and tattered. Companies may schedule regular redevelopments, even though the extent of economic, market, operational, and other changes has been minimal. This reduces emphasis on intensive scorecard monitoring, and ensures that the scorecards are aligned with the current business. This would not have been possible, even in the 1990s. As the costs and hassles of scorecard developments have reduced, especially as companies are moving the function in-house, they are being done more frequently, often every 18 to 24 months. Even so, some lenders will still use the same scorecard for five years or more. By this time, the business should be asking whether scorecards are applicable at all within that environment, or whether some other option could serve its needs better.

3.5 Summary

This section has focused upon some of the mechanics of credit scoring, in order to provide an overview of several topics presented in subsequent modules. The first issue was the scorecard appearance. Most lenders use traditional scoring models that: (i) break various *characteristics* (like 'applicant age') into *attributes* (like 'less than 30'); and (ii) assign *points* to each attribute, such that the total *score* for each case provides a measure of risk relative to other cases. Desirable customers will get high scores, while those less palatable get low scores.

There are a number of different statistical techniques that can be used to develop scoring models. Traditional models are associated with *parametric techniques*, which unfortunately make certain assumptions about the underlying data that are not always true. In the early days, DA and LPM were the primary choices, largely because of computational speed. Unfortunately, they are considered ill-suited to situations with binary outcomes (good/bad, default/not-default), and *logistic regression* is now favoured. Even so, most results comparisons show there is no clear winner. The differences in ranking ability are small, as the manner in which DA and LPM are applied tends to address the assumption violations. It is also possible to use *non-parametric techniques* that require no data assumptions. Machine learning techniques, such as NNs, *genetic algorithms*, and *K-nearest neighbours*, suffer from a lack of transparency, and are prone to overfitting. Of these, neural networks are the most commonly used, primarily for fraud scoring. *Decision trees* are powerful analysis tools, but provide poor results, because they require huge amounts of data.

While a lot of time and effort is spent on obtaining data, developing scorecards, and designing strategies, gremlins can creep into the system. Biases can arise, no matter whether humans or machines make the decisions. There will always be a *flat maximum*, in terms of quality, that no model can exceed, and it can only be hoped that it passes a less stringent 'reasonable-model' test. Bias exists to the extent that a model's quality falls short of this level. While some of it results from assumptions made during the scorecard development process (sample selection, transformation), it is more likely to arise because of data issues (poor data quality, lack of access to key data sources), or misapplication of the final model.

Where the bias is greater than what is considered acceptable, other options are available. First, cases can be *scored for guidance* only. This applies especially where the potential loss or profit is high, and key information cannot be captured in the score. Second, there may be *generic scores* available to supplement internal data. For many small lenders, the cost of developing bespoke scorecards is not worthwhile, and they may rely solely upon bureaux' generics. And finally, underwriter experience can be used to develop an *expert model*. This can provide a viable alternative in instances where no generic exists, or where there are one or more bespoke or generic models—and possibly other information—that can be integrated into a hybrid.

When measuring the results, there are a number of different aspects. Given that credit scoring is used to drive business processes, lenders need to know what value it is providing in those processes, both in terms of *selection* (accept/reject rates), and *performance* (good/bad/default rates). At the same time, they may also wish to use the scores further in risk-based pricing and finance calculations. Credit scores can be used to derive *default probabilities*, and other measures can be derived for *loss severity* (a function of EAD, LGD, and remaining *maturity*). To ensure that they are working, assessments can be done of scorecards': (i) *power*, their ability to discriminate according to risk; (ii) *accuracy*, how close the estimate is to the actual result; and (iii) *drift*, the extent to which the power and accuracy change over time. Power and drift can be measured using *measures of separation/divergence*, the primary ones being the Gini coefficient, Kullback divergence measure, chi-square statistic, and K-S statistic. Accuracy is usually assessed at the portfolio level, such as changes to overall default rates, but it is also possible to use *binomial probabilities*, the *Hosmer–Lemeshow statistic*, or the *log-likelihood measure*.

The scorecard development process is quite a long one, especially for greenfield developments. Lenders must first decide upon their *objectives*, which may include process efficiencies, increased market share, and/or reduced bad debts. A *feasibility study* helps to determine whether or not this is possible, and *key players* need to identified. A critical component of the process is *data preparation*, which includes: data acquisition, good/bad definition, observation and outcome windows, and sampling. Thereafter, *scorecard modelling* requires: data transformation, variable selection, reject inference, segmentation, and training. Once completed, the scorecards can be *finalised*, which requires: scorecard validation and sign-off, calibration, strategy setting, loading, testing, and post-implementation monitoring. *Decision making and strategy* issues include: level of automation, change management, overrides, policy rules, referrals, and strategy enhancement. Finally, there are also *security issues*, relating to: documentation, confidentiality agreements, and change control.

Credit scoring relies on a base assumption that the future will be like the past. It is usually sufficiently true, but drift may occur that affects scorecard and system performance, including changes to: (i) the *economy*; (ii) the *market* being serviced; (iii) lenders' *infrastructure*; and (iv) borrowers' *attitude towards credit*. Changes in available *technology* will also play a role, and lenders may wish to replace scorecards because new or improved data is available. Likewise, new systems may be implemented that improve operational performance. In any case, lenders may nonetheless opt to replace scorecards for no specific reason, other than to ensure they are current, and maintain a competitive edge.

Module B

Risky Business

4 The theory of risk

> The revolutionary idea that defines the boundary between modern times and the past is the mastery of risk: the notion that the future is more than a whim of the gods and that men and women are not passive before nature.
>
> *Peter Bernstein*, in Against the Gods—The Remarkable Story of Risk

Risk is part and parcel of any business endeavour. It arises from different sources that require different data and models to assess it, and different tools to control—and even play—with it. While this textbook focuses specifically upon credit risk, it is only one of many possible risks that companies face. This chapter covers a broader risk framework, which is treated under three headings:

The risk lexicon—The playing field that enterprises operate in (market, economic, social, and political factors), and the risk types (business, market, credit, operational).
Data and models—Tools that are used to determine the extent of the risk.
Control and experimentation—Actions that can be taken to manage risk; and to determine and test changes that might optimise the process.

4.1 The risk lexicon

For 400 years, the only companies actively involved in risk management were insurance companies, both long- and short-term (mostly personal-life and asset insurance respectively). Over the past 150 years, this broadened to the management of credit and market risk, but mostly at the individual transaction level, and not the enterprise level. That has changed over the past 40 years, as the increasing pace of change has brought increasing complexity, and a greater number of unknowns in terms of both potential costs, and further opportunities. Companies have always managed enterprise-level risks subconsciously, albeit perhaps not very well, but in the 1980s, they started to put greater effort into identifying[1] and managing them. Outside of the insurance industry, the idea of having 'Risk' in somebody's title is a new one—a knee-jerk reaction to an increasingly volatile environment that is here to stay. This discipline has developed a language of its own, some of which is covered in the section that follows.

[1] Scenario planning was first used by Royal Dutch Shell in the late 1970s, and has become a crucial element of risk management.

4.1.1 Risk linkages

The focus of this text is credit risk, which is something not to be viewed in isolation. No matter whether lending is the primary or secondary activity, other risks play a role, many of them interconnected (see Figure 4.1).[2] Consider events that have affected the global economy since 1970:

Interconnected risks—a thumbnail sketch

The Vietnam/American War gave rise to massive costs and inflationary pressures in the United States, leading to the demise of the Bretton Woods Agreement and massive increases in commodity prices like oil and gold; OPEC was awash with money. Some of this was lent to developing countries like Brazil, Mexico, and Argentina, only to cause various emerging market crises in the 1980s. When oil and agricultural prices fell, it had a serious effect on the savings and loan industry, which was already suffering from improvements in technology and deregulation in the 1970s. The last 50 years have also been typified by ever-increasing technological change, which has brought new products and rapid obsolescence, while new and cheaper communications tools have fostered outsourcing, both nationally and internationally.

During the 1980s, socialism fell into general disfavour, such that state-run companies were privatised, and previously monopolised markets deregulated. Governments in Eastern Europe and the Soviet Union were toppled, but communism was replaced by Islamist fundamentalism as the world's major threat. The most infamous shock to date is 9/11 and its impact upon American society, but this is only one of many events associated with the volatile religious, political, and leadership situations in the Middle East. Events like the fall of the Shah of Iran (1980), the Iran/Iraq war (1988), and two American wars against Iraq (1990, 2003–), all lead to oil price instability—and Saudi Arabia may be the next powder-keg. At the same time, emerging markets have benefited from spurts of phenomenal

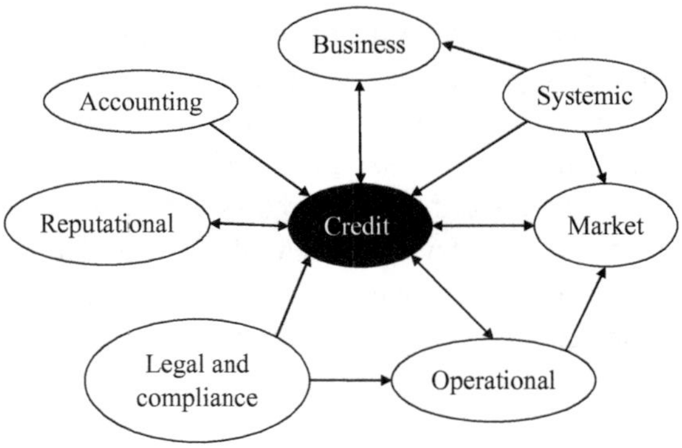

Figure 4.1. Risk linkages.

[2] Olsson (2002).

growth, but have not been immune to shocks (SE Asia 1997/8). Disease and its communicability have also become factors, not only as they affect humans (SARS, West Nile virus), but also animals (mad cow disease, foot-and-mouth, bird flu).

For the twenty-first century, uncertainties will continue to arise from energy (especially if a viable alternative to oil is found), water, and disease. Factors that will continue unabated are: *technology*, continued product obsolescence as new products replace old; *communications*, increased interconnectivity between people and countries, especially as the Internet becomes ubiquitous, and knowledge transfer between individuals is accelerated; *ideology*, Islam will continue to rise, but if not, it could be replaced by something even more fundamental; *power*, economic and political might has shifted between Portugal, Spain, Holland, France, England, and the United States over the last 500 years, and the twenty-first century may belong to China and India, while other emerging markets continue to develop.

Thus, credit risk is just one part of a larger superset of risks faced by the business. It could be addressed in isolation, but many of the concepts are generic.

4.1.2 The playing field

Risk is like a bar of soap—as soon as you think you have got hold of it, it suddenly slips from your grasp and goes in an unexpected direction.

Olsson (2002)

Besides offering the delightful quotation above, Olsson (2002) also describes the myriad of risks that enterprises face, a list that is in no way exhaustive (Figure 4.2). Much of it is generic, and appears in many textbooks on risk. His analysis starts by setting out key areas that affect supply, demand, competition, and so on:

Market opportunities/threats—Arising from the combination of demand, technology, and the current competitive environment.

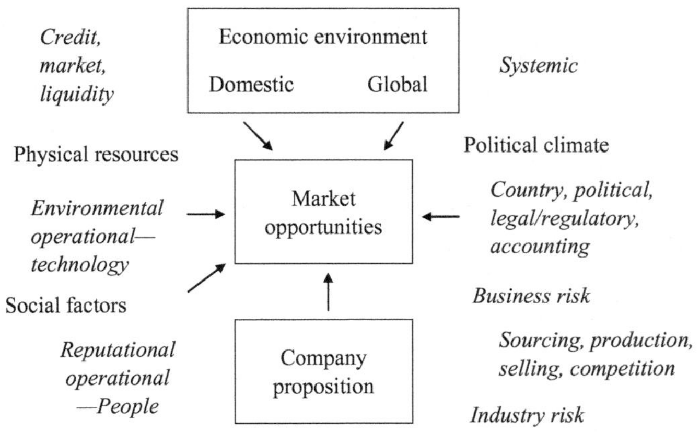

Figure 4.2. Risky Business.

Company proposition—What the company offers in terms of product, price, promotion, package, and distribution.
Physical resources—Relating mainly to primary production by the agricultural and mining sectors, but also applies elsewhere.
Economic environment—Both local and foreign, including factors like interest rates, exchange rates, taxation, economic growth, and commodity prices.
Social factors—Items that affect both the labour market and level of potential demand. This includes population size, education levels, work ethic, religion, and social stability.
Political climate—Ideological stance of government, government involvement in economy, political and economic freedom, and stability versus potential upheaval.

4.1.3 Risk types

The four primary risks in this environment are business, credit, market, and operational risk. *Credit risk*, the primary focus of this book, has already been dealt with quite extensively, but to summarise, it is any risk arising because of a real or perceived change in a counterparty's ability to meet its credit commitments. It covers not only potential non-payment, but also changes in risk grades that affect the market value of traded debt, and the possibility of incurring extra costs to get the money back.

Schönbucher (2003) splits credit risk further into arrival risk (probability-of-default (PD)), timing risk (time-to-default), recovery risk (loss-given-default (LGD)), and market risk (change in market price of defaultable assets). Market price correlation risk refers to the extent to which market risk is correlated with the other three, the balance being made up of economic and other market factors. Arrival and timing risk apply to the obligor, while the other two apply to the specific obligation. He also highlights default correlation (joint arrival) risk, which is the probability that several obligors will default together, and is related to concentration risk.

Operational risk covers any event that can impact upon the operations of the firm, which at the extreme, can lead to failed internal processes. These can result from problems related to staff, technology,[3] fraud, infrastructure, communications, physical security, or internal policies and procedures. It also includes instances where individual staff members fail to disclose potential problems. Examples during the 1990s include Barings Bank and Orange County, where inadequate controls allowed staff members to make huge investment bets that went bad. *Business risk* relates to any enterprise's failure to achieve business targets as a result of misreading the economic or competitive environment, or having inappropriate strategies and/or resources to produce and/or sell its product. And finally, *market risk* is related to

[3] It has been said that, for modern banks, it could be impossible to recover if core systems are down for more than three days.

random changes in market prices, including foreign exchange rates, interest rates, commodity prices, share prices, and even property prices.

While these are the major risks, there are many others, which are grouped here under the headings of business environment, business dealings, extraterritorial, personal, and intelligence.

Business environment

There are several risks that relate to the business environment, including industry, legal/regulatory, environmental, and reputation risk. *Industry risk* refers to factors that affect all players in a given industry similarly, which can arise from product and economic cycles, industry concentration and barriers to entry, technological change, and any other inherent source of volatility. These affect all stakeholders, including investors, lenders, suppliers, labour, and government. *Legal/regulatory risk* arises from potential non-compliance with employment, health and safety, environmental protection, accounting disclosure, insider trading, and other requirements (see Module H, Regulatory Environment). *Environmental risk* refers to the potential for company actions or activities to have a detrimental impact upon the natural environment. Costs may arise because of legal liabilities and reparations, physical repair, or impacts upon public perceptions that affect marketing. And finally, *reputational risk* refers to the possibility of adverse publicity that may affect stakeholders' confidence in the company. The publicity may relate to labour practices, environmental concerns, product pricing, or any other number of issues. It usually relates to crises, and if handled well, the event may be beneficial in the longer term.

Business dealings

Companies are affected by changes in the business environment, but many of these risks only manifest themselves in specific dealings, or functions, within the business. For example, credit risk can arise because of any number of factors, but ultimately results from *counterparty risk*, which refers to any and all risks associated with dealings with a single counterparty (borrower, or other party to a transaction), or parts thereof, whether through one or many transactions. In contrast, *liquidity risk* arises when poor financial planning makes it difficult for an enterprise to meet its obligations, which is aggravated by thin markets, where the true value of investments cannot be realised. For the latter, Chorafas (1990) also refers to *event risk* that arises from an unforeseen change to a debtor, such as a change of ownership, especially from a leveraged buyout where the debtor's debt servicing obligations increase substantially.[4] If one or more inordinately large investments are made, or many investments are subject to the same risks (like same industry or region), the result is *concentration risk*. And finally, there is also an *accounting risk*, which arises from the possibility that an entity's financial statements (balance sheet, income statement, or any projection) are not an accurate representation of its financial situation, whether as the result of errors, or misrepresentation.

---

[4] This applies largely to corporate bonds, and can be offset by 'poison puts' that guarantee investors' capital.

Extraterritorial

Another group of risks is those arising from dealings with entities in foreign countries, including: *sovereign risk*—relates to any national government, government-owned utility, or any loan backed by a government guarantee; *country risk*—events in other countries, including currency and banking crises; *transfer risk*—inability to exchange a currency when required, either because of a refusal, or rationing by the national government; and *political risk*—any change in a country's political framework that has a potential impact upon the economy, or local business environment.

Chorafas (1990) noted that prior to the 1970s, there was very little US bank lending to foreign national governments. This changed with the 1973/4 spike in oil prices, which generated huge amounts of money for OPEC countries while many developing and third-world nations suffered. Banks started lending to the third world, often with the encouragement of their own governments, without giving adequate consideration to fundamentals. Risk was heightened because the loans were not for specific development projects that would aid repayment ability, and many of the borrowing countries viewed it as aid. Much of the money ended up supporting totalitarian ideologies and regimes. Countries like Brazil, Argentina, and Mexico also acquired 'debtor power', meaning that total debt was so large that lenders were scared to call in the loans, and instead lent more.

Personal

Most articles on risk focus on those that affect business operations or asset values and, for the latter, much of the focus is on large loans and traded securities. Little thought is given to the human element, whether in terms of staff or customers. *Employee risk* (loss of key persons, strikes, or fraud) seems to be lumped under operational risk, while *customer risk* is treated under counterparty risk. In retail credit, the personal customer element plays a significant role, and consideration needs to be given to the risks that affect individuals, whether the borrower or a related party. Broadly speaking, there are two classes: (i) *character risks*—including irresponsible borrowing, dispute, moral hazard, and skip/run-away risk; and (ii) *personal distress risks*—including illness/death, domestic dispute, job loss, and personal disaster (loss of assets or livelihood through accident, fire, flood, or other natural disaster). Character risks are more measurable, as they are a function of each individual's own behaviour, and are reflected in past credit performance. In contrast, distress risks are more difficult to measure because of their nature, and because much of the information that would provide value is either unavailable, or cannot be used due to anti-discrimination legislation.

Intelligence

Further risks can arise, related to the enterprise's ability to gather and process information. First, at case level there is a distinction between whether the risk is: (i) *idiosyncratic*, highly unusual and peculiar to specific cases; or (ii) *universal*, common to all cases. For example, in the consumer credit environment universal risks are those that can be ascertained from an assessment of the applicant's character, standing in the community, past dealings, bureau record, and so on. In contrast, idiosyncratic risks include those of the applicant getting hit by a bus, having his place of employment burn down, or winning the lottery.

Second, at process level there is a distinction between whether the risk is: (i) *endogenous*—arising inside the system, where the 'system' could range from a production process to the global economy; or (ii) *exogenous*—arising outside the system, especially random shocks such as earthquakes and other events that cannot be foreseen. Disaster recovery planning is crucial in such instances. Idiosyncratic and exogenous risks will likely result in unexpected losses, whereas there is usually sufficient data for endogenous and universal risks to come up with expected loss (EL) estimates.

Third, there are *systemic risks* where a small event within a system has an unforeseen ripple effect that impacts upon other not obviously connected areas in the financial system. It applies especially where failure of one market participant causes others to fail, but can arise in other circumstances. Olsson cites two examples: (i) the *Kobe earthquake*, which caused a fall in the Tokyo stock market that undid Barings Bank; and (ii) the *Y2K bug*, where so much time and effort was spent to protect against a myriad of possible scenarios. Fourth and finally (and especially relevant in this textbook) is *model risk*, which can arise from: (i) an error in model development or implementation; (ii) a change in the environment where it is used; or (iii) its use in situations for which it was not designed.

If nothing else, this discussion provides a framework that highlights the limitations of risk models. Models are best used for risks that are: (i) universal, (ii) endogenous, and (iii) non-systemic. Even then however, there will always be the possibility that the models will be based upon incorrect assumptions, and pose further risks.

4.2 Data and models

Nothing is constant but change, and the future is certainly more like the recent past than the distant past

Mark Schreiner

Risk assessment is easiest in instances where past experience can be combined with data. The extent to which this is possible depends upon the type of risk. Olsson presents what he calls an 'uncertainty matrix', an adaptation of which is provided in Figure 4.3. According to this, the greater the certainty of the possible outcomes and their probabilities, the easier the risk is to measure. Thus, it is relatively simple to assess market, liquidity, and credit risk, but much more difficult to assess operational, country, reputation, and systemic risks. The primary constraint is the availability of data, and the curse of rare events.

4.2.1 Data types

Risk measurement relies upon having data, and some means of turning it into information. These are covered in Module C (Stats and Maths) and Module D (Data!), but some high-level

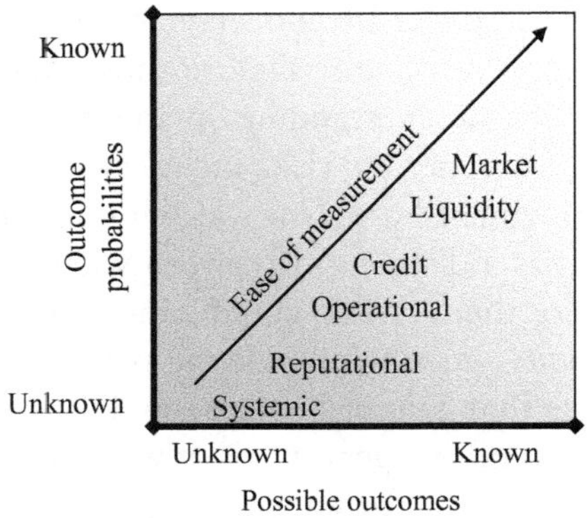

Figure 4.3. Ease of measurement.

concepts can be mentioned briefly here. The types of data used to assess risk can be split out according to source, time, inputs, indicators, and view:

Source—*Internal* to the company, obtained directly from the *customer*, or from *external* agencies.

Time—*Vertical* data, time series reflecting fluctuations in specific variables (such as the prices of shares or publicly-traded debt); or *horizontal* data, multivariate snapshots for each observation, at a given point in time. Credit scoring data is almost exclusively horizontal.

Inputs—*Objective*, can be clearly observed and verified; and *subjective*, based on an individual's assessment, where there may be a level of bias related to a lack of experience with similar situations.

Indicators—*Leading*, gives advance warning, such as stock market prices that indicate changes in the broader economy; and *lagging*, adjusts after the horse has bolted. Both relate to correlations, and not necessarily causation. Scorecard characteristics tend to lag the true causal event (such as illness, job loss, domestic dispute), but lead the default event.

View—*Backward-looking*, based purely upon what has happened in the past; and *forward-looking*, includes a human assessment of what may happen in the future, whether through a direct judgmental assessment (credit rating agencies, credit underwriters), or market prices that summarise the views of market participants (share and bond prices).

On the final point, the ideal is to have forward-looking indicators for case-by-case retail decision-making, but that is infeasible in volume-driven environments. There are no market prices for individual loans, and subjective assessments are inefficient and expensive. Even so, credit scoring models have proven extremely robust, if only because of the inferences that are possible, given sufficient and stable historical data. In need, lenders can adjust strategies to accommodate their assessment of the future.

4.2.2 Model types

Credit risk assessment has undergone its own industrial and technological revolutions over the past two hundred years. Its industrial revolution started with the broader industrial revolution, and evolution of consumer societies. It resulted from the *imposition of structure* upon the process, which was required where the lenders' employees made the decisions. Means had to be found of conveying hard-earned experience from one generation to the next—including the development of policy and override frameworks, concepts like the five C's (Section 6.1), and tools like ratio analysis (Section 6.3.2).

Policy—Rules used to limit the decision, when certain conditions hold true. These are usually based upon past experience, especially where higher than normal losses are associated with those conditions.

Overrides—Decisions can be overturned by other (usually higher) authorities, whether people or policies. Judgmental overrides of policy rules should only be done if they can be motivated by information that is not recognised by the existing framework.

In contrast, the technological revolution is more recent, and has been reliant upon computers to *speed and streamline* the data collection and analysis process. It provided an entirely new dimension to the range of options available. As illustrated in Figure 4.4, pure judgment and policy have given way to the use of different types of models. The choice depends upon the desired level of structure, and the amount of available data:

Pure judgment—Low on structure, low on data. Relies upon subjective assessments, with no model or template.

Expert system—Little data exists, but underwriters have sufficient experience to develop a set of rules or model, that can be used either to drive or guide decisions. They are seldom

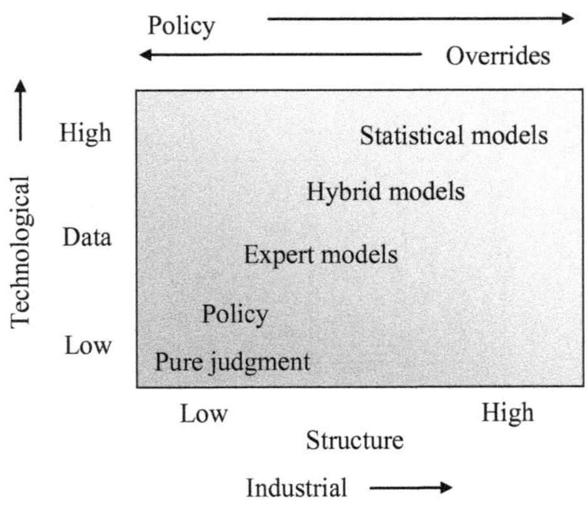

Figure 4.4. Risk assessment revolutions.

reliable enough to be the sole decision driver though, and there will usually be a judgmental overlay.

Hybrid models—Data availability varies. A combination of model types is used, depending upon what can be constructed for different aspects of the risk assessment. Outputs from different statistical, and possibly expert, models are integrated into a single model.

Statistical models—High on structure, high on data. While predictions are more reliable, they have the disadvantage of a data addiction—and only the best and most highly structured data will do.

Expert models are developed either by using input from experts, or by analysing how they make their decisions. Some authors will classify artificial intelligence (AI) techniques—like neural networks (NNs)—under the heading of expert models, because they mimic the way in which people learn. This is a totally different and unrelated concept.

According to Falkenstein et al. (2000), systems that make effective use of both quantitative and judgmental inputs: (i) collapse the inputs into a single measure; and (ii) apply judgment to exceptions, in order to incorporate data outside the model.

An overview of the mix between pure judgment and statistical models for different types of lending is provided in Figure 4.5. The matrix has transaction volumes and potential profit as the x- and y-axes respectively. The upper-left quadrant contains *wholesale credit* sectors where little data or experience exists, and the potential profits are high—especially true for large and/or complex loans, such as sovereign, corporate, and project-finance lending. These areas suffer most from the highly unstructured nature of what little data is available. In contrast, the lower-right quadrant contains *retail credit*—especially consumer and SME lending—where

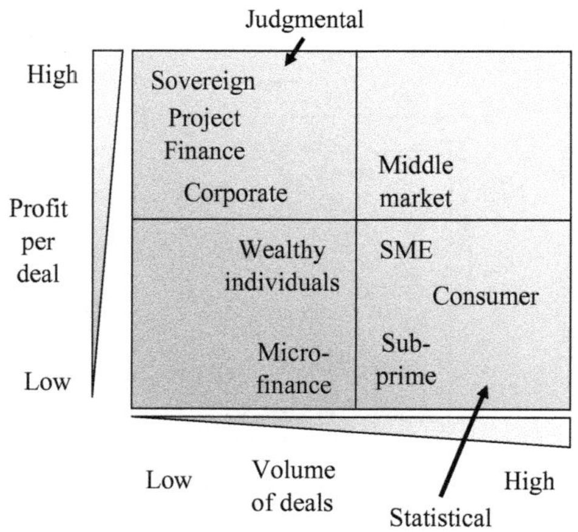

Figure 4.5. Volumes and profits.

statistical models provide the main voice. These have limitations though, and are only the better choice where:

(i) The environment is relatively stable.
(ii) The data is highly structured.
(iii) The loss associated with individual transactions is relatively low.
(iv) Lower costs are perceived to provide a significant competitive advantage.
(v) There are sufficient volumes and potential profits to justify the investment.

According to Chorafas (1990), the Pareto principle also applies to credit. A very small proportion of customers often provides the greatest proportion of profits! Thus, service offerings should be stratified according to potential profitability: 'Emphasis must necessarily be placed on the most lucrative parts of the market—which is often more demanding, more risk, and also requires a steady vigilance in product development, as there is no copyright in the finance business.'

The treatment of the different types of lending is not cast in stone and, over time, the thresholds have been shifting. Risk assessment skills have become fewer and dearer, while data has become broader, deeper, and cheaper. As lenders become more comfortable with credit scoring, it is increasingly being used to assess individuals and companies where ever larger values are involved. The rest of this section looks at pure judgment and expert models in more detail.

4.2.3 Pure judgment

The development of sophisticated mathematical and statistical modelling techniques has not led to the total demise of tried and tested ways of lending. Pure judgment is still widely used, especially for relationship lending, or any lending where little data or experience exists. Some commentators even refer to it as 'judgmental scoring', even though that term would be more aptly applied to expert models. If and when ranking models are first introduced, the grades/scores will either be used to: (i) provide guidance, and not be the sole basis for the decision; or (ii) limit the possibilities for overzealous underwriters.

There is a temptation to refer to 'pure judgment' as 'manual scoring', except manual implies the use of hands instead of minds. The expression 'manual decision-making' and 'manual underwriting' should thus be oxymorons, except where underwriters are doing forced labour as paper shufflers, who apply strict policy rules, and no thought. Irrespective, both terms are occasionally used to refer to the judgmental process.

While the value of pure judgment is downplayed in much of the credit scoring literature, Bunn and Wright (1991) highlight that underwriters can provide much better forecasts in shifting environments with unstructured data, where much relevant predictive information is not—or cannot—be reflected in a historical model:

(i) Data requirements vary from one deal to the next.
(ii) The relationships between the characteristics and risk are uncertain.
(iii) The potential loss arising from individual decisions is sufficiently high to justify the cost of having appropriately qualified individuals and/or committees.

This is most pronounced in wholesale credit, but can also apply to: (i) *emerging retail environments*, where the rules are not well defined, especially larger SMEs and high net worth individuals; (ii) *relationship lending*, where the lender has consciously decided against transactional lending; and (iii) *sub-prime markets*, where data is thin, but the margins or potential pay-offs/benefits are sufficient to justify the investment.

4.2.4 Expert models

While data-driven models have become the norm for retail credit, in many cases it is physically not possible to develop them. That does not automatically imply a lack of knowledge though, and it may still be possible to get many of the benefits of decision automation—especially speed and consistency of decision-making—through the use of expert models. Indeed, consistency by itself may provide a significant improvement.

Expert models can take on different forms. Some are presented as scorecards, others as decision trees, while still others use a combination of the two. They are developed by harnessing the knowledge of people that have the requisite experience. Although these 'domain experts' may not be able to define the exact quantum of relationships between the various factors, they usually have a firm grip on the dependencies.

Chorafas (1990) also commented that the power of domain experts comes from a process of inference, representation, decision, and control. In order for an expert system adequately to mimic individual expert(s), it has to consist of three types of knowledge—factual, judgmental, and procedural—that may each be stored in separate databases, but must work together. The systems are usually based upon probabilities 'to deal with situations that cannot be reduced to mathematical formulae' and, as such, are heuristic in nature.

Domain experts are also capable of ranking individual cases, for use in a scorecard development. The subjective grades can then be used as the target variable in an *ordered logistic regression*. The resultant scorecard can be used to provide rankings, but default estimates are only possible with proper calibration. Alternatively, scorecard developers/vendors with experience in similar environments may be able to develop a *generic*. In either case, the resulting model may be used: (i) as an interim measure, until such time as sufficient data and performance are available to develop a data-driven model; or (ii) as a more permanent solution, which is refreshed occasionally using updated inputs from the experts.

The process can only be automated if the model can be operationalised. The main challenge is to assemble data from different computer systems, or provide a platform for it to be obtained and captured manually. Inputs will be mostly objective data, but can also include subjective evaluations of various factors. The latter is frowned upon, but in some instances can provide a channel for an underwriter's forward-looking view. It is, however, only advisable if they can provide insights that are not already embodied in the objective data.

Businesses' expectations of expert models may be high, but the models cannot be held up to the same standards as statistical models. Validation is difficult in the absence of performance data, and it may only be possible to benchmark against external measures, such as bureau scores or rating agency grades. Ultimately, expert scores or grades are usually used for guidance only, albeit underwriters' latitude to override the scores may be limited.

4.3 Conclusion

This textbook focuses on credit risk, but it cannot be treated in isolation from the many other uncertainties that lenders face. This chapter provided a general risk framework, which was split out in terms of: (i) the *sources* of risk; and (ii) the *data and models* used to assess risk. Threats can arise both internally and externally, whether from the market, economic, social, or political environment. The greatest risk is whether the *business* proposition is appropriate for the market, but *credit* risk is a close second for many companies. The other primary risks are *market* and *operational* risks, with others arising from: *business environment* (industry, legal/regulatory, environmental, and reputational); *business dealings* (counterparty, liquidity, concentration, and accounting); *extraterritorial* (sovereign, country, transfer, and political); and *personal* (character and personal distress). Businesses are also affected by their ability to gather and process *information*, which gives rise to the distinction between idiosyncratic and universal risks (case level), and endogenous and exogenous risks (process level). Systemic risks can also arise from small events that have unforeseen consequences. Finally, and importantly in the context of this book, significant risks can arise if the *models* used for decision-making are ill-defined.

Fortunately, credit risk can be measured more easily than most, if only because there are *known outcomes*, and lenders usually have sufficient data or experience to determine *probabilities*. The data can: (i) come from internal or external sources, including the customer; (ii) be vertical (time series) or horizontal (multivariate snapshots for each observation); (iii) include subjective and objective inputs; (iv) use leading and lagging indicators; and (v) provide either a forward- or backward-looking view. Credit scoring uses all sources, but tends to rely on horizontal data comprised of objective leading indicators, with a backward-looking view.

The type of model that is appropriate for each situation depends upon how well structured the data is, and how much of it there is. *Statistical models* require both, but there are a lot of other models that may be used along the way. *Pure judgment* can be used where both data and structure are lacking, but relies upon the experience of individuals. *Policies* are applied when there is significant collective experience, and can provide greater structure in other instances (they are not always appropriate, and overrides provide a countermeasure). *Expert models* can be used if individuals' experiences can be encapsulated in a model, whether a decision tree or scorecard. And finally, *hybrids* can be used to combine statistical models and judgment where the former are insufficient to provide a decision.

5 Decision science

Man's strength has always been his ability to change his environment to suit himself. He observes the ways things work, and controls disturbances to maintain consistent results, or experiments to improve them. In some instances the risks may be life threatening, but they can be lessened with proper controls. The greater the danger, the greater the need for protection! Business enterprises operate similarly, except the mechanisms involved are more formalised. Control can be viewed as having the following dimensions:

Policies—Rule sets that define qualifying criteria, lending limits of underwriters, risk mitigation requirements, and so on.
Procedures—Series of actions that must be performed: (i) in order to circumvent policies in certain circumstances; or (ii) to mitigate risks once some predefined event has occurred.
Structure—Organisational design relating to the level of centralisation, staff roles and responsibilities, and levels and delegation of authority.
Infrastructure—Resources used for information gathering, computing power, communications, and deployment.

In extremely simplistic terms, the concepts of structure, infrastructure, and procedures could be likened to management, hardware, and software respectively. Management sets the direction, the hardware provides an engine room, and the software is the grease that ensures everything operates efficiently.

A construct often used to describe the ongoing risk-control process is the 'feedback loop', which in more recent years has been changed to the 'adaptive-control system'. The concepts originated in electronics and engineering, respectively, and are very similar, except the latter is more sophisticated, and tailored to industrial environments. Both refer to mechanisms meant to maintain consistent results, by adjusting inputs to counter changes in output. In its simplest form, the feedback loop has four parts:

Monitor—Ensure that things are going according to plan.
Feedback—Communicate any problems that have been identified.
Identify—Determine the source of the problem.
Control—Decide upon and implement a course of action.

The controls involve strategies that fall under four broad headings:

Ignore—Do nothing. This is the easiest and cheapest option, but does nothing to address the potential extra risk.

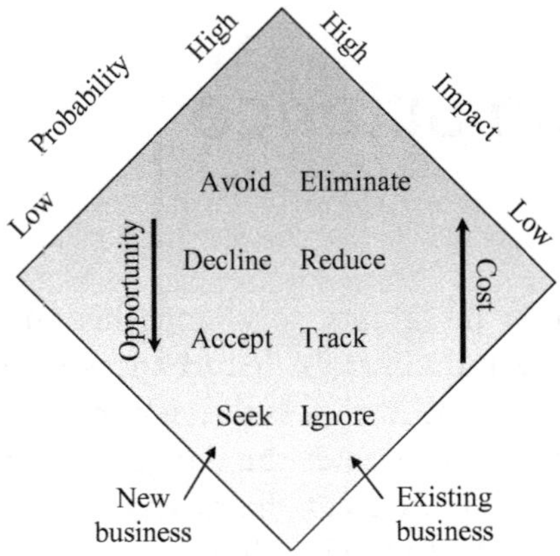

Figure 5.1. Risk strategies.

Track—Require more information, and give greater scrutiny.
Reduce—Take corrective actions to mitigate the risk, while allowing the process to continue running.
Eliminate—Get rid of the risk, by shutting down the process to prevent an even worse situation. It involves high costs/losses, especially when investments are sold at fire-sale prices.

The choice of strategy will be based upon the combination of probability and potential impact. This is illustrated in Figure 5.1, which also provides the equivalents for new business. Other considerations are the ability to respond and cost of response, versus the probability that the action will have the desired effect.

The control process is reactive, and requires a lot of human input when dealing with rare but severe events, such as exogenous shocks (natural disasters, fire, system failure). Disaster recovery plans can be implemented, but just how they are implemented will vary depending upon the situation. In contrast, if the risks are known and recurring, it is possible to set up a proactive risk management system. The controls are designed and implemented as need requires, which is usually the case for retail credit.

5.1 Adaptive control

The adaptive-control system (see Figure 5.2)[1] is a further evolution of the feedback loop. The concept has been hijacked from automated environments where closed systems are used to

[1] Åström and Wittenmark (1995).

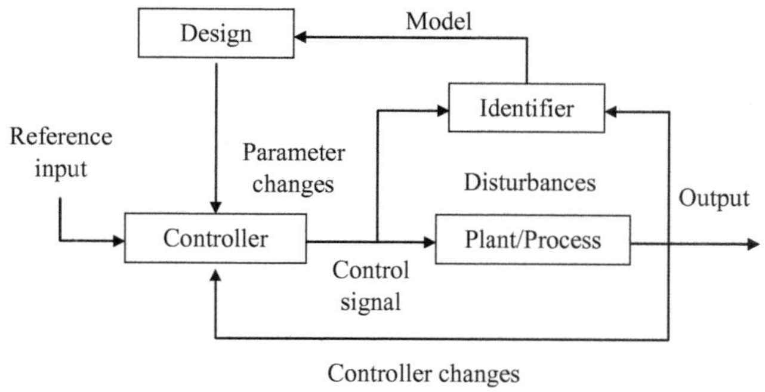

Figure 5.2. Adaptive-control process.

manage highly technical processes, like Boeing 737 autopilots, and NASA robots. Adaptive-control systems have four main building blocks:

Process—The business function being controlled, whether application processing, account management, collections, marketing, etc.
Controller—Governs the process, by executing a set of parameterised instructions.
Identifier—Measures consistency of operation, reports the source and magnitude of any distortions, and where significant, motivates changes.
Design—Determines the changes needed to address the distortions.

A conventional control system only has the first two components, being the controller and the process. It is the last two, identification and design, that are the essential and active ingredients of an 'adaptive-control' system. Within this model, there are also flows of information:

Reference input—Parameters that set the framework.
Control signal—Communicates parameters from the controller to the process.
Output—Results generated by the process, such as bad debts, attrition, number of customers, revenue, and so on.
Disturbances—Distortions that arise both inside and outside the process, which are identified by strict monitoring and comparison against expectations.
Parameter changes—Design changes to controller parameters.
Controller changes—Modifications to the design of the controller itself, like adding new information sources and capabilities.

Consumer credit control systems such as Probe™, TRIAD™, Falcon™, and others use a similar approach, but there is still a lot of human intervention to adjust strategies (controller). Closed systems are as of yet infeasible for credit decision-making.

In general, this is a reactive approach to risk management, which is appropriate for managing existing products and markets. If the economy is slowing and people are starting to default, then it is wise to be stricter in collections. If competitors are stealing customers, or the demand for credit is reducing, then a change in cut-off strategy may be in order. If, on the other hand, the lender wants to be proactive and take on more or less risk, or risks of different types, then changes to the reference inputs are required. There are, however, extra risks that come with fiddling with the system. The best way to determine whether a proposed change will have the desired effect is to apply some scientific rigour, which is covered next.

5.2 Be the master, not the slave

If you want to make enemies, try to change something.
Woodrow Wilson, 28th US president, 1856–1924

In the twenty-first century, people have to deal with ever-increasing amounts of change, both as observer and participant, driver and passenger. Credit scoring was initially developed to model a world that was still fairly steady; men had lifetime employment, women stayed at home, and the kids rushed home to watch the Flintstones at 4:30 pm. That was not enough though—somebody had to bring in the wooden spoon and start stirring things up, with 'What if?'

This question can only be answered if there are tools in place to aid conscious and calculated changes, while ensuring that the proper risk/return dynamic is maintained. Within the realm of business, the term 'science' is commonly used. Arsham (2002) refers to the decision sciences as those commonly known as management science, success science, and operations research. But why use the word 'science'? For many, it is a much loathed subject from their school days, and if they did not hate it, the answer to the above question might make them hate it! The following is a summary of ideas presented by Bala (2001), which is further summarised in Table 5.1.

A brief history of scientific philosophy—an essay

Mankind has spent much of his existence trying to understand and control his environment. True knowledge was gained haphazardly, while myths and legends were used to explain the rest. The first to use a structured approach was Euclid, the fifth-century BC Greek, who developed geometry using an axiomatic system based on infallible truths. Unfortunately, this only works in mathematics, and fails when trying to learn about nature.

It was more than a thousand years after the fall of the Roman Empire before seventeenth-century Renaissance thinkers questioned the dogma of the church. The findings of Galileo and others caused philosophers to search for fail-safe frameworks for discovery. Francis Bacon (1561–1626) proposed deductive reasoning, based on observation and experimentation, what we know as the 'scientific method'. René Descartes (1596–1650), of 'I think, therefore I am' fame, proposed inductive reasoning based on mathematics and pure logic—a deconstructive approach of breaking a problem down into its parts. English and French scientists were divided along these Baconian empiricist and Cartesian rationalist lines for a century. During this period, the concept of science as a structured and continuing search for knowledge arose; a search that uses hypotheses, theories, laws, concepts, and models as tools of discovery.

Neither the empiricist nor rationalist approach worked when Isaac Newton (1642–1727) tried to explain gravity. He instead used a synthesis of both to come up with his mechanistic view of the universe. The Newtonian normativist view relies on expectations that scientific methods will produce principles that accurately predict and explain a variety of phenomena. Theories were measured by explanatory adequacy, predictive accuracy, scope of success, simplicity of assumptions, etc.

All pre-twentieth-century scientific thought was directed at mapping a mechanistic universe, but the map had some wrong turns. In 1905, Albert Einstein (1879–1955) published his special theory of relativity, which showed that many Newtonian rules become invalid as the speed of light is neared, and in 1927, Werner Heisenberg (1901–1976) came up with quantum mechanics, which ruled out certainty and impossibility at the sub-atomic level. This is not to say that these newer models are truths; they are simply the best representations that we currently have. Scientists are still trying to come up with a grand unifying theory, to marry the theories of the very big (relativity) and very small (quantum). One possibility is string theory, which describes everything as infinitesimally small threads, but it is not quite there yet.

Einstein's and Heisenberg's theories contested all knowledge of physics derived using Newtonian normativism, which forced scientists to look out from under their determinist umbrella, and revise much of what was previously thought sacrosanct. They upset the idea that slow accretion of knowledge, using accepted frameworks, was a fail-safe way of developing scientific knowledge. The Newtonian approach is not invalid though; it is a special case within Einstein's bigger and richer description. Experience, reason, and norms still form the basis of most knowledge discovery today.

The scientific philosophies developed by Bacon, Descartes, and Newton are applied mostly to the hard sciences; the natural sciences that we usually associate with the earth and the stars—like physics, chemistry, biology—which can be subjected to the rigours of the scientific method, and the results relied upon for centuries. Since the early twentieth century, these philosophies have also been increasingly applied to the soft sciences; the social sciences that deal with man's relationship with both man and his environment—like psychology, sociology, economics, etc. These relationships are fluid though, and scientifically derived results have shorter shelf lives. Both man and society change, and explanations will change along with them. Within the soft sciences, and even most hard sciences outside of physics, the normative approach should be viewed as sufficient, but any theories must recognise possible changes to the assumptions upon which they were based.

Table 5.1. Philosophies of science

Francis Bacon	René Descartes	Isaac Newton
Baconian empiricism	Cartesian rationalism	Newtonian normativism
Observation/experimentation	Logic/reasoning	Expectations to qualify
Deductive	Inductive	Synthesis
Scientific	Mathematical	Mechanical

Scientific principles were not used in business until the early twentieth century, when they were popularised by Frederick W. Taylor's 1911 'Principles of Scientific Management'. Taylor focused on reducing the amount of time taken for production-line tasks, with the sometimes inhuman use of a stopwatch (Arsham 2002). The field of operations research only came into being after the Second World War, to develop new methods of dealing with complex logistical problems, especially those of managing huge armies and ensuring they are well supplied. The advent of computers then allowed these same tools to be applied increasingly in business.

> [Management Science/Operations Research is] a scientific method of providing executive management with a quantitative base for decisions regarding operations under their control.
>
> *Mores-Kimball*

Arsham (2002) also highlights that decision sciences differ from other sciences, in that there is a decision-maker, who usually is not—and should not be—one and the same person as the analyst/scientist. Analysts thus require communication skills to contextualise the results and express them in terms that laymen can understand.

Credit scoring and the scientific method

What does that have to do with credit scoring? Credit scoring is a tool used for credit decisioning, a decision science that falls within the realm of economics, which in turn falls within the social sciences. Hence, some of the frameworks used in science can be borrowed, but organisations must be able to modify not only their assumptions, but also strategies and resources, going forward. Destination is more important than journey however, and lenders are more interested in 'what' will happen than 'why' (although the latter is definitely a bonus).

The scientific method is usually presented as a process of experimentation involving:

Observe—Viewing and describing the world around us.
Hypothesise—An attempt at explanation of either the observed or related phenomena.
Experiment—Use the hypothesis as a prediction tool.
Decide—Use the results of the experiment to accept or reject the hypothesis.

If the experiment's results support the hypothesis, it is used as the basis for a theory or law, otherwise it is rejected and the researcher tries again. The goal is to come up with a hypothesis that satisfactorily predicts the outcome of the experiment (Wolfs 2002). This is the empiricist

Table 5.2. Experimental design frameworks

Scientific method	Bisgaard et al. (2002)	Arsham (2002)	Generic
Observe	Define	Perceive	Design
Hypothesise	Measure	Explore	Execute
Experiment	Analyse	Predict	Analyse
Decide	Improve	Select	Improve
	Control	Implement	

approach—the only things missing are the deconstructive and repetitive aspects. If the hypothesis is rejected, or even if it is accepted, the problem can be broken down into further parts for greater understanding.

This text's interest is in how the scientific method is used in the retail credit environment. While there have been no frameworks presented specifically for that context, there are several for business applications, which have marked similarities (Table 5.2). In all instances, the goal is to provide a formalised approach for learning and decision-making:

Bisgaard et al. (2002)

Define—The problem statement, presented in measurable and actionable terms.
Measure—Observe variation of performance data from that expected.
Analyse—Determine the sources of variation, or reasons for success/failure.
Improve—Modify process drivers to achieve desired goals.
Control—Hold on to the gains.

Bisgaard's framework is the same as that used as the backbone of Six Sigma, which is a scientific means of achieving continual improvement in business processes, by focusing upon quality and service issues to reduce the costs of poor quality and customer dissatisfaction. See Keller (2005).

Arsham (2002)

Perceive—Realise that there is a problem, a need, or a goal to be achieved.
Explore—Find out the set of possible actions that can be taken.
Predict—Try to determine the outcome for each of the identified alternatives.
Select—Choose the best alternative, based on both effectiveness and risk.
Implement—Put the chosen action into place.

General experimental design

Design—Plan the experiment: define the problem, objectives, and possible solution(s); identify the input and output variables, and the quantity and type of data needed.
Execute—Perform the experiment: select the sample, apply possible solution(s), and collect the data.
Analyse—Analyse the data against all of the output measures. These will include measures relating to cost of the action, and effectiveness at achieving the desired result.
Improve—Determine whether the (or which) challenger strategy is good enough to replace the champion, communicate the results, and motivate to have it implemented.

As indicated earlier, the primary defining feature of the soft sciences and business is how quickly the assumptions change. For the latter, this applies especially in times of rapid growth, when both organisations and their processes evolve through repeated cycles of increasing complexity and simplification to achieve optimal performance.

5.2.1 Champion/challenger

The usual approach to business is very reactive, if only because so much time is spent fighting fires that management and staff members have little energy to be proactive. The problem is, however, that in today's rapidly changing world, companies have to be the architects of change, not the victims. Lenders should continuously be considering questions like, 'What if we phone instead of mail?', 'What if we charge more for NSF cheques?', and 'What if we provide higher limits?'.

Adaptive-control systems in the credit scoring world also have a feature called champion/challenger, a 'what if?' tool that allows performance comparisons of the new and novel against the tried and tested, so that lenders can have a forward view, instead of always looking in the rear-view mirror. The champion strategy is always the *dominant* one currently in place, which has worked in the past, and is trusted. The challenger strategy is the *underdog*—the contender that has to prove itself before gaining acceptance. Only if the challenger wins can it become the new champion.

According to Thomas et al. (2002) the champion/challenger approach is also commonly used: (i) in medicine, for testing drugs and treatments for various ailments; and (ii) by companies looking to introduce new versions of a product, such as toothpaste, to the market.

The use of a champion/challenger approach limits the risks that might arise from hasty implementation of poorly thought out strategies. According to Thomas et al. (2002:164), it can be used at any stage in the risk management process, as long as it is possible to: (i) identify a random sample of cases, say 5 per cent, where the challenger can be applied; and (ii) track their performance separately. The final decision on whether to accept the challenger will then take into consideration:

Marginal cost or benefit—of using the new strategy.
Effectiveness—of providing the desired results, in terms of risk (defaults, losses), value (assets, revenue), or attrition (dormancies, closures).
Positioning—whether it has caused customer complaints, or may impact on the lender's reputation.

While experimentation is an extremely powerful tool, there are limitations: (i) the number of challenger cases is often too small to draw proper conclusions; (ii) it takes time before the challenger's effect on the test group becomes apparent; and (iii) there is always the danger that even though the experiment yielded positive results, the challenger might be disastrous when implemented in full—especially if the business environment changes significantly.

These limitations could, at least partially, be countered by using simulation. This is a process often used in the sciences, which has come to be one of the recent buzzwords in the credit scoring world. It involves the use of computer models to simulate what the impact of different strategies would be on the entire loan book, in different hypothetical scenarios. Existing business is not affected, yet the lender can have some idea of the potential impact almost immediately. Once again though, care must be taken, as the assumptions used may not hold true.

5.2.2 Optimisation

There is little difference between champion/challenger and optimisation, with one exception. Champion/challenger usually looks at two possible options, existing and proposed. In contrast, optimisation tests a number of different proposed challengers that are each rated on a number of different metrics to determine which provides the best results. The illustration in Figure 5.3 provides an example where mailings have been done using different marketing strategies, each of which is rated according to risk and response. The strategies employed varied by type of mailing, wording and duration of offer, interest rate, fees, loan term, etc. The response rates are actuals (not scores), while the risk and revenue items are based on scores at the time of application.

The lender knows certain costs will be incurred, and has set its breakeven revenue score to \$100. There might have been 20 strategies selected for testing, but only 6 beat this benchmark. In the example, *G* provides the best response rate, but *Y* the lowest risk. *G* provides the highest average revenue though, so the final choice will depend upon the lender's risk appetite. Besides revenue, further measures may be considered using similar graphical representations, before making the final decision.

The generation of strategies cannot be random. Choices will be subject to constraints, perhaps related to them not being possible for specific subpopulations or in combination with certain others. For example, mail campaigns could be ruled out for high-value customers.

5.2.3 Strategy inference

A topic that has been discussed in credit scoring circles, but for which there is almost no mention in the literature, is strategy inference. This is a tool used to determine players' strategies,

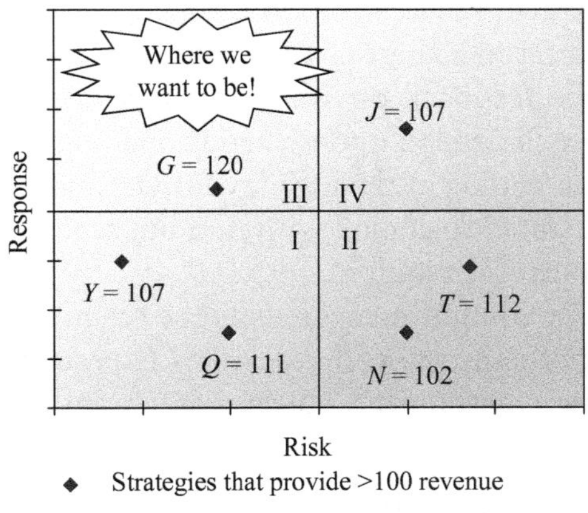

Figure 5.3. Optimisation.

based upon an analysis of moves made in prior games, where the games may be: (i) non-cooperative win/lose *zero-sum* games; or (ii) co-operative games that have *mutual gain.*

John Nash, of 'A Beautiful Mind' fame, pioneered the study of game theory in 1950, for which he received the 1994 Nobel Prize in economics.

Models of how players respond in a variety of situations can be developed by combining data on their moves, with the information used to make the decisions. The task is simplest when analysing moves in chess or similar games, but the same techniques can be applied in war, business, sports, and other competitive endeavours. Unfortunately however, this science is still in its infancy—at least outside of some specialist applications.[2] Irrespective, it needs to be mentioned briefly, because it is happening and some of the same tools are being used.

The difference between normal scoring and strategy inference is the change in direction; normal scoring is used by players to set strategies, while strategy inference is used to guess what the opponent's strategies are. Lenders are most interested in borrowers' strategies, so that they can tailor their own strategies accordingly. An example is a bespoke 'not-taken-up' scorecard (never opened, never used, or used briefly before closure), that could be used to adjust the offering in terms of interest rate, repayment term, collateral requirements, communications mechanism, and so on. These characteristics would also be included in the list of predictive variables.

Although not directly related, an accept/reject model could be considered as a strategy-inference tool, because it allows the scorecard developer to gain some idea of what the lender's past strategies were, without specific knowledge.

5.3 Summary

The defining feature of the late twentieth century was the recognition of risk as a specialist area, which needs to be managed. Policies and procedures are the primary tools, both of which require structure and infrastructure. These can be seen as parts of an adaptive-control process, used to monitor and provide feedback on system performance, and to identify and correct problems. The action taken will depend upon whether the risk is inside the system (eliminate, reduce, track, or ignore), or standing at the gate (avoid, decline, accept, seek). While effective, the process may not be well suited to coping with rare but severe events, in which case human input, and/or higher level control is required.

No system is perfect, but in volume environments the scientific method can be used to seek improvements: observe, hypothesize, test, decide. This experimentation framework takes on different forms in the business environment, but they are essentially the same. In credit scoring,

[2] Even within the literature, the strategy inference analysis is presented as decision trees (Engle-Warnick 2001), albeit this is largely to assist ease of understanding.

the original and still most common approach is *champion/challenger*, where the challenger strategy is tested on say 5 per cent of the population, and only implemented in full if it performs better than the reigning champion. An extension of this is *optimisation*, where several strategies are tested simultaneously. *Simulation* can also be used, which is computationally more intensive, but provides quicker results and a better understanding of the various options. Another powerful, but relatively infrequently used, tool is *strategy inference*, where lenders model borrowers' responses to their strategies, and adjust them accordingly.

Whether or not these methodologies have gained widespread acceptance is another question. Champion/challenger has been known, understood, and applied successfully by many organisations, but a common complaint is that it takes a lot of management time to design, test, and implement new strategies—time that is too often consumed by the day-to-day management of the business. Even so, the successful organisations of the future will most likely be those that can successfully master and capitalise on these approaches, especially optimisation and simulation.

6 Assessing enterprise risk

Credit scoring is usually associated with retail credit, where the rich availability of data makes it the 'gold standard' of credit risk assessment. It was first used for consumer credit, but over time, the same concepts have been applied to SMEs, whose fortunes are closely tied to those of their owners. It may be ill-suited for use in other areas though, especially if more appropriate techniques are available. Commentators such as Allen et al. (2003), who are more familiar with the wholesale market, consider credit scoring a second choice because of its 'portfolio approach', referring to its reliance upon a historical review of similar cases, as opposed to factors specific to each case.

The question must then be asked, 'How should credit risk be assessed once credit scoring becomes inappropriate, and where does this point lie?' Credit scoring has been gaining increasing acceptance, especially for middle-market companies where the amount of data has been growing over time, as has lenders' ability to assess it and use it for pricing, limit setting, or ongoing risk management. This demand comes not only from lenders, but also their shareholders, trade creditors, regulators, and others, and is increasing even further with the Basel II Accord. This trend cannot carry on forever though, as there are several hurdles (data availability, model validation, consistency across organisations) that prevent larger wholesale loans from being managed as portfolios (Basel Models Task Force report, 1999, cited in Ong 2002).

This textbook was initially intended to focus purely on retail credit—especially consumer credit—but at some point, it became clear that some small attempt must be made to provide an overview of what is done for enterprise lending, from SMEs right through into the wholesale arena. The topic is covered under the following headings:

(i) **Risk assessment 101**—An overview of credit risk assessment, covering the 5 Cs, types of risk assessment tools, and rating grades.

(ii) **SME lending**—Covers the shift from relationship to transactional lending for smaller companies, and where relationship lending is still being used.

(iii) **Financial ratio scoring (FRS)**—The use of credit scoring techniques, to determine default probabilities based upon information contained in obligors' financial statements.

(iv) **Credit rating agencies**—Looks at the agencies that assess the largest and publicly traded companies, the letter grades that they provide, and issues that have been raised.

(v) **Modelling with forward-looking data**—The historical, options-theoretic, and reduced-form approaches, which rely upon one or both of rating agency grades and market prices.

6.1 Credit risk assessment 101

The first tool used for providing structure to credit risk assessment was the concept of the 5 Cs, a framework that could be almost as old as credit itself.

Capacity—Ability to repay liabilities out of income.
Capital—Financial resources available to handle unforeseen events, and meet commitments should income not materialise.
Conditions—How the current environment may impact upon the enterprise, whether via competition, economic, industry, or other factors.
Character—The quality of management: Who are they? What experience do they have? Are they well suited to lead the company?
Collateral—Security provided, including the pledge of assets, guarantees from third parties, or other risk mitigation.

This framework still dominates where information is obtained directly from the client in relationship lending. The task now is to determine how this relates to modern environments, where further structure has to be imposed upon the risk assessment. The goal is to minimise subjectivity, which requires data and/or experience. The rest of this section looks at: (i) the data sources that can be used; (ii) risk assessment tools; and (iii) risk grades, which are the end result.

6.1.1 Data sources

Time spent brooding over figures is seldom wasted.
J. H. Clemens

The starting point is the data sources that lenders use to assess the credit risk of business enterprises:

Human input—Employees' eyes and ears are still a primary source of information. The goal is to ensure that their observations are as objective as possible, but in many instances, subjective inputs are required.
Market value of traded securities—This is the gold standard of corporate data. The level, volatility, and buy/sell spreads of market prices provide forward-looking information, which is a summary of market participants' views on obligors' credit risk. Both bond and equity prices may be used. *Traded debt* may be issued by private companies, public utilities, governmental agencies, and others.
Financial statements—A review of obligors' financial positions, as presented in recent balance sheets and income statements.

Payment history—Information on borrowers' payment patterns, which is a loose surrogate for character/management.

Environment assessments—Review of industry and regional factors, whether using economic data and forecasts, or by deriving historical aggregates, based upon internal/bureau data.

Codes must be assigned before industry or region can be reflected in a score. Physical postal codes suffice for the latter, which is important for companies that lack geographical diversification. Industry classifications are trickier, either because there is insufficient information, no appropriate code can be found, or the customer is operating in more than one sector (a 'primary industry' is usually chosen). Different classification frameworks exist, including the International Standard Industry Classification (ISIC). Some of the major industry sectors are: agriculture (farming/fishing/forestry), mining, manufacturing, utilities (electricity/gas/water), construction, trade (retail/wholesale), transportation, real estate, personal and professional services, and community services (education/health/ sanitation).

This list is not exhaustive; other sources, such as application forms and credit evaluations, could also be included. Each provides information on one or more of the five Cs (see Table 6.1). The most far reaching are judgmental assessments and the value of traded securities, but each has its faults: the former because they are expensive and slow to react and the latter because they tend to overreact.

Exactly what data is used also varies depending upon the amount (to be) borrowed, and the size of the obligor being assessed. The former drives what information is requested, while the latter affects what is readily available and relevant. Where the size of the borrower and amount lent is small, less time and effort will be spent on the assessment. The information provided in Table 6.2 provides an indication of the large number of smaller firms in the United States, and the pattern is similar elsewhere.[1] Indeed, it is generally accepted that the bulk of economic growth is driven by SME activities. The 'class' definitions will vary from country to country, but the general pattern of what information is available and used for risk assessments will be similar (see Table 6.3):

Table 6.1. Data versus the five Cs

Data source	Capacity	Capital	Conditions	Character	Collateral
Human input	✓	✓	✓	✓	✓
Traded securities prices	✓	✓	✓	✓	✓
Financial statements	✓	✓			
Environment assessments			✓		
Payment history				✓	

[1] Falkenstein et al. (2000:11).

Table 6.2. USA firms by size (assets) 1996 IRS data

Class	Range	Number
Small	<$100K	2,500,000
Small, middle	$100K–$1M	1,500,000
Middle	$1M–$100M	300,000
Large	>$100M	16,000

Table 6.3. Company size versus data

Company size	Market prices	Judgmental assessment	Financial statements	Payment history	Personal assessment
Very large	✓	✓	✓		
Large		✓	✓		
Middle		✓	✓	✓	
Small			✓	✓	✓
Very small				✓	✓

Very large—The market value of traded securities can only be used for publicly listed companies, and those with traded-bond issues.

Large—Judgmental assessments dominate for larger companies with significant debt. This applies especially to rating grades provided by credit rating agencies, but also to internal grades. Payment histories and personal assessments are not considered relevant.

Middle—Caught in a range where there is neither market data, nor sufficient exposure to justify full fundamental analysis. Analysis becomes backward-looking, focussing upon what has happened to obligors in similar circumstances in the past. The primary data used is financial statements provided by the borrowers, along with industry assessments. Payment histories and personal assessments may feature.

Small—Below a certain level, financial statements may be either unavailable or unreliable (out of date, poor accounting/auditing, or a lie factor). Instead, focus shifts heavily towards obligors' payment histories, often obtained at a cost from credit bureaux, and data on recent revenue inflows that can be confirmed from bank statements.

Very small—Finally, at some point, it becomes difficult or impossible to divorce the individual and the enterprise, especially for sole proprietorships. Lending will be based on, or heavily influenced by, assessments of the borrowers in their personal capacities.

6.1.2 Risk assessment tools

Now that the data sources have been considered, the tools used to assess them can be covered. Falkenstein et al. (2000) mentions the following:

Rating agency grades—Letter grades provided by credit-rating agencies for a fee. These are judgmental assessments of both quantitative and qualitative factors, which may use statistical tools where possible. These apply to large firms only, which numbered approximately 16,000 in the USA in 1996 (see Table 6.2).

Public-firm models—Based upon options theory, the most popular of which is Merton's model. Assuming that markets are efficient, then the equity price and volatility can be combined with the level of liabilities to provide a default probability.

Private-firm models—Provides a probability-of-default (PD) based on companies' financial statements and industry classifications. The approach is very similar to that used for retail credit scoring.

Hazard models—Applies to agency-rated companies with liquid traded debt. It relies upon an analysis of bond prices relative to risk-free securities. This is similar to Merton's model, except bond spreads are analysed instead of default rates. The most well known is that originally presented by Jarrow and Turnbull (1995).

The final three types, mentioned below, relate very closely to the loss probability, loss severity, and bureau models used in consumer credit.

Portfolio models—Attempt to model the loans as a group, using default and exposure estimates for individual loans. This relies upon using correlations and calculating worst-case scenarios at given confidence intervals.

Exposure models—Models that assume the account has defaulted, and are more interested in the magnitude of the loss and not the probability. These include exposure-at-default (EAD) and loss-given-default (LGD) models. The LGD will be a function of the collateral type, seniority, and industry.

Business report scores—Provided by Dun & Bradstreet (D&B), Experian, and other credit bureaux. These scores are based on liens, court actions, creditor petitions, and company age and size, and are used primarily for assessing trade credit.

Most of the data mentioned is publicly available, and the scores are developed to predict bankruptcy, liquidation, or severe delinquency. Trade creditors' data may be used directly as part of the assessment, in much the same way that payment profile data is used in the consumer market. In some environments, the credit bureaux are acting as intermediaries to pool bank data, often for use by banks only.

Table 6.4. Models versus data

Model type	Traded securities	Financial statements	Environment assessments	Payment history	Judgmental assessment
Rating grades	✓	✓	✓		✓
Public-firm	✓				
Private-firm		✓	✓		
Hazard	✓				
Exposure		✓		✓	
Portfolio	✓			✓	
Credit bureau				✓	

Once again, a rather imperfect representation of the relationship between the various model types and data sources is provided in Table 6.4. Credit scores are not specifically mentioned in the list, but two come close: (i) business report scores are effectively bureau scores; and (ii) private-firm models are developed using similar frameworks, but are based mostly upon financial statement data, with no use of company demographics (except industry), or payment behaviour.

6.1.3 Rating grades

Just as in retail credit, wholesale lenders ultimately want two things: (i) a risk-ranking mechanism, and (ii) some indication of loss probability and loss severity. Rating grades are the most common tool used for ranking risk, and for each a default rate can (hopefully) be calculated. Two types of rating grades are referred to: (i) *rating agency grades*, provided by credit rating agencies for bond issues, or on commission for large obligors; and (ii) *internal rating grades*, derived by individual lenders, especially for private and/or smaller obligors that do not warrant the attention of the rating agencies. Standard practice is to rate each counterparty by its default probability, and each transaction by both default probability and loss severity.

According to Schuermann and Jafry (2003a), Standard & Poor's (S&P) and Moody's are the two dominant rating agencies in the United States, and their ratings are used most in quoted studies. They differ somewhat in their approaches however, as S&P ratings are more closely aligned to probability of default, whereas Moody's ratings also take into consideration potential recoveries.

Allen et al. (2002:11) quote Treacy and Carey (2000), who recommend that the grades should cover both the default risk and recovery rate, but reported that only 40 per cent of surveyed banks used both for their internal rating grades. This is largely due to the difficulties associated with obtaining reliable internal loss data for the latter.

The letter grades used by the credit-rating agencies contain up to three characters plus a modifier, such as 'BBB+' or 'Baa1'. There are seven major grades, or more than 19 with modifiers.

Moody's uses higher level grades of the form [Aaa Aa A Baa . . . D], with number [1 2 3] modifiers like Baa3. In contrast, most other agencies use grades of the form [AAA AA A BBB . . . D] with '+' and '−' modifiers. Moodys' Baa3 is roughly equivalent to a BBB- from the other agencies. John M. Bradstreet pioneered the use of credit reference grades for companies in 1851, while John Moody was the first to use similar rating grades for bonds in 1909.

In contrast, internal ratings usually use single letter grades (A, B, C, . . .) or numbers (1, 2, 3, . . .) to indicate increasing risk (B is less risky than C). The number of grades may be between 5 and 13, as individual lenders usually have neither sufficient data nor the breadth of customers to warrant as many grades as the rating agencies. Nonetheless, lenders will usually map their grades onto the agency grades, in order to have a basis of comparison.

If a ratings framework is working properly, it is hoped that the final ratings will have certain properties. Cases within a rating grade should be *homogenous* and their movements *predictable*:

Homogenous—All of the entities within a grade should be of approximately the same risk. Unfortunately however, risk can take on different aspects: default risk, recovery risk, ratings transition risk, and credit spreads. It is difficult to achieve homogeneity on all counts, so the focus is usually upon default risk.

Predictable—Transition rates from one grade to another should be consistent over time. There are, however, areas where the transitions behave differently (industry/country), and they tend to vary within the business cycle.

These issues are usually mentioned with respect to environments, where the ratings are used as a basis for pricing, capital requirements, or other purposes. For individual cases, the grades should be both *stable* and *responsive*, which initially seems like a contradiction:

Stable—Grades should not change drastically from one period to the next, like from 'AAA' to 'C'. When looking at a transition matrix, most cases should remain on or near the diagonal.

Responsive—Grades should respond quickly to new information, and contain all available credit related information. It is relatively easy when data is updated regularly and assessed automatically, but more difficult for irregular judgmental assessments.

The shelf life of the grades varies, depending upon the market (wholesale, enterprises, retail, or consumers), and the nature of the available data. Judgmentally derived grades for large, publicly traded firms will be much more stable than automatically derived grades for SMEs and the middle market. Rating-grade stability will also be affected by whether a point-in-time or through-the-cycle approach is used:

Through the cycle (TTC)—Emphasises conservatism, and stability of the risk estimates and rating transitions. It looks beyond the immediate economic situation, and instead focuses on the expected risk during stress scenarios, like a trough in the firm's business cycle.

Point in time (PIT)—Focuses on obligors' current risk situation, given prevailing industry and economic conditions. This is most prevalent when using 'forward-looking' market prices as the basis for the analysis, and behavioural scores with a short time horizon.

The through-the-cycle approach is the one typically used by the rating agencies, and is favoured by financial regulators. If lenders were to rely solely upon point-in-time estimates, systemic risk would increase; lenders would be tempted to increase their exposures further when times are good, and rein them in when times are bad, which can lead to disastrous consequences when there are shocks. Lenders that have their own internal rating grades typically use something in between, perhaps taking a two to three year view, if only because it is less demanding, and perhaps more appropriate for the loan maturities being considered.

The Basel Committee for Banking Supervision (2003b) has assumed, for the purposes of Internal Ratings-Based approaches, that agency ratings are through-the-cycle, as opposed to point-in-time probability of default estimates.

It cannot be assumed that a grade, no matter how it is derived (rating agency grades, bond prices, equity prices, etc.), will see every default coming. Defaults can occur with no warning, even shortly before the event, as shown by Barings, a centuries-old bank, bankrupted by the dealings of a single trader, Nick Leeson.

6.2 SME lending

It is accepted that SMEs are a major source of economic growth in many economies, but the manner in which their banking relationships are handled varies. According to Allen et al. (2003), a primary pattern that has emerged over the past several years is that larger banks have been moving away from traditional relationship lending, and moving instead towards transactional lending:

Relationship lending—Old-style lending, where the customer relationship and local knowledge are key aspects.

Transactional lending—Focus upon assessing individual transactions, and the use of quantitative assessment techniques such as credit scoring.

In effect, the type of credit evaluations being done has shifted directly from one end of the spectrum to the other. This has accelerated the growth of some banks, especially where mergers and acquisitions have been driven by the economies of scale that can be achieved from transactional lending. Meanwhile smaller banks, at least in the United States, remain focused upon lending into a *niche market that values the relationships*. Allen et al. (2003) quote: (i) Feldman *(1997)*, only 8 per cent of US banks with assets up to five billion dollars used credit scoring; and (ii) Treacy and Carey (2000), quantitative methods were used mostly on larger companies, while for SMEs, loan officers assessed qualitative factors in judgmental assessments. This probably has more to do with the nature of the lenders serving those markets than the markets themselves, and is changing fast.

6.2.1 Relationship lending

Old-style lending, being that based upon the five Cs, is called relationship lending. Risk is assessed by a loan officer or bank manager, who has personal knowledge of the client, his/her reputation, standing in the community, connections, current product holdings, history with the organisation, and so on. This is accompanied by a duty of secrecy, relating to any information obtained directly from the client.

According to Berger and Udell (2001), small businesses are more opaque, and thus have fewer financing opportunities than large companies—both in terms of bank credit and trade finance. Those with strong banking relationships benefit from lower interest rates, reduced collateral requirements, lower reliance on trade debt, greater protection against the interest rate cycle, and increased credit availability. In an earlier study for the United States, they showed that the average SME banking relationship age was nine years, indicating that these relationships have some importance. Concerns were also expressed that:

- Any information that is gathered resides with the loan officer, and cannot be easily passed on to others within the organisation. It does not work well for larger, geographically diversified banks, but this factor may be diminishing as technology improves.
- Relationship lending relies upon soft data about the firm, the owner, and the local community, that is difficult to quantify, verify, and transmit within the organisation.
- Finally, the soft data also leads to an agency problem; the loan officer is contracting on behalf of the bank, but may not be able to communicate through layers of management, and further to shareholders. Agency problems are fewer with smaller lenders, which have flatter hierarchies, especially where the bank owner and president are one and the same.

Relationship lending also suffers from disadvantages related to: (i) less than optimal risk assessments; (ii) potential discrimination against minorities, especially where there is little competition; and (iii) unintentional cross-subsidisation of borrowers. Pricing is based largely upon the length and breadth of the relationship, and there is a tendency for new customers, and those with single-product holdings, to subsidise existing customers. Prices will reduce as the relationship matures, especially in more competitive markets. In the absence of competition, banks are not only more likely to charge higher prices, but also take greater chances.

The problem with relationship lending is the cost, due to the time and effort required to cultivate the relationship. Niche banks can use this as the basis for their competitive advantage, but their capacity for growth will be limited. In contrast, larger banks will focus upon efficiencies to lower costs, grow their loan book, and optimise capital utilisation. Even so, Bassett and Brady

(2000) reported that, over the period from 1985 to 2000, smaller banks enjoyed an interest margin of 1 per cent higher than larger banks, and had a return on assets (ROA) that increased from 0.7 and 1.1 per cent, as opposed that for larger banks, which varied from −0.4 to 1.1 per cent.

This has combined with other factors to increase the larger banks' motivation to move into the retail market. In particular, their base of easy money has been eroded, as depositors have gained access to capital markets. As a result, many of the larger banks have opted to focus their efforts on driving down costs, for loans in the middle and SME markets. This has had the benefit of not only growing the total debt market, but also providing lenders with increased portfolio diversification.

6.2.2 Transactional lending

According to Berger and Udell (2001), the primary difference between transactional lending and relationship lending is the 'hard' versus 'soft' nature of the data being used. Rather than relying upon information known only to the loan officer, transactional lending relies upon other technologies, especially credit scoring. While many lenders use transactional technologies to the exclusion of relationship lending, they can also complement each other.

The problem with lending in the lower end of the market is two-fold. First, obligors' size and transparency are correlated—*the smaller the company, the more opaque*. Financial statements are often not worth the paper that they are printed on, and collateral is often worthless in the event of default. Banks have had to capitalise on data sources that are readily available and trustworthy, and focus on unsecured lending. Second, there are *fixed costs associated with service delivery*, which makes lending more expensive. Hence, the need for the focus on costs, and the large amounts invested in decision automation and delivery infrastructures. Thus, transactional lending has provided a key to the SME market for larger lenders, who use credit scoring to evaluate borrowers' payment histories, both internally and with the credit bureaux. As Allen et al. (2003:19) explain it, rather than focussing upon individual loans, lenders instead rely upon the diversification provided by a large portfolio of small loans.

> Allen et al. (2003:18) refer to the credit scoring models used for transactional lending as 'portfolio risk measurement tools', even though they are used to assess the values of individual loans, and not the portfolio as a whole.

Credit scoring in the small-business market really only started in 1995, with the introduction of a Fair Isaac (FI) model, which was restricted to loan values under $250,000 (Berger and Udell 2001). Two of the most widely used products in the United States are FI's *Small Business Scoring System* (SBSS), and *SMELoan*, which originated in Hong Kong. The FI development identified some of the key factors as: SME, time in business and total assets; and *personal*, age, number of dependents, and time at address. The development also confirmed what lenders already knew—that information on the individual is more important than that for the enterprise. In contrast, the SMELoan model focused exclusively on the firm. Lenders need only 'collect data on sales, cash flow, and accounts receivable', and combine it with the transaction history, to identify problem accounts.

Lenders' own dealings with each SME customer can also be assessed. For banks, the cheque accounts are especially information rich, as practically all transactions pass through them, and can provide an extremely strong indication of the SME's short-term health. Data relating to the principals is particularly powerful, but privacy legislation may restrict what personal bureau data may be used for assessing an enterprise. Without appropriate consents, it may be limited to principals' negative data only. For smaller juristics, lenders will insist not only upon personal suretyships, but also permissions to do bureau searches for payment performance elsewhere.

Impact on the market

Credit scoring's impact on the SME market has been significant. Less face-to-face contact and documentation are required, as lenders instead focus on the wealth of readily-available hard data. Significant productivity gains have been achieved, with borrowers benefiting from improved access and increased choice, while lenders have increased volumes and reach. This results in greater economies of scale, increased geographical diversification, and greater competition. It is dependent upon credit information being readily available via the credit bureaux, where the United States has the lead.

According to Longenecker et al. (1997), the Hibernia Corporation implemented credit scoring in 1993. By 1995, they had already increased their throughput from 100 applications per month to 1,100, and grew their loan book from $100 to $600 million, with reduced bad debts. Cited in Allen et al. (2003:20).

With regard to pricing, there is a correlation between borrowers' credit scores, and the interest rates that they are charged. Many lenders use risk-based pricing, but this is not always the case. Prices for individual products may be fixed, but borrowers that are declined on one product may be accepted on another that is more expensive, and/or has stricter terms.

Allen et al. (2003) note that much greater price differentiation can be achieved with transactional lending. With relationship lending, there are 'concerns about objectivity and consistency'.

One of the fears about decision automation is that it will eventually lead to further consolidation within the banking industry, reduced competition, and higher prices. Experience thus far does not reflect this though. Allen et al. (2003:25) note several studies on the topic:

(i) SMEs have enjoyed greater access to credit where consolidation has been greatest, possibly because the larger banks are better at diversification (Black and Strahan 2002).
(ii) Small-business loans that are displaced during the consolidation process are then picked up by other lenders (Berger et al. 1998).

(iii) The interest rates offered by larger banks are lower (Berger et al. 2001), and there is less cross-subsidisation of established borrowers by new borrowers.
(iv) At worst, bank mergers have had little impact upon the availability or cost of credit (Scott and Dunkelberg 1999).

While credit scoring can provide benefits to almost any bank, there is often a reluctance to move from relationship to transactional lending. Smaller banks in particular believe that their competitive advantage lies in personal service. They may, however, underestimate the appeal of Wal-Mart style prices and convenience.

6.3 Financial ratio scoring

Financial ratios are related to firm failure the way that the speed of a car is related to the probability of crashing: there's a correlation, it's non-linear, but there's no point at which failure is certain.

Falkenstein, Boral, and Carty (2001)

The fact that the companies most likely to experience financial difficulties exhibit similar characteristics has long been known. According to Falkenstein (2002):

> . . . firms with high default risk have very measurable and theoretically straightforward characteristics: low profitability, high leverage, low liquidity, small size, high volatility, high inventories, and extreme growth.

It should thus come as no surprise that for three centuries, assessments of companies' financial health, whether for equity or debt investments, relied upon a judgmental review of their financial statements. Financial ratios were key inputs into the assessment because they normalise the data for size to facilitate comparison against a peer group (which, for business enterprises, is usually others operating in the same industry). Such an analysis can be used no matter what the enterprise, but is particularly important for middle-market companies, where: (i) the debt requirements are not large enough to justify a full fundamental analysis; and (ii) there are no share or bond-market prices available for analysis. The same approach could also be used for SMEs, but unfortunately, their statements' reliability is often suspect (if available at all).

Although this realm does not have the same data richness as consumer credit, predictive statistics can and are being used—only the approaches used differ. The resulting models are often not referred to as credit scores, but instead as 'bankruptcy prediction' or 'business-failure classification' scores (Liu 2001). Some individuals/agencies might refer to them as 'private-firm' models,[2] even though the same models could be used for both private and publicly traded companies.

These 'financial-ratio scoring' (FRS) models are really a form of credit scoring. There are two forms: (i) those intended to predict the rating agency grades, such as the Fitch model used for larger companies; and (ii) those that predict default risk directly, such as Moody's KMV's

[2] This label gives rise to some confusion, as Private Firm™ was a KMV product that is still used by Moody's KMV. Moodys' own product is RiskCalc™.

RiskCalc, which are used primarily for 'high yield' (speculative grade) and unrated private companies (Falkenstein 2002:173), which are 'too large to be considered an extension of its owner' (Allen et al. 2003:30). In both cases, results are best combined with other available information, to provide a single risk grade.[3] The following section covers FRS models under the following headings:

Pioneers—An overview of the research that provided the theoretical framework.
Predictive ratios—A look at the main characteristics that usually feature.
Restrictions—Factors that may impact upon the reliability of the scores, or the extent to which they can be relied upon.
Rating agencies—The role of the credit rating agencies in rating private companies.
Internal risk grades—Some considerations for lenders that are using FRS scores to derive internal risk grades.

6.3.1 Pioneers

It was only during the twentieth century that the use of financial ratios for assessing credit risk started receiving the attention of academics. Most of the original focus was upon public companies, if only because of the availability of information. Fitzpatrick did the first known studies in 1928 and 1932, when he compared the financial ratios of defaulted and non-defaulted companies.[4] Little was done thereafter until 1968, when rapidly evolving technology allowed Edward Altman to develop his z-score model, using multivariate discriminant analysis (DA) (see Shimko 1999:51). The resulting model had five financial ratios: earnings before interest and taxes (EBIT) to assets, retained earnings to assets, working capital to assets, sales to assets, and market capitalisation to book value of debt.

The letter 'z' in 'z-Score' refers to a normalised value, with a mean of zero and standard deviation of one, which effectively expresses the score as the number of standard deviations from the mean. Falkenstein (2002) comments that Altman's was not necessarily the best approach, but he is accepted as the pioneer, because his is the oldest and most well established model.

The z-score model was updated by Altman, Haldeman, and Narayanan (1977), who developed a model using a sample of 53 bankrupt and 58 non-bankrupt firms, more or less evenly split between manufacturers and retailers, that all had over $20 million in assets when first observed. There were a number of ratios and other values tested, but only seven ratios featured in the final model: Return on Assets (ROA), earnings volatility (variance in ROA), debt service (interest cover), cumulative profitability (retained earnings to total assets), liquidity (current ratio),

[3] Where financial statements are available in sufficient quantities, lenders are likely to integrate them with payment histories to create a single risk estimate, whether through data-driven or expert models.

[4] See Falkenstein et al. (2000).

capitalisation (market value to total capital), and size (total assets). The top three predictive characteristics were cumulative profitability (which provided one quarter of the predictive value), followed by earnings stability and capitalisation. Surprisingly, the final model was quite predictive on a holdout sample: 90 per cent accuracy over a one-year window, and 70 per cent over five years (how this was measured could not be determined).

The Altman et al. (1977) article was reprinted in Shimko (1999). It also presented a means of selecting the cut-off score, which is based upon minimising the misclassification errors: $Z_{\text{cut-off}} = \ln(q_1C_1/q_2C_2)$, where q_1 and q_2 are the assumed probabilities of bankrupt and not bankrupt, respectively, for the full population; and C_1 and C_2 are the cost of false positives and false negatives, respectively.

These studies are significant, if only because they were the pioneering works in the field. There have been a number of other academic research papers over the years, all of which were similarly constrained by the small number of defaults. Falkenstein et al. (2000) highlight that in the 30 or so papers since 1970, the median number of defaults was only 40. This may be sufficient to provide a model that is better than random guessing, but is not enough to prove which model is best. Since then though, the rating agencies have been able to assemble significant databases, which have facilitated the development of much more robust models that can be used by lenders—for a price.

6.3.2 Predictive ratios

Even though the number of monetary accounting values that can be used in credit risk assessments is manageable—perhaps 21 for the balance sheet, and 14 for the income statement—the number of possible financial ratios is huge (Table 6.5 is far from exhaustive). Surprisingly though, the number of ratios that typically feature in scoring models is small, because of the correlations. Almost all of the credit-related information will be provided by six or so ratios, many of which will be common across different models. This should not be a surprise, as underwriters would also only use a few key ratios for assessing financial statements, albeit their choice may vary depending upon the industry, and the size of the firm.

Allen (2001:33) provides a list of the predictive ratios, highlighted in about 30 studies. It is almost impossible to pick out common ratios, but certain accounting values are repeated within them:

Income statement—Interest expense, operating expense, depreciation, gross income (sales), operating profit, EBIT, net profit before tax (NPBT) or after tax (NPAT), and/or a cash flow figure (net of depreciation).

Balance sheet—Structure (total liabilities, total assets, shareholders' funds); working capital (inventory, debtors, and current liabilities); debt (total debt, long-term debt) and others (fixed assets, intangibles, cash).

Table 6.5. Financial ratio analysis

Size	Value	Units	Growth	Value	Units
Total assets	3,060	000s	Asset growth	−1%	%
Tangible net worth	1,207	000s	Turnover growth	−11%	%
Total revenue	3,703	000s	Sust. growth rate	1,376%	%
Liquidity	**Value**	**Units**	**Working capital**	**Value**	**Units**
Cash as % of assets	35.0%	%	Stock days	252.36	Days
Current ratio	2.38	Times	Receivables days	69.10	Days
Quick ratio	1.00	Times	Payables days	10.71	Days
Return	**Value**	**Units**	**Operating**	**Value**	**Units**
On equity	12%	%	Gross profit margin	31.0%	%
On net worth	12%	%	Operating profit margin	10.7%	%
On assets	5%	%	Net profit margin	3.9%	%
On surplus	12%	%	Sales/assets	121.0%	%
On net assets	119%	%			
Debt coverage	**Value**	**Units**	**Gearing**	**Value**	**Units**
Finance charge cover	2.12	Times	Liabilities/equity	1.54	Times
EBITDA/interest	2.15	Times	Liabilities/net worth	1.54	Times
EBITDA/current liabs	0.36	Times	Debt/equity	1.42	Times
Total liab. payback	12.69	Yrs	Debt/net worth	1.42	Times
Cash breakeven T/O	2,468	000s	LTD/(LTD + Net worth)	0.38	Times
Margin of safety	3,335.8%	%	Ret. earnings/current liabs	1.07	Times
Debt/operating cashflow	n/a	Times	Ret. earnings/total liabs	0.65	Times

Also, for publicly traded companies, the total market capitalisation is extremely important. Once it reduces below total debt, there is an implied bankruptcy!

Falkenstein et al. (2000) did an analysis of data on Moody's Credit Research Database (CRD), which was again referred to in Falkenstein (2002). The former highlighted 10 values—9 ratios plus total assets—based upon 17 inputs (Table 6.6, based on 2000:61), as being the dominant credit-related factors that could be derived from financial statement data. These were classified under the headings of profitability, capital structure, liquidity, size, growth, and activity. This list was slightly modified in the latter (2002:75), where they are stated as volatility, size, profitability, gearing, liquidity, growth, and inventories.

Allen et al. (2003:31–32) report on the predictive factors for the United States and Singapore models, which are notable both for their similarities and their differences. In the United States, the weights were 23, 21, and 19 per cent for profitability, capital structure, and liquidity respectively (as in Table 6.6), while for 3,400 Singapore companies the weights were 26, 24, and 14 per cent—except size replaced liquidity in third place.

Table 6.6. Moody's credit research database—predictive characteristics

Type	Name	Calculation	Contribution (%)	
Profitability	ROA	Net income/assets	9	23
	ROA growth	(Current ROA-Prior ROA)	7	
	Interest cover	EBIT/interest	7	
Capital sructure	War chest	Retained earnings/assets	12	21
	Gearing	Liabilities/assets	9	
Liquidity	Cash to assets	Cash/assets	12	19
	Quick ratio	(Current assets—inventories) current liabilities	7	
Size	Total assets	Assets/consumer price index	14	14
Growth	Sales growth	(Current sales/prior sales)−1	12	12
Activity	Stock turn	Inventory/cost of goods sold	12	12

Table 6.7. Financial statement characteristics

Pinches et al. (1973)	Falkenstein et al. (2000)
	Assets/consumer price index (Size)
Return on investment	Net income/assets, Net income growth, Interest cover
Gearing	Liabilities/assets
Capital intensity	Retained earnings/assets
Liquidity	Quick ratio
Inventory turnover	Inventory/cost of goods sold
Receivables turnover	
	Sales growth
Cash position	Cash/assets

Factor analysis

As indicated earlier, even though the final model will be comprised of a limited number of characteristics, there may be a large number of candidates at the start. Lenders are often tempted to limit these based upon past experience, but this may miss valuable information. The other approach is to use factor analysis, to identify uncorrelated characteristics or groups thereof (see Chapter 17, on Characteristic Selection).

Pinches, Mingo, and Caruthers (1973) used factor analysis to collapse data for 51 ratios onto 7 factors, shown in the left-hand column of Table 6.7. These explained anywhere from 78 to 92 per cent of the variance in the 51 ratios, depending on the year, and were also shown to be stable over time. Not unsurprisingly, they correspond quite closely to those that were presented by Falkenstein. Other studies were done in subsequent years, and in 1981 Chen and

Shimerda provided a judgmental analysis of the results from 26 of them, of which 5 had used factor analysis. In total, 65 different ratios had been used, of which at least half featured as being relevant in at least one study. The conclusion was that the primary factors are very similar to those provided by Pinches et al. (1973).

6.3.3 Restrictions

Financial-ratio scoring models have the advantage of being a cheap and easy means of extracting maximum benefit from available information, but they are restricted by the nature of financial statements. First, there is a narrow focus. No insight is provided into the business, beyond what is provided in the financial statements. Qualitative factors like competitive position, management quality, market trends, and economic forecasts have to be incorporated through other means.

Second, they are backward-looking. Financial statements focus upon historical performance, and do not provide any representation of future prospects. Income and expense projections cannot be used, and historical balance sheet figures are often much different than the true (market) values.

Assets are usually reflected at cost, net of depreciation, whereas the true values may be much higher or lower. This applies especially to property, plant and equipment, and intangibles, such as trademarks, copyrights, and intellectual capital. Similar problems arise with liabilities, especially where finance was raised at fixed interest rates, in foreign currency, or off balance sheet in some special purpose vehicle. Lenders may try to accommodate revaluations in their risk assessments. There is a trend towards mark-to-market valuations, but it may be infeasible where the assets are illiquid, intangible, or contingent.

Third, the reliability of the data is an issue. Creative and/or lax accounting often means that the financial statements are not a proper reflection of the firm's true position. The most reliable are properly audited statements provided by large companies, while the least reliable are those provided by small companies that might have been drawn up by the owner or an accounting officer.

FRS models typically rely upon the annual financial statements (as opposed to any pro forma statements or management accounts), as they are the most reliable, and are readily comparable. Note, however, that the other types of statements are often requested, and/or used by underwriters for interim reviews.

According to Dwyer et al. (2004:9), 69 per cent of the financial statements for the RiskCalc v3.1 model could not be included due to 'notable errors'. Blume et al. (1998) found evidence that accounting information was more predictive for larger firms than smaller firms.

Fourth, the data is sticky. Financial statements are only provided irregularly, and it may take some time before they are spread. Unless other more dynamic information is included, the resulting assessments are often not representative of the obligors' current situation.

The presentation of results for public/large companies is usually within two to six months after the financial year-end: *listed companies*, because of the demands of listing requirements; and *large companies*, because they receive greater attention from accounting firms, due to the higher fees being paid. In contrast, statements for smaller and unlisted companies may only be received nine or more months after financial year-end, and as a result, lenders may be relying upon information that is two or more years out of date.

Fifth, financial norms and business cycles differ from industry to industry. Credit scoring requires separate models when predictors differ, but this is difficult when data is limited. Fortunately, most industries can be treated together, and be rated up or down according to their relative risk. Certain industries must be treated separately though, such as financial services, real estate, and some others.

Credit underwriters usually do comparisons against peers within an industry. According to Falkenstein (2002:174), it is possible to normalise the financial ratios for each industry, but it would be self-defeating, as it may unwittingly forgive risks inherent within an industry, that are evidenced by say higher gearing.

And finally, financial spreading packages focus upon the most common accounting values, and often some experience is required to interpret the financial statements and spread them correctly. Any management comments, or values presented in the notes to the financial statements, can only be incorporated through a judgmental overlay.

As a result of these concerns, the ratings provided by these models are usually either used as input into judgmental assessments, or other models. If used to drive strategies, there is usually much more room to contest the rating than in the consumer credit market.

6.3.4 Rating agencies

Most of the credit rating agencies provide products that are used to do financial-ratio scoring of middle-market firms, including Moody's KMV, S&P, and Fitch. These agencies have a real advantage in this space, both in terms of data and experience:

Data—Over the years, they have been able to assemble significant databases of financial statements and default data, for different countries.

Experience—Likewise, they have also developed expertise, and methodologies that are specific for analysing financial statements and industry data. This includes knowing which variables are likely to provide value, and being able to build upon previous models.

The rating agencies collect financial data wherever possible. For public companies, the information will be readily available, and need only be captured and assessed. In contrast, data for private companies is somewhat trickier; subscribers will provide an agency with obligors' financial statement data, to develop and track the scoring models. Even so, there is still not as much data as one might expect. Moody's only had 1,500 and 1,400 defaults for the period 1989 to 1999, for their private- and public-firm models respectively (Falkenstein et al. 2000). This is, however, sufficient to provide reliable models.

Over the period from 1920 to 2002, only 3,500 of the 16,000 corporate bond issuers ever rated had ever defaulted (Ong 2002:20).

The way in which the models work varies from one agency to the next. The two primary approaches are: (i) do a *direct estimation* of default risk; or (ii) try to predict the rating agency grades. For the former, estimated default frequencies are derived for a single year, and/or the compounded cumulative frequency for periods of up to five years. These are then mapped onto rating grades. In theory, a BBB provided by a rating agency, or its models, should have the same estimated default frequency, irrespective of how it is derived, whether judgmentally, from the market value of securities, or using an FRS model.

Note here that FRS can also be used for public companies, and the results used as input into the judgmental assessments. Agencies will, however, normally have separate scoring models for public versus private, and/or large versus small, as the credit dynamics (and hence the predictive variables) differ between the groups. In particular, larger companies have greater access to debt funding. The agencies may also develop separate models for different countries, and perhaps different industry groupings within each. One particular grouping that demands separate treatment(s) is financial services and real estate companies, which have much higher gearing than other concerns.

FRS models can be implemented in two ways. First, the data can be provided through to the rating agency, which will return an estimated default frequency. Second, the model can be implemented on lenders' internal systems. The former has the advantage of requiring less of an infrastructure investment, while with the latter, the models are more transparent, and there are fewer networking issues.

Moody's KMV

This is not meant as a plug for Moody's KMV, but because there is so much publicly available information, some of their products must be mentioned. The best known financial-ratio scoring product is Moody's RiskCalc™, which is designed to assess companies of $100,000 and above. According to Dwyer et al. (2004), the V1.0 model was launched in 2000 and was already being used at 200 financial institutions worldwide in 2004. Since the merger with KMV, it has been modified to also: (i) use the structural, market-based approach that was the basis for KMV's Private Firm™ Model; (ii) incorporate general and industry-specific economic trends, at least for the United States; (iii) allow lenders to do stress testing, by assessing default rates under historical economic scenarios; and (iv) provide 'full version' and 'financial statement only' (FSO) modes. The RiskCalc™ results feed into another product, called Credit Monitor, which is used to monitor the efficiency of the models, and the risk in the broader market.

Table 6.8. Moody's KMV—CRD

CRD	Worldwide November 2003	USA and Canada	
		1989–1999	–2002
Defaults	97,000	1,621	3,764
Firms	1.5 mn	24,000	51,000
Financials	6.5 mn	115,000	225,000

RiskCalc™ uses local information for different countries. Its feedstock is the CRD, which contains data collected for countries including, but not limited to, the United States, Canada, United Kingdom, Korea, Japan, Singapore, and the Nordic countries (Denmark, Finland, Norway, and Sweden). The data is supplemented on an ongoing basis.

Dwyer et al. (2004:8) provide details, shown in Table 6.8, that illustrate the CRD's growth. The number of defaults over the last three years to 2002 was disproportionately high, which had the dubious benefit of providing data for a full economic cycle (the periods 1990–1991 and 2000–2002 were recessions). The US/Canada figures were used for both the RiskCalc V1.0 and V3.1 developments, and exclude finance, real estate, and insurance companies. The accuracy ratios for the V1.0 model were 49.5 and 30.7 per cent, for the one- and five-year models respectively, and with the V3.1 FSO model, these values improved to 54.3 and 35.7 per cent (Dwyer et al. 2004:26).

In general, v1.0 was already very predictive, but the improvement provided by v3.1 is significant. Even so, the model results are not as good as the full credit ratings done by Moody's, nor are they as good as many retail credit scores that use both positive and negative information. Two observations can be made. First, Moody's full credit ratings provide substantially better results (see the graph in Ong 2002:23). These come at significantly greater cost, but it proves how much value can be obtained from a *fundamental analysis* that incorporates management quality, market conditions, competitive situation, and so on. Second, given that RiskCalc relies almost exclusively upon financial statement data, it is clear how much value can be provided by *accurate financial information*. Unfortunately however, for retail customers, financial data is often unavailable or unreliable, in which case other data must be exploited—such as account behaviour, credit bureaux, etc.

6.3.5 Internal grades

While credit rating of business enterprises is an area where the rating agencies tend to dominate, many lenders will nonetheless develop their own models, either using their own default data (scoring), or by tapping the knowledge of their own credit staff (expert models), or both (hybrids). There is often a demand for their internal grades to be mapped onto the rating agency grades, but care must be taken. In the absence of empirical default data, it is not easy, and lenders sometimes map the data to achieve the same distribution, and not the same default rates. This is a mistake, as it is unlikely that they will have either the data or experience to derive a model as powerful as those provided by the rating agencies. If mapping is absolutely necessary, then it is best to err on the conservative side.

Lenders do, however, have the benefit of other data sources, such as account behaviour, bureau data, and personal information on the owners/directors. For smaller obligors, this information can be very valuable, if not crucial: (i) it takes time for the financial statements to be provided, and the information is not always reliable; and (ii) the other data sources can provide early warning signs, that indicate financial difficulties long before the next set of financials are received. Something that must also be taken into consideration are the costs associated with the various types of information:

Financial statements—The most costly data source, only this cost is borne by the customer, not the lender. Putting pressure on smaller customers to provide updated financials can jeopardise the relationship, so they are often only requested for loans above a certain threshold.

Bureau data—There is usually a marginal cost associated with obtaining bureau data on a client. The cost of an individual enquiry may seem low, and be considered reasonable for new-business origination, but can become expensive, if done regularly as part of existing-account management.

Own data—The cheapest source of information is lenders' own data on their clients, especially behaviour on transaction accounts, or deals that have been fully paid off in the past. This does, however, demand an infrastructure investment, to capture, assemble, assess, and deliver meaningful data.

Lenders have a tendency to treat the different types of data in isolation, whereas the best approach is to integrate it into a single risk assessment. Exactly how this is achieved is problematic. Hybrid models are often used—objective models wherever possible, with a judgmental overlay to fill in the gaps.

6.4 Credit rating agencies

Many market participants place a lot of trust in the rating agencies' analysis, and the majority of institutional investors are restricted to investments in certain rating classes. Even investors who do not believe in the accuracy of credit ratings use them as a first classification of the riskiness of the obligor.

Schönbucher, P. (2001)

Anybody that follows investment markets will be familiar with the credit rating agencies, whose primary function is to provide credit risk assessments for investors, especially for publicly traded bond issues. Companies that are issuing, or refinancing, debt also have a particular interest in having a credit grade as high as possible, as this affects the interest rate to be paid on their borrowings. In order to avoid any bias, the rating agencies tend to ensure that the people doing the assessment have no contact with companies being assessed, but instead obtain the information through others.

There are three primary rating agencies in the world: *S&P*, *Moody's*, and *Fitch IBCA*. The first two dominate the American market, while Fitch is dominant in many countries outside the United States. Other rating agencies exist that specialise in certain countries/regions, or economic sectors. Rating grades are determined using information from a variety of sources, including financial statements, industry and country assessments, and interviews with management. Also, of late, there have been moves to improve grades' responsiveness to recent events, by including mathematical analysis of bond price movements.

6.4.1 Letter grades

As indicated earlier, rating agencies usually present their assessments of a borrower's credit risk as a rating grade. Examples of the higher-level risk grades, and the associated default risk, are presented in Table 6.9. The 'odds' column refers to the 'not default' to 'default' ratio. The values provided are approximations of the five-year average default rates, calculated using information from S&P.

Grades for bonds can be split into four classes, two that are used for day-to-day assessment, and two exit states:

Investment grade—Some investors, especially larger financial institutions that are investing on behalf of others, may only invest in bonds in lower-risk grades, usually either BBB- or BBB and above.

Speculative grade—Grades below the investment grade definition. These bonds are often highly illiquid, hence the term 'junk bonds', and trade at large discounts. Substantial profits can be made, but the risks are commensurately greater.

Default—For traded bond issues, 'default' refers to any missed coupon or capital payment, as compared to severe arrears for normal lending. Bankruptcy is a default state in each of these. According to Schuermann and Jafry (2003a), if a company goes bankrupt and is rehabilitated, it is treated as a new entity.

Withdrawn—Events may occur that make it impossible to rate the firm: (i) the debt is repaid, either voluntarily or under duress; or (ii) the entity ceases to exist, either because of merger and acquisition activity, or voluntary liquidation. These cases are classed as

Table 6.9. Letter grades

Grade	Default odds	Label	Description
AAA	5000/1	Unquestioned	Extremely high credit quality
AA	2000/1	Excellent	High quality and stable
A	1000/1	Strong	Capable of meeting commitments
BBB	275/1	Satisfactory	Sound, minimum investment grade
BB	50/1	Fair	Good company, but some uncertainty
B	20/1	Speculative	Susceptible to environment changes
CCC	8/1	Doubtful	Early warning!
D		**Defaulted**	
		Unrated	Too small, or no debt outstanding

'NR' for not rated (S&P), or 'WR' for rating withdrawn (Moody's). It is more common amongst riskier grades, which are dominated by smaller firms that do not have the same borrowing capacity. Given that it cannot be determined whether the rating arose because of bad behaviour or other reasons, these cases are typically excluded from any analysis.

An example of a bad transition to NR/WR is where an obligor foregos an agency rating, because of deteriorating credit quality known only to the bond issuer. Even so, according to Schuermann and Jafry (2003a:7), the reasons for ratings being withdrawn are usually benign. They quote Carty (1997), who analysed Moody's data from 1920 to 1996, and found that only 1 per cent of the ratings had been withdrawn due to deteriorating ratings.

Rating agencies typically do not provide descriptions like those shown in Table 6.9, but do provide data to support the validity of their grades, such as the default rates shown in Figure 6.1, and the transition matrices shown later in Tables 6.12 and 6.13. This information is used by lenders to analyse changes in their portfolios' credit quality over time, not only for historical analysis, but also for doing inference regarding future movements when doing pricing.

Figure 6.1 also illustrates the concept of mean reversion, which is associated with information's time decay. When analysed over successive periods from a given observation point, the differences in default rates are greatest in the near future, and over time, will narrow to cluster about the mean. All credit scores and risk grades display this pattern, but the rate of decay differs. Dwyer et al. (2004:24) note that failure to take this into account can result in mispriced loans. If a one-year default rate were used for pricing five-year loans, low-risk customers would be undercharged, and high-risk customers overcharged.

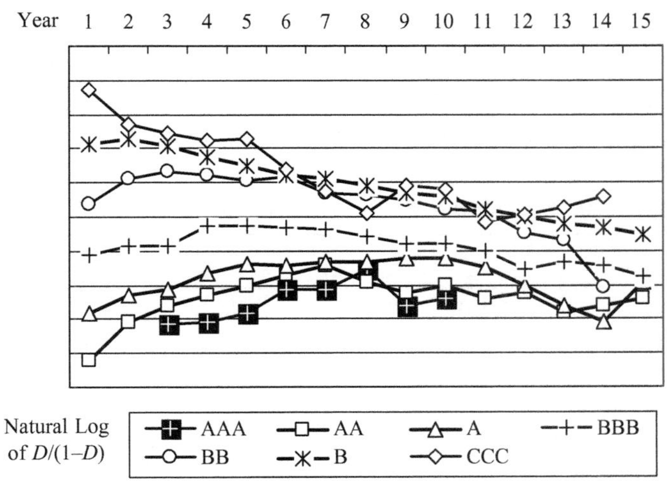

Figure 6.1. Default rates and mean reversion.

6.4.2 Derivation

For large corporate borrowers, the rating grades are a combination of objective analysis of available data, and subjective views. Ratings are done at two levels: (i) for individual bond issues, which are published for consumption by bondholders; and (ii) a composite issuer rating, which is the one typically quoted in the financial press, and is meant for the broader investing public.

According to Delianedis and Geske (1999), the 'fixed income' market in the United States is two to three times larger than the equity market. Bond investors have greater interest in ratings migrations than defaults, because ratings have such a huge influence on bond prices, and defaults are rare.

Ratings for individual bond issues should reflect not only the probability of default, but also the loss severity, which varies with the term structure, seniority, and level of security for each bond issued (see Table 6.10). According to Ong (2002:32), the closer the lender is to an asset, the higher the recovery rate, and hence the higher the grade. For the issuer ratings, individual bond ratings are first converted into long-term senior unsecured equivalents, and any bonds issued by what is substantially the same economic entity are treated together. This accounts for parent/subsidiary relationships, mergers and acquisitions, and contractual no-recourse arrangements.

Credit ratings do not come cheap, and costs vary. According to Schuermann and Jafry (2003a) a rating for a bond issue costs $25,000, or half a basis point for issues over $500 million; they also quote another paper that indicated S&P charged 0.0325 per cent of the face amount. The rating is usually paid for by the issuer at time of issue, but lenders can also commission an agency to investigate a specific borrower.

A contention sometimes made is that with efficient markets, bond prices should not react when issuers' rating grades change, because the prices should already include all available information. The only conclusion that can be drawn is that the markets are not fully efficient, and the rating agency grades do contain information that the market finds of value—most of which comes from the agencies' in-depth fundamental analysis. Individual investors either do not have the: (i) time and resources available to invest in the assessment; or (ii) same level of access to information and insights, much of which is obtained directly from the bond issuer.

Table 6.10. Speculative grade recovery rates—1982–2000

Seniority	Recovery per cent
Senior secured bank loan	67.06
Equipment trust	65.93
Senior secured	52.09
Senior unsecured	43.95
Senior subordinated	34.59
Subordinated	31.88
Junior subordinated	22.48

6.4.3 Issues

Rating agency grades are by no means perfect. Schönbucher (2003:224) and others list several problems, especially when using them as the basis for pricing:

Small numbers—Because of the high cost, only a relatively small number of companies have been rated. This makes proper analysis, using tools like transition matrices and survival analysis, difficult.
Delay and momentum—Rating grades are slow to reflect obligors' credit risk, and tend to adjust in increments.
Population drift—The base of rated obligors has been changing over time, especially as smaller obligors are being rated, and agency activities are expanding outside the United States.
Downward ratings drift—There is more likely to be a downward movement than upward. Further, the average level of credit ratings issued has been getting lower over time.
Business cycle sensitive—The assumption is that rating-grade transitions are cycle neutral, but they have been shown to vary over the business cycle.
Risk heterogeneity—It would be expected that the credit risk and credit spreads within a given risk grade would be constant, but the profiles can be quite different.

Small numbers

The small number of bonds in issue limits the amount of analysis that can be done. Problems are greatest in the riskier grades, especially anything containing a 'C'. For their work on transition matrices, Schuermann and Jafry (2003a) obtained a full set of data from S&P CreditMetrix™ for the period from 1981 to 2001. This included 55,010 obligor years of data, but only 9,178 unique obligors—about 5½ years of data for each company (withdrawn ratings were excluded). The data was for a mix of industrials, utilities, banks, insurance companies, and real estate companies. Sovereigns and municipalities were excluded, as were companies whose ratings were withdrawn. Of these, 71 per cent were BBB or better (investment grade), with almost 30 per cent in the A grade alone (the analysis ignored any '+' or '−' modifiers). The number of defaults was only 840 obligor years, or 1.53 per cent for the entire sample, which limits the possibilities for any statistical analysis.

The 'law of small numbers' typically refers to individuals' tendency to put excessive faith in conclusions drawn based upon a limited number of observations. This is especially apt when modelling using small samples. See Rabin (2000).

Ratings delay and momentum

While the rating agency grades are widely relied upon, many commentators have noted that they suffer from both ratings delay, and ratings momentum. Ratings delay refers to the time lag before new credit related information is reflected, which can lag changes in market prices by several months (Schönbucher 2003:224). This is related to ratings momentum, a phenomenon where if a rating grade changes in one period, the next change is likely to be in the same direction. Perhaps the most extreme examples of ratings delays are highly rated companies that went bankrupt almost without warning. Many of these occurred during the earliest years of the twenty-first century, some aggravated by poor corporate governance, the two most notorious examples being Enron and WorldCom.

According to Ong (2002), in the United States several of these defaults were in recently deregulated industries, especially telephone and electricity companies. In some instances, they resulted from accounting improprieties, perhaps in part the result of the 1990s Gekkoesque 'Greed is Good!' culture that drove much of it.

Such cases often featured prominently in the financial press in months immediately prior to default, without any ratings change, even though their stock and bond prices were falling. At such times, not only accounting standards, but also the entire financial system, are questioned—including banks' practices in structuring off-balance-sheet financing, and analysts' stock recommendations, both of which impacted upon the public's faith in the stock market.

These events also put a spotlight on the reactive nature of rating agency grades, especially with Basel II looming. As a result, all of the agencies have been working to improve their grades' sensitivity to credit related information by: (i) providing more frequent updates; (ii) shifting away from the traditional through-the-cycle approach, so that recent events are reflected; (iii) including an analysis of the price movements of publicly traded securities as part of the assessment; and/or (iv) assessing the effects of off-balance-sheet activities.

Population drift

Another factor to be considered is the changing nature of the population being assessed: (i) the number of graded companies has increased significantly over time; and (ii) the primary growth areas have been amongst smaller companies that are inherently riskier, and companies outside of the United States. Schuermann and Jafry (2003a:28) provide an indication of how great these changes have been. The total number of obligors worldwide, net of 'not rated', was about 1,400, 2,200, and 5,000 in 1981, 1991, and 2001 respectively, with the most dramatic growth from 1993 onwards. Of the total, 70 per cent of the financial statements were from American companies (6,398 of 9,178), but this had dropped from 98 per cent to nearly 60 per cent over the 20-year period. The increase in non-US companies was especially pronounced from 1989 onwards.

Downward ratings drift

The downward ratings drift has also been the subject of some debate. Delianedis and Geske (1999:12) highlighted this trend in data on US corporates obtained from S&P for the period

1986 to 1996. They found that the percentage of AAA, AA, and A grade companies decreased from 44.8 to 33.8 per cent. Further, on average, less than 10 cases per year shifted by more than three grades up or down, and the number of downgrades exceeded upgrades by about three to one. This downward drift was negligible prior to the 1980s, but since then it has been considerable (see Carty and Fons 1994).

Why is this? Blume, Lim, and McKinlay (1998) tried to determine whether the declining ratings were caused: (i) by decreased credit quality, or (ii) increased standards being applied by the rating agencies (read 'rating agency conservatism'). They concluded the latter. Other commentators have also noted an increased use of debt, and better data being available to the rating agencies, which might also affect the ratings.

Strangely, the year of the rating featured in the Blume et al. model. It is possible that this was affected by improved access to financial markets, with the result of a lower cost of borrowing and increased credit appetite from riskier companies. The means of the accounting values used showed little change other than a slow increase in the average market value.

Thus, there appear to be two main factors, as well as several subfactors, that are causing the rating agencies to provide lower grades. First—the growing number and higher risk of rated firms. Three forces are at play: (i) greater reliance upon debt as a funding mechanism, especially at the lower end of the market, as investors' increased risk appetite has reduced borrowing costs; (ii) greater demand for obligor ratings, as investors have recognised their value; and (iii) rating agencies growing their markets, beyond traditional geographic and economic boundaries. And second—higher standards are being applied by the rating agencies. Here too, a couple of forces are at work: (i) greater conservatism, especially where applying through-the-cycle ratings; and (ii) more and better data, and improved assessment techniques.

Business cycle sensitive

How do risk estimates react to changes—peaks, troughs, and points in between—in the business cycle? One of the assumptions when using rating grades is that transition and default rates are stable over time. There are, however, variations over the business cycle, which may not be readily predictable. The generally accepted interpretation is that agency grades should be cycle neutral. There is, however, ample evidence to suggest that credit ratings—and the associated default probabilities, transitions, and survival rates—vary systematically with the business cycle. According to Pesaran et al. (2004), Moody's has changed its rating process in this regard—'Moody's has been striving for some time to increase the responsiveness of its ratings to economic developments'. This is further illustrated by Schuermann and Jafry (2003a), who mention that the default rate for S&P's 'CCC' rated companies increased to 62 per cent in 2001, 'one of the worst years on record from the perspective of corporate bond defaults', far above the long-term average of 40 per cent.

Risk heterogeneity

One of the core assumptions of any risk measure is that all cases with the same grade, or score, should be of homogenous risk. This implies that all possible information has been included in the assessment, and it is not possible to differentiate further. This is an impossible goal to reach, as there will always be factors that are left out of the assessment. For the moment however, this discussion is limited to data that is available and can be assessed. There are two primary instances of within-grade heterogeneity noted in rating agency grades. First, the difference between banks and industrials, as banks tend to be of much lower risk than industrials of the same grade, and are treated separately. Second, differences across countries, in that there are differences in the transitions between rating grades according to geography. As a result, wherever possible, analysis should be focused upon groups of interest, with the hope that there is sufficient data for the numbers to be reliable. Schönbucher (2003:224) also notes that the credit spreads within a rating grade are not constant over time, and while it may be possible to assume a single spread curve as an approximation for a portfolio, it causes problems for individual bonds if the spreads are to be used as the basis for pricing.

6.4.4 Research focus

There will always be questions about credit rating grades, and according to Blume, Lim, and MacKinlay (1998), much of the research to date has focussed on three questions:

- Do they measure what they are supposed to measure?
- Do they contain information not already discounted into the bond price?
- How do the rating agencies use publicly available information?

Do they measure what they are supposed to measure? The studies have confirmed a correlation between the rating agency grades and the probability of default, but qualify it by questioning the reliability of the rating grades. A correlation has also been shown between the grades and the yields of traded securities. Unfortunately, the ability to assess credit risk is not as sophisticated as that for the market risk of equities, due to the lack of liquidity and a shortage of mathematical tools that can be applied. Even so, it is still possible to measure it using variations of the Black and Scholes' (1973) and Merton's (1974) models.

Do they contain information not already discounted into the bond price? Most of the studies indicated that the bond yields tended to adjust to changes in rating grades, but a couple could not confirm this. In general, credit ratings focus on financial and industry data, which makes them very sticky, and raises concerns that the grades are not adjusted downwards quickly enough. A prime example of this was Enron, whose credit risk grade remained high even while its bond prices were falling.

How do the rating agencies use publicly available information? Much of this type of research involved reverse engineering, to determine the composition of the ratings, and in general has shown that publicly available information can be used to predict the rating. Over time, the rating agencies have been working to include other indicators, especially in the United States, where debt issues are publicly traded, and price movements reflect new information (Schuermann and Jafry 2003). This allows the grades to be reassessed daily, based upon the value of the securities.

6.5 Modelling with forward-looking data

Perhaps the biggest disadvantage of credit scoring is its backward-looking nature; any assessment is based purely upon a historical analysis, and it is assumed that any cases with similar characteristics will behave in like fashion. There is, however, forward-looking information available, including the *rating agency grades*—even though they may be sticky—and *market prices*, that reflect investors' views on obligors' future fortunes. According to Yamauchi (2003:16), there are three types of approach that are used to analyse forward-looking data:

Historical method—Relies upon a straightforward analysis of grade movements and default histories, perhaps using Markov chains or survival analysis.
Structural approaches—Attempts to model the structure of the default process, using financial statement data, or some proxy of asset value and volatility, which assumes that the modeller has the same information as the firm's management.
Reduced form approaches—Relies upon the value of publicly-traded debt over time, and assumes that: (i) the same information is available to both the modeller and the market; and (ii) the risk can be determined from price volatility and/or the credit spreads.

The latter two modelling types are usually mentioned together. According to Jarrow and Protter (2004), structural models are best suited for assessing company management and determining capital requirements; while reduced-form models are the option of choice for pricing and assessing market risk (Table 6.11). Note here that the literature available on the topic of structural and reduced-form models does not always agree, and the following section is based upon the author's interpretations.

6.5.1 Historical analysis

Prior to the advent of the more advanced mathematical techniques, the only way for lenders to assess the risk associated with agency rated debt was through a straightforward analysis of past

Table 6.11. Modelling approaches using forward-looking data

Data source / Model type	Agency grades	Financial statements	Market prices
Historical	✓		
Options-theoretic		✓	✓
Reduced form	✓		✓
Forward-looking	Y	N	Y

defaults and rating transitions. Rating agencies publish transition matrices and survival/hazard rates covering different periods of time, and investors use this data in their risk modelling and pricing. When used for pricing, lenders sometimes make the mistake of compensating themselves for the expected loss (EL), but provide for a minimal risk premium beyond that. This suffers because of many of the problems associated with rating agency grades, mentioned in Section 6.2, such as: (i) it ignores default rate volatility over time; (ii) volatility is greatest amongst the riskier grades; (iii) grades are sensitive to the economy; and (iv) differences exist by industry and geography. Even so, it is still a powerful tool for gaining insight into a portfolio.

Survival analysis and Markov chains are both covered in Section 9.2 (Forecasting Tools), which includes an example of the use of survival analysis for analysing rating-grade mortality. Markov chains deserve further attention here though. According to Schuermann and Jafry (2002), they are used for: (i) portfolio risk assessment and provisioning; (ii) modelling the term structure of credit risk premia; and (iii) pricing of credit derivatives. They also comment that under Basel II, capital requirements will be driven largely by default estimates obtained from a ratings migration analysis, whether using S&P, Moody's KMV, or Fitch Ratings—or perhaps even lenders' own internal ratings.

The Schuermann and Jafry (2002) paper notes that the traditional approach presented here, also called a frequentist or cohort approach, is the industry standard. They propose a duration/hazard rate approach (with either a time homogenous or time non-homogenous methods), that can be used to provide better results.

The matrices reflect expected changes in obligors' credit quality over time. A schematic of a credit rating migration matrix is provided in Figure 6.2, which illustrates some typical patterns: (i) it is diagonally dominant, meaning that most companies will stay in the same grade from one period to the next; (ii) jumps of more than two grades are rare, and the further from the diagonal, the sparser the data; (iii) the extent of movement is greater amongst riskier companies; (iv) companies will often jump to the default state without warning; and (v) movements to 'not rated' are most common amongst riskier grades.

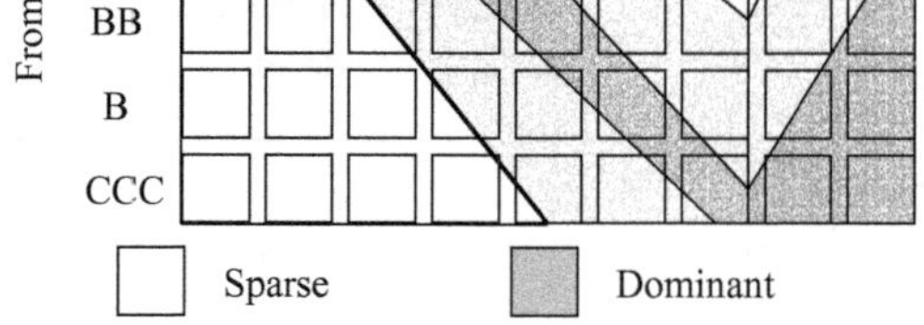

Figure 6.2. Rating migration matrix.

The same type of matrix can be derived for cohorts defined by behavioural scores. These matrices will have many of the same characteristics as credit migration matrices, but have greater variation off the diagonal, if evaluated over the same time period, due to the narrow base and volatility of the information used in the assessments.

Any analysis depends upon having sufficient mass within the various cells; otherwise the results may not be reliable. It follows that the problem is greater in those areas further removed from the diagonal. According to Schuerman and Jafry (2003a) this is why the industry standard is for rating agencies to: (i) only publish migration matrices with higher-level grades, and not split them by the '+' and '−' modifiers; and (ii) to collapse all 'CCC' or worse categories into one. The number of states is thus reduced from 19 to 7, 'which ensures sufficient sample sizes for all rating categories'.

Each agency publishes ratings migration matrices for its published grades, which subscribers use to do their own modelling, like the one-year and five-year tables presented in Tables 6.12 and 6.13 respectively. If the one-year matrix is multiplied by itself five times, the result approximates the five-year matrix. Greater migration is apparent, due to the longer time period that has elapsed.

Table 6.12. One-year transition matrix

	Investment				Junk			
	AAA	AA	A	BBB	BB	B	C	D
AAA(%)	89.35	9.51	1.10	0.00	0.04	0.00	0.00	0.00
AA(%)	1.05	89.10	9.33	0.39	0.08	0.01	0.00	0.04
A(%)	0.07	2.38	90.88	5.47	0.82	0.20	0.03	0.15
BBB(%)	0.04	0.41	5.90	87.27	4.97	1.12	0.10	0.20
BB(%)	0.02	0.08	0.85	5.15	83.02	8.22	0.41	2.24
B(%)	0.01	0.05	0.29	0.62	6.01	82.57	2.99	7.46
C(%)	0.00	0.01	0.13	0.57	2.77	6.44	59.97	30.10
D(%)	0.00	0.00	0.00	0.00	0.00	0.00	0.00	100.00

Table 6.13. Five-year transition matrix

	Investment				Junk			
	AAA	AA	A	BBB	BB	B	C	D
AAA(%)	57.13	31.35	9.98	1.12	0.31	0.07	0.01	0.03
AA(%)	3.35	59.21	30.88	4.97	1.01	0.34	0.03	0.21
A(%)	0.34	8.55	66.52	18.30	4.05	1.35	0.13	0.76
BBB(%)	0.21	2.02	20.30	53.83	15.02	5.82	0.45	2.35
BB(%)	0.11	0.42	4.24	16.12	44.82	20.25	2.13	11.91
B(%)	0.04	0.31	1.55	3.14	16.06	41.70	4.82	32.38
C(%)	0.01	0.06	0.65	2.85	6.61	9.98	7.98	71.86
D(%)	0.00	0.00	0.00	0.00	0.00	0.00	0.00	100.00

The transition probabilities in the two tables are estimates, but are adequate to give an indication of how the rating grades work. They are based upon data provided by Moody's, as presented in Yamauchi (2003:59), but the letter grades are those used by other agencies. Movements into the 'not rated' category were not provided.

These matrices must, however, be used with care. Rating agencies usually only provide one set of numbers that is supposed to be widely applicable. According to Yamauchi (2003), 'there are significant differences between banks and industrials, US versus non-US obligors, and business cycle peaks and troughs'. As of yet none of the credit rating agencies provides separate tables by country, even though the business cycles may vary greatly between them. If US companies dominate, they may not be representative of other environments—especially emerging markets. This is particularly pertinent for speculative grade borrowers, where changes in the business cycle have a greater impact.

6.5.2 Structural models

A part and parcel of the human condition is the constant striving to explain the environment, using logically compelling arguments that make sense. Some attempts have been far off the mark, like having the earth as the centre of the universe, yet we continue. In academia, these explanations are set out in theories and models, which are intended to describe the structure of the environment and its workings. Such structural models are best applied in the physical sciences, and tend to work dismally in disciplines like economics. An exception is some of the models used to assess credit risk, at least if their market acceptance is to be used as the yardstick. According to Falkenstein et al. (2000:17),

> People like structural models. [They are] usually presented in a way that is consistent and completely defined so that one knows exactly what's going on. Users like to hear the story behind the model, and it helps if the model can be explained as not only statistically compelling, but logically compelling as well. That is, they should work irrespective of seeing any performance data. Clearly, we all prefer theories to statistical models that have no explanation.

Here 'structure' refers to the structure of the firm, and the economic process that leads to default. Risk is a function of value and volatility, and is presented as a 'distance to default.' There are two structural models that have been presented for assessing the probability of bankruptcy: Jarrod W. Wilcox's (1971)[5] gambler's ruin model, and Robert C. Merton's (1973) option-based pricing model.

The little known Wilcox gambler's ruin model relies upon information about a company's cash flows; the firm's equity is a reserve, cash flows supplement or drain the reserve, and bankruptcy occurs when the reserve is drained. The name 'gambler's ruin' comes from its initial application to hypothetical gambling problems. The problem is treated as a Markov chain. For example, assuming that each game has 2/1 odds (50:50 probability), what is the probability of losing the initial stake after X games? Wilcox applied this to companies, by treating equity as

[5] See Falkenstein et al. (2000:174).

the initial stake, and cash flows as having two possible states—positive or negative. Lenders' interest is in the probability of the reserve cushion being depleted. Distance to default is calculated as the sum of equity and expected cash flow, divided by cash flow volatility.

In contrast, Merton's model is based on options theory, and has been referred to as an 'options-theoretic structural approach' (Allen et al. 2003). The share price is treated as though it were the value of a European put option (exercisable only at maturity) on the firm's assets, with a strike price equal to the value of the firm's liabilities. Default is assumed to occur when the value of the firm's assets reduces to below the value of its debt. Simply stated, the relationship is:

Equation 6.1. Distance to default $\frac{A-D}{\sigma_A}$

where A is the asset value, D is the amount of debt, and σ_A is the volatility of the asset values. A more comprehensive representation of the formula is given in Equation 6.2, as well as the Black and Scholes model upon which it is based.

Equation 6.2. Black and Scholes', and Merton's models

		Merton's equity valuation	Black and Scholes' option pricing
C	Option value	$C=-M\Phi(-d_1)+Xe^{-rT}\sigma(-d_2)$	$C=+Me^{-qT}\Phi(d_1)-Xe^{-rT}\Phi(d_2)$
D_1	Present value volatility	$d_1=\frac{\ln\left(\frac{M}{Xe^{-rT}}\right)+\left(\frac{\sigma^2}{2}\right)T}{\sigma\sqrt{T}}$	$d_1=\frac{\ln\left(\frac{M}{X}\right)+\left(\frac{r-q-\sigma^2}{2}\right)T}{\sigma\sqrt{T}}$
D_2	Future value volatility	$d_2=d_1-\sigma\sqrt{T}$	

where

M =	Market value	Total assets	Share price
X =	Strike price	Total debt	Exercise price
σ =	Volatility	of ROA	of Share price
e =	Natural log odds		
T =	Time	To maturity	To expiry date
R =	Risk-free rate of return		
Q =	Yield	Not applicable	Dividend

According to Allen et al. (2003:27), Merton's model assumes that 'asset values are log normally distributed', an assumption that often does not hold. As a result, alternative approaches may be used for distance to default and probability of default mapping, such as using historical default rates for calculating distance to default.

KMV's Credit Monitor™ and Private Firm™ models both use an options-theoretic approach. Credit Monitor™ provides default predictions for all major companies and banks, based upon their share prices. According to Yamauchi (2003:20), it uses: (i) market prices of both the firms' assets and their equity; and (ii) the volatility of each. Expected default frequencies are provided for horizons from one to five years, but its greatest power lies within an 18-month window. In contrast, Private Firm™ focuses on middle-market companies, for which no market prices are available. According to Dwyer et al. (2004:7), it instead 'uses a small subset of financial statement data, and a statistical mapping to estimate company value and business risk'.

Most of the research on the structural approach has focused on its ability to predict bond prices and yields, as opposed to default probabilities, even though the models should be equally capable of predicting both. In general, the conclusion has been that the *predicted spreads are much lower than those actually observed*, especially for shorter maturities, where liquidity and incomplete accounting information are much greater issues. There are also other problems that have been highlighted: (i) it requires input on the value of the firm, which may be suspect or not readily available; and (ii) adjustments have to be made to Merton's model, to recognise the interdependence between interest rates and credit risk.

6.5.3 Reduced form models

The other way of measuring credit risk is to analyse the values of borrowers' traded liabilities, which was first proposed by Jarrow and Turnbull (1995).[6] This is called 'reduced form', because it assumes that the companies' structure is represented through the value of these securities. Other names used are 'Intensity Based', and 'Default Correlation' models. Rather than determining the distance to default using information provided by the obligor, reduced-form models instead use exogenous information. According to Yamauchi (2003):

> Recently there has been much development of rating based reduced form models. These models take as a premise that bonds when grouped by ratings are homogenous with respect to risk. For each risk group the models require estimates of several characteristics such as the spot yield curve, the default probabilities, and the recovery rate. These estimates are then used to compute the theoretical price for each bond in the group.

Three components are evident in this statement: (i) bond prices (spot yield curve for each risk grade); (ii) rating-grade transitions and default probabilities; and (iii) recovery rates. Yamauchi (2003) raises several issues:

Rating grades—One of the key assumptions is that the rating grades are an accurate assessment of risk, which has been disputed.

Bond prices—Because of the use of market prices, it is not possible to relate default or recoveries with the underlying characteristics of the bonds or issuers.

Credit spreads—The approach assumes that the risk in each grade is the same, but the credit spreads vary between bonds within the same grade.

[6] Unfortunately, the mathematics behind the Jarrow and Turnbull model is complex, and cannot be treated within the scope of this textbook.

In general, the consensus is that reduced-form models cannot be used to provide direct estimates of default risk, as on average the credit spreads overstate the risk. Reduced-form models are, nonetheless, widely used for pricing debt securities and analysing credit spreads—at least in the United States, where there are well-established markets for traded debt. They cannot be used in environments where these markets do not exist, as is the case in many other parts of the world. Adjustments to the model can, however, be made for illiquid markets.

Credit spreads

The credit spread is a bond's (or other loan's) risk premium, being the difference between its yield, and the risk-free rate (usually the yield on domestic government bonds of equivalent maturity). Yamauchi (2003:13) quotes Schmid (2002), who states that credit spreads compensate for two risk components. *Default risk* is that typically associated with a borrower being unwilling or unable to meet its obligations. In contrast, *spread risk* is associated with changes in the market value of debt securities, usually arising from rating-grade migration.

The market requires this spread not only as compensation for credit risk, but also liquidity risk, market risk, and the call/conversion features of some bonds. According to Ericsson and Olivier (2001), the spread cannot be decomposed into its constituent credit and liquidity components. Liquidity is a function of both the firm's assets and gearing and, as a result, the two risks are highly correlated and interrelated. Yamauchi (2003) illustrates the spread mathematically as:

Equation 6.3. Credit spread $(1+r)^T=(1+r+\alpha)^T(1-q)+q\phi$

Where r is the risk-free rate, T the time to maturity, α the credit spread, q the default probability, and ϕ the recovery rate. The spread will change over time, either in a continuous fashion, or in sudden jumps. *Continuous changes* are usually minor adjustments to the market's assessment of the company and its general risk tolerance, whereas *sudden jumps* will occur with changes in the credit rating, and any generally available news indicating imminent or actual default.

For *investment grade* securities, credit risk is only a small portion of the spread, but is greatest when the security is first issued, and narrows over time. In contrast, for *speculative grade* securities, the spread is wider for nearer maturities, when the market has to assess whether or not the company will be able to refinance. Credit spreads will also be heavily influenced by the business cycle and increase to compensate for higher default rates during downturns.

6.6 Conclusion

In the credit industry, a distinction is made not only between the retail and wholesale markets, but also between consumer and enterprise lending. Consumers are part of the retail space, while enterprise lending is split between retail and wholesale. This textbook focuses primarily on consumer credit, where the masses of data make it ideal for credit scoring. Over time however, credit scoring is being used more and more for the rating of businesses. There are limitations though, and the reader needs to have some understanding of where it can be used, and where not. It may be appropriate for SMEs, but not for large publicly traded companies.

The traditional framework used for rating both business and personal lending is the 5 Cs (capacity, capital, conditions, character, and collateral), which relies upon personal contact with the client. Today's lenders rely upon a variety of data sources, including payment histories, financial statements, share and bond prices, environmental assessments, and human input. Which is/are most appropriate depends upon the size of the firm, with payment histories providing most value for smaller companies, and market prices for larger companies where those prices exist.

The credit risk of businesses, and especially larger enterprises, is typically stated as a risk grade; whether it is provided by a *rating agency,* or is an *internal grade* produced by the lender. The number of grades in the scale may vary from 5 to 25, albeit the standard under Basel II is to have a minimum of 7 grades, with 2 default grades. They may be stated either as letters or numbers, with the most well known being the 'BBB+' style grades used by the rating agencies. Such grades are expected to have certain qualities. First, all cases with a given rating grade are expected to be *homogenous* for risk, and what happens subsequently should be *predictable.* Second, at the case level, the rating grades should be *stable*, but still be *responsive* to relevant new information, as and when it is received. Stability is a function of a variety of factors, including whether a through-the-cycle or point-in-time approach was used; the more fundamental the analysis, the greater the stability. There will, however, always be cases where nobody sees the default coming.

Credit scoring's influence has been greatest in SME lending, which was slow to adopt the new technology. Its defining feature was that less information is available than for the wholesale market because: (i) it resides with the loan officer; and (ii) much of it is difficult to quantify, verify, and transmit. As a result, lenders benefited from higher information rents, as customers found it difficult to establish a banking relationship elsewhere. The sector was dominated by smaller banks that specialised in *relationship lending*, while the larger banks instead focussed on the wholesale market. As the latter's margins were eroded by their loss of cheap funding sources, they started to eye the lucrative returns being made by small players. Technologies used in the consumer space were adapted to develop their *transactional lending* capabilities in the SME market. This not only had the advantage of lower costs, but it also extended their geographical reach. Personalised service may have been lost, but SMEs benefited from greater access to credit and lower interest rates, as well as the flexibility to move their banking relationships without punitive costs.

Credit scoring of middle-market companies is very new, and mostly restricted to financial ratio scoring (FRS). Attempts had been made over the years, but although many of them provided results that were statistically significant, they were not good enough to implement in practice. Today, the best models are developed by the rating agencies, if only because they have more data and greater experience. The first practical model was *Moody's KMV's RiskCalc*, for middle-market and larger companies. Another is a model used by *Fitch*, which tries to predict the rating agency grades instead of defaults, and is used only for the largest corporates, on a par with companies with traded bond issues. Both of these models are quite powerful, but their reliance upon financial statements presents problems in terms of their narrow focus, backward view, data quality, irregularity of updates, issues relating to industry treatment, and problems with interpreting and spreading the statements. As a result, lenders may also develop their own *internal rating grades*, with or without external model inputs, which integrate financial data with other information pertaining to each counterparty.

Agencies such as Moody's, S&P, and Fitch play a major role in providing credit ratings for bond issuers and other larger lenders. Their ratings are typically provided as letter grades, that may be 'investment' or 'speculative' grade, with separate 'default' and 'ratings withdrawn' categories. The grades are powerful, but as with any data, they are subject to decay, and over time the default rates will exhibit *mean reversion*. There are also other issues, including problems with small numbers, ratings delay and momentum, population and a downward ratings drift, business cycle sensitivity, and risk heterogeneity within the grades.

And finally, there are various ways of deriving default probabilities. The standard approach is historical analysis, which rating agencies publish as transition matrices and/or survival rates. The ideal is to use forward-looking data in the assessments, which implies human input, whether into the risk assessments directly, or via market prices. The latter can be achieved by using: (i) structural models like the Merton's model—which is based on options theory, and requires some information on assets and liabilities; and (ii) reduced-form models, which rely upon an analysis of credit spreads.

Module C

Stats and maths

7 Predictive statistics 101

I don't want any of your statistics; I took your whole batch and lit my pipe with it.
Mark Twain, author of Huckleberry Finn, in 1893

Unfortunately for Mark Twain, statistics have become a driving force behind modern economics and business, even if many people share his views. Nonetheless, some understanding is required, even if it might bore some people to tears. What follows is a fairly comprehensive overview of topics to which entire textbooks are dedicated. The focus is on providing the non-statistician some understanding of the terms that are used. As a result, it is of a very high level, and looks at:

(i) **An overview of techniques**—A brief look at all of the modelling techniques, focussing upon which are the most popular, and why.
(ii) **Parametric techniques**—Linear probability modelling (LPM), discriminant analysis (DA), and logistic regression.
(iii) **Non-parametric techniques**—RPAs, neural networks, and genetic algorithms.
(iv) **Critical assumptions**—Different assumptions that accompany the use of many of these modelling techniques.
(v) **Results comparison**—A review of which techniques provide the best results, if any.

Some statistical notation

When writing this textbook, an attempt was made to omit statistical notation in its entirety, but as time went by, bits and pieces crept in regardless. The end result is that a brief overview of this specialist shorthand is required. If nothing else, it should assist the reader when reviewing more academic works. The following relate to datasets and probabilities:

$A \cup B$	= and/union—case is member of either A or B.
$A \cap B$	= or/intersect—case is member of both A and B.
$A \subset B$	= subset—A is a member of B, but B is not necessarily a member of A.
$A \mid B$	= given—refers to both $A \subset B$ and B, that is, A given B.
$A \in B$	= element—A is a single element of B, as opposed to a subset.
$p(A)$	= probability with value 0 to 1 that case is in subset A, also stated as 'p_A'.
$p(A \mid B)$	= conditional probability of A given B.
$\therefore$	= therefore.

Table 7.1. Notation examples

G = 11	P(G) = 11/27
R = 9	P(G∩A) = 9/27
G∪B = 18	P(G∣G∪B) = 11/(11+7)
A∩B = 3	P(A∩B) = 3/27
G∪B∣A = 9+3	P(G∪B∣A) = 12/18

Source table

	A	R	=
G	9	2	11
I	6	3	9
B	3	4	7
=	18	9	27

Table 7.2. Bayes theorem proof

p(A∣G) = p(A∪G)/p(G),
p(G∣A) = p(G∪A)/p(A), and
p(A∣G) = p(G∪A).
∴ p(A∣G) * p(G) = p(G∣A) * p(A)
p(A∣G) = p(G∣A) * p(A)/p(G)

Thomas et al. (2002)

Examples of how this notation is used are provided in Table 7.1, and in the proof of Bayes theorem in Table 7.2. The basis of the proof is that p(A∣G) can be stated as p(A∪G)/p(G), which for the information in Table 7.1 could be stated as 9/11= (9/27)/(11/27). Its purpose is solely to confirm that this conditional probability is the equivalent of p(G∣A) * p(A)/p(G), which would be 9/11 = (9/18) * (18/27)/(11/27).

Some expressions relating directly to credit scoring are:

p(Good)—Usually used in the sense P(G∣G∪B), the probability of an account being good where the set has been limited to goods and bads.

x—An attribute, or set of attributes, that records in a dataset can assume. It is often used to represent all accounts that fall within a given score range, but can refer to any single attribute, or multiple attributes.

P(x)—The probability that an account has attribute(s) x. This is the column per cent for a totals column, covering at least G∪B.

P(Good∣x)—The probability of a case being good, given that it has attribute(s) x. This is the row per cent in a characteristic analysis, G/(G+B).

P(x∣Good)—The probability of a case having attribute(s) x, if it is good. This could be seen as the column per cent for goods in a characteristic analysis, G/ΣG.

Other notation relates more to mathematics and statistics:

Σ = repeated addition, or sum.
Π = repeated multiplication, or product.
α = alpha, significance level used when testing hypotheses.
ϕ = standard normal cumulative distribution, for mean 0 and standard deviation 1.

z	= z-statistic, or number of standard deviations from the mean.
X^2	= chi-square, a goodness of fit measure for frequency distributions.
μ	= mean or average.
σ	= standard deviation, and σ^2 is the variance.
r	= sample correlation coefficient.
x_i	= the value for the variable x for the ith record.
β_i	= beta, the multiplier to be applied to a given variable, x_i, in linear regression.
b_i	= regression coefficient, may be used for linear or other regression equations.
$\hat{s}$	= the caret indicates that this is an estimate of the variable 's'.
e	= an error term, for example $s_i - \hat{s}_i$, the difference between actual and estimate.
λ	= hazard rate, or percentage of cases that do not survive.
$\exp(y)$	= the exponent, e^y.
$\ln(x)$	= the natural log.

As stated, this is a limited set, as the total set of mathematical notation is enormous, and often unintelligible to the layman.

7.1 An overview of predictive modelling techniques

The tools [of credit scoring] are based on statistical and operational research techniques and are some of the most successful and profitable applications of statistical theory in the last 20 years.

Crook, Edelman, and Thomas (1992)

This is where we start getting into some of the meat of modelling techniques. Each has its own strengths and weaknesses, which often vary according to the circumstances. Table 7.3 provides a brief summary of the six main techniques:

Table 7.3. Predictive statistics overview

Method	Main technique	P/NP	Summary
Linear regression	Ordinary least squares	P	Determine formula to estimate continuous response variable.
Discriminant analysis	Mahalanobis distance	P	Classify cases into pre-specified groups, by minimising in-group differences.
Logistic regression	Maximum likelihood estimation (MLE)	P	Determine formula to estimate binary response variable.
Decision trees	RPAs	NP	Uses tree structure to maximise between group differences. Complex for large trees.
NNs	Multilayer perceptron	NP	AI technique, whose results are difficult to interpret and explain.
Linear programming	Simplex method	NP	Operations research technique, usually used for resource allocation optimisation.

P/NP: Parametric (P), Non-Parametric (NP).

Each technique has its own advantages and disadvantages, and when making the choice, certain aspects relating to both the data and the modelling technique must be considered:

Modelling considerations

Suitability—Is the method appropriate for the task at hand? Problems may arise because of violations of one or more assumptions mentioned in Section 7.4. These are not always sufficient to invalidate the model, but may demand that extra care be taken.

Development speed—How easy is the method to learn and apply, and how quickly can the models be developed? Lenders want to avoid long development times, especially in fast changing environments.

Adaptability—How easily can the method be adapted to accommodate problems that are specific to a particular development? Possible problems include small numbers, interactions, ability to stage characteristics, ease of controlling for certain factors, etc.

Output transparency—Is model output easy to understand and explain, and how easy is it to control the amount of complexity? This is vital in situations where the business requires an understanding of the model, or the reason for a decline has to be provided to the customer.

Data considerations

Interactions—Can they be detected and modelled? Some methods can detect the interactions, while with others the users can model known interactions if the data is manipulated.

Skewed target—Are there very few bads? Where the numbers are low, decision trees or NNs might be used instead of regression.

Continuous/discrete—Logistic regression is typically regarded as best suited for modelling a binary outcome. Even so, LPM is still widely used.

Rare events—Where a continuous variable is dominated by zero values, an 'expected value' can be determined by splitting the problem into two parts: (i) the probability of a non-zero value; and (ii) a prediction of that value, if not zero. For example, probability of default (PD), and any of exposure-at-default (EAD), loss-given-default, or time-to-default. Different statistical techniques may be used for each.

This list provides a few of the questions that may be asked, and is not meant to be comprehensive. Logistic regression, LPM, and DA are the techniques most commonly used in credit scoring, even though almost all predictive modelling techniques provide similar rankings (see Section 7.5). Even so, some effort has to go into choosing which technique is most appropriate for a task.

The analyst's main interest should be in providing assistance in decision-making and not in finding methods of solution that are more elegant or marginally faster than existing methods.

Prof. Hossein Arsham

Perhaps the three most important factors to consider are: (i) *ease of calculation*, (ii) *transparency*; and (iii) whether or not it is *statistically suited* for the problem. Ease of calculation dominated during the early days of credit scoring in the 1950s and 1960s, and as a result LPM and linear DA were the primary tools used, even though they are not well suited. Necessity is the mother of invention though, and scorecard developers derived tricks that addressed many of the critical assumption violations.

As time progressed, and computing power increased, MLE became more feasible—first with *logit* (logistic) and then *probit* (Gaussian). Both are less demanding in terms of the statistical assumptions made, but are computationally intensive, and were infeasible at a time when computers were big and slow. Today, the differences in computation time are hardly noticeable, and logistic regression is used by between 80 and 90 per cent of scorecard developers. Most of the rest still use linear techniques, because of their flexibility and relative ease of use.

Most predictive statistical techniques provide similar rankings, and the choice will be driven by some other factor. Linear techniques are often used by organisations where credit scoring has a long history, the existing methodology is well entrenched, and/or a business opts for the tried and tested. In contrast, logistic regression dominates where credit scoring was introduced later, where purists have insisted upon avoiding critical assumption violations, and/or where lenders wanted the score to provide them with a probability estimate (increasingly critical, especially for banks with Basel II).

Ideally, scorecard developers should be familiar with both approaches, and others, and be able to recognise where they can provide the most benefit. According to Falkenstein et al. (2000), the focus of linear regression/DA is to split the population into two groups, and is well suited to identifying a cut-off used in a selection process. In contrast, logit/probit models are more focused upon probabilities, and provide better inputs for risk-based pricing.

Non-parametric techniques have all been proposed for credit scoring, but have not been widely adopted. These include decision trees and AI techniques, such as NNs, genetic algorithms, and K-nearest neighbours. NNs are, however, widely used to identify fraud, as the models can adapt more quickly to changing circumstances, where data is sparse.

This provides a brief summary. All of the techniques are covered in more detail in the following pages, which are concluded with a summary of various studies compiled by Thomas et al. (2002), showing that these predictive modelling techniques provide similar results.

7.2 Parametric techniques

All models are wrong, but some are useful.

George E.P. Box

Our starting point is the parametric techniques mentioned above: LPM, DA and logistic regression. The common factor is that there are certain critical assumptions made when they are used, mostly relating to relationships within the data. While the assumptions are referred to when discussing each of the techniques, refer to Section 7.4.2 for more detailed descriptions.

7.2.1 Linear regression/probability modelling

Some of the simplest possible relationships are linear; as one value increases, another changes at a constant and known rate. These are so simple that people are continually looking for linear relationships, along with other patterns, that can help them to understand changes in the world around them. Such models often come unstuck however, as any investor will admit who has bet on the continuation of past trends and lost.

Even so, many relationships are linear, or close enough for linear regression to be used. The example in Figure 7.1 shows the results of a *simple linear regression* using *ordinary least squares* to find the relationship between the month-end prices of two shares, which have both been increasing over a 36-month period. It is calculated by solving for values of β_0 and β_1 in Equation 7.1, which minimise the sum of the squared error terms, Σe^2, where $e = \hat{y} - y$.

Equation 7.1. Simple linear regression $y_i = \beta_0 + \beta_1 x_{i1} + e_i$

The same formula, without the error term, is then used to provide estimates in future. In the Figure 7.1 example, Share Y has a base value of −789.95, and increments by 15.975 for every one-point increase in Share X. Obviously, X always has a value large enough to make Y positive. Whether or not this information can be used for anything is another question—there are a lot of other factors at play, and it is only based upon past information.

The same relationship could also be described using multiple linear regression (MLR), the only difference being that Share Y's prices would be explained using a greater number of variables—other share prices, economic statistics, financial details, and so on. The final model will then have multipliers allocated to each of the independent variables that have been identified as being relevant.

Linear regression—A brief history

While it is only with the advent of computers that it has become quick and easy to perform, linear regression has a long history. Sir Francis Galton (1822–1911) was an English polymath, amateur scientist, and cousin of Charles Darwin, who championed the now much discredited concept that the human stock could be improved through selective breeding.

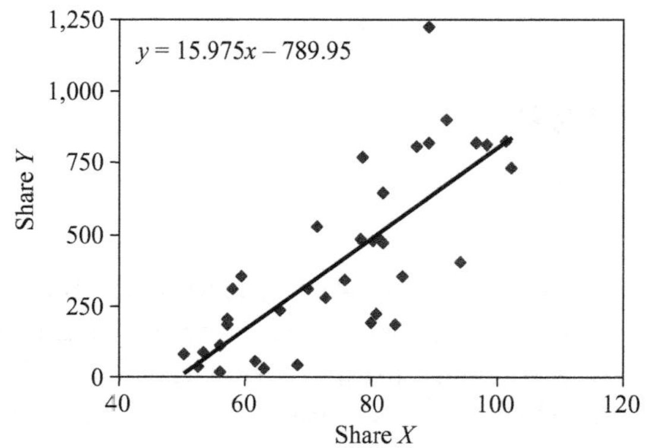

Figure 7.1. Linear regression.

He popularised the concept amongst Victorian intelligentsia and scientists, and coined the term 'eugenics' in 1883. Galton introduced the concepts of regression and correlation, to help put the study of heredity onto a scientific footing. This also marked a major shift in statistics, from the study of averages to the study of differences. He introduced the concept of regression in 1889, building upon Legendre's 1805 concept of least squares, to provide a means of deriving explanatory equations by minimising the summed squares of the error terms. The term 'regression' came from its use to illustrate the tendency for offsprings' quantifiable traits (such as height) to regress towards the population mean, rather than the values for their parents, and his study of heredity was based on an analysis of which deviations of the parents would be passed on to the offspring. The term multiple (variable) regression was first used in 1908 by Karl Pearson, but it was Sir R.A. Fisher (1992/5) who laid the ground work.

The problem with linear regression is that it makes the most assumptions: (i) *linearity*, (ii) *homoscedasticity*, (iii) *normally distributed error term*, which implies a continuous and normally distributed target variable, (iv) *independent error terms*, (v) *additivity*, (vi) *uncorrelated predictors*, and (vii) *use of relevant variables*. In credit scoring, or any instance where there is a binary outcome, linear regression is referred to as linear probability modelling (LPM). The end result is an estimate of $p(\text{Good})$, the formula for which is provided in Equation 7.2.

Equation 7.2. Linear probability modelling $$P(\text{Good})_i = \beta_0 + \sum_{j=1}^{p} \beta_j x_{ij} + e_i$$

The probability for each record i, is the sum of a constant and the products of a series of weights β_j and variable values x_{ij}, where the variables take on different values for each record, and the weights differ for each variable j (the error term e_i is ignored). The problem arises because many of the assumptions mentioned above do not hold true. The most problematic are 'normally distributed error terms' and 'homoscedasticity', because the result only has two possible values, 0 and 1. This is exaggerated further because the predicted values often fall outside the 0 to 1 range. Because of these violations, statistical purists criticise its use for credit scoring.

What seems to be overlooked, is that the scorecard development methodologies used with LPM address the critical assumption violations, effectively turning it into a non-parametric technique.[1] This is done by:

(i) Binning predictors into dummy variables, to address linearity and normality assumptions for predictors.
(ii) Using the scores—at least directly—for ranking and ranking only, which addresses the issue of normally distributed residuals, and—for the most part—heteroscedasticity.
(iii) Working with large sample sizes and sufficient bads, which reduces the standard error caused by any multicollinearity, and where applicable, autocorrelation.
(iv) Using separate scorecards to address interactions, which helps to address additivity.

[1] Fox (2005) presents 'binning' as one of several ways of treating variables to do 'generalised additive non-parametric regression'.

(v) Limiting the number of parameter coefficients and ensuring they make sense, which is a prerequisite to reduce the standard error, where correlated predictors are used.

It helps to have an understanding of these issues, in order to bridge the gap between statisticians and practitioners, but please note that the above is by no means a comprehensive treatment of the topic. Further reading is advised, but ultimately the proof is in the pudding. People using LPM have confidence in their methodology, and find it a fast and flexible tool that provides models with ranking abilities comparable to other techniques. If predictive accuracy is also a requirement, as is required for banks under Basel II, then the resulting scores have to be transformed into default probabilities using some form of calibration technique (see Chapter 20, on Calibration).

Error measures

Before moving on, a brief look should be taken at the error measures used to indicate how well the regression has worked. The two most commonly referred to are the standard error and R^2. The formula for the standard error is shown in Equation 7.3, where: s_e = standard error; $e = \hat{Y} - Y$ = error term; n = number of cases being evaluated; and k = number of explanatory variables being used:

Equation 7.3. Standard error $$s_e = \sqrt{\frac{\sum(\hat{Y}_i - Y_i)^2}{n-1-k}}$$

This does, of course, assume that the error terms are normally distributed, in which case 68.3, 95.5, and 99.7 per cent of the errors will lie within one, two, and three times the standard error, respectively. Before describing the full formula, the denominator—also referred to as the degrees of freedom (d.f.)—should be covered. This is the number of independent variables represented in a statistic, usually calculated as $(n-1)$ (such as for the standard deviation). In the current instance however, the explanatory variables are an added complication; each is treated as an assumption that reduces the d.f. The formula becomes $(n-1-a)$, where a is the number of assumptions.

Besides the obvious result, that the standard error can only be calculated where the number of observations exceeds two plus the number of explanatory variables, there are a couple of factors that must be noted, which are key concepts in statistics. First, the standard error *decreases for larger sample sizes*, but after a certain point the reduction becomes negligible. Second, the standard error *increases as more variables are included* into a model. Basically, the explanatory value added by any new variable must be sufficient to offset the decrease in the degrees of freedom. Where sample sizes are large, the effect of this will be negligible, but it highlights the need for models to be kept simple.

Another statistic commonly used to assess regression results, linear and otherwise, is the coefficient of determination, also called the R^2 statistic, shown in Equation 7.4. Rather than indicating the absolute size of the error though, it indicates how much of the error is explained by the model, relative to simply using the mean as an estimate (naïve model). As can be seen from the equation, as the estimates approach the actual values, the R^2 value approaches a limit of 100 per cent.

Equation 7.4. Coefficient of determination $R^2 = 1 - \frac{\sum(\hat{Y}_i - Y_i)^2}{\sum(\bar{Y} - Y_i)^2}$

7.2.2 Discriminant analysis

A term that causes much confusion for the layman is discriminant analysis (DA), a statistical technique that is used to determine group membership, where there are two or more known groups (cluster analysis tries to identify unknown groups).

> Fisher (1936) used what is sometimes called 'Fisher's linear DA'—the use of MLR to discriminate between groups—to differentiate between different types of irises. The three species were Setosa, Versicolor, and Verginica, and it was done using only the sepal length, sepal width, and petal length. Irises are a type of daylily flower with six sections; petals are the top three sections, and sepals the bottom three.

The DA works by using some classification tool to minimise the distance between cases within a group, and maximise the differences between cases in different groups. It has the following steps:

> (i) Define the groups.
> (ii) Define the model form, usually using some form of regression model.
> (iii) Derive the model, using a chosen statistical technique.
> (iv) Test, using a validation sample.
> (v) Apply, either to assist in explaining or predicting group membership.

The number of models derived will be the lesser of: (i) the number of groups less one, which in the simple two-group scenario only requires one model; and (ii) the number of predictors used in the analysis, which is not a problem in data-rich business environments. The use of DA in credit scoring usually assumes the simple two-group case, ignoring the indeterminates or other groups. In any case, group membership is determined by assessing the score(s) from the discriminant model(s).

An example is where a company wants to determine which product a customer is most likely to take out next, assuming there is a database of what customers have taken out in the past. If there are three products—say cheque, card, and personal loan—then models are developed for say cheque and card, using whatever information is available. Cut-offs are determined for each, and if a case does not fall into either of them, then it is assigned to the personal loan group.

This is where a funny name comes into play (as though there have not been enough of them already, and there are more to come). According to Garson (2005), the Mahalanobis distance is the number of standard deviations between the value for a case, and the centroid (analogous to an average score) for the group. Separate values are calculated for each group, and each case is assigned to the group where the distance is smallest. It takes into consideration correlations within the data, and if the predictors are uncorrelated, it is equivalent to Euclidean distance. These values can be converted to chi-square p-values for analysis. Care must be taken here, as the calculations assume that the variance/covariance matrix is the same for each group.

The DA will suffer from any and all of the assumptions associated with the statistical technique used. The most common form is linear DA, which uses linear probability models. These suffer from high misclassification errors when predicting rare groups, so equal samples for each group are usually used. While linear DA was the original, logistic regression is now preferred because: (i) there are fewer assumption violations, especially as it does not demand normally distributed independent variables; (ii) it works better where group sizes are very unequal; and (iii) many people find the resulting models easier to interpret.

7.2.3 Logistic regression

While linear regression was used to provide the bulk of early scoring models, it was known that there were shortcomings, largely because the target variable is binary. Logistic regression is more appropriate for binary outcomes, and hence most credit scoring. It uses a process called maximum likelihood estimation (MLE), which: (i) transforms the dependent variable into a log function; (ii) makes a guess at what the coefficients should be; and (iii) determines changes to the coefficients, to maximise the log likelihood. It is an iterative and calculation-intensive approach, which may require about six attempts to reach convergence, using one of several convergence criteria (Garson 2005a). The end result is a regression formula of the form:

Equation 7.5. Logit regression $$\ln\left(\frac{p(\text{Good})}{1-p(\text{Good})}\right)=b_0+b_1x_1+b_2x_2+\cdots+b_kx_k+e$$

that is applied to each observation. The value on the left-hand side of the equation is the natural logarithm of the odds (for example, $Z = \ln(80\%/(1-80\%) = \ln(4/1) = 1.386$). This can easily be converted back into a probability, using the formula $p(\text{Good}) = 1-1/(1 + \exp(Z))$. According to Mays (2004), it is common practice for the logistic regression to predict the bad/good instead of the good/bad odds, which results in the same Z-scores, but with the opposite (usually negative) sign. The other transformation formulae are modified to suit.

There are other non-linear means of deriving regression equations, which are often associated, and even confused, with logistic regression: probit—'probability unit', also uses MLE; and tobit—borrowed from economics, and used where all of the characteristics have positive values.

> Probit assumes a normal (Gaussian) distribution, as opposed the logistic distribution assumed by logit. Thomas (2000) provides a regression formula that solves for $\Phi^{-1}(p_i)$, where $\Phi(x)=\left(\int_{-\infty}^{x} e^{-y^2/2}dy\right)/\sqrt{2\pi}$, which provides a point on the inverse normal cumulative distribution curve.

Logistic regression requires the following assumptions: (i) categorical target variable; (ii) linear relationship, but this time with the log odds function; (iii) independent error terms;

(iv) uncorrelated predictors; and (v) use of relevant variables. This is a shorter list than for LPM. While used primarily for binary target variables, it is also possible to use 'ordered logistic regression' for ordinal outcomes, such as subjective risk grades and survey responses.

Logistic regression's primary disadvantage was its computational intensiveness, especially problematic where models have to be rerun countless times for cosmetic changes. Improvements in computers have made this less of an issue though, and today logistic regression has been accepted as the option of choice for developing credit-scoring models, in particular because: (i) it is specifically designed to handle a binary outcome; (ii) the final probability cannot fall outside of the range 0 to 1; and (iii) it provides a fairly robust estimate of the actual probability, given available information.

The origins of logistic regression

Unlikely as it may sound, logistic regression's roots lie in the study of population growth. In 1798, Malthus claimed that, if left to themselves, human populations would increase in geometric progression, which at the time seemed valid, given the then massive population growth in many European countries. According to Cramer (2002), 40 years later, Alphonse Quetelet, a Belgian astronomer turned statistician, realised that such growth could not continue indefinitely, and asked his pupil Pierre-François Verhulst (1804–1849) to work on the problem.

Verhulst defined an S-shaped curve that puts an upper boundary on the pattern (Figure 7.2), and his findings were published in three papers between 1838 and 1847. In the first, he set out the arguments, and used the curve to describe population growth in Belgium, France, Essex, and Russia, prior to 1833. It was only in 1845, that he provided the formula used to define what he called a logistic curve, $P(Z) = \exp(Z)/(1+\exp(Z))$, where Z is the natural log of the odds for a binary outcome.[2]

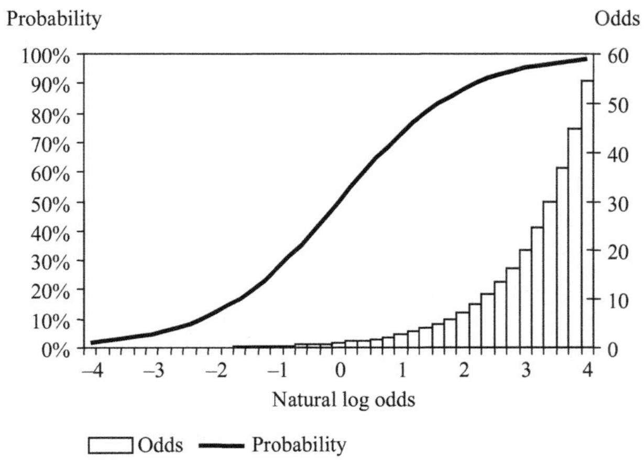

Figure 7.2. Logistic function.

[2] Verhulst gave no reason for the choice of the name logistic. The French term *logistique* is derived from the Late Latin *logisticus*, meaning 'of calculation'.

Later in the nineteenth century, the same function was used to describe autocatalytic chemical reactions, but for the most part it was forgotten and only rediscovered in the 1920s when it was applied to population growth in the United States. It also took some time before some means was derived to develop regression models to explain Z. Although there were several independent discoveries from the 1860s, most of the approaches prior to 1930 relied on ad hoc adjustments of numbers and graphs to improve estimation. Charles Bliss and John Gaddum are recognised for standardising the estimation process in the early 1930s, and it was Bliss who first used the term probit (probability unit) and set out the basis for MLE.

Then, in the 1950s, Joseph Berckson suggested the logit (logistic unit) approach, which uses minimum chi-square estimation. In the era prior to computers, it gained favour quickly due to its ease of use, even if it was heavily criticised by academics. It was only in the late 1970s that the improved processing power of computers made the probit approach feasible for many problems (Bugera et al. 2002). In 1980, the first academic work on its applicability to credit scoring was published (Wiginton 1980), and it has since become one of the core statistical techniques used.

7.3 Non-parametric techniques

This section covers the other side of the parametric/non-parametric dichotomy. While parametric techniques require many assumptions about the underlying data, non-parametric techniques require few, if any. Under this heading fall RPAs, NNs, genetic algorithms, and K-nearest neighbours. Other techniques falling into this category, such as support vector machines, are not covered.

7.3.1 Decision trees/RPAs

There are two types of people in this world: those that can stay focused, and those that . . . hey look, a squirrel.

RossN.com

You are probably familiar with the concept of a 'decision tree'. This is a graphical tool, with a branch- or root-like structure of boxes and lines, used to show possible turns of events that may—or may not—be controllable, even if the name implies that each branch is supposed to represent options available to a decision-maker. Decision trees are also used for data visualisation in classification and prediction problems. The most primitive form is a type of expert system, where a rule-set is defined by people with hands-on experience, which is still done for medical and other diagnoses, where there is insufficient data to do any empirical analyses.

More advanced forms can be derived based upon data analysis. In the example provided in Figure 7.3, the splits are determined from the top down. The top of the tree is referred to as the *root node*, each subsequent level as a *child node*, and at the bottom are the terminal nodes. There will be two or more splits each time, and there may be a different number of levels

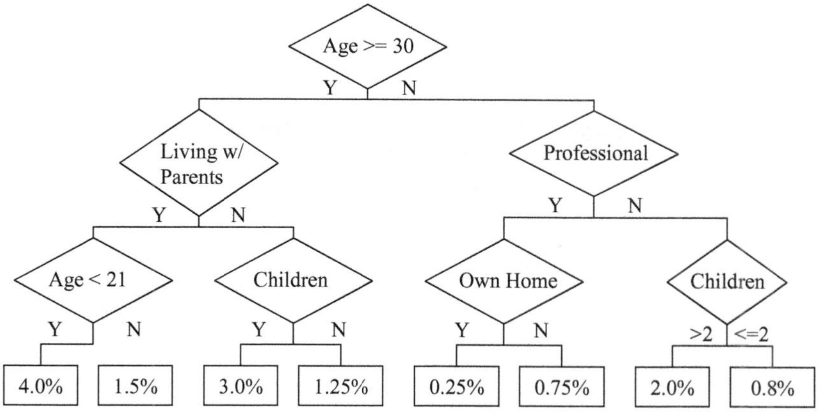

Figure 7.3. Decision tree.

depending upon the branch. When finished, the *terminal node* values could be used either as estimates (scores), or as a grouping tool. For a binary outcome, the value is a probability, say *P*(Bad), and all branches with probabilities beyond a predefined cut-off, say above the average *P*(Bad), would be put into the Bad group.

Early attempts at deriving decision trees used trial and error. According to Thomas et al. (2002), Breiman and Friedman each independently came up with the idea of using analytical tools to determine the rule-set in 1973. It was only in 1984, however, that their further collaboration with Olshen and Stone produced CART (Classification and Regression Trees), a sophisticated, mathematical, and theoretically-sound procedure for deriving decision trees.[3] Their concept was first applied to credit scoring by Makowski and Coffman, in 1985 and 1986, respectively.

The primary technique used is called a RPA, which describes the manner in which the branches are found—through repeated attempts at finding the best possible split. Several rules define the RPA procedure:

(i) **binning,** determining how predictors are to be binned;
(ii) **splitting,** selecting which characteristic is to be used;
(iii) **stopping,** when to stop creating new sub-nodes;
(iv) **pruning,** how to drop nodes to avoid overfitting;
(v) **assignment,** how to classify each node as good or bad.

The splitting rule, required to split the population into different homogenous and mutually exclusive groups, is the most complex. The goal is to minimise the distance between members

[3] The authors were renowned statisticians at the University of California, Berkeley (Breiman and Stone) and Princeton (Friedman and Olshen).

in a group (similar default rates), and maximise the distance between groups (different default rates). Some measure is used to test each possible split, usually the Kolmogorov–Smirnov statistic, but a basic impurity index, Gini index, entropy index, or half-sum of squares, may also be used. Different approaches are available, such as CART, CHAID (Chi-square Automatic Interaction Detection), and QUEST (Quick Unbiased and Efficient Statistical Tree).

The RPAs have a number of advantages and disadvantages relative to other techniques:

(a) The approach is *non-parametric*, and well suited to categorical analysis. Its main strength is its ability to identify patterns, including finding and exploiting interactions. In general however, regression models provide better results, where interactions are not an issue.
(b) The results are very *transparent* and easy to implement, especially for simple trees. These advantages may, however, be lost when extra complexity makes the tree a bush.
(c) As trees become bushier, there are fewer cases in each node, bringing with it the *potential for overfitting*, and unreliable results. Very large datasets are required to provide both detail and reliability.
(d) It is *computationally simple*, using only one measure to choose variables and determine splits, but it is relatively inflexible. In contrast, with NNs the process is opaque and all variables are used, but with training the model can adapt to changing circumstances (see Tanigawa and Zhao 1999).
(e) It allows for *quick and easy identification* of extremely high- and low-risk categories, where policy rules may be in order.

In general, RPAs are not well suited for predictive modelling, but there are instances where they may be considered. An example is where the amount of data available for a scorecard development is limited, such as for a new product. This could possibly be addressed by defining an initial tree structure using available data, and thereafter use bootstrapping (sampling with replacement) to calculate different terminal node probabilities that are then averaged.

In spite of the shortcomings, RPAs are still powerful tools for use within the business. They are best used for quick and dirty data exploration, whether to gain insight into data, describe the data to the business, identify key predictive variables, identify scorecard splits, or act as a benchmark for other models.

7.3.2 Neural networks

Humankind continually strives to improve its situation, and over the past few centuries has endeavoured to replace the efforts of man with machines. In more recent years, this effort has extended beyond manual labour into the domain of thinking and decision-making. While the goal is to make life easier, many people fear that computers will, eventually, gain the ability to think and function in a similar fashion to humans. Indeed, it has been a recurring science-fiction theme for a computer to become self-aware, and decide that its own self-interest takes precedence over, and is at odds with, those of its human creators (as was the case with

HAL, in 2001: A Space Odyssey). In the early days of the third millennium, this still remains the realm of science fiction, but progress on artificial intelligence (AI) is being made. Indeed, the use of any predictive model as part of a decision process could be construed as AI, but there is one technique that has AI at its very core.

These are neural networks (NNs), which can be described as networks of computing elements that can respond to inputs, and learn to adapt to the environment. They are purportedly able to mimic the manner in which the human brain works—especially when it comes to self-organisation and learning. Unlike other statistical techniques, which follow formulaic procedures, NNs are instead trained through the presentation of repeated examples (Chorafas 1990). The end result is something like a decision tree, except the detail is much finer, with decision rules that are much more complex.

According to NeuralT (2002) there are several different paradigms that can be used to develop NNs. The best suited for credit scoring is the *multilayer perceptron* (MLP, or back propagation), which has the advantage of handling both non-linearity and interactions easily. It is also referred to as a 'universal classifier', because it is theoretically able to model any decision process. Other paradigms include *Radial Basis Function* (RBF), *Self-Organising Maps* (SOM), and *Kohonen Networks*.

The NNs have the advantages of being able to: (i) process huge amounts of data; (ii) discover (pattern recognition) and track relationships in the data, especially interactions; (iii) deal with non-linear relationships within the data; and (iv) train themselves, based upon differences between observed and actual results. There are several practical problems with NNs though:

(i) They are data-hungry and computation-intensive, requiring a lot of iterations before a final model is obtained (Fractal et al. 2003).
(ii) They are expensive to implement and maintain, especially as regards the ongoing training, to allow them to adapt to changing circumstances.
(iii) They are opaque, as the relationships detected by the models are very difficult for their creators to interpret.
(iv) There is a significant chance of overfitting.

The NNs are ill-suited for any environment where the decision logic must be understood, especially for consumer credit application scoring, where companies must advise decision reasons to customers, or where the business demands some understanding of the underlying processes. They may, however, be well suited where accurate and adaptive predictions are critical, and transparency is secondary. Note, though, that according to Allen et al. (2003:10), DA outperforms all forms of NNs in minimising Type II errors, where good loans are classed as bad.

The NNs are seldom used for credit scoring. According to Thomas (2000), their primary use has been in areas where there is less data, such as for scoring corporations or credit union customers. They are also well accepted for fraud scoring. As quickly as lenders identify fraud, and put in control mechanisms, *modus operandi* are changed and new weaknesses found. NNs have the ability to adapt to these changing circumstances, but require monitoring and retraining along the way.

7.3.3 Genetic algorithms

Another non-parametric approach is evolutionary computing, which is based upon concepts from biology and Darwinian natural-selection processes, and is usually lumped into the AI camp. It was first proposed in the 1960s, but it was only with the advent of parallel computing, which makes more complex modelling possible, that significant interest was generated. There are two primary approaches:

(i) Evolution strategies, developed by Ingo Rechenberg of the Technische Universität Berlin, which use elitist selection, and vectors of real numbers for object and strategy parameters, to represent individual solutions. They were first used to maximize the thrust provided by a two-phase jet nozzle, with an unexpected final design like a candlestick.
(ii) Genetic algorithms, developed by John Holland of the University of Michigan, which use a more random selection process and bit strings to represent genomes/chromosomes of individual solutions.

Both approaches have been used in engineering, computer science, financial services, and biology. Dawson et al. (2000) note that the evolution strategy approach is quicker, but often finds a local maximum, and can have engineering problems. In contrast, genetic algorithms are slower, more likely to find the global maximum, and may have computational problems. Over time, the proponents of the two approaches have noted the similarities, and today they are almost indistinguishable, to the extent that the terms are used interchangeably. Most of the credit-scoring literature refers to genetic algorithms, which are the focus here.

According to Fractal et al. (2003), genetic algorithms are heuristic search algorithms, which try to find an optimal result within a search space through survival-of-the-fittest evolution. The initial problem is to generate genomes that represent characteristics of different possible solutions, which are each then measured for fitness. Parents are chosen according to their fitness levels, and offspring are generated by combining characteristics from each (cross-overs, or recombination), and throwing in random variations (mutations). The process is repeated through generations, until no further improvement can be achieved. In some instances though, the fitness function may also be varied to simulate a changing environment.

Perhaps the greatest advantage of genetic algorithms is that they may be able to find alternate solutions that would not be readily apparent. In general, there is often more than one possible viable solution for a problem. Most statistical techniques will start from a given point, and follow a course towards a solution, that may be a local maximum. In contrast, genetic algorithms will start with a number of different solutions across the surface of possible solutions, and look for the global maximum.

According to Dawson et al. (2000), the primary use of genetic algorithms is where: (i) there are *many possible solutions*, and an exhaustive search is required; (ii) the interest is in *optimisation*, not necessarily the best solution; (iii) good solutions are easy to identify, but *not easy to find*; and/or (iv) there are *multiple targets*, which require simultaneous optimisation. While extremely computationally intensive, they are well-suited to rapidly changing environments, where they can run continuously in the background to find new solutions.

Within credit scoring, genetic algorithms are almost always presented as a possible tool, but they are seldom used. Like all heuristic methods, they suffer from a lack of transparency, and a high potential for overfitting. Other issues are the scarcity of skills, the high computing requirements, and whether or not credit risk problems suffer from local maxima that differ wildly from the global maxima. The other techniques are probably sufficient, but genetic algorithms should not be discounted, as they may prove to be effective tools for simultaneous optimisation of not only risk, but also revenue, retention, and response.

7.3.4 K-nearest neighbours

The final non-parametric technique covered here stems from the realm of machine learning and data mining. The K-nearest neighbours (kNNs) technique is extremely simple, and is used to determine group membership, by finding cases within a set of training data whose predictors are most similar to an 'unseen' case, or new case for which group membership is not known. The symbol 'K' refers to the number of neighbours that will be used for the imputation, and where K = 1, the target value for the most similar case is used.

The technique works by measuring the similarity between examples. If all of the predictors are numeric, this may be done using: (i) *Euclidean distance*, square root of the sum of the squared differences; or (ii) *City-block distance*, sum of the absolute differences. For non-numeric values, the values have to be converted into numbers, perhaps 1 and 0, for 'match' and 'no match' respectively. It is also suggested that the results should be normalised, so that each variable has the same range of possible differences. An advantage of the technique is that new cases can be easily added to the training dataset.

While kNN is a relatively simple technique, and has been shown to be very powerful for many medical and other problems, in credit scoring it is not very practical: (i) no model or score is provided, only a classification; (ii) the decision is not transparent; (iii) processing times may be slow, as comparisons have to be made against every record in the training set; and (iv) the infrastructure required to do on-line searches is ungainly. Irrespective, it should be kept in mind as a possible solution for tackling new problems.

7.3.5 Linear programming

Linear programming (LP) is a technique that comes from the field of operations research, which also includes tools like dynamic programming, integer programming, network-flow programming, non-linear programming, and queuing optimisation. In general, the original goal of the tools was to aid decision-makers in resource allocation problems. Some of the original research in this area occurred during the 1930s, with studies on transportation problems (Kantorovich), game theory (Morgenstern and von Neumann), and input–output models (Leontief).

While Kantorovich and von Neumann were two of the earlier forerunners in LP, it was only in 1947 that George Dantzig came up with the interior points method, or simplex method, while working with the US Air Force to improve their logistics capabilities. The initial goal was to develop a means of doing dynamic scheduling over time, especially under uncertainty, but this was never achieved. Even so, the simplex method was extremely effective at handling problems in a stable environment, and it was quickly adopted elsewhere. During the 1950s, it was used at the Rand Corporation and the US National Bureau of Standards, and during the 1960s was widely used by oil companies. In general, the use of LP has grown along with computers, and has sought to make best use of their power.

As a broad generalisation, LP is a means of solving resource allocation problems that have constraints. For credit scoring, it would work by solving for the β values in a problem that is presented in the form:

Minimise $\sum e_i^2$ subject to:

$$y_i=\beta_0+\beta_1 x_{i1}+\beta_2 x_{i2}+\ldots+\beta_k x_{ik}+e_i$$

$$\beta_1 < \beta_3$$

$$\beta_2 > 0$$

etc.

In other words, it attempts to come up with a regression equation that minimises some error term, which can vary, while ensuring that individual point allocations fall within given constraints. The primary advantage of LP is that the scorecard developer has greater control over the final scores, by being able to include required biases in the 'subject to' statements. For example, to specify that the points for 'Age > 62' must be equal to the maximum for all the other scores applied to age. While it is technically possible to use this technique for credit scoring, it is seldom, if ever, used in practice. It is computationally instensive, and the statistical significance of the point allocations cannot be tested. The actual performance of the resultant models may suffice, but lenders can achieve better results elsewhere. As a result, it is only covered briefly here.

7.4 Critical assumptions

The above discussion focused on the predictive modelling techniques, without providing any understanding of the critical assumptions made when they are used. This section covers these assumptions, and—at least in some cases—the extent to which they are important. It is treated under three headings:

(i) **Data factors**—Goal is to predict, not to explain; more data means better models; and treatment of missing data, whether predictors or performance.
(ii) **Statistical assumptions**—Predictors—normally distributed, uncorrelated, linearly related to the target function, and additive for the final model (regression only); error terms—independent, normally distributed, and homoscedastic.
(iii) **Addressing violations**—Data transformations; treatment of multicollinearity; and variable selection.

7.4.1 Data factors

Data issues are covered in great detail in Module D (Data!), including factors relating to the required quality and quantity, target definition, observation and outcome windows, and sample construction. A couple of points can be made here though. First, credit scoring's goal is to predict performance, not explain it. Science looks to find explanations of man's environment, usually to facilitate greater control. Statistics have become a key tool in this realm, but statistical analysis usually only provides insight into distributions, correlations, and interactions—not causes. Extra effort is required to prove causation, starting with ensuring that the true causal variables are amongst the predictors. In credit scoring, causes often fall into the job loss, illness, domestic upset, and financial irresponsibility camps. Missed payments are only a symptom, but are still highly predictive, even if they do lag, the true causal event. That said, as data improves and lenders are able to get closer to data relating to the root cause, the better their predictions will be.

Second, more data means better models. A key concept in statistics is the *standard error*, the calculation of which varies depending upon which statistic is being considered. For a sample mean, it is the standard deviation divided by the square-root of the sample size. Thus, as the sample size increases, the standard error and associated confidence interval reduces, but at an ever-decreasing rate. The concept also applies to the results of credit-scoring developments, whether the regression coefficients or the resulting predicted values. If an assumption is violated that is known to increase the standard error, increasing the sample size can offset its effect—but not entirely. Extra effort must still be taken to ensure that the final model makes practical sense.

And third, there will be instances of missing data, which can refer to predictors, or outcome performance. As regards predictors, whether one or many, there are several ways to address the problem, the simplest of which are:[4]

Listwise deletion—The entire record is deleted; used in instances where the number of records with missing data is small enough that they will not be missed. In credit scoring however, the number of variables used in the calculations can be huge, and there are often so many missing variables that a significant part of the dataset would be lost.

Mean imputation—Populate all missing values with the mean of those records that have values. This can be used in credit scoring, but care must be taken when using the arithmetic mean of a value that is not normally distributed. Where weights of evidence or probabilities are used, this would refer to those values relative to the entire population.

Dummy variable—When binary dummies are used, missing cases may be excluded, or a separate dummy used to represent them. In most instances though, missing data dummies will not enter a model, as they are usually associated with average performance.

[4] Other ways of dealing with missing data are pairwise deletion, and full-information maximum likelihood.

Missing outcome performance

As regards outcome performance, the missing data is usually the result of a selection process—such as new business origination—where reject inference is required to address the selection bias, and develop a model for the full population. The most commonly quoted works on this topic, are Rubin (1976), and Little and Rubin (1987), who describe three types of missing data scenarios:

Missing completely at random (MCAR) $P(A) = P(A|X, y_{obs}) = P(A|X, y_{mis})$
Acceptance is independent of both the data and outcome performance.
Missing at random (MAR) $P(A|X) = P(A|X, y_{obs}) = P(A|X, y_{mis})$
Acceptance is dependent upon the data but independent of outcome performance.
Missing not at Random (MNAR) $P(A|X) \neq P(A|X, y_{obs}) \neq P(A|X, y_{mis})$
Acceptance is dependent upon both the data and outcome performance.

Where: A is a binary accept/reject flag; y is the outcome performance; obs and mis refer to observed and missing respectively; and X refers to a set of data for each record, which may include a score, or reasons for a policy decline. In both the MCAR and MAR cases, data is said to be 'ignorably missing', and analysis can be based solely upon observed performance. In contrast, in the MNAR case, the performance data is 'non-ignorably missing', selection bias is evident, and reject inference is required.

To explain further, MCAR applies to cases where choices were made totally at random, like using a coin toss. For a statistical analysis, this is the ideal situation, as the available data can be used as is, with no reweighting. This does occur in practice, but is something to be avoided where there are associated costs, varying benefits, and other possible options.

As a result, selection criteria are derived using X, which results in two scenarios that have one crucial difference: (i) can the accept probability be determined using the X dataset only (MAR)?, or (ii) has it been influenced by extraneous factors that are also related to the outcome performance y (MNAR)? The MAR case is best illustrated by a selection process that is totally objective and/or fully automated, and all of the data used is represented within X. For any X, the outcome distribution should be the same, irrespective of the selection status A. Here it is also possible to use augmentation, which involves reweighting the 'accepts', to derive a model that applies to the entire population.

In contrast, with MNAR, the decision was influenced not only by X, but also by other extraneous factors related to perceived outcome performance, including characteristics no longer available on the system, the experience and prejudices of underwriters with the power to override system decisions, and declined customers persistence. In this case, the outcomes are said to be 'non-ignorably missing', and there may be substantial differences in the risk of the 'accept' and 'reject' populations that cannot be captured in a model developed using observed performance only. Selection bias arises, such that performance for the rejects has to be inferred using available tools and information, making adjustments where necessary. Reject inference is covered further in Model D: Scorecard Development.

According to Smith and Elkan (2004), this framework has been referred to not only in the context of credit scoring, but also epidemiology, econometrics, clinical trial evaluation, and sociology. It arises when analysing the performance of any selection process, where the resulting selection bias must be considered; the observations are not random, but are limited only to those that have been selected in the past. Most researchers represent it in terms of a Bayesian network, where there is conditional independence. Smith and Elkan refer to it in the context of 'active learning', where labels have to be assigned to observations, but label assignment is costly, and as a result the learner will be picky, and only assign labels to those from which the most information can be gleaned. For credit scoring, and other related business problems, the end goal is not information, but profit.

7.4.2 Statistical assumptions

When applying any statistical technique, assumptions are made about the population being analysed, which will vary depending upon the technique being used. The scorecard developer must ensure that the model has the correct functional form, and where the assumptions are violated for a given dataset or problem, it may be necessary to change the technique being used. These assumptions are of two types: (i) variable assumptions, and (ii) residual assumptions.

Variable assumptions

The first set of assumptions relates to the form of the data provided, whether the distribution of individual variables, or relationships between different variables. These assumptions are:

Normally distributed variables—Many statistical tests demand a normal distribution. Linear predictive-modelling techniques assume that the response variables are normally distributed.

Uncorrelated predictors—Independent variables used in a regression should not be correlated with each other, otherwise there may be a multicollinearity problem. This inflates the variance of the resulting coefficients, and 'leads to correlated errors in the regression coefficients themselves, . . . ' (Vaughan and Berry 2005). Resulting models may be difficult to interpret, perhaps even presenting a 'wrong-sign problem', and less reliable when applied in practice.

Linear relationship with target function—The use of parameter coefficients within a regression assumes a straight-line relationship between the independent predictor variables (observations), and the dependent response function (target variable or function thereof, such as logit or probit). Any variable where this does not hold true should be transformed into a new variable which has an approximate linear relationship; or where sufficient data exists, into dummy variables.

Additivity of final model—Applies to regression models, where it is assumed that a given change in a predictor affects the target similarly, irrespective of the values of the other predictors. It will not hold true where there are significant interactions, which may demand separate regression models for different subgroups, or a non-parametric technique.

The relationship between correlation and standard error for two characteristics, x_1 and x_2, can be stated as sterr $= 1/(1 - \text{corr}(x_1, x_2))$. As the correlation approaches 100 per cent, the standard error approaches infinity, that is assuming both characteristics are included in the final model.

Residual assumptions

The second set of assumptions relates to the residuals, meaning the difference between the predicted (expected), and observed (actual) values. This is represented in regression equations as the letter '*e*', an error term often added as the last item in a regression equation. There are three assumptions in this camp, violations of which are either symptoms of variable-assumption violations, or use of an inappropriate modelling technique in a given situation:

(i) **Normally distributed**—The distribution of error terms has the shape of a bell curve. This assumption is most important where samples are small.
(ii) **Homoscedasticity**—The residuals have a constant variance across the range of the estimates, meaning that no matter what the estimate, it will be wrong by approximately the same amount. The opposite is called *heteroscadisticity*, which is not fatal, but implies that the model's reliability varies across the range of possible estimates.
(iii) **Independent**—There should be no *autocorrelation*, which arises where multiple observations of the same cases are included at different points in time, as the status in one period usually has a significant bearing on the status in the next.

7.4.3 Addressing violations

Assumption violations are not always deal breakers. This section takes a brief look at: (i) addressing the non-linear relationship between a predictor and target variable, through transformations; and (ii) some issues related to multicollinearity.

Linearity—transformations

Where characteristics are not normally distributed, new variables can be created that are transformations of the originals, and can be used in their stead. The most common transformations quoted in statistics textbooks involve either *square roots* $(\sqrt{y})$, *natural logarithms* ($\ln(y)$ or e^y), or *inverses* ($1/y$). These are seldom used in credit scoring though, either because they cannot be implemented in lenders' delivery systems, or because other methods are more appropriate—especially in data rich environments. Instead, model developers address non-linear relationships with the target variable, by first classing the predictors, and then transforming them using either: (i) the *weight of evidence*, used with logistic regression, covered further in Section 8.2.1; (ii) *probabilities*, used with linear regression; and (iii) *dummy variables*, which can be used anywhere. The first two options both provide measures of relative risk for each range of the classed characteristic. In contrast, the latter requires that binary 0/1 variables be created, for all but one of the specified ranges.

Multicollinearity

Credit systems provide a wealth of relevant data, but much of it is highly correlated, which increases the standard error, and may make the model less reliable in practice. Perhaps the greatest symptom of multicollinearity is a large number of variables with small point allocations, that nonetheless provide a seemingly powerful model. Multicollinearity can be addressed though: (i) *factor analysis* can be used, to summarise predictors into uncorrelated factors; or (ii) the model developer can take extra effort to ensure that the point allocations make *common sense*, and that the number of terms is limited to those that truly add value. If the second option is chosen, increasing the sample size helps to reduce the standard error.

Surprisingly, factor analysis is most commonly mentioned with regard to financial ratio scoring (FRS), and only receives brief mention in the retail credit-scoring literature. This is primarily because the resulting factors are difficult or impossible to implement in retail delivery systems (which tends to be overlooked in academic environments). Instead, developers may choose one or two of the most predictive charcteristics to represent each factor.

Variable selection

Both linear regression and logistic regression have iterative procedures that are used to automate the variable-selection process, and will influence the amount of multicollinearity in the final model:

(i) **Forward selection**, which starts with no variables, and selects variables that best explain the residual (the error term, or variation that has not yet been explained).
(ii) **Backward elimination**, which starts with all of the variables, and removes variables that provide little value in explaining the response function.
(iii) **Stepwise**, either forward or backward, which are combinations that have the same starting points, but consider moves in the other direction at each iteration.

Ultimately, the goal is to find the set of characteristics that best explain any variance in the target variable, and different models can be compared using an adjusted R^2 measure. With forward-stepwise regression (FSR), once there are three or more variables, it will always try to remove variables, before considering another one for entry. Backward-stepwise regression (BSR) operates in reverse. FSR is the more popular of the two options, if only because the resulting models are simpler, and the results more interpretable. In contrast, BSR is less interpretable, but will include variables that may not seem predictive in isolation, but provide value as part of a set. Note that each of the options has three parts (Statsoft 2003):

(i) **Initial model**—No variables or all variables. Users may also opt to exclude the intercept, and/or specify variables that must be included in the final model.
(ii) **In-/exclusion criteria**—At each iteration, forward entry will select the characteristic with the lowest p-value (based on an F- or other statistic), and backward removal will reject that with the highest.
(iii) **Stopping criteria**—The routine will stop once no new variables can be found that meet the criteria, say a p-value of 0.05, or once a specified number of steps has been reached.

If multicollinearity exists, the resulting set of variables will depend on the method used, and starting point for the model. Where FSR is used, those variables that have the least correlation with each other are usually included in the model first, and correlated variables will feature later. While the stopping criteria will usually prevent spuriously correlated variables from being included, this is not a given. In retail scorecard developments, modellers will spend a great deal of time ensuring that the regression coefficients make logical sense, and methods are used to ensure that variables truly do add value as predictors.

These automated variable-selection techniques have both fans and critics. The fans rightly refer to the speed with which the assessments can be done, especially with modern computers, and argue that further in-depth evaluation has a minimal effect on the choices. Even so, critics see them as imperfect tools,[5] and argue that: (i) they use statistics meant for *hypothesis testing*, and it is not possible to validate a hypothesis using the same data that generated the hypothesis; (ii) there are problems with the *statistics being used*, like the R^2 values being inflated, or the F- and chi-square statistics not having the required distributions: (iii) where there is multicollinearity, characteristics may be selected because of *chance features* of the dataset being assessed, and can change with the introduction of new observations. Another possibility is to create a large number of models and assess them using another statistic, such as the Gini-coefficient, while simultaneously ensuring that each makes logical sense.

Critics instead argue in favour of having a proper understanding of the data. As stated by Albright et al. (2003:647),

> There should always be some rationale, whether based on economic theory, business experience, or common sense, for the variables that we use to explain a given response variable. A thoughtless use of stepwise regression can sometimes capitalize on chance to obtain an equation with a reasonably large R^2 but no useful or practical interpretation.

In the current context, the amount of effort employed to choose characteristics should be a function of how the model will be used. If it affects business decisions, involving aggregate values in the millions, some extra care and attention is required.

---

[5] See comments made by Frank Harrell, Ira Bernstein, and Ronay M. Conroy on www.stata.com, or search on 'stepwise variable selection' and 'problems'.

7.5 Results comparison

If you want to inspire confidence, give plenty of statistics. It does not matter that they should be accurate, or even intelligible, as long as there are enough of them.

Lewis Carroll, a.k.a Reverend Charles Lutwidge Dodgson (1832–1898).

An issue often debated by lenders and scorecard developers is which statistical technique is best suited to credit scoring. Given the nature of the problem, logistic regression should be the most obvious choice, but linear probability models and DA are still used.[6] The other methodologies have been proposed, but generally have not received wide acceptance.

One would think that the results from the various techniques could be readily compared, but the research has been inconclusive. Thomas et al. (2002) provide a summary of several studies, shown in Table 7.4, which indicates more similarities than differences. Similarities occur, because for any given predictive modelling problem using a given dataset, there is a 'flat maximum' percentage that cannot be exceeded by any statistical technique, but each of them can come quite close. According to Falkenstein et al. (2000), once the most important explanatory characteristics have been identified and normalised, there are a large number of possible coefficient combinations which provide solutions approaching the flat maximum.

Although no firm conclusion can be drawn, there seem to be three general areas of consensus. First, where people have significant experience with a given technique, they usually also have a significant *bag of tricks*, that allows them to get the best possible results from it. Second, greater value can usually be obtained from *improving data quality*, and bringing in *new data sources* (thus pushing up the flat maximum), than the use of any new and exciting statistical technique. Third, what was once a key consideration is *now minor*, and it might even be possible to use several techniques as part of a single development.

While no specific recommendation can be given, some advice can be offered. If an organisation already has a significant investment in a given scorecard methodology, it should stick with

Table 7.4. Comparison of results—percentage of cases correctly classified

Author	Linear	Logistic	RPA	Linear programming	NN	Genetic algorithm
	Regression					
Henley (1995)	43.4	43.3	43.8			
Boyle et al. (1992)	77.5		75.0	74.7		
Srinivisan and Chakrin (1987)	87.5	89.3	93.2	86.1		
Yobas et al. (1997)	68.4		62.3		62.0	64.5
Desai et al. (1997)	66.5	67.3			66.4	

[6] Experian still uses linear probability modelling for much of its credit scoring.

it. If, however, it is looking at developing expertise in credit risk scoring for the first time, or an opportunity to change presents itself, then either logit or probit should be used, if only because: (i) they are statistically more acceptable; and (ii) the resulting scores can provide estimates. At the same time, many long, frustrating, and fruitless discussions filled with much intellectual onanism will be avoided. There may, however, be instances, where non-parametric techniques should be considered, especially where there are a lot of interactions, or it is a fast changing environment.

8 Measures of separation/ divergence

> These measures of goodness of fit have a fatal attraction. Although it is generally conceded among insiders that they do not mean a thing, high values are still a source of pride and satisfaction to their authors, however hard they may try to conceal these feelings.
>
> *J.S. Cramer* (1987)

On a recent visit to Paris, my partner and I visited the Eiffel Tower, and opted to go to the top, level 3. This is 295 metres above the ground, and once at the top, I automatically felt slightly giddy, as though there were a slight sway in the tower. I had never felt this way before though, in spite of having looked over precipices of more than a kilometre on hiking trips in South Africa. When I mentioned the sway to my partner, she said, 'Nonsense!' I insisted that there must be some sway in the tower, whether because of wind (it was a calm day), the movement of lifts up and down its centre or legs, or the shifting of the many and gawking tourists on any of the three observation platforms. She conceded that there might be some sway, but thought it imperceptible, and insisted I was suffering from vertigo. I swore the contrary, but had no means of proving my point. No doubt it can be measured, but we had neither the tools nor access to information to settle the argument.

Credit scoring also needs tools to measure movements. The tools are called measures of separation, measures of divergence, or power/divergence statistics, which are used to determine differences in data distributions, whether during: (i) coarse classing; (ii) variable selection; (iii) segmentation; (iv) result evaluation; (v) post-development validation; or (vi) post-implementation monitoring. These are *bivariate statistics*, that originated in a variety of different disciplines, including mathematics, economics, psychology, and electronics. In some cases, like the Pearson's correlation coefficient, they have little application in credit scoring, but are covered here as part of a larger conceptual framework. For the most part, these statistics are used to assess:

Power—Measure of ranking ability, or dependence between a characteristic, score, or grade and a binary outcome. It shows the extent (random↔perfect), and usually the direction (positive/negative), of the correlation. Measures used include rank-order correlation measures, as well as the information value, KS statistic, and chi-square.

Drift—Measure of variation between expected and actual results. In this instance, direction will seldom be of interest, as the primary concern is deviation from expected. Measures used include the stability index, KS statistic, chi-square statistic, binomial test (and its normal approximation), and the Hosmer–Lemeshow statistic.

According to Thomas et al. (2002:155), in credit scoring the primary distinction is where the statistics are used. *Power* is of greatest interest when scorecard performance is being measured, and is used to prove that the scorecard will add value; big is good, and implies better risk ranking ability. In contrast, when monitoring *drift*, the hope is for minimal change; small is good, and means less variation from the baseline.

The value of these tools comes from their ability to collapse information, whether into a single number, graphic, or table. Furthermore, the statistics often provide a scientific basis for hypothesis testing, which requires the formulation of a null (H_0) and alternate (H_A) hypothesis, and use of a statistical test, to determine whether to reject the former.[1] Care must, however, be taken, because specific ranges are sometimes of interest (especially when dealing with cut-offs), and not the entire distribution. As a result, the KS curve, Lorenz curve, Receiver Operating Characteristic curve, Misclassification graph, and other graphical tools are presented, to aid visualisation of what is happening across the range of values.

As shown in Table 8.1, at a high level, the types of divergence statistics can be described by what is being compared: (i) *frequencies*, of classed characteristics; (ii) *rankings*, of raw or classed characteristics; and (iii) *cumulative percentages*, the percentage of the total that falls below each ranked value, also called an empirical cumulative distribution function (ECDF). Each of them has different strengths and weaknesses, depending on the situation, and although valuable, they should not be used in isolation. Cognizance must always be taken of: (i) peculiarities specific to each situation; (ii) other measures and tools; (iii) the scorecard developer's intuition; and (iv) any insight that can be provided by experts within the business.

Measures of separation—use and interpretation

When used to measure ranking ability, all of the tools mentioned here share the common trait, that the flat maximum that can be achieved depends upon the problem; in particular the portfolio being assessed, and its relationship to the outcome. If a *risk-heterogeneous* group is being assessed, whether using a single characteristic or the final scorecard, the results will always be

Table 8.1. Measures of separation

	Frequency	Ranking	Cumulative percentage
Chi-square	✓		
Kullback divergence	✓		
Spearman's rank		✓	
Gini coefficient		✓	✓
KS statistic			✓

[1] A null hypothesis can never be accepted. Either the evidence is sufficient to reject it at a given confidence level, or it is insufficient.

higher than for a *risk-homogenous* group. There are four primary sources of the homogeneity:

(i) It may be an inherent feature of the population. Anderson (2003b) and Mays (2004) both refer to the seemingly poorer results that are obtained in *sub-prime markets*, which results because the individuals are clustered at the high end of the risk spectrum (this is compounded by the next point, as they have traditionally been financially excluded, and hence not credit active).

(ii) It may result from data deficiencies, either because of poor data quality or lack of relevant data. The greatest leaps in predictive power occur when new data sources become available, that: (i) provide greater transparency, and new insight into the behaviour of individuals (like shared-performance data); and (ii) have a low correlation with that already available.

(iii) It may be the result of truncation by a selection process. Two unrelated examples are: (i) admission-test scores, versus academic grades in universities; and (ii) application scorecard developments, versus post-implementation monitoring of accepts. Assuming that rejects would have performed worse than accepts, any measure of separation based solely on accepts, will be lower than that for the full through-the-door population.

(iv) It may be a by-product of the segmentation, especially when separate scorecards are applied at different levels in the risk spectrum. As in the maxim, 'the sum of the parts is greater than the whole', so too may the apparently poor performance of an individual scorecard be more than offset by its contribution as part of a scorecard set.

As a final note, measures of separation should be used with caution. Falkenstein (2002:184) warns, 'one should not be focused on any single statistical test (of ranking ability), as this indirectly encourages overfitting'. It must also be ensured that 'apples are compared with apples', especially as regards the binary outcomes (good/bad definition), time frames (outcome window and censoring), and truncation (score cut-off and policies). Where firm conclusions have to be drawn, then appropriate statistical tests should be used, as not all measures are suited for hypothesis testing.

Divergence statistic

Perhaps the most straightforward summary measure of separation is the divergence statistic. It is a parametric statistic assumes the values for both groups are normally distributed, and is calculated as the squared difference between the means of two groups, divided by their average variance. The greater the spread of possible values, the greater the difference has to be before the two distributions are considered different. For the instance where the two groups are goods and bads, the formula for a given characteristic can be presented as:

Equation 8.1. Divergence statistic $D^2 = \dfrac{(\pi_G - \pi_B)^2}{(\sigma_G^2 + \sigma_B^2)/2}$

It can be applied to any continuous characteristic, including scores. It suffers, however, because it says nothing about the shape of the distribution. According to Mays (2004), this statistic is closely related to the information value, and it is also mentioned in Siddiqi (2006). It is only covered briefly here, because it is seldom encountered in practice, probably because of: (i) its limited focus on continuous characteristics; (ii) the assumption that the two distributions are normally distributed; (iii) the potential distorting effect of outliers; and (iv) other measures are more common. Even so, it is probably appropriate for many, if not most, score distributions provided by logit and probit models.

8.1 Misclassification matrix

A very simple way of evaluating how well a predictive model has worked, or at least those with binary outcomes, is to calculate the percentage of accounts that have been correctly classified. This was used for Table 7.4, to compare the results from various predictive modelling techniques. The percentage correctly classified is derived from a misclassification matrix that is created by:

(i) choosing a score cut-off;
(ii) marking all accounts below the cut-off as expected bads, and all those above as expected goods;
(iii) cross-tabulating the expected goods and bads against the actuals, using the development definition, or any other definition of interest;
(iv) determining the percentage of accounts that fall into each cell;
(v) calculating the various ratios that can be derived from the model.

The correctly classified cases are the *true positives* (bads) *and negatives* (goods). If they do not correspond, they are labelled false positives (type I error, expected bad that is good) and negatives (type II error, expected good that is bad). The use of 'negative' for goods and 'positive' for bads is consistent with other tests used to identify rare events, such as in the medical profession; positive means that a patient has been diagnosed with the disease (e.g. HIV positive) whether correctly or incorrectly. In some cases, such as for the construction of the ROC curve discussed in Section 8.4.5, the focus is restricted to those predicted as bad; both true positives (hits), and false positives (false alarms).

The example in Table 8.2 presents the results for a through-the-door population of 150,000 applicants, with a bad rate (admittedly very high) of 32.6 per cent. The cumulative percentage of applicants by score was then reviewed, and it was found that a cut-off of 500 provides almost the same percentage. When split out into the four groups, the false positives and negatives (11.9 and 11.1 per cent respectively) become evident, indicating a misclassification rate of 23.0 per cent. It did, however, have 77 per cent correctly classified.

An issue is the choice of cut-off. The most common choice is one where the number of accounts below the cut-off is equal to the sample bad rate. Lenders may, however, wish to measure changes at different reject rates, and can construct a graph like that in Figure 8.1.

Table 8.2. Misclassification matrix

Actual	Predicted		Totals
	Negative/good	Positive/bad	
Negative/good	83,275 55.5%	17,850 11.9%	100,125 66.4%
Positive/bad	16,700 11.1%	32,175 21.5%	48,875 32.6%
Totals	99,975 66.6%	50,025 33.4%	150,000 100.0%

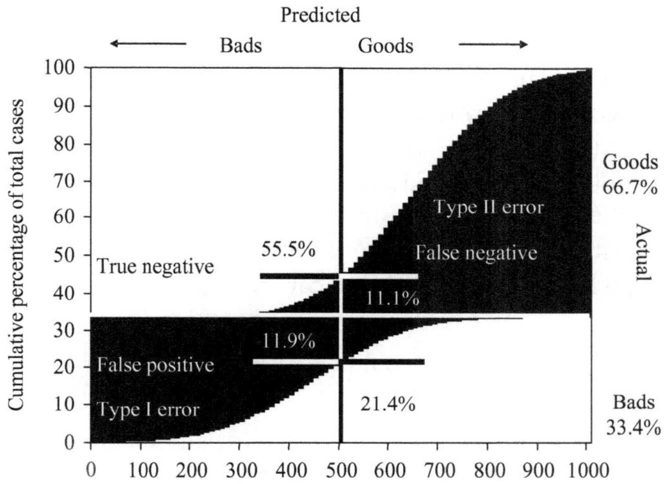

Figure 8.1. Misclassification graph.

Here it can be seen that at a cut-off of 500, 64.2 per cent of the bads (21.4/33.4) have been identified, at the expense of misclassifying only 16.7 per cent of the goods (11.1/66.7).

It must also be noted that although the misclassification rate is commonly used to measure model accuracy, it is seldom sufficient, unless the total misclassification costs can also be calculated. Unfortunately, 'per case' figures are difficult to derive, and analysts often use a significant degree of latitude when assuming Type I and II error costs. A similar type of analysis can also be used for comparing scorecards against each other, except in that instance, swap sets between the two are identified. This is particularly useful for comparing recently developed scorecards against those currently in place.

8.2 Kullback divergence measure

Kullback's divergence measure is used to gauge the difference between two frequency distributions. It is used in many disciplines, but as with many other statistics, it has become masked under other names, that do not give proper credit to its author. In credit scoring, it is referred to either as the: (i) *information value*, when measuring power, to compare good and bad distributions; or (ii) *stability index*, when measuring drift, to compare a distribution at two points in time. It is based upon the *weight of evidence* (WoE), which provides a simple, and theoretically well-grounded, tool for assessing relative risk based upon available information. The WoE is covered first here, before delving into the information value and stability index in more detail.

8.2.1 Weight of evidence

Each and every day we make decisions based upon the probability of some event occurring. We decide on whether or not to cross the street, based on how much traffic there is, and how fast it is going; and on whether or not to take a raincoat, based on the morning weather report, or a look outside to see if it is sunny or cloudy. The probability is far from empirical, as it relies upon personal experiences, or information gained from others.

In 1950, Irving John (Jack) Good, an Englishman who was a Second World War code-breaker, published a book that addressed these personal and subjective probabilities. For any decision, one assesses the circumstances and determines a weight of evidence. Basically, this converts the risk associated with a particular choice onto a linear scale that is easier for the human mind to assess, which for credit scoring can be expressed as:

Equation 8.2. Weight of evidence $$W_i = \text{Ln}\left(\left(\frac{N_i}{\sum N}\right) \middle| \left(\frac{P_i}{\sum P}\right)\right)$$

where P = occurrence (positive), N = non-occurrence (negative), and i = index of the attribute being evaluated (such as 'Income $< X$'). The precondition is, of course, non-zero values for all N_i and P_i (small adjustments can be made to ensure this).

> The WoE formula above is that most often used. It can be restated as: $W_i = \text{Ln}(N_i/P_i) - \text{Ln}(\Sigma N/\Sigma P)$, which illustrates two components: a variable portion for the odds of that group, and a constant portion for the sample or population odds. The WoE for any group with average odds is zero. Note that the two natural log odds values are both restatements of $\ln(p_N/(1-p_N))$. For a characteristic transformation, the WoE variable has a linear relationship with the logistic function, making it well suited for representing the characteristic when using logistic regression (logit).

The WoE is used: (i) to assess the relative risk of different attributes for a characteristic, to get an indication of which are most likely to feature within a scorecard; and (ii) as a means of transforming characteristics into variables. Some software packages provide it as a value or graph (see Figure 8.2), as it is a very useful tool for binning. Attributes with similar relative risks are usually merged. Unfortunately, the WoE does not consider the proportion of accounts with that attribute, only the relative risk. Other tools are used to determine the relative contribution of each attribute, and the total information value (covered next).

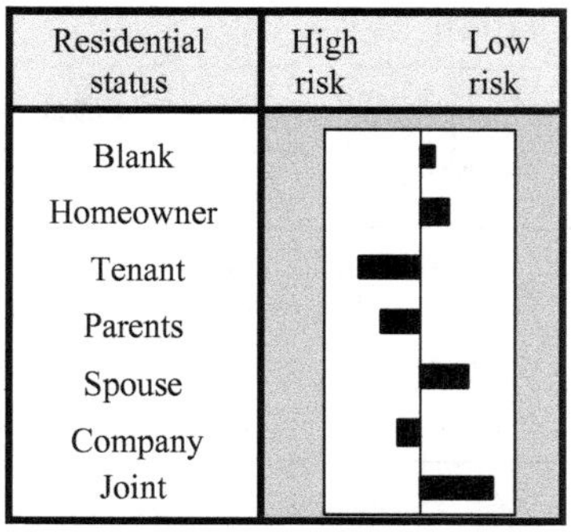

Figure 8.2. Weight of evidence.

With the advent of Basel II and the PD/EAD/LGD framework, there is a trend towards speaking in terms of probabilities, instead of risk grades. While this should not pose a problem for the cognoscenti, it could cause confusion for banks' rank and file. Jack Good's work effectively showed that people can relate to risk grades of 3, 6, 9, and 12, better than percentages of 0.04, 0.32, 2.5 and 17, as the human mind processes risk on a log scale.

8.2.2 Information value

In the early days of credit scoring, Fair Isaac (FI) adopted a measure that they dubbed the information value, to measure the predictive power of a characteristic. Very few people give proper credit to Solomon Kullback, who first published it in 1958, at the time when FI was first finding its feet. It is technically referred to as the Kullback divergence measure, and is used to measure the difference between two distributions. When applied to test results it is expressed as:

Equation 8.3. Information value $$F=\sum_{i=1}^{n}\left[\left(\frac{N_i}{\sum N}-\frac{P_i}{\sum P}\right)\times \text{WoE}_i\right]$$

where N = negative identification (goods), P = positive identification (bads), WoE = the weight of evidence, i = index of the attribute being evaluated, and n = total number of attributes. The result for each attribute reflected between the square brackets is called the 'contribution'.

Values for F will always be positive, and may be above 3 when assessing scores provided by highly predictive behavioural scorecards. Characteristics with values of less than 0.10 are typically viewed as weak, while values over 0.30 are sought after, and are likely to feature in scoring models. Table 8.3 provides a very simple example for applicants' income. The predictive power is marginal; and if included in a scorecard, the point allocations would be low.

Please note, that *weak characteristics* may: (i) provide value in combination with others; or (ii) have individual attributes that could provide value as dummy variables. They should

Table 8.3. Information value calculation

Income	Outcome		Good/bad odds	Column (%)		WoE	Contribution
	Good	Bad		Goods	Bads		
Low	5,000	2,000	2.5	14.3	33.3	−0.847	0.161
Middle	10,000	2,000	5.0	28.6	33.3	−0.154	0.021
High	20,000	2,000	10.0	57.1	33.3	0.539	−0.074
Totals	35,000	6,000	5.8	100.0	100.0	Info value =	0.109

thus not be discarded indiscriminately. Further, even if not considered for the model, they should still be retained for future analysis (Mays 2004).

Like the Gini coefficient, the information value is also sensitive to how the characteristic is grouped, and the number of groups. Unlike the Gini coefficient, however, the information value will provide the same result, irrespective of how the attributes are ordered. It can, however, be difficult to interpret, because there are no associated statistical tests. As a general rule, it is best to use the information value and/or chi-square to assess individual characteristics, and the Gini coefficient (in combination with other measures) for the final scorecard.

8.2.3 Stability Index

The Kullback divergence measure is also used to measure drift, in much the same way as the chi-square statistic. In this instance however, it is called a population stability index, as shown in Equation 8.4, which measures the difference between the development sample and a more recent distribution:

Equation 8.4. Population stability

$$F = \sum_{i=1}^{n}\left[\left(\frac{O_i}{\sum O} - \frac{E_i}{\sum E}\right) \times \mathrm{Ln}\left(\frac{O_i}{\sum O} \Big/ \frac{E_i}{\sum E}\right)\right]$$

where O and E are the observed (recent population) and expected (development sample) frequencies. The result will always be positive, and a traffic-light approach is used to provide warnings: (i) *green*, less then 0.10, no cause for concern; (ii) *yellow*, between 0.10 and 0.25, some cause for concern, and (iii) *red*, greater than 0.25, concern! It can be used on the final score to provide a measure of score drift, or on individual characteristics. Please note that, once again, the precondition is positive values for all O_i and E_i.

According to Thomas et al. (2002:155), the stability index lacks sophistication and consistency, but it can be considered in combination with other measures, like the Gini coefficient and K-S statistic, which allow significance testing.

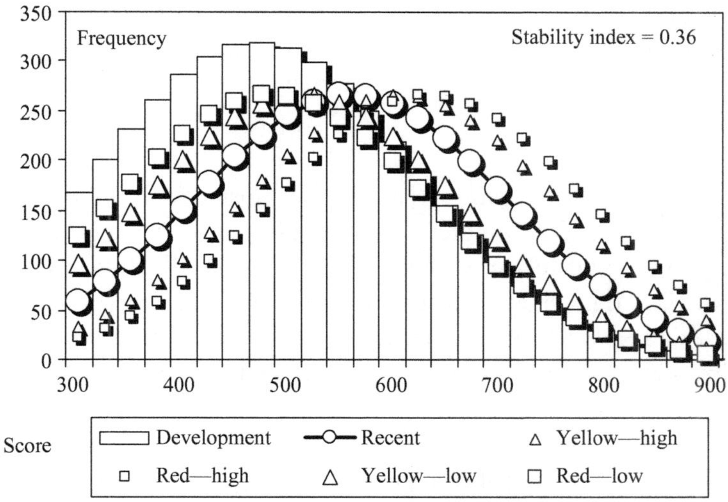

Figure 8.3. Population stability.

The graph in Figure 8.3 illustrates the score distributions for a recent through-the-door population (columns), a development sample (circles), and the warning lights either side of the development sample. The recent and development distributions indicate that not only volumes have increased since development, but also the applicant risk as measured by the score. The population stability index of 0.36 has gone past both the yellow and red lights to the left, indicating that special attention is required (the yellow and red lines are shown for illustration purposes only, and are hypothetical, assuming that neither the volume nor variance changes, only the mean).

Please note, that the scorecard could still be working well, regardless. The only way really to know is to have sufficient performance to assess its ranking capability directly. Otherwise, all that can be done—other than immediately fine-tuning or redeveloping the scorecards—is to get a better understanding of what is causing the shifts. That is the purpose of the score shift report (see Section 25.3.2).

8.3 Kolmogorov–Smirnov (KS)

It seems no conversation on credit scoring is complete without mention of the KS statistic. Even senior executives of financial services companies are familiar with it and in fact try to impress each other by claiming their scoring model has a bigger KS than the next guy's.

Mays (2004:121)

One of the statistics commonly used in credit scoring, as well as countless other disciplines, is the KS statistic. This was developed by two Soviet mathematicians, A.N. Kolmogorov and N.V. Smirnov. Kolmogorov first proposed it in an Italian actuarial journal in 1933.[2] Smirnov

[2] See also the biographical sketch later in this section. No biographical information could be found for Smirnov.

built upon it 1939, and tabulated it in 1948. It is one of several statistics, like the Gini coefficient, that is built upon an analysis of the empirical cumulative distribution function (ECDF). It is a better non-parametric measure for assessing error (or 'goodness of fit') in curve fitting than many others.

According to Mays (2004), the KS statistic is the most widely used statistic within the United States for measuring the predictive power of rating systems. This does not appear to be the case in other environments, where the Gini or AUROC seem to be more prevalent. In any event, it is dangerous to use any single measure in isolation. Mays also indicates that as a measure of ranking ability, KS values range from 20 per cent, below which the model's value should be questioned, to 70 per cent, above which it 'is probably too good to be true'.

The KS Curve (also known as the fish-eye graph) is a data-visualisation tool used to illustrate scorecard effectiveness. It charts the ECDF percentages for goods and bads against the score. In the left-hand chart within Figure 8.4, it can be seen that 56.6 per cent of the bads fall under the score of 470, but only 12.2 per cent of the goods.

The statistic of interest is where the difference is greatest.[3] This is the KS statistic, being the maximum absolute difference between the two curves: $0 < D_{KS} < 1$. In the right-hand graph of Figure 8.4, the distance at the score of 470 is 44.4 per cent (56.6 less 12.2 per cent), but this increases to 49.4 per cent at a score of 550.

Equation 8.5. KS statistic $D_{KS} = \max\{\text{abs}(cpY - cpX)\}$

While this is a very simple measure to understand, it may be too simple. The KS statistic often applies to a point on the curve that has no relevance to the problem at hand, especially where

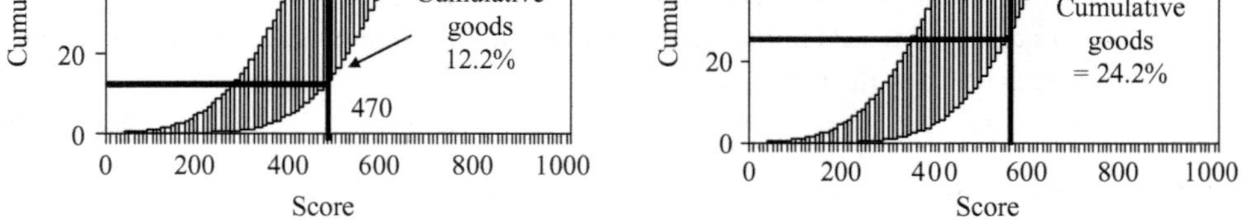

Figure 8.4. Kolmogorov–Smirnov.

[3] The treatment differs depending upon whether one or two samples were used to generate the values.

it is a long way above or below an application scorecard cut-off. It is thus usually used in conjunction with other measures.

The most common uses of the KS statistic are as a measure of predictive power, and to determine whether or not two distributions differ. Hypothesis tests can also be applied, by comparing it to $\mathrm{KS}_{\mathrm{critical}}$. If it is less, then there is a good chance that the two distributions are the same. $D_{\text{KS-critical}}$ is calculated as $c/\sqrt{n}$, where c varies according to the significance level, and type of distribution being assessed, and n is the sample size. In most instances, it is sufficient to assume that the distribution is normal, in which case 'c' is 1.36 at a 0.05 significance level.

For an exponential and Weibull distribution, the values would be 1.08 and 0.874 respectively at the 0.05 significance level. If the type of distribution is not known up-front, this can be determined using the expected probabilities for different possible distributions, and applying a chi-square test to determine which is most appropriate. If that is not feasible, the more conservative 0.874 should be used.

Note however, that $D_{\text{KS-critical}}$ is very sensitive to changes in the value of n. If two samples of 10,000 each are being compared, then $D_{\text{KS-critical}}$ is 0.0136 (1.36%). The logic is that as sample size increases, sampling error decreases, and the test has to be more stringent.

Note that this formula assumes that the two populations are the same size. Mays (2004) provides a formula for two sample sizes, where $D_{\text{KS-critical}}$ is $c/\sqrt{(n_1+n_2)/(n_1 n_2)}$, '$n$' is the size of the respective populations, and 'c' is 1.22 at the 95 per cent confidence interval, for a one-sided test of significance. She does not indicate how the value for 'c' was obtained.

Like with the other statistics, care must be taken when using the KS statistic for comparisons. Mays (2004) warns that when assessing application scorecards, the KS statistic for the accept population will differ depending upon the cut-off, because of the truncating effect of rejects, especially if compared to the full development sample. In the particular example she provides, the KS statistic for high risk, low risk, and the full sample are 27, 36, and 49 per cent respectively. Likewise, when assessing variation in scorecard performance over time, it must be ensured not only that the good/bad definition and outcome windows are the same, but also that the same score cut-offs and policy rules are applied, or are at least similar.

Andrei Nikolaevich Kolmogorov (1903–1987)—Biographical sketch

A.N. Kolmogorov was a renowned Soviet mathematician, who wrote over 300 papers on practically every aspect of mathematics, and spent much time advancing mathematics in their school system, especially for the gifted. His father was the agronomist son of a clergyman, and his mother was of aristocratic stock. Kolmogorov was born out of wedlock in Tambov, during a delay while his mother was returning from the Crimea, and she died in childbirth. He was brought up by her sister, a woman of high social ideals, on his grandfather's estate in Tunoshna, near Yaroslavl, and took his maternal grandfather's surname.

As a teenager, Kolmogorov worked briefly as a railway conductor, prior to starting at Moscow University in 1920.

Besides mathematics, he also had interests in poetry and history, and briefly spent time researching fifteenth and sixteenth century manuscripts on agrarian relationships in ancient Novgorod. He stuck with mathematics though, and before graduation in 1925, he had already published eight papers. His interest in probability theory started in 1924, and in 1929—besides completing his doctorate—he published a paper titled *A general theory of measure and the calculus of probabilities*, that started providing foundations where none had existed before. His 1931 paper, *Analytical methods in probability theory*, built on Markov's work to develop the modern theory of Markov processes (diffusion theory). In 1933, his *Foundations of the Calculus of Probabilities* was published in German, which became the definitive work on probability theory, and established it as a formal branch of mathematics. He is also renowned for several articles on the theory of poetry and the statistics of text, where he analysed Pushkin's poetry. Kolmogorov was highly recognised by the Soviet state, receiving seven orders of Lenin, the Order of the October Revolution, and the Hero of Socialist Labour. Of all Soviet mathematicians, he was the most recognised outside of the USSR, and besides receiving a number of honorary doctorates, he was also elected as full or honorary member of many foreign mathematical and other societies.

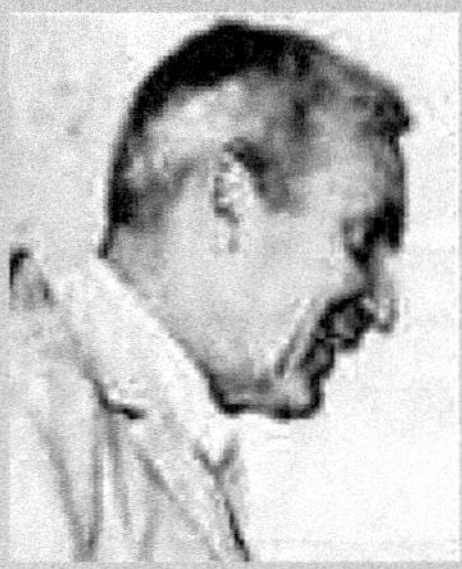

8.4 Correlation coefficients and equivalents

It is part and parcel of human nature to notice and strive towards an understanding of patterns and relationships, whether because of idle curiosity or an innate desire to control the environment. If X varies with Y then maybe, just maybe, Y can be influenced if X is controlled. The problem, of course, is that correlation does not imply causation—the observed relationship is, more often than not, related to one or more other factors. Even so, knowledge of these correlations provides direction, and much time is invested in studying them.

As a result, the workhorse of modern quantitative analysis is the correlation coefficient, which measures the degree of association, or covariance (ratio of shared variation to independent variation), between two variables X and Y. Here the two most commonly used types of correlation coefficient are discussed: product-moment, which measures how well a linear formula, of the form $Y = a + bX$, can describe the relationship, and for which both X and Y must be scalar; and rank-order, which measures the extent to which the relationships are monotonic, where X and Y can be either scalar or ordinal. Where it is a linear relationship, it is assumed that both values are normally distributed. If the assumption is seriously violated, then a rank-order correlation may be the better option.

The strength of the correlation is usually represented by a coefficient, in the range $-1 \leq r \leq +1$. The sign indicates whether the two variables move in the same, or opposite directions. If the value is at or near zero, the variables are statistically independent; but at the

Table 8.4. Correlation

Value range	Strength & direction	
$r = +1.0$	Perfect	Positive
$+0.9 < r < +1.0$	Strong	
$+0.5 < r \leq +0.9$	Moderate	
$0.0 < r \leq +0.5$	Weak	
$r = 0$	Uncorrelated	
$-0.5 \leq r < 0.0$	Weak	Negative
$-0.9 \leq r < -0.5$	Moderate	
$-1.0 < r < -0.9$	Strong	
$r = -1.0$	Perfect	

two extremes, they are statistically dependent. The labels used in Table 8.4 are highly subjective, whereas in truth the strength of each measure will vary according to circumstances. A key use of correlation coefficients is for hypothesis testing, which also requires:(i) the formulation of a null and alternate hypothesis to be tested (ii) calculation of the standard deviation or variance; and (iii) degrees of freedom (see sections following). This section covers all correlation measures that provide values in the –1 to +1 range, or equivalent, and associated tools:

Pearson's product-moment—Assesses linear relationship between continuous variables.
Spearman's rank-order—Similar, but modified to assess monotonic rank orders.
Lorenz curve—Data-visualisation tool, that plots the cumulative percentages of *X* and *Y* based upon some rank ordering, whether on *X*, *Y*, or a third variable.
Gini coefficient—Calculates the area between the curve and diagonal in the Lorenz curve; the higher the absolute value, the greater the rank correlation.
Receiver operating characteristic (ROC)—Like the Gini coefficient, except it calculates the full area below the curve.

8.4.1 Pearson's product-moment

Although Galton first proposed the concept, it was Karl Pearson who came up with the original and most widely-used formula, which bears his name—the Pearson product-moment correlation coefficient. It is related to ordinary least squares regression, but rather than deriving a beta coefficient, it instead measures the extent to which there is a linear relationship between the two variables. The correlation for a population is usually represented using the Greek character ρ (*rho*), but for a sample the Roman letter r is used. The formula varies from textbook to textbook, but the one most commonly used is:

Equation 8.6. Pearson's correlation
$$r = \frac{N\sum XY - \sum X \sum Y}{N\sqrt{(\sum X^2 - (\sum X)^2)(\sum Y^2 - (\sum Y)^2)}}$$

It does have some restrictions. Both X and Y must be continuous characteristics that are approximately normally distributed, which makes it inappropriate for use with binary, ordinal, and discrete characteristics. Furthermore, if there are outliers, the results can be highly distorted. If the product-moment coefficient proves infeasible, it is still possible to consider a rank-order correlation statistic as an alternative.

The formula can be restated as $r = \left(\sum z_X z_Y\right)/N$ if the two variables are represented by standardised Z-scores. These are obtained by substituting X and Y for V in the formula $z_V = (V - \overline{V})/\sigma_V$ to obtain new values with a mean of zero ($\overline{z}_V = 0$) and standard deviation of one $(\sigma_{z_V} = 1)$. This requires that X and Y are normally distributed, or can be transformed into something that is normally distributed.

A related statistic is the coefficient of determination, or r-squared (r^2) which can be interpreted as the proportion of variance in Y that is contained in X. Thus, if $r_{x,y}$ has a value of 0.9, then $r^2_{x,y}$ indicates that 81 per cent of the variance in Y is explained by changes in X, and vice versa. It is recommended that r-squared be used as a measure of association, as the correlation coefficient overstates the relationship, especially at lower values of r.

Karl Pearson—biographical sketch[4]

Karl Pearson (1857–1936) could well be called the founder of modern quantitative analysis. He was an English mathematician, renowned for dealing in symbols and formal truths. His interest was in the analysis of large samples to determine correlations, and he is credited with the product-moment correlation coefficient, the chi-square statistic, and the 1894 coining of the term 'standard deviation'.

After graduating from Cambridge University in 1879, he dabbled briefly in German literature and law, but by 1883 he was a professor of mathematics at University College, London (UCL), where he spent the rest of his working career. In his book, *The Grammar of Science* (1892), he anticipated some ideas later proposed by Einstein's relativity theory. He also developed a keen interest in heredity and evolution, and over the years 1893 to 1918, he wrote 18 papers, that are collectively referred to as the *Mathematical Contribution to the Theory of Evolution*. Although he claimed to be a socialist, and supported their causes, he was a believer in eugenics, and openly advocated 'war with inferior races'.

---

[4] Department of Statistical Science, University College London, England.
http://www.ucl.ac.uk/Stats/department/pearson.html
School of Mathematics and Statistics, University of Saint Andrews, Scotland.
http://www-history.mcs.st-andrews.ac.uk/Mathematicians/Pearson.html
O'Connor J.J. and Robertson E.F.
http://www-groups.dcs.st-and.ac.uk/%7Ehistory/Mathematicians/Pearson.html

In 1901, Pearson, Weldon, and Galton co-founded the journal Biometrika to develop statistical theory further. In 1911, Pearson founded the UCL's Department of Applied Statistics, the first university statistics department in the world, and was the first Galton Professor of Eugenics, chairing from 1911 to 1933. He had serious theoretical disputes with Ronald Fisher, who focused on small samples and finding causes. Fisher refused an offered post of Chief Statistician at the Galton Laboratory in 1919, because he would have reported to Pearson.

Hypothesis testing

While it is very nice to have a measure of association, in many cases it is needed to draw a conclusion. With correlation coefficients, tests can be done for dependence, independence, or direction in the correlation. A null and alternative hypothesis, H_0 and H_A respectively, are stated of the form:

H_0—The two variables are not linearly related, $\rho = 0.0$.

H_A—They are linearly related, $\rho <> 0.0$

The other possibility is to compare the correlation coefficients that have been calculated for two samples. For example, if the lender wants to determine whether the correlation between income and age for both loan and card applicants is the same, then the hypotheses would be stated as:

H_0—The two variables have the same correlation, $\rho_{\text{LOAN}} = \rho_{\text{CARD}}$.

H_A—They do not have the same correlation, $\rho_{\text{LOAN}} <> \rho_{\text{CARD}}$.

In either case, if the value of r calculated for the sample does not fall within the range demanded by the test, then the null hypothesis must be rejected. There is a problem however, because the t-tests and z-tests demand that the values tested are—at least approximately—normally distributed, and r is not. If multiple samples are taken from the same population, the distributions will only be approximately normal if the correlations are relatively low, say less than 0.5. For higher values, the distribution tends to be skewed to the right, with the skewness increasing as *rho* approaches 1.0.

In order to correct this, *rho* is converted using Fisher's z' transformation, which if applied to the r from repeated samples, would provide a value for z' that is normally distributed, and has a standard error $\sigma_{z'} = 1/\sqrt{(N-3)}$, where N is the number of observations. The transformation is:

Equation 8.7. Fisher's z' transformation $$z' = 0.5 \ln\left(\frac{1+r}{1-r}\right)$$

The transformation has minimal impact for values of r below 0.4, but starts growing thereafter. Hypothesis testing can be done using a Student t-test to: (i) determine whether or not the correlation is dependent, independent, or equal to a certain value; or (ii) to compare correlations for data taken from independent samples.

> Example: There is a correlation of 0.2 between income and age in a sample of 1,000. We wish to determine with 95 per cent confidence that they are not independent:
>
> H_0: The two variables are independent, $\rho = 0.0$.
> H_A: The two variables are not independent, $\rho <> 0.0$.
>
> The value for z' is also 0.2, so the next task is to obtain $z_{critical}$ from the Student *t*-test table in Appendix B. This can be tricky, because this is a two-tail test—H_0 states that the variables are independent, so ρ must fall in a restricted range around zero. The p-value used is then 0.975, and the associated $z_{critical}$ is 0.3. Given that $[-z_{critical} < z' < z_{critical}]$, the null hypothesis can be rejected.

This example was quite simple, as hypothesis tests often require much more complex calculations to come up with a *z*-score that can be used in the Student's *t*-test. To look for help on the Internet, do a search on (Pearson correlation Fisher transformation).

8.4.2 Spearman's rank-order

While using a product-moment correlation is first prize, it is often precluded because the relationship is not linear, or the distributions are not normal. Approximations are still possible however, if the two characteristics are at least ordinal. It was Charles Spearman who came up with a formula, to measure the monotonic relationship between the rank-ordering of two variables. This is effectively the same as Pearson's correlation coefficient, except it is a non-parametric test. It has the distinct advantages that it: (i) can assess non-linear relationships; and (ii) is not affected by outliers. The formula used is:

Equation 8.8. Spearman's rank-order correlation $$r_S = 1 - 6\,\frac{\sum(x_R - y_R)^2}{N^3 - N}$$

Within the formula, the term $(x_R - y_R)$ refers to the difference in the respective ranks for the same observation, and N refers to the total number of cases that are being ranked. There is a complication here for ties, in which instance the average rank for the tied cases should be used.

How would this statistic be used in credit scoring? Its primary use is for the comparison of different scores or grades, provided for the same group of cases. Comparisons can be made of new versus old, option A versus option B, or internal versus external. Its use is especially prevalent for benchmarking a lender's own internal credit-risk grades against those provided by a rating agency, or a model developed to assess the same cases. In any case, if the two are perfectly correlated, then the extra one provides no value, but there is usually a much less-than-perfect correlation.

Charles Spearman—biographical sketch

Charles Spearman (1863–1945), a British behavioural psychologist and statistician, spent 20 years in the army before completing his Ph.D. degree in 1904, at age 41. Besides being known for the rank-order correlation coefficient and factor analysis, both introduced within months of each other in the same journal in 1904, he also formulated the classical mental tests, and came up with a two-factor theory of intelligence, distinguishing between 'general intelligence' and 'specific factors of intelligence'.

8.4.3 Pareto principle and Lorenz curve

Surprisingly, several of the tools used to analyse and illustrate the result of scorecard developments stem from the field of economics. These include the Lorenz curve and the Gini coefficient. The late 1800s and early 1900s saw the emergence of both Marxist and Fascist ideologies in Europe, and a focus on income distributions in different countries by various academics. Vilfredo Pareto (1848–1923) was an Italian engineer, who later became an economist, and later yet took to sociology. In 1896, he noted how 80 per cent of the land in Italy was owned by 20 per cent of the population, and saw that this ratio also applied to land ownership and income in other countries. The ratio also applied in many other instances, and is now known as the 'Pareto principle', or '80/20 principle'.

In 1905, the American mathematician Max Otto Lorenz (1876–1959) went further, to develop what is today called the 'Lorenz curve', as a data-visualisation tool for displaying income inequality within society. The income data is sorted in decreasing order, and the cumulative percentages of both income and population are calculated as $\mathrm{cp}V_i = \sum_{j=1}^{i} V_j / \sum V$, where i refers to the rank in an ordered list.

The results are both then plotted on an XY graph, like that shown in Figure 8.5 (the graph will be inverted if income is sorted in ascending order). From this, it can be seen that

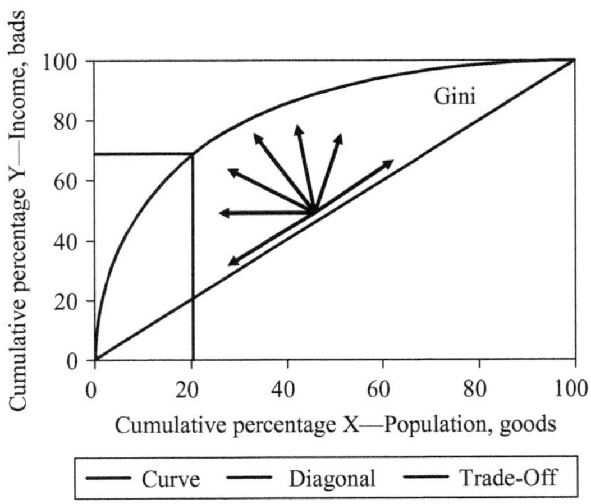

Figure 8.5. Lorenz curve.

20 per cent of the population earns 70 per cent of the income. The space within the bow above the diagonal represents the extent of inequality. Perfect equality lies exactly on the diagonal, and perfect inequality would cover the entire space above (or below) the diagonal.

This same curve is applied in the scoring world, to illustrate a model's ability to separate good and bad accounts, and may be called a *power curve*, *trade-off curve*, *efficiency curve*, or *Receiver Operating Characteristic* curve. Cumulative bads are plotted on one axis, and cumulative goods on another. A model that has no predictive power implies perfect equality, and a model that is perfectly predictive implies perfect inequality.

8.4.4 Gini (rank correlation) coefficient

One of Pareto's contentions was that income inequality would reduce in richer societies. In 1910, Corrado Gini proved him wrong by comparing income inequalities between countries, using what is today known as the Gini coefficient, which is the area between the curve and the diagonal, as a percentage of the area above the diagonal (for interest, the average value for most developed countries is about 40 per cent, and the greatest inequalities are in Brazil and South Africa, with values around 60 per cent). It is calculated as:

Equation 8.9. Gini coefficient $$D = 1 - \sum_{i=1}^{n}((cpY_i - cpY_{i-1})(cpX_i + cpX_{i-1}))$$

where cpY is the cumulative percentage of ranked income, and cpX is the cumulative percentage of people. The result is a rank correlation coefficient, which is exactly the same as the Somer's D statistic provided by many statistical software packages. The Gini coefficient is not used for hypothesis testing, but does provide a powerful measure of separation, or lack of it. Table 8.5 provides a highly simplified example.

The same calculation has been co-opted into a lot of other disciplines, including credit scoring, where it is often referred to as an accuracy ratio or power ratio. The Gini coefficient is used as a measure of how well a scorecard is able to distinguish between goods and bads, by having the bads as P and goods as N as in Table 8.6. The end result is again a value representing the area under the curve (also see Section 8.4.5).

The Gini coefficient does have some sensitivity: (i) it can be exaggerated by increasing the *indeterminate range*; and (ii) it is sensitive to the category definitions, in terms of contents, number, and ordering. As a result, some care must be taken in how it is used and interpreted.

Table 8.5. Income inequality

Income class	Totals		Per capita income	Cum (%)		$cpX_i + cpX_{i-1}$ (%)	$cpY_i - cpY_{i-1}$ (%)	Z (%)
	People	Income (mn)		People	Income			
Rich	1,000	60	60,000	20.0	50.6	20.0	50.6	10.1
Middle	1,500	36	24,000	50.0	81.0	70.0	30.4	21.3
Poor	2,500	22.5	9,000	100.0	100.0	150.0	19.0	28.5
	5,000	118.5	23,700	Gini coefficient = 1 − Sum (Z)				40.1

Table 8.6. Scorecard effectiveness

Score	Outcome		G/B odds	Cum (%)		cpN_i+ cpN_{i-1} (%)	cpP_{i-1} cpP_{i-1} (%)	Z (%)
	Goods	Bads		Goods	Bads			
Low	5,000	2,000	2.5	2.0	33.3	2.0	33.3	0.7
Middle	45,000	2,000	22.5	20.0	66.7	22.0	33.3	7.3
High	200,000	2,000	100.0	100.0	100.0	120.0	33.3	40.0
	250,000	6,000	41.7	Gini coefficient = (1−Sum(Z))				52.0

When measuring scorecard power, the exaggeration associated with a potentially oversized indeterminate range can be avoided by assessing bads versus not bads, possibly even using a stricter bad definition. The same applies to any other correlation measure, when used to rate the final model's predictive power.

The Gini coefficient provides a single value that represents predictive power over the entire range of possible values. There are a lot of instances though, especially in application scoring, where lenders' greater interest is in the model's power around the cut-off. It is thus always wise also to consider some other measure when comparing model results, such as the percentage of bads at different accept rates.

What is an acceptable Gini coefficient? There are no hard and fast rules, and those rules of thumb that do exist vary, depending upon the development type. In most retail application scoring, a Gini coefficient of 50 per cent plus is more than satisfactory, while less than 35 per cent is suspect, and 30 per cent possibly unacceptable. In contrast, for behavioural scoring with a one-year outcome window, Gini coefficients of over 80 per cent are possible, while anything below 60 per cent might raise suspicions. In all cases, these values apply to resulting scorecard sets, and not individual scorecards.

Corrado Gini—biographical sketch

Gini's association with a measure of income disparity might make one fallaciously think his interest was in social welfare, but in truth, Corrado Gini (1884–1965) was a keen fascist theorist who published *The Scientific Basis of Fascism* in 1927, and was the leader of Italy's eugenics movement under Mussolini from 1934. He was born into a family of landed gentry in the Treviso area of Italy. He started studying law at the University of Bologna, but his interests directed him into the social sciences (especially demography, sociology, and economics), and statistics—the latter being used to complement and support other research. He took over the Chair of Statistics at the University of Cagliari in 1910 and at the University of Padua in 1913, and provided several significant contributions to the field of statistics before the end of First World War. His ideas were not, however, well accepted in the statistical arena, because he did not explore their mathematical basis.

In 1920, Gini founded the journal *Metron*[5] (which he directed for the rest of his life), the focus of which was restricted to ideas that could be practically applied. He was active politically, and highly regarded within Italian political circles. In 1923, he moved to the University of Rome, where he later became a professor, founded a sociology course, set up the School of Statistics (1928), and founded the Faculty of Statistical, Demographic, and Actuarial Sciences (1936). In 1926, he became president of the Central Institute of Statistics, and founded the journal *La Vita Economica Italiana*. In 1929, he also founded the Italian Committee for the Study of Population Problems, and in 1934, its journal *Genus*. The committee survived the Second World War and the fall of fascism, primarily due to the quality of its work. Over the following years, he was president of many professional societies, and received several awards before his death in 1965.

8.4.5 Receiver operating characteristic

While many of the statistics used in credit scoring have their origins in the social sciences, one statistic's origin is totally different. The Receiver Operating Characteristic (ROC) was developed in the 1940s, to measure radar operators' ability to distinguish between a true signal and noise. In the 1950s and 1960s, it was adopted in the field of psychology, for the study of behavioural patterns that were barely discernible, and could not be explained using existing theories. Today, the ROC is used widely in medicine, engineering, and other fields—including credit scoring. It falls under the heading of 'signal detection theory', two key concepts of which are: (i) *sensitivity*, ability to mark true positives; and (ii) *specificity*, ability to identify true negatives. Using the example illustrated in both Table 8.2 and Figure 8.1, at a score of 500, the sensitivity and specificity are 83.3 and 64.2 per cent respectively.

Building upon this, the ROC curve is the plot of $X = \Pr[S_{FP} \leq S_{Cut-off}]$ against $Y = \Pr[S_{TP} \leq S_{Cut-off}]$ as the cut-off is varied, where $X=$ sensitivity, the true positive rate, or hit rate; and $Y = 1-$ specificity, which is the false positive rate, or false alarm rate. The resulting curve looks much like that in Figure 8.6, which uses the same data as Figure 8.5 for the Lorenz curve.

In most of the literature, especially for medicine and psychology, the ROC curve is very jagged, as it is usually used in instances where the signal is weak or non-existent. A concave curve above the diagonal occurs only where the likelihood ratio $LR_i = p_i^+/p_i^-$ has a monotonic relationship with the measure being evaluated. If the curve goes below the diagonal, the model is getting it wrong, but a reversal of sign will correct it.

[5] Most of this information was obtained from Metron's web page, http://www.metronjournal.it/storia/ginibio.htm

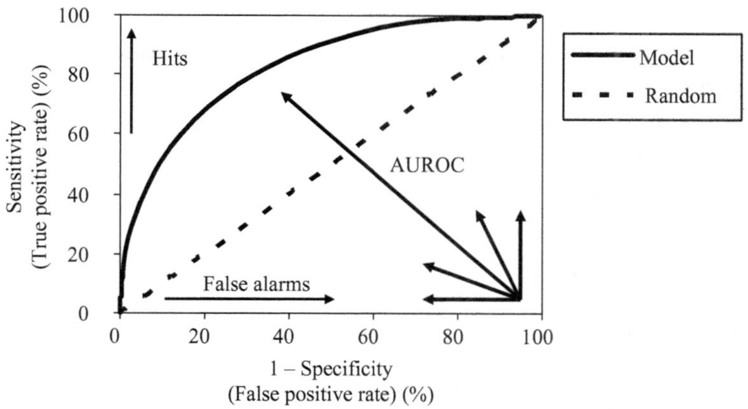

Figure 8.6. ROC curve.

There is also a summary statistic, much like the Gini coefficient, except it provides the percentage of the total area under the ROC curve (AUROC), as opposed to the area above the diagonal. This may either be called AUROC, or the c-statistic. The relationship between it and the Gini coefficient is $c \approx (D+1)/2$; for example, a Gini coefficient of 52 per cent equates approximately to an AUROC of 76 per cent. The formula usually used is:

Equation 8.10. AUROC $c_{P,N} = \Pr[S_{TP} < S_{TN}] + 0.5 \Pr[S_{TP} = S_{TN}]$

In English, it states that the area under the curve is equal to the probability that the rating for a true positive will be less than that for a true negative, plus 50 per cent of the probability that the two ratings will be equal. An AUROC of 50 per cent implies that the model is no better than making a random guess; and a value of 100 would indicate the unlikely occurrence of perfectly correct predictions. Likewise, any value less than 50 per cent implies that the model is getting it wrong with some consistency, and 0 per cent means the predictions are perfectly wrong.

This value can then be used for confidence testing, but the maths are complex and seldom used within credit scoring. With some patience, anybody who is interested will be able to find the formulae on the Internet. One possible source is Engelmann et al. (2003), who use the Mann-Whitney U test, and present several formulae for the variance that differ according to the hypothesis. The simplest is for testing whether a model has any predictive power: $\sigma^2 = (N_D + N_{ND} + 1)/(12N_D N_{ND})$.

As a final note, when measuring scorecard performance: (i) the Lorenz curve and ROC curve are almost exactly the same; and (ii) the values for the Gini coefficient and AUROC are extremely highly correlated. Yet, in spite of Lorenz and Gini being first on the block, they have been supplanted by ROC and AUROC. The reason is that Messrs. Lorenz and Gini were economists, who developed tools for univariate analysis of a ranked variable (income) versus its count, which tools are only by chance being applied more broadly. In contrast, ROC and AUROC were developed by radio-analysts, for bivariate analysis of a ranked signal-detection measure versus a binary outcome, 'Was there a signal or not?' Credit scoring is a form of signal detection mechanism that fits neatly into the latter camp, and as a result, the ROC/AUROC concepts have taken precedence. The Gini coefficient is still often referred to, but as another way of measuring the area under the ROC curve.

8.5 Chi-square (χ^2) tests

You can predict nothing with zero tolerance. You always have a confidence limit, and a broader or narrower band of tolerance.

Dr Werner Karl Heisenberg, German physicist and Nobel laureate (1901–1976)

And now, another ancient Greek character! This time their equivalent of 'x', which is called 'chi', pronounced like 'sky' without the 's'. The chi-square (χ^2) test looks for a linear relationship between two characteristics, and the resulting p-value provides a measure of reliability—the probability that the *similarity (goodness of fit)* or *difference (independence)* between them is not a chance occurrence. It is usually used to evaluate a theory or hypothesis, by comparing *observed (actual)* and *expected (estimated)* distributions. There are several variations of the chi-square calculation, but the original and most commonly used is Pearson's chi-square test:

Equation 8.11. Pearson's chi-square $$\chi^2 = \sum_{i=1}^{n}\left((O_i - E_i)^2 / E_i\right)$$

where O = the observed frequencies and E = the expected frequencies for each class i. Basically, a χ^2 value of zero indicates a perfect fit, and χ^2 increases as the distributions become dissimilar, eventually becoming so high that one can only conclude that the two distributions bear no relationship to each other (independent).

> Expected frequencies should be large. A rule is that no E value should be zero, and no more than 1/5th should have values less than 5 (some people insist on at least 10 or 20), otherwise the test can be considered invalid. Categories can be collapsed to accommodate this. If there are only two categories (d.f. = 1), Yates' correction should be applied, which deducts 0.5 from each $(O-E)$.

The chi-square is then converted into a p-value—a percentage that indicates whether or not the fit is a random occurrence: as χ^2 approaches 0, the p-value approaches 100 per cent; and as χ^2 increases, the p-value approaches zero. The task of converting this into an exact probability is where the test starts getting complicated, if not painful. For those less technically inclined, the rest of this section should be skipped.

The conversion depends upon the degrees of freedom (d.f.), meaning the number of independent pieces of information contained in a statistic. In most instances, the d.f. is calculated as $(n-1-a)$, where n is the number of classes, and 'a' is the number of assumptions, if any, made in the null hypothesis.

What usually happens is that null and alternative hypotheses, H_0 and H_A respectively, are stated of the form:

H_0—The observed distribution fits a certain distribution.

H_A—The observed distribution does not fit a certain distribution.

A test is performed to determine whether the null hypothesis is true at a given threshold significance level (SL); the higher the SL, the less the chance that the null hypothesis will be wrongly rejected. For the above instance, the threshold p-value is equal to the SL, α or $p_{critical}$; but if the hypothesis is turned around, then α is equal to one minus the SL. Nowadays, most spreadsheets and statistical packages are capable of calculating the p-value directly from the two distributions, or alternatively, can calculate the p-value using only the χ^2 and d.f. The null hypothesis is rejected if $p < \alpha$ for goodness of fit, or if $p > \alpha$ for independence.

In the absence of such tools, it is necessary to revert to the dark ages, and the use of tables. The p-value and d.f. are used to select $\chi^2_{critical}$ from a table, like that in Appendix A, and $\chi^2_{critical}$ is then compared against the calculated χ^2 value. The null hypothesis is rejected if $\chi^2 > \chi^2_{critical}$ for goodness of fit, or $\chi^2 < \chi^2_{critical}$ for independence.

Problem: Complaints have been received from the application processing area that high application volumes during certain quarters are affecting service levels. If this is true, then management will have to assign extra resources for peak periods. Application volumes per quarter (000s) have been averaged over several years, and the results are provided in Table 8.7, which yields a χ^2 value of 2.33. Management wishes to ascertain with 80 per cent certainty that these fluctuations imply meaningful differences before doing anything.

H_0—The number of applications is evenly spread over the quarters.
H_A—The number of applications is not evenly spread over the quarters.

From Appendix A, it can be found that $\chi^2_{critical}$ for d.f. = 3 and α = 20% is 4.642, which is more than the calculated χ^2 of 2.33. The hypothesis is rejected, and things are left alone. If the confidence level was reduced to less than 50 per cent (unlikely!), the decision might change.

Table 8.7. Chi-square calc

Group	Actual	Expected	Chi-square
Q1	292	300	0.21
Q2	320	300	1.33
Q3	285	300	0.75
Q4	303	300	0.03
Totals	1200	1200	2.33
N = 4, d.f. = 3			p = 50.74%

These functions are a bit easier to work with, if they can be visualised. Figure 8.7 illustrates the relationship between χ^2 and the associated probabilities, where there are 11 categories, and independence is being tested at an 80 per cent SL ($p = 0.2$). The points to the right are those above the $\chi^2_{critical}$ value of 13.442. As can be seen, the test becomes more demanding as $\chi^2_{critical}$ increases, and p decreases.

A similar pattern exists for other d.f. values, but the shape of the distributions changes. In Figure 8.8, it can be seen that as the number of categories increases, so too does $\chi^2_{critical}$ for each confidence level. Also, where the d.f. is low, the probability distribution is highly skewed to the left; but as it increases, the distribution starts looking like a normal distribution.

In credit scoring, the most obvious uses of the chi-square test are: (i) to measure the drift in score or characteristic distributions over time; and (ii) to measure the differences between the good and bad distributions. In these cases, contingency tables are being compared, and the rules for calculating the d.f. changes. The number of assumptions being made is affected by

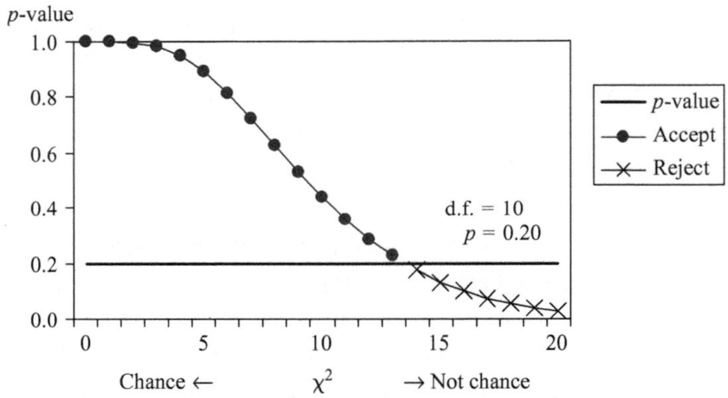

Figure 8.7. Chi-square distribution.

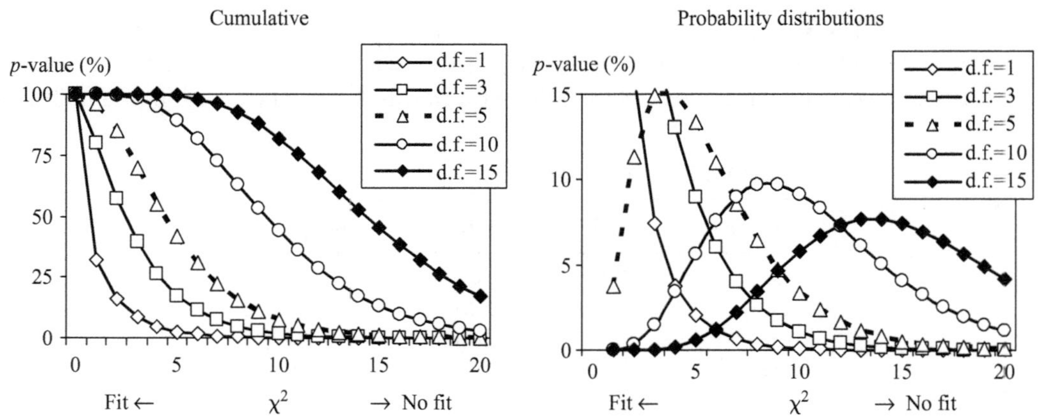

Figure 8.8. Degrees of freedom.

whether or not the row and column totals (expected and actual) have been anchored:

If both	d.f. = $(n_{rows} - 1)*(n_{columns} - 1)$
If rows only	d.f. = $n_{rows}*(n_{columns} - 1)$
If columns only	d.f. = $(n_{rows} - 1)*n_{columns}$
If neither	d.f. = $n_{rows}*n_{columns} - 1$

Most academic works on the use of the chi-square statistic limit themselves to the 'both' formula, since it is the most common case. The following provide various examples of different tests done, using the historical and recent data in Table 8.8:

(a) Differences between historical and current odds (illustrated): estimates are the recent row totals, apportioned according to the historical good/bad odds. At a χ^2 value of 22.9 with d.f. = 4 = 4 * (2 – 1), the differences cannot be considered random at any probability level—$\chi^2_{critical}$ is already 13.3 at a p-value of 1 per cent.
(b) Differences between historical and current good/bad distributions: estimates are the recent total (14,190), apportioned by the historical cell percentages. The resulting χ^2 value is a massive 495.1, indicating that the variations are not random, no matter what the degrees of freedom (which in this instance is 7 = (4 * 2 – 1)).
(c) Characteristic's current predictive power: estimates are the current row totals, apportioned by average good/bad odds (8.5). In this case, χ^2 is 5.5 and d.f. = 3 = (4 – 1)*(2 – 1). This time there is some question about the result, since there is still at least a 10 per cent chance that these differences are random.

While this provides an understanding of the mechanics, software packages can often calculate the p-value, given a specified confidence level. Care must still be taken to ensure that the degrees of freedom used are appropriate for the problem.

There are two particular instances where the chi-square statistic can be used as part of the scorecard development process: *coarse classing* and *variable selection*. In both cases, the calculation is the same as '(c)' above. It can also be used to audit whether a model's scores adequately represent its characteristics' predictive power, which would use approach '(b)'

Table 8.8. Chi-square—good/bad odds test

Group	Historical			Current			Expected			Chi-square	
	Good	Bad	Odds	Good	Bad	Odds	Good	Bad	Odds	Good	Bad
A	2,313	252	9.2	1,805	220	8.2	1,826	199	9.2	0.25	2.01
B	3,773	443	8.5	4,950	592	8.4	4,960	582	8.5	0.02	0.16
C	2,565	304	8.4	2,888	303	9.5	2,853	338	8.4	0.43	4.07
D	2,999	295	10.2	3,051	381	8.0	3,125	307	10.2	1.78	14.23
Totals	11,650	1,294	9.0	12,694	1,496	8.5	12,763	1,427	8.9	2.47	20.48
	12,944			14,190			14,190			22.95	

above.[6] Scores are calibrated onto probability estimates, which are then used to 'fuzzy parcel' (see Section 19.3) each record into good and bad portions (irrespective of what the actual status is) that are then used as expected values.[7] A chi-square test is then done to compare actual versus expected for each characteristic, and low p-values may reflect a problem.

8.6 Accuracy tests

After one has tortured the data until it confesses, get a clean set of data and see if the confession was valid.

cretog8 (a.k.a. *John Ashcroft*), www.everything2.com

Most of the tests that are used in credit scoring test the models' power, or ranking ability. This is entirely appropriate when scorecards are used only for ranking, but they are being used increasingly to provide default probability estimates, whether for pricing, forecasting, or capital allocation purposes. There are also accuracy tests available, which can be applied: (i) in-sample, out-of-sample, or out-of-time; (ii) to the raw or calibrated scores; or (iii) to the entire risk spectrum, or a part thereof. The types of tests covered here are:

(i) **Binomial test**—Used to compare observed and estimated success rates for a single group.
(ii) **Hosmer–Lemeshow test**—Based upon the binomial test, but is applied across groups.
(iii) **Log-likelihood**—Provides measures of both power and accuracy for ungrouped cases.

Please note that these tests must be used with caution. No matter how accurate the scorecard is, the moment it is implemented the business tries to break it! Customers are punted or shunned according to their deemed risk, with high-risk cases receiving collections priority when problems arise. Also, credit is a dynamic environment, influenced by infrastructural, competitive, and economic changes. Many of these tests assume that the phenomena being observed are independent, whereas in truth the outcomes are correlated—both within the population, and over time.

8.6.1 Probability theory

The basic approach used for testing estimates' accuracy is the binomial test, which is applicable only to dichotomous outcomes—bad versus good, default versus not default, bankrupt versus not bankrupt. In order to understand the mathematics behind it, a brief review of probability theory is required. This is the realm of Bernoulli trials, which have three properties: (i) there are only *two possible outcomes*, success and failure, where the label is arbitrary; (ii) there is the *same success probability* for all trials; and (iii) the *results are random*, and each trial is independent of other trials.

---

[6] The audit may be done on all cases, or alternatively only on accepts, to ensure that the reject inference has not adversely affected the model's applicability to the accept population.

[7] It is easiest where the model provides a value that either is, or can be easily converted into, a reliable probability.

Jacob (Jacques) Bernoulli (1654–1705)

Jacob Bernoulli originally studied philosophy and Calvinist theology, and it was some dismay to his parents when he turned to mathematics, but he went on to become Professor of Mathematics at the University of Basel, in 1687. He was the first in his family of mathematicians to become famous, before his younger brother Johann (1667–1748), and his nephews Nicolaus (1687–1759) and Daniel (1700–1782). Both Jacob and Johann were renowned in Europe for their contributions to calculus, yet their relationship in later years was acrimonious. Today, Jacob is even more famous for his *Ars Conjectandi* (The Art of Conjecturing). He had formulated most of the ideas between 1684 and 1689, but because of the work's ambitious scope it was never completed, and was published as an *opus posthumus* by Nicolaus, in 1713.

Part I started with a review of Christiaan Huygen's 1657 tract, 'On Rationalisation in Dice Games', which complemented the theory of equity for pricing gambles, presented by Blaise Pascal in 1665 (see box below) and Jan De Witt in 1671. Parts II and III looked at combanitorics and games of chance respectively. Only Part IV was unfinished, which covered the possible application of the theory of equity to probabilities, and probabilities' practical uses in politics, law, and business. Unfortunately, he could not find the data that would provide him with real life examples outside of gambling. Johann was asked by other academic luminaries to finish and publish the work, but was prevented by Jacob's widow and son, who distrusted his intentions. As a result, some of the mathematical ideas were already obsolete by the time it was published. In 1708, Pierre Rémond de Montmort published the first edition of his *Essay d'analyse sur les jeux de hazard*, and later in 1713, the second edition was published with some of the latest ideas from Montmort, and both Johann and Nicolaus Bernoulli.

Ars Conjectandi was the first substantial work on probability theory, and covered the general theory on permutation and combination, the law of large numbers, and the binomial and multinomial theorems.[8] Such works were novel, in an era when mankind and science were searching for deterministic and mechanistic answers for everything, and believed that chance could only exist because of human ignorance. Even Bernoulli's theorem (law of large numbers) states that certainty can be determined with sufficient trials. Probability theory was not of scientific interest, but was used as a means of pricing uncertain future outcomes (economics, gambling, and contracts).

Pascal's original pricing problem related to the apportionment of monies from an unfinished gamble, but the concept was extended to explain reasonable expectations and behaviour. He eventually returned to the church, and in 1669 he published *Pensées*, which contained three wagers, one of which is today known as Pascal's Wager. People's belief in God was supported through purely logical argument—'If you gain, you gain all; if you lose, you lose nothing. Wager, then, without hesitation that He is'.

The first step in understanding probability theory is to understand factorials, which involves repeated multiplication of an incrementing integer, as shown in Equation 8.12. Note that

[8] Wolfram Research, scienceworld.wolfram.com

factorials only work for non-negative integers and, when working with fractions, only the integer portion is used.

Equation 8.12. Factorial $n! = \begin{cases} 1 & \text{if } n = 0 \\ 1 \times \cdots \times n & \text{if } n > 0, n \in 1, 2, 3, \ldots \end{cases}$

Thus, $2! = 2$, $3! = 6$, $4! = 24$, $5! = 120$, and so on. The increases are exponential, and beyond a value 170! most spreadsheets fall over. Even so, it is sufficient for many problems. This is then used to calculate the number of possible combinations that can be created from a set of unique items, as shown in Equation 8.13:

Equation 8.13. Number of combinations $_nC_k = \binom{n}{k} = \frac{n!}{k!(n-k)!}$

Where C is the number of possible combinations, n is the total number of cases, and k is the number selected. Thus, if there are nine unique items in a box, and three are selected, there are

$$_9C_3 = \binom{9}{3} = 9!/(6!\ 3!) = 3{,}62{,}280 \times 720 \times 6 = 84$$

possible combinations. No matter what the value of n, the possible values of $_nC_k$ will always be highest at, or immediately around, $n/2$; and will appear normally distributed, but lumpy, for small values of n. The probability of any one particular combination is then the inverse, $1/_nC_k$.

8.6.2 Binomial distributions

The number of combinations is only of interest because it is a key component of calculating the binomial distribution, meaning the frequency distribution of successes for any given number of Bernoulli trials, and a success rate estimate. For any given situation, the number of observed successes X will have a binomial distribution, denoted as $B(n, \hat{p})$, where n is the number of trials, and $\hat{p}$ the estimated probability of success. If there are nine items that fall into two categories, with an estimated proportion of 30 per cent for one of them (success), then the counts can range from 0 to 9, with a distribution represented as $B(9, 30\%)$. The probability of observing exactly k successes is then:

Equation 8.14. Binomial probability $\Pr(X = k) = {}_nC_k \times \hat{p}^k \times (1-\hat{p})^{n-k}$

Thus, if the expected success rate is 30 per cent, then the probability of exactly 3 successes out of 9 trials is $\Pr(X = 3) = {}_9C_3 \times 0.33 \times 0.7^{9-3} = 84 \times 0.007412 = 26.68\%$. In similar fashion, the probability of having at most three successes, is the cumulative probability for all integers 0 to 3, or $\Pr(X \leq 3) = \sum_{i=0}^{3} \Pr(X = i) = 72.96\%$.

The binomial test is typically associated with medical trials, where researchers are trying to determine if the observed and expected rates are consistent, especially for rare events. The problem is stated as a null and alternative hypothesis, of the form:

H_0—the observed and expected probabilities are the same, $p = \hat{p}$;

H_A—the observed and expected probabilities are not the same, $p \neq \hat{p}$;

For equality, a two-tailed test is used, where both upper and lower bounds are determined. For a confidence level of 99 per cent, critical values at SLs of both 0.005 and 0.995 are required. If the observed value lies outside the resulting range, then the null hypothesis is rejected. A one-tailed test is used to determine whether the observed value is greater or less than an estimate.

The example in Table 8.9 illustrates not only the two-tailed test, but also the HL statistic (covered next). The goal is to determine a range of observed default rates that would be considered acceptable, given previous estimates and a confidence level. To do this, the minimum number of successes that would provide probabilities at the lower and upper bounds for the significance level are calculated, as per Equation 8.15:

Equation 8.15. Critical binomial $k_a = \min(k \mid \Pr(X \leq k) > \alpha)$

where k_α is the critical value for k, and α is the significance level. For the score range 'Low to 302' in Table 8.9, the number of trials was 2,600 with 600 observed successes (defaults), as compared to an estimated 594 at a rate of 22.87 per cent. At a confidence level of 99 per cent, the confidence interval is between $k_{0.5\%} = 540$ and $k_{99.5\%} = 650$,[9] or alternatively, a bad rate of between 20.77 and 25.00 per cent.

Table 8.9. Binomial accuracy tests

Score	Counts			Bad rate (%)		Boundaries			z-Stat	HL-Stat
	Total	Good	Bad	Obs.	Est.	0.5%	99.5%	Critical		
Low–302	2,600	2,000	600	23.08	22.87	20.77	25.00		0.25	0.06
303–348	4,100	3,500	600	14.63	14.85	13.44	16.29		−0.39	0.15
349–380	5,600	5,000	600	10.71	9.30	8.32	10.32	X	3.63	13.20
381–406	9,600	9,000	600	6.25	5.69	5.09	6.31		2.36	5.59
407–430	18,600	18,000	600	3.23	3.43	3.09	3.77		−1.52	2.30
431–454	22,600	22,000	600	2.65	2.05	1.81	2.29	X	6.47	41.91
455–479	50,600	50,000	600	1.19	1.21	1.09	1.34		−0.57	0.32
480–510	80,600	80,000	600	0.74	0.72	0.64	0.80		0.91	0.83
511–555	140,600	140,000	600	0.43	0.42	0.38	0.47		0.20	0.04
556–High	200,600	200,000	600	0.30	0.25	0.22	0.28	X	4.47	19.94
Total	535,500	529,500	6,000	1.12	1.06	1.02	1.09	X		84.35
		Gini coefficient = 51.5%					z-stat critical = 2.58			

[9] These values were calculated using the CRITBINOM function in Microsoft Excel.

The same calculation was repeated for all of the score ranges, and the default rates were found to be significantly different from the estimates for three of them, as well as the overall estimate. The latter is especially disconcerting, as the difference between 1.06 and 1.12 per cent is small. Remember though, that these tests are conservative, and often fail to recognise the dynamic environments in which lenders operate.

Normal approximation to binomial distribution

While the binomial formula provides exact probabilities, it is unfortunately not computationally feasible for larger values. The alternative is to use the normal approximation, which can be used as long as both the expected number of successes and failures are greater than 10 ($10 \leq n\hat{p} \leq (n - 10)$). Its base assumption is that the probabilities are normally distributed. The number of standard deviations that a value lies away from the mean is represented by the z-statistic, which is usually calculated as $z = (X-\mu)/\sigma$, where X is a value that lies somewhere within the distribution, μ is the mean, and σ is the standard deviation.

In this case: (i) X and μ are the observed and expected number of successes respectively, where $X = k$ and $\mu = n\hat{p}$; and (ii) the variance for a binomial distribution is $\sigma^2 = n\hat{p}(1-\hat{p})$. Thus, the z-statistic calculation changes to:

Equation 8.16. Binomial normal approximation $$z = \frac{k-n\hat{p}}{\sqrt{n\hat{p}\,(1-\hat{p})}}$$

As an example, for the score range '349 to 380' in Table 8.9, there are 600 successes out of 5,600 trials, whereas the estimated success rate of 9.3 would have provided 521.0 successes. The z-statistic is:

$$z = (600 - 5600 \times 9.3\%)/\sqrt{5600 \times 9.3\% \times (1-9.3\%)} = (600 - 521.0)/21.7 = 3.63$$

If the expression $\Phi^{-1}(\alpha)$ is used to denote the inverse standard normal cumulative distribution (NORMSINV function in MSExcel), which is used to provide the critical z-statistics given the required significance levels (α), then for a confidence level of 99 per cent, the threshold values of $\Phi^{-1}(0.005)$ and $\Phi^{-1}(99.5\%)$ are -2.58 and 2.58 respectively. In this instance, the hypothesis that the observed and expected success rates are equal must be rejected.

Rather than testing hypotheses, the task is often to calculate specific probabilities. If $\Phi(z)$ is used to denote the standard normal cumulative distribution function (NORMSDIST function in MSExcel), and $z_{n,\hat{p},k}$ the z-statistic (for a number of trials, success probability estimate, and number of successes), then the probabilities are:

Equality	$\Pr(X = k) = \Phi(z_{n,\hat{p},k+0.5}) - \Phi(z_{n,\hat{p},k-0.5})$
Less than	$\Pr(X < k) = \Phi(z_{n,\hat{p},k-0.5})$
Less than or equal to	$\Pr(X \leq k) = \Phi(z_{n,\hat{p},k+0.5})$
Less than	$\Pr(X > k) = 1 - \Phi(z_{n,\hat{p},k+0.5})$
Less than or equal to	$\Pr(X \geq k) = 1 - \Phi(z_{n,\hat{p},k-0.5})$

Note the ±0.5 adjustments to the observed successes, made because it is a continuous distribution. This adjustment should also have been reflected in Equation 8.16, but its omission has little impact where the number of observed successes is large.

For the case where 50 trials are performed with an estimated success rate of 20 per cent, the probability of exactly 5 successes using the normal approximation is: $\Pr(X=5) = \Phi(z_{50,\ 20\%,\ 4.5}) - \Phi(z_{50,\ 20\%,\ 4.5}) = \Phi(-1.591) - \Phi(-1.945) = 5.58\% - 2.59\% = 2.99\%$ In contrast, if the exact binomial test, as per Equation 8.14, were used, the result would be 2.95 per cent.

8.6.3 Hosmer–Lemeshow statistic

Both the binomial distribution and its normal approximation focus on individual groups or ranges, but an estimate's accuracy can also be tested across the entire risk range. The most commonly known approach is the Hosmer–Lemeshow statistic shown in Equation 8.17.

Equation 8.17. Hosmer–Lemeshow statistic $$\mathrm{HL} = \sum_{k=1}^{g}\left(n_k \times \frac{(p_k - \hat{p}_k)^2}{\hat{p}_k \times (1-\hat{p}_k)}\right) = \sum_{k=1}^{g} z_k^2$$

where k is an index for each group, and g the total number of groups. Note that all the authors did was sum the squared z-statistics for each of the ranges, much like squaring the error terms, only in this instance it is being used like a goodness of fit measure. The hypothesis is:

H_0—The observed and expected probabilities are not the same;

H_A—The observed and expected probabilities are the same;

The resulting values fit a chi-square distribution, but with a d.f. of g–2.

Although the effect is small, it may be wise to enforce the $10 \le n\hat{p} \le (n-10)$ constraint prior to calculating this statistic. For risk grades, those with insufficient numbers are collapsed with their neighbours. For scores, it usually sufficient to split the range into deciles using the number of trials, but it may be problematic for groups where the estimated success rates are very low. The number of successes should instead be used, as has been done in Table 8.9 (p. 215). For that example, the final HL statistic is 84.35, and for the 10 categories the d.f. is 8. For a significance level of 1 per cent, $\chi^2_{\mathrm{Critical}}$ is 20.09, which leads to the conclusion that the estimated and observed rates are inconsistent even at that lax level. Indeed, the hypothesis could not be accepted at any significance level.

It is worth noting that the HL statistic was originally presented in a textbook on logistic regression, as a means of testing the estimates provided by the resulting models. When used to assess model performance on out-of-time data, it should only be used to test recalibration, as in dynamic environments, it is unlikely that the estimates will be reliable to the levels expected by these tests.

8.6.4 Log-likelihood

Model reliability can be broken down into two dimensions, power and accuracy. Power is more important than accuracy, because the latter can be provided after the fact, through calibration. As a result, most of the measures focus solely upon model power. Irrespective, there may be times when lenders wish to compare models in terms of both. One means of doing this is to use a log-likelihood measure.

These measures are usually used in hypothesis testing, in much the same way as chi-square tests, except the focus is to determine the distribution that best fits a known distribution. It also forms the basis for maximum likelihood estimation (MLE). In the current instance, however, the error term needs to be represented in a manner similar to the chi-square statistic. The result shows how far away an expected distribution is from the actual, but has the added advantage that it can work for individual cases—not just a classed distribution. It is shown in Equation 8.18, in a form that can be applied to both positive and negative test results.

Equation 8.18. Total log-likelihood $$\text{TLL} = \sum_{i=1}^{T} \begin{cases} P_i \times \ln(P_i/\hat{P}_i) & \mid P_i \neq 0, \hat{P}_i \neq 0 \\ N_i \times \ln(N/\hat{N}_i) & \mid N_i \neq 0, \hat{N}_i \neq 0 \end{cases}$$

where P is a positive indicator (0 or 1), N a negative indicator $(1-P)$, $\hat{P}$ and $\hat{N}$ are probability estimates, T is the total number of cases being evaluated, and i is the index for the case being evaluated. The results can then be converted into a likelihood figure as:

Equation 8.19. Likelihood $$L = \exp\left(\frac{\text{TLL}}{T/2}\right)$$

The final result is an error term, which indicates the likelihood that the probability estimates are NOT a proper representation of the values; the lower the value, the better the fit. Unfortunately, it provides no indication of whether the error comes from problems with power or accuracy. In order to split it out, likelihoods are calculated for two naïve models; one using the estimates, and another using the actuals. For naïve models, the total log-likelihood formula in Equation 8.18 simplifies to:

Equation 8.20. Naïve TLL $$\text{TLL}_{\text{Naïve}} = P \times \ln\left(\frac{T}{P}\right) + N \times \ln\left(\frac{T}{N}\right)$$

where P, N, and T are the total number of positives, negatives, and records respectively. The value of T will be the same for each naïve model, but P and N will almost always be different for the expected and observed totals.

Equation 8.21. Accuracy $$\text{Accuracy} = 1 - \frac{(L_{\bar{E}} - L_{\bar{O}})}{L_{\bar{E}}}$$

where: $L_{\bar{E}}$ is the likelihood for a naïve model using estimates, $L_{\bar{O}}$ is the same, but for observed values. According to this formula, if the likelihood figures for the two naïve models happen to be equal, the accuracy is 100 per cent, irrespective of the model's power. Note here that for most models this figure should be very high, and the minimum required threshold may be 95 per cent or higher. The counterpoint to this is power, which can be calculated as:

Equation 8.22. Power $$\text{Power} = \frac{L_{\bar{E}} - L_E}{L_{\bar{E}} - 1}$$

Table 8.10. Log likelihood

	Actuals		Estimates		Log Lik
#	P	N	P	N	
1	1	0	0.90	0.10	0.105
2	0	1	0.10	0.90	0.105
3	1	0	0.80	0.20	0.223
4	1	0	0.70	0.30	0.357
5	0	1	0.50	0.50	0.693
6	0	1	0.40	0.60	0.511
7	1	0	0.70	0.30	0.357
8	0	1	0.20	0.80	0.223
9	1	0	0.80	0.20	0.223
Total	5	4	5.10	3.90	2.797
Model		TLL	LL	Lik	Power
Test		2.797	0.622	1.862	70.8%
Naïve		6.185	1.374	3.953	99.9%
Naïve		6.183	1.374	3.951	Accuracy

where $L_{\bar{E}}$ is the likelihood for a naïve model using estimates, and L_E the likelihood for the model being tested.

In this case, power will approach 100 per cent as L_E approaches 1, which represents a perfect model, and will also provide accuracy of 100 per cent. At the other end, power will approach 0 per cent if the same estimate has been applied to every case—as happens with any naïve estimate—and will be negative, where the estimates are tending towards randomness. Please note that this makes no reference to the actual rankings, and as such is not a rank-order correlation coefficient.

An example of this calculation is provided in Table 8.10, for a small group of nine cases. The total log-likelihood for the sample is 2.797, which provides a log-likelihood of 0.622 and likelihood of 1.862. In contrast, the likelihood calculated using the observed average, provides a likelihood of 3.953, and using the estimated average, provides 3.951. The accuracy is 99.9 per cent, which anybody would be comfortable with. In contrast, the power is 70.8 per cent, which means that the model being tested explains 70.8 per cent of what its naïve equivalent cannot. Please note that this approach should be used with caution. While it can be used to compare models, no confidence intervals can be provided for use in hypothesis testing.

8.7 Summary

While Chapter 7 (Predictive Statistics 101) focused on the predictive-modelling techniques used to develop credit scoring models, this chapter has moved on to the mathematical formulae used both as part of the scorecard development process, and to assess the results. Lenders' primary interests are in prediction (high power) and stability (low drift), and as a result there are several power/drift statistics that focus on these aspects. Power measures are used throughout the model development process, including coarse classing, variable selection, segmentation, and final result evaluation. The results can, however, be affected by real or apparent homogeneity, whether it is: (i) an inherent feature of the population; (ii) the result of data deficiencies; (iii) selection process truncation; or (iv) the product of segmentation. In contrast, drift (or divergence) measures are used primarily for post-development validation and post-implementation monitoring, albeit they could also be calculated against a recent sample and used as part of the scorecard development. A further aspect is accuracy, to ensure that the overall probabilities are more or less in line with those expected.

The measures used to assess power and drift are not mutually exclusive, and many can be used for both. The tools presented were: (i) *misclassification matrix*, the most basic tool, a 2×2 contingency table detailing true/false and positive/negative for both predicted and actual, which also provides the 'per cent correctly classified'; (ii) *Kullback divergence measure*, which is based upon the weight of evidence, and used to calculate the information value (power) and stability index (drift); (iii) *Kolmogorov–Smirnov curve* and *statistic*, the former displays two ECDFs in a 'fish-eye graph', and the latter provides the maximum percentage difference; (iv) *correlation coefficients*, including Pearson's product-moment, Spearman's rank-order, Gini coefficient (area between Lorenz curve and diagonal), and AUROC (area under receiver operating characteristic); (v) the *chi-square test*, used to examine contingency tables, where the number of cells affects the d.f.; and (vi) *accuracy tests*, including the chi-square test, binomial test (and its normal approximation), Hosmer–Lemeshow test, and log-likelihood.

Exactly how these measures are used is covered in Module E (Scorecard Development Process). A brief summary can be provided here though (Table 8.11 provides a high-level overview, but is for guidance purposes only, and should not be interpreted strictly). When doing *coarse classing* (binning characteristics in an optimal fashion), *variable selection* (choice of those that will provide value in a predictive model), *segmentation* (determine whether and which separate models are required), and *final performance assessment* (to rate the models' predictiveness out-of-sample and/or out-of-time) the goal is to extract maximum power. The most commonly used measures are: (i) for *predictors*, the information value and chi-square statistic; and (ii) for *scores*, the AUROC, Gini coefficient and KS statistic. Drift assessments, whether of characteristics or the final score, rely mostly on the stability index and chi-square statistic. The latter's advantage is that there are specific confidence thresholds, which do not exist for the stability index. When assessing the stability of the final score, both the stability index and KS statistic may be used, and there are guidelines for each.

The Gini coefficient—also called an accuracy ratio, Somer's D, or power statistic—is widely referred to, and often suggested for broader use, but assumes a rank ordering. As a result, it cannot be used to assess non-monotonic (especially categorical) characteristics, which applies to many of the characteristics used in retail credit. It is also not possible to use it for hypothesis testing. Its primary use is to assess the rank ordering of the final score or grade. The KS statistic is also commonly used for that purpose, and can be used for statistical tests, but unfortunately focuses upon a single point in the score range. In either case, care must be taken, because too heavy a focus upon a specific measure may lead to an overfitted model, and poor

Table 8.11. Use of statistical measures

	Predictive power		Stability	
	Predictors	Scores	Predictors	Scores
Chi-square	✓		✓	
Kullback divergence	✓		✓	✓
AUROC/Gini coefficient		✓		
KS statistic		✓	✓	

out-of-sample performance. Hence it is wise to use more than one measure, and perhaps also data-visualisation tools, like the misclassification graph or strategy curve.

Other measures were covered, but they tend to have more specific uses. First, *Spearman's rank-order correlation* is used primarily to assess the differences between different scores or grades calculated, or available, for the same set of cases (often for benchmarking internal versus external grades). The *chi-square test* is used to assess both power and drift, through a comparison of contingency tables, and is heavily influenced by the number of classes (d.f.). The *binomial test* is used to assess predictive accuracy for a single group, with a binary outcome. An extension of its normal approximation is the *Hosmer–Lemeshow statistic*, which can be used to assess the full model. And finally, the *log-likelihood* calculation forms the basis of MLE and logistic regression, but can also be used to assess both power and accuracy at the same time. Unfortunately, it is not possible to use it for hypothesis testing.

9 Odds and ends

This section is used for topics that do not fit neatly elsewhere in this module. These are treated under three headings: *descriptive statistics*—covering cluster analysis and factor analysis; *forecasting*—covering Markov chains and survival analysis; and *concepts*—which include correlations, interactions, monotonicity, and normalization.

Credit scoring is almost exclusively associated with predictive modelling, but there are other techniques that are sometimes encountered. Descriptive techniques are used to describe the data, and are sometimes used as part of the scorecard development process, either to gain a better understanding of the data, or to create new variables. In contrast, forecasting techniques are used to do modelling at the portfolio level, to aid in various finance functions. Both categories each have two main techniques associated with them, as shown in Table 9.1, which are covered next.

The final part of this section looks at various statistical concepts encountered during the scorecard development process: *correlations*—extent to which variables move in tandem; *interactions*—instances where the relationships between variables change because of a change in another variable; *monotonicity*—where the value for one variable increases or decreases consistently across the range of possible values for another variable, even if at different rates; and *normalisation*—bringing values into line with a standard.

9.1 Descriptive modelling techniques

Descriptive statistical techniques are used to gain a better understanding of data, and for data reduction. Two techniques are commonly encountered: (i) *cluster analysis*, which identifies groupings of records that share common traits; and (ii) *factor analysis*, which identifies groupings of correlated characteristics. If thought of in terms of a spreadsheet, with observations as rows and characteristics as columns, cluster analysis and factor analysis provide summaries of the rows and columns respectively. Factor analysis is sometimes used in credit scoring, as part of the characteristic selection process, whereas cluster analysis is seldom encountered except where it has been used to create lifestyle indicators.

Table 9.1. Descriptive models and forecasting tools

Function	Technique	One-line overview
Description	Cluster analysis	Finds record clusters with similar characteristics.
	Factor analysis	Creates uncorrelated factors comprised of correlated variables.
Forecasting	Markov chains	Analysis of changes in states, using transition matrices
	Survival analysis	Analysis of mortality rates over extended periods.

9.1.1 Cluster analysis

Cluster analysis is used to identify groups of cases that share common features, by defining homogenous clusters such that the distances within a cluster are as small as possible, and distances between clusters are as great as possible (if this seems familiar, the same concept is used in recursive partitioning algorithms, discriminant analysis (DA), K-nearest neighbours (KNN), and use of various measures for coarse classing). Perhaps the main example is lifestyle codes, characteristics that are generated using a person's residential address, and perhaps even just the postal code, if it is sufficiently detailed (UK and Canada).

Lifestyle codes are based upon the 'birds of a feather flock together' principle, and are derived using census and/or property register data. Cluster analysis looks for similar patterns in household income, marital status, number of dependants, accommodation status (own, rent, board), accommodation type (house, flat, trailer), property value and age, number of bedrooms, etc. The end result is a series of mutually exclusive clusters. For example, cluster 1 might be (on average) high income, living in properties more than 30 years old, worth more than one million; group 2, less than 35 years old, with two dependants, and living in a house; group 3, people with three or more dependants, in low-rent flats; and so on. Each of these would then be given a label like 'old money', 'happy families', and 'urban decay', and the rule sets would then be applied to existing and new data. These codes will often appear in credit scoring models, as they contain information that is predictive, and uncorrelated with other available data.

The task of obtaining data and identifying clusters is not easy. Where lifestyle codes are used, the task of developing them, and ensuring that they are kept up-to-date, is usually outsourced. The end deliverable is a mapping table or algorithm, based on the postal code and/or address details, used to determine the cluster to which each address belongs. Assignments are usually quite stable, and will only be revisited when new census data becomes available.

9.1.2 Factor analysis

> . . . additional information adds linearly to one's confidence, even though after a certain point the inflated standard errors from collinearity worsen out of sample prediction. It may seem prudent to 'look at everything', and this definitely allows better ex-post anecdotal explanation, but it is a statistical fact that additional information, after a certain point, just adds confusion in the form of worse predictability.
>
> *Falkenstein et al.* (2000)

Factor analysis is a descriptive multivariate statistical technique, used to analyse a matrix of correlation coefficients in order to derive a set of latent composite characteristics—or 'factors'—that are uncorrelated with each other. When considered together, the resulting factors describe the dataset with a *minimal loss of information* (in an example provided by the Napier Business School in Edinburgh, an analysis of 10 questions from a survey on soft drinks provided three factors with an information loss of only 15.1 per cent).

> Factor analysis was first proposed by Charles Spearman in 1904. He used it to analyse boys' test scores, and came up with his *two-factor theory* of test results: *general intelligence* (abbreviated as 'g'), and something specific to the test. More recently, general intelligence is being spoken of in terms of rational, practical, emotional, spiritual, and other factors.

In predictive modelling it is used as a data reduction technique, in order to address potential multicollinearity, but rather than using the factors directly, scorecard developers will instead choose one or two of the underlying characteristics to represent each factor. Two primary approaches are used: *principal-component analysis*, which is the most well known; and *common factor analysis*. They could be used as part of any predictive modelling, but most uses in credit scoring deal with the assessment of financial ratios, due to the large number of characteristics that can be generated from a set of financial statements. Mays (2004) and Siddiqi (2006) each refer to it only briefly with respect to retail credit scoring. The topic is covered in much greater detail in Section 17.3.3, as part of characteristic selection during the scorecard development process.

9.2 Forecasting tools

There are a couple of tools used in forecasting that must be covered: Markov chains and survival analysis. Although some academics have proposed them as possible credit scoring techniques, they have not been used in practice. Instead, they are widely used for pricing, provisioning, and capital-allocation purposes.

9.2.1 Markov chains

Credit scoring uses snapshots of historical information, observation, and outcome, to develop a risk-ranking tool. For behavioural-risk scoring, most models will use a one-year outcome. To then determine the probability of an account going bad, or defaulting, over any given period, historical information is again used to work out the rates. But what if percentages are needed within, or beyond, the one-year period, or whatever period was used for the scorecard?

Before the advent of behavioural scoring, the primary indicator of default probabilities was the past-due status, perhaps combined with other attributes. These could quite easily be modelled using a Markov chain, which allows the business to predict the future distribution, using only the current distribution and a transition matrix indicating the expected movements between states. Once again, the origins of this tool lie far outside the realm of business, as illustrated below:

Andrei Andreyevich Markov (1856–1922)

Andrei Markov was a mathematics professor in St Petersburg from 1893 to 1905. According to Basharin et al. (1989), he continued his work on large number and probability theory after his retirement, and published a series of papers from 1906 to 1913. His 1907 paper presented the general concept of a chain, and the 1913 paper provided its now famous first application; Markov had a keen interest in poetry, and did a study of the sequence of 20,000 letters in A.S. Pushkin's poem 'Eugeny Onegin' to determine the distribution of vowels and consonants. Also in

1913, the 3rd edition of his book *Calculus of Probability* was published, which included both the full 1907 paper and the illustration. It was not until 1926, however, that the term 'Markov chain' was used for the first time by S.N. Bernstein. Surprisingly, prior to his death in 1922, Markov found few uses for his own brainchild, yet it has since found applications in physics, biology, linguistics, economics, engineering, medicine, and elsewhere (Thomas et al. (2001) make specific mention of road maintenance, bridge repair, and health care expenses).

The matrices will always have certain properties: (i) the number of possible states is both comprehensive and finite; (ii) the matrices are always square, with the same states along each axis; (iii) the cells will all have values between 0 and 1; (iv) cells with values of 1 are exit/absorption states that cases enter, never to return; and (v) the total of the 'from' cells will always equal 1. They are similar to behavioural scorecards, in that they: (i) are based on probabilities; (ii) are derived from an analysis of historical information; and (iii) rely upon both observations and outcomes. The primary differences are that they: (i) can have more than two possible outcomes; (ii) are not applied to individual accounts, but to groups of accounts with similar characteristics; and (iii) can also be applied to monetary values.

In the simple representation in Figure 9.1 there are three states, with no exit state. The current distribution is stage 0, each subsequent distribution is calculated as part of a sequence, and the nine links between each stage represent the transition matrix.

Table 9.2 shows a situation sometimes used for illustration—the change in the electorate's voting patterns over time, whether at individual, constituency, or any other level. The first table provides the transition probabilities between Conservative/Republican (C), Liberal/Democrat (L), and Independent (I), from one election to the next, and the second shows the

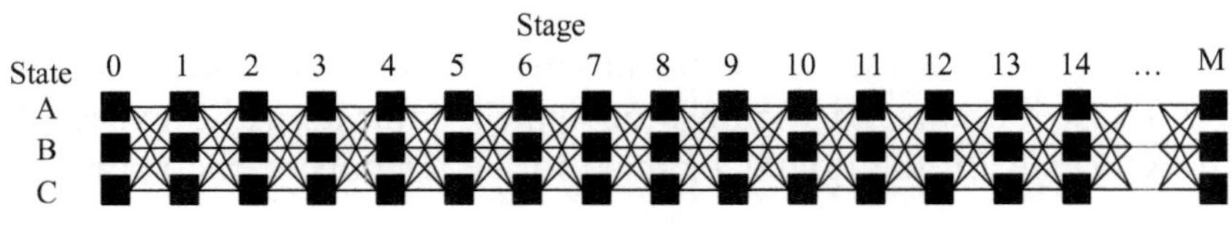

Figure 9.1. Markov chain illustration.

Table 9.2. Transition probabilities

		From T_n		
	State	C (%)	L (%)	I (%)
	C	70	20	20
To	L	20	60	40
T_{n+1}	I	10	20	40
	Total	100	100	100

Table 9.2. continued

State	Year												
	0	1	2	3	4	5	6	7	8	9	10	11	12
C	1,000	700	550	475	438	419	409	405	402	401	400	400	400
L		200	300	350	375	388	394	397	398	399	400	400	400
I		100	150	175	188	194	197	198	199	200	200	200	200
Total		1,000	→										

transitions for an assumed 1,000 individuals who currently vote Conservative. It is assumed that 70, 20, and 10 per cent will vote for Conservative, Liberal, and Independent respectively.

To calculate the distribution after one period is easy. But what about two or more? This is done through repeated application of the transition matrix, which can be expressed mathematically as:

Equation 9.1. Matrix multiplication $\pi_m = \pi_0 * \prod_{i=1}^{m} P_i$

where π_m is the distribution after m periods, π_0 is the current distribution, and P is the transition matrix. If the same matrix is used for every period, which is the norm, then $\Pi_{i=1}^{m} P_i = P^m$.

The Π symbol indicates repeated multiplication, as opposed to 'Σ', which indicates repeated addition. A similar concept is used when discounting future cash flows. The discount factor is $d = 1/\Pi_{i=1}^{t} (1+r_i)$, which converts to $d = 1/(1+r)^t$ if the same interest rate is used in every period.

The calculation of the individual cells is a bit tricky to express, but for the example in Table 9.2 could be stated as:

$$S_{i+1,k} = S_{i,\text{"C"}} \times p_{\text{"C"},k} + S_{i,\text{"L"}} \times p_{\text{"L"},k} + S_{i,\text{"I"}} \times p_{\text{"I"},k}$$

where, $S_{i,k}$ is a count in state k at time i, which may be the number of cases or some other measure; and $p_{i,k}$ is the expected percentage that will move from state j to state k, which is calculated as $a_{j,k}/a_j$, based upon an analysis of one or more historical periods.

In matrix mathematics, the notation $x_{i,j}$ refers to a specific cell in matrix 'x', that is in the 'ith' row and 'jth' column. The rows and columns of the transition matrix in Table 9.2 have been transposed, so that the example can fit on the page; the notation in the equations is correct.

By year two, the voting pattern would be:

$$S_{2,\text{"C"}} = 700 \times 70\% + 200 \times 20\% + 100 \times 20\% = 490 + 40 + 20 = 550$$

$$S_{2,"L"} = 700 \times 20\% + 200 \times 60\% + 100 \times 40\% = 140 + 120 + 40 = 300$$

$$S_{2,"I"} = 700 \times 10\% + 200 \times 20\% + 100 \times 40\% = 70 + 40 + 40 = 150$$

This looks quite complex, but is relatively simple ('relatively' being the key word). For each successive period i, the distribution can be determined by applying Equation 9.2 to each state k:

Equation 9.2. Transition cell calculation $$s_{i+1,k} = \sum_{j=1}^{m}(s_{i,j} \times p_{j,k})$$

where $s_{i+1,k}$ = the number of cases in state k in the next period.
$s_{i,j}$ = the number of cases in each state in the current period.
$p_{j,k}$ = the probability that a case in state j will move into state k.
i = current period; k = state being evaluated; m = number of states.

As can be seen, after 11 periods, the voting patterns of voters that voted Conservative in the latest election reaches a 'steady state' of 40, 40, and 20 per cent, for Conservatives, Liberals, and Independents respectively. At first glance, this seems extremely odd, but if there are enough periods: (i) every distribution will find its steady state; and (ii) for a transition matrix with no absorption states, the steady state will be the same irrespective of the initial distribution.

According to Jafry and Schuerman (2003), the amount of time required for a matrix to decay within a given percentage of the steady state can be calculated. For a two-state matrix, where $p = \begin{pmatrix} 1-p_1 & p_1 \\ p_2 & 1-p_2 \end{pmatrix}$, the calculation is $k_x = \log(x)/\log(|-p_1-p_2|)$. Thus, if p_1 and p_2 are both 5 per cent, it will take $k_{10\%} = \log(.1)/\log(.9) = 21.85$ periods to come within 10 per cent of the steady state.

Ideally, the end result would provide a model that is 'memoryless' (the Markov property), meaning that the transition matrix contains all of the information required to provide a reasonable estimate of the future, using information only about the present, and not the past. If the Markov property holds true, then the predicted future distribution should equal, or at least approximate, the actual distribution. If the property is not sufficiently strong, it can be improved by changing the segmentation.

The goal of segmentation is to improve how well a model represents the dynamics of the population. Thomas et al. (2001) state the problem, as being 'to define a set of subpopulations $r \in R$ and states S^r for each subpopulation, r, so the process for each subpopulation is Markov'.

Assuming that there is appropriate data, and the matrix is not too complex, it is fairly easy to experiment and assess the impact of changes, to ensure that this is true, or nearly true. The Markov property is elusive though, especially when statistical tests are applied, and the matrices can become very complex very quickly:

(i) The number of possible states can become very large.
(ii) Second, or even third or fourth order states may be defined, that use movements not over a single period, but two, three, or four periods, in a single matrix.
(iii) Separate matrices can be defined: (i) to accommodate seasonality; or (ii) for different subgroups and migrations between groups.
(iv) Different measures may be used in each, like the number of cases versus monetary value.

This is not an exhaustive list, but provides an indication of the potential complexity. Also, note that when the number of states is high, many of them will be sparsely populated, and the resulting probabilities will be unreliable.

Thomas et al. (2001) refer to a χ^2 test for Markovity, proposed in Anderson and Goodman (1957). This compares the transitions of a→j→k to b→j→k, and if the Markov property holds, the j→k transitions would be the same for all values of a and b. In most credit scoring cases, the χ^2 values will fall outside of the acceptable range of values. This may, however, be lessened by using a second-order matrix that combines the states from two periods. It will be large, and many of the cells will be empty, because the transitions are impossible; yet 'it is surprising how often this second-order state system is almost Markov'. Opinions regarding the benefit of doing higher-order modelling differ, but seem to indicate that little benefit is provided by third- and fourth-order states.

That was the academic part. The obvious question now is, 'How does this relate to the credit environment'? Cyert et al. (1962) first proposed the use of Markov chains in the consumer credit environment, using monetary values. According to Thomas et al. (2001) though, even though they have been suggested as an alternative to behavioural scores, there 'have been few commercial systems based on the ideas'. Instead, Markov chains are widely used for doing bad-debt provisioning and forecasting, in two environments:

Account level—Focuses primarily upon movements between arrears statuses, usually over periods of one or three months. It may also take into consideration credit scores, account age, outstanding balances, or other factors. The uses are for bad-debt provisioning, and for estimating resource-allocation requirements in collections and recoveries. Profit modelling has been suggested, but this is not widely done in practice.

Enterprise level—Focuses upon annual movements between risk grades assigned to businesses. The grades may be provided by rating agencies, or derived by lenders internally, and are used not only for provisioning, but also pricing, risk management, and portfolio valuation.

More information is provided about the practical use of Markov chains in: (i) Section 6.5.1, on the historical analysis of credit ratings; and (ii) Section 25.1, on portfolio analysis reporting.

9.2.2 Survival analysis

Another tool used in credit scoring, that has also been borrowed from another discipline, is survival analysis, which is used in fields like life insurance (human mortality), engineering (component failure), and medicine (malady incidence). It is similar to Markov chains, except the sole focus is whether cases stay within the system, and do not go into an exit state. A population is segmented into groups, where survival rates are known to vary, and rates are determined for each at different points in time.

The end result is a survival (distribution) function, whose calculation is illustrated by Equation 9.3. It effectively determines the probabilities that the life of a unit, (T), will be greater than the stated time period (t), which is the ratio of surviving units (S_t) to the starting population (S_0). This equals the repeated product of one less the hazard rate (λ_t), for each year.

Equation 9.3. Survival function $$s_t = \Pr(T > t) = \frac{S_t}{S_o} = \prod_{n=1}^{t}(1-\lambda_n)$$

When assessing risk grades, survival functions are needed for loans/companies of different credit qualities, as illustrated in Table 9.3. These values are typical of historical defaults, and may be smoothed to be more meaningful when used for future projections.

From these figures, it is possible to calculate the average hazard rate over any period, using the formula shown in Equation 9.4; $\lambda_{t,t+\Delta t}$ is the probability of unit failure between periods t and $t+\Delta t$, given that failure has not yet occurred. It is not possible to do any analysis beyond the furthest observation point, at which point data is considered to have been 'censored'.

In the life insurance industry, there is some academic research being undertaken to derive the hazard function for those with ultra-long lives, like past 100 years of age. The analysis is, of course, complicated by a lack of data.

Table 9.3. Credit quality survival function (S&P)[1]

Year	AAA	AA	A	BBB	BB	B	CCC
0	1.0000	1.0000	1.0000	1.0000	1.0000	1.0000	1.0000
1	1.0000	0.9999	0.9996	0.9978	0.9902	0.9470	0.7806
2	1.0000	0.9996	0.9989	0.9950	0.9703	0.8872	0.7075
3	0.9997	0.9991	0.9981	0.9921	0.9465	0.8412	0.6563
4	0.9994	0.9984	0.9968	0.9870	0.9256	0.8090	0.6176
5	0.9990	0.9975	0.9951	0.9820	0.9078	0.7856	0.5787
6	0.9982	0.9963	0.9935	0.9771	0.8889	0.7680	0.5638
7	0.9974	0.9947	0.9917	0.9727	0.8773	0.7523	0.5560
8	0.9960	0.9937	0.9899	0.9690	0.8665	0.7399	0.5518
9	0.9955	0.9930	0.9879	0.9661	0.8571	0.7301	0.5426
10	0.9949	0.9921	0.9859	0.9632	0.8500	0.7212	0.5347

[1] The information was provided by Standard & Poor, and quoted in Galil (2003).

Equation 9.4. Hazard function $\lambda_{t,t+\Delta t} = 1 - \left(\frac{S_{t+\Delta t}}{S_t}\right)^{\left(\frac{1}{\Delta t}\right)}$

From Table 9.3, the survival rates for 'B' grade customers in years four and eight were 82.71 and 72.54 per cent respectively. Using the formula, it is found that the average cumulative compounded annual default rate was 3.33 per cent per year. From this can also be derived an instantaneous rate of default, or default intensity, which is the probability of default at time t, assuming the loan has survived to time t, and Δt (change in t) approaches zero.

This example is relatively simple, and focuses solely upon corporate risk grades. Survival analysis can also be used in other types of forecasting, in particular loss and profitability forecasting—inclusive of recoveries, recoveries costs, risk mitigation, and so on. Forecasting can take survival into consideration, not only from the credit risk point of view, but also account attrition. Two or more survival functions could be used as part of a single forecasting model—one for each exit state. The challenge is to determine what these survival functions are.

9.3 Other concepts

When dealing with data, there are a lot of concepts that one comes across, and assumptions are often made about people's level of understanding. The following sections try to clarify several topics that are often encountered:

Correlation—Extent to which variables vary in tandem with one another. In predictive models, correlations between predictors can result in multicollinearity, an increase in the standard error and, potentially, a wrong-sign problem.

Interaction—Instance where the relationship between a predictor and the response function varies depending upon the value of another variable. It can be addressed using generated characteristics, and possibly separate scorecards.

Monotonicity—Instance where the relationship between two factors always moves in the same direction, even if at different rates. It is usually expected of bivariate relationships between the response variable and numeric predictors and, especially, the final score.

Normalisation—To bring into line with a standard! It may be done by conversion to z-scores, rescaling, partitioning, or using ratios to standardise for size, or some reference value.

9.3.1 Correlations

Characteristics are correlated if they vary in tandem with one another. For example, time at address and time at employer are highly correlated, because many people move house when they change jobs. Likewise, a correlation between 'Occupation = Student' and 'age < 22' is expected. In credit scoring, statistical techniques are used to identify the attributes that best explain the difference between a good and bad customer. If, say, students are identified as being higher risks, then it is likely either to ignore, or provide fewer points to, 'age < 22', as the two overlap.

When using regression techniques, high levels of correlation (collinearity) can cause problems. Adding new characteristics increases a model's fit, and improves its ability to explain the past, but simultaneously increases the standard error, and thus decreases the predictive value of the model. This is sometimes referred to as 'factor overload', and can be avoided by ensuring that each term makes a significant contribution to the model's predictive power.

Scorecard developers must also beware of the 'wrong-sign problem'. Where two variables are correlated with the target variable, and with each other, then there is a distinct possibility that the weaker of the two will also be included in the model, but with a sign opposite to that expected, while the coefficient of the stronger variable is exaggerated. It may seem trivial statistically, but affects the robustness of the model, and can lead to wrong decisions that are very embarrassing to explain to management. As a result, wrong signs must be identified and addressed, usually by removing the offending variable from the candidate list.

9.3.2 Interactions

In contrast, interactions arise when different predictive patterns exist for different subgroups within the population. The picture presented by two characteristics viewed in isolation, and in combination, can be totally different.[2] Two examples are:

Age and residential status—Homeowners are usually better risks, but this can present financial burdens for young applicants. Likewise, living with parents is often positive for the young group, but negative for older applicants. If an applicant is 35 and still living with parents, then questions about future prospects arise.

Marital status and number of dependants—Married couples with children are normally better than average risks, but this changes for singles.[3] A simple set of possible combinations are: married and none, married and some, single and none, single and some. Single might also include divorced and separated, if analysis shows they are similar.

The example in Table 9.4 illustrates the good/bad odds for different combinations of age and residential status, in a hypothetical sample. The interaction is highly pronounced, because the predictive pattern changes between the young and old groups. If a single regression model,

Table 9.4. Interactions

G/B Odds Age	Residential status			
	Own	Rent	LWP	Total
Young	10.0	12.5	18.0	12.6
Old	20.0	14.3	10.0	14.3
Combined	14.3	13.2	12.7	13.5

[2] Falkenstein et al. (2000) use the term 'conditionality' for the same concept.
[3] Marital status is an illegal characteristic in the United States, and may be contentious elsewhere.

$y = a + (b_{age}x_{age}) + (b_{res}x_{res})$, were developed to represent the risk, it would rate the old and owner groups as better risks, and the young and live with parents (LWP) as worse risks, and ignore the interaction. In order to address it, the lender would either have to: (i) generate another characteristic for the different age and residential status combinations (Young and Own, Old and LWP, etc.); or (ii) develop separate models for each group.

In either case, scorecard developments require foreknowledge of the interactions, or significant effort to identify them. Where the delivery systems allow the use of generated characteristics, scorecard developers can bring 'domain knowledge' into the development process. Some scorecard developers maintain this is a 'skill'—or even an 'art'—that separates the best from the mechanical modellers. In contrast, non-parametric methods (covered in Section 7.3), such as classification trees and neural networks (NNs), are well suited to identifying interactions, and dealing with them. Some scorecard developers will use classification trees to identify the interactions, and then use generated characteristics to model them in a regression (Thomas et al. 2002).

9.3.3 Monotonicity

Characteristics like 'Age', 'Time at Address', and 'Time at Employer', are numeric variables, but it is unwise to use their raw values. Regression formulae assume that there is a linear relationship between the predictor variables and the response function, but this is seldom the case. Indeed, in many instances the relationship is not even monotonic, let alone linear, where 'monotonic' means that it increases or decreases, and never changes direction.

In credit scoring, many of the relationships are non-monotonic, and the scorecard developer has to decide whether this should be recognised in the model—the answer to which is usually a firm 'No'! An example is 'Customer Age', as illustrated in Figure 9.2. One might expect the risk to reduce as applicants become older, but this is not the case: as people get to their late 20s, their incomes increase, and they do not yet have the expenses; into the 30s, they are having

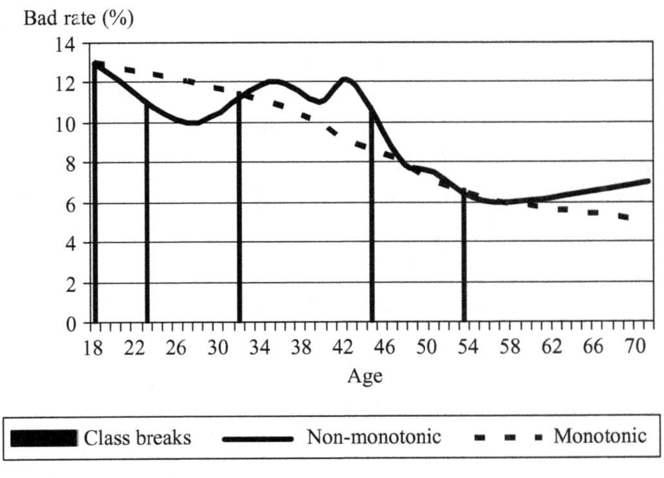

Figure 9.2. Monotonicity/classing.

babies and buying houses; by the late 30s, careers have been firmly established and kids are growing; in the 40s, parents are hit with the cost of university, or buying the kid a car; in the 50s, the children are less of a burden, and people have more money to spend; and finally, into the 60s and beyond, there might be a slight increase in risk, because some people have not provided for retirement.

Although the general pattern of risk reducing with age is recognised, lenders are loath to recognise the minor variations. To address it, a monotonic pattern is forced by the coarse classing, to ensure that the average point allocation increases with age.

9.3.4 Normalisation

One of the problems encountered in any type of data analysis is that comparison can be difficult, because of the various influences at play. It is, however, possible to normalise data, meaning 'to bring into line with a standard', so that comparisons are easier. There are a number of different approaches (also see Saisana 2004), as set out below:

Truncation—Removal of cases that are considered abnormal from the dataset. In credit scoring, this would occur for cases that are not typical of normal business, especially where they fall outside of the control of that business unit.

Ranking—Determining the relative position within a ranked dataset, which is often provided relative to the total number of cases. It loses information contained in the differences between the raw values, but is simple, and eliminates the effect of outliers.

Rescale by range—Raw value less the minimum for a group, divided by the range. This provides a value between 0 and 1. It recognises the relative differences in the raw values that would be missed by ranking, but may be adversely affected by outliers.

z-score standardisation—Calculated as the raw value less the mean value, divided by the standard deviation. The end result is a new variable, with a mean of zero, and standard deviation of one. This often requires some data manipulation beforehand, either truncating extreme values, or transforming the functional form, to provide a normal or near normal distribution (e.g. raising to a power or taking a root, to adjust skewing).

Partitioning—Splitting characteristics into various bands or groups, and assigning a new value according to the group. The new values may: (i) provide a simple expression of rank (-1, 0, $+1$); (ii) be a reference value obtained elsewhere; or (iii) be the result of a calculation, whose result is specific to that group.

For local means—Calculated as the raw value, times the mean for the sample, divided by the mean for the group. This is used to adjust for local abnormalities, such as when there are known differences over time (seasonality, maturity, trends), or across groups (industries, market segments, geographical region). Once the data is normalised, it is easier to compare across groups.

For reference means—Calculated as the raw value, divided by a reference value that has been supplied. This may be a historical average, the most recent value, or a target value.

For size—Refers to any instance, where ratios are used to aid comparison, whether across entities of different sizes, or over time in inflationary environments. This applies especially to currency values, and ratios like the 'assets to liabilities ratio' or 'return on assets'. In financial ratio scoring (FRS) models, one characteristic will usually be left to represent size, such as total assets or revenue.

Falkenstein et al. (2000) make the distinction between: (i) levels, being the value at a point in time; and (ii) trends, being the change in those values over time. Data on levels is always much more predictive. Trends will be secondary within the models, and possibly not feature, but lenders will still see them in the changing credit quality of the book over time, as represented by the aggregate scores or grades.

While this list is quite comprehensive, there are others that have not been mentioned. The key point is that data often has to be normalised, in order to make sense of it and use it. The most common cases are: (i) the use of *ratios*, to normalise for size; (ii) the use of *z-scores*, in instances where the amount of data is limited—in particular for financial ratio analysis; (iii) the use of a *weight of evidence*, to represent the risk associated with each characteristic; and (iv) the use of *dummy variables*, either for categorical characteristics, or instances where there is sufficient data to justify it.

9.4 Basic scorecard development reports

Up until now, the focus has largely been on statistical techniques, and calculations that can be done at the touch of a button. Credit scoring also relies upon various reports to monitor the process, whether as part of: (i) the scorecard development; or (ii) scorecards' use within the business. There is a significant overlap between the reports used in the two areas, but this section looks only at the former, and leaves the latter for Chapter 25, Monitoring. The reports covered here are:

Characteristic analysis report—Provides details about the relationship between predictor variables' attributes and the target variable, and is used as by the scorecard developer to make decisions regarding the appropriate binning.

Score distribution report—Provides detail about the score and its relationship with the target variable, whether as counts, a bad rate, or good/bad odds.

New business strategy table—Provides details of trade-offs between accept and bad rates at different cut-off scores, which is used to assess cut-off strategies.

9.4.1 Characteristic analysis report for scorecard development

During the scorecard development process, the characteristic analysis (CA) report is primarily a binning tool, which is also used to make sure the *point allocations* make sense. It can take on a number of guises, depending upon the function being performed, with the focus always being upon the distribution of different characteristics in the development sample, by: (i) the *decision* that was made (for selection processes); (ii) the *subsequent performance* of the account; and (iii) the *points* or *coefficients* being allocated.

Report formats will vary, depending upon: (i) the scorecard development methodology being used; (ii) type of development; (iii) stage in the development process; and/or (iv) preferences of the scorecard developer. With standardised software packages, templates may be fixed, or have limited facilities for tailoring. In contrast, many scorecard developers prefer packages that allow tailoring to their own preferences. Extra columns may be included for 'Not Taken Ups' and/or 'Inactive' accounts, and there may be separate sections for 'Accepts' (known), 'Rejects' (inferred), and 'Total' (known plus inferred). Other details may also be included to aid the analysis, such as: (i) a comparison of the development sample distribution versus a recent sample, to assess potential drift; and/or (ii) summary statistics, to indicate power or drift.

The example in Table 9.5 summarises the final results for Graduate (Y/N) from an application scorecard development, with the points provided for the 'All Good/Bad' model after reject inference, and the coarse classes assigned to dummy variables. The columns show detailed values/ranges (fine classes), bin assignments (coarse classes), weights of evidence, information values, points, counts (total, good, bad, indeterminate), ratios (odds, rates), distributions (sample and recent), and the information value. When using this type of format, the scorecard developer must follow a distinct process:

(i) Characteristics are *fine classed*, to provide as much detail as possible for the analysis.
(ii) The fine classes are binned into *coarse classes*, that either have the same relative risk, or logically belong together.
(iii) In instances where there has been *significant drift*, the classes may be grouped further, or the characteristic moved to a later stage in the analysis.
(iv) The characteristics are *transformed* into variables, and the model is run.
(v) Checks are done to *guard against overfitting*, and ensure that the coefficients make logical sense.
(vi) If any issues are identified, the characteristic selection, coarse classing, stopping rules, and other factors will be *revisited*, and the model rerun.

Table 9.5. Characteristic analysis report

							Graduate Y/N								
Fine class	Coarse class	Woe	Info value	Points	Total	Odds	Good	Bad	Indet.	Reject	Reject rate	Sample dist	Recent	Recent dist
N	Null	−0.08	0.415		50,824	15.6	28,841	1,852	3,295	16,836	33.1	59.6	25,867	72.0
Y	Grad	0.12	0.575	18	34,441	19.1	22,977	1,206	2,276	7,982	23.2	40.4	10,055	38.9
Total			0.991		85,265	16.9	51,818	3,058	5,571	24,818	29.1	100.0	35,922	100.0
Information value *		100			Fine classed		0.991		Coarse classed		0.991			

Fine classing is typically for the full population, and may be adjusted for different scorecard splits, if any are identified. Coarse classing is similar, except it may also be adjusted at different stages in the scorecard development process (known good/bad, accept/reject, all good/bad), albeit greater emphasis should be placed on known performance. The key factors to be considered when assessing the final coarse classes are that: (i) the groupings make logical sense; (ii) there are sufficient goods and bads in each; (iii) the points are consistent with the relative risk; and (iv) the frequency distribution remains relatively constant over time.

9.4.2 Score distribution report for scorecard development

Once a scorecard has been developed, it can be applied to any record, whether in- or out-of-sample. Like the characteristic analysis report, the score distribution report also takes on different guises, but with a focus upon frequencies within score bands, across the range of possible scores. In its simplest form, it will only provide details for the overall score distribution—especially for a recent sample or the current population, with no performance. Indeed, these can both be compared to the development sample distribution using a population stability report. During the scorecard development though, and when post-implementation performance is available, the analyst has to delve deeper, into how the scores relate to the processes where they are being used. Hence there will be an interest in the distributions of the performance statuses—accept/reject, good/bad/indeterminate—by score.

If the development sample was constructed without knowledge of the full population, it will not be possible to provide exact odds or bad rates, but some assumptions can be made to provide an indication. This often occurs for greenfield developments in emerging environments, especially where there is a branch network with paper-based files.

The report is much like a characteristic analysis report, except the characteristic being analysed is the final score. It can be used both as part of the scorecard development process, and for post-implementation monitoring. The resulting rank ordering means that summary power statistics (AUROC, Gini, KS) can be calculated, to assess how well the scores are working.

One must consider the possible effects of truncation where it occurs, as this will influence the apparent predictive power of a model (see p.189). For selection processes, the censoring of rejects means that a like-for-like comparison with the all good/bad model is not possible. Any performance comparison must be done against historical accepts only, and even then should be done with caution.

Table 9.6 provides an example of an 'All Good/Bad' score distribution, split into 10 bands with (approximately) equal numbers of records. Other possibilities are to have: (i) equal numbers of bads or goods; (ii) minimum row percentages for bads and goods; (iii) a defined change in risk between bands; or (iv) defined risk for each band. The latter is a primitive form of calibration, used to ensure consistency of meaning across scorecards and/or over time.

Table 9.6. Score distribution report

	Total	Good	Bad	Bad rate (%)	GB odds
Low–363	42,929	30,926	12,003	28.0	2.6
364–411	42,516	29,806	12,710	29.9	2.3
412–446	42,548	31,781	10,767	25.3	3.0
447–477	43,100	34,417	8,683	20.1	4.0
478–507	43,312	36,782	6,530	15.1	5.6
508–538	43,063	38,563	4,500	10.4	8.6
539–573	43,207	40,413	2,794	6.5	14.5
574–617	43,249	41,829	1,420	3.3	29.5
618–684	42,712	42,257	455	1.1	92.9
685–High	43,427	43,405	22	0.1	1973.0
Total	430,063	370,179	59,884	13.9	6.2
			Gini coefficient = 50.7		

This example only has goods and bads, but it can also be done using goods, indeterminates and bads, defaults versus not defaults, or other classifications.

The scorecard developer may map the scores onto an existing scale, or create a new scale (see Chapter 20, on Calibration). Ultimately, the goal is to have a set of grades that can be used for strategy setting, portfolio assessment, and financial/regulatory reporting. The number of bands will vary depending upon the circumstances. Even where a large number of bands are possible, it may be difficult to design strategies appropriate for that level of detail. The end result should be that which makes the most sense in the business.

9.4.3 New business strategy table

There are two types of men, those who want to be somebody and those who want to do something.

Dwight Morrow

Now that we have it, what do we do with it? At this point, the score cut-off has to be set—a threshold above which applicants will be accepted, and below which they will be rejected. If everything has gone according to plan, the scorecard should show promise of providing an improvement upon the current process. In the ideal world, this decision should be based upon marginal profitability—choose a cut-off at the breakeven point, where the losses from the marginal bads offset the profits from the marginal goods. Sounds simple enough, but scorecards are from Mars and marginal profits are from Venus. When it gets down to the finer detail at the margin, companies usually do not understand it. Irrespective, there is still a strong move towards profit-based cut-off analysis (see Section 26.4).

This section focuses on the traditional means of setting score cut-offs, where possible strategies are expressed as cumulative percentages of accepts/rejects and goods/bads at different scores. These are easily understood, and can be readily compared to what has happened in the past. They are either presented in a strategy table or strategy curve, as presented in Table 9.7 and Figure 9.3 respectively. The latter is also an effective tool for visual comparison of

Table 9.7. Strategy table[4]

Cut-off score	Accepts	Bad	Reject rate (%)	Bad rate (%)
Current	58,418	4,323	32.1	7.4
0	86,036	1,197	0.0	13.9
460	69,655	5,564	19.0	8.0
465	68,640	5,282	20.2	7.7
470	67,597	5,005	21.4	7.4
475	66,523	4,735	22.7	7.1
480	65,421	4,473	24.0	6.8
485	64,289	4,218	25.3	6.6
490	63,128	3,971	26.6	6.3
495	61,940	3,732	28.0	6.0
500	60,725	3,502	29.4	5.8
505	59,481	3,280	30.9	5.5
510	58,211	3,066	32.3	5.3
515	56,913	2,863	33.8	5.0
520	55,591	2,668	35.4	4.8
525	54,245	2,483	37.0	4.6
530	52,875	2,307	38.5	4.4

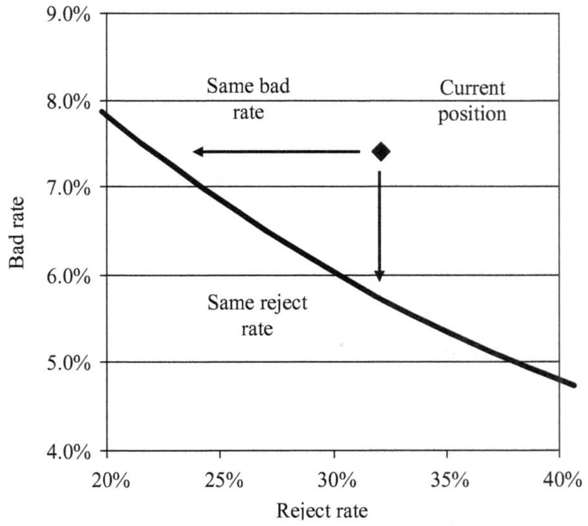

Figure 9.3. Strategy curve.[5]

[4] Table 9.7 has 5-point increments over most of the score range, but in practice 1-point increments would be used in the range where the most likely cut-off strategies would be found—in this instance probably from 460 to 510.

[5] The strategy curve in Figure 9.3 is very similar to, but should not be confused with, the trade-off curve (see Figure 8.5), which has cumulative goods along the bottom.

competing scorecards; the lower the bad rate at any given reject rate, the better. The graphs may cross though, in which instance, the best scorecard is the one that performs best in the expected cut-off region (see Figure 18.2), or wherever the greatest value is expected to be obtained.

Before choosing a cut-off, it should always be remembered that: (i) the scorecard is a *credit tool*, that must fit in with the company's marketing and other general strategies; and (ii) there are *lots of assumptions* made during every scorecard development. A degree of conservatism is often prudent, and it is best that this exercise is done using a holdout sample.

With those caveats in mind, there are several broad strategies that can be chosen. The company can be aggressive, by choosing to maintain the same bad rate. If the historical bad rate was 7.4 per cent, a logical score cut-off would be 470. If used going forward, the reject rate would reduce from 32.1 to 21.4 per cent and the number of accepts would increase 15.7 per cent to 67,597. This option would be chosen in a competitive environment, where the company wishes to grab market share, or grow the total market. In either case, a large increase in the acceptance rate puts heavy reliance upon the reject inference done during model development, and care must be taken in post-implementation monitoring.

Another possibility is to maintain the same accept rate. In the same example, the reject rate was 32.1 per cent, which lines up with a cut-off of 510 on the new scorecard. This provides a 29.1 per cent reduction in the number of bads, to 3,066. This is a conservative strategy, which is best where the primary goal is better risk management in a tough market, where market share is less of a concern.

Further possibilities lie between these two points, and possibly outside. Whatever the cut-off chosen, there will still be further issues of setting terms and conditions—loan amounts, interest rates, repayment periods, etc. These rely upon grouping scores in the accept region into risk bands, and applying common strategies within each.

9.5 Conclusion

Credit scoring is primarily associated with predictive models, and not descriptive or forecasting techniques, which nonetheless can play a supporting role in the scorecard development process, or in the finance function. Descriptive statistical techniques focus on describing the data, and can be used for data reduction. There are two main techniques: (i) *cluster analysis*, which identifies groups of similar cases that could possibly be treated on a like basis (e.g. lifestyle codes); and (ii) *factor analysis*, which condenses correlated characteristics into uncorrelated factors, and is used either to simplify the dataset, or address multicollinearity, where it is an issue.

In contrast, forecasting techniques are predictive tools, but are based upon the analysis of aggregated data—perhaps using the output of predictive models as inputs. Once again, there are two main techniques: (i) *Markov chains*, which are derived using transition matrices that represent changes in states, and require the rather elusive Markov property (memorylessness); and (ii) *survival analysis*, which focuses upon mortality associated with specific groups, relative to a specific exit state (which in credit scoring will either be related to credit risk, or account attrition). Both of these techniques are used primarily in the finance function, whether for pricing, provisioning, or capital allocation purposes.

Several important concepts were covered that are important in scorecard developments. First, *correlation* relates to the extent and direction of two variables tandem movements, usually defined on a scale from −1, through 0 to +1. The goal is to choose predictors that are correlated with the target, but uncorrelated with each other. This is not always possible though, and care must be taken to ensure multicollinearity does not introduce unacceptable errors. Second, *interactions* are instances where the relationship between variables changes, based upon the value of another variable. They can be addressed either by using generated characteristics, or developing separate scorecards. Third, *monotonicity* means that the relationship between two factors always moves in the same direction, across the range of possible values for each. It is a requirement for the bivariate score to risk relationship, and is often also required of models' constituent predictors. And fourth, *normalisation* refers to instances where data is brought into line with a standard. It may be done by converting characteristics' values into z-scores, rescaling, partitioning, or using ratios to standardise for size or some reference value.

Finally, credit scoring is a discipline that is very report intensive, whether during assembly, development, validation, implementation, or post-implementation monitoring. In many cases, the same reports are used at different stages, perhaps with various modifications to tailor them to a task. This section limited itself to three reports, used during the scorecard development process: (i) *characteristic analysis*, used to evaluate characteristics' relevance; (ii) *score distribution*, which focuses on score ranges and frequency distributions; and (iii) *strategy tables and curves*, which focus on cumulative bad and reject percentages, by score, in order to evaluate possible strategies.

10 Minds and machines

> A lot of business applications, such as financial modelling, are actually scientific in nature and even though the nature of the problems may be different in business, the mathematics is very similar to tasks we have already tackled in science.
>
> *Chorafas* (1990)

When the above is read, one must understand science as also including fields like economics, engineering, psychology, and others. It is also not only the maths that is similar, but also much relating to the people and computers that are employed. By extension, this also applies to credit scoring. While it tries to bring science into lending decisions, it will always be both science and art. Creativity can be used at all stages in the process, but some developers wield their paintbrushes better than others. This section focuses on both the people and the software:

People and projects—Scorecard developers (internal versus external), and project participants (project team and steering committee).
Software—(i) applications software required for developing scorecards; and (ii) decision engines that are used to combine scores and strategies.

10.1 People and Projects

> Most model developments are undertaken for one of three reasons (1) There is no model in place, (2) the current model is not effectively ranking risk, or (3) while the model is ranking risk, newer technologies are available to build a model that does better.
>
> *Wiklund* (2004)

Significant resources are required to take the project all the way through, from feasibility study to post-implementation monitoring. As a result, scorecard developments are usually run as projects, and managed as such. Depending upon the role, individuals' involvement in a project can take a variety of forms:

- **direct**—part of the project team;
- **indirect**—called upon in need, to perform one or more functions;
  - **active**—performs project-related tasks, outside of the team meetings;
  - **advisory**—provides advice on an area of expertise;
    - **delegator**—will involve others in the task, but is responsible for delivery;
    - **doer**—performs the required task personally.

This is provided solely as a useful framework, and not much more. It is up to the reader to decide which labels apply, when the various role-players are considered:

(i) **Scorecard developer**—staff members or consultants employed to develop scorecards;
(ii) **Project team**—responsible for ongoing execution of the project;
(iii) **Steering committee**—makes higher-level decisions affecting the business.

These are not all of the people involved in credit scoring. There are also those involved in implementing, validating, monitoring, strategy setting, forecasting, dealing with business units, and other functions, to ensure that everything is working according to plan, and that optimal benefit is being obtained. They are not covered here, even though their roles are just as important.

10.1.1 Scorecard developers

A decision-maker who thinks that she or he can turn the analyst loose without guidance and expect to get relevant information back that can be applied directly to the problem and then forgotten will not make the best use of quantitative inputs. Instead, the interaction between the decision-maker and . . . analyst must be open, interactive, and focused on the ultimate goal of the effort: to develop and make the best use of the quantitative input to a decision problem.

Prof Hossein Arsham

Credit scoring's evolution can be likened to that of the automobile. At first, the drivers also had to be mechanics, or have one in their employ. The first four-stroke engines were simple, and for years basic motor mechanics was taught to teenagers in school. From the 1970s onwards though, there was increasing design complexity, including the incorporation of computers to perform and monitor many functions. The teenagers have now grown up, but unlike their dads of old, they are no longer able to tinker with the car in the garage.

Likewise, when credit scoring was first implemented, it involved decision-makers at the highest levels, and required inputs from them. Over time, the tool has become widely accepted, and responsibility for signing off scorecards is often delegated. Rather than becoming involved in the mechanics of scorecard developments, company directors just want their vehicles to work, with dashboards to monitor them. The big question is, 'Who will develop the scorecards?' There are two main possibilities: (i) outsource the task to a vendor or consultancy; or (ii) develop the skills in-house. Each has its own advantages and disadvantages.

10.1.2 External vendors/consultancies

Fair Isaac (FI) was the first vendor to propose and develop credit scores, and over the years the number of vendors grew, as skills became more widespread. At the time, scorecard vendors would approach lenders to sell the new tool, which was especially beneficial to the fledgling credit card companies that were experiencing rapid growth, and whose underwriters had little experience. Vendor-supplied business analysts and scorecard developers would work together with the business, not only to develop scorecards, but also the infrastructure required to automate back-office decision processes. Over time, the services were presented to the entire retail-credit industry, as a way to streamline their origination processes.

The limiting factor in the 1960s was not only a skills shortage for the breakthrough technology, but a lack of appropriate computer systems and software. Back then, everything was bigger, slower, hotter, and much more expensive. A mainframe with the power of today's palmtops could take up a large, specially air-conditioned room, and still only handle a single regression.

For 20 years or more, there were only a handful of companies with the resources necessary to develop scorecards, the prime examples being FI, Experian, Equifax, and TransUnion. Each developed their own methodologies and software, and gained significant expertise in their use, including being able to recognise the pitfalls that may be encountered during developments of different types. Since then, the number of agencies and boutiques offering these services has been growing, such that lenders can pick and choose between them. There are a variety of factors to consider when evaluating them:

Experience—How much exposure have the consultants had to scorecard developments, especially those of a similar nature to that being proposed?

Availability—Do they have the available resources to develop the scorecards within the required time frames? Or to revisit the scorecards, if any issues arise?

Technology—Does the vendor have a formalised methodology that is consistently applied? Do they have the necessary hardware and software for data processing?

Support—How extensive is the vendors' network, if further assistance is required? Are they also providing the delivery infrastructure?

Costs—What is the financial impact? Charges may include consultancy services, back-office support, and travel (often international).

Flexibility—Will the project schedule provide sufficient latitude to test different scorecard splits and characteristics? Does their methodology allow for features desired by the lender?

Transparency—Are they willing to inform the lender regarding the internal workings of their methodologies?

Of these, experience and availability are the key factors. While the cost of external consulting can be extremely high, it is often worth it—especially for lenders that do not have the time or inclination to develop and manage an internal team. Scorecard development resources can be scarce though, and lenders may struggle to find a vendor with any available when required. As a result, many will develop internal teams, but even then, relationships will still be maintained with external vendors to: (i) provide support, when internal resources are scarce; (ii) provide fresh insights into the process; and (iii) keep lenders apprised of best practice within the industry, such as technological developments, legislative changes, and other environmental factors.

The 'transparency' aspect should be highlighted further, as it has become particularly important where good corporate governance demands that businesses—especially banks in the Basel II environment—need to understand their key processes. Consultancies are very protective of their methodologies, and many of them use proprietary software packages that greatly aid consistency, but are black boxes whose results cannot always be explained or improved upon. Indeed, in many cases their own staff may not have a comprehensive understanding of the process. This opacity can be dangerous for the scorecards' end users.

10.1.3 Internal resources

As indicated earlier, one of the main impediments to developing scorecards was access to skills and technology. During the mid-1980s, technology started becoming more accessible, and some lenders started investing in their own in-house scorecard development capabilities. According to Wiklund (2004), internal teams tend to:

(i) Be cheaper!
(ii) Be smaller and more flexible than those used by vendors.
(iii) Be more flexible, in terms of the techniques that they employ.
(iv) Have greater knowledge of the lenders' business, including data and processes.
(v) Offer quicker turnarounds, when any rework is required.

In terms of flexibility, Wiklund (2004) makes special mention of internal developers not being 'locked into scalable modelling processes for business reasons'. These can put limitations on what can or cannot be done within the scorecard development process. Also, internal developers may work and consult in the areas where the scorecards are being used, including providing assistance when setting strategies, doing ongoing monitoring, and determining where else the scores may provide value. They are also more likely to develop several competing models, and choose the one that has a better fit with the lenders' business and processes.

This does not, however, mean that having in-house scorecard developers is a panacea. It comes at the cost of considerable management time and risk. Learning curves are long, and skills are in high demand, for: (i) experienced scorecard developers; (ii) the people that manage them; and (iii) the people that know how to apply credit scoring within the business. McCahill (1998) notes that staff will always wish to advance their careers, and there are very few people who wish to remain professional scorecard developers. Indeed, lenders may struggle to keep people in their roles long enough to benefit from the investment. Care must be taken to ensure that: (i) salaries are kept in line with the market; (ii) there is a challenging environment, where they have plenty of exposure to, and input into, the business; (iii) there is career planning, and advancement opportunities; and (iv) that they get outside exposure, whether through conferences or relationships with vendors, as internal teams are small and isolated.

Choosing the right people

Scorecard development is a specialist quantitative analysis (quant) function. The labels of rocket scientist and propeller-head are sometimes jokingly used, but these overstate the situation. The skills levels have become lower over time, especially as the processes have become better known and understood. Model risk can, however, still arise from the complexity of the scorecard development process, and is affected by the personality, experience, and background (mathematician, statistician, engineer) of the scorecard developer. It is greatest where:

(i) The scorecard development process has not been standardised, or well documented.
(ii) Developers do not have an adequate theoretical knowledge of the tools they are using.
(iii) There is experimentation and deviation outside of the formal process, especially by inexperienced developers.
(iv) Adjustments are made for special circumstances, and their impact is not well understood.
(v) There is no independent validation by people outside the development team.
(vi) The values at risk are large, especially for new business origination for banks and credit card companies, where margins are low.

The final factor plays a major role, and influences who is best suited to the task. High-volume low-value environments can be equated to industrial processes, where small errors and poor-quality inputs can go unnoticed for a long time, with potentially disastrous results. A formalised scorecard development process is crucial for these key projects, and the best developers are those that: (i) can focus on detail; (ii) are less likely to experiment with the process; and (iii) have an interest in how the scorecards are influenced by, and affect, the business. Misfits can become highly frustrated, as the learning curves are long, and they may bore of the task before they are fully competent.

This is complicated even further if the predictive modelling techniques used are proven in practice, but frowned upon in academic circles, and questioned by new initiates. Lenders may wish to limit their modelling techniques to accepted methodologies, like the use of logistic regression instead of LPM, in order to facilitate the training of new initiates with statistical backgrounds (and placate any concerns that might be raised by consultants and auditors).

In contrast, for non-key projects—and those where the formal process is in its infancy or under review—there is more latitude to experiment, but the level of theoretical knowledge required is greater. Non-key projects include: (i) smaller portfolios, especially where the type of business is new; and (ii) scorecards used for marketing and retention. These projects may also provide a valuable testing ground for potential changes to the formal process.

10.1.4 Project team

Where scorecard developments are run as formal projects, it usually occurs at two levels: (i) at *steering committee* level, where the major decisions are made; and (ii) *project team* level, which is responsible for execution of the project. In both cases, decisions are typically reached by consensus, and rely heavily upon the expertise of members, both inside and outside the group. At times there may be disagreements, where it is necessary for somebody further up the line to make a call.

While projects are underway, the project team must make technical decisions that affect the end result. Issues that cannot be resolved have to be referred to the steering committee, especially where they relate to ensuring adequate resource availability, getting co-operation from different areas of the business, technical issues around the development, or others. Gaining co-operation is crucial, but in most organisations, the scorecard development has to compete for resources with other projects, and the day-to-day running of the business. The roles on the project committee can be summarised as:

Project manager—Reports to the project champion, and is responsible for project execution.
Scorecard developer– Person responsible for developing the model.
Internal analysts—Employees responsible for data assembly, understanding the data, and determining potential changes in the pipeline.
Functional experts—Individuals with a comprehensive knowledge of the business, especially markets and processes past, present, and (expected or required) future.

Some of the internal analysts and functional experts will not be part of the committee per se, but will be called upon in need. Other parties in a similar position are:

External vendor—Much of the technical scoring expertise may be provided by scorecard vendors or consultancies, which provide a variety of different services.
Credit bureau—Where lenders need external data, the parties providing it may be included, especially where retrospective data or a greater understanding is required.

Any other constituencies that may be impacted by the scorecard development, whether inside or outside the organisation, should also be kept apprised of what is happening. An example is a dealer network that will be affected by the development.

10.1.5 Steering committee

Over the years, scorecards have become well accepted in organisations, and for many they are second nature. This section covers the steering committee, which may be dealing with multiple projects over time. It plays the initial role, to decide whether scorecards should be (re)developed, commission a feasibility study, and put the project team together. Committee members will have a higher-level view of the organisation than the project team. Once a project is underway, the steering committee will be called upon 'when things go wrong or when factions within an organisation are setting up roadblocks . . . ' to resolve issues so that the project can move forward (Wiklund 2004). Issues addressed by the steering committee would include:

responsibility at executive level;
scope, timing, budget, and deliverables of the project;
composition of the **project team;**
choice of **external vendor** and contract terms (if applicable).

One of its primary goals is to ensure effective communication within the organisation, and it must include people with the power to make things happen. Key roles and areas represented include:

Project champion—An individual who believes in the project, and its benefits; whose normal responsibility might be credit risk, decision support, or information technology.
Sponsor—The person who signs the cheques, and has an interest in the time, cost, quality, and benefits to be derived. This could be the CEO, or a director from the affected area.
Targeted CRMC function—Business unit that will implement the scorecard, whether credit risk, marketing, collections, or elsewhere.
Affected CRMC functions—Changes at one stage in the cycle can have significant impacts downstream, so other areas should be kept informed.
Strategy and marketing—People familiar with existing policies, who can be tasked with ensuring that strategy and marketing will take best advantage of the new tools.
Sales and distribution channels—Individuals responsible for product and/or service delivery, possibly including dealer/broker networks.
Engineering and technology—Technical areas that will be responsible for building the system, or providing programming or networking resources.
Compliance and legal—To ensure that the decision system complies with relevant company policy, code of practice, and legislation.

Steering committee members will be responsible for ensuring that adequate resources and co-operation are provided by their areas, and for communicating any factors that may affect them. Meetings would be held: (i) monthly to ensure that goals are being met, and appropriate co-operation is being obtained; and (ii) whenever the project team has to report back on milestones that have been reached, to ensure that the steering committee is comfortable with the results, and the assumptions that have been made. These milestones may be further communicated to the affected divisions. Where there are disagreements, they are referred to the company executive.

Many worthwhile projects fail due to either lack of a champion, or the champion not having sufficient clout within the organisation. In many banks, the decision-support function is important enough for the departmental head to have executive status, and this individual will be the champion for new projects and updates.

10.2 Software

Computers are useless. They only give you answers.

Pablo Picasso (1881–1973)

The next aspect to consider is machines, but with a focus on software instead of hardware. The software is comprised of the scorecard development packages and the decision engine that is used to determine and deliver the appropriate decision for each case. Scorecard development, monitoring, and strategy design also rely upon databases (which may be referred to as electronic data warehouses, analytical data marts, or other labels), which fall outside the scope of this text. They are, however, just as important, if not more so, as the networking required to bring the data together and communicate the results.

10.2.1 Scorecard development

There are two types of software that are available for scorecard developments: generalist statistical software such as SAS and SPSS; or specialist credit scoring software, such as Experian's Sigma™, Scorex's Toolbox™, SAS Credit Scoring, and others. There are a variety of factors to consider when making the choice:

Cost—Whether up-front, or as an annual license fee.
User-friendliness—Ease with which staff are able to learn and use it.
Service support—Access to qualified technicians, who can assist when problems are encountered.
Flexibility—Allows processes to be modified to handle special situations.
Transparency—Extent to which underlying processes can be understood.

We will not compare individual packages, but the broad groupings of black, grey, and white box approaches (Figure 10.1).

The totally black box approach is to rely on scorecard vendors to do the developments, or use bespoke credit scoring software. This can be very effective, but has its disadvantages:

Opacity—The results may not be well understood by those using it.
Cost—The software can be expensive, especially when developed by scoring consultancies that would rather lenders continue to use their services.
Inflexibility—Having to fit the problem to the software (instead of vice versa), which limits the available options.

At the other end of the spectrum is the white box approach. This uses generalist statistical packages, and demands that scorecard developers know, and hopefully comprehend, every element of the process. Software costs are lower, but the learning curve is steeper—not only learning how to use it, but also how to apply it effectively to the credit scoring problem. The

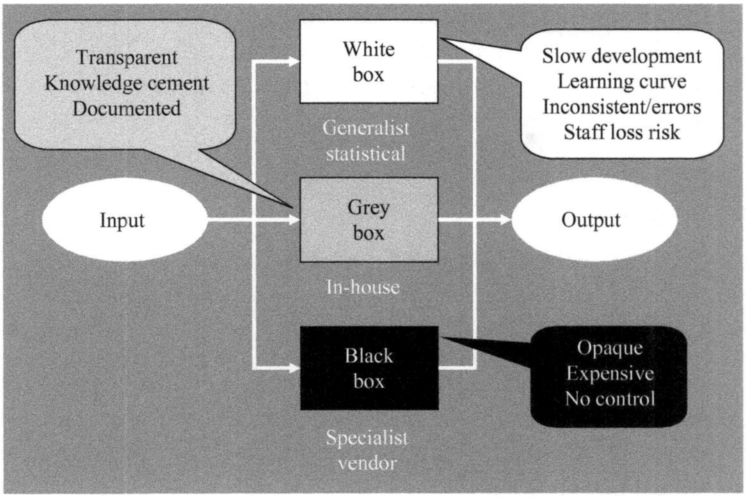

Figure 10.1. Software strategies.

curve can be expedited by buying skills, either by poaching experienced scorecard developers, or by using consultants to train staff. Potential problems with this approach are:

Risk of staff loss—Skills investment is high, and loss of a staff member is costly—especially if lost to a competitor.

Staff assimilation—New staff will always question the accepted way of doing things, often to minute detail, distracting them from getting the job done.

Errors—Small mistakes may be made, either because of incorrect assumptions, a small error in coding, or use of unapproved procedures.

Slower development times—Software may not be user friendly, or staff may be distracted by trying new things, instead of focusing on getting the job done.

Inconsistency across developments—Each developer will have developed his own bag of tricks, making the task of auditing the process more difficult.

Poor development documentation—May have been limited to key assumptions, while seemingly inconsequential but critical decisions have not been noted.

Poor process documentation—A living document is needed, to set out best practice learnt to date, which may be a difficult task in its own right.

Companies sometimes develop their own specialist software, to standardise and speed up scorecard developments, while reducing the required skills levels and chance of human error. This grey box approach acts as knowledge cement for whatever has been internally accepted as best practice, by putting it into program code, hopefully without losing too much flexibility.[1] A necessary and useful by-product is documentation of whatever has been learnt to date, which is insurance against key people leaving, and reduces the cost of staff turnover.

[1] Some vendor scorecard development software packages are able to combine both standardisation and flexibility by acting as program code-generators, especially those built using a SAS backbone.

Something that must be warned against, is that software vendors tout the flexibility of their software, but these packages often make significant assumptions. This is fine, as long as the lender is comfortable with those assumptions. Otherwise, the grey or white box approaches are more appropriate.

10.2.2 Decision engines

. . . a complex piece of computer software designed . . . to make decisions about appropriate actions to take in any customer situation, or any delivery channel through which contact is made.

McNab and Wynn (2003)

Once the scorecards and strategies have been defined, they need to be implemented for use in the company's operations. While it may be possible to do this in the existing system, these are often highly inflexible, incapable of providing the required information, and very difficult for lenders to maintain on an ongoing basis.

Many companies will instead use a decision engine, software that is independent of lenders' core systems, that compiles the required information, makes the decision, and returns the result back into the operational environment. Decisions could be based solely upon a score and cut-off, but in most volume-driven environments will involve complex rules and multiple scorecards to determine the product offering, limits, pricing, cross-sell, and other features. Furthermore, lenders also require the capability to do experimentation using champion/challenger strategies.

Decision engines have become key in the field of customer relationship management, where the goal is mass customisation. Two of the best known are Triad® (FI) and Strategy Manager® (Experian). They are especially crucial where there are several different products that all require similar capabilities, as they provide a standardised platform that can be tailored to each.

10.3 Summary

The final section of this module focused on the minds and machines aspect of scorecard development. Credit scoring may be a statistical process, but the scorecard development process will always have a human element. There were three aspects covered: (i) the *scorecard developers*; (ii) the *project team*; and (iii) the *steering committee*. Lenders may rely upon external consultancies for scorecard development expertise, but there is an increasing trend towards in-house developments. The former can provide a quality product, but can also be expensive and inflexible, while the latter seeks to provide more tailored solutions at reasonable cost.

The project team is comprised of the *project leader*, *scorecard developer*, *internal analysts*, and *functional experts*. The project leader is the communications link to the steering committee, usually reporting to the project champion. Internal analysts are required to obtain and understand data, while functional experts will be called upon to explain past and planned events in the business.

In many respects, this section was written in reverse, as it is the steering committee that has true control over the project. It decides when it is required, commissions the feasibility study, and organises the resources. It is comprised of the *project champion* and *sponsor*, as well as representatives from the *target and affected functions*, *strategy and marketing*, *sales and distribution*, *information technology*, and *compliance and legal*. They must have the power to make things happen, and obtain resolution where it is required, especially where there are competing projects.

As regards software, there are two aspects: (i) the software used to do the *scorecard development* directly; and (ii) *decision engines* used as delivery platforms. For scorecard development software, lenders can use three approaches: *black box*, specialist vendor-supplied software, that may be poorly understood; *white box*, generalist software that is flexible, but may be more prone to errors; and *grey box*, software developed in-house, to formalise the scorecard development process. The latter can sometimes be done by modifying black or white box software to meet lenders' needs.

And finally, decision engines are used to calculate scores, and apply the lenders' strategies. In the early days, lenders would develop their own software, but today there are parameterised packages that can be easily modified, without the involvement of computer-programming staff. Decision engines may be expensive, but the benefit that can be gained across different business processes is significant.

This concludes the tools module. It may seem strange that tools have been presented as a separate section, especially the statistical and mathematical tools. This format was chosen because many of them are used at different stages in Module E (Scorecard Development Process) and Module F (Implementation and Use).

Module D

Data!

11 Data considerations and design

> Neither sophisticated software nor statistical techniques can overcome the inherent limitations of the raw data that goes into them.
>
> *McNab and Wynn* (2003)

While there may be a variety of possible data sources that can be used, they are not always readily available, or may be inadequate for current purposes. This section looks at various data issues that must be considered, many of which could lead to potential pitfalls during the design stage, if ignored. The topics covered are:

Transparency—The extent to which there is sufficient information available to do an adequate risk assessment.

Quantity—The depth and breadth of data, which can be a function of the data's accessibility, and the homogeneity of the group being assessed.

Quality—Ability to meet specified needs, which can be split into relevant, accurate, complete, current, and consistent.

Data design—Types of data used, in practical or statistical terms, including special cases such as missing data and division by zero; and form design issues, to maximise data's value.

11.1 Transparency

The end goal is to provide a 'measure of creditworthiness', the appropriateness of which will depend how transparent borrowers' circumstances are, to whoever is doing the assessment. The word transparent normally refers to a substance's ability to transmit light, but in this instance, it means 'easy to understand and analyse'. This refers to the bulk of application form data, data on past dealings, and data provided by the credit bureaux. In contrast, something is opaque if it is either 'impervious to light', or 'not easily understood'. A (potential) borrower's creditworthiness is opaque if:

No credit history—Credit seekers may be creditworthy, but without current or past credit activity, nobody can tell—for example, first-time home buyers who have no credit records, yet have flawless records at paying rent and utilities. The influence is greatest in youth, immigrant, subprime, and micro-finance markets.

Intelligence unfriendly—The information required for an assessment may be elusive, because the data is poorly structured, and/or the technology is poorly developed or non-existent. This is especially so in emerging environments.
Highly complex—Credit scoring needs structured data, which can only be achieved through experience. The challenges are much greater where borrowers' circumstances are complex, and non-standard. This applies especially to wholesale lending, multiple counterparties, and complex deal structures.

Credit bureau data is criticised in the subprime market, because files are thin, and much data that could aid the assessment is unavailable. In August 2004, FICO introduced its Expansion™ score, designed specifically for 'credit-underserved' individuals—as many as 50 million Americans. It uses data from non-traditional sources, such as cheque verification agencies, rent-to-own stores, payday lenders, utility and rental payments, and others. Initial results were positive, with over 50 per cent scoring above the subprime ranges (Interact, July 2005).

Stanton (1999) defined four categories, based upon individuals' creditworthiness and transparency, to illustrate their influence on the probability of a correct prediction, as set out in Table 11.1. Correct decisions are more likely when individuals' creditworthiness is transparent, and problems when it is opaque. Transparency will always be greatest for a credit-active population, especially where there is one or more credit bureaux in operation. Problems with opacity will occur with: (i) cultures with an aversion to credit; (ii) rural and other areas with few shopping opportunities; (iii) youth and immigrants who may be seeking credit for the first time; and (iv) inner city, emerging market, and other groups that have traditionally had little access to credit. Lenders should be aware of the extra risks, and design their processes and strategies accordingly.

Where creditworthiness is known to be opaque, lenders have two options: (i) increase interest rates charged, to offset the extra risk; or (ii) put extra effort into determining what information can add value, and how to obtain and assess it. Transactional lenders usually strive for the latter; especially where there are competitive pressures, and the potential profits warrant the extra cost. Where borrowers' financial situation is highly opaque, credit scoring cannot be used. There is, however, a lot of middle ground where it can provide value into judgmental decisions.

Table 11.1. Effect of opacity

Data	Credit quality	
	High	Low
Transparent	True negatives	True positives
Opaque	Type II errors	Type I errors

Micro-finance markets are now using credit scoring, but have issues because of low credit activity, high illiteracy, and poor infrastructure. Local knowledge is necessary, and lending in these markets is highly relationship intensive. The loan officer's own observations and opinions usually play a major role in the decision, and the credit score will be just one input. The human input may be costly, but this market is distinct, because borrowers are highly interest rate insensitive, and are more concerned with the affordability of the repayments. Even seemingly usurious rates may be fair for very small loans. If a stallholder can sell goods purchased for $100 for $200, then a $20 p.m. working capital charge can be justified.

Middle-market businesses. Traditional credit scoring focuses upon personal details and credit behaviour, and dominates for high-volume low-value lending to SMEs, especially one-man operations that can be treated as extensions of the individual. Other information is needed as enterprises' sizes increase, and their circumstances become more complex. In the middle market, credit scores (derived using account behaviour, bureau information, and/or company financial statements) may form part of a subjective assessment. This can provide a forward view, by recognising factors that cannot be reflected in a purely empirical grade, for example market conditions and management ability to adapt to changing circumstances.

The term 'transparency' does not appear in the normal credit scoring vocabulary, even though it is one of the basic assumptions. It acts more as an unspoken constraint, which limits: (i) the ability to develop scorecards; (ii) their predictive power; and (iii) the extent to which they can be relied upon for credit decisions. Instead, concerns are usually voiced in terms of the quantity and quality of available data:

Quantity—is there enough of it, both depth and breadth?
- **Access**—is it legal and available for use?
- **Ease of collection**—how easy is it to collect and process?

Quality—is it accurate and consistent?
- **Relevance**—is the data relevant to the task at hand?
- **Age of data**—is it too new, or too old?
- **Manipulability**—can somebody fool the system?
- **Transparency**—can the data be readily interpreted?

Such concerns apply to all scoring, whether: (i) application, behavioural, or transaction; (ii) risk, response, retention, or revenue; (iii) credit, collections, fraud, or marketing.

11.2 Data quantity

As a rule . . . he who has the most information will have the greatest success in life.

Benjamin Disraeli (1804–1881), twice Prime Minister of England

Credit scoring is primarily associated with the retail credit arena, where volumes are high, values are low, and data is plentiful. These are the little fish in the financial ocean, where one need only cast a net both to catch and study them. In contrast, when hunting whales (wholesale credit), there are fewer of them, making them more difficult to catch, observe, and understand. Over time however, new information sources are being harnessed to allow broader application of credit scoring, at least to smaller whales.[1] The precondition for any analysis is having sufficient data, without which the results can be questionable. Data quantity is discussed under the headings of:

Depth and breadth—Number of cases, and variables available for each.
Accessibility—Limitations on availability, caused by infrastructure, access, or legal limitations.
Homo-/heterogeneity—Diversity, which affects whether cases can be treated together.

11.2.1 Depth and breadth

Any analysis, including predictive model developments, is dependent on having sufficient data, in terms of both: *depth*, the number of cases; and *breadth*, the amount of information available for each. In retail credit, the commonly quoted minimum numbers for depth are 1,500 goods, 1,500 bads, and 1,500 rejects per scorecard (Lewis 1992; McNab and Wynn 2003, amongst others). No logic is provided for the choice of these numbers, but they: (i) have worked in practice for many years, to ensure representative samples; and (ii) are sufficient significantly to reduce the effects of multicollinearity and overfitting, when working with correlated variables.

The primary constraint is almost always the bads. In many instances, it may be difficult to get 1,500 for a single scorecard, let alone multiple scorecard splits (see Section 15.4, Sample Selection, and Chapter 18, Segmentation). Smaller numbers may suffice, but extra care is required. Fewer cases are required for validation, perhaps a minimum of 300 of each, as there are no weights being applied (Mays 2004).

As for breadth, no guidelines exist. There may be hundreds of characteristics at the start, yet the final model(s) will only have between 6 and 25 characteristics that: (i) make logical sense; (ii) are predictive; and (iii) must be available within the business process, where the scorecard is to be used.

11.2.2 Homogeneity/heterogeneity

To use a credit-scoring system cost-effectively [to create portfolios for securitisation], a lender must also make its small-business loans fairly homogenous. . . . Using a scoring system to rate heterogeneous loans would be like using the same machine to process many differently shaped and sized widgets.

Ron Feldman (1997)

[1] This applies especially to companies' financial statement data, where lenders have pooled data to develop models for small and middle market enterprises.

The diversity of the group being assessed also plays a role. To develop a model, the group must be homogenous (similar) enough for it to be treated together, yet heterogeneous (dissimilar) enough, in terms of credit risk or whatever is being measured, that a scorecard can provide value. Where the group is highly heterogeneous, interactions can arise that make it difficult or impossible to treat the group as one (also see Chapter 18, Segmentation). Some key factors to consider are:

Different target definitions—If there are different good/bad definitions, and transactions are put through substantially different processes, then they should not only have separate scorecards, but also be monitored separately. This applies especially for different products or markets.

Different data sources—If groups are characterised by different data sources, it may be indicative of substantial interactions. This will be evidenced where substantial groups lack data on some sources. Questions should be asked, as it will influence the development.

Significant interactions—In many instances, relationships between predictors and the target variable change from one group to another. Separate scorecards are developed if the extra effort improves the predictive power, but the scorecards are still monitored together.

In any of the above cases, the number of available records will be limited to those in the subgroup, and any rules relating to quantity (1,500 goods, etc.) would apply to each. At the other end of the spectrum, a group may be highly homogenous for risk, or at least according to the information available. This will be evident if a large proportion of cases fall into a narrow risk range; for example, if 30 per cent fall within a single risk range/grade/band, where on average the odds double from one to the next. Unfortunately, in this case little can be done in terms of depth! The only option is to increase breadth, by looking for new data sources, and even that may not help.

11.2.3 Accessibility

By 'accessible', it is meant that the data can be obtained and used as part of a credit assessment. A lot of data may be predictive, but is inaccessible, because:

Data collection—It is available, but in an inappropriate format—like paper forms in files, cupboards, or dusty warehouses. If so, the forms have to be extracted, software written, and data captured, before the scorecard can be built. Today, this is less of an issue, as most volume-driven lenders have sophisticated data-gathering processes. It does, however, still arise for greenfield developments in emerging environments.

Communications—Can the data be provided on an ongoing basis to the business process where it will be used? This is an issue for bureau and internal characteristics, where infrastructure must be built, or updated, for data transmission, acceptance, and storage.

Anti-discrimination—Details like gender, race, religion, and others may be prohibited by legislation (see Chapter 34). They may be banned outright, but are sometimes allowed if their influence is small, and they form part of a holistic assessment.

Data privacy legislation—Legislation may demand that information only be used for the purposes for which it was collected. As a result, credit grantors may be restricted from using shared-performance data for marketing, or voters' roll data in credit decisions.

Information sharing agreements—In order to access pooled data, lenders must meet membership requirements, by: (i) subscribing to the reciprocity agreement; and (ii) possibly, being in a specific line of business (retailing, banking, mail order). Most private credit bureaux offer their services to all credit grantors, and are quite relaxed when providing information for scorecard developments. They can, however, stop data feeds to subscribers that breach the reciprocity agreement.

Retail credit providers have been able to develop relatively rich data sources, whether from application forms, credit bureaux, or internal systems, especially for new business origination, account management, and collections. In contrast, marketing suffers, not only because of data privacy legislation, but also concerns about poaching.

11.3 Data quality

At the most basic level, information for managing risk is usually a by-product of a processing system designed for a completely different purpose and there is often insufficient policing of the quality of data going in at the front end of the system so the value of information produced is undermined.

Olsson (2002)

One of the maxims applied to scorecard developments is 'garbage in, garbage out'. There is no statistical manipulation that can turn manure into mink! Indeed, just a few pieces of quality information can be crucial to a decision-making process. This section discusses the concept of data quality under the following headings:

Relevant—Has a bearing on the outcome. In credit scoring, it is relevant if it could provide meaningful input into a score and/or decision.

Accurate—Is a true reflection of the situation, which implies that it is correctly completed, captured, and stored.

Complete—Contains all of the information that is required. Individual fields may be missing, or even whole records.

Current—Has been updated recently. Data can age very quickly, and after a period of time becomes worthless.

Consistent—Has the same meaning over time; if it is wrong it will always be wrong, and can be relied upon to be wrong in the future.

11.3.1 Relevant

When developing credit scoring models, the primary interest is in correlation, not causation; yet it is important to be conscious of potentially spurious correlations, and to ensure that the characteristics are relevant to the problem at hand. Several questions can be asked about each characteristic, to ensure that it is relevant:

(a) If it can be measured, how predictive is the characteristic?
(b) If not measurable, is there any evidence of it having provided value elsewhere?
(c) Will it be readily available when needed, and if not, can it be obtained?
(d) Does it make logical sense?

For many greenfield developments, lenders and their underwriters will already have significant experience with lending to that market, and have a good feel for what the relevant inputs should be. If the data is not already stored electronically, it can be captured from application forms. In contrast, where lenders lack experience in a market or function, the task may not be so simple. What has been used for credit risk assessments may not be as appropriate, for retention, revenue, or response. For example, risk assessments typically focus on monetary debit and credit values, but retention may be better served by monthly transaction counts.

Relevance is also an issue, especially in terms of the data privacy (or fair credit reporting) legislation governing the credit bureaux. In the days of index cards and filing cabinets, personal character information was used, sometimes based solely on hearsay and gossip, which is today no longer permitted. It might have been valid when viewed in the context, but is now considered irrelevant. This put pressure upon the bureaux to limit their data to that which can be shown to be credit-related.

11.3.2 Accurate

A key component of data's relevance is its accuracy. Does the data provide a true representation? If not, it becomes irrelevant, no matter how much is available. Given the amount of money invested to collect it, it makes sense for lenders to invest that little bit extra to ensure its accuracy. This applies not only to credit scoring, but any business process.

In credit, incorrect data can result in 'perceived' customer misbehaviour, with a significant adverse impact upon service levels (see Table 11.2, which was compiled based upon an outline provided by McNab and Wynn 2003). This is particularly true where account, contact, and/or

Table 11.2. Inaccuracies and their effects

Data type / Process	Address	Phone and email	Payment account	Other
Payment processing			✓	
Collections	✓	✓	✓	✓
Recoveries and tracing	✓	✓		✓
Recording of judgments	✓			
Fraud detection				✓

address details are loaded incorrectly. If debit-order details are wrong, 'she' (the customer for the purposes of this paragraph, which is interchangeable with 'he') will be in default, once it fails. If she does not receive a memory-jogging reminder, she will not know until she notices the larger than expected balance in her bank account—if she has not already spent that. If the phone number is wrong, she will not receive collections' friendly telephonic reminder, telling her the payment is overdue. If she runs away, saying she cannot pay, then recoveries and tracing will not have the correct previous address details to track her down. And so on . . .

The scoring aspect is of primary interest, as it is affected by a much broader range of characteristics, than just contact and bank account details. In either case, the inaccuracies can stem from a number of sources, described here under the headings of:

- **Poor process design**—Problems that arise from form design, data capture, system errors, and matching problems.
- **The lie factor**—Details may be manipulated, in order to improve the probability of a request being accepted

Poor process design

Process design plays a major role in ensuring data quality, which if done poorly can result in two types of errors: *errors of commission*, where data is incorrect, inconsistent, or duplicated; and *errors of omission*, where data is missing, either blank fields or missing records. Such errors can arise from a variety of different sources:

- **Form design**—Forms may be long and confusing, and questions unclear. Thomas et al. (2002) provide an example, where when asked for telephone numbers, many applicants replied 'Yes' or 'No', but once a ☎ graphic was included, actual numbers were provided.
- **Data capture**—Poor equipment, staff training, or checking procedures. This is relevant primarily where paper forms are submitted, and processed centrally.
- **System errors**—There may be incorrect rules, or calculations, used to derive certain fields. This may be a design fault, or result from changes to upstream systems.
- **Matching**—Problems linking customers and their records. This applies especially to credit bureaux, who manage data provided by many subscribers, over whom they have little control.

Errors can significantly influence both the accept probability, and the terms offered. Their effect will vary depending upon the type of error, and possibly the borrower characteristics. A CFA/NCRA (2002) report noted that errors have the greatest impact on thin files, especially individuals struggling to establish a credit record—students, immigrants, and subprime markets. It may result from the incorrect inclusion of derogatory information, or omission of positive performance. Where data is totally insufficient, no matter what the cause, it becomes impossible to provide a rating.

The lie factor

A significant factor in credit scoring (and any other selection process where the subject has a significant personal interest in a favourable outcome), is the temptation to cheat; which might range from simple embellishment to outright fraud. Embellishment is not only a temptation to the applicant, but also other interested parties, including staff members, who receive incentives for new business done; and dealers/agents, who earn a mark-up or commission on sales.

Whenever applicants provide information, there is a possibility of misrepresentation, no matter who is behind it. Steps should be taken to ensure that the system cannot be defeated or manipulated (Wiklund 2004; Thomas et al. 2004). Lenders can: (i) implement separate *fraud checks*; (ii) request *supporting documentation*, especially for key fields; and (iii) stress characteristics that are *less manipulable*. For the latter, the focus is put on data from automated sources, where there is no human intervention. Indeed, the combined power of credit bureau and internal transaction data is so great, that it has reduced lenders' reliance upon application forms, to the extent that risk assessments can often be done without them. Unfortunately however, there are limitations. Details obtained directly from the customer are crucial for: (i) *no or thin bureau*, where there is insufficient information to assess the risk properly, especially for subprime and credit-inactive groups; and (ii) *large loan amounts* (home loans, business loans), which demand extra input, especially where financial data is required (income, expenses, assets, and liabilities).

11.3.3 Complete

Data collection is like assembling a jigsaw puzzle—it is never finished until all of the pieces are in place. Unlike a jigsaw puzzle, however, it may be impossible to tell that a piece is missing, and very easy to carry on blissfully unaware. Lenders can only ensure that there is as much data available as is reasonably possible. Missing data must be minimised, which can be done at two levels: (i) *characteristic*, individual pieces of data are missing, such as income or occupation; and (ii) *sub-record*, records of existing credit facilities, or court records.

At the characteristic level, the score may be suppressed if key fields are missing, or if too many scored fields are missing. In contrast, if one or more non-crucial fields are missing, they may either be ignored, or meaning can be ascribed to their missingness. As pointed out by Lewis (1992), if a missing field expected a 'Y/N' response it might mean 'No', or simply that that applicant would not answer. This can be determined by comparing the three categories; if the good/bad odds for 'N' and blank are close, then they should be treated together.

At the sub-record level, the problem is much more difficult. The records may be missing because they were not received (associated divisions, bureau subscribers), or because of matching problems (especially if there is an incorrect or missing personal identifier). In either case, values will be unknowingly understated. If the level of missingness is constant, or improves, it forms part of the base assumptions; but if it deteriorates, the data quality can be seriously compromised.

11.3.4 Current

When circumstances change, so does the most appropriate decision, especially in competitive environments. As a result, decision-makers need up-to-date information, whether engaged in war, business, marriage, or elsewhere. Without regular updates, data decay sets in. In credit, it can result from changes to customers' own circumstances (house move, job change, divorce, financial standing), or the data that defines them (place names, postal codes, phone codes). The update frequency will vary though, depending upon the data's acquisition cost, and benefit derived. The primary reason lenders develop complex application processing systems is because so much of the risk can be controlled at point of entry. Lenders are in a power position, because they have something applicants want; and applicants comply because they acknowledge lenders' need for information.

Once the facility is in place though, the task becomes more difficult. For existing customers, the treatment will vary depending upon the type of information. Lenders wish to minimise customer contacts, so *customer-supplied* data should be maintained centrally, and disseminated to business units that need it. Occasional courtesy calls may then be used to update it, as necessary, if the opportunity does not arise from other customer contacts. This data changes irregularly, but when it does change, its impact can be significant. In contrast, *automatically generated* data, whether from credit bureaux or internal systems, will be updated regularly, with the frequency determined by what can be economically justified. This is especially crucial for transaction products, where this data is used for ongoing account management.

There is also an added complication when developing scorecards. Credit scoring is used to make future decisions, based on information available at that time. It follows, that the predictive models must be based upon data that was, or would have been, available when decisions were made in the past. Thus, care must be taken to ensure that the data is not too recent! A common error is to confuse outcome data with predictors, when they are provided in the same file. Fortunately, this error is usually quickly identified. More problematic are instances where there is no application processing archive, and data is instead sourced from a customer file or billing system, which houses customers' most recent details. If the data is relatively static or seldom updated, like occupation or education, it will not be a concern. In other instances, the data may be rendered unusable.

Application/behavioural trade-off

A knee-jerk reaction to data decay is to exclude data beyond a certain age, an example being application scores. When prospective customers first apply, a fairly comprehensive picture can be obtained from application and bureau details. Thereafter, there is a period where the application data decays, with little or nothing to replace it. As the relationship becomes entrenched, more and more reliance can be put upon internal behavioural data, especially for transactional products and multi-product relationships. While some lenders may do their account management using internal data only, it helps also to include recent bureau data, if cost-justified. This is happening more and more, as bureau data becomes ever more integrated into lenders' own processes. Failing that, the application scores may still provide value.

The trade-off between application and behavioural scores is illustrated in Figure 11.1. The curve's shape varies from product to product, and lender to lender. Pure behavioural scores for non-transaction products, such as home loans, motor vehicle finance, and fixed-term loans, are information weak, and application scores could still add value even after a long relationship. In contrast, transaction products provide powerful information, and supplant application data more quickly.

11.3.5 Consistent

Lenders' processes seldom remain static, especially in fast moving environments, where innovation and modernisation are the norm. Someone, somewhere, is always trying to improve something or other that can have unintended consequences downstream. The end result is inconsistencies arising from:

- **Forms**—Different forms may be used with different questions, different layouts, or different wordings that might attract different answers.
- **Systems**—Different systems may be used, with slightly different treatments in terms of processes, calculations, or who is included.
- **Controls**—Different levels of rigour or different types of checks may be applied (like dual capture versus programmed consistency checks).

This 'operational drift' may impact upon: (i) the probability of adverse selection; (ii) business process efficiency; and/or (iii) variances in customer service levels. For any lender trying to steer a clear course, inconsistencies can only be addressed as they are identified. Some drift will be so minor that it does not require any action, but in other instances, a realignment/redevelopment may be in order.

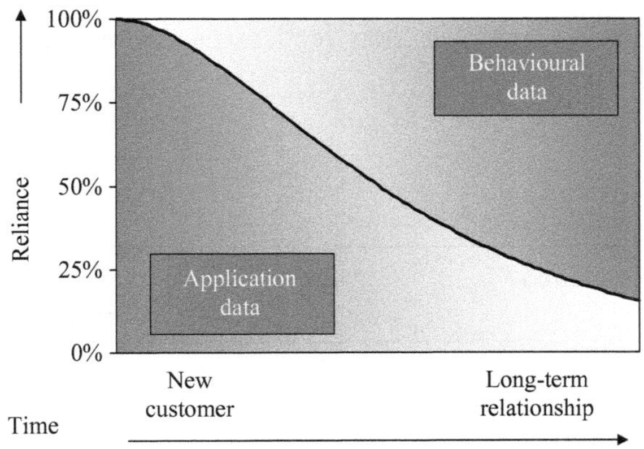

Figure 11.1. Application/behavioural trade-off.

Besides inconsistencies in individual characteristics, changes in relationships between characteristics are also a concern. Credit scoring depends on the future being like the past, and if there have been any changes—or significant changes are foreseen—then the reliability of any calculated scores will be affected, as will lenders' ability to base decisions upon them; perhaps subtly, perhaps greatly.

11.3.6 Impacts on credit bureaux

Data quality is not only an issue for lenders, but also credit bureaux, and borrowers who may be incorrectly prejudiced by bureau reports. A 1992 report, commissioned by the Credit Bureau Association and conducted by Arthur Anderson in the United States, was almost positive in comparison to the negativity of later studies. Of 15,703 applicants rejected based upon bureau information, 7.8 per cent asked for a credit report, 1.9 per cent disputed the information, and 11.8 per cent of those resulted in an overturned decision. Less complementary was a study undertaken by the NCRA in 1994, which showed significant percentages of duplicates, missing information, stale data, and incorrect matching.

More recently, a 2002 report by the Consumer Federation of America (CFA) and National Credit Reporting Agency (NCRA) summarised several studies that highlighted the high levels of errors in credit bureau reports. The few that can be presented in a common format are summarised in Table 11.3, and although not very scientific (and now dated), they remain indicative. The columns are: 'Year'—when study was conducted; 'Ind'—number of individuals surveyed; 'Rep'—number of credit reports that were reviewed; 'Total Errors'—inaccuracies of any kind; 'Major Errors'—inaccuracies that could cause a refusal of credit.

The CFA/NCRA 2002 report indicated that errors put 20 per cent of the American credit-active population at risk of being misclassified as subprime, and many more were being mispriced. Lenders may benefit from statistical averaging, but this matters little to affected consumers, especially those paying 3.25 per cent more than they should be on a 30-year home loan. A common factor throughout was problems with matching individuals to their data. In a comparison of lender- versus consumer-requested credit reports, it was noted that consumers were usually presented with fewer errors, because the match criteria (name, address, etc.) were better. The CFA recommendation was that lenders use information from more than one bureau for decision-making.

Table 11.3. Bureau report inaccuracies

Year	Organisation	Ind	Rep	Total errors (%)	Major errors (%)
1991	Consumers union	57	161	48	20
1998	Public interest research group	88	133	70	29
2000	Consumers union	25	63		50

Ind = Individuals, Rep = Credit reports

11.4 Data design

The design of a credit scoring system must also cater for how data will be provided and stored, which imposes certain constraints when developing a scorecard. The following discussion considers several aspects of data design: *data types*, terms used to describe data, both statistical and practical, and treatment of special cases; and *form design*, some guidelines for the design of forms used to gather data.

11.4.1 Data types

In statistics, many terms are used to describe data elements, usually to indicate what can and cannot be done with them. Credit scoring is a relatively specialised field though; the number of types is limited, and more straightforward labels are used. There are also a couple of terms relating to data manipulation that should be mentioned, being 'converted' and 'generated' characteristics.

Characteristics, attributes, and variables

Data contained in company databases has two dimensions: *records*, which contain details for individual cases; and *fields*, containing individual pieces of information about each case. These equate to the rows and columns within a spreadsheet. In credit scoring, the fields are more commonly referred to as either 'characteristics' or 'variables'. The two words are practically synonymous, but *characteristic* stresses that it contains distinguishing qualities for each record, while *variable* suggests their random nature (Thomas et al. 2002). Credit scoring practitioners favour the term 'characteristic', while 'variable' is reserved for transformed inputs into the modelling process. In turn, each characteristic has possible values called *attributes*. For example, 'gender' is a characteristic, while 'male' and 'female' are attributes. Characteristics are of different types, the most obvious distinction being between numbers and text, each of which has different subcategories. This was initially not intended as a statistics textbook, yet it is almost impossible to avoid some concepts. A characteristic can be:

Statistical classifications

CATEGORICAL—Groupings based on a common qualitative characteristic, such as gender (male, female), or colour (yellow, red, blue).

Binary—Consisting of only two possible categories, usually true/false or other opposites (also called 'dichotomous'). Most target variables in credit scoring are binary.

Nominal—Distinct categories that are: (i) represented by labels (names) or codes (letters/numbers); and (ii) provide no indication of relative rank.

Ordinal—Indicate relative position in a sequence, but not distance from those occurring before or after, making it inappropriate for use 'as is' in calculations. It is usually associated with subjective grades, such as excellent, very good, good, fair, and poor.

NUMERICAL—Stated as numbers, either whole or real. The relative differences have meaning, which enables their use in mathematical calculations.

Continuous—Occurring in an unbroken series, with an infinite number of possible values, between high and low. Associated with real numbers, and especially measures such as temperature, weight, distance, and time.

Discrete—Distinct and separate, or not continuous. Associated with whole numbers occurring in a sequence. Discrete numerics that are sufficiently granular are often treated as continuous.

Cardinal—Discrete, but specifically refers to counts within a set. Georg Cantor proposed the concept in 1873 as a part of 'set theory'. It is often considered synonymous with discrete.

These descriptors will almost always be associated with nouns like 'scale', 'data', 'variable', and 'characteristic'—which are often used interchangeably. In credit scoring, variables are also often referred to using more readily understood labels:

Practical classifications

Code (nominal)—Characters or numbers used to designate specific categories.

Currency (near continuous)—Any monetary value, whether: (i) balance or limit; or (ii) transaction values. May be provided as a total, average, current, minimum, maximum, range, or limit.

Count (discrete)—Number of occurrences, whether already provided (number of dependents), or derived by the lender (dishonours) or credit bureau (enquiries, judgments).

Ratio (continuous)—Result of dividing one numeric value by another, which is most commonly used to normalise currency values for size—for example, assets to liabilities ratio.

Time (discrete)—Period elapsed since the last occurrence of a given event (account opening, account active, judgment), usually in days or months.

Score (near continuous)—A value indicating the probability of a future event. In some instances, one score will be used as an input into another.

Grade (ordinal)—Like a score, except it either: (i) refers to a score range; or (ii) is assigned subjectively.

The currency, count, and ratio categories are often restricted to a specific time period. For example, 'maximum borrowing last six months', 'value of deposits last three months', and 'number of enquiries current month'.

Manipulating variables

Rather than just limiting the characteristics to those provided, new characteristics can be created. Lewis (1992) highlights two more types. First, *converted characteristics* are obtained from a single characteristic, which was inappropriate for use in a scorecard. For example, age can be obtained as the difference between the birth date and a reference date; 'Home Phone

(Y/N)' by checking whether the phone number is blank; 'Number of Credit Cards Held' by doing a count; and 'Region' from a postal code. Second, *generated characteristics* are obtained by combining two or more characteristics in: (i) logical combinations that use an 'and' statement, and are used to address interactions; or (ii) calculations, especially ratios used to normalise data for size.

Special cases

There are many instances where a single characteristic has two data types, especially numerics where special cases are represented by codes. They could be treated by creating separate nominal variables, but that creates extra processing and archiving overheads. Possibilities that are often catered for are:

Missing data—Data may be missing for different reasons, each of which might have different implications: *not found*, no reference to the individual could be found; *no record*, the individual was positively identified, but there has been no credit activity; and *none on record*, that no occurrences of the item of interest (accounts, judgments, enquiries) were found.

Statuses—If a variable becomes irrelevant because of some status that has been assigned, it might be replaced with that code. Examples are arrears statuses, like legal or written off, that replace the number of months in arrears.

Division by zero—When calculating a ratio or percentage, any division by zero may cause the computer program to crash, so these cases must be isolated. If at all possible, a value other than zero should be assigned, to avoid confusion with a zero numerator.

Division by negative—In like fashion, most ratios are intended to have meaning only when the denominator has a positive value. If both numerator and denominator can take on positive and negative values, then it is impossible to tell which one caused the negative ratio, and a double negative causes even more confusion, by providing a positive.

In these instances, a non-zero value should be used for each special case. To maintain consistency, the same codes should also be used across different characteristics with similar issues. For example, for numeric variables codes in the range 99980 to 99999 could be used, and any valid values capped at 99979. Care would have to be taken to ensure that these codes do not influence any downstream calculations.

11.4.2 Form design

In credit, application forms are designed primarily to determine applicants' needs, and to ensure that they can be properly identified, and contacted in future. Other information is still required though, to aid the credit decision (see Table 12.1). The primary challenge is to ensure that as much relevant information as possible is obtained, without going overboard. Treatment will vary, depending upon whether the responses are qualitative or quantitative.

Numerical (quantitative) responses

For quantitative responses, the choice is often between: (i) the value itself; (ii) calculation inputs; or (iii) ranges into which the value can fall. Many forms are designed so that applicants will choose a class of age, income, or other numeric characteristic. The logic is usually either to make the form more user-friendly, or less invasive. In general, the guiding principle should be to demand as few inputs as possible, but ensure that they are those which provide both maximum value, and maximum flexibility. Rather than an applicant providing an age of 28, or choosing the 27 to 30 group, the birth date is instead requested, so that the age can be calculated at any date in the future. Rather than providing an income in the $100 to $125 range, it would be provided as $113, so that ratios and averages can be calculated. And rather than providing a debt to income ratio of 3.21, the values of $321 debt and $100 income would be provided, so that they can be combined with other inputs for other ratios.

Categorical (qualitative) responses

When designing a process, it can be a significant challenge to define appropriate categories for a given qualitative characteristic. For items like gender, the choices are straightforward; but for others, they may be vast. While free-form fields are a possibility, they present problems, because: (i) of problems reading applicants' handwriting, at least until handwriting recognition software is available; (ii) of different names and spellings, for what are substantially the same categories; and (iii) grouping is still necessary. As a result, lenders try to present most qualitative questions as multiple-choice, perhaps with an 'Other' option and associated free-form field for further elaboration. These will be presented either as *tick-boxes* (paper-based and electronic capture), or *drop-down boxes* (electronic capture only). Even then however, care must be taken; applicants can struggle, both when there are too many choices, and too few.

Where the number of possible options is large, a possibility is to split the problem into more dimensions, assuming that such data can be requested on a form. This is illustrated by the residential status and occupation/education examples in Table 11.4. Typical classes of residential status are own, rent, living with parents, and other. It can however, be expanded into combinations based on type and ownership. When dealing with paper applications, the simpler option is more likely, but with drop-down boxes on computer screens, greater detail becomes feasible.

> Some instances can be quite complicated, especially if there is a detailed standard framework. An example is International Standard Industry Classification (ISIC) codes, which are used to classify industries within which businesses operate. The number of possibilities is large, and often confusing. To aid data quality, the system should provide some means of guiding the capturer to an appropriate classification.

The same applies to applicants' occupations, whether doctor, banker, plumber, nurse, student, clerk, railwayman, security guard, etc. The possibilities are endless, and changing. In 1980

Table 11.4. Categorical variables—increasing dimensions

Residential status		Occupation and education			
Type	Ownership	Occupation industry	Employment level	Highest education	Current employment status
House	Fully owned	Manufacturing	Manager	University	Employed
Complex	Mortgage	Transport	Supervisor	Technical institute	Student
Apartment	Rent—furnished	Finance	Clerical	College	Unemployed
Trailer	Rent unfurnished	Service	Salesman	Trade school	Unemployed w/Income
Dormitory	Live with parents	Construction	Tradesman	High school	Retired
Other	Shared	Medical	Trainee	None	N/A
	Other	Government	Grunt[2]	N/A	
		Military	Owner		
		N/A	N/A		

there were only a few computer programmers, and health and fitness could hardly have been considered an industry. In contrast, TV repairmen became rare creatures as repair costs approached the price of a new box. A better picture may be possible using several characteristics with fewer choices, such as occupation industry, employment level, highest education, and current employment status.

11.5 Summary

Credit scoring is totally dependent upon data, and there are many data considerations during scorecards' development and use. This section covered aspects relating to transparency, data quantity, data quality, and data design. Improving *transparency* has been a huge factor behind recent credit growth, as lenders have gained better views of their customers. *Opacity* occurs in markets where customers have little or no credit histories, with poor infrastructures, or where there are restrictions on access to data. Such markets have higher risks, and rather than avoiding them, lenders will either charge a premium, or spend a bit extra to improve their transparency. Transactional lenders prefer the latter, with examples being subprime and middle-market lending.

Related to transparency is data quantity and quality. Data quantity is defined by its *depth* (cases) and *breadth* (characteristics), which may be limited by *access restrictions*, *ease of collection*, and *group heterogeneity*. Credit scoring works best in high-volume low-value markets,

2 The term 'grunt' is an informal term for infantrymen that originated during the Vietnam/American War, first in the U.S. Marines, and then the Army. Today it is often used to denote unskilled or low-ranking workers. In business settings, it is suggested that this not be used as a choice on an application form, as most people see themselves in this light, even at managerial level. Some other alternative should be found.

where data is plentiful. Scorecard developments need a minimum of 1,500 goods and 1,500 bads, but are possible with less. If a portfolio is sufficiently diverse to require more than one scorecard, these limits apply to each. There are no real guidelines for breadth, other than that there should be sufficient relevant data to provide a scorecard comprised of anywhere from 6 to 25 characteristics.

Data quality is greatest if it: (i) has a bearing on the outcome (*relevant)*; (ii) is a true reflection (*accurate*); (iii) contains everything required (*complete*); (iv) has been recently updated (*current*); and (v) has the same meaning over time (*consistent*). It is heavily influenced by process design, which may cause *errors of commission and omission*, whether resulting from form design, data capture, system errors, or matching problems. Even attempts at correcting problems can cause problems, as system changes can create inconsistencies with unintended consequences downstream. There is also a lie factor, whether by the applicant, staff members, and other interest parties.

Finally, significant effort must be put into data design, including deciding on data types, and general form design. In any data mining endeavour, 'records' and 'fields' are referred to. A large number of the former is a volume issue, and of the latter is a design issue. In credit scoring, there are also distinctions between *variables* (implies a random nature), *characteristics* (elements used to describe), and *attributes* (defining features). *Data types* relate to characteristics/variables, which may be stated in statistical or practical terms. *Statistical classifications* include categorical (binary, nominal, ordinal) and numeric (continuous, discrete, and cardinal), but scorecard developers will think in *practical terms* of codes, currency, counts, ratios, time, scores, and grades. *Special cases* must also be accommodated, such as missing data, division by zero, negative values, and special statuses. Improper form design can also limit the value of data collected. A general principle is to strive for maximum value from as few questions as possible. Less emphasis is being put on forms today though, because of the increased power of credit-bureau and internal-performance data.

12 Data sources

In primitive credit environments, no application forms were used. Lending was based upon personal knowledge of the borrower, and either the interest rates were high enough to counter the risks of incomplete knowledge; or the penalties for non-payment were high. Indeed, these features still apply to loan-sharking today. In modern times, pressures are different: (i) personal knowledge of customers is less, especially in instances of high staff and customer mobility; (ii) interest rates and other fees are limited by legislation, and competitive pressures; and (iii) exorbitant and unreasonable penalties have become infeasible. The result is that lenders have had to become much more adept at gathering information about people they, at least initially, know nothing about.

When making their decisions, credit providers try to access as much relevant information as economically justifiable. Applicants are asked questions, either in person or via an application form, about their identity, contact details, income and expenditure, employment and residential stability, and so on. Reviews are made of past dealings, and calls are made to credit bureaux, to enquire on account performance elsewhere. Lenders then assess the information, and base the decision on past experience with similar applicants. The data sources used during this process are:

Customer—Any information provided by the customer, whether via an application form, or supporting documentation, including financial statements.
Internal—Compiled from application processing, accounting, customer contact, and other systems maintained by the lender.
External—Available from outside the company, whether credit bureaux, voters' roll, phone books, or other sources.
Environment—Economic or aggregate information relating to a country, region, or industry.
Staff—Any input provided by staff members, especially where the analyst provides subjective views on elements deemed relevant to the decision.

While these five categories apply generally, only the first three are widely used for consumer credit scoring. Factors relating to the environment and analysts' views play a relatively minor role, albeit with exceptions. For example:

Economy—Will have a significant impact upon the strategies used with the scorecards.
Industry—Plays a significant role in the assessment of business customers, especially where financial statements are requested.
Staff—Can play a significant role when assessing micro-finance and middle-market businesses. Staff members have the power to override the scores, or answer questions that are scored.

Each data source has a cost associated with it. McNab and Wynn (2003) split data sources out into two generic types: *free*, anything that the applicant willingly reveals, whether on an application form, via a survey, over the phone, in an interview, or in identity and financial documentation provided; *cost*, fees paid to credit bureaux or monies spent developing and maintaining systems and obtaining data from internal databases. This does, however, ignore the high cost of interacting with customers.

12.1 Customer supplied

The starting point for any credit relationship is the 'through-the-door' customer, who expresses an interest in applying for credit. Given the risks involved, at least in the absence of third-party guarantees, credit grantors have to ask for information from the customer that: (i) identifies who they are dealing with; (ii) advises what is being requested; and (iii) provides an indication of creditworthiness. The amount of information demanded will vary according to the circumstances. If the loan size is small, or potential profit is high, then there is less pressure to have a comprehensive picture. The two types of customer-supplied documentation are:

Completed—Forms that have to be filled in by the customer. This will include: (i) an application form detailing the customer's identity, contact, demographic, banking and financial details; and possibly (ii) a template for the customer's income, expenses, assets, and liabilities.

Supporting—Documentation already in the customer's possession: payslip, identity document, utility statements, and detailed financial statements where these have already been completed for other purposes. Lenders may demand to see the originals, or certified copies (identity documents/payslips), to avoid potential fraud, money laundering, and other criminal activities.

It is the 'completed' documentation that causes customers the most aggravation, albeit the demands of 'financial intelligence and control' (anti-money laundering) legislation have also increased the requirements for supporting documentation. The rest of this section focuses on two items:

(i) **Application form**—Which may be completed either by the customer or a third party, whether on paper or electronically.

(ii) **Financial statements**—Which will be demanded whenever greater scrutiny is required, but are not requested lightly, because of the extra costs and inconvenience.

12.1.1 Application form

A major tool in this exercise is the credit application form, which most people are familiar with, in one form or another. They are commonly available at banks, retailers, etc., and most look surprisingly similar. Indeed, people may tire of answering the same questions. Perhaps one day there will be smart cards, such that customers can pre-populate their details with a quick swipe through a machine, but in the meantime, they will have to endure the tedious task of filling in the forms. Given the general public's level of familiarity, rather than presenting a hypothetical application form, this section provides a walk through its purposes and parts. Most applications will have the following sections (see Table 12.1):

Contact details—Used primarily by the billings and collections areas, but they may also feature in scorecards as 'Phone Number Given (Y/N)'.[1] Personal identifiers are also used to match records on past or current loans, both internally and externally.

Loan details—Requested loan amount, repayment period and frequency. There may be instances where the requested loan may not be feasible, but alternative offers can be made.

Loan purpose—Reason for which the loan is being sought, or a description of the goods being purchased. Some lenders may ask what the borrower uses credit for generally.

Table 12.1. Application form characteristics

	Personal:	Work:	Financial:	Bank:
Contact	Name, Title, ID number	Employer name	Income:	Name and branch Account type Account number References Credit/store cards held
	Address/Previous Address		Own Spouse Other	
	Phone numbers: Home, Work, Cell, Fax		Expenses:	
	Email		Rent/Bond Motor vehicle Other credit	Request:
Stability	Time @ address	Time @ employment		Loan amount Repayment: Method Period Frequency Insurance Opt out clauses Other
	Time @ previous address	Time @ previous employment	Balance sheet: Assets Liabilities	
Demographics	Gender	Type of business/ Industry	Security:	
	Age/ Date of birth		Goods: Age Type Surety Guarantor	
	Marital status/ #Dependants	Employment level		References: Credit Personal
	Accommodation type	Level of education	Signature:	Conditions:

[1] Contact details are particularly important in subprime lending and other instances where borrowers are particularly transient. In some instances, details of the applicant's boss, relatives, or friends may also be requested.

Security—If not the asset being purchased, then some other assets, or the name of an entity willing to stand surety, or provide a guarantee. Outside home loans and motor vehicle finance, the use of collateral is not widespread, except for very high-value loans.
Applicant stability—Time at address and employment, current and possibly previous; and possibly industry and employment level. These factors continue to play a role, even where high job mobility is associated with high-paying jobs.
Demographics—Level of education, employment level, age, accommodation status, etc. These provide insight into applicants' stability, and future professional and income prospects.
Financial—Income, expenses, assets, and liabilities. These are indications of applicants' ability to repay, but in retail (especially consumer) environments they are often unreliable.
Bank details—Required to set up debit orders, but may also be included in the scorecard as 'Other Bank', 'Type of Account Held', or 'Credit/Store Cards Held'.
Financial sophistication—Credit and store cards held, and any credit references provided by the applicant.
Options—Repayment insurance and other options, which the borrower may select. Insurance can be a very profitable income source, but applicants requesting it usually know that they are higher than average risk.

Application forms have to cater for the needs of at least four different business areas, whose interests are sometimes at odds with each other:

Account management—Address details, to send regular statements; bank details, to set up debit orders; and contact details, to communicate with the customer if there are problems, especially if collections actions are required.
Credit—Data needed to profile good versus bad customers, including affordability, stability, financial sophistication, and security details. These classifications are often not clear-cut, and the layout provided in Table 12.1 is not an exhaustive list of possible questions.
Marketing—Lenders must know who is coming through the door, if they want to tailor their marketing effectively. Campaigns aimed at young families could easily attract dinkies (dual income, no kids), and it would never be known, in the absence of 'Number of Dependants'.[2]
Legal—The application form is documentary evidence that can be used in court to support the existence of a contract. The alternative would be to have a contract separate from the application form, but this is often more paperwork and unjustifiable overhead.

The information required by each area does not always overlap, and most areas have a tendency to want more information about customers. Lenders are trying to move towards more customer-friendly application forms—meaning shorter and easier to complete—albeit some lenders will try to fit as many questions onto a single page as possible (causing eye-strain for some).

[2] Lenders would also be looking for cross sales, and retailers want information for future marketing campaigns. If a clothing retailer is having a special on women's undergarments, it may wish to limit the mail shot to female customers in order to avoid offending the men.

12.1.2 Financial details

When making loan decisions, a key factor is the customers' financial strength, and ability to repay. This is not only to ensure that lenders get their money back, but also to ensure that customers do not over-indebt themselves. Many application forms will ask for a few key items, but often this is not enough, especially for asset finance and large loans to high net-worth individuals, companies, governments, and so on. At the extreme, this may include the full financial statements: balance sheet (assets/liabilities) and income statement (income/expense).

Comprehensive and reliable financial data is extremely powerful, because it provides the best possible picture of a borrower's financial strength. It has its costs though, because it takes time for customers to produce, and lenders may find it difficult to collect, capture, and keep up-to-date. It is also often treated with suspicion, because most individuals and small businesses do not have a good feel for their own finances. Lenders often design special templates to help them. If nothing else, by working through the form, prospective borrowers can work out whether they can afford the loan or not. Thus, lenders often only ask for detailed financial information where it is probably produced for other purposes, or where the loan value is large enough to warrant it. In the consumer market, this might be for home loan purchases, and large loans to high net-worth individuals. In the enterprise market, it would be for larger exposures, say greater than $100,000, but the threshold will vary from lender-to-lender.

The consumer and enterprise markets differ substantially in what information is required—as set out in Table 12.2. In both cases, the analysis is done by reviewing a variety of financial ratios, such as debt to equity, and repayment cover. Note, however, that any analysis of the financial statements should take into consideration special circumstances, such as industry, or the use of the funds to finance productive assets.

While detailed financials are often requested in the consumer market, only key items would be used in credit scoring models—like repayment to income, or disposable income—largely because so much value is obtained from borrowers' credit histories. As a result, policy rules are often applied, or the data is assessed judgmentally by an underwriter, in particular to assess

Table 12.2. Financial statement items

	Consumers	Enterprises
Balance sheet		
Assets	Property, motor vehicles, households goods, jewellery, unit trusts, insurance policies, and investments	Non-current: fixed, moveable, intangible Current: cash, inventories, debtors
Liabilities	Home loan, motor vehicle loan, overdrafts, credit cards, revolving credit	Equity: share capital, reserves Non-current: long-term debt Current: creditors, overdraft, current portion of long-term debt
Income statement		
Income	Wages, interest, dividends, rentals	Trading revenue, finance income
Expenses	Taxes, rent/bond, utilities, school, groceries, transport, subscriptions, clothing, entertainment	Cost of sales, depreciation, lease expense, taxation, extraordinary items, dividends

affordability. In contrast, larger enterprises prepare financial statements as a normal part of business, for both shareholders and creditors. Lenders request this information not only when asked for new or increased credit limits, but also as a part of regular credit reviews. The information is not always reliable though, and there are several factors that should be considered:

(i) **Who produced them?** Financial statements may be produced by: (i) the customer, in which case it may be by the accounting department, an accounting officer, or even the owner; or (ii) a third party, which may be an accounting firm or bookkeeper.
(ii) **Type of accounts?** The statements may be annual accounts, detailing results for the full period; interim accounts, for a part period; or management accounts, to provide an indication of company performance. Creditors rely mostly on annual accounts for model developments, but might consider others in a judgmental review.
(iii) **Have they been audited?** Having financial statements audited costs money, and many smaller, privately held companies will not bother.
(iv) **Who is the auditor?** Auditing firms vary in reliability, and at times even audits done by reputable firms are suspect, as evidenced by Enron and other accounting scandals.
(v) **Is the audit qualified?** Qualified accounts indicate that the auditor has a concern, which may or may not be material. The lender may have to investigate further.
(vi) **Age of accounts?** There can be a significant delay between a company's financial year-end and receipt of the financial statements, especially where there is a dispute between the auditor and the enterprise. Even where there is no dispute, borrowers may be loath to provide their financials to creditors if the news is bad.
(vii) **Size of company?** The larger the company, the better the data. More time and effort is put into preparing the accounts, and greater priority is given by the auditing firms. The turnaround time for large firms may be as little as two months, while for smaller companies it may be six months or more—and they still have to be forwarded to creditors.

Financial statement information was once the cornerstone of lending, to both individuals and enterprises, but has been receiving less emphasis, as the potential value of readily available performance data has increased. Care must be taken before discarding financial information though. Performance data is a powerful short-term measure of creditworthiness, but it may not provide a good indication of debt capacity over the longer term. Financial statement information is the best for assessing financial strength and affordability, and generally has a low correlation with performance data. It may still provide significant value, especially for larger SMEs and middle-market companies. In these markets, skilled credit evaluation managers put a lot of faith in financial statements, and will acknowledge performance scores, only where it conveys information on poor performance that might require remedial action. Good performance is viewed as neutral or marginally positive, as it is considered the norm.

12.2 Internal systems

Even though back-office functions, like billing and accounting, were the first to be computerised, it took about thirty years before the technology evolved far enough, for their data to be interrogated for any type of customer intelligence. Today, internal data is a key resource, especially where there is a large amount of repeat business, or a broad credit product range. It plays two major roles:

Predictors—The data can be used either directly, or as a part of scores, to aid in decision-making. It may be used in isolation, or integrated with other data at any stage in the risk management process.

Performance—Provides an indication of how the account performed subsequently, which is used as the target variable when developing models to rank applicants' risk, whether the predictors are obtained from the customer, internal systems, or external agents.

The rest of this section will take a look at two aspects of internal data: (i) data types, and (ii) internal database types.

12.2.1 Data types

The primary source tapped for credit information is the account management system. McNab and Wynn (2003) classify its data into two types: (i) *static*, details that hardly ever change; and (ii) *dynamic*, those that change regularly (see Table 12.3). Dynamic characteristics relate to transactions, and will often be calculated for different periods, like last 1, 3, 6, and 12 months. The key characteristics for credit risk scoring are those relating to historical arrears, however defined. In most good/bad definitions, the dominant characteristic is either 'Months in Arrears' or 'Worst Arrears'.

Many of these characteristics are highly correlated, especially when they relate to the same aspect—such as arrears—over different time periods. This can provide confusing scorecards, if all of them are used as predictors. As Thomas et al. (2002) states 'Deciding which to keep and

Table 12.3. Account management data

Static	Dynamic
Product type	Outstanding balance
Open date	Payment due
Market segment	Credits/repayments
Original loan/account limit	Debits/purchases
Loan term	Available credit
Cycle/billing date	Interest income
Interest rate	Fee income
Repayment method	Date last payment
Settlement value	Date last purchase
Date closed	Arrears amount
Date in recoveries	Arrears in months
Lost/stolen/fraud/deceased indicators	Times in arrears

which to ignore is part of the art of scoring'. This is covered further under Characteristic Selection, in Chapter 17.

12.2.2 Credit risk management cycle

Lenders' internal data sources are not restricted to account management systems; there are several others—most of which are generated by, or for, a specific stage in the risk management cycle. Many of these are covered in greater detail in Module G (Credit Risk Management Cycle), but are summarised briefly here.

Customer contact

Contains information relating to interactions with the customer, both inbound and outbound, including the nature and outcome of the contact: *inbound*—the customer contacts the business, with enquiries and complaints; *outbound*—the lender contacts the customer, through telemarketing and direct mail campaigns. One might also include purchased marketing lists within this category, as they are used to determine whom to contact.

New business origination

Application form details, and any information obtained as part of the process; in particular, credit bureau and account performance details. The data is used primarily for the initial decision and application process monitoring, but can also aid account management during the early part of a customer relationship.

Account management

Summarised details for existing accounts, including minima, maxima, averages, ratios, and counts, for various bits of information over different time periods, like 'Days over Limit last 3 months', 'Number of Dishonours last 6 months (L6M)', 'Current Month's Average Balance', 'Maximum Delinquency L6M', 'Current Value Delinquent', 'Ratio of Payments to Payments Due L6M', and others. The database may also contain: a few key characteristics obtained from the application (date of birth, gender, marital status, phone number given, but at the extreme may contain all); a geographic lifestyle indicator; and possibly some limited bureau data.

> Although the use of certain personal characteristics may be either illegal or qualified in credit risk assessments, they can often still be kept on the behavioural system and used for marketing, reporting, or other purposes. In the early 1990s, some South African lenders removed racial classifications from their databases in anticipation of expected political and legal changes, only to find that they could not report on their lending activities to 'previously disadvantaged individuals', when requested to do so by government.

Collections/recoveries

Contains a copy of the accounts' details at the time they hit collections, updates on account activity in the interim, and other information relating to collections contacts with the customer (by phone or in writing), and their outcome.

12.2.3 Operations and customer relationship management

Over and above the data sources maintained specifically for managing credit risk, there are others used for the ongoing operational management of the account, managing operational risk (fraud), and managing the customer relationship.

Customer management

Used to summarise the entire customer relationship, in order to drive customer-level strategies. The extent of product-level detail will vary from organisation to organisation. Some rely on summarised information to reduce data archiving requirements, while others may use full account-level data. Some companies may also supplement this information with marketing (date last contact) and financial (lifetime value) information.

Transaction and payments

Transactional data is the ultimate level of detail, containing details on payments to and from an account: when, how much, who, and why. 'When' and 'How much' will be included for both payments and receipts, while 'Who' and 'Why' may vary. An account number and name/reference may appear for electronic transactions and debit orders, a merchant code for credit card purchases, or a cheque number on a cheque payment.

Authorisations

Credit card transaction details are temporarily held on a separate database while they are pending approval. After the decision is made, approved transactions are posted to the main account, and declines are (hopefully) stored in a separate database. This database is unnecessary for non-transactional products, or where transactions are processed immediately, and reversed later if declined.

Local knowledge

Contains personal details about customers, sometimes those that may be considered unrelated to the lending relationship. This will often have a bearing upon customer risk, but cannot be captured in conventional scorecards, because of its diverse nature. One UK lender maintains a database of customers' personal relationships, like links between family members (father, sister, aunt, etc.), acquaintances, business associates, companies, etc., and notes about personal circumstances. This is used for managing overrides, and brings old style local-knowledge lending into the modern credit world.[3]

Financial spreading

Contains details of customers' financial details, in particular income statement and balance sheet. This usually applies only to middle-market companies, but could also be applied to SMEs and individuals. The problem at the lower end of the market is: (i) obtaining the data; (ii) data quality issues; and (iii) spreading financials into a consistent format. The latter can be made easier if spreading is delegated to the customer, especially with specially designed accounting software and communications links.

Security

Details of any risk mitigants put in place to secure the loan, including pledges (guarantees, suretyships), and collateral (fixed, movable, and liquid assets). For consumer credit, the security is usually the asset being financed, perhaps with a guarantee from a parent, or similar party. When lending to enterprises, the situation becomes more complicated. In general, transactional lenders put little reliance on collateral, due to the costs and risks involved.

Fraud

Contains information on known and suspected frauds, including names, identification numbers, phone numbers, and addresses. It is searched for every new application, and all matches are referred. If the applicant is genuine, and has unknowingly acquired a fraudster's address or phone number, the database is updated accordingly.

12.3 Credit bureaux data

The credit bureau is an institution with no direct dealings or relationship with consumers, largely unknown and misunderstood, maintaining large databases of information which may or may not be accurate. It has the power to determine whether or not an individual is given credit to consolidate his or her bills, buy a home or start a business; it is the epitome of the remote database, in its size and potential for harm equalled only by the comprehensive records of taxation authorities.

Owens and Lyons (1998)

[3] The legal treatment of local-knowledge databases is less clear. A strict interpretation of the legislation may deem it excessive information, but it will probably not get the same attention as long as the data is not used in the application process, and is not disclosed to outside parties.

Many retailers wish to promote sales by offering deals with 'SIX MONTHS TO PAY', but when doing so they are taking a risk. Rather than pushing up prices or interest rates to cover credit losses, they instead tried to prevent serial uptown shoppers from taking everyone on the downtown block for a ride. Credit bureaux (also called 'credit reference agencies') were established for retailers to pool their experiences, and gather data from other sources. Today, the bureaux are a critical information source for the retail-credit industry, and are perhaps rivalled only by certain government agencies for their intelligence-gathering capabilities. While specifically intended for use by credit providers, bureau data is also often used by service providers (phone, utilities, etc.), and for employment screening, tenant checks, and other purposes. The bureaux are well established in developed countries, but in emerging markets they may be non-existent, or the data may be thin, and delivery slow (see Chapter 14, Information Sharing).

Credit bureaux bring together information from subscribers (banks, finance houses, retailers, credit card companies, service providers) and public sources (courts, public authorities, tax registers, etc.) and collate it into a bureau record using name, address, birth date, and/or a personal identifier as a linking mechanism(s). It is all kept in a giant data pool that subscribers can dip into, but not everybody likes the same water temperature. Searches and account performance in one segment should provide much greater value within that segment than elsewhere (mortgages, bank cards, revolving credit, instalment credit, etc.), so the bureaux will facilitate separate analysis. Thomas et al. (2002) split the information provided into a variety of different types:

Publicly available—court records, voters roll.
Previous searches—counts of enquiries made.
Shared performance data—credit data pooled from different lenders.
Aggregated information—data at post-code level.
Fraud warnings—invalid address or personal identifier.
Bureau added value—bureau scores.

These data types are quite diverse, and can be described along several dimensions: (i) *purpose*—credit, fraud, verification; (ii) *source*—public, subscriber, and generated; (iii) *risk management cycle stage*—origination, management, recoveries, fraud; and (iv) *bureau function*—assembly, data pool, and value add. These are illustrated in Table 12.4. The bureaux will package this data in a variety of different forms. Some of the services offered, amongst others, are:

Customer monitoring—Advise lenders of new negative data against their customers.
Identity verification—Ensure applicants are who they say they are, whether through comparing details provided against those held by the bureau, or by asking them to confirm.
Fraud detection—Maintain a known fraud database, and/or use sophisticated routines to match and compare details from different credit applications.
Marketing—Provide the capability for lenders wishing to pre-screen campaigns to check for negative performance.
Tracing—Assist lenders with finding defaulters who have moved address, by looking out for new credit applications with updated contact details.

Table 12.4. Bureau data types

Type	Data purpose	Data source	Generating risk cycle	Bureau function
Court records	Credit	Public	Legal	Assembly
Voters' roll	Credit	Public	N/A	Assembly
Enquiries	Credit	Generated	Origination	Value add
Shared performance data	Credit	Subscriber	Management	Data pool
Aggregated data	Credit	Subscriber	Management	Value add
Bureau scores	Credit	Combined	Combined	Value add
Personal identification	Verification	Generated/ public	Origination	Assembly/ data pool
Application	Verification/ fraud	Subscriber	Origination	Data pool
Fraud warnings	Fraud	Subscriber	Origination/ management	Data pool

The rest of this section provides greater detail on each of the data types.

12.3.1 Enquiries/searches

When a watering hole is dug, it does not take long before animals come to drink. Once it becomes known as a regular source of water, a drought's intensity can be judged by the pitter-patter of passing paws and hooves. In credit, the pitter-patter is the footprint of enquiries made whenever consumers apply for new credit, or credit-related facilities. Lenders can listen, because the bureaux keep count and provide details of prior searches. Items, such as the date, type of search, who made the search, and type of industry, are recorded for future reference. Enquiry data is unique, as it is the only data that the credit bureaux create entirely by themselves. The more established and accepted the bureau is, the more likely it will have richer and more predictive search information on each individual.

Purpose of enquiries

Enquiries are the gateway into the credit bureau. They need to be considered from two angles: (i) purpose; and (ii) means of access. The main purposes are:

Marketing solicitations—Where lenders pre-screen customers prior to making offers.
Application processing—The enquiry is customer initiated as the result of a request for credit.
Account management—Any enquiries done as part of ongoing management of existing accounts, whether for risk management, collections, or fraud.
Scoring and analytics—Enquiries made that have no bearing upon the individual customers, but are used solely to develop scorecards or monitor portfolio performance.

In the United States, the credit bureaux keep separate counts of marketing and account management enquiries, which may be used in other assessments, and are provided to individuals when doing online enquiries regarding their own bureau activity (Mays 2004). No record is maintained of scoring and analytic enquiries. Application-processing enquiries are the only ones relevant for assessing new credit applications, and it is crucial that the others do not contaminate this count. Both lenders and bureaux must guard against the others leaving a footprint, because the system cannot distinguish between them. There is also an issue of double-counting, when staff members make manual enquiries over and above the automated call.

Care must be taken when interpreting this data, as it is impossible to determine the circumstance or outcome of every enquiry. As regards circumstance, an example is individuals that have recently moved house, or are doing renovations. There may be a large number of enquiries as they purchase furniture and fittings, which would otherwise be highly negative. Even so, a single month with 10 or more enquiries should provide a significant warning sign, and 15 may be sufficient reason for a policy reject. As for outcomes, if the enquiry cannot be directly associated with a new line of credit, it is impossible to determine why no link can be found. The application could have been rejected by the lender, but often the applicant decides not to proceed, because: (i) personal circumstances have changed, and the credit is no longer required; (ii) a better offer was obtained elsewhere, which often happens with high-ticket items like houses and motor vehicles, where people shop for the best deal; or (iii) the enquiry was made without the knowledge of the individual, by a dealer/broker who forwarded the application to several lenders that each made their own enquiry.

Means of access

Information collated by the credit bureau is provided to subscribers as a credit report. The means used to access it vary depending upon the purpose:

Telephonic—A staff member phones the bureau to be provided with details over the phone, or possibly by fax. This is seldom done today, except where communications are poor.

Manual online—A staff member gets a direct connection to the bureau's system, and views a bureau record on a computer screen. Done for ad hoc enquiries, or instances where an automatic link is not available.

Automatic online—A computer-to-computer connection is made. Usually implemented as part of an application process, where response times are critical.

Current batch—A group of records is processed simultaneously, to get a view of their current credit standing. Usually conducted for marketing or account management purposes.

Retrospective batch—Ditto, but records are obtained for past dates specified by the subscriber. Usually undertaken for application scorecard developments, based on the application date.

The online and batch enquiries may return comprehensive details (of every enquiry, account, judgment, etc.), or summary details. Comprehensive details are the norm for manual enquiries, while summary details are required to build and apply scoring models. New bureau-manager technologies allow lenders to obtain and store comprehensive details for future reference, whether done online or in batch.

12.3.2 Publicly available information

A lot of information is freely available to the public, but to access it, it is necessary to visit the public library, local courthouse, or local voter-registration authority. Credit bureaux add value by bringing the data together into a single repository, and better yet, putting it into an easy-to-access electronic format. The physical collection and capture may be done either by the bureau, by the entity that controls the source, or by contractors that act as intermediaries. This applies to: *court records*, details on bankruptcies, judgments, liens, and other court orders; and *voters' roll*, the register of voters, which is something specific to the United Kingdom.

Court records

Court records are a crucial source of information on severe past defaults, whether for insolvencies, judgments, or other court orders. In each case the court record will include: defendant's name, date of birth, personal identifier (if available), address, judgment amount, complainant's name, and reason for the judgment. Care must, however, be taken with these records:

Data retention—There are usually legal restrictions, or public relations considerations, regarding how long the data may be retained, or used as part of the credit decision. Periods may range anywhere from 3 to 10 years.

Matching—Many countries either do not have a personal identifier, or the courts do not record it in their records. If so, customers can only be matched with court records using name, address, and perhaps date of birth. In this case, the bureaux' data-enhancement capabilities can provide a key competitive advantage.

The court records refer primarily to two types of legal actions, bankruptcies and judgments. No lending is allowed to bankrupts, while judgments are extremely prejudicial. Literally interpreted, 'bankrupt' means that an individual's bank has ruptured, or rather they are financially ruined, and are not able to meet the claims of their creditors. Its synonym is 'insolvent', which literally means 'being incapable of providing a solution', or in this instance being unable to clear one's debt. Bankruptcy/insolvency[4] is something that is adjudged by the courts, who:

(a) Legally prevent the individual or legal entity from entering into any new debt.
(b) Release them from direct demands from creditors, making them instead answerable to the court.

---

[4] Edelberg (2003) cites research indicating that as income increases the probability of default decreases, but the probability of bankruptcy is higher once default occurs, because individuals have greater incentive to shield it from being garnisheed.

(c) Provide liens against assets, which effectively allows debtors to take possession, pending settlement of the debt.

Bankruptcy may be voluntary, where the individual or legal entity wishes to escape the demands of creditors; or involuntary, where debtors foreclose in the hope of preventing any possible further claims against assets, which is their only hope of recovering the debt. A by-product of the legal process is a court record, the existence of which should cause an automatic decline of any request for credit, otherwise the lender can be deemed to be acting contrary to the orders of the court.

Judgments are *legal orders* for the borrower to repay, which can cover debt, rent, service charges, or any other obligation. They are credit providers' last recourse prior to foreclosure, and their ultimate whipping tool. The claimant must show that the defendant has been given a default notice or letter of demand to: (i) advise of the proceedings; (ii) offer a chance to repay or come to an arrangement; or (iii) provide the opportunity to file a defence. If nothing has been done within a specified time, a court order will be issued that gives the defendant a limited time to repay, for example, one month. If the debt is not repaid, a judgment is registered. While judgments do provide valuable information to a lender that is considering a new loan application, there are some faults:

Collections usage—Judgments should only be taken as an option of last resort, but collections areas sometimes use them as a powerful method of persuasion; a judgment is registered, and then lifted when the customer repays. It has the dubious advantage of providing more data to lenders that do/can not share performance data, but results in a very disgruntled public. According to McNab and Wynn (2003), from the early 1990s the number of judgments in the United Kingdom reduced by half, as companies became more reluctant to use them.

Legal and admin costs—Some lenders will take out a judgment for every bad debt as a matter of principle, to protect both themselves and the general public, but legal services are not cheap. Others may choose not to incur the cost, if the loan value is small and there is little chance of recovering the debt.

Judgments and other adverse details also have a very high public profile. A 2003 survey by the South African Department of Trade and Industry [Notice 1249] on the role of credit bureaux indicated that of 300 respondents 32 per cent had not been informed of the listing, 24 per cent had paid or had goods repossessed, 14 per cent had suffered accident, illness, or job loss, and 9 per cent had informed the lender of financial problems. While these numbers may be extreme, they are probably indicative of public perceptions in other countries. The report also highlighted that many employers do credit checks, and these sometimes result in people being refused employment (similar effects occur where credit checks are used to screen prospective tenants).

Voters' roll

Many countries have a personal identifier that allows companies to track individuals, especially when they change addresses. The United Kingdom does not, and as a result, UK lenders are at a significant disadvantage. Voters' roll data (VRD) is used, at least partially, to offset this shortcoming. Unlike many other countries, it is not obligatory to vote in the United Kingdom, and as a result this information adds value to credit risk scoring. Characteristics like 'Same Surname on Voters' Roll? (Y/N)', 'Years on Voters' Roll', and a check of 'Years on Voters Roll' against 'Years at Address', all provide valuable stability measures, and an indication of civic-mindedness that is correlated an with individual's attitudes towards debt, and still adds value over and above all other data.

Local councils supplement their income by converting their voters' rolls into electronic format, and selling it to the credit bureaux for a not insignificant price (Thomas et al. 2002). According to Wilkinson (2003), VRD was used in manual underwriting for years, and was then built in to automated processes. In 2002 however, a disgruntled voter, who protested against the sale of voters' rolls to credit bureaux, challenged its use.[5] The Data Protection Act states that information should only be used for the purposes for which it is provided, and the VRD is compiled for voting. The credit industry countered and argued for its retention, not because it was required for credit scoring, but because it was crucial for lenders to comply with new money-laundering legislation that requires financial institutions to know their customers. Rather strange that there are two competing and contradictory forces: 'Data Privacy' and 'Know Your Customer'.

In order to solve the problem, there are now two registers maintained: one that is open to the entire public and another that may only be used for credit checks, money laundering, and other limited purposes. About 25 per cent of people opt for the second. There is, however, still the possibility that the United Kingdom may adopt their National Health Insurance number as a personal identifier, in which case VRD will become redundant. A motivating factor is the United Kingdom's proposed integration into the European Community, as most member states—who are critical of the United Kingdom in this regard—not only have personal identifiers, but also local authorities where all residents are obliged to register.

12.3.3 Shared-performance data

Shared-performance data is constructed from details (provided by bureau subscribers) of the various tradelines that can be linked to an individual or other legal entity. At the core, these details include, but are not limited to:

Balance details—The outstanding debt and facility limit. High levels of debt and limit utilisation are a greater risk.
Account type—Revolving, instalment, bank credit card. Greater risk is associated with heavy utilisation of unsecured debt.

[5] *Robertson v. Wakefield Council*. (McNab and Wynn 2003).

Arrears—Too new to rate, current, late payment, 30/60/90/120/150+days in arrears, arrangements made, repossession, bad debt.
Activity—Date open, closed, or last active. Closed accounts may be excluded or ignored, and the date last active is used to determine when records are deleted. Date opened can be a key field to provide the time since the first and last accounts were opened.
Relationship—Primary accountholder, joint account, unknown, surety. This purportedly allows joint accountholders to get the benefit of good performance, but is not included in scoring models (Mays 2004).
Industry codes—Bank, retailer, card issuer, finance company, credit union, cellular operator. This is shown, and is used to deliver generic industry solutions and bespoke bureau scores.
Subscriber code—Identifier for credit provider. This is not shown with the tradeline record, but can be used to aggregate, or exclude, details for subscribers' own customers.
Consumer statement—Possible explanation for dispute.

Of these, the most important variable is the arrears status, which is presented as a 'payment profile' for each tradeline. These are represented as strings of numbers and/or letters detailing accounts' delinquency history, where the latest month's status takes the left-most position. For example, a string of '002213210100' contains a full year's history, and indicates that, although the account is currently up-to-date, it was three months in arrears six months ago, and it took some effort to bring it right. It is also relatively simple to evaluate, to determine the worst delinquency status over the past 3, 6, or 12 months. Such strings also provide collections staff with a quick overview of account performance on a computer screen.

Collections and recoveries agencies

Collections and recoveries agencies may also subscribe to the credit bureaux, and provide information in return—like 'Date Received for Collections', 'Original Creditor', 'Original Amount', 'Balance Outstanding', 'Repayment Terms', and others. Such information can be very valuable, especially if lenders write off debts without taking judgment; or the quality of bankruptcy and judgment matching is suspect. Lenders could create links to the agencies' systems directly, but it is usually simpler, and cheaper, to route this through a credit bureau that has a relationship with one or more agencies.

Medical collections

A special case is medical collections, which are usually noted as a separate class. The reason is the number and duration of disputes that can arise between consumers, health care providers, and medical aids. Privacy concerns have also been raised by consumer bodies (CFA 2002), as the type of malady can often be inferred from the service provider's name. This is of most concern where bureau reports are used for employment screening, as a potential employer may be able to deduce personal- or family-health problems (fertility, mental health, Aids), that may demand time off. To counter this, bureaux may restrict the amount of detail that employers may view.

Householding

The credit histories of associated individuals have also proved predictive of individual creditworthiness, but this presents ethical problems. Why should your housemate's creditworthiness affect yours? This could relate to the relatives, other people in the same house, same street, etc. The use of related-party information was rife a few decades ago, when credit bureaux were using index cards and filing cabinets. In the United Kingdom, this persisted into the 1990s, as the bureaux used to return information not only for the applicant, but anybody else living at the same address, irrespective of whether the people were related (same surname) or not. Over time, the facility was limited to same surname only, but both public outcry and data-privacy legislation subsequently removed even this. Since then however, changes in legislation have allowed the use of householder information for low-score overrides, where the applicant has little or no credit history (Taylor 2004).

Business dealings

There is a strong correlation between entrepreneurs' personal and business dealings, and when accounts are in the name of legal entities, it can be difficult to link them. If at all possible, they should be combined in a single assessment. As the size of the enterprise increases though, especially where there are a large number of shareholders, the behaviour of the individual becomes less relevant.

12.3.4 Fraud warnings

Credit bureaux may also provide 'fraud warning' services. The warnings can come from three main sources: (i) *known fraudulent activity* reported by subscribers; (ii) use of *third-party information*; and (iii) *data-pooling arrangements* to screen applications for potential fraud or embellishment. These warnings do not mean that the application is fraudulent, but that the lender should be more diligent when validating details. The application should only be rejected if it is proved fraudulent.

Known frauds

For known fraudulent activity, the bureaux can act as central repositories for data provided by subscribers—in particular, any available identifying characteristics, such as name, address, phone numbers, and personal identifier. A fraud warning is then returned for applications that match on any of them. Other characteristics returned might include the contributor, fraud type, date loaded, and loss amount. False positives can arise, because the fraudster has moved address, changed phone numbers, or the applicant was the victim of identity fraud. If subsequent checks indicate that the identity details are correct and the application is genuine, then the details must be removed from the known-fraud database. An exception is 'Protective Registration', a CIFAS facility where applicants intentionally request that their details be loaded, whether because of known or suspected identity theft.

Third-party information

Fraud checks can also be done using information obtained from third parties other than credit bureaux, including government departments, voter-registration authorities, telephone companies, the post office, property registers, and others. What is possible will vary depending upon technology, and regulations for each country. Where allowed, a search against a government database containing a personal identifier (Social Insurance Number, Social Security Number, Identification Number), can indicate that the identifier does not exist, the individual is deceased, there is an inconsistency in the date of birth, or that documentation has been reported as lost or stolen. A search of a telephone database can: (i) confirm the residential address; (ii) ensure that the home phone area code and residential postal code correspond; and (iii) ensure that the phone number is not a phone booth. Care must be taken because, in many cases, phone booths are valid numbers, especially for shared accommodation, and educational, medical, and other institutions, where they must be used for private calls. Likewise, a search of a post-office database can indicate addresses known to be mail drops, correctional facilities, or other high-risk addresses.

Application data sharing arrangements

Applicant details can also be verified by comparing the current credit application to prior applications made by the same individual elsewhere. This is a data-pooling arrangement that includes not only personal-identification and contact details, but also details on income and employment. If the comparison indicates that details—such as income or employment—have been manipulated, the application may be rejected or further checks undertaken, to determine if there is fraud or embellishment.

12.3.5 Bureau scores

The amount of information provided by the credit bureaux can be massive, and many companies are neither able, nor interested, in trying to assess it by themselves. Instead, they subscribe to generic bureau scores that summarise available bureau data, which are often obtained at a price over and above the normal enquiry. The most well known are risk scores: *FICO®*—a general default risk score 'designed to rank-order consumers as to whether they will pay as expected' (Mays 2004) that is tailored to information on the different credit bureaux; *BEACON®*—FICO® score provided by Equifax; *EMPIRICA®*—FICO® score provided by TransUnion; and *DELPHI®*—bankruptcy score developed by Experian.

> According to CFA (2002), FICO scores are produced for between 190 and 200 million Americans—almost every adult—and are major determinants of interest rates and loan terms. The FICO website says that there are six pricing ranges, whose meaning is fairly standard: two subprime ranges—500–559 and 560–619—and four prime ranges, starting at 620, 675, 700, and 720. Consumer ignorance of their own scores, and aggressive sales tactics by subprime lenders, mean many consumers with scores 620+ have subprime loans.

According to Mays (2004), FICO scores are 'designed to rank the likelihood that an applicant will go 90-days delinquent on any consumer credit loan or account, within the next two years'. This is effectively a pooled-data behavioural score. Other types of bureau scores are: (i) *generic industry scores*, that target a specific industry or product type; (ii) *revenue, retention, repayment* (collections) and other scores, used at various stages in the risk management cycle; (iii) *generic application scores*, a rare breed that focuses on new business enquiries only, and includes application form details; and (iv) *bespoke bureau scores*, that target a specific subscriber's customers, and are either delivered by the bureau, or calculated by subscribers upon receipt of bureau data.

According to Princetich and Tobin (1998), American credit bureaux can provide separate scores for 'auto loans, bank cards, installment loans, personal finance loans, mortgages, insurance policies, retail cards, and cellular accounts'. The good/bad definition used for each may vary, and for mortgages, a 'bad' is typically associated with 'foreclosure, write-off, or bankruptcy'.

Mays (2004) also comments that, 'more credit decisions today are affected by generic than by customised scoring models'. The comment refers to the United States, where many lenders rely upon bureau scores, rather than developing their own models. They may be developed using very rich data sources, and be highly predictive, but this presents a risk. Most lenders will not solicit any individuals with FICO scores less than 660, which results in everybody chasing the same customers. Thomas et al. (2002) provides both criticism of, and justification for, the use of generics. Criticism, because of their generic nature, especially given that the resulting score does not relate to a specific lender's experience, market position, or product. Justification, because there are context-specific instances where they should be used:

(a) By small lenders, who have neither sufficient data to develop their own scorecards, nor sufficient staff to manage the process.
(b) By new entrants, with little or no experience with the market or product.
(c) By credit providers who are new to credit scoring, and wish to focus on using scores, as opposed to developing and managing them.
(d) As an independent measure of application quality over time, to provide a benchmark for internal scoring systems.

12.3.6 Geographic indicators

Birds of feather flock together.

Proverb

A well-known maxim touted by professional property punters is to focus always on the three Ps—position, position, and position. Well, 'position' can also provide value in credit scoring. Borrowers' creditworthiness is affected by their environments, which include industry and geography. Unfortunately, industry is seldom captured in consumer credit (except perhaps under profession), but geographical location is immediately obvious from the physical address (postal is second choice). Its use for credit risk scoring may be contentious though, because of sensitivities about 'red-lining' suburbs, especially because adversely affected suburbs are often largely inhabited by minorities. If used, it must be only one of many elements in a decision, and play a minor role. To use it, lenders must decide on whether they will: (i) use post-code regions, or lifestyle indicators; and (ii) use the classifications directly, or calculate geographic aggregates. Lifestyle indicators are almost always used directly.

Geographic aggregates

Most credit bureau data relates directly to individuals, but they can also calculate new variables to aggregate values for different geographic regions. Thomas et al. (2002) provide several examples at postal-code level: percentage of houses with judgments; percentage of accounts up to date; percentage of accounts three or more months in arrears; and percentage of accounts written off in the last 12 months. The value of this information varies, depending on the size of the postal-code footprint. This ranges from a couple of city blocks in the United Kingdom and Canada to entire suburbs or towns in the United States and South Africa. If the footprint is too small, there may not be enough cases for the aggregated data to be reliable; if it is too large, it may encompass so many different types of residences that the data is too general to be of value.

A similar concept is the use of regional economic data (cross-sectional data), such as unemployment rates and GDP growth rates. This would not be done by the bureau though, but by lenders themselves. While the values could be used directly, another possibility is to normalise them relative to the national average.

Geographic lifestyle indicators

Rather than grouping by geographical region, addresses can instead be grouped by lifestyle indicators or, alternatively, demographic or sociographic indicators. Their greatest use is where other customer data is not available (marketing), or where it can provide some risk insight not available at the individual level (credit). Lifestyle indicators are derived using cluster analysis, which identifies distinct groups, such as 'old money', 'happy couples', 'urban squalor', etc. Exactly which addresses are included in each group will be affected by the types of data used: *demographic*, individual details, which can be stated as facts or figures; *attitudinal*, relating to general opinions and perceptions; *lifestyle*, an indication of local culture, which combines both demographics and attitudes; *preferences*, usually relating to products. The main sources of this information are:

National census—Primarily demographic data gathered by government, for example income, number of children, level of education, etc. It has the advantage of being some of the most comprehensive, in terms of number of people covered, but dates quickly, and does not give much of a feel for how people think.

Market research—Relates to attitudes and preferences, but also includes demographics to determine correlations. This data is usually based upon samples taken within regions, and while it may be updated more frequently than census data, it may not be as reliable.

The size of the postal-code footprint can again affect the results. Where it is large, there may be many socio-economic groups living within the same area, and street-address details are required to provide more accurate classifications. This can be complicated where there are different spellings for the same street (or different languages), shantytowns, and informal addresses (shack-lands). One instance that cannot be countered is live-in employees, which is a fact of life in some developing countries.

There are a number of different lifestyle code products that vary according to country, including: ClusterPlus (First Data Solutions), PRIZM (Claritas), MicroVision (National Decision Systems), Mosaic (Experian), and Spectra Grid (Spectra). With each, there is the risk of using outdated information, and of favouring or prejudicing the wrong people if not used wisely. To obtain the codes, lenders either have to: (i) load the necessary tables and software on their own systems; or alternatively (ii) obtain them via the credit bureau, as a value-added service.

12.3.7 Miscellaneous sources

There are other sources of information that may be provided via the bureaux, a third party, or directly from source:

Motor vehicle registrations—Provided by a company called Polk in the United States, and used for marketing. It is a source of individuals' ages and addresses, and little more.

Telephone and city directories—For name, telephone and/or address details. It may not be possible to use this for credit decisions, but is invaluable for collections, tracing, fraud, and marketing. Provided by First Data Solutions and Metromail in the United States.

Property services—Provide lists of valid addresses, and registered owners.

In the United States, lenders may not discriminate against people with unlisted phone numbers. TransUnion offers a 'Phone Append' facility, which searches telephone number databases purchased from the phone companies. If a number is confirmed as unlisted (as opposed to 'No Phone'), a 'private number' indicator is returned.

12.4 Summary

Retail credit risk assessments can be done using a variety of information sources, which can be treated under the headings of: (i) *customer-supplied*; (ii) *internal systems*; and (iii) *external agents*. There are different advantages, disadvantages, and costs associated with each. Customer-supplied data is obtained as part and parcel of the account-origination process, and although provided willingly, it still has its costs—not only in terms of collection, capture, and storage, but also inconvenience to the customer, and potential damage to the customer relationship. Its shelf life is also limited, if no mechanisms are in place to keep it up-to-date. As a result, lenders strive to minimise the amount of information required from customers, and instead make optimal use of other sources. The exception is larger loan values, which fall outside of lenders' comfort zones for their automated processes.

Lenders' own internal databases can be the cheapest source of information, especially when they are a by-product of other processes. They play a key role when assessing repeat business or multiple products; and in account management, collections, and marketing. The expensive part is putting the appropriate systems in place, especially behavioural and/or customer scoring. Once in place, lenders try to extract maximum value out of their own data, but there are some disadvantages: (i) they often only reflect problems when it is too late; and (ii) the scores are volatile, with shelf lives shorter than those of Christmas toys, so they must be updated at least monthly.

Finally, lenders will access as much information as economically feasible from external sources—much, or most, of which is channelled via the credit bureaux. Their most powerful data relates to judgments, payment profiles, and enquiries, but value can also be extracted from geographical aggregates, lifestyle codes, and other data. Bureaux also offer generic *bureau scores* that summarise their data, which many lenders rely upon instead of developing their own bespoke scores. External vendors charge for the privilege, usually on a per enquiry basis; however, fixed charges are sometimes negotiated for larger customers where volumes are predictable. The benefits usually justify the cost when used for account origination but may be questionable thereafter. Over time, bureau costs in most countries have been dropping, and when sufficiently low, it should also be feasible to use the data for other functions, such as account management and marketing. Implementation of bureau-manager software can facilitate this further.

13 Scoring structure

The first two chapters of this module covered data considerations, and the various data sources. This section moves on to some of the practical issues encountered when bringing data together and making it work. It is an awkward chapter, in that many of the topics are not normally treated together—if treated at all—in other texts. Even so, each topic provides some insight into the scoring process:

Customisation—The broad types of scoring solutions available to lenders, especially generic solutions that can be applied across a spectrum, and customised solutions developed for a specific customer, product, and process.

Hosting—Bespoke and generic solutions are usually hosted on internal and external systems respectively, but exceptions exist. The choice depends on the organisation's ability and willingness to invest in the required infrastructure.

Data integration—Merging of data from different sources for use in strategies. Separate scores may be derived for each, and applied sequentially or in a decision matrix. Alternatively, the scores, raw data, or some combination of the two, can be integrated into a single score.

Credit risk scoring—Types of credit risk scorecards, including application, behavioural, customer, and collections. The delineations relate mostly to the credit risk management cycle (CRMC) stage where used.

Matching—Whether dealing with individuals or companies, the correct records for each case have to be found. This requires one or more matching keys, or complex matching algorithms.

13.1 Customisation

Lenders wanting to use credit scoring must decide on the required level of customisation (the decision will, of course, only be made after the company has decided that it wishes to use scores at all). According to Mays (2004), the choice will be affected by several factors: *Feasibility*—Is it possible for the lender to develop a scorecard that provides added value? The big issue is expectations, and how the scores will be used in decision-making. *Development*—Are the resources available to do the development? There may be issues with staff and scheduling. *Implementation*—Can the solution be effectively implemented? Whether bespoke or generic, lenders have to consider how the data will be gathered, scores calculated, and decisions delivered to where they are needed. The following sections will look at customisation under the following headings:

(i) **Generic scorecards**—One-size-fits-all, which can be bought off the shelf, ready for use by any lender that fits the profile. The developmental data requirements are nil or minimal, and although usually easier to deploy, they are not as predictive as bespoke models. Generics are most appropriate for small lenders, new market entrants, and those not wishing to invest in a bespoke development. A subcategory is generics developed using pooled data.

(ii) **Bespoke scorecards**—Custom made and specifically tailored for the company, product, and process at hand. They are more predictive, but also more costly, demanding in terms of data and management time, and more difficult to implement and maintain. They are most appropriate for larger companies and banks, for which lending is the main business, or a significant part thereof.

(iii) **Expert models**—Both generic and bespoke scorecards are associated with data-driven models. There are instances where there is so little data that even a generic model is not possible. The alternative is to base models upon inputs provided by experts in the field.

13.1.1 Generic scorecards

The term 'generic scorecard' applies to any scorecard that has been developed using borrowed experience, whether data or judgment. The most widespread and well-accepted type is bureau scores, which provide a generalised measure of creditworthiness, based upon consumers' (or businesses') credit-reference data.

The game and the players

Generic scorecards are usually developed where there is a broad need in the market, but individual lenders are unable to invest heavily in providing their own solutions. Although generic scorecards are 'one-size-fits-all' by definition—some tailoring can be done around a number of different dimensions:

Purpose—Risk, response, revenue, or retention.
Process—Marketing, application processing, account management, collections, recoveries, fraud/verification.
Place—Credit bureaux, internal systems, customer supplied.
Producer—Store credit, bank lending, service providers.
Product—Credit card, cheque, revolving credit, vehicle finance, home loans, utilities, trade credit.
Population—Low-income, subprime, thin-file, small businesses, middle-market businesses.

Mays (2004) lists over 70 different generic scores available from various vendors in the United States, also including bankruptcy, insurance, income prediction, and segmentation models. The range of available options differs from country to country, is greatest in developed countries, and is increasing in all environments, as data and the technology to exploit it improve. Bureau scores are considered 'generic', because they cover a broad spectrum of consumers and businesses listed on bureaux, and are used for a variety of purposes. They are often used on a stand-alone basis for new business assessments, but this ignores other information that may be available, especially that supplied by the customer. This can be partially solved using generic application scores.

Generic application scores

Which came first, the chicken or the egg? For newly developed application-processing systems and new markets, lenders want scorecards to drive decisions, but sometimes there is no data to develop them. Generic application scorecards can provide an interim solution, to integrate customer-supplied, internal, and credit bureau data. Vendors were providing this type of generic long before bureau scores gained broad acceptance, and some research has been done into their effectiveness. In one case, an attempt was made at developing a single European scorecard (Platts and Howe 1997), and in another, a scorecard that used combined data for American credit unions (Overstreet 1992). In both cases, it was shown that the generic application scorecard was better than no scorecard at all, but fell far short of bespoke scorecards. This can either be taken as proof that: (i) there is a lot of information loss when developing a generic, either because of different infrastructures, or the large number of assumptions being made; and/or (ii) there are significant differences between populations that affect credit scorecard performance (Thomas 2000).

Who develops generics?

A 'P' that falls outside of the above list is 'provider', or who develops the generic scorecard. There are a number of different organisations that develop, market, and use generic scoring solutions:

Credit bureaux—Information services that provide details on customers' credit histories, and provide scores as a value-added service (Equifax, Experian, TransUnion). Section 12.3 covers credit bureaux in much more detail.

Scorecard vendors—Companies with areas that specialise in developing scorecards (Fair Isaac (FI), Experian-Scorex). When providing generic solutions to *lenders*, they rely upon data they have been able to accumulate and pool, and/or experience from previous developments. When providing solutions to *bureaux*, the nature of the data makes any resulting scorecard a generic by definition, unless there is special tailoring for an individual subscriber.

Co-operatives—Special arrangements that assemble data pooled from several lenders. This is typically used where lenders do not have sufficient data to develop their own solutions, and no appropriate generic exists.

Credit rating agencies—A special case, applied to scorecards used to assess companies' annual financial statements. These may be developed to model actual defaults (Moodys KMV), or to mimic the rating-agency grades, with no link to default rates other than for validation (Fitch ratings).

All of the above relate to companies who develop scorecards for use by others. There are still others that develop generics to aid their own ends:

Franchisors—Companies that have an interest in quality control, and wish to protect their interests. The primary examples are Visa and MasterCard, who may provide fraud-scoring products for franchisees.

Recoveries agencies—Companies that purchase defaulted portfolios, whose profit is recovered amounts, less the purchase price and recoveries costs.

Securitisers—Companies such as Fannie Mae and Freddie Mac, that use scores to value loan portfolios prior to purchase and transformation into marketable securities.

Securitisation is a relatively new feature of the American market, which is less than 20 years old, and quickly spreading to other countries. Kitchenman (1999)[1] concluded that because of securitisation, US mortgage rates are perhaps 2 per cent lower than in Europe, a significant saving on their home-loan balances of $6 trillion. Cate et al. (2003) ascribe this not only to the liquidity provided by securitisation, but also to the better risk assessment made possible by the data quality and efficiency of the credit bureaux.

Advantages and issues

There are both advantages and disadvantages associated with the use of generics. The advantages are fairly straightforward, whereas the disadvantages will vary depending upon the situation.

Data—Lenders do not have to wait until they have sufficient history with their own customers. They can either use an existing generic to borrow from other lenders' experiences or work with other lenders to pool data for a new one.

Cost—Generic scorecards are cheaper, because costs are spread over a larger number of lenders. Where hosted externally, the charge is usually on a per enquiry basis for smaller subscribers, but can be fixed for larger players with predictable volumes.

Management—Credit scoring can be very demanding on management time, and generics can allow management to focus on the business instead of the scorecards.

[1] Quoted in Cate et al. (2003).

Lenders have to consider whether these benefits are sufficient. They can be substantial, but there are also a lot of potential issues that may undermine their use, to the extent that lenders may be forced to do a bespoke development, or do without:

Applicability—Is the generic score appropriate for the proposed product, market, and process? For example, if most bureau data comes from retailers and service providers, its generic score may not be as appropriate for banks, finance houses, and mortgage lenders.

Stability—Is the data stable? Changes to a credit bureau's infrastructure, and subscriber base, can impact upon the reliability of their scores.

Omissions—Has highly valuable data been left out of the generic solution? For credit risk assessment, credit providers often have extensive internal data from broad customer relationships that will be rich information sources, if correctly harnessed.[2]

Data fields—Are the appropriate data fields available? This applies to generics intended for internal implementation, and compounds issues regarding scores' reliability. Lenders must consider how closely their data fields correspond to those intended in the generic's design.

Communication—How will the generic score be obtained? If it is provided by an external agency for each individual enquiry, then a stable communications link is crucial, if response times are important.

Transparency—Is clarity required regarding score drivers, and is it provided? Most bureau scores are proprietary, but there are increasing demands for transparency—either to provide customers with decline reasons, or to ensure that lenders have an adequate understanding.

13.1.2 Bespoke scorecards

At the other end of the spectrum are bespoke scorecards, which are tailored for a lender, product, market segment, and process. The primary advantages are: (i) *predictive power*, their ranking ability is greater than generics'; (ii) *control*, lenders have greater control over the scoring process; (iii) *sources*, a greater number of data sources can be included, in particular lenders' own past history with clients, some of which may not be fully reflected by the credit bureau. Once again though, there are a lot of potential issues that must be considered, which may make a bespoke scorecard infeasible:

Data—Is there sufficient relevant data to develop a scorecard, and is it in an appropriate form? Problems can arise because it is a new product, new market, or greenfield development.

Target—Can observation data be matched with outcome performance, and has sufficient time elapsed for accounts to mature? Problems arise where charged-off and/or closed accounts have been purged.

---

[2] There is a vendor in the South African micro-lending market that provides combined account management, application processing, and bureau services. In this case, lenders have the potential to obtain the best of many worlds.

Development—Are the required human and other resources available? There may be a problem with finding the right people, and scheduling the development.
Cost—Will the benefits justify the development and implementation costs? Mays (2004) indicates that the cost of developing and implementing a 'customised system' (bespoke) ranges from $40,000 to $100,000.

If any of these issues are considered significant, a generic scorecard may provide a more viable alternative.

13.1.3 Expert models

Credit scoring is almost always associated with data-driven models, but in some instances, even generics may be impossible to find, and infeasible to develop. All is not lost though! Many of credit scoring's benefits arise purely because it provides consistent results, and experts' experience can be tapped to develop a preliminary scorecard. In the consumer market, scorecard developers may have enough experience to construct judgmental scorecards that have surprisingly good results (Mays 2004). Failing that, underwriters' experience can be tapped directly, by developing a decision tree to mimic their thought process, or by developing a scorecard to predict a judgmental grade.[3]

Similar concepts can also be used for new markets, and greenfield systems. These models will not be as predictive as bespoke models, but experts are very adept at identifying the most relevant factors for a decision. Once the system has been up and running for a period of time, and empirical performance becomes available, the model can be adjusted. While expert models should only be used as stop-gap measures, there are instances where other options will never be possible, because of the small size of the group being considered.

13.2 Hosting—internal versus external

When discussing bespoke versus generic, a distinction should also be made between hosting types. This defines the deliverable, and delineates who will be responsible for sourcing the data, calculating the scores, and ongoing of monitoring of the scorecards (Table 13.1). Hosting on lenders' own internal systems provides the greatest control and flexibility, but involves huge management hassles. In contrast, external hosting removes management hassles, but the frustration then becomes lack of control. Internal hosting is done for most bespoke developments, and external hosting for most generics. Externally-hosted bespoke scorecards are sometimes called 'hosted solutions'.

The primary difference between the two options is that the internal hosting allows maximum use of lenders' own data, but this has very high infrastructure costs. The number of

[3] Models developed to predict judgmental grades are rare in retail credit, but more common in wholesale. For example, Fitch Ratings has developed models that are used to predict the average rating that would be provided by Fitch IBCA, Moodys KMV, and Standard & Poor.

Table 13.1. Hosting—in-house versus vendor

	Internal	External
Deliverable	Scorecard	Score
Provision	In-house	Delivered
Data source	Provided by the customer and/or sourced from in-house systems, often supplemented by vendor data.	Details on dealings with many lenders, general financial standing, and other factors.
Cost	Fixed, relating to developing, implementing, and managing the scoring infrastructure.	Variable, cost of enquiries on a per transaction basis.

transactions that lenders must process to justify it can be huge. In contrast, the costs for external hosting are (usually) variable, and will be cheaper where transaction volumes are low, but this comes at the expense of ignoring potentially valuable internal data. As the cost per enquiry and volumes increase however, so too does the DIY (do it yourself) motivation. There are four possible combinations of generic/bespoke and internal/external options:

Generic/external—Sit-down fast food. Avoids the expense of scorecard development, and the hassle of managing the scoring process. This label can be applied to most bureau-supplied generic solutions, in particular FICO scores. These are often used stand-alone by smaller retailers and service providers, for whom credit is a secondary function.

Generic/internal—Takeaways and TV dinners. Lenders want an in-house solution, but do not have the data for a bespoke development. The scorecard is based upon the vendor's own experience, data, or the pooling of data across many lenders. It is most common for new and emerging markets that have a heavy reliance upon data, obtained directly from the customer.

Bespoke/external—Fine-dining restaurants. A tailored scorecard is built, but is implemented on external systems. Most of the benefits of bespoke developments are achieved, without the hassles of managing the process. This is most appropriate for lenders that do not wish to develop their own scoring infrastructure, but believe they need tailored solutions. It is rare, but several credit bureaux have the capability of delivering bespoke scores. Note, however, that they may not have the same richness of data as the bespoke/internal route.

Bespoke/internal—Mom's home cooking. A customised solution, that makes best use of data obtained from lenders' internal systems, and is implemented in-house. It is expensive, but is cost justified in high-volume environments. It is most commonly used by large and geographically diversified banks and retailers. If internal data quality is suspect, then greater weight can be put on external data (and vice versa).

The bespoke/internal (in-house scorecards) and generic/external (bureau scores) are at opposite ends of the spectrum, and variations exist. Many lenders using bespoke application scorecards will also use generic bureau scores as part of the process.

13.3 Integrating data

The earliest credit scoring models focused on application-form information, and whatever data could be readily obtained from internal systems. External information might have been limited to 'Clean on Bureau', and certain other bureau statuses that had been written on the application form, after having received the information by phone or fax. Obtaining bureau data was a crucial step that often lengthened the time-to-decision by hours, if not even days, where there were (or still are) problems with telephone lines. As automated communication links were established, it became easier to access this information, which dramatically increased the wealth of external data that could be economically included in a risk assessment. As a result, external information is now being used in processes where high costs used to make it infeasible—like account management for existing customers. Whenever information is obtained from diverse information sources—whether internal, external, or both—lenders will have to determine how it will be combined. Broadly speaking, the three main approaches are:

(i) **Independent**—Focuses on scores that provide the most immediate value, and ignores the other data sources. The choice depends upon each lender's own constraints.
(ii) **Discrete**—Calculation of scores that summarise all of the data from each data source, which are then integrated in a decision matrix, or considered sequentially.
(iii) **Consolidated**—Calculation of a score that summarises the data from all sources, possibly using source-level scores as inputs into the final model.

13.3.1 Independent

When presented with information from a variety of different sources, the easiest option is to use only those that provide the greatest value, at least cost (Figure 13.1).

External scores—Scores based upon data obtained from external sources, which ignore internally held data that is not reflected on the external source.
Internal scores—Any scores that have been developed specifically for that process, which ignore data from other sources. These are usually bespoke scores, but may be generics.

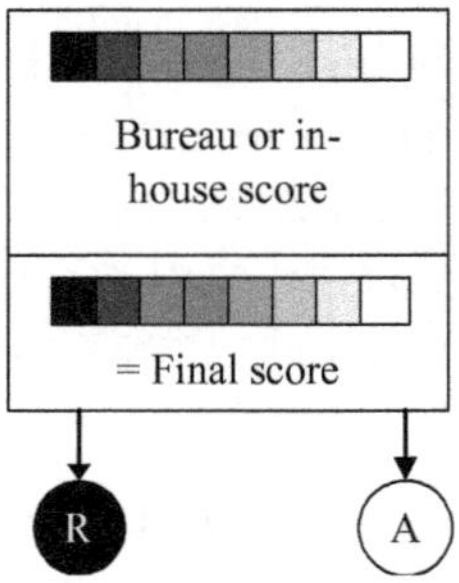

Figure 13.1. Independent scores.

A huge overlap exists between these stand-alone options and the hosting concepts mentioned earlier. The difference is that the assessment is now limited to data readily available from that source. External scores are appropriate when almost all of the relevant data on internal data sources has a second home on the external source. Retailers and service providers can rely exclusively upon bureau scores, because their own data is also resident there. These subscribers can instead focus on managing their businesses of selling clothing, furniture, or cell-phone contracts.

Internal scores are relied upon where: (i) internal data provides a rich information source that can add significant value without any other inputs; or (ii) external data is unavailable, difficult and/or costly to access, or of suspect quality (especially in some developing countries). For the former, banks have a wealth of information that cannot be reflected by the bureau, especially for transaction products like cheque accounts and credit cards. For the latter, banks sometimes decide not to share data on certain key products (cheque accounts, home loans), and may be blocked from using some or all bureau data for those products.

13.3.2 Discrete scores

Credit scoring relies upon having relevant data, but it is not always available all at once. Discrete scores may be used for different blocks of data—usually one per data source—within the decision process. A rule of thumb is that source-level scores should be predictive enough to provide value on a stand-alone basis, for some or other purpose. If not, then different data sources should be combined until this is achieved. Which data sources are included will be a function of availability and cost. A common approach is to integrate internal data, before bringing in external sources. The scores can then be integrated either through: (i) *sequential evaluation*, score and policy are used as filters, to prevent undeserving cases from passing through to the next stage, or perhaps divert them into another channel; or (ii) a *decision matrix*, scores are combined via a table that allows them to be used simultaneously (Figure 13.2).

The most well-known use of sequential evaluation is where pre-bureau scores are calculated using internal data only, and bureau data is called for only if there is a good possibility that it may change the decision. This is used primarily to keep bureau costs down. When cases do

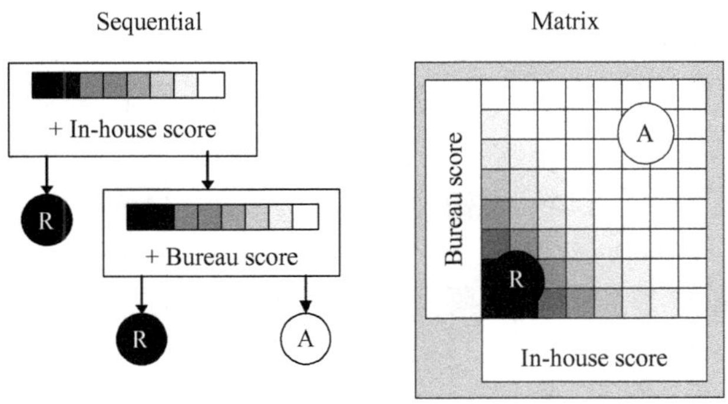

Figure 13.2. Discrete scores.

pass to the next stage, a decision matrix can still be used to make the accept/reject decision and determine pricing.

Decision matrices are the more common approach, and are fairly simple where there are only two scores, but the complexities compound exponentially as the number of data sources increases. According to Mays (2004), this approach provides more accurate results, but is more difficult to design, implement, and manage. In particular, the provision of decline reasons becomes a problem. Matrices are covered in more detail in Section 13.3.4.

13.3.3 Consolidated scores

The final approach is to consolidate all of the data into a single measure, which may be done in one of three ways:

(i) **Scores only**—Data from each source is summarised as a single score, and these scores are then integrated into a final score.
(ii) **Data and scores**—Combines one or more scores with other information.
(iii) **Data only**—Data is integrated directly without the use of any source-level scores.

A modular approach using source-level scores simplifies the process, but: (i) will not recognise all of the correlations within the data across the different sources; and (ii) should only be used if it can be assured that the source scores' meaning will be stable over time, even if the scorecards used to calculate them change. Use of source-level scores is recommended where it is infeasible to host all of the data on a single system (Figure 13.3).

In contrast, using the underlying data from each source recognises all correlations within the data, and provides the most powerful scorecards, but suffers because: (i) there are significant data communication, management, and storage requirements; and (ii) the entire scorecard has to be redeveloped if there is significant drift. It is recommended where the resulting risk assessment capabilities provide a key competitive advantage, volumes are high, and margins are low.

These concepts apply irrespective of whether the data sources are internal, external, or both. They apply mostly: (i) in the new business process, where application, internal account

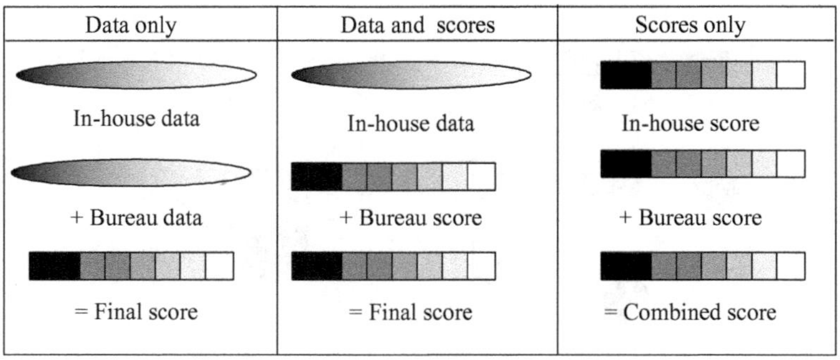

Figure 13.3. Integrated scores.

performance, and bureau data are brought together; (ii) for account management, where other products and/or credit bureaux are included; (iii) customer scoring, where data from different accounts is combined; and (iv) collections, which combines its own data with internal performance and bureau data.

13.3.4 Decision matrices

A little more attention is paid to the decision matrix, due to the extent of its use. Scores are split into ranges and cross-tabulated, to find the good/bad odds (or bad rates) for each combination, which are then used to decide customers' fates. In the Table 13.2 example, if the two scores are considered in isolation, the odds range from 0.9 to 7.8, but when considered in tandem, they range from 0.1 to 24.5. If any cases with odds of less than 2.0 are to be rejected, the swap set can be seen in B2/L5, B4/L2, and B5/L1&L2.

This is based on the most common type of representation, where the two axes are risk measures that are highly correlated with the outcome, represented in the cells. A similar format can be used to combine different types of scores, like risk and retention, and have a number of different performance measures represented within the cells. This provides a better understanding of the correlations between the scores and key outcomes.

The greatest advantage of a decision matrix is that the lender can still make decisions when data from certain sources is missing—for example, where there is a relationship with only one bureau that is not always available. This simplicity also provides extra flexibility when setting strategies; lenders can shift emphasis between scores, especially if their validity has deteriorated.

There are two disadvantages. First, there may be *correlations* between the underlying characteristics that are not properly reflected. For example, when using bureau data, details for products of the same type, say credit cards, are more predictive than those from other products. This may be overlooked, if there is no industry-specific generic. Second, the number of *possible combinations* can make analysis difficult, and complicate implementation. In the example, there are 25 cells just for credit risk. What if the lender wishes to combine this with an early settlement score? Some means would be required to assess each of the combinations.

Irrespective, decision matrices are commonly advised as the best choice by credit bureaux and others. Given that bureau scores are value-added services for which there is an extra

Table 13.2. Decision matrix

GB odds	Bureau					
Lender	B1	B2	B3	B4	B5	Total
L1	0.1	0.3	0.5	1.3	2.2	0.9
L2	0.5	0.9	0.6	2.1	3.0	1.5
L3	0.5	1.2	1.5	3.7	7.5	2.7
L4	1.0	1.5	2.0	3.8	8.5	3.2
L5	1.9	3.0	8.3	8.1	24.5	7.1
Total	0.9	1.5	2.2	4.0	7.8	3.2

charge, it follows that it would be in the credit bureaux' own best interests to tout their benefits. Consolidated data-only scores would probably provide better results, but require greater investment. Unfortunately, there appears to be no academic work to provide a comparison of the approaches, or to describe who uses them.

13.4 Credit risk scoring

Lenders do different types of credit scoring, and each will address the task in different ways. This section takes a brief look at customisation and integration, with respect to specific types of scorecard developments.

Application scoring

The greatest benefit from credit scoring comes from guarding the front door, and the motivation for 'bespoke' scoring systems is greatest where: (i) the overall values at risk are high; (ii) the profit margins are low; and (iii) bureau data lacks the required richness. Thus, those most likely to make the investment are larger banks, and those in countries where credit bureaux' data is insufficient for the task. In contrast, where the bureaux are data rich (developed countries), and lenders are small or margins are high (retailers, service providers), much greater reliance can be put on the bureau scores, especially if the lenders' own data is well represented. As regards integration, smaller lenders are more likely to keep bureau and internal scores separate, while larger lenders are more likely to integrate the data into a single score. Funnily, it also seems that US lenders rely on decision matrices, while UK lenders focus on integrating the underlying data.

Behavioural scoring

Initially, almost all behavioural scores focused exclusively on performance details for the product in question. Over time, it has become feasible to incorporate demographic, other product, and bureau data into the assessment. It is also possible to use the application score as one of the inputs, but the lender must have some means of recognising application scores' information decay. In general, most behavioural scoring is done using bespoke, internally-hosted scores, and if other scores are used, they are usually kept separate.

Customer scoring

During the 1990s, customer scoring was touted as the next revolution in credit scoring. Unfortunately, the much-touted benefits have been elusive for most lenders, who still rely heavily upon product-level scores for most of their account management. The one area where customer scores are used is for cross-sales, as they can provide an indication of a customer's overall risk profile before an offer is made (Figure 13.4).

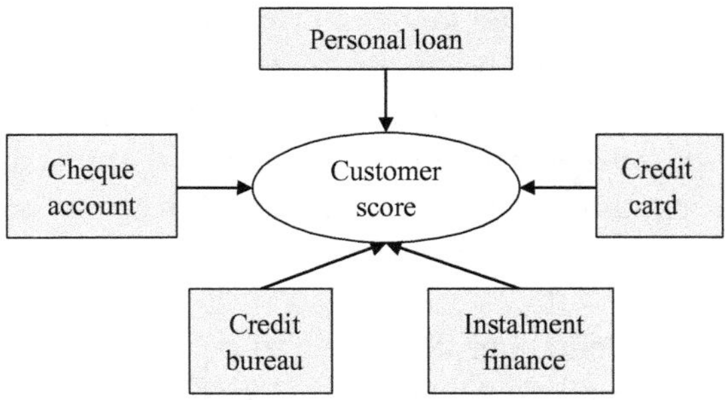

Figure 13.4. Customer scoring.

Customer scoring presents a challenge when bringing information together, because of data volumes and issues with product-level scores. The most predictive customer scores are obtained if product-level detail is used, but the overheads for storing and processing the data are significant. In contrast, integrating product-level scores has lower overheads, and provides results that are almost as good, but the customer score's stability will deteriorate each time an upstream score is redeveloped.

Collections scoring

The collections function has traditionally been driven by the arrears status—in particular, the number of months in arrears. As application and behavioural scores were developed, they were brought in to refine the process. Collections generates a wealth of valuable information by itself though, and collections scores came about as a means of harnessing collections data (like promise-to-pay, and contact history), and combining it with other product and bureau data, specifically for use in collections. Indeed, the credit bureau may even provide a repayment score, to assist the assessment of later-stage delinquencies. If a lender does develop a bespoke-collections score, it is probably best to keep the scores separate, as its useful life may be short.

13.5 Matching!

Like anything, data only provides value if you can find it when you need it! This maxim applies whether we are talking about paper filing systems, electronic data storage, or children's toys. In credit scoring, the issue relates to the databases used to store or archive borrowers' details, which may lie on any number of diverse systems. Means are needed to bring it together, and match new cases to data already on file. This is the domain of relational databases, and the use of keys:

Primary key—Data field with unique values for every record in a database, which can be used as an identifier. There is usually only one, but it is possible to use several in combination.
Matching key—Any data field used to link records in two databases.
Foreign key—Data field in another database that is being matched.

There are two broad types of matching keys used in credit scoring. First, internal matching keys are those used internally within a company, to assist them with managing their data. The most obvious are account and application numbers, but there are others. Application numbers are required to match newly-opened accounts to the original applications. Customer numbers will be used, where there are multi-account relationships. Care must be taken, however, because the customers may be unaware of these numbers, and be assigned several of them as new accounts are opened.

Second, nationally-accepted matching keys refer to personal and company identifiers used within a country. Some of the best known personal identifiers are the Social Insurance Number (USA), Social Security Number (Canada), or National Health Number (UK). These are not always allowed for use outside of social services though, as in the United Kingdom. Other countries, such as South Africa, have an Identification Number (ID Number) that is widely used for all purposes. Tax numbers have been used in South America, but there may be a large pool of people that are not registered for tax, either because their income is insufficient, or registration is difficult or expensive.[4] For companies, and other juristic individuals, the situation is much easier because privacy issues are fewer. Company registration numbers are the primary identifiers, while VAT, taxation, and other numbers may be used in other instances.

Failing the existence of identifiers, lenders have to resort to other less efficient means, and sophisticated matching software is required. Problems can arise with: names—common names,[5] diminutives, nicknames, and misspellings; and addresses—different spellings, different languages,[6] street name changes,[7] and the manner in which addresses are presented.[8]

Credit bureau

The greatest challenge for matching lies with the credit bureaux, which receive data from different subscribers, and have to collate it, so that they can create reports for every credit user. Mistakes can be made, and some bureaux can return the probability of a right party match, either as a percentage or some description (exact, close, possible). Alternatively, the lender can specify the match level required. In some cases, legislation or regulators may require stricter matching criteria.

Some credit bureaux have also invested heavily in data enhancement, especially where personal identifiers are not available, whether nationally or on a given data source. This practically causes them to act as automated sleuthing agencies, to help maintain links between individuals and their credit history. Personal information provided by a subscriber is supplemented by the bureaux' own databases, and automated routines check for logical associations. This

---

[4] In Brazil, tax numbers can be allocated by banks to low-income clients at the time of account opening, solely for the purposes of identification, and do not imply a tax obligation.

[5] Such as S. Andersons in Sweden, D. Jones in Wales, P. Govenders in India, B. Nguyen in Vietnam, or K. van der Merwes in South Africa.

[6] Canada, South Africa, Belgium, Switzerland, and others.

[7] South Africa and any other societies undergoing substantial political changes where street, town, and city names are being changed to reflect current political realities.

[8] Thomas et al. (2002) use the UK example where a customer includes a property name in the address, like 'The Old Mill, 63 High Street, London', but the 'Old Mill' portion is not on the database record.

includes: maintaining details on both current and past addresses; including a full first name, if a nickname had been used; recording a new surname, if forename, birth date, and contact details match (women recently married or divorced); or tagging a personal identifier onto judgment records, where a name and address match is found. In any case, the original details provided must be maintained, and any supplementary information kept separately. This not only provides business benefits, but also aids compliance with data protection legislation.

13.6 Summary

The first two chapters in this module focused on data considerations and data sources. This section has moved on to some of the issues relating to scoring structure: *customisation, hosting, integration, and matching*. For greenfield developments, the first decisions to be made are: (i) the appropriate level of customisation—whether to buy one-size-fits-all (generic), or invest in a tailored solution (bespoke); and (ii) whether hosting will be on in-house or external systems. The most common types of generic scores are bureau scores hosted on external systems, while bespoke scorecards are usually hosted in-house. The latter provide better results, but are expensive, data intensive, and require significant management time. In contrast, generics are cheaper, but the results are not as good. Generics are used (i) by small lenders, who wish to focus upon their own business, and (ii) for greenfield developments and new markets, where there is little experience.

When obtaining data from different sources, lenders have to develop means of integrating it, for use in their decision-making. Broadly speaking, they can: (i) use the scores *independently*, with a focus either on internal or external scores; (ii) use them *discretely*, either sequentially or via a decision matrix; or (iii) *consolidate* the scores, data, or both, into a single risk score. Smaller lenders seem to favour using a decision matrix, while large banks favour consolidation using data or scores.

There are a variety of different types of credit risk scoring, and the available choices vary with each. Application scoring is the oldest form, but behavioural scoring, customer scoring, and collections scoring have evolved quickly as the resources have become available. Each of these will have different data sources and types of data, which can be combined in different ways.

Finally, the crucial element for bringing all of this data together is being able to match a customer to the relevant data, from different data sources. Matching requires keys, the most obvious of which are account and application numbers. Use of customer numbers is another possibility, but this assumes that there will always only ever be one customer number per customer. Customer-level matching is easiest where there is a widely accepted personal identifier, such as a Social Insurance Number, Social Security Number, National Health Number, or Identification Number, that is recorded against every account or customer. If these are non-existent or unreliable, then names and contact details may be used, but this presents challenges because of problems with common names, misspellings, and name changes.

14 Information sharing

In many competitive environments, people play with cards very close to their chests, and any evidence of collusion is considered cheating. In a game like credit however, there are risks that can cause all players to lose, which can be reduced if lenders show at least some of their cards. The real antagonist is not the competing lender, but the borrower; and the information being shared relates not to one's own strategies, but the customers'. In explaining the emergence of the credit bureaux, and their ability to lower the cost of collecting and evaluating data, Barron and Staten (2003) state:

> Lending markets almost always display an information asymmetry between borrowers and lenders. Borrowers typically have more accurate information than lenders about their likelihood of repaying a loan. Lenders have an obvious incentive to evaluate a borrower's creditworthiness, and the outcome will affect whether to approve the loan, as well as its price. Borrowers have an incentive to signal their true risk (if it is low), or disguise it (if it is high). Given the amount of loan principal at stake, both parties have incentives to incur costs (often large ones) to reduce the information asymmetry.

Lenders' primary tools are information sharing arrangements, which have two main forms: (i) *private registries*, either independent agencies (credit bureaux), or co-operative arrangements run by an industry association or chamber of commerce; and (ii) *public registries*, meant either to protect consumers, or to protect banks against themselves. This chapter covers the topic under the headings of:

(i) **Credit registries**—The types of registries and data that is shared, with particular focus on public (government) versus private (for profit) registries, and positive (shared-performance data) versus negative (publicly available and default) data.
(ii) **Do I or don't I?**—A look at: (i) the 'Principles of Reciprocity' that govern information sharing arrangements; (ii) the benefits that can be achieved from sharing; and (iii) the concerns that might inhibit lenders from participating.

14.1 Credit registries

The types of data provided by credit bureaux were covered in Section 12.3, but little was done to describe their *raison d'être*. Such information sharing has led to a massive change in the way retail credit-risk assessments are done, by reducing: (i) the amount of information required directly from customers; and (ii) reliance upon collateral, allowed by the shift towards cash-flow based lending. The growth of credit bureaux has not only increased access to credit, but also lowered the cost, and made it easier for people to move from less formal (micro-finance, store credit, prepaid phones) to more formal (banks and credit card issuers) credit markets. This section has the following headings:

Private versus public—The two major types of credit registries, being private credit bureaux and public credit registries (PCRs).
Positive versus negative—The types of information typically available via the credit bureau. 'Positive information' typically refers to shared performance data (between lenders), and 'negative information' to that publicly available, especially from the courts.

14.1.1 Public versus private

Most of this textbook relates to English-speaking countries, where private credit bureaux are the norm. In contrast, PCRs play a role in much of continental Europe, as well as in many developing countries where there are no private players.

Public registries—Statutory governmental agencies created to fill some social need. Participation is mandatory for affected lenders (and/or loans above a given threshold), and is governed by regulation.
Private registries—Driven by a profit motive, and (usually) subject to competition. Participation is optional, and is governed by contract.

Over the past few decades, the growth in credit registries around the world has been phenomenal, in particular in countries that were previously unserviced. Indeed, the World Bank, IFC, and USAid have all been active in facilitating the development of credit bureaux in developing, underdeveloped, and transition economies. Japelli and Pagano (2005:8) highlight that between 1950 and 2000, the percentage of countries with private registries grew from 20 to 60 per cent, while those with public registries grew from 5 to 50 per cent. According to Miller (2001), at that time, the median age of private credit bureaux around the world was 10 years, and 30 per cent had been established since 1995. Djankov et al. (2004) indicated that, as at 2003, private bureaux were operating in 55 of their sample countries, including all OECD countries except France. Experian, Equifax, and TransUnion, owned or were affiliated with half of the bureaux sampled. Jentzsch (2005) states that 'by 2004, 49 countries had public registries, 46 had private credit bureaux and 17 countries had both'. Jentzsch stresses that the two types are not mutually exclusive, and 'can play a complementary role if properly designed . . . otherwise they might get into direct competition on very unequal terms which would be undesirable'. Better risk assessments should result if both sources are accessed.

An observation made at the Cape Town Micro-finance workshop in 2005, was that the maximum number of 'major' credit registries that can seemingly be supported within any one country are three private and one public.

Information sharing provides an *ex ante* substitute for *ex post* creditor protection; it is easy to guard the back door where creditor protection is good (collections enforcement); but where it is poor, greater effort must be put into guarding the front (account origination). According to Djankov et al. (2004), even higher levels of private credit to GDP (PC/GDP) result when both are in place, albeit protection of creditor rights is more important in rich countries. The introduction of public and private registries accelerated the growth in PC/GDP over the following three years by 2.2 per cent and 4.2 per cent respectively.

World Bank researchers class legal origins into five country groups: English, French, Germanic, Nordic, and Soviet. Countries of French legal origin (FLO) tend to have the greatest protection of creditor rights, but the treatment of insolvents is so harsh that it inhibits entrepreneurial activity (di Martino 2002). Prior to 1978, public registries were limited to FLO countries, yet by 2003, half the countries in the world had them. Much of the growth was in Germanic legal origin countries, which includes many in Eastern Europe; they are still much less common in Nordic, Soviet, and English common-law countries (see Table 14.1). Over the same period, creditor protection in all groups has remained relatively constant.

The French system is also referred to as the Napoleonic legal system, and includes those of France, Spain, Portugal, and any colonies that adopted their legal systems. Spain and Portugal adopted the French legal system while under Napoleon's rule.

The rest of this section is based on work by Miller (2001), Love and Mylenko (2003), Jentzsch (2005), and Japelli and Pagano (1999, 2005).

Public registries

Public registries are created to serve some social need. They usually fall under the central bank, and are set up as statutory governmental agencies responsible for banking supervision. Depending upon their goals, some minimum or maximum reporting threshold will be set, and all banks must report on any loans falling within those thresholds. Goals may include, amongst others, to act as a means of controlling credit extension (large loans); and to ensure access to credit, without over-indebting consumers (small loans).

Table 14.1. Registry penetration in 2003

Legal origin	Countries			
	#	Public (%)	Private (%)	Both (%)
English	35	25.7	48.6	5.7
French	64	76.6	35.9	21.9
Germanic	18	61.1	55.6	22.2
Nordic	4	0.0	100.0	0.0
Soviet	11	18.2	0.0	0.0
Total	132	79.5	40.9	15.2

According to Djankov et al. (2004), the public registries in Germany and Saudi Arabia focus on large loans and banking supervision, while those in Belgium, Ecuador, Malaysia, and Taiwan distribute extensive information that matches or exceeds that provided by private bureaux.

Where a private registry exists, the public registry tends to focus on larger loans, and control of the financial system. Other functions may be to monitor credit quality, validate lenders' credit scoring models, and monitor potential systemic risk within the economy. Where the goal is consumer protection, it is usually to fill a gap where no private bureaux exist, their scope is limited, or consumer protection is poor. This applies whether for the whole market or a particular market segment, especially in poorer countries. The focus is usually negative data, but exceptions exist.

As is befitting of governmental agencies, public registries are monopolies that are usually unresponsive to lenders' needs. Few value-added services, such as bureau scores, are offered. Because participation is mandatory, coverage will be high within their specified mandate, but the number of types of institutions and breadth of data included will be lower than for private registries. In most cases, the focus is on banks, and non-banks do not contribute or have access to data. Given the prevalence of non-bank lending (leasing, store credit) in some markets, it follows that this will reduce the predictive power of available information. The data may also be inappropriate for use in lender decision-making, because: (i) its relevance is poor; (ii) the focus may be on current outstanding debt, with little attention paid to payment history; and (iii) it may be disseminated in consolidated form, and not at individual customer level. Where public registries are cost effective, they may keep private registries out of the market, which is not necessarily the intention. Upping the minimum reporting threshold can increase the level of private participation within an economy.

According to Japelli and Pagano (2005), the type of data reported varies from country to country. Germany requires reports on loan exposures and guarantees, Belgium on defaults and arrears, and Argentina on all those, plus interest rates.

Private registries

Private registries are driven by a profit motive. In most developed countries, they are independent concerns, but may also be bank owned, or the initiative of a bank association or chamber of commerce. Some are sector specific (banks, retailers, micro-finance), but the ideal is to share performance data across sectors. Like most private concerns, these registries are competitive, and responsive to subscribers' needs. They are technology driven, and innovative in the services that they offer (bureau scores, fraud checks). They will be subject to regulations relating to data privacy and credit access, which may be sectoral (USA) or general (Europe and elsewhere). Where two or more bureaux compete, the coverage of each may vary.

Several researchers have shown that there is a strong correlation between the existence of private credit registries, and the ratio of private credit extension to GDP (or GNP). Miller (2001) takes this further, to show a correlation with the number of years that a credit registry has been active within an economy. Love and Mylenko (2003) highlighted the benefits for SMEs from the reduction in information asymmetries, allowing them greater access to affordable credit, especially from banks, and especially for younger firms (even so, the extent is insufficient to meet the financing needs of those firms). They suggest that the existence of a private registry makes public registries redundant in terms of facilitating access to credit, and that the latter's focus should then shift to banking supervision. Where there is strong co-operation between the regulators and private bureaux, the need for a public registry may disappear entirely. A further factor mentioned is that there tends to be a higher-quality legal system ('stronger rule of law'), where there are private registries.

Competitive advantages between private registries

Lenders can often choose between different bureaux, and need to determine which is best suited for their needs. The bureaux will return different results, because of differences in their market, geographical, and technological dominance:

Market dominance comes from having a large number of subscribers and a rich enquiry database; first player advantage also helps. The dominance may be broad-based, or limited to certain subscriber types.

Geographical dominance means having market dominance in a given region. In the United States, the major bureaux started out as small operations that served a geographical area and grew, either organically or through acquisition (Furletti 2002).

Technological dominance refers to the ability of the systems to match people to profiles, to recreate applicant profiles at time of application, to provide timely data suited to the task, etc. This will help to offset the first player advantage of a competitor.

The decision of which bureau is best will be affected by the cost per enquiry, perceived quality of the information, geographic location of the customer, ability to swap between bureaux, and so on. If there are regional considerations, the subscriber may decide to use the bureau that is dominant in the area where the customer lives. If not, it will either stick to one bureau, or allocate a percentage of enquiries to each. Where the value at risk (VaR) is large, it may even be viable to use information from multiple bureaux. In the United States, there was a move by a mortgage securitiser, Fannie Mae, to use only a single bureau, but this was met with public resistance. Public opinion favours them having access to more, as otherwise unlucky customers could be declined, or be charged higher interest rates based upon a spin of the wheel (CFA 2002).

14.1.2 Positive versus negative

There are only two types of people in this world—those that split everything into two types, and . . .

Anonymous

While it is almost a given that a credit bureau or registry will be available, the types of data allowed may vary from country to country, and subscriber to subscriber. This typically relates to a dichotomy—positive/negative, white/black, or non-default/default. The former of each pair refers to good account performance for a customer, and the latter to bad. The polarising labels can be misleading though, for two reasons. First, they are used to distinguish between shared performance versus publicly-available data. The bias against the latter exists because there is usually little in the public domain that can work in customers' favour. Second, some confusion arises because there are a lot of shades of grey, and shared-performance data can sometimes be as bad, or worse, than default and judgment information. This is because: (i) lenders often do not take judgment due to the expense involved, especially for small fish, and defaults may not be reported separately; (ii) the match rate may be better on positive data, due to the ready availability of personal identifiers; and (iii) some lenders may be using judgments as a collections tool.

In some environments, shared-performance data forms an integral part of the credit culture, while in others it may not be available at all, due to real or perceived legal restrictions, or poor infrastructure. In still others, some bureau subscribers may opt not to use it. In any event, most countries started by sharing negative information only, and have moved towards sharing both. Barron and Staten (2003) mentioned moves by Brazil, Argentina, and Chile, but stressed that the bulk of information in their credit files was still negative.

Forgiveness period

Associated with the concept of positive and negative data is the concept of a blacklist. Most credit scoring practitioners will argue that no such thing exists, and that anybody wanting access to credit can find it—at a price! The affected public sees it differently though; they are only aware that they are declined, or have to accept unacceptable terms, and possibly be forced towards unscrupulous operators, because of past misdemeanours. Well, perhaps they have a point!

Credit bureaux act as lenders' collective memory, and decisions have to be made regarding the forgiveness period—the amount of time before defaults, judgments, dishonours, and other transgressions are excused. Japelli and Pagano (2005:18) point out that: (i) if it is extremely long, it can *ex ante* discourage potential borrowers from incurring debt, and *ex post* make it impossible to resume normal economic activity, with no incentive to repay once defaulted; and (ii) if it is too short, it would not act as a deterrent, and the data would be of little value. The happy medium is somewhere in between, and the period tends to be longer in countries where there are greater problems with enforcing creditors' rights.

There will, however, usually be some absolute maximum for different types of records, set by legislation. Most countries have consumer-protection legislation that limits the data-retention period, and/or the age of adverse data that may be used in credit processes. According to Cate et al. (2003), in the United States the FCRA prevents the credit bureaux from holding negative information (delinquencies, charge-offs, judgments) for longer than 7 years, and bankruptcy for 10 years (changed from 14 years in 1979). Japelli and Pagano (2005) note the rather interesting treatment by the Belgian public registry office (Central Office for Credit to Private Individuals), which only records default information where '"punishment" is stricter for more serious misconduct'. Arrears may be kept on file for 1 year after repayment, and defaults for 2 years after repayment, but no record can be kept for longer than 10 years.

14.2 Do I or don't I?

Even though credit bureaux are active in many countries, not all lenders participate. In spite of the many potential benefits, lenders often have concerns. This sections looks at:

Principles of reciprocity—The set of rules that govern the scheme—in particular, what information should be provided, how often, and how it may be used.
Motivators—A look at the benefits provided by sharing data, which include reductions in adverse selection and information rents, and its function as a borrower-discipline device.
Inhibitors—Factors of concern to lenders that are considering sharing data, including fears about poaching, data-quality issues, potential legal issues relating to data privacy, and their focus on relationship banking.

14.2.1 Principles of reciprocity

Credit registries require some compulsion. Borrowers and lenders must agree to participate, with appropriate penalties if they do not.

Djankov et al. (2004)

All information sharing arrangements are governed by 'Principles of Reciprocity' (PoR) agreements, whose primary premise is that 'only those that give shall receive'. This is only fair where the data is provided for free, meaning that contributors do not receive any payment for the data they provide. The scheme will be run either by a private third-party company that charges for its services (such as a credit bureau), or a co-operative in which the participating lenders each take a share of the costs. In both cases, the charges are usually based on usage.

Shared performance data is a powerful tool, but contributing subscribers do not have carte blanche to use it as they will. The PoR will normally define when it may or may not be used. For credit bureaux' PoR, the restrictions are greatest for marketing solicitations to non-customers, and least for through-the-door application processing and existing account management. Its use for marketing to existing accountholders lies somewhere in between. If its use is prohibited, lenders are still able to use public, enquiry, and other information provided by the credit bureau. Rules vary from country to country. In the United Kingdom, data ownership lies with subscribers, while in the United States, the bureaux have ownership of at least some of the data. Bureaux in the United States thus have more freedom to offer positive data for use in marketing campaigns.

Such information sharing arrangements are used not only for credit, but also fraud and tracing. Each will have its own PoR, and in some instances credit bureaux will host the scheme. Some of the schemes are:

Function	Acronym	Country	Name
Credit	CAIS	UK	Credit Account Information Sharing
Fraud	CIFAS	UK	Credit Industry Fraud Avoidance System
	SAFAS	RSA	South African Fraud Avoidance Scheme
Tracing	GAIN	UK	Gone Away Information Network

All of these schemes rely on having a critical mass, and are thus dependent upon the number of contributors, and the volume of data provided. Some of them may be criticised by the general public for 'invasion of privacy', but ultimately, they are crucial for keeping credit losses at a minimum, and hence ensuring affordable finance for all.

14.2.2 Motivators

Economists are increasingly conceding that data sharing (especially about consumers) and free-flowing information has been a key to U.S. economic flexibility and consequent resiliency. It contributes to our mobility as a society, so that structural shifts within the economy cause temporary disruptions but without crippling long term effects.

Barron and Staten (2003)

Credit information sharing benefits not only lenders, but also borrowers and the broader economy. According to Djankov et al. (2004), there are six major factors that significantly add to the private credit to GDP ratio: (i) sharing of positive and negative information; (ii) access to information for both firms and individuals; (iii) access to information from banks, retailers, and/or utilities; (iv) access to five or more years' data; (v) all loans greater than 1 per cent of income per capita are included; and (vi) laws allow consumers to inspect their own data. Japelli and Pagano (1999, 2005) summarise the benefits as:

- Reducing the probability of adverse selection, as extra knowledge improves bad rate prediction, which further helps to improve loan targeting and pricing, and to increase the total lending book. In its absence, lenders are likely to redirect funds to where they can better assess the risks. For regulators, this may also lead to lower systemic risk.
- Acting as a 'borrower discipline device'[1] that: (i) motivates borrowers to maintain a clean credit record; (ii) limits their ability to over-indebt themselves across multiple lenders; and (iii) cuts insolvents, and those with severe repayment problems, off from credit.
- Reducing 'informational rents', or lenders' potential gains from asymmetric information. This levels the information playing field, by reducing lenders' (pricing) power over their customers, which in turn leads to a reduction in borrowers' potential for moral hazard.[2]

Under what circumstances are lenders most likely to insist on sharing data? The incentives are greatest when:

- Borrowers are highly mobile, both geographically and between lenders.
- Lenders are small, and are dealing with a heterogeneous population.
- The credit market is growing, and potential demand for loans is high.

[1] This is a double-edged sword. Japelli and Pagano (2005) point out that low-risk borrowers may engage in riskier behaviour, based purely on the knowledge of their good credit standing.

[2] Gehrig and Stenbacka (2002) were two of the few authors who argued that information sharing is uncompetitive, but their arguments are highly theoretical, mathematical, and unconvincing.

- It is expensive to obtain, store, and maintain customer information.
- Collateral is not available, or its value is difficult to realise.
- Advances in technology make sharing cost effective.

The benefits do not accrue to lenders only. In a world without sharing, borrowers are tied to their banks, because that is who knows them best, and the costs of switching are high. Barron and Staten (2003) quote William McDonough (New York Federal Reserve Bank president), who said:

> . . . the portability of information makes us more open to change. There is less risk associated with severing old relationships and starting new ones, because objective information is available that helps us to establish and build trust more quickly.

A measure of benefits

In some cases, as in Australia, lenders are still not allowed to share performance data. There are, however, pressures to change this, and as a result, some studies have been done to illustrate the benefits of including it in credit scoring models. According to Fair Isaac (FI), where used, it may provide as much as 65 per cent of the predictive power of a model.[3] It also allows bureaux to provide effective customer monitoring services, to alert lenders to significant changes in the credit quality of existing customers.

Japelli and Pagano (1999) highlight that information sharing also increased available credit within the economy, in particular bank lending to the private sector. They developed a model to predict the total 'bank debt to Gross Domestic Product ratio' (debt to GDP ratio) for about 30 different economies. It considered characteristics such as negative data only, positive and negative data, GDP growth rate, rule of law, protection of creditors' rights, and the historical origin of the national legal system (English, French, German, Scandinavian, etc.) The analysis indicated that the availability of negative information added 10 per cent to the debt to GDP ratio, and the use of positive information a further 13 per cent.

Other research that provided support for sharing positive information was conducted by Barron and Staten, which was mentioned in IFC (2001) and published in Miller (2003). They used bureau data provided by Experian for May 1997, to test what would happen in the United States if it had: (i) Australian legislation, which only allows negative data; or (ii) fragmented industry-specific reporting, as in several Latin American counties that allow bank- or retailer-only reporting. The results are provided in Table 14.2, which shows the results from a strategy that maintains an accept rate at 75 per cent, and a target bad rate of 4 per cent. For example, a lender that targets a 4 per cent bad rate would see its accept rate reduce from 83.2 to 73.7 per cent, which implies a loss of about 11,500 customers per 100,000 that would otherwise qualify.

---

[3] Japelli and Pagano (2005). The comments refer primarily to the power of positive information in credit bureau scores.

Table 14.2. Benefits of positive data (Barron and Straten (2003))

Compare	Negative-only		Retail-only		Bank card-only	
Sample size	312,484		67,130		110,633	
Constant	Accepts = 75%	Bads = 4%	Accepts = 75%	Bads = 4%	Accepts = 75%	Bads = 4%
Measure	Bad rates	Accept rates	Bad rates	Accept rates	Bad rates	Accept rates
Limited	4.07%	73.7%	2.97%	80.6%	1.95%	75.4%
Full info	3.04%	83.2%	2.13%	90.6%	1.69%	83.4%
Difference	1.03%	11,500	0.84%	11,000	0.34%	2,500

In general, from the table it can be concluded that: (i) being limited to negative-only is detrimental; and (ii) ring-fencing prejudices both retailers and banks, but retailers much more so. The conclusions published in the IFC report indicated that where credit information is restricted: (i) credit losses will be higher; (ii) the cost of credit will be higher; (iii) early warning of, and preventive action against potential losses, becomes more difficult or impossible; and (iv) consumer credit will be less available, especially for applicants that are young, lower income, financially vulnerable, or have recently changed address or job.

While information sharing is usually associated with increased competition, this is not always the case. An exception is closed-user groups that act as a potential barrier to entry for new entrants into a market, including firms that are trying to expand their range of services by offering credit. It is a particular problem where the first player is a co-operative arrangement between banks. Japelli and Pagano (2005) provide the example of the *Buro Credito*, a registry set up by the Mexican Bank Association, Dun & Bradstreet (D&B), and TransUnion. Two subsequent attempts at setting up private bureaux in Mexico were foiled, because they could not get bank participation. Other Latin American countries with similar arrangements are Argentina and Brazil. It is in the best interest of any country for the bureaux to be open access, offering services to any and all firms wishing to offer credit.

14.2.3 Inhibitors

In spite of the potential benefits, there are still factors that inhibit data sharing—especially amongst banks and other formal credit providers. In general, their concerns about information sharing usually relate to: (i) poaching; (ii) data quality and consistency issues; (iii) potential legal concerns; and (iv) their own risk management capabilities.

Poaching

> Credit bureaux are exposed by a potential conflict of interest, especially when they are owned by the lenders themselves; each lender would like to exploit the information provided by other lenders without exposing his own.
>
> *Japelli and Pagano* (2005:7)

Many lenders are reluctant to share positive information, because they believe it will aid poaching by competitors, especially new entrants into the market trying to target the most profitable segments. Concerns fall into two camps: (i) unsolicited marketing offers; and (ii) through-the-door customers. Unsolicited offers may be received as part of marketing campaigns, and credit bureaux typically do not allow the use of positive information for this purpose—only negative information and the bureaux' own enquiries. In contrast, for through-the-door customers, no such restrictions apply. If a customer applies as part of a marketing campaign, then any lender is entitled to access the positive information.

Such concerns inhibited many banks in the United Kingdom and Commonwealth countries, and some are still loath to share. Even when they do participate, they may provide data for all but a key product, usually the cheque account or home loan portfolios. In contrast, American banks were until recently limited to operating within state boundaries, which acted as a barrier to entry and limited competitive risks. As a result, their concerns about poaching have been fewer, and their credit bureaux play a greater role in the economy.

Such concerns are not limited to competitors of the same type—there may also be potential cross-sector competition. An example is subprime lenders, as illustrated in a speech made by John Hawke Jr., the US Comptroller of the Currency, to the Neighborhood Housing Services of New York on 5 May 1999:

> Subprime loans cannot become a vehicle for upward mobility if creditors in the broader credit market lack access to consumer credit history. Yet, a growing number of subprime lenders have adopted a policy of refusing to report credit line and loan payment information to the credit bureaus—without letting borrowers know about it. Some make no bones about it: good customers that pay subprime rates are too valuable to lose to their competitors. So they try to keep the identity and history of these customers a closely guarded secret.

Another group that has been mentioned is credit-card issuers, who have been accused of reporting lower credit limits to reduce the credit scores. Such practices may not be illegal, but it is considered an unfair lending practice. As regards illegal use of positive information for targeted poaching—it has been known to happen, in spite of strict credit-bureau rules against it. Where contraventions become known, actions should be taken against offending subscribers.

Data quality

There may often be issues regarding the quality of data provided by the credit bureaux, and its consistency. Problems can arise because:

Number of subscribers—While the growth in the number of subscribers has been large, some bureaux may not be perceived to have sufficient mass to add value, especially in developing countries with thin credit markets.

Subscriber profile—Besides new retailers, cell phone companies only started emerging in the 1990s, and banks only started contributing in the United Kingdom and RSA in the mid-1990s and early 2000s respectively. This affects the data consistency! The information provided currently may not be representative of what was available for past applicants.

Regularity of updates—Subscribers may provide updates on an irregular basis, and data quality may vary. Analysis of the data may indicate a problem that must be fixed before it is loaded, which can cause significant delays.

Unstable technology—Where the bureaux are recently established, the infrastructure may not yet be bedded down. This might be because of changes within their own systems, or problems with communications technology linking lenders with the bureaux.

Potential legal concerns

One of the primary pillars of legislation governing the sharing of data between lenders, is data privacy legislation. This was particularly acute in the United Kingdom, Commonwealth, and elsewhere, where the 1924 Tournier case precedent was applied to bank lending. This is covered more fully in Chapter 33 (Data Privacy and Protection). In other countries, it can be even worse! Legislation may either prohibit the sharing of positive information (Finland, Australia), or even prevent the establishment of any private credit bureaux (France, Israel, Thailand).

Focus upon relationships

Traditional banking was based on trust, and relied upon a customer relationship to obtain information, and upon collateral for security. Even today, many banks may believe that this is sufficient, and are reticent to invest in new technologies. This is especially true where the banks have dominant market positions, or for smaller banks that believe their relationships provide a competitive advantage. In contrast, small retailers and other non-bank lenders have little interest in forging a relationship, or managing collateral, as credit is merely a means to a sale. They are usually the most eager to share information, and often spearhead the formation of credit bureaux in their countries.

According to FI's Interact magazine, in South Africa there were three stages as different classes of lenders bought in. These were: 1993/4—small retailers; 1997/8—large retailers; 2000/2001—some banks. The concept of sharing data is totally alien to most emerging economies though. In Shanghai, it only came about through pressure from the People's Bank of China, and is still only a regional initiative.—'Sharing positive data: the next big step for emerging consumer economies', *Interact*, January 2004, p.19.

14.3 Summary

While Section 12.3 covered some of the mechanics of external data, this section has focused on the theory—especially why lenders would wish to share performance data, and why it works in the best interests of both the consumer and the economy. The greatest benefits come from: (i) *improved risk assessment capabilities*, which implies a reduced probability of adverse selection; (ii) *reduced information asymmetries* between lenders and borrowers, which lowers the required risk premium and cost of borrowing; and (iii) *borrower discipline*, whether by preventing new lending to people already in trouble, or discouraging consumers from over-indebting themselves.

The primary information sharing arrangements for retail credit are operated by credit registries, which may be public or private, both of which have grown dramatically since 1950. Private credit bureaux were the forerunners, and even today tend to be the innovators. Their primary goal is to facilitate the provision of data, to aid lenders in their decision-making. In contrast, public registries were latecomers, and are usually meant to fulfil some social good, like aiding in monitoring the financial system to protect against systemic risk, or facilitating public access to credit.

Credit registries can provide two major types of data: *negative*—information on insolvencies, judgments, and other defaults; and *positive*—shared-performance data that includes accounts that are currently up-to-date. While the former is useful as a borrower-discipline mechanism, the latter provides even greater value, and usually leads to greater credit availability within an economy. There are concerns about how long the forgiveness period should be for bad behaviour. Blacklists may not exist in truth, but as far as the public is concerned, negative information has that effect. Where the memory is too long, it may limit economic activity, because: (i) potential borrowers may be discouraged; and (ii) defaulted borrowers cannot recover. The happy medium lies somewhere between.

This does not mean that lenders have accepted information sharing blindly. Concerns exist due to: (i) the potential for *customer poaching*; (ii) *data quality issues*, relating to the size and stability of the subscriber base, update reliability, and technological stability; (iii) *data privacy concerns*, which may either limit or preclude sharing; and/or (iv) a belief that *relationship lending* is sufficient, or provides a competitive advantage. Customer poaching is usually the greatest concern, but the use of shared data is usually governed by a contractual PoR agreement that prohibits the use of shared-performance data for marketing purposes. Poaching can happen, but there are sanctions. Such principles apply not only to credit information used for new business assessments, but also fraud and collections initiatives.

15 Data preparation

This module's focus has been on data and, so far, data considerations, data sources, scoring structure, and information sharing have been covered. This chapter covers data preparation, the first practical stage in the scorecard development process. It is a crucial stage, as any mistakes can have a severe impact upon the results and may be impossible—or very expensive—to redress. It is covered under the headings of:

Data acquisition—Obtain data from the different sources.
Good/bad definition—Determine exactly what is to be predicted.
Observation/outcome window—Determine the months to be used for the development.
Sampling—Choose a manageable number of cases, to be representative of the population.

Most of the focus is on application scorecard developments, as that is where the issues are greatest.

15.1 Data acquisition

The first task is to obtain data from the identified sources, and bring it all together. Previous chapters described the sources, but no guidance was provided on what was to be done with each. Data acquisition is covered under the headings of:

Application forms—Collection and coding.
Credit bureau—Retrospective bureau searches.
Internal systems—Behavioural data obtained from internal systems.
Performance data—Outcome data for each case.
Matching—Bringing it all together.

Some of this material will be viewed with nostalgia by old-timers, but be totally foreign to newcomers—especially those whose only experience is with environments where credit scoring is well entrenched, and highly automated. Most relates to greenfield developments with no application processing system, which applied to every company at some point, and still applies to some. It should, however, also highlight many of the issues encountered with brownfield developments, where systems are well established.

15.1.1 Application data

People take much for granted! Just as they used to get by without television sets, automobiles, cell phones, and washing machines, lenders also used to get by without credit scoring. And just like the others, credit scoring has become indispensable—bordering on addiction—for many. Processes have evolved so that applications are captured, data is gathered, scores are calculated, decisions are provided, information is stored away, and performance is tracked—automatically. The task of developing a new scorecard becomes easy, because all or most of the required data is readily available. In the not-so-distant past, data acquisition was much more painful. Some of the less fortunate credit providers are only just starting this journey, especially in certain subprime, micro-finance, and small-business markets, as well as in many developing and third-world countries.

Data collection

Lewis (1992) covers the most important questions to be asked for first-time developments: (i) Are the application details for both accepted and rejected applicants available, in any form of database? (ii) If either of these two groups is not available, then where are the application forms? And (iii) once the application forms have been obtained, how will the details be captured? In the worst case scenarios, little or no historical information has been stored away. The task of obtaining and capturing it, at least by today's standards, is onerous. Cupboards, drawers, filing cabinets, and dusty warehouses are searched to find boxes containing yellowing forms, which can bring back memories for some. It is further complicated by having to ensure sufficient numbers of good and bad accounts (especially the bads), and problems determining how many applications each sampled case is supposed to represent.

Lewis (1992) indicates that some companies he dealt with did not have a 'master billing file', meaning that accounts administration was decentralised. This seldom happens today in first-world environments, but at times still applies in emerging markets. In these cases, it is impractical to sample every branch—whether due to the distances involved, or the disruptions caused. Instead, a number of branches, thought to be representative of the market as a whole (income, lifestyle, geography, etc.), must be carefully selected and sampled.

In other scenarios, some information may be available, but not all. Quite often, the person in charge of application processing will try to cut corners, by: (i) pre-screening applications, to focus data-capture efforts only on those applications that have a higher probability of being accepted, or (ii) not bothering to store the details of any rejects, in order to save on manpower or disk space. These practices may save money in the short term, but will cause difficulties later—especially if the applications are discarded.

Coding

Once the application forms have been gathered, the next step is to get the data into computer usable form—a process referred to as coding, because many of the application details are transformed into codes. This requires: (i) determining the codes to be used; (ii) writing the software to aid the capture process; (iii) ensuring that differences in application forms are accommodated; and (iv) employing data-capture operators to do the physical capture. The codes chosen should be as logical and consistent as possible. Residential statuses such as 'Own', 'Rent', and 'Live with Parents' may be coded as 'O', 'R', and 'L'. For cases like 'time at address', a distinction might be made between 0 and blank, by using a special code like 999 for the latter. Throughout the process, it is essential to document what these codes mean. Codes assigned for the first development often become entrenched when a formal application processing system is developed. It is very frustrating for people that come later on, to try to interpret codes with little documentation.

Further complications arise if there are significant differences between past, current, and expected future application forms. Assuming that only minor modifications have occurred, or are expected in future, the variations and their implications must be assessed. Do some forms ask for marital status using tick-boxes with 'Married' and 'Single', while others have an extra field for 'Separated', and still others, a free-form field? Is the customer's age asked for as 'Age', or is it derived from the date of birth? The current application form is the best starting point. Even if it looks very different from old ones, the information has probably changed little. Where variations exist, either: (i) the data capturer must be instructed about how to treat the differences; or (ii) special program code must be written to convert the details into a consistent format.

15.1.2 Bureau data

And of course, lenders need to use the information provided by their spies—the credit bureaux. Hopefully, it has been stored in-house in the appropriate format, as it otherwise comes at a cost. There are two possible situations: (i) use the bureau data obtained at time of application; or (ii) do a retrospective bureau search. Given the power of this data, any effort expended to ensure its quality will be well spent.

Bureau data stored

When credit scoring systems were first developed, the only bureau field was often 'Bureau Clean (Y/N)', which was probably noted in the 'Office Use Only' space. It might have been based on: (i) the underwriter's judgmental evaluation of the bureau data; or (ii) one or more rules like, 'If no judgments, then clean, else dirty'. As technology evolved, credit providers could receive and store these details electronically, and the volume of readily available data grew. Even so, there are some instances where retrospective bureau extracts are required:

(i) **Data instability**—Arising because of rapid growth or contraction in the bureaux' subscriber base—either whom they are serving, or how many. A change in subscribers will affect enquiries; a change in contributors, will affect shared-performance data.
(ii) **New characteristics**—Very often, new bureau fields are made available to lenders, whether totally new (geo-codes), existing bureau characteristics that were previously out-of-bounds (banks gaining access to retailers' shared performance data), or expanded relationships (moving from using one to multiple bureaux).

Retrospective bureau searches

In the absence of historical bureau data, lenders can request a retrospective bureau search. All it requires are personal identification details—like name, address, SSN/SIN/ID number—and the application date for each. The date requested should be immediately prior to the application date, otherwise the data may mistakenly include that enquiry, and/or loan. The retrospective search is then done as a batch run, which unfortunately may take days or weeks, before the results are provided. There are two types of retrospective searches, choice of which depends on the bureaux' infrastructure.

Archive approach—Summary data is stored each month for every individual/company, and is available for the next several years for analysis. Such snapshots are ideal if the bureau data is stable, but can cause problems if not.

Split-back approach—Every bureau enquiry, judgment, and payment profile is retained as a separate dated record, and applicants' retrospective data is reconstructed based on the parts. New bureau contributors are asked to provide historical data, and data provided by old contributors is removed. Although this is theoretically the ideal approach, there may be practical problems, like obtaining historical data from new subscribers, in the required format.

In either case, retrospective searches cost money. The charge is often calculated on a per-enquiry basis, but lenders may be able to negotiate a cap or flat fee for the service. If there are hundreds of thousands of applications, it is not necessary to search for all of them—costs can be cut by sampling (see Section 15.4).

15.1.3 Lenders' historical (observation) data

The expression 'a leopard does not change its spots' is sometimes used to state that behaviour, whether of humans or animals, is unlikely to change. Likewise, data on existing or past dealings will show a strong correlation with performance on new business, often more so than bureau data, and much more so than pure demographic data. This relationship has always been known, so it should be no surprise that knowledge of past dealings had a strong bearing upon judgmental decisions. When application processes are automated, this 'Own Data' should be seamlessly obtained, assessed, and stored, along with externally-sourced data.

There may also be other product areas for which data is not readily available, either because: (i) it was a separate company previously; or (ii) the computer systems could not talk to each other. In such cases, it may still be possible to use their data, assuming the necessary communications links are put in place. First, policy rules could demand more conservative strategies, if the other data indicates potential problems. It may not be very scientific, but allows lenders to extract maximum immediate benefit, and simultaneously to collect data for future analysis.

Second, as with the credit bureaux, a retrospective search can be done—assuming the data has been stored somewhere, and is available. For example, a bank whose primary product is a cheque account (A) acquires a credit card operation (B). It wants to redevelop B's application scorecards, and a link will be created to bring A's performance details automatically into B's account-origination process. This has never been done before, but A's performance files for the past three years are available. The bank can search for performance on A, at a date immediately prior to the application date, using whatever match key is available.

15.1.4 Lenders' performance (outcome) data

In most organisations, the first computerised back-office functions were accounting and billing (this also applies to newly emerging companies), as these were quick wins that obtained the most value out of an expensive technology. As a result, lenders almost always have outcome-performance data available in electronic form. Exactly which fields are required will vary, depending upon what is being predicted: 'Delinquency Status' for default risk; 'Open/Closed Status' and 'Months since Last Active' for attrition/dormancy; 'Account Balance' for revenue; and so on. While the billing system is the most obvious source of information, there may be problems, because:

(a) It usually only contains the most recent information, and does not have an archive.
(b) The data is not provided in an easily analysable form.
(c) Observation and performance details must be matched, and the matching key (such as an application number), may not be recorded on the billing system.

Thus, most lenders will maintain a performance archive separate from the billing system, to ensure that data is readily available in an appropriate format for scorecard development and other analyses.

Outcome window

When developing scorecards, lenders need to decide upon the outcome window (covered in more detail in Section 15.3). Before that though, a choice has to be made between: (i) a *static* outcome point, where the same date is used no matter what the observation date; or (ii) *staggered* outcome points, where the period in between is constant (Figure 15.1). Most application scorecard developments use a static outcome period, as it allows lenders to extract

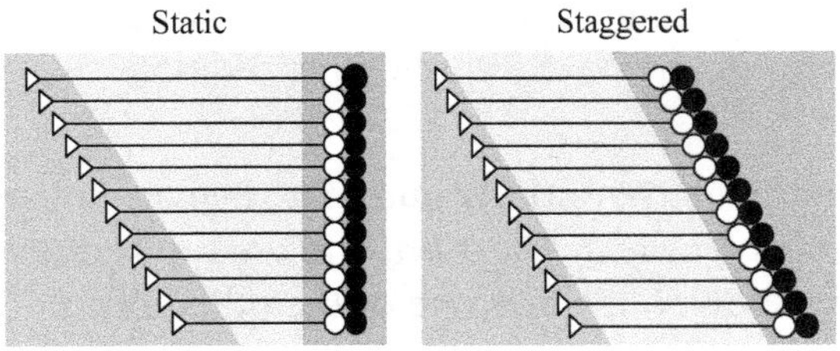

Figure 15.1. Outcome points.

the most value out of available data. Accounts that go bad early may have profiles much different from those that go bad later on, and censoring will affect scorecard effectiveness. In contrast, the staggered option may be used for behavioural scorecard developments. In both cases: (i) sufficient time is needed for the cases to mature, meaning that it should include the peak period for going 'bad', by whatever definition is chosen; and (ii) the different windows are stacked, prior to sampling.

Thomas (2000) states that there is a lot of literature available on the treatment of censored data, but that as of yet, there is no approach that is appropriate for the type of censoring encountered in credit scoring.

Information format

The billing system may also not present the data in a format that can be used in a good/bad definition (see Section 15.2). For example, it may have 'Date Last Payment' as a single field, and 'Amount [30 60 90 120 150] Days in Arrears' as five separate fields. For current purposes, these are inappropriate, and have to be translated into another form. Thus, 'Date Last Payment' is transformed into 'Days/Months since Last Payment', and the 'Amount [X] Days Delinquent' fields are converted into a single 'Months Delinquent' field. Information from the current and preceding months is then combined, to derive a maximum delinquency over the past 3, 6, and 12 months. These same characteristics must be stored, for use in monitoring.

15.1.5 Initial data assembly

The final stage of data acquisition is data assembly, a task much like putting together a Lego or Meccano model. Application, performance data, other product, and bureau data all have to be merged, and presented as a single file (see Figure 15.2):

(i) Decide upon the good/bad definition to be used.
(ii) Obtain historical observation data, and match it to performance.
(iii) Decide upon the observation and outcome window, based on an analysis of the historical outcome-performance data.
(iv) Using the historical data: (i) determine initial possibilities for segmentation; and (ii) create a sample that has sufficient goods and bads from each segment.
(v) If required, submit the sample to the credit bureau for a retro search.
(vi) Merge the results back to create the development sample.

This can be tricky in unautomated or poorly automated environments, especially if there was no account number recorded on the application, and no application number recorded for the account. In most instances, however, either one or the other exists.

There are also cases where lenders want to use not just the performance on accepted accounts, but applicants' worst performance on any account of the same type. Besides the

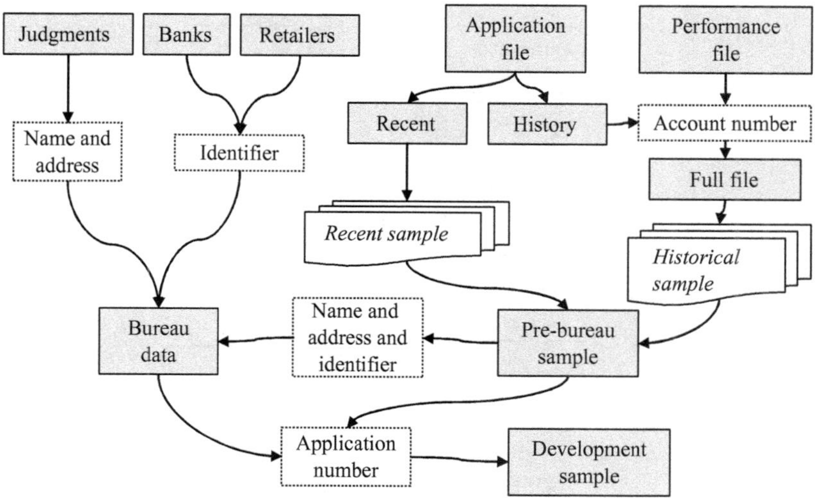

Figure 15.2. Data assembly—flow diagram.

extra benefit of increasing the number of bads, this also addresses cases where the same application leads to multiple account numbers over time. This applies especially to credit cards that are reissued after being lost or stolen, and to repeat loans made to the same individual over the period.

15.2 Good/bad definition

A part and parcel of the human condition is that people try to generalise the world around them, to pigeonhole what they see and experience into a limited range of possible categories that make life easier to understand. Things are seen as yes or no, good or bad, black or white, alive or dead, up or down, left or right, open or closed, pregnant or . . . not. Credit scoring tries to differentiate between the good, the bad, and the ugly . . . uhhh, indeterminate, but how does one tell which is which? That is the job of the good/bad (performance) definition, the second most crucial element—after data—of any scorecard development, which is used to create the target variable for modelling. The 'good/bad' label is often a misnomer though, as it is often better described as a 'bad/not bad', or 'good/bad/indeterminate/exclude' (GBIX) definition.

Most of the credit scoring literature focuses on the good/bad definition's performance elements. Figure 15.3 illustrates the definition for a selection process, like application processing. The arrows indicate the logical order in which categories could—or should—be assigned, while the dashed lines indicate that assignment into the other categories is, or may be, inferred. The 'reject', 'not taken up', and 'mutual accept' categories do not apply to non-selection processes, such as behavioural or fraud scoring.

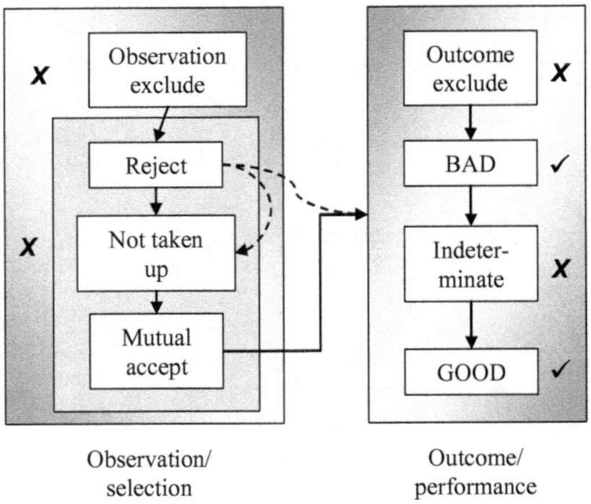

Figure 15.3. Good/bad definition.

15.2.1 Selection statuses

The statuses at observation have nothing to do with account performance, but with the selection process, and the search for mutual attraction between customer and lender:

Observation exclude—Not to be scored, neither in the development nor in practice.
Reject—Not selected; lender says 'No'.
Not taken up (NTU)—Selected, but not used; lender says 'Yes', but borrower says 'No'.
Mutual accept—Selected and used; both parties say 'Yes'.

Observation excludes

If there are certain sub-groups where the score will not influence the decision-making, then it is wise to exclude them. Inclusion will distort the scorecard, decreasing its validity for those accounts where the lender wishes actively to apply strategies. The most common reasons for excluding entire groups are:

Other business area—Even though the data resides on the system, other business areas are responsible for those accounts. This may apply to entire market segments, channels, or accounts that have passed on to another stage in the CRMC.

Policy—The scores will not make any difference. Cases may be accepted regardless, because of employer agreements and/or guarantees (policy accepts), or are rejected outright, because of a statutory decline rule (policy rejects). Treatment may change, if the policy changes.

Insufficient data—Thin-file cases, where it is impossible to provide a score with any meaning.

Discontinued—Products, markets, or origination channels that the lender has either exited, or has plans to do so.

Out of scope—The group is thought to be significantly different, and/or is not to be managed using the same strategies. A separate scorecard may be in place, or planned. If not, the group may be excluded from the development, but scored for guidance in production.

Hard bad—Cases that have exited the system, or are so close to the point of no return that their classification is indisputable. This applies especially to behavioural scoring for accounts already in recoveries, tagged as non-performing loans, or say four or more months in arrears.

Ideally, these groups should be excluded from the data assembly in their entirety, but some of them might only be identified during data analysis. They should also be excluded from (or treated separately within) any scorecard monitoring.

Rejects

This is the most problematic category, because lenders only have performance for accepted accounts. Leaving rejects out results in scorecard bias though, so reject inference is used to make educated guesses about how they would have performed if accepted (see Chapter 19, Reject Inference). Of primary interest is whether the accounts would have been good or bad. Rejects can also be inferred into the not-taken-up and indeterminate categories, but failure to do so should not have a huge effect, unless they constitute a substantial portion of the through-the-door population. No reject inference is done on hard-core policy rejects, irrespective of whether the policy is related to unacceptable risk (extremely high reject and/or bad rates), legal constraints, or product rules.

Not taken up

It is not only the lender that has the right of refusal; accepted applicants can also refuse, by never opening or activating the account, whether because they did not like the terms, they took up an offer elsewhere, or the loan was no longer required. This status may be inferred for rejects, especially where the NTU rate amongst accepts is large.

Mutual accept

This is the category of most interest—instances where the customer has taken up the product, and has used it (or abused it). It includes all accounts that have an outcome-performance status, and is treated next.

15.2.2 Performance statuses

The other part contains the four mutually exclusive outcome-performance statuses that are traditionally associated with the good/bad definition:

Good—Desired state, something to be welcomed in future.
Bad—Unwanted state, something to be avoided in future.
Indeterminate—In between, greeted with mild reluctance and not revulsion (optional).
Excludes—Any outcome outside of the intended purpose of the scorecard, like an operational-risk event (fraud) in a credit risk scorecard.

Of these four states, only two—good and bad—are used in the final scorecard development. The other two, indeterminates and excludes, are omitted, so that the model provides a much clearer picture of whatever is being predicted.

Goods and bads

Medical practitioners refer to patients as being either 'positive' or 'negative' with respect to the existence of some malady: 'Positive' means true, the patient is sick; 'negative' means false, the patient is free of disease. Diagnoses are based upon test results that are not always foolproof—leading to the possibility of true positives, false positives (type I errors), true negatives, and false negatives (type II errors). This terminology is widely used in science and research when dealing with rare events, and is sometimes carried over into business.

In this case, credit scores are the measurements used: 'positives' are bads, that involve either a loss, or lost opportunity; and 'negatives' are goods, that are desirable accounts. False positives are rejects that would have been good if accepted, and false negatives are accepts that fared poorly. Definitions will vary depending upon the type of scorecard, but the general interpretation is: 'bad'—the lender is struggling to get its money back, or has already given up (risk); the account is dormant (retention); or the customer does not go for the bait (response); 'good'—the account has either been repaid, or the agreed repayments are being made (risk, collections); it is open and active (retention); or the customer has taken up the offer (response). Just as medical maladies have an 'incubation period', accounts have a maturity period, before the 'negative' call can be made. In contrast, positives, especially 'hard bads', might qualify for inclusion within three or four months.

Indeterminate

In many instances, the definition of good and bad is straightforward. Bankruptcy is like pregnancy—either you are, or you are not. In other cases there is a grey area, where the classification is less obvious, such as mild delinquencies, where there may have been a few extra phone calls (risk), or sporadic activity (retention). The logic for the indeterminate range is that: (i) seemingly bad behaviour may be the result of technical arrears, or company strategies; and (ii) good and bad are more clear cut, which should hopefully aid identification of truly problematic accounts.

Not everybody is a convert though. Beardsell (in Mays 2004) presents results showing that the use of an indeterminate range provides no extra value, but this is only one test, and there have been others showing the contrary (the difference might lie in whether a worst-ever or current-status definition is used). Care must also be taken because an incorrectly specified indeterminate range can provide seemingly better results, but substandard scorecards. Neves (2003) suggested that for credit risk scorecards, the appropriate indeterminate rates are from 5 to 15 per cent for application developments, but 10 to 20 per cent for behavioural developments, because of the shorter timeframes.

Another potential indeterminate group for risk scorecards is early settlements. This may be contentious, as they are good accounts, but may be justified, because their inclusion can bias scorecards in favour of applicants that are often unprofitable. The alternative is to treat them as good, and use a separate early-settlement score as part of the assessment.

Outcome excludes

The final group is similar to indeterminates, except instead of falling between good and bad, it falls outside. Outcome excludes are statuses that may compromise the model with respect to its given objective. For example, where a customer dies, or it is a known fraud, the loss is not a credit loss for the purposes of a credit risk scorecard.

'Deceased' cannot always be identified on lenders' systems, and some may choose to leave them in. Also, while the payments may cease, the lender still has a claim against the deceased estate, and the cases may still be treated in EAD and LGD modelling.

Other potential outcome exclusions are closed, dormant/inactive, and insufficient experience. Thomas et al. (2002) describe 'insufficient experience' as any account that has not had enough activity to assign one of the other three labels—not just new accounts, but also those with little activity, prior to dormancy/closure. Such cases should not be included in: (i) the scorecard development; (ii) the population to which the scorecard is applied; and (iii) performance monitoring.

Note, that many lenders lump indeterminates and excludes into a single indeterminate category, for ease of reporting. This has little effect where exclude rates are low, but can distort the results as they increase. Mays (2004) suggests ensuring like-for-like comparison by treating frauds and mortalities the same way, in both the scorecard development and ongoing monitoring, especially if they comprise more than 2 to 3 per cent of bad loans.

15.2.3 Current-status versus worst-ever definitions

Lenders have a choice between: (i) a *current-status* definition that focuses upon the account status at outcome point only; and (ii) a *worst-ever* definition that uses the worst status over the outcome period. In both cases, the intention is to specify something near the 'point of no return', where the chances that the account will recover are low. Which one is appropriate depends upon: (i) the type of development; (ii) what the definition is being used for; and (iii) factors relating to the business.

For behavioural scoring, Basel II requires 90-days worst-ever, but if the scores are to be used to manage early-stage delinquencies, current-status is more appropriate. Otherwise, the model's ability to recognise self- and easy cures will be limited, unless extra effort is expended to determine an expected loss, or a harsher definition is used. In any case, it should be possible to map the results for regulatory purposes.

For application scoring developments, either type of definition can be used. A current-status definition ensures that *technical arrears* and *self-cures* are not unfairly prejudiced, but a worst-ever definition is necessary if *distressed restructurings*—like re-aging of credit card arrears—are common practice, and difficult to detect. Data availability also plays a role, as worst-ever definitions are only possible if account performance is available over the full outcome window. Siddiqi (2006) mentions current-status only as a means of deciding upon an appropriate worst-ever days-past-due threshold, yet many lenders still prefer the current-status approach.

In general, for credit risk assessments, 60 days-past-due dominates current-status definitions, while 90 days dominates for worst-ever. The latter seems to be more common for application scorecard developments, and is the basis for Basel II reporting. Even so, lenders should choose that which makes most business sense. Unfortunately, nobody seems to have done any research on the merits, or lack thereof, of the two approaches. This might be a worthy research topic, for anybody so inclined.

15.2.4 Definition setting

A question often asked is, 'What is an appropriate bad rate?' One subject expert retorted with the question, 'How long is a piece of string?' The answer depends entirely upon the circumstances, and lenders need definitions that are appropriate for them. Given that the good/bad definition is so crucial for the scorecard development, a good deal of attention should be given to developing it. Several approaches can be used:

Consensus—Based upon judgmental inputs from experts within the company.
Empirical—Based upon empirical analysis of the lenders' own data.
Prescribed—Set by an external agency, to ensure consistency.

Consensus

In early developments, the definition was derived based on consensus between scoring experts, experienced underwriters, and company management, and varied in complexity depending upon the product and purpose. It could be as simple as 'repaid/not repaid' for a short-term loan product, or a highly complex set of rules for cheque accounts. It may be driven by the lender's accounting policies, or be the result of meetings that resemble the Mad Hatter's tea party. The end result is a series of statements like:

If Account Status = NPL	then 'Bad' else
If Current_Delinquency >90 days and Current_Arrears >50	then 'Bad' else
If Times_60d_L12M >3 and Maximum_Arrears >50	then 'Bad' else
If Account_Age <10	then 'Excl' else
If Months_Since_Last_Active >6	then 'Excl' else
If Purchases_L6M <50	then 'Excl' else
If Current_Delinquency >0	then 'Ind' else
If Times_60d_L12M >0 and Maximum Arrears >10	then 'Ind' else
	'Good'

These are applied as a filter, meaning that the status associated with the first true statement is used. While not invalid, this approach is difficult to validate, and can suffer from unwarranted complexity.

Prescribed

Many lenders will not develop their own definitions, but instead use a standardised definition demanded by a third party. First, a *securitiser* or *collections agency* may prescribe one for use when valuing loan portfolios. In most instances, they will apply their own scorecards to standardised data fields, especially bureau data. Second, *technology vendors* may insist upon a specific definition for use with their systems. This achieves a great deal of consistency, making analytics much easier. Finally, *regulatory bodies* may demand a specific definition for reporting. This applies especially to banks under Basel II, which has specified a definition of 90 days-past-due on any material obligation (or 90 days-in-excess of agreed limit for cheque accounts), or whenever it is known that there is a significant probability of loss. Note, however, that most banks will still use a definition considered best for business purposes, and calibrate those scores onto the statutory definition for reporting.

Empirical

The final alternative is to derive the definition, based on an empirical roll-rate analysis. A 'hard bad' or 'super bad' definition is used to define soft bad (high probability), indeterminate (medium probability), and good (low probability), using either a tabular approach (see below) or recursive partitioning algorithm (see Section 7.3.1). Hard bads are cases where lenders have given up on getting all or part of the money back, or expect to experience serious costs in doing so. It may be non-performing loan (NPL), 120 days-past-due, or some other status. An example is presented in Table 15.1, which assesses the percentage of accounts rolling to NPL after one year. Both old and new definitions are shown. The lender had treated accounts at '30 days' as indeterminate and '60 days' as bad. The analysis shows, however, that the roll rate for accounts at '60 days' is 17.5 per cent, and that for 'late but not yet 30 days' is 5 per cent.

Table 15.1. G/B definition—roll rates to NPL

	Total	Old definition	NPL	NPL rate (%)	Definition
Guaranteed/VIP		Exclude			Exclude
NPL	4,000	Hard bad	4,000	100.0	Hard bad
4 down	500	Soft bad	300	60.0	Soft bad
3 down	1,000	Soft bad	350	35.0	Soft bad
2 down	2,500	Soft bad	438	17.5	Indet.
1 down	5,000	Indet.	625	12.5	Indet.
Late payment	8,000	Good	400	5.0	Indet.
Up-to-date	283,000	Good	2,830	1.0	Good
Total	300,000		4,943	1.6	

There is no predefined threshold for these developments, but lenders will develop rules of thumb. For credit risk scoring, this will usually be in the 25 to 30 per cent range for bad (definite loss probability), and the goods should be at least better than the portfolio average (or better yet, be profitable). In the example, bads would be limited to accounts 90-days-past-due or more, and goods to those up-to-date.

The example is very simplistic, especially as it is restricted to using one delinquency status characteristic. Other characteristics may be used, including those from other products and the credit bureau, but lenders should ensure they are available for ongoing scorecard monitoring. In need, preliminary models can be developed using competing definitions, and compared by evaluating how well they work, using a separate 'hard-bad/not hard-bad' definition. Some general words of caution though: (i) use common sense; (ii) err on the side of simplicity; and (iii) beware of definitions that can be affected by small changes to lender strategies.

15.2.5 What a good/bad definition should be!

When engaged in the dating game, people think in terms of the desired attributes of a potential partner. Terms used by a man to describe his ideal lady may be petite, caring, and intelligent, while the lady may be looking for somebody who is tall, strong-willed, and adventurous. Good/bad definitions also have ideal attributes, but much different from those of ideal partners. The following list of required attributes was presented by Jes Freemantle (unpublished, but used with permission), and is split into three groups: (i) *relevant*, well-suited for meeting the models' goals, based upon customer behaviour, objective, polarising, and focused on the recent past; (ii) *adequate*, sufficient, robust, comprehensive, and implementable; and (iii) *transparent*, capable of being understood by anybody that reviews it.

Relevant

When lost in London, a map of Sydney helps little. Likewise, the definition must be suited to the type of scorecard being developed. For example, risk scorecards would use data on bad statuses, arrears, and over-limit excesses, while retention scorecards use closure/dormancy statuses, months since last active, and transaction counts. In order for the definition to be relevant, it must be: (i) *focused*, capable of providing a model that will serve the purpose for which it is being developed, whether controlling credit risk, account retention, fraud, or some other dimension; (ii) *polarising*, 'goods' should be good (profitable, productive), and 'bads' bad (loss, or lost opportunity); (iii) *objective*, it should be possible to substantiate the classifications through empirical analysis; (iv) *based on customer behaviour*, and ignore or minimise the effect of lender strategies; and (v) *current*, focus on recent behaviour, and ignore data that is dated, or might be the result of teething problems. The final point may be contentious, especially given the popularity of worst-ever definitions.

With respect to customer behaviour, one rather confusing case is overdrafts on cheque accounts. Besides credit turnover drying up, the two key factors are over-limit excesses and NSF cheques—both of which are linked to lender strategies. Customers are able to minimise the duration of excesses, but can do little about bounced cheques. It is advisable to ignore the latter, or use it sparingly, in any definition. Assuming that the lender is at least trying to enforce the agreed limits, the duration of excesses has almost the same meaning as days-past-due, and should suffice.

Adequate

Just as the definition should be relevant to the task, it should also be: (i) *robust*, it should remain valid for longer than the scorecard(s), and not suffer from small changes in customer behaviour, business environment, or company policy and practice; (ii) *comprehensive*, it must be valid for all cases in the sample, otherwise a separate definition and scorecard is required; and (iii) *sufficient*, it should provide enough goods and bads to develop a model, which may only be achievable by adjusting the hurdles.

On the 'comprehensive' point, Siddiqi (2006) notes that there may be demands for definitions to be consistent across different products within a company. This aids decision-making and monitoring, and minimises training and programming requirements. Beware, however, as there may be instances where a consistent definition is inappropriate, in particular where the repayment culture varies across different segments, and it is accepted as part and parcel of the business. A possible solution is to have separate definitions for scorecard development and management reporting, so that the scorecards can focus on serving the purpose for which they are developed. The same applies to regulatory reporting, and could extend as far as scorecard monitoring—but the latter may be dangerous territory.

Transparent

One of the problems with credit scoring is getting the business to trust it, and if they do not trust the good/bad definition, then they are unlikely to trust the final model. One of the keys to getting business buy-in is ensuring that the definition is transparent: (i) *simple*, expressed using as few characteristics as possible (in some instances, however, it is difficult to identify sufficient bads, and use of other attributes, such as '3 times 30 days', allows their numbers to be supplemented (Mays 2004)); (ii) *logical*, it should make common sense, and possibly be based on accounting, write-off, or other policies; and (iii) *implementable*, scorecard performance must be monitored, which means the definition—either exact, or with only minor deviations—has to be implemented in the monitoring system.

15.3 Observation and outcome windows

If you are a coach looking for football players, you first have to put out the word, and then go through some form of selection process. Some prospective players will be suited to the task, some not. Likewise with credit scoring, just because the data exists does not mean all of it is used. The observation and outcome windows are restricted to ensure an optimal period is chosen, and the data is sampled to speed the development process. In both cases, lenders must ensure that the data is representative of the expected future through-the-door population.

First, some more definitions: *observation*, pertaining to data, to be used as predictors; *outcome*, pertaining to data, used to define the target; *date/month*, time when information is collected; and *window*, a period of opportunity. These terms are combined to come up with things like: *observation date*, date when observation data was collected for a record(s); *outcome date*, date when performance data is collected for the same record(s); *observation window*, time period over which the observations are collected; and *outcome/performance window*, time period between observation and outcome date. The outcome window could be anywhere from nanoseconds in the realm of physics, to decades for things like genetics. When choosing the appropriate window for credit scoring, there are three factors to consider:

Maturity—As more time goes by, the bad rate will increase, but at a decreasing rate. Accounts are considered mature when the bad rate curve is almost flat (if that happens).

Censoring—Exclusion of cases that go bad outside the chosen window, which results in information loss.

Decay—Changes in the business—whether the through-the-door population, infrastructure, policies, economy, or some other factor—that makes older information, or models, less valid.

According to Siddiqi (2006), the typical outcome window for application scoring is between 18 and 24 months for credit cards, and from three to five years for home loans. In contrast, behavioural scorecards will use a window of between 6 and 12 months, while collections scorecards may use a month or less. In general, lenders tend—often unwisely—towards shorter periods, to recognise the rapidly changing nature of the business.

Application scoring

The illustration in Figure 15.4 illustrates the bad rates for applications processed over a three-year period. Although not shown, there were on average 2,500 applications per month, of

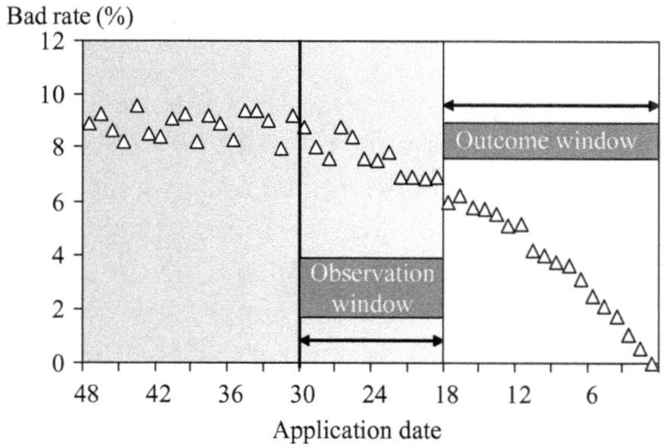

Figure 15.4. Application scoring—sample window.

which about 15 per cent were rejected. The bad rate levels off at about 9 per cent, but it takes about 30 months to get there. Beyond that point, the applications are not thought to be representative of today's business. Even so, after about 18 months, the bad rate starts levelling at about 7.25 per cent, and increases only slowly after that. If the period from 18 to 30 months is used, there are just over 2,200 bads, which will at least allow the construction of a single scorecard.

This example is realistic, and lends itself to a scorecard development, but other cases are more problematic. First, it may not be possible to get the required 1,500 bad accounts. Where there is a limited number of bads, it is still possible to work with lower numbers, perhaps as low as 400 to 500, but this makes the results more suspect, and requires greater care and diligence once implemented.

Hoyland (1995) mentions two commonly accepted means of increasing the number of bads: (i) include more recent bads, as the passage of time will not change their status; and/or (ii) lengthen the original sample window to include older accounts, perhaps three or more years old. For the latter, tests should be conducted to ensure that the original and older groups have similar profiles, at least for the most influential characteristics used in risk assessment.

Second, the time effect may not flatten out as quickly, and still be strong within the observation window, which poses a problem where the nature of the applicants has changed over the period. For example, if there is an increasing number of younger applicants, then the scorecard may erroneously associate lower age with lower risk.

There are two ways of controlling for the time effect. First, the 'Months Since Application' can be used as a control variable (Mays 2004). This requires no *ex ante* knowledge of the bad rates, and will not mask any real trends within the data, but does not assist with the coarse classing or any other bivariate characteristic analysis.

Control variables

Control variables are included as part of a regression, but are not included in the final implemented model. 'Age of . . . ' variables can be used not only to control for the time effect, but also seasonality, cyclicality, once-off marketing campaigns, changes to strategies, and other historical blips. Christmas and other holidays are of particular concern, as well as periods immediately before and after. Borrowers overspend, ignore account payments, and may receive bonuses over this period. According to Lacour-Little and Fortowsky (2004), the use of banned variables (race, gender, etc.) as control variables could also be used to protect against disparate impact.

Second, the time effect can be controlled by reweighting the bads, so that the bad rate for any month is the population average (for more on weights, see Section 15.4.3 on Stratified random samples). The reweighting formula is:

Equation 15.1. Time-effect reweighting $$w_i = W_i\left(\frac{B/T}{B_m/T_m}\right) \Bigg|\; \begin{matrix} m_i = m \text{ and} \\ y_i = 0 \end{matrix}$$

where W and w are the original and modified weights; B and T are the total number of bads and applicants respectively; m is the application month or period number; y is the good/bad flag ($y_i = 0$ being bad); and i is the index of the record being assessed. A very simplified example is presented in Table 15.2.

This approach is simple and easy to understand, and will assist subsequent bivariate analysis. It has disadvantages though, as it may mask real trends within the data, and it requires *ex ante* knowledge of the bad rates. The effect of the masking should be small, as the time effect usually dwarfs natural drift. As for having no knowledge of bad rates, this occurs where there is a sample, but it is not known how many each case should represent. It occurs primarily in emerging environments, where data is obtained manually from paper forms. The resulting models may be just as effective for ranking risk, but cannot be used to provide estimates.

Finally, if two or more scorecards are required, sufficient observations are needed for each. Again, for any single scorecard there should be a minimum of 1,500 goods, bads, and rejects, and it cannot just be assumed that 1,500 times the number of scorecards will suffice. As shown

Table 15.2. Reweighting for time effect

Age in Qtrs	Actual bads			Controlled bads	
	G+I+B	B	Odds (%)	Weight	Count
5	7,895	174	2.2	2.49	432
6	8,542	393	4.6	1.19	468
7	7,532	467	6.2	0.88	412
8	8,221	575	7.0	0.78	450
9	7,963	589	7.4	0.74	436
Total	40,153	2,198	5.5		2,198

Table 15.3. Uneven sampling

	Total	Old	Young
Total	30,000	22,500	7,500
Rejects	6,000	3,375	2,625
Accepts	24,000	19,125	4,875
Bads	3,000	1,688	1,313
Goods	21,000	17,438	3,563
Reject rate (%)	20.0	15.0	35.0
Bad rate (%)	12.5	8.8	26.9

in Table 15.3, sampling 3,000 cases provides an uneven split between the Old and Young scorecards, and insufficient bads for the latter. This can be solved by over-sampling both groups, to provide the required 1,500 bads for each.

There are other issues that can make the development a veritable minefield, in particular if processes have changed over time: (i) a subgroup of applicants may have been processed manually, but were then put through the system; (ii) rejected applications may have been discarded, but are now captured; or (iii) collections strategies may have been lax, but then became more aggressive. All of these present some form of bias that is not the result of any customer behaviour, and efforts should be taken to prevent them from impacting upon the scorecard results.

Behavioural scoring

The task with behavioural scoring is similar, yet it is both simpler and bigger: simpler, because rather than looking at a window where a number of consecutive months will have the same outcome point, the time period between observation and outcome can be kept constant; bigger, because there may not be one but many of these windows. A single window may be sufficient, but provide only barely enough bads. By stacking multiple 6- or 12-month windows, the number of available bads can be increased, such that even the bads can be sampled.

15.4 Sample design

Managers take little buckets and row their boats over an ocean of data, dipping the buckets here and there. By examining what they find in the buckets they will be able to draw firm conclusions not only about what is in the buckets, but also about the entire ocean.

George S. Odiorne, Management Decisions by Objectives, p. 196
paraphrased in Groebner and Shannon (1989)

Have you ever noticed how so many toothpaste adverts show the toothbrush with a healthy load of toothpaste, more than you will ever use? This has been a ploy used by marketers since the 1960s. Only a little bit of toothpaste is required, but if they can convince consumers that more is better, then sales increase. It is similar with credit scoring; there may be a lot of information available, but as with Brylcreem (hair gel), a little dab will do ya. These dabs are called samples—cases that are selected and analysed, to draw conclusions about a much larger population. Sampling is essential in statistics, to make the gathering and processing of data cheaper, faster, and easier.

If there is no data, then the problem lies in acquiring it—and the cost of acquisition can be extremely high. If there is too much data, the problem is to reduce it to a manageable size. In either case an unbiased random sample is required, meaning that it must be truly representative of the population under consideration. An obvious example is that scorecards are biased towards past customers, but there are countless others. One of the most common instances of bias is marketing surveys, where only people with home phones are contacted.[1] A rather light-hearted hypothetical example is provided below.

Biased samples—the country mouse

A mouse-hole salesman wants to sell country mice mouse-holes in the city, but country mice have a concern about cats. A study is required on the risk of mice becoming supper for a cat in the city or country. He phones mice listed in the mouse telephone directory, but forgets to recognise that only 5 per cent of country mice have telephones, as compared to 80 per cent of city mice.

The resultant model will probably miss out on a lot of things. For one, most of the mice with phones will live in houses with cats, and city mice living nice compete with city cats for scraps, and often come second in more ways than one. In contrast, most country mice do not have phones and feed in the fields and only the foolish dare to cohabit with cats even if they can get fat, so they are more likely to stay out of cats' way. Mice choose not between city and country, but between phone and no phone.

If only mice with phones are phoned this will never be detected. It should not be condoned if the survey is defective. Mail shots could be used to sample mice with no phones, but then only erudite mice who read and write will respond. Sigh! It's hard to reduce bias for country mice, who have no phones at home.

By random, it is meant that cases are selected without any regard to their attributes. It is usually sufficient to choose every Xth case, but the use of coin tosses or random numbers is even better. For the latter, a random number generator can be used to assign a value between 0 and 1 for each record in the dataset.

A seed value is used to provide a starting point for the random number generator, and the same series of numbers will be returned as long as the same seed is provided. To obtain a different series every time, the time of day (hhmmss = hour, minute, second) returned by the system clock, or some variant, can be used as the seed.

[1] This has become less of an issue in first-world countries as telephones have become ubiquitous, but still applies to developing and under-developed countries.

To create the sample, each record is processed, and the sampling criterion for each is:

Equation 15.2. Random sampling $$s_i = \begin{cases} 1 & if\, R_i < C_i \\ 0 & if\, R_i \geq C_i \end{cases} \quad \text{where} \quad C_i = \frac{S - \sum_{j=1}^{i-1} s_j}{N - i - 1}$$

where *s* is a binary sampling flag [0 or 1], *S* is the total sample size required in units, *N* is the total number of records in the group, C is the selection criterion threshold, where $[0 \leq C \leq 1]$, *R* is a random number, where $[0 < R < 1]$, and *i* is the index for the record being assessed. A very simple example is where one case is to be selected out of a group of 10. The numerator for C starts with a value of 1 and reduces to 0 once a case is selected, while the denominator starts at 10 and reduces by 1 each time. Thus, the criterion threshold will be 0.10 for the first record, and increase by 0.10 until a record is chosen and it becomes zero. If no record has been chosen by the 9th record, the threshold becomes 1/1 or 100 per cent, and the 10th record must be selected. This logic applies whether the sampling is being done for the entire population or strata therein.

15.4.1 Sample types

A development sample is not always one homogenous whole. Different parts are required for different purposes. Some of the terminology used is summarised in Table 15.4, which also indicates where the terms overlap.

Development sample—All of the records to be used for the scorecard development, both historical and recent, training and validation. The term is sometimes used synonymously with 'training sample', but the latter is a subset.

Historical sample—Records from the observation window that have performance associated with them. This includes both the training and holdout samples.

Validation samples—Records that are used to check the results. This includes both the recent sample, and out-of-sample cases (holdout/out-of-time).

Training sample—Records from the sample window that are used to develop the predictive model. This is the most relevant part of the dataset, which requires the greatest number of observations.

Table 15.4. Sample types

	Development	Historical	Validation
Recent	✓		✓
Training	✓	✓	
Holdout	✓	✓	✓
Out-of-time		✓	✓

Holdout sample (In-time)—Remaining records from the historical sample. These are held back to be used for model validation. Statistical models can be overfitted, meaning that they are so finely tuned to the training sample that they do not work for other accounts.
Out-of-time sample—Data from a time period different to that used for the training sample, usually just after. It is used to ensure that the model will work across different time periods, and is preferable to a holdout sample, but not always feasible.
Recent sample—Records from the three or so months immediately before the development, which will have no performance. It is used to: (i) ensure stability of the characteristics used in the scorecard; and (ii) evaluate the stability of the final scorecard.

15.4.2 Maximum and minimum sample sizes

As mentioned elsewhere, almost all of the literature on credit scoring recommends a minimum of 1,500 bads, 1,500 goods, and 1,000 rejects for a scorecard development. These minima were derived in the 1960s, when the task of collecting data was backbreaking (and computers did not have much backbone), and have lived on. Why collect more data than is needed?

While this textbook focuses upon cases where the amount of available data is significant, there are still instances where there are very few bads. Models have been developed with as few as 40 bads, but these were done in academic environments and were never used in practice. It is, however, quite common for scorecards to be developed using as few as 400 bads, but these are subject to greater validation and monitoring.

In the meantime, technology has evolved such that modern Nintendo games have as much computing power as NASA used to put the man on the moon in 1969 (this may be an exaggeration, but makes a point), and modern application-processing systems make data collection easy and cheap. If data is readily available, then why not take advantage? With this in mind, the first sampling step should be to decide upon the maximum sample size. This will be a function of: (i) *data collection*, obstacles to assembling both observation and performance data; (ii) *computing power*, larger sample sizes can be used if it is sufficient; and (iii) *extra costs*, incurred to obtain information from outside sources, especially credit bureaux.

When faced with large amounts of data, lenders must realise that beyond a certain point, the extra information does not add much value. In particular, it serves no purpose for the number of goods to be disproportionately large relative to the bads (assuming bads are fewer than goods), and in rare cases there may even be more bads than needed.

Another issue is the cost of obtaining external data, in particular retrospective bureau enquiries. Tariffs are usually on a per enquiry basis, but lenders may be able to negotiate a fixed price up to a certain maximum number of enquiries. If the bureau does give a substantial discount, it may be sensitive regarding how many records can be sampled. The information they hold is powerful, and very large samples can allow lenders to use the information for purposes beyond scorecard development. This is not in the bureaux' best interests, because: (i) it goes against the grain of data reciprocity agreements; and (ii) it reduces the potential income they could otherwise earn.

15.4.3 Stratified random samples

The use of totally random samples can pose difficulties, as there may be small but important subgroups that must also be adequately represented. This is addressed by over-sampling the rare groups in a stratified random sample, where the root 'stratum' is used in the same sense as in chemistry—a layer that possesses similar qualities. With gold and oil, different strata are sampled to determine the feasibility of mining or drilling respectively. In credit scoring, the main concern is ensuring that there are adequate bads at outcome, but lenders may also wish to ensure proper representation, by observation GBIX status (behavioural only), market segment, new versus existing customers, or other potential scorecard splits.

Each sampled record is assigned a weight, so that the statistical process can be tricked into believing it is working with the full population. For example, if the sampling rate for a stratum is 1.0 per cent, then the weight for each record within the stratum is 100, effectively causing them to be counted 100 times. While this is the norm, and is necessary to provide probability estimates, models with the same or similar ranking ability can be derived using unweighted data.

Weights can also be adjusted and used to reflect misclassification costs, in which case the sample and population proportions no longer equate. There is also scoring literature that maintains that weighting up the bads in this fashion can provide better results, but nobody bothers to explain how or why.

Assuming that the number of bads is known, should the goods be matched 1-to-1 (a 'balanced sample'), or should the population's good/bad odds be maintained? According to Thomas et al. (2002), the ratio used in practice varies between the two, but the benefit of having extra goods diminishes as the imbalance increases.[2] For an individual observation stratum, a rule of thumb is to restrict the number of sampled goods to three times the number of bads. Other generalised rules that could be used are that:

- The required numbers should be increased to ensure that there are enough for a holdout sample, which may range from 10 to 50 per cent of the total.
- In the rare instances where there are more than 5,000 bads for the training sample, the extra numbers are unlikely to add much value.
- Where applicable, fewer numbers of rejects are required. These could be limited to between 67 and 100 per cent of the total number of bads.

[2] Thomas et al. (2002) also cited Makuch (2001), who indicated that little extra value is obtained from having more than 100,000 goods.

- Fewer numbers are required for indeterminates and excludes. They are not used in the development directly, and could be limited to between 40 and 60 per cent of the number of bads.
- Consideration must be given to accounts that may be dropped after the sample has been analysed. A common occurrence is where 'Exclude' groups are only identified through analysis, especially groups with extraordinarily high or low bad rates.

There is no universally applicable optimal sampling strategy; what works in some cases may not work in others. An incorrect approach can be avoided if high-level issues are considered—in particular, that the primary directive is to ensure a representative sample for all groups of interest. Given that the cost of making incorrect assumptions at this stage can be significant, the sampling strategy should be reviewed before any data is extracted, especially from external sources.

Application scoring example

The most basic case is to obtain the absolute minimum required for a single scorecard, as illustrated by the application scoring example in Table 15.5. Here there are 100,000 records, but the lender only wants to use 5,000 at most, in order to keep bureau costs down. If a totally random 5 per cent sample were used, it would only yield 200 bads. If the minimum required numbers are sampled for each status, the resulting sampling weight is not 20.0, but 2.7, 8.0, 48.7, and 15.0 for bads, indeterminates, goods, and rejects respectively.

Behavioural scoring example

This can be taken one step further for a behavioural scoring development, as illustrated in Table 15.6, which shows the population and sample counts, on the left- and right-hand sides respectively. If the scorecards are to be representative of accounts that are already problematic

Table 15.5. Stratified sample—application

		Selection		Performance		
	Total	Reject	Accept	Bad	Good	Indet.
Population	100,000	15,000	85,000	4,000	73,000	8,000
Rates (%)	100.0	15.0	85.0	4.7	85.9	9.4
<u>*Totally random*</u>						
Size	5,000	750	4,250	200	3,650	400
Weight	20.0	20.0	20.0	20.0	20.0	20.0
<u>*Stratified random*</u>						
Size	5,000	1,000	4,000	1,500	1,500	1,000
Weight	20.0	15.0	21.3	2.7	48.7	8.0

Table 15.6. Behavioural—no oversampling

Obs G/B status	Population counts					Sample counts—simple				
		Outcome G/B status					Outcome G/B status			
	Total	Bad	Good	Indet.	Exclude	Total	Bad	Good	Indet.	Exclude
Soft bad	2,000	800	700	400	100	80	30	30	8	12
Indet.	8,000	400	4,600	2,000	1,000	320	120	120	32	48
Good	90,000	2,800	67,700	5,600	10,900	3,600	1,350	1,350	360	540
Totals	100,000	4,000	73,000	8,000	12,000	4,000	1,500	1,500	400	600

Table 15.7. Behavioural—with oversampling

Obs G/B status	Sample counts					Weights				
		Outcome G/B status					Outcome G/B status			
	Total	Bad	Good	Indet.	Exclude	Total	Bad	Good	Indet.	Exclude
Option 1: Minima										
Soft bad	1,013	400	400	107	107	2.0	2.0	1.8	3.8	0.9
Indet.	1,013	400	400	107	107	7.9	1.0	11.5	18.8	9.4
Good	1,773	700	700	187	187	50.8	4.0	96.7	30.0	58.4
Totals	4,000	1,500	1,500	400	600	25.0	2.7	48.7	20.0	20.0
Option 2: Realistic										
Soft bad	1,820	700	700	336	84	1.1	1.1	1.0	1.2	1.2
Indet.	1,040	400	400	160	80	7.7	1.0	11.5	12.5	12.5
Good	7,280	2,800	2,800	570	1,110	12.4	1.0	24.2	9.8	9.8
Totals	10,140	3,900	3,900	1,066	1,274	9.9	1.0	18.7	7.5	9.4

at observation, then there is a problem. There are only 80 bads and 320 indeterminates, yet these may be the accounts where the greatest benefit can be obtained from scoring.

In this case, accounts that change status have to be adequately represented. There are enough goods that go bad, but nowhere near enough bads and indeterminates that become good. Table 15.7 shows two options. Option 1 keeps the sample size down by obtaining the minimum number of accounts for each group, while Option 2 maximises the value obtained out of available data, without going to extremes. As can be seen, the sample size is 2.5 times larger for Option 2, but at just over 10,000 cases this is still highly manageable, especially if there are no bureau costs involved.

15.5 Summary

Data is perhaps the most crucial and time-consuming aspect of any scorecard development. This module's first four chapters covered data considerations, data sources, scoring structure, and information sharing, while this chapter moved on to the first part of the physical scorecard development—data preparation. It consists of four main activities: (i) *data acquisition*; (ii) setting the *good/bad definition*; (iii) determining the *observation and outcome windows*; and (iv) *sampling*.

The data acquisition aspect is a process of obtaining data from various sources—application processing, billing, credit bureaux, and performance files—and bringing it together using one or more matching keys. The challenges have changed over the years, as lenders have developed automated processes to collect and manage the data. Even so, there are still times—especially in emerging environments—where data collection (acquisition from primary sources) is required, which can include digging through boxes in dusty warehouses to find rejected applications that are then captured.

A good/bad definition is applied at the outcome, to derive the target variable used for statistical modelling. The 'good/bad' label is a misnomer, as there are also other outcome-performance classifications (indeterminates and outcome excludes), as well as selection classifications (rejects, not-taken-ups, and observation excludes). Only the goods and bads are used in the scorecard development, but for application scoring, rejects' performance is inferred, to try to ensure that the scorecard is also relevant for them (see Chapter 19, Reject Inference). The definition may be based upon consensus, be prescribed by an external agency, or be empirically derived. A *consensus* definition is agreed by the business, with little empirical analysis. In contrast, an *empirical* definition is based upon analysis, and a *prescribed* definition is set down by company policy, or an external agency. The definitions will either be 'current-status' or 'worst-ever', with 60 and 90 days past due being the most common choices for each respectively. Each approach has different advantages and disadvantages, and lenders should use a definition that best meets business needs.

The next step is to determine the observation and outcome windows, and three competing forces influence the appropriate time periods for each: (i) *maturity*, allowing sufficient time for bad behaviour to become apparent; (ii) *censoring*, omission of performance on relevant cases, because the period is too short; and (iii) *decay*, inclusion of data that is so old, that it is no longer relevant. Lenders must guard against factors that are unlikely to recur (like marketing campaigns), or which they do not want reflected in the model (seasonality, cyclicality). If necessary, such factors can be guarded against by removing records, including control variables, or modifying weights.

The final step is sampling, and the minimum sample size usually suggested for credit scoring developments is 1,500 goods, 1,500 bads, and 1,000 rejects, or thereabouts. Where scorecard splits are identified, these minima should apply to each, but it is possible to create scorecards with less. Sufficient data is needed not only for training, but also for validating the model, albeit fewer cases are required for the latter. Issues also exist with having too many cases, as there are costs, and anything beyond 5,000 bads in a single scorecard is probably superfluous. Important subgroups should be properly represented by doing stratified random sampling, whether based upon performance statuses or potential scorecard splits.

This is the end of the most physical and frustrating part of predictive modelling, which can take up more than half the time required for the project. The sexier aspects are covered in Module E (Scorecard Development Process), coming up next.

Module E

Scorecard development

16 Transformation

When first presented with a dataset, many budding statisticians think that they are able to use the data directly, but this is unfortunately not the case. The first step is to analyse the data, and transform it into something usable. The following section on data transformation is covered under the headings of:

(1) **Transformation methodologies**—Univariate and bivariate approaches, with a focus on the latter (dummy variables and risk measure substitutes).
(2) **Classing**—Characteristic analysis report, fine classing, and coarse classing.
(3) **Use of statistical measures**—Information value, chi-square, and rank-order correlations.
(4) **Pooling algorithms**—Adjacent, non-adjacent, and monotone adjacent approaches.
(5) **Practical cases**—Judgments, industry, and occupation.

16.1 Transformation methodologies

For the moment, it is assumed that a traditional regression model will be developed. There are two broad classes of transformation methodologies that can be used:

(i) **Univariate**—Only refer to the predictor, and focus on providing a normally distributed substitute.
(ii) **Bivariate**—Refer to both the predictor and response variables, and try to capture the non-linear relationship between them.

Univariate methodologies

Regression equations assume a linear relationship between each predictor variable and the response function, and the results provided by statistical techniques are usually more robust if the predictor variables are normally distributed. There are a number of univariate transformation algorithms that can be used to achieve this, including *logarithmic*, *exponential*, *standardisation* (using the z-statistic), *polynomial expansion*, or some combination of the above (Falkenstein et al. 2000). The same technique need not be used for every variable, so different combinations can be used. Possibilities vary for numeric and ordinal characteristics, while categorical characteristics have to be represented using dummy variables, or risk substitutes.

Univariate methodologies are more appropriate for smaller samples. They are relatively easy to use, and come standard with many statistical packages, but are seldom used for retail credit scoring, where the larger sample sizes justify bivariate approaches. Other shortcomings are that: (i) some non-linear relationships may be impossible to represent; (ii) certain key relationships within the data may be missed; and (iii) many delivery systems cannot support them. The last point is crucial! Greater flexibility exists where models are applied on the same computer systems used for their development, which often occurs with marketing and attrition models in a PC-based environment. In contrast, many retail credit-risk models are applied on mainframe and/or networked systems, with limited flexibility.

Bivariate methodologies

While univariate methodologies create (approximately) normalised substitutes for numeric characteristics, bivariate approaches create substitutes that best represent the relationship between the predictor characteristics and the response function. There are two approaches, *dummy variables*, and *risk-measure substitutes*, both used for traditional regression models. They have the same first two steps: (i) create fine classes for each characteristic; and (ii) bin these into coarse classes (attributes), that are logical groupings and/or have similar risk (see Section 16.2). The approaches have particular advantages, in that the results are: (i) transparent, which eases explanation to and acceptance by the business; and (ii) easy to implement within lenders' production systems. Some people frown upon their use for continuous characteristics because of the discontinuities they create, but they probably provide the best way of dealing with the non-linear relationship between the raw characteristics and the target function.

What does retail credit scoring use?

Bivariate approaches dominate retail credit scoring, for several reasons. First, many of the characteristics are *categorical*, and those that are not can be easily converted. This is unlike FRS for enterprises, where most characteristics are continuous. Second, there are demands for *transparency*, to aid understanding and acceptance by management, because the models are driving something akin to industrial processes, to aid speed and quality of decision-making. This is unlike marketing and other throw-away scorecards, which do not have the same demands. And third, the retail credit environments that have used credit scoring have *enough data* to make bivariate approaches feasible. The rest of this section covers the two bivariate approaches in more detail:

(i) **Dummy variables**—Creates binary variables for all but one attribute of each characteristic.
(ii) **Risk measure substitutes**—Creates new transformed variables for each characteristic that manufacture linear relationships with the target function.

16.1.1 Dummy variables

Dummy variables are not stupid, they are more like mannequins used in place of the real thing. Rather than using a characteristic in its raw form: (i) it is binned into coarse classes, of which at least one is assigned to a 'null group'; and (ii) separate binary 'dummy' variables are created for all coarse classes not in the null group. The *null group* should contain classes that: (i) are closest to average risk; (ii) contain the greatest number of observations; (iii) are blank or missing (if not treated separately); and/or (iv) have insufficient data, or insight, confidently to group them with others. This 'null group' treatment aids the interpretation of the results, as the points assigned to each dummy should then agree with the risk, both for that attribute (wrong-sign problem), and relative to adjacent classes (if the risk is higher, then the points should be lower, and vice versa). When the regression is run, coefficients are assigned to the dummies (attributes) that best explain the target function, and by definition the null groups automatically get zero points.

> Failure to omit at least one of the coarse classes may cause problems in the regression. The last attribute adds no extra information, and if included would provide a set of variables with perfect collinearity—meaning that they have an 'exact linear relationship amongst themselves', because $X_1 + X_2 = 1$ (Mays 2004).

When the regression technique is applied, greater emphasis is put onto predictors' extreme values, or attributes that are rarer, and associated with the poles of the risk spectrum. Correlations between individual attributes are considered, to provide more targeted results, but the extra variables cause it to suffer from two problems. First, reduced degrees of freedom and an associated potential for overfitting. This should not pose a problem for most retail credit scoring developments, especially where: (i) the requisite 1,500 each of goods and bads are available; (ii) the number of dummies featuring in the final model are limited, using a stopping rule based on the model's out-of-sample predictive power (see Section 17.4.1); and (iii) the scorecard developer ensures that the final model makes logical business sense. Second, it is more labour intensive. Problems arise because: (i) extra variables mean larger datasets and longer run times; and (ii) many iterations are required to collapse the coarse classing, or exclude characteristics, until the point allocations make logical sense. Individual runs are quite quick, but repetitions are tedious. The model may be rerun many times, with changes to characteristics (including removal) and coarse classing, until the model makes sense.

An example is provided in Table 16.1, which could be for any 'Y/N' characteristic, but in this instance is 'Home Phone Given' for an application scorecard. The bulk of applicants have a home phone, and their bad rate is closest to the sample average. Thus, the 'Y' class is assigned to the null group, and a dummy variable is created for the 'N' class.

The missing group presents a complication though. It may be labelled as blank, no hit, or something else. Most statistical texts or software packages demand that they be assigned values, whether the sample average, a value inferred from other details, or another. For most credit scoring, the treatment is best summarised by Lewis (1992): either it contains predictive information, or it does not. For the former, it can be treated as a valid class, either separately or

Table 16.1. Dummy variable example

Coarse class	Count	Bad rate (%)	Dummy name	Coefficient
Yes	30,000	5.0	Null	(0)
No	3,000	12.0	PhoneN	−10.2
Missing	2,000	10.0	PhoneN	−10.2
Total	35,000	5.9		

in combination with others. For the latter, it is excluded from the analysis; if no information is there, then no conclusions are drawn. In most cases, missing values have a profile similar to the general population and are put into the null group; in this instance however, there is an obvious structural difference; they are more like the 'No' category, and are binned together with them.

16.1.2 Risk measure substitutes

The other approach is to create a single transformed variable to represent each characteristic, which captures the risk represented in each of the classes, effectively manufacturing a linear relationship with the target function. The two main substitutes that can be used are: (i) the *weight of evidence*; and (ii) the *probability of good*. In retail credit, the former is the de facto standard for use with logistic regression, and provides similar results to dummy variables. In contrast, the probability of good may be used with discriminant analysis (DA) and linear probability modelling (LPM), but the results are often questionable, and it tends to be avoided. The rest of this discussion will be limited to the weight of evidence (WoE), which was also discussed in Section 8.2.1.

To start, a WoE is calculated for each coarse class, and a single transformed variable is used in the regression. The process might involve: (i) generating models, using different characteristic combinations; (ii) dropping characteristics with wrong signs, and iterating until no wrong signs occur; (iii) ranking the models by some predictive measure; and then (iv) choosing one of the top-ranking models that makes logical business sense. It might provide less targeted results, but has some advantages: (i) the degrees of freedom is greater, which should enhance the models' robustness; (ii) the relationships between a characteristics' attributes can be fixed; and (iii) it is easier for less experienced modellers to understand.

The regression will provide coefficients for selected variables, which are multiplied by the WoE to determine the points. Table 16.2 revisits the Table 16.1 example, showing the result if a coefficient of 19 is assigned to the calculated WoE. It is only for the purposes of the illustration that the examples' results are similar, as in practice the two approaches would provide different results. As for the missing values, they can also be excluded from the calculation, and either included as a dummy variable or ignored (especially where the same cases are missing across multiple variables).

16.1.3 Which should be used, when?

Which of the two approaches are used will vary, depending upon a number of factors. First, the *choice of statistical technique*. For LPM, the choice is limited to dummy variables, because it is nearly impossible to use any other transformation methodology to provide either: (i) a normally distributed variable; or (ii) a manufactured linear relationship. Flexibility is greater with logistic

Table 16.2. Risk measure substitute

Coarse class	Count	Bad rate (%)	G	B	WoE	Coeff = 19
'Y'	30,000	5	28,500	1,500	0.172	3.28
'N'	3,000	12	2,640	360	−0.702	−13.33
Missing	2,000	10	1,800	200		
Total	35,000		33,000	2,000		

regression, where other issues prevail. Second, *data factors*. Although risk measure substitutes manufacture linear relationships with the target function, correlations and interactions within the data may make a non-linear representation, and hence dummy variables are the better choice. Third, *development importance*. If industrial-strength decision-making is required at a critical stage in the credit risk management cycle (CRMC) for a major high-volume low-margin portfolio, then dummy variables might be more appropriate (assuming sufficient data). Fourth, the *amount of available data*. As indicated, use of dummy variables reduces the degrees of freedom, but this may be ameliorated by scorecard developers' efforts to ensure parsimony. As a result, it is more data hungry, and the weight of evidence should be used if there are any concerns about data quantities. Fifth, *available resources*. Unfortunately, not everybody is familiar with both approaches, including many of the available software packages. That which is known and available should be used. Sixth, *development deadlines*. Assuming that sufficient computing power is available, and deadlines are short, then the weight of evidence is the better choice. Seventh, and finally, *is it worth it?* If the dummy variables provide improved predictive power, and associated business benefits, the benefits must be sufficient to justify the extra costs and complexities. In general though, the weight of evidence's ease of use and general availability tends to dominate for logistic regression. If there are doubts about any of these, and the software and scorecard developer have the capabilities, it might be possible to use both simultaneously and see which variables prevail. Alternatively, an initial model could (hypothetically) be developed using weights of evidence, before dummy variables are considered in another stage.

16.2 Classing

We have already alluded to the concept of classing, the purpose of which is to determine how each characteristic will be represented in the model. There are a number of different terms that are used to describe the process, which will vary from one environment to the next. In general, there are two stages to the classing process:

(i) **Initial enumeration,[1] or fine classing**—Done at the outset, to provide the finest level of detail possible for analysis.
(ii) **Binning, grouping, or coarse classing**—The fine classes are binned into a smaller number of coarse classes, primarily to achieve larger pools of similar risk.

[1] Lewis (1992).

Within this text, the terms fine and coarse classing are used. For example, 'Customer Age' might have one year per fine class, and an analysis would then be done to group them into a smaller number of meaningful coarse classes. Treatment of the characteristics will differ, depending upon their type:

Continuous/discrete/ordinal—There is an implied ranking, and the characteristic to risk relationship is usually expected to be monotonic. Failing any special codes or values, any groups where this does not hold true will be merged with their neighbours.

Categorical—There is no implied ranking. The groups can only be assessed based upon what is expected from experience or available data. Where grouping is required, and numbers are small, some judgment may be used.

The age example is a continuous characteristic, which is usually presented to a scorecard developer as discrete values. In contrast, marital status has categorical values, whose relative risk may vary depending upon the environment.

These are not tasks that are done once and forgotten. Fine classing may be done at the outset for the full sample, but some tailoring may be required for each of the different scorecard splits. The same applies to coarse classing, except with dummy variables it may have to be revisited many times, before the classes are considered acceptable.

16.2.1 Characteristic analysis report

Whether using dummy variables or risk substitutes, a characteristic analysis (CA) report is used to aid binning, albeit with some variations. CA reports can have a number of different formats in credit scoring, that vary depending upon the type of scorecard development (application, behavioural, etc.), methodology being used, (dummy variable, weight of evidence), and stage in the development process (initial enumeration, known good/bad, all good/bad). The reports will always have characteristics' attributes as rows, and during the scorecard development process the columns would include: (i) the *decision* that was made (for selection processes); (ii) the subsequent *performance* of the account; and (iii) the *points or coefficients* that have been allocated.

With *standardised software packages*, the report templates may be fixed or have limited flexibility. In contrast, many scorecard developers prefer packages that allow tailoring to their own preferences, which may vary across developments or stages within the development process. Extra columns may be included for not-taken-ups and/or inactive accounts, and there may be separate sections for accepts (known), rejects (inferred), and total (known plus inferred). Other details may also be included to aid the analysis, such as: (i) a comparison of the development-sample distribution versus a recent sample, to assess potential drift; and/or (ii) summary statistics, to indicate the level of power or drift. Both the fine and coarse classing may be revisited at each stage in the scorecard development process, and for each scorecard.

The example in Table 16.3 summarises the final results for marital status from an application scorecard development, with the points provided for the 'all good/bad' model after reject

Table 16.3. Characteristic analysis report

Marital status												
Fine class	Coarse class	WoE	Points	Total	Odds	Good	Bad	Indeter-minate	Reject	Rej. rate	Sample dist.	Recent dist.
Missing	Null	−0.33		346	12.2	159	13	25	148	43.0	0.4	0.2
Single (S)	Sing	−0.15	−4	27,255	14.6	15,679	1,074	1,795	8,706	31.9	32.0	31.3
Married (M)	Null	0.10		48,231	18.7	30,245	1,617	3,189	13,180	27.3	56.6	57.7
Separated (P)	Null	0.04		318	17.6	166	9	22	121	38.0	0.4	0.5
Divorced (D)	Sing	−0.20	−4	5,282	13.9	3,425	246	303	1,309	24.8	6.2	6.7
Widowed (W)	Wido	0.23	10	1,365	21.3	773	36	105	451	33.0	1.6	2.2
Unclassified (U)	Wido	0.27	10	2,468	22.2	1,370	62	133	903	36.6	2.9	1.6
Total				85,265	16.9	51,818	3,058	5,571	24,818	29.1	100.0	100.0

Info. value * 100 Classing: Fine = 1.830 Coarse = 1.747 Stability index = 1.140

inference, and the coarse classes assigned to dummy variables. The columns show detailed values/ranges (fine classes), bin assignments (coarse classes), WoE, points, counts (total, good, bad, indeterminate), ratios (odds, rates), distributions (sample and recent), and the information values (fine and coarse).

16.2.2 Initial enumeration/fine classing

Perhaps the most tedious task is the initial enumeration, which requires that every characteristic be reviewed, to determine the maximum number of classes that can be used to represent it. In most cases, the software being used will provide tools to assist. The first use for the fine classes is to check for errors and possible changes in the infrastructure, which can arise in any number of ways. Possible errors that must be guarded against are:

Capture errors—Incorrect capture, because of sloppiness, improper interpretation of the capture instructions, or failing to provide for certain instances in the system design.

Program errors—Errors in the computer program used to perform any calculations.

Infrastructure changes—Changes in business processes that have changed the meaning of one or more characteristics, or caused new fields to be added or old fields dropped.

Outcome characteristics—Identify any characteristics containing outcome data, and ensure that they are not mistakenly used as predictors in the final model (beware of information values far above the norm for the sample).

Policy overrides—There may be certain groups with extremely high or low reject rates, where policy rules drive the decision. If this is expected to continue, the scorecard developer may exclude those groups from the analysis.

Where any discrepancies are identified, it is wise to correct them at the outset. A great deal of time and effort has already gone into assembling the data, and at this stage it should—hopefully—still be relatively easy to rectify problems. Once the scorecard development process begins, it is very difficult to backtrack. Two initial enumeration examples are provided here: 'Residential Status', as an example of a discrete characteristic (Table 16.4); and 'Time at Employer', for a continuous characteristic (Table 16.5). These are fairly typical, and do not reflect any errors in the underlying data.

Categorical characteristics

For discrete characteristics, it is usually best to create separate fine classes for every possible value. There may be instances, however, where the number of possible values is extremely large, and logical groupings have to be found upfront—such as for industry or postal codes. Residential status is not one of these; the number of values is usually manageable, perhaps limited to those presented in Table 16.4. This is a characteristic that appears on most application forms, and the business should have a feel for the odds of each attribute, at least relative to the others.

The first step is to consider the odds of each fine class. Does it make sense that the best odds are for applicants whose homes are either jointly owned, or solely owned by the spouse? Does it make sense that the worst odds are for those that rent, have company accommodation, or are living with parents? For this particular case, it does! For 'Joint', there are probably two incomes, and the 'Spouse' probably earns enough to support them both. Tenants, and those staying in company-owned houses, are less stable than the average homeowner. And finally, the living with parents group will be younger than average, with issues of affordability, stability, financial literacy, and so on.

In this example, the primary thing to be wary of is the large number of blanks, nearly 25 per cent of the total, especially given that the most comparable group is homeowners. This should be investigated further, and it could transpire that these cases come via a specific channel, where that data is not collected.

Table 16.4. Fine class—categorical

	Residential status					
	Good	Bad	Indet.	G/B odds	WoE	Info. value
Blank	12,878	698	1,252	18.5	0.09	0.113
Homeowner	13,856	703	1,401	19.7	0.16	0.158
Tenant	9,461	777	1,262	12.2	–0.33	0.000
Parents	7,234	527	919	13.7	–0.21	0.005
Spouse	1,595	73	138	21.8	0.26	0.026
Company	599	45	119	13.3	–0.24	0.000
Joint	6,173	245	493	25.2	0.40	0.146
Total	51,797	3,069	5,584	16.9		0.448

Continuous characteristics

A common approach for the initial fine classing of continuous characteristics is to create a standard number of equally sized groups, usually anywhere from 20 to 50. It is possible to go beyond 50, but it seldom offers any value, given that there will usually be at most 10 coarse classes. There may also be problems where values cluster on a single value—which often happens with values like zero, 100 per cent, and codes indicating missing information—leaving insufficient cases to spread over the balance. Another possibility is to include a minimum of 1 or 2 per cent of both goods and bads in each group, which ensures that the good/bad odds for each is meaningful. Both of these approaches are dependent upon having a sufficiently large sample size, and still usually demand that the scorecard developer ensures that the attribute breaks are meaningful—such as using a range of 125 to 200, instead of 128 to 197.

The data in Table 16.5 illustrates fine classing for a continuous characteristic, in this case 'Time at Employer', expressed as years and months. From this it can be seen that: (i) there is no clustering; and (ii) as a general pattern, the good/bad odds increase with length of employment, but the relationship is non-monotonic. Subsequent coarse classing will ensure that the end result is monotonic. There may, however, be characteristics where the non-monotonicity should be acknowledged, such as income or asset values that exhibit higher risk, at both ends of the spectrum.

A word of caution here: Be patient! Both fine and coarse classing can be extremely tedious, especially when the number of characteristics is large. In some instances there may be well over a hundred, or perhaps even two hundred, depending upon the type of development.

Table 16.5. Fine class—continuous

			Time at Employer			
Value	Good	Bad	Indet.	G/B odds	WoE	Info. value
000	136	5	11	29.2	0.54	0.001
001	582	62	75	9.4	–0.59	0.005
to 003	686	60	59	11.5	–0.39	0.002
to 006	1,113	120	136	9.3	–0.60	0.011
to 100	2,778	218	355	12.8	–0.28	0.005
to 106	1,810	161	144	11.2	–0.41	0.007
to 200	2,856	246	350	11.6	–0.38	0.010
to 206	1,265	89	163	14.2	–0.18	0.001
to 300	3,974	328	414	12.1	–0.33	0.010
to 400	4,217	271	497	15.5	–0.09	0.001
to 500	3,883	249	486	15.6	–0.08	0.001
to 700	5,132	287	532	17.9	0.05	0.000
to 1,000	6,323	282	623	22.4	0.28	0.008
to 1,500	6,912	277	665	25.0	0.39	0.017
to 2,000	4,281	142	449	30.1	0.57	0.021
to High	5,873	262	611	22.4	0.28	0.008
Total	51,820	3,058	5,571	16.9		0.107

16.2.3 Binning, grouping, coarse classing

Once the fine classing is done, the next stage is coarse classing, also referred to as binning or grouping. Each characteristic must be considered individually and, if at all possible, automated pooling algorithms should be used to aid the process. Coarse classes may also be revisited during the development process, whether: (i) for the known good/bad, accept/reject, and known+inferred models; (ii) for each of the identified subpopulations; and (iii) to ensure the final models are parsimonious. The primary goal is to develop models that are both predictive and robust. Thus, there are two overriding and contradictory goals when defining the coarse classing:

(i) **Keep it simple**—Try to represent the characteristic with as few coarse classes as possible, especially when using dummy variables. The greater the number of classes, the greater the complexity and potential for overfitting. Remember that the greatest benefit comes from having the right characteristics, and not a large number of coarse classes.

(ii) **Minimise information loss**—Having too few coarse classes can lose valuable information. Note, however, that the more powerful the characteristic, *ceteris paribus*, the greater the number of potential coarse classes. Even so, in most cases, the number is unlikely to exceed 10, and in many instances there will only be two (especially for 'true/false' and 'zero/not zero').

Larger numbers of coarse classes are possible, but only when powerful scores are included within a model. Two examples are: (i) inclusion of a risk score (bureau or behavioural) within an application scorecard; and (ii) composites, that combine bespoke, bureau, and possibly also behavioural scores. In such instances, the original score's granularity will be lost, unless it can be used directly (such as a logistic score in a logistic model).

Thereafter, there are a number of other factors that have to be taken into consideration:

Logical groups—Each group should either be: (i) of similar risk; or (ii) have some logical connection. The former is the primary driving factor, but there may be insufficient numbers to draw firm conclusions, in which case some judgment must be used.

Logical breakpoints—For continuous characteristics, the breaks used to define the classes should also make sense. For example, zero, one thousand, and fifty thousand should be used instead of −5, +997 and +43,950. Codes and natural barriers (like zero, 100 per cent, or the number of days within a period) should also be recognised.

Logical relationships between groups—The difference in risk between the coarse classes should make intuitive sense. For continuous characteristics, it is usually wise to enforce monotonicity, albeit there are some exceptions.

Logical relationships between points—Likewise, when a model is run, the differences in points should correspond to differences in relative risk—at least in terms of direction. A problem common to all types of transformations is wrong signs (a symptom of multi-collinearity), while with dummy variables there may also be gaps where no points have been assigned, or there are inconsistent points.

Sufficient numbers—If the number of cases within a class is too small, it is difficult to draw firm conclusions, and there is a distinct danger of overfitting the model to the development sample. In such cases, the group should be collapsed into a neighbouring or other group, possibly the null/average group. For categorical variables this will require some judgment.

Relevance—This is related to statistical significance, and is especially an issue when a variable is weak, or the number of cases is small. Statistical packages also provide a *p*-value for measuring statistical significance, and it is usually recommended that any variable with a *p*-value greater than 0.03 be removed.

Stability over time—The characteristics and attributes used should be stable, which can be checked by monitoring the relative frequencies for each group over the time (see 'stability index'), failing which, the coarse classing should be revisited.

For 'sufficient numbers', the definition of 'too small' varies. Lewis (1992) states that there should be at least 40 each of both good and bad per coarse class. In contrast, Hoyland (1995) and others state that each coarse class should comprise at least 5 per cent of the applicant population.

As a final comment, exactly what the groups consist of is up to the scorecard developer to decide, with the caveat that the business will always have the final say. It is not necessary to get them completely right the first time, as they can be revisited. By the time it is presented to the business however, there should be some degree of certainty.

16.3 Use of statistical measures

While most of the coarse classing can be done by a scorecard developer just looking at the data, there are times when the choices are less obvious. In such instances, some means is required to assess the predictive power under different classing options. Possibilities are: (i) the chi-square (χ^2) statistic; (ii) the information value (F-statistic); and (iii) the Gini coefficient (Somer's D).

16.3.1 Measures of predictive power

These statistics are used not only for coarse classing, but also to assess characteristics' predictive power, as part of the characteristic selection process (see Section 17.2). Each has a different origin, usual application, and suitability:

Information value—Specifically intended as a measure of divergence, and takes into consideration the relative frequencies in each group.
Chi-square—Used to test for significant differences between observed and expected values, but has been co-opted here to compare observed and expected frequencies, where the latter assumes the average bad rate applies to each attribute.
Rank-order correlations—Statistics such as the Gini or Spearman's coefficients, that assume a rank order, and a monotonic bivariate relationship.

The information value is the most commonly used, because it: (i) can be applied to all classed characteristics, irrespective of the underlying data type; and (ii) can assess the overall power of a characteristic. Its disadvantage is that its 'weighted average' nature causes it to overlook small niches of extremely high (or low) risk accounts. The chi-square statistic can also be applied to all classed characteristics, but is slightly less effective at measuring the overall power. It can however, highlight the small niches that the information value misses. And finally, rank-order correlation coefficients will work for monotonic characteristics, and even near-monotonics prior to fine-classing, but they are not well suited otherwise (even the existence of a single code precludes them). Given the predominance of categorical characteristics in retail credit scoring, and the use of codes within numeric characteristics, the other statistics are favoured. If this author were to make a recommendation, both coarse classing and characteristic selection should use the information value, but the chi-square statistic should also always be checked, to make sure nothing is being missed.

Mays (2004:92–100) highlights the difference between the three measures (see Table 16.6) according to: (i) the type of relationship they are meant to assess; (ii) whether they penalise variables whose values vary widely; and (iii) whether they assess the direction of a relationship. Please note, she is referring to the 'score chi-square', and the *p*-values generated by software packages (especially SAS's PROC LOGISTIC) to test whether a regression parameter is non-zero. She also refers to Spearman's rank-order correlation, and not the Gini coefficeint, albeit both are rank-order correlation coefficients with similar properties.

Table 16.6. Characteristic comparison

Measure	Relationship type	Penalises homogeneity	Assesses direction
Chi-square	Linear	No	No
Information value	Any	Yes	No
Spearman's	Monotonic	No	Yes

16.3.2 Coarse classing example

When coarse classing, characteristics' predictive power will usually reduce every time classes are collapsed. The goal is to reduce the number of classes as far as possible, but with minimal information loss. Thomas et al. (2002:132) provide an example of these calculations. Table 16.7 shows the calculated values for the original set of fine classes, and Table 16.8 shows three different possibilities for coarse classing.

In this case, the statistics concur on the choice of classing option, but this is not always the case. Even so, it would be overkill to provide all three for every characteristic. In later examples, only the information value is provided; the chi-square statistic would have been second choice, and could still add value.

A couple of other observations can be made from Table 16.8. First, although the option with four coarse classes provides higher measures, the number of cases for 'Other' is low, and it is wise to collapse it further. As for the other two options, the choice is already fairly evident

Table 16.7. Coarse classing—input

Accommodation status	Total	Actual			χ^2	F	D
		Good	Bad	Odds			
Owner	6,300	6,000	300	20.0	192.1	0.2928	0.2000
Live w/parents	1,050	950	100	9.5	0.3	0.0003	–0.9510
Rent unfurnished	2,000	1,600	400	4.0	222.2	0.1802	0.8083
Rent furnished	490	350	140	2.5	187.8	0.1295	0.2714
Other	140	90	50	1.8	102.9	0.0644	1.0450
No answer	20	10	10	1.0	35.6	0.0195	0.0200
Totals	10,000	9,000	1,000	9.0	740.7	0.6867	0.3937

Table 16.8. Coarse classing—analytics

Accommodation status	Total	Actual			χ^2	F	D
		Good	Bad	Odds			
Other	160	100	60	1.7	134.4	0.0824	0.0007
Renter	2,490	1,950	540	3.6	377.9	0.2953	0.1290
LWP	1,050	950	100	9.5	0.3	0.0003	0.0561
Owner	6,300	6,000	300	20.0	192.1	0.2928	0.4000
Totals	10,000	9,000	1,000	9.0	704.6	0.6708	0.4142
Other	2,650	2,050	600	3.4	470.5	0.3605	0.1367
Live w/parents	1,050	950	100	9.5	0.3	0.0003	0.0561
Owner	6,300	6,000	300	20.0	192.1	0.2928	0.4000
Totals	10,000	9,000	1,000	9.0	662.9	0.6536	0.4072
Renter	2,490	1,950	540	3.6	377.9	0.2953	0.1170
Other	1,210	1,050	160	6.6	14.0	0.0137	0.0880
Owner	6,300	6,000	300	20.0	192.1	0.2928	0.4000
Totals	10,000	9,000	1,000	9.0	583.9	0.6017	0.3950

Table 16.9. Non-adjacent pooling algorithm

Accommodation status	Owner	LWP	Rent unf.	Rent furn.	Other	No answer
Owner		0.245	0.027	0.125	0.215	0.275
Live w/parents	0.293		0.123	0.049	0.009	0.000
Rent unfurnished	0.473	0.181		0.295	0.229	0.192
Rent furnished	0.422	0.130	0.310		0.191	0.145
Other	0.357	0.065	0.245	0.194		0.082
No answer	0.312	0.020	0.200	0.149	0.084	

from the good/bad odds, but the statistics make it even more obvious—'Owner', 'Live with Parents', and 'Other' are the logical choices.

16.4 Pooling algorithms

Pooling algorithms are routines that can be used for automated coarse classing, in order to ensure minimal information loss. There are three types: (i) *non-adjacent*, for categorical characteristics; (ii) *adjacent*, for ordinal, discrete, and numeric characteristics; and (iii) *monotone adjacent*, for instances where a monotonic relationship is assumed with the target variable. In each case, the scorecard developer should be able to review, or control, the classing. The following examples focus solely on the information value (an *ex post* review of the chi-square values is also recommended).

16.4.1 Non-adjacent pooling algorithms

Non-adjacent algorithms require no assumptions about the fine classes' ordering. An example is provided in Table 16.9. The cells either side of the diagonal reflect contributions to the information value for all possible attribute pairs: those below are the sums if the two attributes are treated separately; those above are the values if the two are treated together. The pair that should be collapsed is that with the least difference; in this case 'No answer' and 'Other'.

The process would then be repeated with the remaining five, then four, and then three groups. In this particular instance, the results would be the same as the subjective classing achieved in Table 16.8, being 'Owner', 'Live with Parents', and 'Other'. This is not always the case though. Care must also be taken to ensure that the resulting groups are logical, especially where numbers are small.

16.4.2 Adjacent pooling algorithms

In contrast, adjacent pooling algorithms assume that only neighbouring attributes can be grouped, which applies to any ranked characteristic, whether ordinal, discrete, or continuous. The example in Table 16.10 again uses 'Time at Employer', but with different numbers than those presented earlier. Categories with less than 100 bads have already been grouped (which is more easily done

Table 16.10. Adjacent pooling algorithm

Time @ Employer—fine class								
Class	Total	Outcome			Information value*10E4			
		Good	Bad	Odds	F_i	$F_i + F_{i-1}$	$F_{i,i-1}$	$(F_i+F_{i-1})-F_{i,i-1}$
to 3	518	410	107	3.8	31.2			
to 6	469	368	101	3.6	37.9	69.09	68.75	0.34
to 100	1,152	957	194	4.9	2.3	40.25	21.96	18.29
to 106	590	485	105	4.6	5.1	7.43	6.54	0.89
to 206	1,386	1,144	242	4.7	8.0	13.14	13.03	0.12
to 300	1,335	1,086	248	4.4	24.7	32.74	30.40	2.35
to 400	1,349	1,106	243	4.6	14.9	39.66	39.05	0.61
to 500	1,486	1,217	270	4.5	18.9	33.79	33.75	0.04
to 700	2,090	1,736	354	4.9	4.9	23.77	20.30	3.46
to 1,000	2,531	2,184	347	6.3	44.6	49.52	11.02	38.50
to 1,500	2,382	2,062	320	6.4	53.2	97.77	97.40	0.36
to 2,000	1,623	1,399	225	6.2	25.6	78.77	78.18	0.59
to High	1,195	1,045	150	7.0	49.0	74.61	70.19	4.42
Total	18,107	15,199	2,908	5.2	320.3			
Resulting Coarse Classes								
to 6	986	778	208	3.7	68.7			
to 500	7,299	5,996	1,303	4.6	68.0	136.7	113.4	23.3
to 700	2,090	1,736	354	4.9	4.9	72.9	69.3	3.6
to 2,000	6,537	5,644	892	6.3	122.7	127.6	70.7	56.9
to High	1,195	1,045	150	7.0	49.0	171.7	167.0	4.7
Total	18,107	15,199	2,908	5.2	313.3			

than for the non-adjacent equivalent). There are separate information value columns for: (i) the current fine classing [F_i]; (ii) current and prior classes, if treated separately [F_i, F_{i-1}]; (iii) current and prior bins, if combined [$F_i + F_{i-1}$]; and (iv) the marginal difference between separate and combined treatment. The best pair to collapse is the one where the least information is lost, which in this instance is the 0.04 lost by combining the 'to 400' and 'to 500' groups.

Like the non-adjacent algorithm, this also relies upon iterations. One pair is collapsed at a time, until the groups make sense. In this instance however, it is quite obvious that 0 to 6, 13 to 30 months, 37 to 60, and 85 to 180 months can each be collapsed. Monotonicity, although not a goal (see Section 16.4.3), can be achieved after reducing the number of classes from 13 to 5, with an information loss of only 0.0007 (reduced from 0.0320 to 0.0313). It could even be taken down to 3 groups, with a further loss of only 0.0008 (to 0.0305).

16.4.3 Monotone adjacent pooling algorithms

In many instances, monotonicity in the resulting coarse classes is required. Thomas et al. (2002:138) present a 'maximum likelihood monotone coarse classifier', which is also called a 'pooled adjacent violators algorithm', but is perhaps better described as a 'monotone adjacent pooling algorithm' (MAPA). It can be applied to any ordinal characteristic, but in Table 16.11 is being applied specifically to scoring results. The starting point is the calculation of the cumulative bad rate for each score, as shown in Equation 16.1, which can be done: (i) at record level; or (ii) based upon some preliminary fine classing:

Equation 16.1. Cumulative bad rate $C_{k,v} = \sum_{i=V_{k-1}+1}^{v} B_i \Big/ \sum_{i=V_{k-1}+1}^{v} (B_i + G_i)$

where C is the cumulative bad rate; G and B are the good and bad counts; V is a vector containing the series of score breaks being determined; v is a score above the last score break; and i and k are indices for each score and score break respectively. Cumulative bad rates are calculated for all scores above the last breakpoint, and the score with the highest cumulative bad rate is identified, and assigned to the vector as shown in Equation 16.2.

Equation 16.2. MAPA $V_k = \max\{v | C_{k,v} = \max\{C_{k,v}\}\}$ for all $v > V_{k-1}$

This is an iterative process, which is repeated until the maximum cumulative bad rate is the one associated with the highest possible score. For the example, the breaks of 183, 187, 193, 196, and 199 provide bad rates of 13.8, 11.8, 9.7, 8.5, and 7.7 per cent respectively.

16.5 Practical cases

Such analytical tools are extremely useful, but tend to be used only if provided in vendor software. The following practical examples illustrate what scorecard developers might encounter in practice, and work through some of the decision process manually. Each example assumes that

Table 16.11. Monotone adjacent pooling algorithm

Score	G	B	T	B/T (%)	ΣT	ΣB/ΣT (%) 1	2	3	4	5
180	361	52	413	12.6	413	12.6				
181	359	55	414	13.3	827	12.9				
183	827	140	967	14.5	1,794	<u>13.8</u>				
184	437	54	491	10.9	2,285	13.2	10.9			
185	988	135	1,123	12.0	3,408	12.8	11.7			
187	1,184	160	1,344	11.9	4,752	12.5	<u>11.8</u>			
189	1,137	115	1,252	9.2	6,004	11.8	11.0	9.2		
191	1,386	145	1,531	9.4	7,535	11.4	10.6	9.3		
193	1,790	202	1,992	10.2	9,527	11.1	10.5	<u>9.7</u>		
195	1,971	179	2,150	8.3	11,677	10.6	10.0	9.3	8.3	
196	1,959	185	2,144	8.6	13,821	10.3	9.8	9.1	<u>8.5</u>	
197	1,329	110	1,439	7.6	15,260	10.0	9.5	8.9	8.3	7.6
199	993	83	1,075	7.7	16,335	9.9	9.4	8.8	8.2	<u>7.7</u>

the resulting coarse classes will be transformed into dummy variables, and that the scorecard developer will assign one of the resulting coarse classes as a 'null' class (see Section 16.1.1).

16.5.1 Number of court judgments

While many lenders will rely upon the bureau score, others will bring raw bureau data into their assessment; items like 'Number of Enquiries', 'Value of Judgments', or 'Number of Payment Profiles'. In this particular instance (Table 16.12), the number of judgments over a period is being considered.

Two issues need to be highlighted. First, there may be one or more 'missing data' categories, such as 'No Match' or 'No Record', for each characteristic. This will apply to all, or a section, of the record, and it may be best to create a separate binary characteristic for 'Missing/Not Missing', that is only treated once. If weights of evidence are being used, the missing category should be dropped from that calculation. And second, there is a problem with rare events: (i) very few individuals have two or more judgments; (ii) those that do are unlikely to be applying for credit; and (iii) the probability that they will be accepted is very low, which is why they do not bother to apply; and (iv) the number of known bads is tiny—only 17. If a minimum of 40 goods and 40 bads were required in each coarse class, then the resulting groups are No Match, No Judgment, and Judgments.

This characteristic presents another quandary! The reject rate amongst cases with two or more judgments is 80 per cent plus. If a group of accounts has been rejected by a policy that is effective, and is likely to be retained, should they be included in the reject inference process? Many people question the entire process of reject inference, and once the number of accepts becomes this small, the results become even more dubious. A good practice is to review all fine

Table 16.12. Court judgments—coarse class

Attribute	Coarse/ dummy	WoE	Total	Odds	Good	Bad	Indet.	Reject	Rej rate	Sample dist	Recent dist
No match	NMatc	−0.26	4,381	13.0	2,452	188	348	1,393	31.8	5.1	3.1
= 0	Null	0.03	76,738	17.5	48,421	2,773	5,060	20,483	26.7	90.0	89.6
= 1	Judg	−0.49	3,093	10.3	819	79	128	2,067	66.8	3.6	5.3
= 2	Judg	−0.93	670	6.7	85	13	23	549	82.0	0.8	1.3
= 2+	Judg	−0.48	386	10.4	43	4	12	326	84.5	0.5	0.7
Total			85,268	16.9	51,820	3,058	5,571	24,818	29.1	100.0	100.0
Fine class			Information value = 0.01218						Stability index = 0.02045		
No match	NMatc	−0.26	4,381	13.0	2,452	188	348	1,393	31.8	5.1	3.1
No judgment	Null	0.03	76,738	17.5	48,421	2,773	5,060	20,483	26.7	90.0	89.6
Judgments	Judg	−0.54	4,150	9.9	948	96	164	2,942	70.9	4.9	7.3
Total			85,268	16.9	51,820	3,058	5,571	24,818	29.1	100.0	100.0
Coarse class			Information value = 0.01171						Stability index = 0.02034		

Table 16.13. Industry—coarse class

Attribute	Group	WoE	Total	Odds	Good	Bad	Indet.	Reject	Rej rate
Missing	Null	0.53	918	28.9	576	20	69	253	27.6
Agriculture	Null	0.25	1,236	21.7	761	35	63	376	30.4
Catering and accom.	Temp	−0.62	1,263	9.2	695	76	103	390	30.8
Civil Services	Null	0.23	4,914	21.4	2,753	129	341	1,692	34.4
Community services	Qual	−0.03	4,411	16.5	2,252	136	310	1,713	38.8
Construction	Temp	−0.39	1,828	11.5	970	84	155	619	33.9
Education	Null	0.46	9,911	27.0	5,462	203	496	3,751	37.8
Engineering	Qual	0.09	5,102	18.5	3,291	178	262	1,371	26.9
Finance	Null	0.33	11,508	23.5	8,753	372	685	1,697	14.7
Importing and exporting	Indus	−0.15	1,311	14.5	756	52	95	407	31.1
Industrial	Indus	−0.14	13,983	14.7	7,761	527	1,070	4,626	33.1
Information technology	Qual	−0.10	4,020	15.4	2,827	184	270	738	18.4
Legal	Qual	0.08	1,075	18.3	695	38	74	268	24.9
Media and advertising	Qual	0.17	1,358	20.1	938	47	84	289	21.3
Medical	Qual	0.01	4,617	17.1	2,785	163	295	1,375	29.8
National forces	Temp	−0.19	1,263	14.0	681	49	65	468	37.1
Natural resources	Indus	−0.02	1,579	16.6	885	53	100	541	34.3
Personal services	Temp	−0.21	5,841	13.7	3,626	264	439	1,511	25.9
Science	Qual	−0.25	225	13.2	141	11	21	51	22.7
Security	Temp	−0.43	1,066	11.0	537	49	61	420	39.4
Selling	Temp	−0.41	4,837	11.2	2,808	250	313	1,466	30.3
Sports and entertainment	Temp	−0.77	393	7.8	229	29	29	105	26.8
Transportation	Temp	−0.15	2,470	14.6	1,541	106	163	660	26.7
Welfare	Temp	0.41	138	25.5	96	4	8	31	22.1
Total			85,268	16.9	51,820	3,058	5,571		29.1
Stable	Null	0.35	28,488	24.13	18,306	759	1,654	7,769	27.27
Qualified	Qual	0.01	20,808	17.10	12,930	756	1,316	5,805	27.90
Industrial	Indus	−0.13	16,873	14.87	9,402	632	1,265	5,574	33.03
Temporary	Temp	−0.32	19,099	12.28	11,183	910	1,336	5,670	29.69
Total			85,268	16.95	51,820	3,058	5,571	24,818	29.11

Information value × 100 Fine classed 7.80 Coarse classed 6.69

classes to identify attributes where both reject rates and bad rates are high (usually relating to poor past dealings, or adverse information on bureau). The rejects in these categories would then be tagged as 'policy rejects', and no attempt would be made to infer their performance. If, however, there are sufficient bads, reject inference is a possibility.

16.5.2 Industry

Most of the examples presented thus far have been fairly straightforward, relatively speaking. There are, however, characteristics where the number of fine classes is large, and the groups are not so obvious. One of these is the industry that the applicant is employed in. Here, there are a large number of possible coarse classes, and it is up to the scorecard developer to come up with groupings that make sense. A manual review of the data in Table 16.13 indicates that, for this development, there are four broad industry groupings: (i) *stable*, very low-risk industries with stable employment, including financial services, education, and civil service; (ii) *qualified*, low-risk industries, where qualifications and/or high skill levels are required, but employment is not secure, including information technology, law, medicine, and engineering; (iii) *industrial*, moderate risk, where skills levels are low, but employment stable, including general industry, natural resources, and import/export; and (iv) *temporary*, high risk, where employment is insecure, including sports and entertainment, catering and accommodation, selling, security, and construction. The labels may not be perfect, but make logical sense given that job insecurity is a major factor in consumer credit.

16.5.3 Occupation

Usually, the patterns highlighted by the characteristic analysis are totally logical, but there are times when they are counterintuitive. This is especially true where reliance is put upon how applicants present themselves on an application form, whether they believe it themselves or not. When it comes to 'Occupation', there are a variety of different possible answers that might be provided, ranging anywhere from building janitor to company owner.

In the example provided, Table 16.14, the fine classing was already hard coded into the source system (the original categories were even more detailed). A review highlights something very odd in those classes normally associated with high-income earners; their risk is higher than is logically expected! The weight of evidence for directors is -0.11, and for managers 0.00, which is contrary to the expectation of better than average risk. The odds get even worse as the position in the company gets higher! Office staff are better than supervisors, who are better than managers, who are better than directors.

There may be a variety of different reasons for this. First, the amount of job mobility amongst the higher income earners may be greater, leading them to higher-risk levels as they change employment and residence. Second, where people give their position as director or partner, the size of the firm may be small and unstable (if 'Firm Size' were available, it might add value). Third, applicants may be embellishing the application, trying to ensure that they are not refused credit, and it is showing up in the results. There are other possibilities, and there is no one right answer. Interestingly, the lowest risk category is 'Professional', which included qualified individuals in private practice.

After review, the end result is five different coarse classes. The null category is comprised of the small number of missings, and all of the categories hovering around the average good/bad odds of 16.9—manager, supervisor, technician. The other groups are professionals, office workers, temporary staff, and labourers. The latter three labels are purely subjective, and are being used to explain the coarse classes.

Table 16.14. Occupation—coarse class

Attribute	Group	WoE	Total	Odds	Good	Bad	Indet.	Reject	Rej rate
Missing	Null	0.53	918	28.9	576	20	69	253	27.6
Director	Temp	−0.11	1,358	15.2	918	60	83	297	21.9
Manager	Null	0.00	8,206	16.9	6,027	357	391	1,431	17.4
Supervisor	Null	0.01	5,441	17.2	3,321	193	367	1,560	28.7
Consultant	Temp	−0.14	4,423	14.8	3,081	208	352	782	17.7
Clerk	Midd	0.16	14,489	19.9	9,469	476	937	3,607	24.9
Secretary	Midd	0.08	3,052	18.4	2,013	110	187	742	24.3
Labourer	Work	−0.51	15,212	10.1	6,636	654	1,388	6,533	42.9
Apprentice	Null	0.01	947	17.1	448	26	53	419	44.3
Professional	Prof	0.37	16,616	24.5	10,634	434	913	4,636	27.9
Semi-professional	Midd	0.12	6,030	19.1	3,404	178	345	2,102	34.9
Technician	Null	0.00	4,610	17.0	2,944	173	259	1,233	26.7
Salesman	Temp	−0.42	1,813	11.1	1,066	96	90	562	31.0
Comm. officer	Midd	0.23	1,484	21.4	900	42	86	455	30.7
Non-comm. officer	Temp	−0.25	667	13.1	382	29	50	206	30.8
Total			85,268	16.9	51,820		5,571		29.1
Professional	Prof	0.37	16,616	24.52	10,634	434	913	4,636	27.90
Office	Midd	0.15	25,056	19.59	15,787	806	1,556	6,907	27.57
Manager	Null	0.02	20,122	17.29	13,316	770	1,140	4,896	24.33
Temporary	Temp	−0.20	8,261	13.84	5,448	394	574	1,846	22.34
Labourer	Work	−0.51	15,212	10.14	6,636	654	1,388	6,533	42.95
Total			85,268	16.95	51,820	3,058	5,571	24,818	29.11
Information value × 100				Fine classed 8.3				Coarse classed 7.8	

It is never certain whether any points will be assigned to a characteristic, or with dummies, to any of its parts. If this characteristic features, then office workers will probably get more points than supervisors and management. This could pose a problem when explaining the model to management and other end users, as they may find it difficult to accept that office workers are allocated higher points than them. Either they will have to have some faith in the explanation, or the 'Manager' and 'Office' classes will have to be combined, and the model rerun. Alternatively, some effort can be made to add data that might explain the inconsistencies.

16.6 Summary

Once the data for a scorecard development has been assembled, the first step is the transformation of characteristics into a form that can be used in the modelling process. Both *univariate* and *bivariate* approaches are possible, the latter being the norm in retail credit scoring, where the large number of observations makes them feasible. The primary *bivariate* approaches are dummy variables and weights of evidence. With *dummy variables*, a null class is created for average or uncertain classes, and separate binary variables for each of the rest. It is the most suitable approach for use with discriminant analysis and linear probability modelling. Where there is sufficient data, it could provide better results for logistic regression, but is labour intensive. With *weights of evidence*, a separate variable is created, containing the values calculated for each coarse class. It is most suitable for use with logistic regression, and requires much less effort. For cases where lenders are uncertain of which is most appropriate, models can be developed using both simultaneously (but WOE should dominate).

The classing process has two stages: *fine classing* and *coarse classing*. The former may already be hard coded, but if not, it is required to make sense out of categorical or numeric characteristics with a large number of possible values. The latter is required to create fewer groups of similar risk (or that logically belong together), with minimal information loss. Both depend on characteristic analysis reports to provide details of frequencies and/or rates for good, bad, indeterminate, and reject.

Coarse classing can be a tedious process, but there are tools that can assist. Statistical measures include the information value, chi-square statistic, and rank-order correlations. The *information* value is most commonly used, but if possible should be used in combination with the *chi-square statistic*, to make sure that small but highly relevant groups are not missed. Grouping can also be made easier through the use of pooling algorithms: *non-adjacent*, for categorical characteristics; *adjacent*, for numeric characteristics; and *monotone adjacent*, for numerics, where monotonicity is required. In all cases, the goal is to reduce the number of classes as far as possible, but with minimal information loss.

A few examples were provided to highlight the manual process. An analysis of *court judgments* indicated problems with: (i) *matching issues*, where no data is found; and (ii) *rare events*, where certain classes must be treated as policy rejects. The classes for *industry* were logical, but it took some time to come up with logical groupings that were related to employment stability within each industry. In contrast, the groupings for *occupation* were counterintuitive, as certain categories that one would expect to be lower risk were not. This can create problems when explaining scorecards to their end users. If nothing else, it highlights the need for some human supervision in the process.

Once the classing has been done, it is possible to develop the first true model—either a known good/bad model in application scoring, or a first pass at a behavioural model. For the latter, or those brave, and possibly foolish, enough to forego reject inference, this might even be the final model. In either case, it is still likely that the coarse classing, and perhaps even the fine classing, will have to be revisited.

17 Characteristic selection

The product of the transformation process is a set of transformed variables, which are potential candidates for inclusion in a model. These may represent about 40 characteristics for a simple application scorecard, and increase to 200 or more if data is used from other internal sources (existing product holding and performance) and/or external sources (credit bureaux). For behavioural scorecards, there may be 60 or more for simple fixed-term repayment products and over 120 for transaction products. Obviously, the number has to be cut down; otherwise, the task can be overwhelming. This is especially true if: (i) numerous iterations are required to finalise a model; or (ii) the predictive modelling technique is calculation intensive. According to Thomas et al. (2002), the latter applies especially to logistic regression and decision trees. Thus, this is another instance where the complexity needs to be reduced. This chapter covers several aspects of characteristic selection, or reduction:

(i) **Considerations for inclusion**—Significance, correlation, available and stable, logical, compliant, and customer-related.
(ii) **Measures of significance**—Information value, chi-square, and rank-order correlations.
(iii) **Data reduction methods**—Treatment during development, correlation assessment, and factor analysis.
(iv) **Variable feed**—Stepping and staging.

17.1 Considerations for inclusion

When reviewing characteristics for possible inclusion in a scorecard, the primary factors to be considered are whether they: (i) are *logical*, and can be explained to the business; (ii) have a significant degree of *predictive* power; (iii) have a *low correlation* with each other; (iv) are *stable, and available* for use; (v) are *compliant*, in that there are no legal, or ethical, restrictions on their use; (vi) relate to the *customer*, and not the lender's strategies; and (vii) result in unacceptable *information loss*, if excluded. Each of these is treated in greater detail in the paragraphs that follow.

Logical

Something that has been stressed on several occasions in this text, is that simpler is better. Ultimately, the goal is to provide a robust model that works not only when implemented, but also for a significant time period thereafter. This is aided if both the characteristics and point allocations make logical sense, which also makes the results easier to explain and sell to the business.

According to Falkenstein et al. (2000), one of the most controversial findings of their variable selection process was a 'preference towards relationships that are intuitive', meaning those that are commonsensical and customary. He also quotes Bunn and Wright (1991), who in the context of expert models, highlighted that 'the usefulness of experts is often more in the set of variables to which they refer, rather than how they actually use them', and that 'the judgmental challenge is therefore the creative one of eliciting key variables from experts'. Thus, one of the first stops when doing a scorecard development should be to consult with underwriters, and other subject experts. The consultative process will not only improve the final model, but also improve buy-in when it is presented. Note, however, that many scoring systems will be able to source, or generate, many characteristics that underwriters are unfamiliar with, so there may be a much broader field to evaluate than what has traditionally been available to the business.

Predictive

The characteristics of interest are those with a significant degree of predictive power, which can potentially add value in a model. Once again, statistics like the chi-square, information value, and rank-order correlation coefficients can be used to rank and compare candidates. The information value is the most commonly used, and can indicate anything from a total lack of correlation to too good to be true.

Uncorrelated

In many instances, a lot of characteristics will be highly correlated with each other, especially those calculated using the same, or similar, base inputs (e.g. financial ratios, and characteristics calculated for different time periods using the same underlying variable). This gives rise to potential multicollinearity, which can lead to poor out-of-sample performance. Only one or two characteristics out of a group can suffice, but groups have to be defined before the characteristics can be selected (see Section 17.3.3 on Factor Analysis). If not done, multicollinearity can only be guarded against by addressing inconsistencies during the development process, either by removing characteristics, or collapsing coarse classes further.

Instances where coefficients have the 'wrong sign' (+/−) are a sure sign of problems, as other coefficients will more than likely have been improperly exaggerated in the opposite direction. This results from multicollinearity, with the possible result of poor out of sample performance. According to Falkenstein (2002:178), if not addressed, 'it places great demands, by requiring that the correlations between the predictor variables remain stable'.

Available and stable

Characteristics should only be used if they: (i) will be available in the production system in future; (ii) have been stable since the sample was taken, and are expected to be stable in future. They may instead be: (i) *discontinued*, and no longer populated; (ii) *new*, and poorly populated; (iii) *unstable*, because of infrastructure changes, or problems; (iv) *inflation sensitive*, like income and asset values; or (v) *manipulable*, either by customers or staff. A major risk arises where one set of software was used to generate characteristics for the scorecard development dataset, and another is in place for the live system (such as retro data extracts versus results from live feeds).

Compliant

If scorecards are to support decision-making, their developers must ensure that the characteristics used are compliant with any legal, policy, or ethical restrictions on their use. Those most commonly forbidden are race, culture, gender, religion, nationality, and sexual preference. The situation varies from country to country, and some will allow their use, if: (i) they form part of a broad-based assessment; (ii) their influence is relatively minor; and (iii) it can be empirically shown that the characteristic is required in the assessment.

There are also instances, as highlighted by Mays (2004), where the rank ordering is required for planning, forecasting, provisioning, or other uses that will not affect individual decisions. If the forbidden characteristics are known to have an influence, they could still be taken into consideration. It need not be via a separate scorecard, but rather, by including those characteristics in a final stage, used only for that purpose.

Customer related

When rating individuals' risk, the characteristics should relate to them, and not lenders' strategies. Lenders' interest is in customer risk, independent of the decision made, so that a decision can be made. In application scoring, customer demographics, indebtedness, loan purpose, and payment behaviour are fair game, but the product offered, loan term, and credit limit granted are not. There are two ways of addressing this. First, *control variables* can be used to neutralise the strategies' influence, and put all cases onto a common footing. A strategy table, using one or more scores (risk, retention, revenue, response), is then used to choose the most appropriate option. Second, if the strategies are mutually exclusive, and only one dimension is being considered: (i) *separate models* can be developed for each strategy; (ii) *new cases* scored using each; and (iii) a *strategy* chosen according to the highest score.

Minimum information loss

Finally, when trimming the number of characteristics, it should be done with the minimum possible information loss. There may be characteristics that are seemingly contentious, weak, or highly correlated with other characteristics, whose exclusion reduces the final model's

Table 17.1. Measures of predictive power

		Good	Bad	Total	G/B
X1	= 1	2,000	3,000	5,000	0.67
	= 0	4,000	1,000	5,000	4.00
	T	6,000	4,000	10,000	1.50
X2	= 1	4,000	3,000	7,000	1.33
	= 0	2,000	1,000	3,000	2.00
	T	6,000	4,000	10,000	1.50
X3	= 1	5,800	3,600	9,400	1.61
	= 0	200	400	600	0.50
	T	6,000	4,000	10,000	1.50

	χ^2	F	D
X1	2,083	0.7466	0.4167
X2	83	0.0338	0.0833
X3	204	0.0780	0.0667

power. In some environments, this factor may even swing decisions regarding the inclusion of otherwise forbidden characteristics, like gender.

17.2 Statistical measures

In Section 16.3, three different measures were presented as tools for assessing characteristic' predictive power: *chi-square* (χ^2), the *information value* (F), and the *Gini coefficient* (D). They are used again here, but this time to rank characteristics by their predictive power, to assess whether they might feature in the final model. Thomas et al. (2002:140) provide an example using binary characteristics, which is illustrated in Table 17.1. As can be seen, *X*1 is clearly the most powerful across the board, but *X*3 is only marked as second choice by two of the three measures—the Gini coefficient contradicts the other two. The following expands upon use of the information value, and chi-square.

Information value

The most commonly used statistic is the information value, possible interpretations of which are provided in Table 17.2. These are based upon a SAS e-Intelligence document, on the use of their Enterprise Miner for an application scorecard, which recommends using characteristics with information values of greater than 0.20, but this is a bit too limiting.

As can be seen, any characteristics with information values less than 0.01 should be dropped, and values below 0.10 are unlikely to provide much value. Final scorecards will be comprised primarily of characteristics with values between 0.10 and 0.50. Care must be taken

Table 17.2. Information value (F-statistic) benchmarks

Value range	Strength	Description
$F \leqslant 0.01$	Uncorrelated	Drop, unless sound reason to consider further.
$0.01 < F \leqslant 0.05$	Very weak	May add value, but can be dropped with minimal loss.
$0.05 < F \leqslant 0.10$	Weak	Possible spurious correlation, unlikely to feature.
$0.10 < F \leqslant 0.30$	Medium	Known correlation, which could appear.
$0.30 < F \leqslant 0.50$	Strong	High information content, and sought after.
$0.50 < F \leqslant 1.00$	Dominant!	Very powerful, but investigate why.
$1.00 < F$	Warning!	Possible error, perhaps outcome posing as observation.

with higher values! First, use of related scores, such as bureau scores and other highly-predictive characteristics, will diminish the value provided by other data, and should be integrated later using a matrix, or subsequent regression. And second, outcome variables are often mistaken for predictors. Only information available at the time a decision is to be made should be used, other than for inference purposes.

Chi-square

The other primary statistic is the chi-square, and its associated *p*-values. No literature could be found relating to its use to compare classed characteristics. They are provided by many scorecard development and statistical software packages, but for the transformed variables—not the frequencies associated with the classed characteristics.

> Mays (2004:94–95) refers to the 'score chi-square' values provided by SAS's PROC LOGISTIC, which tests whether or not the 'regression parameter for the variable in question is different from zero', and comments that it has the disadvantage of being a 'test for a linear association between the candidate predictor variable and the log-odds of the outcome variable'. This is a different usage than presented here.

Mays (2004:94) quotes Rud's Data Mining Cookbook, which recommends that any characteristics with a *p*-value greater than 0.5 be dropped from the analysis and then goes on to recommend 0.3 (for binary characteristics, this equates to dropping all characteristics with a χ^2 value less than 1.07). Other statistical texts recommend an upper limit of 0.1. At the higher levels, other checks must be performed to guard against overfitting (see Section 17.4.1, on testing each step using a hold-out sample).

Combined usage

Rather than using one of the statistics to do the selection, it is also possible to use all three: (i) calculate each statistic for all characteristics; (ii) rank the characteristics' predictive power using each; and (iii) plot the rankings using the information value as the X-axis. If either of the other two measures provides a rank ordering significantly better than the information value, then they should be investigated further before being discarded (see Mays 2004:93 for an example).

In general, care must be taken when using these measures, as the rankings cannot always be relied upon. To quote Mays (2004),

> . . . bivariate analysis can lead us astray in certain cases if proper care is not taken. The best approach is always to further analyze any variable whose statistics run counter to expectations. On further analysis we often find that the relationship we expected between predictors and the outcome is borne out if we account for the influence of other variables that are correlated with the predictor.

17.3 Data reduction methods

As indicated earlier, at the outset there may be a lot of characteristics, and it helps to reduce their number. Even if steps are taken to remove those that will clearly add no value, there will still be a lot of candidates. The next step is to treat potential multicollinearity, which can be done in three ways:

(i) **Treat during development**—Ignore it, and instead deal with it during training.
(ii) **Manual assessment**—Review the correlation matrix manually.
(iii) **Factor analysis**—Use a statistical analysis to identify the correlated variables.

Correlations, factor analysis, and other calculations or procedures referred to in this section are intended for use with numeric characteristics, and any analysis should be based upon the transformed variables (see Chapter 16, Transformation), not the original characteristics. It is, however, possible to do some analysis on the raw characteristics, but care must be taken regarding missing values, outliers, and values that are meant as codes.

17.3.1 Treat during development

As indicated earlier, one consideration when choosing characteristics is possible information loss. In order to get around this, some scorecard developers will use all of the available characteristics, except those that are contentious, obviously not predictive, or would not make sense in that context (like existing account number). This approach is feasible where the number of characteristics is small, the computing resources are significant, and the scorecard developer has sufficient knowledge to counter any statistical vagaries that arise. The logic is that the statistical techniques being used—whether linear probability modelling (LPM), logistic regression, or others—have variable selection routines, to choose the variables that best explain the response function (see Section 17.4.1).

The scorecard developer's task is then reduced to quality control, to ensure that the resulting point allocations for each attribute: (i) *make sense* relative to those for neighbouring attributes, especially if monotonicity is required; (ii) are *consistent* with the risk, and do not have 'wrong signs'; and (iii) are not the result of overfitting. The first applies when using dummy variables; for risk substitutes, it was already addressed during coarse classing.

The extremely oversimplified example presented in Table 17.3 illustrates four highly correlated characteristics, all derived using counts of bureau searches. This is an example of autocorrelation, as all of these values are constructed out of the same base characteristic, but over adjoining time periods, prior to the credit application.

There are three ways that the number of enquiries could be presented: (i) use cumulative totals, as in the example; (ii) treat each period separately; or (iii) generate a new characteristic that captures any interactions.

At first glance, the choice is obvious (assuming highly correlated characteristics are to be removed); the 360-day characteristic has the highest information value! A closer look at the weights of evidence (WoE) reveals something interesting though—a large number of enquiries in the recent past is particularly indicative of higher risk, especially where there are six or more enquiries in the last 90 days, or three or more enquiries during the last month. It is not a statistical aberration, but makes common sense, and should be modelled—the most recent information is the most valuable, especially where it pertains to rare events. Thus, only the value

Table 17.3. Information value comparison

	Number of enquiries last . . .							
	30 days		90 days		180 days		360 days	
	Total	WoE	Total	WoE	Total	WoE	Total	WoE
No match	3,317	0.188	3,317	0.188	3,317	0.188	3,317	0.188
= 0	30,581	0.080	22,083	0.189	14,291	0.336	6,474	0.506
= 1	5,288	−0.161	8,293	−0.036	9,349	0.113	7,692	0.285
= 2	2,155	−0.353	4,522	−0.141	6,174	−0.065	6,574	0.153
= 3	754	−0.516	2,147	−0.414	3,752	−0.166	5,357	0.044
= 4	348	−0.649	1,112	−0.489	2,341	−0.423	3,859	−0.155
= 5	119	−0.723	561	−0.619	1,339	−0.517	2,962	−0.190
= 6	51	−1.265	277	−1.082	820	−0.510	1,937	−0.416
= 7	33	−0.548	155	−1.004	485	−1.005	1,326	−0.391
to High	49	−0.922	227	−1.256	825	−0.989	3,196	−0.767
Total	42,694		42,694		42,694		42,694	
F-stat		0.0354		0.0797		0.1172		0.1328

for 180 days can be safely dropped, and care must be used with the rest. Alternatively, a generated characteristic could be used to combine them.

17.3.2 Correlation assessment

As indicated, one of the risks is multicollinearity, which occurs when correlated variables are used. Note, that correlated 'variables' are being treated, not 'characteristics'. Thus, when using dummy variables, risk measure substitutes, or the results of univariate transformations, the concern should be with correlations between the transformed variables. Unfortunately however, this can be a tedious task, and many people will instead review the raw correlations.

Table 17.4 is a correlation matrix, for the same four enquiry counts illustrated in Table 17.3. The autocorrelation is evidenced by the high correlations, especially between adjacent characteristics (note, the correlations are exaggerated, because each count includes all cases represented in the prior count). The one month and one year values have the lowest correlation, making them the most likely candidates for inclusion. This example is highly abbreviated, as in most cases all of the candidate variables would be included—possibly numbering in the hundreds. In such instances, variable reduction can be done using factor analysis.

17.3.3 Factor analysis

The concept of factor analysis was already touched on briefly in Section 9.1.2, where it is described as: (i) a descriptive statistical technique that can be used to gain a greater understanding of data; and (ii) a tool that can be used for variable reduction prior to model development. Its power lies in its ability to take a set of inter-correlated characteristics, and transform it into a smaller set of uncorrelated characteristics, which is then used to develop a regression model. Many readers of this text will be familiar with SAS statistical programming software, so this section will touch on a couple of the SAS procedures available, in particular PROC VARCLUS and PROC FACTOR. The resulting formulae could be used to create new variables for the scorecard development, but in retail credit scoring this is seldom done in practice.

Table 17.4. Correlation matrix

	Number of enquiries last . . .			
	30d	90d	180d	365d
30d	1.0000			
90d	0.7851	1.0000		
180d	0.6529	0.9215	1.0000	
365d	0.6038	0.8765	0.9712	1.0000

Mays (2004) mentions the use of SAS's PROC VARCLUS, which uses 'oblique principal component cluster analysis'. It is a very powerful data analysis tool that can: (i) identify '*variable clusters*', where each variable is assigned to a distinct cluster; and (ii) provide an *audit trail* of how the clusters are derived. Rather than using the resulting clusters directly, however, one or two variables with the highest potential predictive power (information value, chi-square, or another bivariate statistic) are chosen from each. The output of PROC VARCLUS includes the following:

Standard Output:
For each subsequent stage in the splitting process,
number of variables represented per cluster, and the extent and proportion of variation explained by the clusters, both individually and collectively;
variables represented in each cluster, the *R*-squared values for within the cluster, and with the neighbouring cluster, and the 1 *R*-squared ratio.
standardised scoring coefficients, which can be applied to a new dataset using PROC SCORE to derive the cluster values;
correlation matrices showing the correlations between each;
For the final set of variable clusters,
a table of the intercluster correlations; and
a summary of the results from each step in the analysis.
CORR option:
a correlation table for all of the variables.

The default stopping rule is to stop splitting, when the maximum eigenvalue of the remaining groups is less than one. It is, however, also possible to specify either another maximum, or a maximum number of variable clusters.

An example of PROC VARCLUS's output is provided in Table 17.5, which is for behavioural scoring variables for a cheque account (these should have been derived using transformed variables, but raw values were used instead). It can be seen that there are certain logical clusters: (i) turnover; (ii) borrowing status; (iii) time since; (iv) balances, but not minima; etc. There are also some that are not so clear, like the next group containing debit turnover, fee income, and the value of inter-account credit transfers, which probably relates to activity.

The *R*-squared values show the correlation between the variable and both that cluster, and the next closest, as well as a ratio $(1 - R^2_{\text{Own}})/(1 - R^2_{\text{Next}})$, which indicates how well suited the variable is for inclusion in that cluster—the lower the ratio, the better. As can be seen, limit utilisation has a very poor fit with cluster 2, but is left there, only because the next best choice is even worse.

Table 17.5. Variable clusters

13 Clusters		R-squared with		1-R^2 Ratio
Cluster	Variable	Cluster	Closest	
1	Credit turnover	0.8684	0.2760	0.1818
	Interest income	0.2354	0.1035	0.8529
	Net turnover	0.8635	0.2350	0.1784
	Value automated credits	0.8615	0.2701	0.1898
	Value inter-account DR Tfrs	0.5367	0.1033	0.5166
2	Limit utilisation (%)	0.0071	0.0006	0.9935
	Days in credit	0.9560	0.2435	0.0581
	Days in debit	0.9579	0.2398	0.0554
	FLAG—no borrowing	0.8327	0.2537	0.2241
3	Months since open	0.6257	0.1590	0.4450
	Months since limit increase	0.7519	0.0648	0.2653
	FLAG—never increase	0.6957	0.1187	0.3453
4	Balance	0.7506	0.1384	0.2894
	Maximum balance	0.8399	0.3556	0.2485
	Average balance	0.9386	0.1919	0.0760
	Average credit balance	0.9111	0.2740	0.1225
5	Debit turnover	0.6875	0.1378	0.3625
	Fee income	0.3579	0.0194	0.6548
	Value inter-account CR Tfrs	0.4115	0.0224	0.6020
6	Minimum balance	0.7472	0.6704	0.7670
	Average debit balance	0.7472	0.0904	0.2779
7	Maximum excess	0.8284	0.4846	0.3329
	Balance range	0.8284	0.3076	0.2478
8	Months since last active	0.6841	0.0285	0.3252
	FLAG—never credit	0.6841	0.0427	0.3300
9	Months since last credit	0.1054	0.0154	0.9085
	Days in excess	0.8355	0.3345	0.2472
	Number of days dishonours	0.0992	0.0238	0.9228
	FLAG—no limit	0.7985	0.1689	0.2425
10	Months since last dishonour	0.8178	0.1592	0.2167
	FLAG—never dishonous	0.8178	0.1020	0.2029
11	Credit T/over to Min Baln (%)	1	0.0532	0
12	Baln range to Min Baln (%)	1	0.0446	0
13	Debit int to CrTO L3M (%)	1	0.0006	0

PROC FACTOR

Another possibility is to use PROC FACTOR, which can use a number of different factor analysis techniques to determine the groupings (the default being principal component analysis). The results should be similar to those provided by PROC VARCLUS, but it has the advantage that the results are more robust. Unfortunately though, they are also more difficult to interpret, and to implement. Rather than each variable being assigned to a distinct factor, they are spread across all of them in different proportions. PROC FACTOR's output includes:

Standard Output: (i) Eigenvalues of the correlation matrix; (ii) factor patterns giving an overview of which variables are in each factor; (iii) details on the amount of variance explained by each factor and by each variable.
CORR option: A correlation table for all of the variables.
SCORE option: Standardised scoring coefficients that can be used to create the factors, either from the original dataset, or other datasets that are presented to it.

17.4 Variable feed

We have now decided which characteristics will be considered within the modelling process, but need to determine how this will be done. Each statistical technique has some means of evaluating the order in which variables are included in the model, and the scorecard developer has varying levels of control. There are two concepts that will be touched on in this section: (i) *stepping*, an iterative procedure used for variable selection, that is driven by an automated algorithm, which comes standard with most statistical packages; and (ii) *staging*, a manual process of creating logical groupings that are considered in blocks, with stepping used within each.

17.4.1 Stepping

One of the beauties of modern computing is the speed and ease with which it can be done; a far cry from what was available when many modern statistical techniques were first developed. This also applies to variable selection when developing statistical regression models. The three basic selection approaches were touched on in Section 7.4.3, being forward, backward, and stepwise (either forward or backward). With each, coefficients are re-estimated every iteration. Most practitioners use the term 'stepping' to refer to all three, primarily because: (i) it aptly describes the incremental nature of the process; and (ii) the most popular is forward stepwise.

According to Falkenstein et al. (2000), there is a dilemma, because the variance/imprecision of the regression parameters, which are later converted into point allocations, increases as more variables are added into the model (associated with the increased degrees of freedom). Thus, there is a point where the added imprecision outweighs the value added by the additional variable. Most statistical packages provide a stopping rule, like where the p-value of the next candidate is below or above a given threshold, say 0.05 for a confidence level of 95 per cent. Stated in other words, the procedure will continue until as many variables have been included in the model as possible, and each must make a statistically significant contribution to explaining the response function. Although commonly used, this approach has the disadvantage of doing hypothesis testing using the same data that gave rise to the hypothesis.

There are variations that use an out-of-sample group, usually a holdout sample. All of them demand that a training sample be used to determine the regression parameters. Thereafter, the contribution at each step is assessed out-of-sample, whether by assessing confidence levels, or by determining where the predictive power levels off, based upon another statistic like the KS-Statistic, Gini coefficient, or misclassification rate at a given reject rate. Bailey (2003) provided an illustration using the percentage of bads rejected, which is the basis for Figure 17.1 where it levels out at about step 13.

17.4.2 Staging

For most people, statistical software is usually a black box, whose inner workings are neither well-understood, nor easy to control. As indicated, stepping is an automated and iterative process, used to select the variables that appear in the final model, and all coefficients are recalculated with each new variable. Instances arise, however, where scorecard developers want greater control over how characteristics are introduced, and their eventual influence. One way of achieving this is to sort them into 'stages', meaning characteristic groupings that are each treated separately. There are two types:

(i) **Independent**—Separate scorecards are developed for each group, which are then integrated via a further scorecard, or a matrix. This is used primarily when data is obtained from different data rich sources.
(ii) **Ordered**—Coefficients for each group are fixed, and used as inputs into the next stage, which puts greater emphasis upon characteristics included in the earlier stages.

Please note that the term 'staging' is typically only used by practitioners, with respect to the latter. The two labels used here are not used in practice, but are presented as a means of distinguishing between two different ways of treating characteristics in blocks. There is little or no literature on this aspect of variable selection, let alone a proper name, much in contrast to the sexier automated techniques.

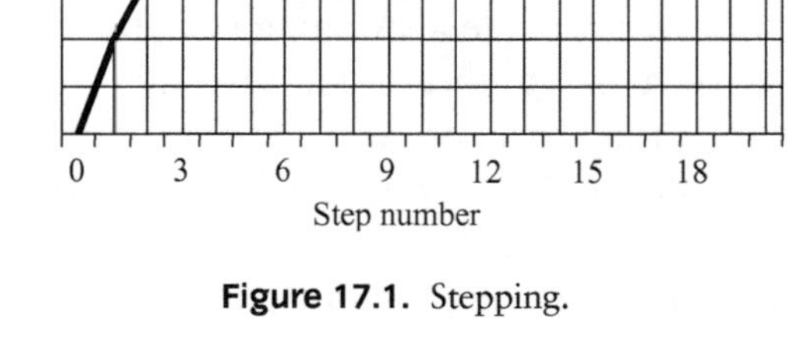

Figure 17.1. Stepping.

Common reasons

The primary reasons for staging, whether ordered or independent, relate to the incorporation of bureau data, or data from other information rich sources. The simplest and most common example of independent staging is where lenders develop bespoke application scorecards for their own data, and integrate these with generic bureau scores via a strategy matrix. Decisions can be made without bureau data if so required, assuming that the 'own data' scorecard has sufficient power for it to be trusted. For many smaller lenders this is the preferred route, if only for its simplicity. The same, or similar, can be done using bespoke bureau scores, but these are much more costly.

In either case, there are a number of reasons why lenders require separate treatment of the different data sources. First, there may be service disruptions, where decisions must be made without it, whether because of busy phone lines (old world), computers not communicating (new world), systems problems at the bureaux, or even because of a dispute over bureau charges. Lenders are under extreme pressure to cater for these possibilities, because they are of little interest to waiting customers. In some instances, the issue may simply be the order in which data becomes available, even if the difference is only milliseconds.

Second, intermediate decisions may have to be made prior to obtaining bureau data. Lenders may opt not to use it, where: (i) the pre-bureau score is so low, or high, that the new data is unlikely to change the decision; and (ii) bureau costs are high relative to the potential income. This can also operate in reverse, where bureau data is used as a pre-screening mechanism prior to making marketing offers, and application data is included only upon receipt.

And third, there may be multiple credit bureaux. Lenders may switch between them for any of the above reasons, or perhaps even use more than one simultaneously, while internal data remains constant. Also, in some cases, the different bureaux' strengths may lie at different ends of the risk spectrum, and the pre-bureau score can be used to decide upon which bureau to use.

For the multiple bureau case, separate scorecards will be required for each bureau, and possibly also for different combinations. Say, for example, that: (i) a lender normally obtains data from two credit bureau, A and B; (ii) knows that there will be circumstances where only one is available; and (iii) data from at least one bureau must be available for any decision to be made. Three different scenarios have to be catered for—A, B, and A&B. Rather than developing totally separate scorecards for each scenario, the lender can instead develop one scorecard for application and existing account performance details, and then use either ordered or independent staging to incorporate the two credit bureaux. This has the significant advantage of: (i) reducing the scorecard developer's workload; and (ii) making the resulting scorecards easier to explain to business. If independent staging is used, the matrix approach is feasible to integrate the own score and a single bureau score, but a further regression should be considered to integrate the three scores (own, A, and B) into one.

Dependent staging

The above relate to instances, where there is a large amount of data from other data sources that is known to be very powerful, but lenders may simply wish to downplay certain characteristics. This level of control is achieved by: (i) sorting the variables into numbered blocks; and

(ii) developing a separate model for each, while controlling for the variance that has already been explained in prior stages. Coefficients are fixed after each stage, and characteristics entered in earlier stages will get greater emphasis (higher coefficients) than if an automated stepping algorithm were used throughout. The model output from each stage could be used as is, assuming it has sufficient power for the lender to put faith in it. Most characteristics will be treated in the first and second stages, but there may be special cases that are treated in one or two further stages.

Most academic statisticians will probably ask, 'Why?', and question the statistical soundness, primarily because correlations with earlier stage variables cannot be properly assessed. Irrespective, many lenders use it and consider it sound. It is possible that in this area, statistical theory is yet to evolve to recognise practical issues, associated with use of predictive statistics in production processes. There are a number of reasons why this level of control may be needed.

First, with external data sources, especially the credit bureaux, there may: (i) be an issue of trust, and; (ii) the lender may be uncomfortable with the amount of power that the bureau data wields over their decisions. For the latter, lenders' pride may play a role, especially where they have invested much in their own internal systems, but the concerns more often relate to quality, thickness, appropriateness, and stability of the bureau data. Lenders can instead opt to put greater emphasis on their own data, and include bureau characteristics in a second stage. This has the further advantage that the point allocations for own data, treated in the first stage, need only be documented and explained to the business once.

Second, credit scoring makes the assumption that the future will be like the past, which is not always valid. There are often instances where significant changes have occurred, or are expected, especially where a comparison of historical and recent distributions highlights significant drift for certain characteristics. In these cases, it would be foolhardy to allow stable and unstable characteristics to compete on an equal footing. While it is possible to drop the unstable characteristics, that is a drastic way of dealing with mistrust. The alternative is to consider them last, allowing stable characteristics to take precedence.

Third, the scorecard should focus on borrowers' characteristics, and pay less attention to those that may be influenced by lenders' strategies, whether credit or marketing. According to Bailey (2003), some sensitive characteristics may be 'marketing-driven'. An example is Card/Payment Protection, which normally indicates a much higher risk, but is sometimes used as a marketing sweetener. In order to prevent it from dominating the scorecard, it can be included last. A similar treatment might be given to other characteristics, relating to past strategies employed, the asset being purchased, and/or loan officer making the decision.

> The point relating to the loan officer is purely hypothetical, and might apply in emerging and micro-finance environments where judgmental decisions still play a large role—especially if the same person is also responsible for collections. The dearth of appropriate data and need for an effective tool may make it palatable to some lenders. Characteristics might include the loan officer's age, years' experience, and number of children.

Fourth, there may be compliance issues for one or more characteristics. For example, lenders are under pressure to reduce reliance upon sensitive demographic characteristics, such as 'gender', and instead use information specific to each customer's past borrowing behaviour. In some countries, the use of such characteristics is *verboten*; in others, allowed; and in still others, the treatment is uncertain. If treated in a later stage, they: (i) can be easily removed in need, with little or no effect upon the rest of the model; and/or (ii) can still be used for forecasting and capital allocation purposes, irrespective of whether they are allowed for case-by-case decision-making.

And finally, the lender may wish to ensure that less predictive characteristics have a greater opportunity to play a role in the assessment generally. This aspect is mentioned in Siddiqi (2006), who suggests that it be used to create a risk profile 'to mimic the thought process of a seasoned, effective adjudicator or risk analyst'. A part of the goal is to make the assessment more broad-based, and more stable. By ensuring a broader spread of characteristics, the assessment is less susceptible to changes in one or two very powerful characteristics. He also suggests that this approach improves the stability of the scorecard over time, as compared to 'limited variable' scorecards. His approach differs though, and it cannot be ascertained from the text whether or not the point allocations are fixed at each stage.

Regression treatment

With *independent staging*, model C can be developed, using scores A and B as predictors. In contrast, for *dependent staging*, model A is developed first, and is included as a fixed value when deriving model C, with no intermediary B. The tricky part is how to control for the results from prior stages. With LPM, and any other technique well-suited to continuous response variables, this is done by modelling the prior stage's residual (see Equation 17.1):

Equation 17.1. Residual modelling $e_{K-1,\,i} = y_i - \hat{y}_{i,\,K-1} = S_{K,i} + e_{K,i}$

where e is the error term/residual, K is the current stage number, S is the score derived for the current stage only, i is the index of record being assessed.

In contrast, *logistic regression* is best suited for binary outcomes, or ordinal outcomes with a limited number of possible values. Coefficients have to be fixed directly, assuming that this capability is provided by the software package being used. With SAS's PROC LOGISTIC, a variable can be forced into a regression with a given coefficient, using the OFFSET option in the MODEL statement. In this case, the total of any prior stage scores would be included as a variable, with an offset coefficient of one. After all of the stages are completed, the final scorecard is constructed by combining the elements from each regression formula, inclusive of the intercepts.

17.5 Summary

Most credit scoring developments, irrespective of type, have a huge number of characteristics that could potentially be used, yet there are usually only between 6 and 15 characteristics that, hopefully, best explain customer behaviour. The large number that comes out of the starting blocks can create significant overheads for the scorecard developer, but the field can be narrowed before the race starts by doing variable selection to limit which ones can be bet on.

For greenfield, and many other developments, the first port of call should be to ask underwriters, and other domain experts, which data has provided value in the past. There are, however, often cases that underwriters are not familiar with, especially: (i) new and unfamiliar data fields; and (ii) where the development is for new markets or functions. Thus, even where such sage advice is available and must be considered, scorecard developers should still start with the broader field. Several questions need to be asked about each characteristic: (i) Does it make sense, and can it be explained to the business?; (ii) Does it have even the slightest hint of predictive power?; (iii) Will it be available in future?; (iv) Have there been any major changes in the characteristic's distribution, or are such changes expected?; (v) Are there any legal or ethical issues surrounding its use?; (vi) Does it reflect customer details, as opposed to lender strategies?; and (vii) Are there conflicts, or overlaps, with existing policy rules?

Measures such as the information value, chi-square statistic, and Gini coefficient can be used to weed out irrelevant characteristics. Potential multicollinearity can then be addressed during the scorecard development, but pre-build variable reduction is also possible using factor analysis. Variable clusters are identified, and either; (i) one or two of the most predictive characteristics are chosen from each; or (ii) the factors are used directly within the scorecard development. Implementation issues make the latter problematic.

Most regression techniques have automated variable selection routines, such as *forward, backward,* and *stepwise*. These make the process easy, but scorecard developers may want greater control. *Independent staging* is used, where separate models are developed for each data source, either because of problems with availability, or to deal with intermediate decisions. In contrast, *dependent staging* is used to reduce the influence of: (i) data from external sources; (ii) unstable characteristics; (iii) lender strategies; (iv) potentially non-compliant characteristics; and (v) characteristics that might otherwise dominate the assessment. It can be done by modelling the prior-stage residual (linear probability modelling), or by including the prior-stage score as an offset variable (logistic regression).

At the end of the process, there may be more than just one set of possible characteristics. Some scorecard developers will develop several models using different combinations, and then assess each of them, based upon: (i) whether or not they make logical sense; and (ii) the amount of value that they will provide, in terms of reducing risk, or increasing business.

18 Segmentation

Segmentation is something usually associated with marketing, where companies adjust the marketing mix to improve the appeal to different groups of customers. Similar concepts are used in credit, but not always for the same reasons. At this point, it is assumed that the purpose of the scorecard (risk, response, revenue, or retention) and the objective (default, bankruptcy) is known. Most of the concepts are common, but to aid understanding this section focuses on credit risk assessment and default prediction, where segmentation's primary goals are to: (i) improve the assessment; and (ii) ensure an appropriate product offering, in terms of loan rate, repayment term, collateral, and other requirements (see Section 26.5, on Risk-Based Pricing). It must be stressed that the resulting segmentation will often not agree with that used for marketing, and may impact upon marketing.

If at all possible, efforts should be made to minimise the number of segments. Every extra scorecard brings with it extra complications, and there will always be the risk of complicating the situation unnecessarily. This applies not only in terms of the scorecard development, but also validation, monitoring, and strategy setting. Indeed, this is an area where scorecard developers and vendors can keep themselves employed, and even pad their packages, especially where the cost of the development is done on a 'per scorecard' basis.

18.1 Segmentation drivers

According to Thomas et al. (2001), there are three factors driving the choice of scorecard splits: strategic, operational, and interactional. While these three factors are valid, an alternative framework is proposed here, including: *marketing*, *customer, data*, *process*, and *model fit*. These may overlap, and ultimately only the 'model fit' factors are valid, but the others are usually indicative of serious interactional forces. Lenders may insist upon specific high-level segmentation, based purely on their own knowledge of the business. Even so, the scorecard developer should test any proposed splits and suggest alternatives, one of which is not to segment.

Marketing factors (strategic) arise where lenders require greater confidence in specific market segments, especially: (i) where they have little past experience; (ii) where they perceive themselves to be weak, or believe that they must make competitive inroads; or (iii) to ensure the ongoing health of the business. They apply especially to new markets and product offerings, where the dynamics are not well known or understood.

Customer factors arise where certain characteristics may not logically apply to certain customers. For instance, lenders may wish to treat customers with no borrowings, and/or thin files, separately from the bulk of applicants, where there is much more data. Even if a single scorecard is used, lenders may be well advised to be more conservative in their strategies with them. Examples are new customers, youth, immigrants, non-borrowers, and underserved markets.

Data factors (operational) relate to what is available, and how and when it becomes available. Different application forms may be used for different channels—say Internet and branch—where the form layouts and details requested are so substantially different that they cannot be treated on a like basis. The appropriate internal and external data can also vary by market segment, and lenders may opt to develop separate scorecards to summarise each source, that are then integrated.

Process factors (operational) relate to the differential treatment given by the account-management and collections functions, which may result in different good/bad definitions being required for each segment. This applies especially where the lender has different areas that are responsible. For example, many lenders will make the distinction between transactional (low-value) and relationship (high-value) lending, where credit scoring rules are applied strictly to the former, but with a lot of latitude for the latter. In such cases, it may be inappropriate to treat the two groups on a like footing. The same applies to collections, where vastly different strategies are being employed, or different areas are responsible; and also where there are substantially different product offerings.

The advent of the Euro raised the possibility of having European cross-border scorecards. It is only speculation, but it is probable that the level of protection afforded creditors' rights in each country causes interactions. If so, countries under the Napoleonic legal system (France, Spain, and Portugal) should be treated separately, as they provide little protection, and different back end processes are necessary. This is a potential area for further research.

And finally, model fit factors (interactional) relate to interactions within the data, where different characteristics predict differently for different groups. The most obvious examples are products with substantially different features (secured/unsecured, revolving/fixed, transactional/non-transactional), but there are many others (new/used, young/old, high/low income, large/small, existing/new customer).

While some interactions can be addressed using generated characteristics (like creating 'married with children' out of 'marital status' and 'number of dependents'), they are often insufficient, and separate scorecards are required instead. The most common example used for *application scoring* is 'Customer Age', as there are substantial differences between how certain characteristics predict risk for 'young' and 'old' groups. In most first-world countries, people are expected to have moved away from home and have established a credit record by the age of 30 (or thereabouts),[1] and there is usually extra risk associated with 40 year olds still living with mother. In contrast, living with parents is typically associated with lower risk for younger applicants, as what would otherwise be spent on rent or bond repayments can instead finance their lifestyles.

Another common split is between new and existing customer because the latter's risk assessment is heavily influenced by current/past performance. Indeed, such information can be so

[1] The Customer Age split usually lies in the 26 to 30 range, which brings on some chuckles when a 32 year old is being called old.

powerful that it contaminates the new customer assessment. For *behavioural scoring*, the splits often relate to the extent of existing risk, like 'Recent Arrears' versus 'No Recent Arrears', for much the same reason—potential contamination by an extremely powerful predictor, which is irrelevant for that group.

18.2 Identifying interactions

How is the scorecard split determined that best addresses interactions? There are different methods used in practice. Many scorecard developers rely upon past experience and guidance from the business, but there are also analytical approaches: (i) a *manual review* of changes in how the most powerful characteristics predict across a number of segments; (ii) *cluster analysis*, to inform of any natural clusters of observations in the data; and (iii) use of chi-square automatic interaction detection (CHAID), or another recursive partitioning algorithm, to derive a decision tree. In each case, analysis should be based only on known performance.

The end result will be a list of possible scorecard splits that can be tested, with the goal of determining whether the split can provide any lift (increased predictive power). Cluster analysis provides nothing in itself, because at no point does it assess the relationship with the target (good/bad) variable. In contrast, CHAID provides a table showing which splits provide the most predictive power, but each branch only indicates the best split at that point, and not overall.

The example in Figure 18.1 presents such a tree, where the first and most obvious split is the Customer Type, 'new' versus 'renewal'. The arrears status plays a significant role under renewals,

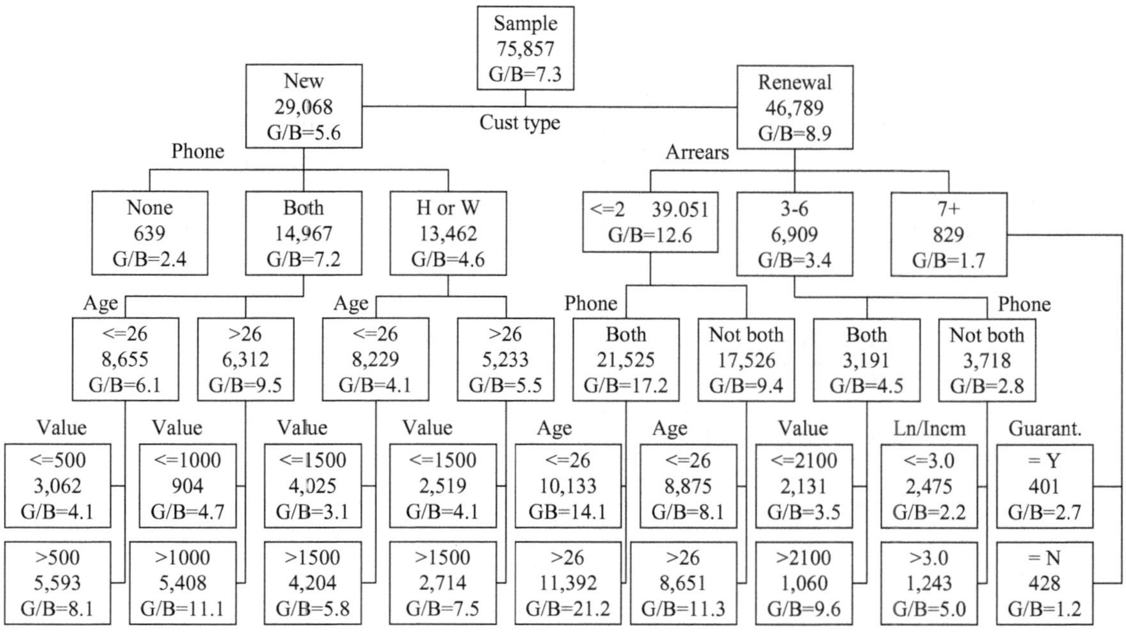

Figure 18.1. Segmentation—classification tree.

whereas contact details are more important for new customers. There are other possible splits though. In particular, Phone and Age recur at various levels, and should be considered.

18.3 Addressing interactions

While separate scorecards are often used to address interactions, many scorecard developers prefer to use generated characteristics within a single scorecard, the combination of age and residential status being a primary example. This avoids segmentation, but does not necessarily provide the best possible solution. Interactions may exist with many characteristics, not just one or two, which makes implementation and analysis tedious. Furthermore, it may not be possible to implement the generated characteristics within the delivery system.

Where separate scorecards are used, there are two possible approaches, independent and mother/child. Use of *independent scorecards* is common and straightforward, but reduces the amount of data available for each scorecard, and hence the reliability of the coefficients. With the *mother/child approach* the coefficients' reliability is enhanced by using all possible cases for the mother scorecard, and then including its score as a predictor in each child. Only those variables that provide a statistically significant contribution over and above the mother scorecard are included in the child.

The mother/child approach is also referred to as a master/niche approach. At one point, a certain scorecard vendor referred to this as a master/slave approach, but changed their terminology when they realised that lenders did not like using 'slave' scorecards.

In either case, it is necessary to make comparisons to determine which split is best. The example in Table 18.1 presents the results for four alternatives: a single scorecard, and then three possible two-way splits (the sub-pop scores should be aligned prior to the comparison, as they may not have the same meaning, especially with linear probability modelling). The results for each are compared using the Gini coefficient. According to this analysis, the 'new' versus 'renewal' split

Table 18.1. Segmentation—Gini comparison

Split	Full pop	Sub pop A	Sub pop B
None	54.2		
		New	Renewal
Cust type	57.3	34.9	51.2
		<= 26	>26
Age	56.5	42.2	49.4
		Not both	Both
Phone	53.9	39.3	47.7

on customer type provides the greatest lift in the Gini—from 54.2 to 57.3 per cent. This is in spite of the 'New' scorecard's seemingly low 34.9 per cent, which is only because it does not have the benefit of performance data on existing or past accounts. People new to credit scoring may criticise scorecards with such low Gini coefficients, without realising that stand-alone performance is much less relevant than performance as part of a scorecard set.

While the Gini coefficient provides a good starting point, it does not recognise that the optimal choice may vary with the cut-off strategy to be employed—what is good for 'same bad rate' may not be good for 'same reject rate' or others. This is illustrated by the strategy curves for the 'Phone (Y/N)' and 'Customer Type (New/Renewal)' splits in Figure 18.2. In general, the best split is that which provides the lowest bad rate at the chosen cut-off (the choice is clear-cut if the bad rates for one of the splits are lower at all possible cut-offs). In the example, the strategy curves cross; the phone split is better only at reject rates below 14 per cent. Given that the most likely cut-off strategies will be in the 14 to 25 per cent reject rate range (same bad rate and same reject rate respectively), the customer type split is preferable. The choice might change if the lender plans to be even more aggressive in its strategies, which is a possibility when economic conditions are benign, or the lender wishes to capitalise on recent improvements to account-management and/or collections processes.

18.4 Summary

Population segmentation is usually associated with marketing, where it is used to improve company sales and profitability by ensuring that the strategies employed are those best suited to the customers being targeted. In credit scoring the purpose is similar, except the customer's propensity to default is of greater concern than sales. There are several different drivers behind the choice of segmentation: (i) *marketing*, to improve confidence in specific sectors; (ii) *customer*, relating to whether certain data items are applicable for given subgroups; (iii) *data*, relating to data availability in different areas; (iv) *process*, where significantly different treatment is received from account management or collections; and (v) *model fit*, or interactional

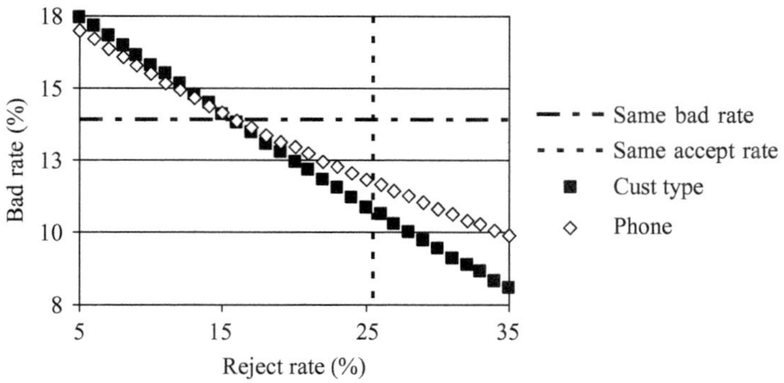

Figure 18.2. Segmentation—strategy curve comparison.

factors, where predictors' values vary across certain groups. Ultimately, all of the above fall into the latter camp, but that label is reserved for those cases where the choice is less obvious.

Very often, splits are based purely on business intuition, and should be tested to ensure that they truly add value. When dealing with a relatively homogenous portfolio, extra analysis will be done to identify potential splits, including the use of *cluster analysis*, *decision trees*, or a *manual review* of a giant matrix, containing bad rates across different variables and subpopulations. This provides a short-list of potential splits to be further tested. Comparisons are then done by developing simple scorecards for the full population, and each of the splits. The master and each of the sets are then compared using a *Gini coefficient*, and by comparing the bad rates at each reject rate. It must be stressed that the greatest concern with segmentation is how the scorecards work as a set, and not individually, as they will never be used alone.

19 Reject inference

When dealing with selection processes, statisticians are faced with a predicament. If scorecards are developed using historical performance, how can rejects with no performance be assessed? The problem is best, and often, illustrated by applicants with derogatory information on file, which violates policy reject rules, such as court judgments recorded on the credit bureaux. Some cases still make it past the gatekeeper, either because there is extra information not captured by the score, or the applicants try harder, because they have better knowledge of their own circumstances. Lenders can 'cherry pick', such that they get the best of the lot from the reject region.

Thus, some means of making educated guesses is required; of using available information to infer reject performance, for use in the final model. It is the next step after segmentation, and is applied to each of the resulting segments. This was one of the biggest puzzles in application scoring in the 1960s. Early developments were limited to known performance, but there was a perception that they were deficient in the cut-off region, where they were needed most. The best approach was random supplementation in that region, and it took several years to come up with some rudimentary solutions (Lewis 1992).

Today, several techniques exist, but their benefits are contentious, and the mechanics are not always clear. The approaches are sometimes embedded within generic scorecard development packages, whose inner workings are secrets guarded with the passion of fourteenth-century cartographers, who protected their bits of knowledge of world geography. In other instances, scorecard developers are quite open about their approaches, but honest regarding their results' reliability. This section covers the topic under the headings of:

Why reject inference?—A look at the logic behind reject inference, differences in accept/reject performance, intermediate model types, and potential gains, or lack thereof.

Population flows—A tool for illustrating the starting and ending positions, and changes brought about by reject inference.

Performance manipulation—Means used to manipulate performance at record level, such as: (i) *reweighting*, of known or cohort performance; (ii) *reclassification*, using either a rule- or score-based approach; and (iii) *parcelling*, done either on a random or split basis.

Special categories—Categories that require special treatment, such as: (i) policy rejects; (ii) not-taken-ups NTUs and indeterminates; and (iii) limit increases.

Inference techniques—Approaches used in reject inference, including: (i) random supplementation; (ii) augmentation; (iii) extrapolation; (iv) cohort performance; and (v) bivariate two-stage approaches.

19.1 Why reject inference?

Application scoring relies on historical data, but outcome performance is missing for many applicants, either because: (i) the application was declined (rejects); or (ii) the applicant did not open, or use, the account as envisaged (NTUs). Rejects should perform worse than average, but there is no way of definitively saying what would have happened to each, had they been accepted. Just as some of the accepted applicants have defaulted (false negatives), so too, would many of the rejects have paid as agreed, had they been accepted (false positives).

The result of this 'partial observability'—meaning, the resulting inability to observe outcome performance for a significant portion of the population (Poirier 1980)—is two-fold: (i) the likely possibility of selection bias; and (ii) a reduction in the number of cases available for analysis. Both can affect the coefficients derived by any modelling processes; the former because of subtle differences between the known- and no-performance groups, and the latter because of small numbers. Most commentators focus on the former, and the potential bias that can result.

Differences in known/inferred performance

The need for reject inference varies, depending upon the current process, data, reject rate, and diversity of the group being assessed. It is greatest where: (i) it is a risk-heterogeneous population; (ii) reject rates are high; (iii) the current process is effective; and/or (iv) there are significant differences between the data used in past decisions and what is available for the model. A statistic oft used to assess whether the rejects' inferred risk is appropriate is the known-to-inferred odds ratio, as illustrated in Equation 19.1.

Equation 19.1. Known-to-inferred odds ratio $\text{KI} = \dfrac{(G_K/B_K)}{(G_I/B_I)}$

where G_K, B_K, G_I, and B_I are the total counts of known goods, known bads, inferred goods, and inferred bads respectively. The higher the value, the greater the risk associated with the inferred group. Both academics and practitioners agree that the appropriate ratio lies between two and four, albeit two tends to be very optimistic, and practitioners would rather err towards the latter. If the current process (a combination of scores, policies, and human judgment) is weak, then the value provided by reject inference will be limited, and a value of two may be appropriate. If, however, the current process is effective, then the need for reject inference in any new development is greater, and the appropriate ratio may be four, or even more.

Intermediate model types

Scorecard developments for selection (application) processes have extra dimensions that do not exist for non-selection processes. Other models may be required, including: (i) a *known good/bad model*, limited to known performance only; (ii) an *accept/reject model*, to provide an accept probability estimate; and (iii) an *all good/bad model*, which uses both known and inferred performance. The first two are intermediate stages, while the latter provides the final model. The accept/reject model may be replaced by an old or bureau score, where it is available and known to be effective.

For the intermediate models, there are fewer modelling restrictions, largely because they will not be used in day-to-day decision-making that affects customers. As a result, scorecard developers may: (i) use other statistical techniques; (ii) use future account statuses and characteristics that are otherwise unavailable or illegal; and/or (iii) ignore the cosmetic appearance of the scorecards.

Potential gains

There is very little theoretical support for reject inference. Most of the literature focuses upon presenting reject inference techniques, and very few attempts have been made to quantify the benefits. Part of the difficulty arises because to test their hypotheses, researchers need datasets with few or no rejects, so that ersatz rejects can be created using different selection strategies. Even then, numbers-driven analyses will never be able to simulate the human elements, especially overrides. From the little bit of research available, there seems to be general agreement, that the potential gains from reject inference are modest. According to Crook and Banasik (2002),

> Useful implementation of reject inference seems to depend on accurate estimation of the potential good-bad ratio for the population of all applicants. Simple application of that ratio then seems indicated. More elaborate tweaking of a vast set of coefficients does not seem to promise much potential benefit.

While that makes logical sense, even coming up with the all-applicant bad rate may be tricky.

Most quantitative studies have focused upon measures like the Gini coefficient, AUROC, and other measures of fit, that may understate the potential benefits. The real issue is the size of the swap set, and potential reduction in misclassification costs. Unfortunately however, these costs are difficult to quantify, but such an analysis should show that the benefits are greater where misclassification costs are high. The little available research is discouraging; Verstraeten and van den Poel (2004) had the luxury of profitability figures for a Belgian mail order company, and could only find a very modest gain of 1 per cent from perfect reject inference. Mail order is, however, a marketing channel where misclassification costs are low.

It should then come as no surprise that reject inference is regarded with a significant dose of suspicion. Most see it less as science than art of dubious value, but have more confidence where random supplementation and/or cohort performance are incorporated (see Sections 19.5.1 and 19.5.4 respectively). For the rest, the benefits may be meagre, but it is a brave scorecard developer that will fore go it, for fear of missing something. Indeed, there are instances where reject inference is absolutely necessary, and the pitfalls may be overlooked without careful scrutiny.

19.2 Population flows

A tool used to illustrate the impact of reject inference is the population flow diagram, which is a form of classification tree, used to show the distribution of cases across different performance categories and changes as reject inference is applied. A simple example is provided in

Figure 19.1, where the starting point is the pre-inference distribution, including accepted goods and bads. There are certain areas though, where the flows may become a little bit more complex. For example, what about the indeterminates? And what about the NTUs that were shopping for credit, but either got a better offer elsewhere, the need went away, or they opted on the side of prudence and forewent the purchase? It is also possible to take the extra effort, and accommodate these in the reject inference process.

Table 19.1 shows the before (known) and after (all) scenarios, and the shifts between the categories (inferred). The greater portion—21,757—of the rejects have now been assigned a performance status: good, bad, indeterminate or NTU. The only accounts that would still be treated fully as rejects, are those deemed as policy rejects, as described in Section 16.2.3.

Such changes affect the relationships between the characteristics, and the good/bad odds. Indeed, it may be wise to revisit the coarse classing, albeit with some caution, before proceeding further with the all good/bad statuses. Tables 19.2 and 19.3 show the effect of these changes on the characteristic analysis (CA) report, for the worst-arrears status on bureau (the numbers refer to 'months in arrears').

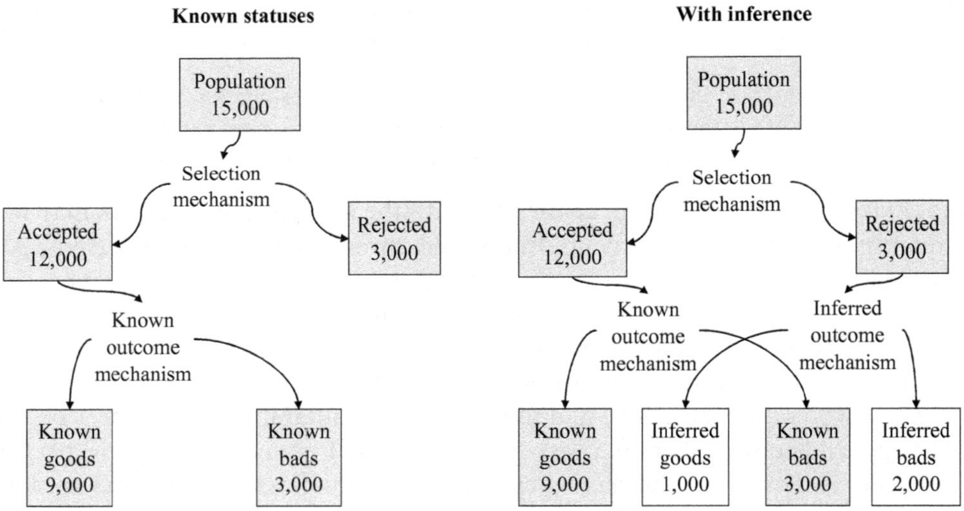

Figure 19.1. Population flows.

Table 19.1. Reject inference

	Accept/known	Reject/inferred	Total/all
Good	49,961	13,818	63,780
Bad	2,948	2,643	5,592
Indet.	5,372	3,319	8,691
NTU	2,169	1,977	4,145
Reject	24,818	−21,757	3,061
Total	85,268	0	85,268

Table 19.2. Reject inference—characteristic analysis

Pre-inference

Attribute	WoE	Total	Odds	Good	Bad	Indet.	Reject	Rej (%)
0–1	0.39	27,878	25.1	21,374	852	1,552	4,101	14.7
2–3	0.04	21,502	17.6	14,716	834	1,483	4,470	20.8
4–5	−0.22	10,353	13.6	5,434	400	821	3,699	35.7
6–8	−0.42	11,968	11.1	4,539	409	723	6,296	52.6
9+, Legal	−0.86	7,028	7.2	1,674	233	396	4,725	67.2
Missing	−0.35	6,538	11.9	4,052	340	619	1,527	23.4
Total		85,268	16.9	51,820	3,058	571	24,818	29.1
		Information value × 100			12.7			

Post-inference

Attribute	WoE	Total	Odds	Good	Bad	Indet.	Reject	Rej (%)
0–1	0.59	27,878	20.6	24,338	1,181	2,037	323	1.2
2–3	0.25	21,502	14.6	17,904	1,227	2,029	342	1.6
4–5	−0.15	10,353	9.9	7,922	803	1,344	285	2.8
6–8	−0.58	11,968	6.4	8,166	1,273	1,887	642	5.4
9+, Legal	−1.09	7,028	3.8	3,687	959	1,115	1,267	18.0
Missing	−0.18	6,538	9.5	4,981	525	830	202	3.1
Total		85,268	11.4	67,166	5,888	9,152	3,061	3.6
		Information value × 100			28.7			

Table 19.3. Known versus inferred

Attribute	No. of Accounts			G/B odds		
	Known	No Perf	All	Known	Inferred	All
0–1	32.7%	17.4%	32.7%	25.1	9.0	20.6
2–3	25.2%	19.0%	25.2%	17.6	8.1	14.6
4–5	12.1%	15.7%	12.1%	13.6	6.2	9.9
6–8	14.0%	26.0%	14.0%	11.1	4.2	6.4
Legal	8.2%	15.9%	8.2%	7.2	2.8	3.8
Missing	7.7%	6.1%	7.7%	11.9	5.0	9.5
Total/Overall	54,878	21,757	82,207	16.9	5.2	11.4
	Information value × 100			12.7	16.9	28.7

Besides the stark reduction in the rejects, it should be noted that the information value has increased substantially from 0.127 to 0.287. Further delving shows that the value for the inferred rejects alone is 0.169 (see Table 19.3), which implies that this characteristic is playing a major role in the reject region.

Please note, that the relationship between the information values for the known and inferred groups will vary from one characteristic to another, and will be reversed for characteristics that play less of a role in the reject decision. Care must also be taken because reject inference only provides an estimation of account performance, had it been accepted. A bit of conservatism is always wise, and the swap set between accepts and rejects from the old and new processes should be kept within reasonable bounds. It can be dangerous to change the entire profile of accepted applicants overnight.

19.3 Performance manipulation tools

Of all the sections in this textbook, this section on reject inference was one of the most difficult to write. Problems arose because of: (i) a lack of literature on the topic; (ii) a lack of consensus on the terminology used; and (iii) confusion between the performance manipulation and reject inference techniques. This section focuses first on some of the performance manipulation tools, before moving on to special categories and then, reject inference techniques. There are three basic types of performance manipulation: *reweighting*, *reclassification*, and *parcelling*. The choice will depend upon the circumstances, and/or the preferences of the scorecard developer.

Reweighting

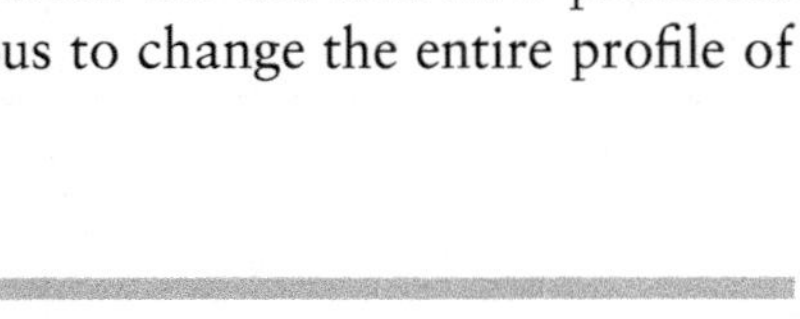

Reweighting has already been mentioned in Section 15.3 (Observation and Outcome Windows), where it was used to control for the time effect. In reject inference, it is primarily associated with augmentation, where accepts' weights are increased, so that they can represent both accepts and rejects. For example, if there are 400 cases in a group, and 500 are required, then the statistical process is fooled by weighting each case up by a factor of 1.25 (see Equation 19.4, p. 411).

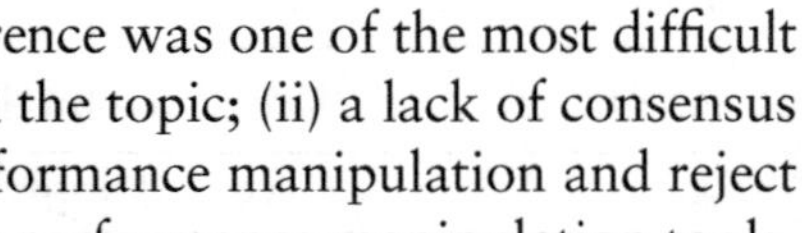

There are also instances where there is known or cohort performance for rejects, and their weights can be changed to provide a desired good/bad odds ratio, using Equation 19.2.

Equation 19.2. G/B reweighting $$W_i' = W_i \times (R + 1) \times \begin{cases} (1-1/(R'+1))/R \mid P_i = 0 \\ (1/(R'+1)) \mid P_i = 1 \end{cases}$$

where W and W' are the current and modified weights, R and R' are the current and required odds ratios, P is the good/bad status (0 = Good, 1 = Bad), and i is the index for each record. For example, to change the good/bad odds ratio from three to four, multiply the goods' and bads' current weights by factors of 16/15ths and 4/5ths, respectively. Note that this approach assumes that the current good/bad odds ratio is reliable, which may preclude its use where numbers are low.

Reclassification

In some instances, scorecard developers want to severely prejudice certain rejects. This is called reclassification, which assigns selected cases to bad only (a variation is 'iterative reclassification', see p. 413). It is done by either using a rule-based approach to simulate policy reject rules, or, alternatively, by using a score:

Rule-based—Cases with serious derogatory information, like two or more judgments on bureau, are labelled as bad. It is sometimes done prior to augmentation, parcelling, or score-based reclassification.

Score-based—Rejects are ranked using an accept/reject, known good/bad, or bureau score, and the worst-scoring cases are assigned to bad.

In each of these cases, it is assumed that all identified cases would have performed as inferred, had they been accepted, and taken up the product. This is an extreme assumption and may result in significant model bias. As a result, reclassification tends to be used only for very high-risk rejects, with other approaches, if any, used for the rest.

Parcelling

Another approach is similar to reclassification, and may be confused with it. The difference is that parcelling assigns cases to both good and bad, and possibly other categories. It is typically associated with extrapolation approaches (see Section 19.5.3), and there are three main options:

Polarised—Use a score to rank the rejects, and then assign those above and below a given cut-off probability to good and bad respectively (see Crook and Banasik 2002).

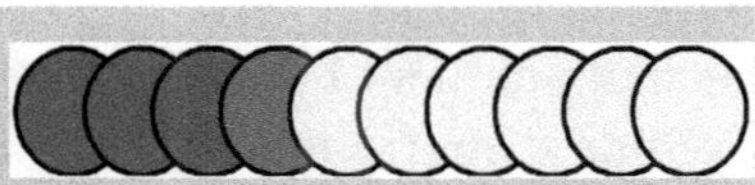

Random—The assignment is done at random, either for the entire reject region, or on a stratified basis. For the latter, it may suffer because of the vagaries of the random selection process.

Fuzzy—Also called duplication, or partial reclassification. Rejects are replicated, and their statuses and weights are adjusted, to apportion them between performance categories. It may also be done for the entire reject region, or on a stratified basis. Its advantage is that the required odds ratios can be achieved exactly, which is especially appropriate where there are specific estimates associated with each score.

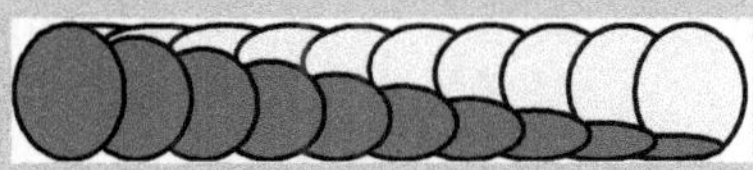

For fuzzy parcelling, a simple example may assist. In order to assign *P*(Good) values to rejects: (i) create two copies of each reject; (ii) reassign one as good, and the other as bad; (iii) modify the weights by *P*(Good) and 1-*P*(Good) respectively; and (iv) include both with known goods and bads for any subsequent modelling. If the original reject has a weight of 5, and a *P*(Good) of 40 per cent is required, then good and bad records are created with weights of 2 and 3 respectively. The total number of accounts being represented is unchanged.

Fuzzy parcelling can also be used to audit whether characteristics are properly represented within the final, or an existing scorecard: (i) calibrate the scores onto *P*(Good) values; (ii) replicate the records, and set 'expected' *P*(Goods) for each; and (iii) compare expected and actual good/bad distributions using a chi-square test. For scorecards already implemented, this can only be done for cases with known performance, and it may be wise to exclude overrides.

19.4 Special categories

Thus far, the focus has been on goods and bads, but there are certain categories that require special treatment: (i) policy rejects; (ii) NTUs, inactives, and indeterminates; and (iii) limit increase applications.

Inferred policy rejects

Amongst the rejects, a special category is inferred policy rejects (IPRs). These are attributes, or combinations thereof, where all or most applicants were rejected, and the bad rate for accepts was much higher than average. The most logical example is two or more legal judgments on bureau, but it applies to any instance where lenders are loath to rely upon the performance of a small number of dubious accepts. Scorecard developers can either reclassify rejected IPRs, or exclude them from any future analysis. Accepted IPRs may still be treated as accepts.

Reclassification will severely prejudice IPRs, effectively causing the policy rule to become embedded within the scorecard. It may also distort the coefficients given to other characteristics and attributes. In contrast, with exclusion, it should still be possible to put some faith in the coefficients, but the IPR rule must be formalised. The concerns are then: (i) whether the lender is still as good at cherry picking as in the past; and (ii) potential instability in the estimates, because of the small numbers.

NTUs and indeterminates

Most good/bad definitions have more outcome-performance statuses than the common garden-variety goods and bads. The most obvious, are indeterminates that lie in between, but there are also

NTUs, being invited customers that do not come to the party. Either the account is: (i) never opened; (ii) opened but never used; or (iii) used briefly before going dormant.

Are these categories considered when doing reject inference? Very little mention is made of their treatment in the literature, but some scorecard developers do address them as part of the process. The necessity depends upon the size of the NTU and indeterminate groups; the larger their size, the greater the need for special treatment. Care should be taken not to invest too much time and effort in this—there may be benefits, but they are not huge.

Each of the categories is treated one at a time. First, rejects are parcelled into NTU and not NTU based upon either a bespoke NTU score, or one of the scores already available. This makes sense, given that true outcome performance can only be obtained for those taken up. Second, the 'not NTU' group is parcelled between good, bad, and indeterminate, based upon either a known good/bad score or old score. It can be achieved directly using three-way fuzzy parcelling, but the scorecard developer has greater control—especially if the good/bad odds are being manipulated—if it is done in two stages: (i) parcel into indeterminate and not indeterminate; and then (ii) parcel the not indeterminates into good and bad.

Limit increases

There are a lot of instances where the application processing system is used for both new business and limit increases. The latter differ in that rejects have known performance. In an ideal world, separate scorecards should be created for each, but it may be infeasible due to insufficient numbers. Even so, reject inference can still be done separately for the two groups. Had the limit increases been granted, the rejects would have had higher bad rates (theoretically), which can be reflected by reweighting the known performance.

19.5 Reject inference methodologies

In Section 7.4.1, Little and Rubin's (1987) missing data framework was covered. There were three possible scenarios: missing completely at random (MCAR), missing at random (MAR), and missing not at random (MNAR). For MCAR and MAR, the performance data's missingness is said to be 'ignorable', and a predictive model can be developed using accepts' data only. With MCAR, missingness is totally random, but it replaces a reject inference problem with a credit quality problem (see Section 19.5.1). In contrast, with MAR, missingness is not random, but accepts' performance can be used to represent the full population (see Section 19.5.2). Finally, with MNAR, selection is also not random, but the performance is 'non-ignorably missing'. Acceptance is somehow correlated with outcome performance, and there may be problems deriving a model that is applicable to both accepts and rejects (see Sections 19.5.3 onwards). It applies if: (i) data used in the original decision is no longer available; (ii) underwriters use external data to do system overrides, or there were other influences related to the risk; and (iii) the data required for rejects does not exist, or is poorly represented amongst the accepts. The latter two points are exemplified by those customer-motivated contests that result in reject overrides, whose better performance cannot be considered typical of equivalent rejects.

There are five basic reject-inference techniques, which can both affect and address missingness, and which are not mutually exclusive:

(i) **Random supplementation**—Accept random rejects, to get extra performance.
(ii) **Augmentation**—Increase the accept weights, for them to represent both accepts and rejects.
(iii) **Extrapolation**—Use known good/bad scores, to infer individual rejects' performance.
(iv) **Cohort performance**—Use of data from other products or lenders.
(v) **Bivariate**—Combines both known good/bad, and accept/reject scores.

For each, there is a fair amount of flexibility in how the technique can be applied. Other techniques have been suggested, but it is impossible to say how extensively they are being used in practice. Siddiqi (2006) mentions cluster analysis and machine-learning techniques, which reassign rejects based upon similarities to known goods and bads, but no mention has been made of their use elsewhere.

19.5.1 Random supplementation

Where selection is totally at random, with no reference to available data or past experience, the performance is 'MCAR'. There are only rare instances where this occurs naturally, but it can be manufactured using random supplementation to 'buy data'. Some high-risk cases, which would otherwise be rejected, are accepted. If taken up, their performance is then known, not inferred, and can be used directly in a known good/bad model.

According to Hand and Henley (1993/4) this is the ideal; the lack of selection bias ensures a more robust scorecard. It comes at a price though; the greater the risk, the higher the cost. The resulting credit quality problem can only be commercially defensible if the benefits—either information or profit—offset the costs incurred. As a result, lenders try to minimise the time period over which it is used, and/or put greater focus on the region immediately below the cut-off. There are two main scenarios:

(1) **New players**—New market entrants that accept all comers, barring obviously problematic cases identified using policy reject rules. As performance data becomes available, policy rules are updated to exclude the worst cases, and scorecards are developed once sufficient data has accumulated.
(2) **Established players**—Lenders accept some cases that would otherwise be rejected, focusing primarily on the region immediately below cut-off, and less in the higher-risk regions.

In both cases, the amount of time required before a scorecard can be built will depend upon accept volumes and bad rates. Rather than waiting, new entrants may opt to develop preliminary scorecards using a more relaxed definition, like any missed payment, to provide a temporary solution until a proper risk scorecard can be built.

Retailers and mail-order houses can use profits from other areas to offset the credit losses, and are in a better position to use this random supplementation approach. In contrast, pure credit providers, such as banks and finance houses, usually find it infeasible. Even so, it was done by some credit card companies during the 1990s, when entering new markets.

19.5.2 Augmentation

Another approach is augmentation, often called *reweighting*, because it relies upon changes to the weights of cases with known performance. First proposed by Hsia (1978), it can be used to address the 'MAR' case, where acceptance is dependent only upon the given data, and there are no extraneous factors. It assumes, as shown in Equation 19.3, that the probability of default Y given a dataset X is independent of whether the observation is an accept A or reject R.

Equation 19.3. Augmentation assumption $P(Y|X) = P(Y|X, A) = P(Y|X, R)$

The process is as follows . . . After deciding how to treat policy rejects, choose a score to assess the accept probabilities. It could be a known good/bad score, bureau score, or old score, but most of the literature leans towards a bespoke accept/reject score. Use it to determine score ranges, and acceptance rates in each. Finally, adjust accepts' weights to represent the entire population, using Equation 19.4.

Equation 19.4. Augmentation reweighting $W_i' = W_i \times \frac{A_j + R_j}{A_j}$, where $S_i \in L_j \ldots U_j$

For each record i, S is the score, and W and W' are the original and modified weights; and for each score range j, L and U are the lower and upper bounds, and A and R are the total accepts and rejects.

Thus, the weights of known performance cases are increased, so that they can represent both accepts and rejects within their score ranges. For example, if the score indicates a reject probability of 80 per cent, then the accepted record would be weighted up five times. A more realistic example of this calculation is provided by Weldon (1999), shown in Table 19.4.

The major advantage of augmentation is its simplicity—there is no reassignment. It does have several disadvantages though, which Ash and Meester (2002) enumerate as: (i) it assumes that the performance of the rejects can be directly imputed from that of the accepts; (ii) it may rely on the performance of a relatively small number of accepts to represent a large number of rejects (in the example, there were only 16 accepts in the lowest score range); and (iii) it assumes that at least some of the accepts have bureau profiles the same as, or similar to, rejects. In general, most commentators indicate that augmentation does nothing to address selection bias, but is otherwise harmless, and will at least allow the modeller to estimate reasonable probabilities for the full population.

Table 19.4. Augmentation

Score	Known				Accept weight	Total
	Accept	Reject	Acc (%)	Rej (%)		
Low–199	16	214	7	93	14.375	230
200–299	81	262	24	76	4.235	343
300–399	632	665	49	51	2.052	1,297
400–499	1,556	933	63	37	1.600	2,489
500–High	1,992	285	87	13	1.143	2,277
Totals	4,277	2,359	64	36	1.552	6,636

19.5.3 Extrapolation

Another reject inference approach is extrapolation, which relies upon known performance to infer what might otherwise have transpired to rejects. It is often called parcelling, because that is the main performance manipulation technique used. Unlike random supplementation and augmentation, it can accommodate different known and no performance distributions (allowing the creation of *mixture models*). According to Hand (1998), it assumes that the same set of information will be available for both old and new scorecards.

Like augmentation, the first step is to determine how inferred policy rejects will be treated. Thereafter, an appropriate ranking score is either chosen, or derived. The required bad rate(s) are then determined, or decided upon, before any one of polarised, random, or fuzzy parcelling is applied (Crook and Banasik (2002) used polarised parcelling based on a known good/bad model, in a comparison of extrapolation and augmentation). For stratified approaches, the values are based on historical data, which may then be adjusted upwards to achieve some desired known-to-inferred odds ratio. In the absence of any adjustments, stratified approaches should provide results similar to augmentation. Of the different approaches, stratified-fuzzy parcelling is the most complicated, but provides the greatest flexibility.

In instances where there is a target known-to-inferred odds ratio, the required number of inferred bads may have to be calculated, using Equation 19.5.

Equation 19.5. Required bads $$B_I = N_R \times \frac{1}{1 + \frac{(G_K/B_K)}{KI}}$$

where B_I is the number of inferred bads required, N_R is the total number of rejects, G_K and B_K are number of known goods and bads respectively, and KI is the required known-to-inferred odds ratio. If polarised parcelling were used for the example shown in Table 19.5, and a KI ratio of 4.0 were required, the worst scoring 333 of the 2,359 rejects would be assigned to bad.

Table 19.5. Extrapolation

Score	Known					Inferred	
	Bad	Good	Reject	Rej (%)	*P*(Good) (%)	Bad	Good
Low–199	16	0	214	93	0.0	214	0
200–299	13	68	262	76	84.0	42	220
300–399	68	564	665	51	89.2	72	593
400–499	44	1,512	933	37	97.2	26	907
500–High	28	1,964	285	13	98.6	4	281
Totals	169	4,108	2,359	36	96.0	358	2,001
G/B odds		24.3					5.6
			Known/inferred ratio				4.3

Source: Weldon (1999)

Weldon (1999) used stratified-random parcelling for his illustration, shown in Table 19.5. The KI odds ratio is already 4.3, which implies an extremely strong model that is being strictly adhered to. In this case, it is questionable whether any further adjustment is necessary, but if the value were too low, the rejects could be further prejudiced by increasing the inferred bad rate for each band.

Iterative reclassification

According to Verstraeten and van den Poel (2004) one of the most renowned reject inference techniques is iterative reclassification, which was first proposed by Joanes (1993/4). This procedure involves the following steps: (i) develop a known good/bad model; (ii) use it to reclassify the rejects using extrapolation; (iii) develop an all good/bad model; (iv) determine a cut-off that provides the same reject rate; (v) repeat the process, assuming that the above and below cut-off cases are known and no performance respectively; and (vi) continue repeating the process until convergence is achieved, where convergence means that the model is predicting the same in both the accept and reject regions. No illustrations of this approach could be found.

19.5.4 Cohort performance

Performance data is not limited to accepts' outcome performance for the product in question; it can be broadened to include performance elsewhere. Applicants often have other credit, and credit-related accounts, either with the same lender, or the one across the street. This 'cohort performance' may be provided as a score, delinquency status, or good/bad flag. After random supplementation, it is the most powerful data that can be used for reject inference. It comes at a cost though, and there may be time, data quality, and match rate issues. There are two possible ways of using it: (i) as the *sole basis for reclassification*; or (ii) as a predictor in a *super known good/bad model.*

Straight reclassification

The easiest way of incorporating cohort performance is to use it as the sole basis for reclassifying rejects. It has several potential pitfalls though. First, the lender must have *access to the performance*. If available in-house this is not an issue, but if credit bureau data is used, there will not only be costs, but also potential delays, data quality issues, and limitations resulting from reciprocity agreements, or data privacy legislation. The bureaux are usually very accommodating, as the data will be used for a model build only, and will not prejudice those consumers. As always, care must be taken to ensure that no footprints are left.

Second, there is a *limited universe of potential cohorts*. Loan performance is required for accounts that are similar, in terms of both opening date and type, which significantly narrows the field. If the match rate is too low, then the field can be broadened on both dimensions. For cases where no suitable performance can be found, some other technique must be used. Third, it *relies upon a loan having been obtained elsewhere*. The applicant may have been so bad that nobody would touch him/her, or the loan's terms so usurious that it is not directly comparable.

Many applicants have credit histories so bad that getting approved is either nearly impossible or extremely expensive. Lenders charge applicants based on risk, so the biggest credit risks will pay the highest rates, have the most collateral requirements, and be the first borrowers the collection agencies go after in the event of default. Any credit behaviour they have may be skewed because of these stringent requirements.

Gregg Weldon

Alternatively, the applicant may not have applied elsewhere, either because the need went away, or the hassle factor became too great. Rejection can be a wake-up call for applicants, to indicate how stretched they have become, and force them to reassess their financial situation. For this pocket of rejects, cohort performance may exhibit lower risk, because the applicant will not have incurred the extra financial obligation.

Fourth and finally, the cohort should have a *comparable outcome definition*. This should be simple where raw performance characteristics are provided, but may become complicated if there are only good/bad statuses or scores.

Super known good/bad models

Cohort performance data can also be combined with observation data to create super known good/bad models (SKGB), so called because of their seemingly extremely high predictive power. In this instance, the time constraint is relaxed and any performance may be used. Use of future data in a predictive model might seem to present a problem, but here it is used solely as a reject inference tool, and nothing more—normal rules do not apply, because the model will never be implemented. Any and all data, both at observation and outcome, can be used as long as it: (i) is correlated with known performance on the product in question, (ii) is available for both accepts and rejects, and (iii) comes at a reasonable price.

For SKGB models, cohort performance can be provided in any form, but is best presented as a performance indicator, or score. It is then used in combination with one or more of the other techniques described in this section. Where customer or bureau scores are being used, please ensure that the performance for the product being assessed is excluded, otherwise a

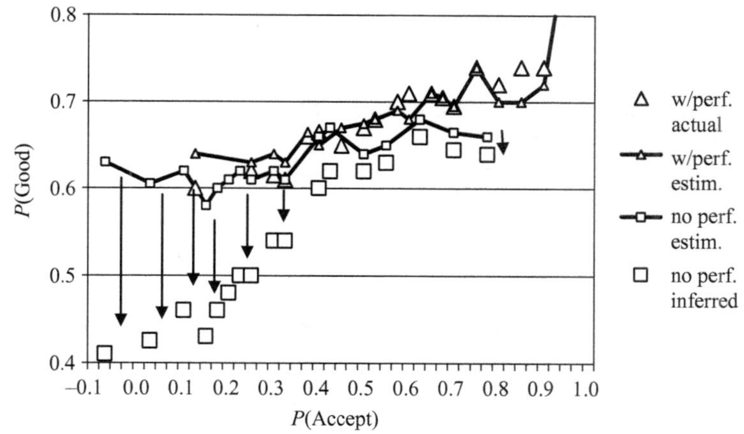

Figure 19.2. Bivariate visualisation.

severe case of double counting and confusion will result. If this capability does not exist, the use of an SKGB model may be precluded, as lenders and credit bureaux seldom foresee the need for such precision excisions.

19.5.5 Bivariate inference

Thus far, it has been shown how an accept/reject score and known/good bad score can be used in augmentation and extrapolation respectively. The question now is, 'Can they be used at the same time?' Given that the process being modelled consists of both selection and outcome performance, it would make sense to acknowledge both dimensions—first accept/reject, and then good/bad. This is the basis for Heckman's (1979) correction, as well as Ash and Meester's (2002) extrapolation approach.

This section focuses on the latter, as it is currently being widely applied in practice.[2] It differs from the other approaches in that it relies upon the use of a data-visualisation tool, such as that shown in Figure 19.2, which is a rough copy of Ash and Meester's illustration. The *actual* values are determined by a cross-tabulation of the score against known performance, whereas the *estimates* are imputed from the known good/bad scores, whether applied to accepts or rejects. There are a number of steps required:

(i) Develop the (super) known good/bad and accept/reject models.
(ii) Apply the known good/bad model to the full population.
(iii) Create a graph, with *P*(Accept) and *P*(Good) on the *X*- and *Y*-axes respectively.
(iv) Use the accept score as *P*(Accept).

[2] It is believed that this two-stage technique was already being used by MDS when it was purchased by CCN (today's Experian) in 1982.

(v) Determine *P*(Accept) and *P*(Good) bands, that capture sufficient cases from which to draw conclusions.
(vi) Plot the average bad rate estimates, for both the known and no performance groups.
(vii) Plot the known bad rates.
(viii) Manually set adjustments to provide higher inferred bad rates; the lower the accept/reject score, the higher the adjustment.

Several patterns can be seen in the illustration (Figure 19.3). First, the *P*(Good) estimates for no performance cases are lower for all possible *P*(Accept) values, indicating that performance is 'missing NOT at random'. Second, for known performance, the actual *P*(Good) values are consistent with the estimates, except for lower *P*(Accept) values, where there is higher risk. Third, the *P*(Good) estimates flatten, and then ramp up for lower *P*(Accept) values, which is probably the result of cherry picking. And finally, the *P*(Good) values for the no performance cases—which will be used to do stratified-fuzzy parcelling—are set at values lower than the estimates, with the prejudice increasing as *P*(Accept) decreases. This will not only reduce the size of the swap set, but also ensure that the relative change is least for those cases that were most likely to be rejected in the past.

This bivariate approach may be a better and more sophisticated technique, but is unlikely to provide a huge amount of benefit over the others. It does, however, provide greater control over the reject inference process. Ash and Meester (2002) also suggest that a SKGB model be used, to incorporate cohort performance into the process.

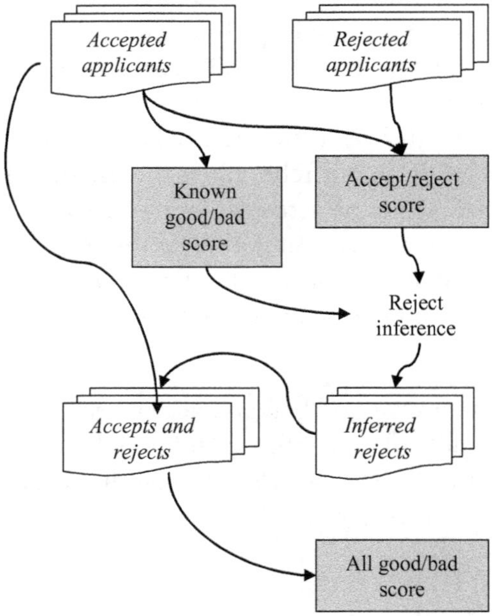

Figure 19.3. Bivariate inference process.

19.6 Summary

When developing performance-prediction models for selection processes, there is a predicament. Historical performance is needed to develop a risk-ranking model, but there is no performance for rejects! Use of known performance only may result in selection bias, especially where there is extensive cherry picking in the reject region. As a result, some educated guesses are required. Several reject inference techniques are available, but many academics and practitioners view it less as science than art of dubious value, and most research indicates that the potential benefits are limited. The best possible approach is to incorporate actual performance, whether for high-risk cases selected at random (expensive!), or cohort performance from elsewhere.

Reject inference is usually discussed using Little and Rubin's (1987) 'missing data' framework: MCAR, where the reject inference problem becomes a credit quality problem; MAR, where it can be addressed through a simple adjustment, using accept data only; and MNAR, where more sophisticated techniques are required. In this chapter, the topic was treated under three headings: (i) the performance manipulation techniques that are used; (ii) categories that require special treatment; and (iii) the reject inference techniques themselves.

Performance manipulation refers to modifying individual records to achieve some desired effect. It may be done using either reweighting, reclassification, or parcelling. *Reweighting* is usually associated with augmentation, but can also be used to adjust the good/bad odds for known or cohort performance. *Reclassification* assigns some cases to bad, and is either: (i) *rule-based*, to simulate policy reject rules; or (ii) *score-based*, where low scoring cases are assigned to bad. And finally, *parcelling* assigns cases to both good and bad and may be: (i) *polarised*, where cases above and below some cut-off are assigned to good and bad respectively; (ii) *random*, where a random number generator is employed; or (iii) *fuzzy*, where the rejects are replicated, and the weight apportioned between categories, to provide a desired odds ratio for each. Random and fuzzy parcelling are usually done on a stratified basis, based on known bad rates.

While most of the focus is on inferring good and bad performance, certain other categories require special consideration. *Policy rejects* may either be reclassified as bad, or alternatively be excluded from the analysis. NTUs and *indeterminates* can be inferred separately, before addressing inference of goods and bads. And finally, for *limit increases* treated as part of a new business process, there is known performance that can be adjusted using reweighting.

Several reject inference techniques are available. *Random supplementation* can be used to buy data and manufacture a missing completely at random scenario. *Augmentation* requires the calculation of reject rates for different accept/reject score ranges, and accepts are reweighted upwards to represent both goods and bads—albeit making the unlikely assumption that accepts and rejects at each score will perform the same. *Extrapolation* requires parcelling between good and bad, using some risk score (known good/bad, bureau, or old). *Cohort performance* may be used as part of the process, either by incorporating it directly, or alternatively by including it in a SKGB model. And finally, there are more sophisticated *bivariate approaches* that use both known good/bad and accept/reject scores as part of the process. While all of these may provide some value, there is nothing that can beat having full observability for the entire population. Failing that, random supplementation and/or cohort performance can improve the results.

Intermission

At this point, a break is in order. The scorecard developer should have a dataset (inclusive of reject inference, if applicable), and can start training the final model. Strangely though, there is nothing that can be said specifically about training that has not already been covered elsewhere. Most of what is now required is running the dataset(s) through one of the predictive modelling techniques presented in Chapter 7 (Predictive Statistics 101), and transformation (Chapter 16) and characteristic selection (Chapter 17) may have to be revisited. Thereafter come the final stages, scorecard calibration and validation, covered next.

20 Scorecard calibration

A scorecard has now been developed for each of the identified segments, but the job is not yet done. Like any measuring instrument (e.g. a weigh scale), scores must be calibrated, to give readings in the appropriate units. This ensures consistency of meaning across scorecards, and provides an opportunity to determine probability estimates for use in finance and elsewhere. It can be done not only as part of the scorecard development, but also at various points thereafter (recalibration). Calibration is required whenever the scores cannot be directly associated with the required probability estimates, whether as the result of: (i) the statistical modelling technique used; (ii) the passing of time, or (iii) differences between the model's target variable and the metric to be estimated. The latter can arise from differences between the definitions (like good/bad versus default) or time frames (short- versus long-term).

The primary concern is consistency of meaning across scorecards; a score of 190 can mean a bad rate of 5 per cent in one instance, but 7 per cent in another, making it difficult to use it confidently to drive strategies. As stated by Falkenstein et al. (2000), 'A model may be powerful, but not calibrated, and vice versa'. Calibration can be achieved by:

Banding—Identify score ranges of common risk that will be used as the basis for risk groups, grades, buckets, pockets, indicators, or whatever label is used by the lender. The number of groups will usually be between 5 and 25, depending on the circumstances.

Scaling—Transform the scores onto a consistent scale, to be used across scorecards and over time. This has the advantage of allowing maximum granularity, such that the final scores act directly as probability estimates.

For both of these, the calibration can be done using the scorecard development's good/bad definition, or another definition required for some specific function. Indeed, just as temperatures can be measured in Celsius and Fahrenheit, so too can scores be mapped onto different scales. The most obvious example is Basel II, which has its own definition, but its 'use test' demands that estimates be based on models used to drive decisions within the business. Most banks have opted to use scorecards built using good/bad definitions to drive their decision-making, and calibrate their scores onto the Basel definition. Likewise, lenders may develop scorecards using relatively short observation windows to represent the current business, but calibrate using a one-year window, to provide estimates for use in financial calculations.

Care must be taken with probability estimates, as they may not live up to expectations, especially where the environment is volatile. They will, however, serve a purpose where there has been a structural shift and the situation has stabilised; or if the lender can make a blanket up or down adjustment, to provide a forward view. Also, there are problems with backtesting any calibration's accuracy. The chi-square, binomial, and Hosmer–Lemeshow statistics assume independence, whereas defaults are correlated over time, whether because of the economy, strategy, or other factors.

20.1 Score banding

A very straightforward way of addressing the scorecard alignment problem is to band the scores into risk grades. The most primitive approaches will either force some desired account distribution (dangerous!), or change in risk between classes. In general, most lenders opt for an exponential increase in risk from one grade to the next, which for retail portfolios typically results in distributions that are normally distributed, but skewed to the right (lower risk), possibly because of data limitations for low-risk accounts.

At one time, five risk bands were considered appropriate for a given portfolio, but there has been a trend towards increasing granularity. Lenders and regulators now have up to 30 grades, covering the entire risk spectrum (bond rating agencies have about 21 grades, including the default grades). Ultimately, the appropriate number will be a function of the portfolio's risk-heterogeneity, and the amount of information available to evaluate the risk. For portfolios that are homogenous, or where data is limited, the number of practical grades may still be limited to between 5 and 10.

In the following section, several techniques are presented, that can assist with determining the optimal number of grades, and/or the breakpoints for each:

(i) **Calinski–Harabasz statistic**—An algorithm usually associated with determining the optimal number of clusters, which can also be used to compare different grouping options.
(ii) **Benchmarking**—Map the scores onto a predefined set of grades and probabilities, which requires the identification of breakpoints that provide the best possible fit.
(iii) **Marginal risk boundaries**—Similar, except the upper and lower boundaries for each risk range are set, and not the average.

20.1.1 Calinski–Harabasz statistic

There are times when the lender has no idea of how many groups there should be. In this instance, a clustering technique provided by Calinski and Harabasz (1974) is generally accepted as the best possible means of determining the optimal number of groups. The goal is to define clusters that maximise within group similarities, and between group differences, using their variances, as in Equation 20.1.

Equation 20.1. CH-statistic $$\mathrm{CH}(g) = \frac{\mathrm{BSS}/(g-1)}{\mathrm{WSS}/(n-g)} = \frac{\sum_{k=1}^{g} n_k(p_k - p)^2 / (g-1)}{\sum_{k=1}^{g}\sum_{i=1}^{n_k} n_k(P_{i,k} - p_k)^2 / (n-g)}$$

where g is the total number of groups, n is the number of observations, p is an observed probability, P is a default 0/1 indicator for each record, i and k are indices for each record and group respectively, and BSS and WSS are the between and within group sums of squares respectively. The CH-statistic is calculated for groups numbering from say 1 to 25, and the optimal number of groups is that with the highest result (this is the traditional presentation, which returns very small numbers; it may make more sense to instead minimise its inverse).

How are the groups to be compared defined? The starting point is to rank all of the observations by risk, and partition the records into equally sized groups, or as close as possible. It may make more sense to do this at score level, in which case the within group sum of squares becomes:

$$\text{WSS} = \sum_{k=1}^{g} \sum_{s=b_{k-1}+1}^{b_k} n_k \left(p_{s,k}\, p_k^2 + (1-p)(1-p_k)^2\right)$$

where s is the score, b is the score break for each range, and $p_{s,k}$ is the observed probability for the score.

While almost all of the available literature refers to this formula as a means of determining the optimum number of groups, it could also be used to compare competing banding options. Testing all possible score breaks would be computationally onerous though, so some shortcuts are required. A possibility is to use a monotone adjacent pooling algorithm (see Section 16.4.3, MAPA) to provide an initial set of score breaks, and then adjust or delete the breaks in search of a set that maximises the CH-statistic.

20.1.2 Benchmarking

In many cases, lenders want to map scores onto a set of risk grades, each of which is associated with a given level of risk. The grades may be internal or external benchmarks, which are either targets, or historically observed values. Lenders' own internal benchmarks can be either, while those for external rating agency grades are usually published historical default rates, and those for regulatory classifications are predefined targets. A good example is where scores, used to rate companies, are mapped to an equivalent Moody, S&P, or Fitch letter grade, such as those

Table 20.1. Rating agency grade benchmarks

Investment grade	AAA	AA+	AA	AA−	A+	A	A−	BBB+	BBB	
Default (%)	0.01	0.03	0.07	0.10	0.14	0.20	0.28	0.44	0.66	
Odds	10,000	3,500	1,500	1,000	700	500	350	225	150	
Ln(Odds)	9.21	8.16	7.31	6.91	6.55	6.21	5.86	5.42	5.01	
Change	1.05	0.85	0.41	0.36	0.34	0.36	0.44	0.41	0.41	
Speculative grade	BBB−	BB+	BB	BB−	B+	B	B−	C+	C	C−
Default (%)	0.99	1.41	1.96	2.78	4.3	6.3	9.1	12.5	16.7	25.0
Odds	100	70	50.0	35.0	22.5	15.0	10.0	7.0	5.0	3.0
Ln(Odds)	4.61	4.25	3.91	3.56	3.11	2.71	2.30	1.95	1.61	1.10
Change	0.36	0.34	0.36	0.44	0.41	0.41	0.36	0.34	0.51	

shown in Table 20.1:

> Some lenders have made the mistake of trying to match the distributions, and not the default rates. Falkenstein et al. (2000) stress that models built on smaller and less reliable data are unlikely to be as predictive as rating agency grades, in which case it is not possible to achieve the same spread of cases. If the distribution is forced, then bias will result.

The default probabilities are idealised, but are close to what is found in practice. If the score ranks risk, then the score bands can be determined using an optimisation approach, determine the breakpoints that minimise the sum of the squared differences between the natural log odds of the benchmark and banded default rates. This can be expressed mathematically as:

Equation 20.2. Benchmark breakpoints $$\min_{s_1,\ldots,s_{k-1}} \sum_{k=1}^{g} n_k \left(\ln\left(\frac{1-p_k^b}{p_k^b}\right) - \ln\left(\frac{1-p_k}{p_k}\right)\right)^2$$

where p_k^b and p_k are the benchmark and calculated odds, s the score breaks between the groups, n_k the total cases in each group, g the number of groups, and k the index for each class. The process is as follows:

> (i) Determine the number of classes required.
> (ii) Determine the score breaks that provide groups of almost equal size.
> (iii) Calculate the sum of the squared differences between the benchmark and banded natural log odds for each.
> (iv) For each break, calculate the sum of squares for a shift up or down one point.
> (v) Choose the modification that provides the greatest reduction.
> (vi) Repeat from (iv), until no further reductions can be achieved.

> The benchmark breakpoints approach is based upon a methodology employed by Fernandes (2005:36), who used the default probabilities, and not the natural log odds. The latter is better suited if there is an exponential increase in risk between grades.

20.1.3 Marginal risk boundaries

Another case often encountered is where the grades are again prescribed, but upper and lower limits are set for each band. It is most common when targets are set either by the lender, or a regulatory agency. In this case, the marginal risk must be determined, meaning the change in risk (good/bad odds, bad rate, default rate) associated with a small change in score. These values are then used to map each score into the appropriate class. The threshold is the score S_k for the last record i, with a marginal risk r_c less than or equal to the limit R_k.

Equation 20.3. Threshold score $$S_k = \max\{s_i \mid r_i \le R_k\}, \quad i \in N$$
$$\text{assuming } r_i \le r_{i+1} \text{ for } i \text{ to } N-1$$

While the concept is a very easy one, the marginal risk calculation is more problematic, because the score-to-risk relationship is seldom, if ever, totally monotonic across the range of scores. Conceptually, the simplest way is a nearest-neighbours approach, where each record is treated with its neighbours. The formula in Equation 20.4 is a modified version of that provided by Banasik et al. (2001b).

Equation 20.4. Marginal bad rate $b_i' = B_i' / (u_i - l_i + 1)$ where $B_i' = \sum_{j=l_i}^{u_i} B_j$

$$l_i = \begin{cases} i-k & \text{if } i-k \geq 1 \\ 1 & \text{otherwise} \end{cases} \quad \text{and} \quad u_i = \begin{cases} i-k+n & \text{if } i-k \leq N \\ N & \text{otherwise} \end{cases}$$

where b' is the marginal bad rate, B' the marginal bad count, B a 0/1 bad status flag, i an index for the record being assessed, l and u indices for the upper and lower records, n the number of neighbours included, k the number of prior records included, and N the total number of cases in the dataset. Values for k and n may be adjusted according to preference and circumstances. They will, however, usually hold the relationship $k = \frac{n}{2}$, so that the values straddle the current record. When testing around an application scorecard cut-off, $k = 0$ or $k = n$ might be appropriate.

Although conceptually simple, this approach fails when being applied in practice. The number of neighbours required to enforce monotonicity may be large, and infeasible. It is possible to do some pre-processing using MAPA, but the problem then becomes one of dealing with large flat ranges with sudden moves up or down. Thus, other approaches are necessary. Glößner (2003) developed a means of fitting a Lorenz curve, but this involves extremely complex mathematics, and does not cater for all situations.

Record-level interpolation of natural log odds

Marginal risk can also be determined, by first determining a series of score pools that ensure monotonicity, and then interpolating values for each score in between. It is done not at score level, but at record level; it is not using the probabilities, but the equivalent natural log odds (NLO). The steps are:

Score breaks—Identify the record index, for the lower bound of each pool.
Midpoint indices—Calculate the average index value for each pool.
Midpoint NLO—Associate the log odds for the pool with each midpoint.
Increment NLO—Calculate the per record increment between midpoints.
Extrapolation—If midpoint log odds are unavailable, then use neighbouring increment.
Record NLO—Interpolate log odds at record level.
Convert—Calculate equivalent probabilities for each record.
Average—Calculate the average probability for each score.

The calculations used are as follows:

Midpoint index: $M_k = (L_k + L_{k+1})/2$

where M is an average index for all records in pool k, calculated using the lower indices L, for that pool and the next. This is the benchmark index for each pool.

Midpoint NLO: $m_k = \ln((1 - p_k)/p_k)$

where m is the NLO for the observed probability p of pool k, which applies only to M_k.

$$\textbf{Increment NLO:}\quad d_k = \begin{cases} d_{k+1} & \text{if } k=1 \text{ or } p_k=0 \\ (m_{k-1}-m_k)/(M_k-M_{k-1}) & \text{if } k>1 \text{ and } p_k \neq [0,1] \\ d_{k-1} & \text{if } p_k=1 \end{cases}$$

where d is the slope between the prior and current midpoints. If it cannot be calculated ($k = 1$, $p_k = 0$, or $p_k = 1$), then the neighbouring NLO increment is extrapolated through that space. This assumption may prove problematic, but the effects can be lessened if the pools are sufficiently large.

$$\textbf{Record NLO:}\quad \ln((1-\hat{p}_i)/\hat{p}_i) = \begin{cases} d_k \times i & \text{if } i \leq M_0 \\ d_k \times (i - M_{k-1}) & \text{if } M_0 < i \leq M_{\max k} \\ d_k \times (i - M_k) & \text{if } i > M_{\max k} \end{cases}$$

$$\text{Subject to: } i \in L_k, \cdots, L_{k+1}-1$$

where i is the record index, $\hat{p}_i$ is the probability estimate for that record, k is the index, of the pool of which that record is a member, and L is the lowest index for each pool. Each value is an interpolation between midpoints (or extrapolation, where one of the required midpoints does not exist).

MAPA-based marginal risk example

Table 20.2 provides an example of the above for a small selection of 20 records. A monotone-adjacent pooling algorithm (MAPA) is first used to determine the score breaks. There is a small error (13 − 12.96 = 0.04), which has been spread evenly over the bads. Of note is that the MAPA algorithm also maximises the score's power, but at the cost of overfitting. With real life cases, further clustering is advisable! Although some power is lost, it is negligible, and offset by the increased robustness of the resulting estimates.

This approach can be used not only to identify marginal-risk breakpoints, but also to calibrate directly from scores to estimates. To do so though, the lender needs to implement a mapping table, to translate the raw scores onto their associated probabilities or scaled equivalents (covered next). It may be difficult in a production system, but is relatively easy for off-line calculations.

20.2 Linear shift and scaling

Banding is an extremely effective way to align scorecards, but many lenders loathe the loss of granularity. There are two opposing views. On the one hand, it provides little value to have a

Table 20.2. MAPA-based calibration

MAPA				Bounds			Ln(Odds)			Results		
Index	Score	p_i	MAPA	Lwr	Mid	Upr	Lwr	Mid	Upr	Ln	p_i'	p_i''
1	602	0	0.00	1	2	3	−0.348	−0.232	−0.116	−0.348	0.414	0.420
2	603	0	0.00	1	2	3	−0.348	−0.232	−0.116	−0.232	0.442	0.448
3	604	1	0.50	3	4	5	−0.116	0.000	0.116	−0.116	0.471	0.471
4	606	0	0.50	3	4	5	−0.116	0.000	0.116	0.000	0.500	0.506
5	608	1	0.60	5	7.5	10	0.116	0.405	0.638	0.116	0.529	0.529
6	612	0	0.60	5	7.5	10	0.116	0.405	0.638	0.232	0.558	0.564
7	612	1	0.60	5	7.5	10	0.116	0.405	0.638	0.348	0.586	0.586
8	614	1	0.60	5	7.5	10	0.116	0.405	0.638	0.452	0.611	0.611
9	615	0	0.60	5	7.5	10	0.116	0.405	0.638	0.545	0.633	0.639
10	617	1	0.71	10	13	16	0.638	0.916	1.195	0.638	0.654	0.654
11	618	1	0.71	10	13	16	0.638	0.916	1.195	0.731	0.675	0.675
12	618	1	0.71	10	13	16	0.638	0.916	1.195	0.823	0.695	0.695
13	620	0	0.71	10	13	16	0.638	0.916	1.195	0.916	0.714	0.720
14	621	1	0.71	10	13	16	0.638	0.916	1.195	1.009	0.733	0.733
15	622	1	0.71	10	13	16	0.638	0.916	1.195	1.102	0.751	0.751
16	625	0	0.71	16	18	20	1.195	1.381	1.566	1.195	0.768	0.774
17	628	1	1.00	16	18	20	1.195	1.381	1.566	1.288	0.784	0.784
18	629	1	1.00	16	18	20	1.195	1.381	1.566	1.381	0.799	0.799
19	629	1	1.00	16	18	20	1.195	1.381	1.566	1.474	0.814	0.814
20	630	1	1.00	16	18	20	1.195	1.381	1.566	1.566	0.827	0.827
	Total	13	13.00								12.96	13.00

large number of bands, if it is not possible to use the greater detail for setting strategies. On the other hand, the greater detail allows greater flexibility, especially when improved data and models have increased the amount of differentiation that can be achieved, and scores have been co-opted into the finance function.

While it is always possible to have more bands, more is often not enough. Thus, lenders have moved away from banding as a one-size-fits-all approach, and have instead moved towards aligning the scores themselves—even if there is some false accuracy, or it is done solely as a precursor to banding and strategy setting. There are two aspects to this process: (i) *linear shift*, which is used to bring the scores onto a common definition; and (ii) *scaling*, which refers to a transformation of the scores and/or point allocations, to ensure that they have a set of desired properties.

20.2.1 Linear shift

The primary means of aligning the scores is to do a linear shift, which requires a regression, using the score as the only predictor for some function of the target variable, as shown in Equation 20.5.

Equation 20.5. Score alignment $f(\text{TARGET}) = b_0 + b_1 \times \text{Score}$

This has two aspects: (i) the type of regression technique used; and (ii) the target representation. The regression technique defines the relationship between credit quality and the final score. Linear probability modelling (LPM) provides a linear function, while logit and probit both provide an exponential function. Care must be taken when the function used differs from that used to create the scorecard, as the results will be less than perfect, especially when linear scores are transformed onto a log scale. Other more complex non-linear alternatives are possible, but they only allow modification of the final score, and not the underlying points.

In contrast, the target representation refers to whether the target variable is the raw binary variable (good/bad, default, or other flag), or the averages calculated for a set of monotonic score bands. The former is usually sufficient, but the latter recognises the raw score's inherent imperfections. Also, if the bands' probabilities are scaled at this stage, no further scaling is required.

Please note that for new scorecard developments done using logistic regression, no alignment should be necessary, as long as the lender is comfortable with deriving estimates using the scorecard's good/bad definition, as opposed to a bad/not bad or default/not default definition. In contrast, linear regression is famed for the unreliability of its estimates, even though the rankings are usually just as powerful.

20.2.2 Scaling

Once the score has been aligned, the next step is to adjust it—and often the point allocations—onto a scale desired by the business. The most well-known example is Fair Isaac's (FI's) practice of scaling application scorecards, so that a score of 200 refers to the average or benchmark odds, and 20 points implies a doubling of odds. This is not the only way though. Banasik et al. (2001b) surveyed lenders to determine what properties they require, which unfortunately cannot all coexist:

(i) the points for each attribute are positive;
(ii) the points for each characteristic's attributes are monotone;
(iii) the total points are always positive;
(iv) the final score must lie within the range 0 to 1;
(v) there is a reference final score(s) associated with a specified credit quality;
(vi) differences in the final score imply a specified change in credit quality, however measured.

Questionnaires were sent to 172 people who had attended the University of Edinburgh Credit Scoring and Control Conferences in 1999 and 2001, and 64 were returned. A summary of the article can be found in Thomas et al. (2002). Of the respondents that indicated that they required differences in score to imply a specified change in credit quality, 36 used the log of odds, and only 6 used *P*(Good). This probably gave an indication of the preference for logistic and linear regression respectively.

As can be seen in Figure 20.1, the most demanded feature ('Always' or 'Usually') is that the final score must be positive (third from top). Next is having a reference score and specified

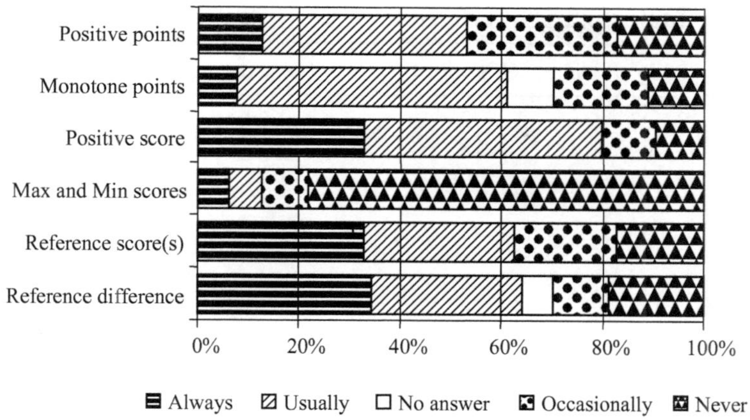

Figure 20.1. Scorecard features.

change in odds between scores, which tend to go hand-in-hand (the two bottom bars). At the other end of the spectrum, the least demanded feature is that scores fall within the range 0 to 1 (Bar 4). Most of these properties are required to make acceptance and implementation within the organisation easier:

(i) **Positive points**—Reduces errors where customers' scores are calculated manually, and removes certain sensitivities if they are advised of how the scores are determined. It is achieved by: (i) finding the highest negative points for each characteristic; (ii) adding it to the points for all of its attributes; and (iii) deducting it from the constant.

(ii) **Monotone points**—Aids understanding and acceptance. Best achieved as part of the scorecard development process, and not after the fact. It is easiest where characteristics' weights of evidence are used as predictors in the scorecard development process, as monotonicity is then ensured at the coarse classing stage.

(iii) **Final score is positive**—Systems may not be able to handle negatives. This can be addressed by increasing the constant by the highest negative score from the scorecard development.

(iv) **Final score falls within range of 0 to 1**—The question may have caused confusion, as it relates to ensuring that associated probabilities fall in the 0 to 100 per cent range. Most lenders will scale the points up, so that most final scores are in the hundreds. Some systems limit final scores to between 0 and 999.

As for the last two properties, relating to reference scores and score increments, they can usually be achieved by applying formulae to the models' results. There are two main possibilities: (i) there is a probability, or odds estimate, that can be converted onto the desired scale directly; and (ii) there are two reference points, for which there are odds estimates that can be used as the basis for interpolating points in between.

Table 20.3. Log reference

Bad rate(%)	G/B odds	LN score	Score
33.33	2	0.69	555
5.88	16	2.77	600
0.78	128	4.85	645

Odds to scaled score conversion

If a reliable odds estimate already exists, whether because the statistical technique provides it directly, or some algorithm was used, scaling can be done using Equation 20.6. It provides constant c' and variable increment i' portions, that are combined with the NLO of the probability estimate to compute a scaled score s'.

Equation 20.6. Log reference $$c' = \frac{S \times \ln(D \times G) - (S + I) \times \ln(D)}{\ln(G)}$$

$$i' = \frac{D}{\ln(G)} \qquad s' = c' + \ln(D_{\text{Orig}}) \times i'$$

where S is the reference score, D is the required good/bad odds at that score, I is the score increment, G the required odds increment, and D_{Orig} the odds provided by the model. An example for a reference odds of 16 to 1 at a score of 600, with odds doubling every 15 points, is provided in Table 20.3. The scaled score equating to 128 to 1 is then 645, calculated as:

$$c' = \frac{600 \times \ln(16 \times 2) - (600 + 15) \times \ln(16)}{\ln(2)} = 540$$

$$i' = \frac{15}{\ln(2)} = 21.64 \qquad s' = 540 + \ln(128) \times 21.64 = 645$$

If the scorecard was developed using logistic regression, the formula can be applied to the underlying point allocations. The points are substituted for $\ln(D_{\text{Orig}})$ in the scaled score s' calculation, and the constant from the original scorecard is replaced by the scaled constant c'. It achieves the same result, barring some rounding errors.

Reference points

The other possibility is where probabilities are not available for each score, but for reference scores towards the upper and lower ends of the score range. If the score to credit quality

Table 20.4. Linear transformation

	Odds	s_x	s'_x
s_1	2.065	0.725	555.69
s_2	125.71	4.834	644.61
S		1.705	576.90

function between those points is linear, or close enough to it, then Equation 20.7—provided by Thomas et al. (2002:148)—can be used to interpolate points in between:

Equation 20.7. Linear transformation $c' = \frac{s'_1 s_2 - s'_2 s_1}{s_2 - s_1}, \ i' = \frac{(s'_2 - s'_1)}{s_2 - s_1}, \ s' = c' + s_i{}'$

As before, the formula has been split into the constant and variable portions. The new variables are the original-score reference points, s_1 and s_2; and scaled-score reference points, s'_1 and s'_2. This equation can be used to adjust the attributes' point allocations, or be applied to the final score directly. For three or more reference points, it must be applied on a non-linear or piecewise linear basis.

An example of how this is applied is provided in Table 20.4. A scorecard has been developed where the odds increase exponentially with score, and it has been determined that the scores of 0.725 and 4.834 have odds of 2.065 and 125.71 respectively. The lender wants this placed onto the same scale as above (reference score of 600 with doubling every 15 points). Scaled scores of 555.69 and 644.61 are calculated for the upper and lower points respectively. These are then used to determine the constant and variable portions, such that a score of 1.705 equates to 576.9 on the new scale. The calculations for this are:

$$c' = \frac{555.69 \times 4.834 - 644.61 \times 0.725}{4.834 - 0.725} = 540$$

$$i' = \frac{644.61 - 555.69}{4.834 - 0.725} = 21.64 \quad s' = 540 + 1.705 \times 21.64 = 576.9$$

If there are only two reference points, rather than modifying the final score, the point allocations can be adjusted. The constant will be the same, but 0.50 points becomes 10.82, which is usually rounded upto 11.

20.3 Reconstitution using linear programming

Banasik et al. (BCT 2001b) also present linear programming as a means of providing a model that has many of the required qualities simultaneously—including being able to match on multiple reference points—not just one or two. Please note, before reading any further, that this approach is highly academic, and has not been applied in practice. The primary problem is that it becomes computationally intensive, and even infeasible, as the number of records and attributes increase. It also affects the rank ordering provided by the original model, even though the primary constraint is that the change in the rankings must be minimised. Some loss of ranking ability will occur, but just how much is not known.

Equation 20.8. Scorecard normalisation using linear programming

	All	Nearest Neighbour
Minimise	$\sum_i \sum_r e_{ir}$	$\sum_i e_i$
Subject to	$s_i < s_r + e_{ir}$	$s_i < s_{i+1} + e_i$
	where $s_n = w_0 + \sum_{j=1}^{p} x_{nj}\, w_j$	
	$e_{ir} \geq 0$	$e_i \geq 0$
	for $1 \leq i < r \leq N$	for $1 \leq i < N$

BCT (2001b) present two possible linear programming approaches (see Equation 20.8). Both determine a new set of weights to provide a new score for each record, as well as rank-error terms. In the 'All' case, the score for every record is compared against all prior records, and the goal is to minimise the rank-error terms. Unfortunately, however, this is a mammoth task—the number of constraints is $N(N-1)$, and the number of variables being solved for is $N(N-1)/2 + (P+1)$, where N and P are the number of records and attributes respectively. For a problem with 1,000 records and only 10 attributes, this translates into 999,000 constraints and 499,511 variables. The other possibility is to restrict the comparisons to the immediately prior records, or 'nearest neighbours', which cuts the number of constraints down to $2(N-1)$ and variables to $N+P$, or 1,998 and 1,020 respectively. The former is ideal and provides the least degradation; the latter results in more degradation, but is more practical.

Thereafter the other properties can be achieved by including further constraints:

Positive Points	$w_j \geqslant 0$ for $j = 1$ to p
Positive final score	$w_0 \geq \sum_{j=1}^{p} \min(w_j x_{ij} \mid j) \times -1$
Monotone points	$w_j \leq w_{j+1} \leq w_{j+2} \leq \cdots \leq w_n$, j and n are the first and last weights for a characteriestic
Specified range	$s_i \geqslant S_{\min}$, $s_i \leqslant S_{\max}$

Table 20.5. Odds doubling

	Lower bound	Baseline	Upper bound
P(Good)	70.0%	98.0%	99.8%
R = Good/bad	2.33	49.00	499.00
Relative	0.05	1.00	10.18
L = Log_2	−4.39	0.00	3.35
M = Trunc	−4	0	3

Reference scores

If part of the linear programming problem is to have either the odds or bad rate increment by a fixed amount for a given score increment, then it is necessary to: (i) identify how many reference records are necessary; (ii) identify the records themselves; and (iii) specify whether it needs to be an exact match or not.

Assuming the odds must double every X points, then the first task is to specify the baseline good/bad odds—often the sample's average odds—and determine the sample's minimum and maximum marginal odds. Thereafter, Equation 20.9 is used as the basis for finding the reference points.

Equation 20.9. Relative $\log_2$ odds $L = \mathrm{LOG}_2(r/R)$

$$T = \mathrm{INT}(\mathrm{ABS}(L)) \times \mathrm{SIGN}(L)$$

where r is a breakpoint marginal odds, R the baseline marginal odds, L the relative $\log_2$ odds, and M the truncated L. The highest and lowest integer values of T must be found such that $R \times 2^T$ still falls within the min and max odds. For the example in Table 20.5, there is a baseline odds of 49, and marginal odds ranging from 2.33 to 499. Thus, eight records have to be found for $T = -4$ to 3 where $r = 49 \times 2^T$. These reference records can also be expressed as:

$$i_V = \max\{i \mid r_i < R_v\}, \quad \text{for } V = 1 \text{ to } 8 \text{ where } R_V = \exp(R \times 2^{(-4+V-1)})$$

Reference scores are then set up as constraints:

Equation 20.10. Reference score $s_i = S_V + d_V^+ - d_V^-$

$$\text{Subject to: } d_V^+, d_V^- \geq 0 \quad \text{for } V = 1 \text{ to } 8$$

where s_i is the calculated score for the reference record, S_V the desired score, and d_V^+ and d_V^- the error terms for each reference point. If any of these records require an exact match on score, then it can be included without the error terms. The error terms are then included in the minimise statement, along with a multiplier for each error component. Equation 20.11 uses the nearest-neighbour rank-error terms, and those from two sets of reference scores:

Equation 20.11. Linear programming:

$$\text{Minimise } a\sum_{i=1}^{N} e_i + a_1 \sum_{V=1}^{M1} (d1_V^+ + d1_V^-) + a_2 \sum_{V=1}^{M2} (d2_V^+ + d2_V^-)$$

Just from the maths presented here, it can be seen that the use of linear programming for score calibration would be onerous. For the moment, the approach is being included in this text as a possible alternative, if only because someone may be able to build upon it further in future.

20.4 Summary

Calibration refers to the fine tuning of a measuring instrument, to ensure that it provides accurate readings in the required units. In credit scoring, it is the same! In some instances, its purpose is solely to ensure consistency across scorecards, but it is increasingly required to provide probability estimates, for use in finance functions. It may be done in two ways: (i) by *banding* the scores to create risk groups; or (ii) by *scaling* the scores, and perhaps even the point allocations, so that the final score provides the measure.

There are three main banding approaches. First, if there are *no prescribed groups*, the CH-statistic can be used to determine an optimal number of groups. Second, if there is a *benchmark* set of grades and required average bad rates, then an optimisation approach can be used, to minimise the sum of squared differences between the benchmark and observed NLO. And third, if *upper and lower bounds* are specified, the marginal risk at each score must be calculated, which was done here using a monotone adjacent pooling algorithm.

When scaling the points or final scores, there are a number of different properties that lenders may require: (i) positive attribute points; (ii) monotone characteristic points; (iii) positive total points; (iv) total points within a given range; (v) baseline score for a given credit quality; and (vi) specified change in quality between scores. Unfortunately, these cannot all exist at the same time. If the scores can be confidently associated with probability estimates, even if only for two points on the scale, then certain formulae can be applied to do the conversion. These assume a linear relationship between the scores and the target function, and vary according to: (i) the *regression technique* applied, usually linear or logistic regression; and (ii) the *target representation*, whether the raw binary target variable or the probabilities for a series of monotonic score bands. For the final scaling, the score can be *converted* directly, if it is already linearly associated with the required function. Otherwise, *reference points* must be chosen, and scores interpolated for points in between. The final approach presented was linear programming, but it is very academic, computationally intensive, and may be infeasible for many of the problems that are encountered.

21 Validation

It is important that we not let models make the decisions, that we keep in mind that they are just tools, because in many cases it is management experience—aided by models to be sure—that helps to limit losses.

Susan Schmidt Bies, Federal Reserve Governor; in Geneva, 7 December 2004.

This is now the home stretch, but some work still needs to be done regarding validation, and presentation of the scorecard for sign-off, which are not mutually exclusive. Best practice is independent scorecard validation, whether by a specialist team, internal audit, or external consultants, but developers must use many of the same frameworks to ensure business acceptance of the model, and to facilitate subsequent review. The primary issues are to ensure: (i) that the model meets business needs; (ii) regulatory compliance; and (iii) when errors are identified, the new knowledge can be used to improve the process. The first is most critical, as the costs of model risk and the resultant adverse selection can break the bank—literally. Management must make informed decisions, and if there is any cause for concern, it must be made known.

Much of the model risk results from the use of theoretical and statistical tools by inexperienced modellers, and policies should be implemented to formalise the 'who' and 'how' of the validation process, which may vary in different circumstances. It is not solely a mathematical exercise, as judgment also plays a role. The major issues are:

Qualitative

Conceptual soundness—A review of the data inputs (nature, quantity, accuracy), statistical methodology and assumptions, and model limitations. Inputs must be appropriate for the job, and the methodology and assumptions should comply with best practice. Documentation must be prepared to help identify the source of any errors or inconsistencies, if they arise.

Quantitative

Predictive power—Ability to separate good and bad. It is assessed using power statistics including the Gini coefficient, AUROC, KS statistic, misclassification matrices, and efficiency curves. Results from similar developments can act as benchmarks, if they can be found (old scorecards, other lenders, advice from vendors).

Explanatory adequacy—Accuracy of estimated probabilities relative to actual rates. It is measured using measures, such as the chi-square statistic, binomial test, Hosmer–Lemeshow statistic, and others, possibly using a traffic-light approach.

Stability—Comparison of development and recent samples, with no reference to account performance. It is not a big issue in its own right, but can be indicative of changes that will cause power/accuracy loss. Movements in the stability index and score shifts can indicate deterioration.

For the 'traffic-light' approach, green, yellow, and red refer to confidence levels of say under 95 per cent, between 95 and 99.9 per cent, and above 99.9 per cent—the higher the value, the greater the latitude for error. Colours are observed over time, to detect whether calibration is out of kilter.

These can be applied as a hierarchy, optimising each in turn before proceeding to the next. For the quantitative aspects, the tools used will depend upon the circumstances. Accuracy need only be validated if the estimates are being used somewhere within the business; otherwise, the focus need only be on power and stability. The stage in the scorecard lifecycle also plays a role:

Post-development—Use of a holdout sample (preferably out-of-time) to measure predictive power and accuracy, and a recent sample to measure stability. Bootstrapping can be used in the absence of a holdout sample.

Implementation—Ensure that scores are those expected. Stability measures may assist, but this is best done in detail, using test datasets.

Monitoring—Regular comparison of actual and predicted outcomes, once the scorecard is in place. Stability measures, score shifts, and vintage analysis may be used. When there is sufficient history, backtesting becomes possible.

According to Burns and Ody (2004), credit risk is assessed using two types of models: (i) *scoring models*, used to distinguish between good and bad accounts; and (ii) *loss forecasting models*, used to predict monetary losses. Each has different limitations, which must be recognised when doing the validation. There is, however, a trend towards using scoring models as the basis for both, with macroeconomic factors included as an overlay. Most of this textbook focuses on credit scoring models; loss forecasting is discussed briefly in Section 26.1 (Loss Provisioning), sans macroeconomic factors. Basel II also requires stress testing under different economic conditions, which is also not covered in this textbook.

While validation is considered critical, there are complications, and no set guidelines on how to do it. In a summary of forum proceedings, Burns and Ody (2004) quote several people on the topic, including: Dennis Ash—(i) scorecards are already old when implemented; (ii) they are not constructed to handle changing economic conditions; (iii) generic scorecards are applied across a wide spectrum of products and lenders, that each act differently; David Hand—(i) it is not possible to validate how well application scoring models perform for rejects; and (ii) the metrics should take into consideration the model's use, especially as metrics meant for continuous distributions should not be applied to accept/reject selection processes; and Nick Souleles—stability issues may be partially addressed by including cross-sectional macroeconomic variables in the models, like unemployment rates and house prices in different regions.

To these must be added the new toy effect; lenders try to break the scorecards as soon as they are implemented. If scores drive strategies, that in turn influence outcomes, then use of the scorecards speeds their obsolescence.

As at time of writing (2006), most of the literature on validation relates to Basel II guidelines for banks, which are focused on capital adequacy. Its requirements are general enough though, that they can be applied more broadly. According to the BCBS's (2005:4) paper on internal ratings validation:

- Lenders are responsible for their own validation.
- Validation's fundamental purpose is to assess the validity of risk estimates, and their use in business processes.
- It is an iterative process, that cannot just be done once.
- There is no one validation method.
- It should cover both quantitative and qualitative elements of the ratings.
- The validation process, and its results, should be subjected to independent review.

21.1 Components

According to the BCBS (2003b) there should be ongoing efforts to ensure the validity of model results, irrespective of the mix of judgment and data. This has three dimensions:

(i) **Parameters**—PD, LGD, EAD, and M. Almost all of this textbook's focus has been upon the probability of default (PD), or some other binary good/bad outcome. LGD and EAD are possible, but difficult because of low numbers, long time horizons, and data problems.
(ii) **Components**—Data, estimation, mapping, and application. *Estimation* refers to the model development process; *mapping*, to calibration onto the Basel II default definition; and *application*, to the models' use within the business.
(iii) **Actions**—Review of developmental evidence, ongoing validation, and backtesting.

The rest of this section focuses on the actions to be done as part of the validation, but treats it more broadly—in particular, to use it as a framework for: (i) presenting the model to the business; and (ii) providing comfort that it will work, upon implementation and at regular points thereafter. The topic is covered under the three Basel headings:

Development evidence—Documentation of all aspects of the development process.
Ongoing validation—Verification and benchmarking.
Backtesting—A comparison of actual versus expected results.

For all of these, lenders should define: (i) the validation process; (ii) event triggers, and actions to be performed to test for model obsolescence; and (iii) accountability! These aspects should be documented, and approved by appropriate levels of management.

21.1.1 Developmental evidence

Scorecard development documentation should cover all critical aspects of the scorecard development process, including any assumptions made along the way, and the results at the end of each step. A full enumeration of what should be documented would include much of what has been presented elsewhere in this book, including, but not limited to:

Reason for the development.
Data design, including observation and outcome windows, sample size, and the good/bad definition.
Model development, including scorecard splits, reject inference and population flows, characteristic analyses for all variables seriously considered for the model, and calibration.
Specifications, including point allocations for each scorecard, score banding or mapping, and recommended strategies.
Validation, any tests performed to assess the power, drift, or accuracy of the model.

Scorecard presentation

While this section focuses on validation, the documentation should also aid acceptance by the business. Consultations would have been held at various points, but this provides an opportunity to clarify and correct. In particular, the business might highlight past, or proposed, changes to infrastructure, marketing, processes, collections, or the economy, that may have affected the model or its applicability in future.

An example of a scorecard presentation report is provided in Table 21.1, which is effectively a characteristic analysis (CA) report, detailing the point allocations, average scores, and bad rates (or alternatively the good/bad odds) for each coarse class. The scorecard was developed using dummy variables, hence there is no fixed relationship between points and risk. Warning bells should ring when the average score or points are inconsistent with the attribute's relative bad rate, which may require that the model be reviewed.

A column that may cause confusion, but assists clarification, is the index—the difference between the average score for that attribute, and the average score overall. It is influenced mostly by the point allocation, but there are instances where the index and the points can be quite different.

For example, only applicants over 54 years of age get positive points, yet the average score increases consistently across each of the age bands. Also, even though the 54+ attribute only gets 16 points, its index is 30 points above average—the balance comes from elsewhere. In contrast, customers that have only had accounts for three months are much higher risks, that get docked 74 points, yet the band is only 55 points below the average. It is obviously getting

Table 21.1. Scorecard presentation

Attribute	Points	Index	Avg score	Count	Bad rate
Constant	987	0	939	48,649	5.3
		Age of customer			
to 23	0	−31	908	2,818	8.6
to 28	0	−13	926	9,778	6.0
to 42	0	2	941	24,937	5.2
to 54	0	11	950	9,027	4.6
>54	16	30	969	2,089	2.6
		Age of relationship			
≤ 3 months	−74	−55	884	3,825	10.6
≤ 1 yr 6 mths	−28	−29	910	9,837	8.0
≤ 2 years	−18	−14	925	2,902	6.7
≤ 4 years	0	8	947	9,102	4.9
≤ 10 years	0	15	954	14,527	3.8
>10 years	10	30	969	8,456	2.7
		Time @ Employer			
≤ 6 months	−27	−40	899	2,928	9.1
≤ 2 years	−16	−24	915	8,125	7.6
≤ 7 years	0	1	940	17,788	5.7
≤ 10 years	0	9	948	5,922	3.8
≤ 15 years	0	14	953	6,450	3.5
>15 years	0	22	961	7,437	3.2

points back elsewhere. The opposite applies to applicants that have been with their current employer for less than two years.

21.1.2 Ongoing validation

Ongoing validation includes both verification and benchmarking, and would be applied to both the mapping and application components. Verification ensures that the process is working as intended, including: (i) checks immediately after implementation and thereafter, to ensure that the test results and those provided by the production system are the same; and (ii) override monitoring, to ensure staff acceptance and check for potential design flaws. Measures include score shifts and population-stability indices, both of which are used to monitor system stability. Implementation issues can arise, especially where the meaning of certain fields has changed because of code or infrastructure changes.

In contrast, benchmarking refers to comparisons against other measures, provided by say rating agencies, credit bureaux, or other banks. It usually involves other data sources and estimates, or 'alternative tools to draw inferences about the correctness of ratings before the results are known', such as judgmental assessments, old or specifically designed models, or external rating grades (BCBS 2003a, para. 73).

21.1.3 Backtesting

Backtesting refers to a comparison of actual versus expected results. It is unfortunately easier said than done, because it takes time before accounts mature enough for them to be compared with the development sample on a like-for-like basis. Prior to that point however, lenders can track preliminary performance and scorecard stability, to see if there are any radical changes, and where they occur. The main tools are: (i) *vintage analysis reports*, that indicate performance at regular intervals after acceptance (see Section 25.2.2); (ii) the *population stability index*, a summary statistic calculated for each scorecard, and the scorecard set (see Section 8.2.3); and (iii) *score shifts*, which indicate the direction of the shift, and where it occurs (see Section 25.3.2). Score shifts are covered in a bit more detail here, to illustrate how monitoring can be done without performance.

Score shift report

Table 21.2 provides an overview of the score shifts in all scorecards used for a particular product. Of all the scorecards, 'Age $\geqslant$ 26 and Renewal' exhibits the greatest change, of about −13.9, followed closely by 'Age $\geqslant$ 26 and New Customer', at −13.3. Care must be taken though, as score shifts within the scorecard can offset each other and mask significant differences.

This is illustrated in Table 21.3. The cheque risk indicator shows a positive shift since implementation, yet the payment profile instalments have been playing an increasingly negative role that has been offsetting it. Further analysis of these characteristics will provide a better insight into changes in the risk profile, and quite possibly, changes in the market generally.

Table 21.2. Score shift—scorecard drift report

Scorecard	Q1 CCY6	Q4 CCY5	Q3 CCY5	Q2 CCY5	Q1 CCY5	Q4 CCY4	Q3 CCY4
Age $\leqslant$ 26 and Time @ Empl $\leqslant$ 2	−6.3	−5.8	−4.8	−3.7	−1.4	1.2	0.6
Age $\leqslant$ 26 and Time @ Empl >2	−12.3	−10.8	−9.8	−8.4	−5.7	−3.9	−5.0
Age >26 and New Customer	−3.3	−3.8	−2.1	0.3	4.2	8.1	9.0
Age >26 and Renewal	−19.5	−16.2	−14.2	−15.3	−11.1	−7.3	−5.6

Table 21.3. Score shift—characteristic drift report

Quarter Year	Q1 CCY6	Q4 CCY5	Q3 CCY5	Q2 CCY5	Q1 CCY5	Q4 CCY4	Q3 CCY4
Bureau:							
PP instalment value	−6.3	−6.1	−5.8	−4.4	−2.7	0.1	1.2
PP worst status ever	−0.9	−1.2	−0.7	−0.6	−0.4	−0.3	−1.0
Time since last judgment	−2.7	−2.8	−2.6	−2.3	−2.4	−2.8	−3.2
Total number of enquiries	0.5	0.7	0.5	−0.4	−0.2	−0.1	0.1
Non bureau							
Instalment to income	−1.7	−1.1	−0.1	0.7	1.9	3.8	3.7
Combined monthly income	0.4	0.4	0.4	0.4	0.4	0.4	0.4
Bank status	−0.2	−0.9	−0.8	−0.8	−0.7	−0.6	−0.4
Residential postal code	−1.0	−1.2	−1.3	−0.9	−0.9	−0.6	−1.1
Cheque risk indicator	6.9	6.8	6.6	6.9	7.4	6.8	7.7
Funding account type	1.7	1.6	1.7	1.7	1.8	1.4	1.6
Score shift	−3.3	−3.8	−2.1	0.3	4.2	8.1	9.0

21.2 Disparate impact

While Basel II focuses on capital adequacy, there may also be concerns regarding fair access to credit. This is especially the case in the United States, where lenders must show not only that the final model is predictive, but also that neither the model nor process result in overt—or avoidable covert—discrimination against a protected group. Regulation B of the Equal Credit Opportunity Act (ECOA) (1974) recognises credit scoring as a powerful fair-lending tool, but one with the potential for abuse. Characteristics that define protected groups are banned (race, religion, gender, national origin, etc.). Age is only allowed as long as older applicants do not get negative points. Lenders must also protect against disparate impact upon those groups, whether by the scorecard or override process. Where it does arise—say because of the postal code, or home ownership status—the characteristic can be retained if the lender can show sound business reasons, and that there are no reasonable alternatives. In other countries this may be practised, even if not required by law.

In order to use a scorecard, the ECOA requires lenders to show that the credit scoring system is an 'empirically derived, demonstrably and statistically sound system'. Witt (1999) expands this to mean that a scoring system must be:

(i) **Empirically derived**—Based upon data for recent applicants.
(ii) **Credit focused**—Developed for the purposes of assessing credit risk.
(iii) **Statistically sound**—Developed and validated using accepted statistical methodologies.
(iv) **Updated regularly**—Periodically adjusted or replaced, to maintain its predictive ability.

Failure to meet these criteria can result in a discrimination claim if a system decision is contested. As regards validation, Witt (1999) goes on to say:

> If the [lender] is developing its own system, [it] should obtain an expert to assure that the system, in its application and construction, is consistent with accepted statistical principles and methodology. Where a system is obtained from an outside vendor the [lender] should obtain a written assurance or a warranty of validation from the vendor.

Thus, even though capital adequacy and fair access to credit are substantially different ends, both put similar demands on the scoring models that are used. Witt's article was written specifically for credit unions, where monitoring is also a major challenge, in particular as regards the difficulties and costs that smaller lenders must incur to implement appropriate systems.

21.3 Summary

While validation is often viewed as an onerous and unrewarding task, it is a crucial part of the scorecard development process. It ensures that the model meets business needs, and provides a means of documenting results, so that the sources of errors and inconsistencies can be traced. Model risk can be high, and failure to detect inadequacies can result in huge losses. There are four major factors that must be shown: (i) *conceptual soundness*; (ii) *predictive power*; (iii) *explanatory adequacy* (accuracy); and (iv) *stability*. Validation is not only done after the development, but also upon implementation, and as a part of ongoing monitoring thereafter.

Much of the literature available on validation relates to Basel II and banks, but is general enough to be applied more broadly. While most of the focus is on the PD or good/bad model, there are also demands for LGD and EAD validation. It must cover data, estimation, mapping, and application of the models. The actions required include the production of: (i) *developmental evidence*, documenting the scorecard development data, process, techniques, and assumptions; (ii) *verification*, to show that the system is working as designed; (iii) *benchmarking*, to show that the risk estimates are similar to those obtainable by others; and (iv) *backtesting*, to compare actual against expected results, once performance is available.

Most of the focus is on corporate governance, but there may also be validation requirements to guard against discrimination and ensure fair access to credit. This applies primarily to the United States, whose ECOA requires that scoring models be *empirically derived*, *credit focused*, *statistically sound*, and *updated regularly*. It is also wise to obtain independent expert opinions, and request vendors to provide written assurances of compliance for their generic models.

22 Development management issues

The scorecard development process has now been covered, and the next steps are strategy setting, implementation, and use—all covered in the next module. For the moment though, some development management issues need to be highlighted, which are covered under the headings of: (i) *scheduling*, prioritisation, when there are multiple developments; and (ii) *streamlining*, means of speeding up the development and implementation process.

22.1 Scheduling

Many credit providers, especially banks, have a number of different portfolios where credit scoring is applied—cheque accounts, personal loans, credit cards, home loan and motor vehicle finance, small-business lending, and so on. They may also be using different types of scorecards for each—application, behavioural, attrition, etc. When scheduling scorecard developments, consideration should be given to:

Portfolio size/importance—Key products should receive priority, whether assessed by portfolio value, revenue, or importance as part of the product offering.

Potential benefits—Address those where the greatest benefits can be achieved first. Credit risk scorecards, especially for new business, will usually take precedence over others.

Resource availability—Are there individuals available to perform the required tasks? Once there are sufficient resources, it becomes possible to address lower-priority developments.

Data availability—In many instances, lenders have to wait until sufficient time has passed for accounts to mature, or make special arrangements to obtain data (retro extracts).

Lenders must also determine how often to redevelop scorecards already in place, especially those in areas of rapid change. While effective monitoring can guide the decision, lenders may have some foresight of economic, infrastructure, and marketing shifts. Behavioural scorecards tend to be the most stable, because there are fewer information sources, and most of the factors relate to company infrastructure and strategy. In contrast, application scoring uses more data sources, and is more susceptible to changes in the economy and the through-the-door population.

Lenders typically rely upon scorecard monitoring to flag when scorecards need to be redeveloped. This approach was developed in an era when the entire scorecard development process—including data assembly, model build, and implementation—was much more onerous than it is today. Nowadays, data is often at lenders' fingertips, skills and technology are readily available, scorecard development methodologies have been formalised, and parameterised systems have made hard coding obsolete. Thus, relying upon monitoring may be a case of the tail wagging the dog, especially when these reports can be difficult to interpret, and regular redevelopment is more feasible.

22.2 Streamlining

Scorecard developments are not easy, but it is possible to speed them up. Such streamlining for 'rapid scorecard developments' can be done in a number of different ways, but in general, it relies upon applying a few shortcuts (see Table 22.1):[1]

Keep it simple—This applies not only to the statistical technique, but also how it is applied.
Piggyback—Reuse what has been done before. This applies not only to code and infrastructure, but also the good/bad definition, fine classing, variable selection, and segmentation.
Formalise—Document the scorecard development process, such that it can be applied by new entrants to the area. Otherwise, much time is spent relearning.
Standardise—Use foresight to make the next development easier. For example, if the delivery system is set up with finer classes, implementation is simply a matter of populating points.
Modernise—For hardware and software, focus on speed and ease of use. Some software packages may be efficient, but difficult to learn for new employees.
Borrow or buy—In need, use generics, and keep a shortlist of external consultants who can be used in need, either to do the developments, or provide advice.
Shorten—If possible, compress the time frames for the observation and outcome periods. If there are insufficient bads, bootstrapping may assist development and/or validation.

Table 22.1. Streamlined redevelopment

Development stage	Redo
Data assembly	✓
Good/bad definition	✗
Scorecard splits	✗
Classing	✗
Reject inference	✓
Model development	✓
Implementation and Testing	✓

[1] See also Bailey (2003), who documented some of the experiences of IKANO.

22.2.1 To piggyback

Piggybacking upon prior work can speed up redevelopment.[2] If the task is onerous, and the results are unlikely to have a significant impact on the final result, then why endure the pain? By changing as few assumptions as possible, and focusing only upon the essential, new models can be developed based upon more recent data in one-third to one-quarter of the time required for the original build. This does not mean that a full rebuild will not be required, only that it will be done much later.

Data assembly—Needs to be redone, albeit most of the program code used for the initial build can be reused. No retrospective searches or calculations will be performed.
Good/bad definition—Leave unchanged. The intention when developing the good/bad definition is that it should last longer than any subsequent scorecard.
Scorecard splits—Leave unchanged, otherwise a full rebuild is required.
Classing—Both fine and coarse classing from the prior development can be reused, but there may be scope to change the coarse classing.
Reject inference—A labour-intensive part of the development, which must be redone.
Model development—Use the same statistical technique as before.
Implementation and testing—If the coarse classing is unchanged, then there are fewer opportunities for mistakes.

It is not a foregone conclusion that the redeveloped scorecards will be implemented; a comparison with the existing scorecards is required, using Gini coefficients, strategy curves, and/or other measures as appropriate. If there is little or no improvement, the existing scorecards should be left intact.

22.2.2 Or not to piggyback

While it would be ideal to piggyback on the previous development, there are instances where the changes in the environment are so great that it is either infeasible or impractical, and a full scorecard rebuild is the better option:

Data sources—New links to the credit bureau or other product areas. The latter can occur with mergers, and other instances where companies are integrating their networks.
Infrastructure—Small but crucial changes in: (i) calculations; (ii) coding of upstream computer systems; or (iii) the layouts used to communicate data.
Procedures—Modifications to policies and processes in different business areas that affect account performance; in particular, channelling of applications, documentation requirements, communications with the customer, collections policies, and so on.

[2] These ideas are largely based upon a presentation given by Jes Freemantle during 2004.

Markets—Entry into totally new markets, especially where the customer profile is radically different from the past.

Organisational—Mergers and acquisitions where portfolios are merged, or divestitures where portfolios are split.

Environment—Substantial changes to the economy or legislative environment, which impact upon the business.

When dealing with new data sources, rather than waiting until sufficient time has passed to accumulate performance on new cases, lenders may opt to obtain retrospective data. This applies primarily to credit bureaux that keep and market such records.

22.3 Summary

This very brief section is not really a chapter, but instead acts as a finale for the module on the scorecard development process. It covered some items relating to management and prioritisation, and ways of streamlining the process. As regards prioritisation, lenders have to consider: (i) portfolio size and importance; (ii) potential benefits that will arise from that sort of development; (iii) resource availability, in terms of money and people; and (iv) data availability.

As regards streamlining, the types of methods that can be used to speed up the process include: (i) keeping it simple; (ii) piggybacking on prior developments; (iii) formalising the scorecard development process; (iv) standardising the implementation; (v) modernising the hardware and software for data management, scorecard development, and implementation; (vi) borrowing or buying skills from elsewhere; and (vii) shortening the observation and outcome windows. Piggybacking can provide a lot of value within the scorecard development process directly, as the good/bad definition, fine/coarse classing, and segmentation can be reused. Eventually, however, changes in the environment will force a redevelopment, whether as the result of new data sources, or changes to infrastructure, markets, policies and procedures, or the organisational structure.

Module F

Implementation and use

23 Implementation

How will the scorecards be applied? How will the results be monitored? How will the customer be advised? These questions should have been considered before the scorecard development started, but if not, the answers are now critical, and will involve varying degrees of technical complexity. This chapter covers some of the primary considerations, under the headings of: (i) *decision automation*, level of automation, responsibility, employee communications, and customer education; and (ii) *scorecard implementation*, data considerations, available resources, migration, and testing.

23.1 Decision automation

When lenders use credit scoring for the first time, it brings about a profound change in the way they do business, to the extent that it influences the corporate culture and relationships with customers. This section covers several aspects relating to decision automation, and although they apply primarily to greenfield developments, some of them must also be considered for brownfields:

(i) **Level of automation**—What hardware, software, and networking is appropriate? Technology is not cheap, and the costs may not be justified.
(ii) **Responsibility**—Who will be responsible for the system? Where will it be hosted? The answer will depend upon the technical competency of the organisation, and the amount of resources available to be invested in it.
(iii) **Employee communications**—How will employees be affected? How will changes be advised? A significant investment needs to be made in change management, to allay their fears and get their buy-in.
(iv) **Customer education**—How will customers be informed of the changes? How will they be advised of declines? Customers should not only be advised of changes, but also of how to contest the system decision.

23.1.1 Level of automation

Credit scoring's primary benefits are consistency and lower cost of decision-making. Technology has provided the added benefit of increased speed, but this can come at a very high cost. In many instances, slower low-tech solutions are more appropriate. It could be as simple as manual tabulation using a score sheet, or use of a hand-held calculator with the scorecard programmed into it.

Some of the low-tech solutions will be totally foreign to new entrants to the credit scoring field, but may have been encountered by those with exposure to emerging environments. Programmable calculators and spreadsheets are still a viable and commonly used option, especially if a scorecard is available but the delivery infrastructure is yet to be developed. They pose challenges though, as it may be difficult or impossible to consolidate the data for monitoring—especially if they are being used across a branch network.

More sophisticated solutions are likely however, especially in an age where computers can be found in previously inaccessible regions. Possibilities range from spreadsheets that are completed by underwriters, to high-tech decision engines with automated data feeds (past account performance, bureau data, etc.). Data storage and data management capabilities are also an issue, as data is required for reporting and analysis. Scorecards can be implemented without any storage or monitoring initially, but over time the issue must be addressed.

Operational aspects of credit scoring must be considered, as bottlenecks can easily undo everybody's best efforts. It matters naught that a computer can assess the data in a nanosecond, if it takes a week to assemble it, capture it, and then advise the customer. In some instances, a letter to the customer may be sufficient, but many environments now demand near instantaneous communications—whether from a sales representative, an ATM, or an Internet interface. Lenders have to automate data feeds, both into and out of the decision system. Indeed, they are trying to reduce reliance on application forms, in order to speed up the process and improve customer service levels.

Finally, how will the customer be advised of the decision? Will it be over the counter when he/she next visits, over the phone, or by mail? It is possible to implement systems that will automatically generate an appropriate letter, and/or an electronic message via email or mobile phone. This is especially important if there is a high probability that the customer (or dealer/agent) has submitted applications to a number of lenders. In some cases, customers are price-insensitive, and will take the first offer received.

23.1.2 Responsibility

War is too serious a matter to entrust to generals.
—*George Clemenceau*

Lenders must decide upon who will be responsible for the implementation and ongoing maintenance of the decision infrastructure—computer hardware, software, and programming staff. There are a number of different possibilities, and the choice is often determined by the organisational structure, culture, and available technology.

Technology

Mainframes were the technological workhorses of the 1960s and 1970s, and were the only option for the first fully automated application-processing systems. These were the responsibility of the data processing department, or what is today more often referred to as information technology (IT). Unfortunately however, where credit scoring falls under their control, it is only one of many functions—along with billing, accounting, human resources, and others—that are clamouring for scarce resources. IT's reliability depends upon: (i) their available manpower; (ii) commitment to, and interest in, the project; and (iii) ease with which they can integrate it, within the existing infrastructure. If their capabilities are limited, then a separate computer system may be required. This has become ever more feasible, as parameterised software and networked/PC-based solutions have become available.

Consideration must be given to how flexible the final system has to be, whether scorecards, strategies, or reports. Flexibility will be a function of the available technology, which often drives who is responsible. In the 1960s, bespoke software was developed using COBOL, or other programming languages, for each computer and each company, and this traditional-systems approach is still used by many lenders. When the programs had to be hard-coded into mainframes, it meant that: (i) IT was probably the only area that had the required access and skills; (ii) modifications were problematic, because they did not get priority; and (iii) the stringent controls demanded by IT lengthened implementation time and increased costs.

According to Wiklund (2004), the implementation cost for a 12-characteristic scorecard might be US$24,000, on top of other costs, to cover the 400 man-hours required to program, implement, and test. It is wise to: (i) have an IT representative on the project steering committee, (ii) use people, whether business analysts or programmers, with prior experience in scorecard implementations; and (iii) limit the characteristics used in the model to those already catered for in the system. The cost of a bespoke implementation at a credit bureau would be $10,000 to $25,000, subject to negotiation.

Functional areas like application processing, account management, and collections have significant motivation to develop and manage their own systems. Going it alone has its own problems though, including, but not limited to: (i) *high skills requirements*, which have to be managed; (ii) ensuring proper *change management*; and (iii) potential *problems communicating* with other computer systems.

Systems co-ordination is especially problematic if the application processing and account-management systems are not networked, to: (i) facilitate automatic account opening; and (ii) ensure a link to subsequent account performance.

As technology has become more sophisticated, this has changed. Today, there are *networked systems*, and perhaps even *distributed computing*, that: (i) obtain data from remote computers; (ii) use parameterised decision engines to calculate scores, apply policies, and provide decisions; and (iii) return the results to where they are needed. Extensive testing is still required, but implementation costs are lower, and there is much more flexibility. An analyst with no programming skills—just competency in using the software provided—can specify characteristics, attributes, scorecard splits, policy rules, and other elements. IT may still handle the hardware and networking issues, but the business unit will have the flexibility it needs.

External agents

Rather than developing and managing these systems themselves, lenders can instead opt to avoid such complexities by outsourcing the scorecards' installation, operation, and maintenance to an external company. This frees up time and resources to focus on managing other aspects of the business, but has a cost implication. Charges may be variable (according to volume), fixed (monthly charge), or some combination of the two. Even entire stages of the risk management cycle can be outsourced, especially application processing and collections. Generic bureau scores, if they are adequate for the task, can be used to minimise the technology requirements. Lenders then need only worry about strategy setting, finance, marketing, managing a branch network, and ensuring that the decisions are executed. This exists in at least one micro-lending environment, where the lenders do not have sufficient resources to develop their own systems.

23.1.3 Staff education

When credit scoring is first implemented, underwriters are usually a fundamental part of lenders' decision-making processes, and may remain so. The introduction of new technologies gives rise to insecurities, relating to how the changes will impact upon their work, or whether they will still have a job. Lenders usually try to redeploy them into other functions, but some cannot make the adjustment. If a change management program is not implemented to allay their fears, implementation of the new system may backfire.

Extensive employee communication is crucial, including staff education to answer questions like, 'What changes are being made and why?', 'How does it work?', 'How does it impact upon the business?', 'How does it impact upon me?', 'What do I tell the customer?', and 'What if I disagree and wish to contest the system decision?' This applies to all staff involved in the decision process, whether management, underwriters, or front-line staff. There may not be a Luddite uprising, but uncertainty can nonetheless be highly disruptive, to the extent that customer service is affected. At the extreme, employees may even sabotage the new system.

According to Wiklund (2004), a major challenge with first-time implementations is to convince underwriters that an intelligent decision can be made using a *score derived using limited data*, often excluding demographic and deal-specific information, which previously weighed heavily in their assessments. Their training should include: (i) some background on the scorecard development process (including sample selection, the good/bad definition, and variable selection); and (ii) a review of the model validation, to demonstrate that the model works. Individual examples may be presented to illustrate what decision would result in future. The primary message is that the scorecards have been empirically derived, and while the underwriters may differ on some specific cases, the scorecard(s) should provide better results overall. It may also assist to apprise them of some of the characteristics used, as the most powerful ones are usually those that they have always put a heavy reliance upon, or are closely related.

Not only the underwriting manager, but also one of the more lead underwriters, should be included when the final scorecards are being presented, as they can assist with getting broader acceptance and support from the underwriting team. Scorecards would still be confidential, kept even from the rank and file in order to avoid manipulation. Where dealer networks are significant new-business channels, Wiklund (2004) also suggests that lenders make investments in keeping dealers' sales staff apprised of changes. Indeed, they may demand as much attention as the internal underwriters, to the extent that in some cases, key players in the dealer network should be included in the project steering committee.

Wiklund (2004) also highlights that the indirect dealer (agent, broker, etc.) channel has two customer sets: (i) the dealer, and (ii) the purchaser. Extra risks arise because the lender is removed from the final customer, and may not understand the forces driving either the dealer, or customer decisions. Dealers wish to ensure their own sales, and they are primarily interested in which lender is most likely to provide a speedy decision and accept the deal, and will channel their applications accordingly. Given that dealers do not take any risk in the lending transaction, the profit on their sales goes straight to the bottom line.

For individuals that will be directly affected, the communications should be as personal as possible. Classroom training and audio/visual presentations may be appropriate. For others, written communications may be sufficient, whether in the form of circulars, articles in company newsletters, brochures, email, and/or Internet publications. Subsequent changes will also require special attention, but this becomes less as the systems become more stable, and decision automation becomes accepted as part of the company culture. Eventually the changes will become minor events, and may require little or no communication, especially once scoring education is integrated as part of the broader ongoing company training.

23.1.4 Customer education

In Stanley Kubrick's '2001: A Space Odyssey', the shipboard computer, HAL, systematically eliminated the crew because it perceived them to be a threat. The year 2001 has come and gone, and both HAL and the space station it commandeered are still some time off, but such science fiction has nonetheless heightened people's perceptions that computers can destroy lives. The truth may be less dramatic, but sufficient for the public to be suspicious. Thus, besides educating staff, some investment in customer education is also warranted, the extent of which varies depending upon the environment and the level of existing acceptance. Over the past few decades, transactional lending has become well accepted in first-world consumer environments. In emerging environments however, whether countries, products, or markets, extra effort may have to be put into customer communications. Customers used to relationship lending are often very unaccepting of the new technology, and may take their business elsewhere.

Most crucial is that front-line staff have knowledge of the process, and are able to communicate it. Providing written materials such as brochures, handouts, and web pages to the branch network will assist greatly to explain: (i) how the system works; (ii) the expected customer benefits (reduced costs, consistent delivery, increased availability); (iii) the appeals process, if a customer wishes to contest a decision; and (iv) how to get further information in need.

Extra effort is required for first-time implementations, including staff training and possibly communications through the media. The worst thing that front-line staff can do is to blame a declined application on the big bad computer (whose feelings are, fortunately, not easily hurt). They must realise that any decision generated by the computer is based upon company policy, as decided by management. The computer does not make the decision, it is just a tool! Extra training may be required to help them deal with disgruntled customers.

Decline reasons!

Customer service will be greatly enhanced if staff members are empowered to communicate what influenced the decision, whether for an outright reject, a higher interest rate than expected, or a different product offered. For rejects, the system must also provide reasons, which in many instances is demanded by law. There are two possibilities for lenders:

Minimum possible information—This may be as cryptic as 'score', 'policy', or 'statute'. Legislation may, however, demand that customers be given greater detail, like the exact policy or statute reason.

Scorecard contributors—The attributes that contributed most to a 'decline on score' can be detailed, but this requires greater infrastructure, and sensitivity. Some of the reasons may be 'no home phone number', 'short time at address', or 'too many bureau enquiries', and the customer may wonder, 'Why does that matter?' The way reasons are communicated will have a significant impact upon customer perceptions.

Appeals process!

Depending upon the environment, the credit provider may also provide an appeals process, that allows customers to contest the system's decisions. It is most common for products like home loans, motor-vehicle finance, and overdrafts, where the decisions have a greater effect on customers' lives. Given the information asymmetries, borrowers usually have better knowledge of their own financial situations, and lenders may find that: (i) some of the information used in the decision was wrong; (ii) there is a fault in the scoring process; or (iii) there is other data not captured by the score. In contrast, where the product's utility to the customer is low, and especially where the appeals process cannot be economically justified (petrol cards, store cards, and Internet-based applications), it may be unnecessary.

In these cases, the appeal is customer driven; but it may also be staff driven, especially if: (i) the potential profits are high; (ii) significant amounts are invested in relationships; (iii) there is information not captured by the score; and/or (iv) there are incentives for new business done. In either case, and especially for volume-driven environments, an override process is required (see Section 24.2).

23.2 Implementation and testing

At this stage, there are: (i) scorecards and strategies that have been signed off by the business; (ii) an implementation platform; and (iii) somebody to take responsibility for implementing and managing the system. The next step is physical implementation and testing. Several factors have to be considered:

(i) **Data, resources, and migration**—Is the data available, and have there been any significant changes? Are the people needed to do the work available? How will cases already within the system by treated?
(ii) **Testing**—How will the scorecards be tested before going live?

23.2.1 Data, resources, and migration

Most of the issues presented in Section 23.1 were very high level, and while they may be very relevant for greenfield projects, most credit providers today are faced with brownfield implementations, where the task is to update an existing system. Whether greenfield or brownfield, the following issues need to be considered:

Data—Have there been any changes that might affect how well the scorecards work?
Resources—Are the required people and equipment available?
Migration—Have there been appropriate communications to those people that will be affected? Is there a policy in place for cases that are work in progress?

Data

Given that data is the feedstock of credit scoring, it is also one of the greatest concerns when implementing new scorecards. There are three major factors to guard against: (i) *characteristic not available*, either because the data source is unavailable, or the characteristic has been dropped; (ii) *operational drift*, which has caused a characteristic's meaning to change; or (iii) *population drift*, which implies that the customer base has changed in some way. In each case, efforts must be made to determine what has happened, its impact, and whether the model can still be used. These problems can usually only be identified when running the model in test.

Some scorecards will have one or more new characteristics/attributes, which have been included in anticipation of them becoming available on the production system. If not yet available, then adjustments must be made. In the simplest case, the points for the missing items are set to zero, and it is assumed that the risk ranking provided by the rest is sufficient. If, however, greater accuracy is desired, then the points can be recalculated for all scorecard characteristics. The former is sufficient if the missing characteristics play a minor role; otherwise, the latter is more appropriate.

Resources

Qualified staff members are required to do the physical implementation and testing of scorecards, which may include writing program code, or modifying software parameters. Coding and testing in mainframe environments can be extremely time consuming, whereas parameterised systems with well-structured test environments can speed up, and reduce the cost of, implementation. In either case, the time and effort required for implementation should not be underestimated.

Migration

When implementing scorecards, consideration must be given to: (i) how the scorecards might impact on the through-the-door population, or other aspects of the business; and (ii) having a policy in place to handle work in progress, where the decision changes. As regards the former, lenders often still insist upon implementing conservative policy rules that, unfortunately, reduce the scorecards' effectiveness. It is useful to sample a swap set, to highlight some accepts that will now be rejected and vice versa, and explain these. The more confidence they have in the system, the better!

As regards work in progress . . . If the customer, or dealer, has already been advised of acceptance, it is best to stick with that decision, in the interest of good public relations. Such reworks occur where applications have been accepted in principle, but some item is still outstanding, and the application has to be reprocessed once everything is in place. This includes instances where: (i) documentation is outstanding, such as a signed agreement or personal identification; or (ii) final processing is required to open the account or draw funds. In other instances there is more latitude, and the lender may opt to use the new system decision.

23.2.2 Testing

Once the scorecard and associated strategies have been implemented on the delivery system, the next step is to test them, prior to first use. Exactly how this is done varies, depending upon the level of automation. The greater the human input, the more testing has to consider human foibles.

Test versus expected

In general, testing ensures that the system is working according to design, and relies upon comparisons of test versus expected results. The analysis should be done: (i) at a *high level*, to see whether the scores and/or decisions are those expected; and (ii) on a *case-by-case* basis, especially where differences are highlighted. If errors are identified, the code/parameters are corrected, and the system retested. Differences may be easy to identify at a high level, but their causes may be elusive. Possibilities to be considered are:

Scorecard parameters—Scorecard splits, attribute codes and ranges, and point allocations.
Strategy parameters—Score cut-offs, limits, and policy rules that determine the decision.
Operational drift—Characteristic not populated, or the calculation or process has changed.

For the first two, it should be easy to fix the problem, but for the latter, someone has to determine and report on how much the operational drift has undermined the results. Operational drift is the most difficult to identify and measure, and can arise from: (i) changes to the questions presented in application forms; (ii) changes to pre-capture screening criteria that are applied, whether sanctioned or unsanctioned; (iii) differences in the calculations used for development and operational data (e.g. between retrospective and online data); and (iv) differences between the development and operational processes.

Thomas et al. (2002) highlight that in October 2001 the UK credit bureaux implemented changes that reduced the number of searches being recorded against individuals, which affected any scorecard that had 'Number of Searches' amongst its characteristics.

With regard to the customer responses, the differences may be the result of: (i) how the *questions* are phrased and ordered, whether in writing, in person, or over the phone; (ii) *cultural background* of the respondent, which can vary by geography, religion, customer age, or a number of other factors; (iii) *environment* in which the question is being asked, which would cover location (at home, at work, at a sales point) and state of mind (chilled, hung over, stressed out); (iv) *form structure*, including how options are ordered on forms, changes between free-form and categorised, and changes in categories; and (v) *interpretation* of the responses by the person capturing the data or asking the question, which might require special training and rule-sets. Any changes in wording, the through-the-door population, marketing channels/ strategies, response formats, or policies for interpreting responses, can have subtle influences upon responses.

Testing processes

Exactly how testing is done will vary depending upon the environment. A low-tech solution is a manual review of individual cases, to ensure that the system result is exactly the same. This can be very time consuming though, and will probably only test a relatively small set of possible scenarios. The high-tech solution is dual processing, but there are less demanding alternatives. In any event, a test system is required that can be run without impacting upon operational processes. It may be a totally independent system, or involve test runs done after hours.

The expected results are usually generated by a computer program set up specifically for the task, often running on the same platform used for the scorecard development (SAS, SPSS, etc.).

Lenders should also provide some means of handling individual queries after scorecards are fully tested and implemented. Staff members often recognise scores and decisions that are out of the ordinary, and question the results. Providing them with a channel to query these cases will increase their confidence in the system, and may be of aid in identifing and rectify implementation errors, and perhaps even design problems.

23.3 Summary

When implementing scorecards, the issues to be addressed vary, depending upon whether it is a greenfield or brownfield project. Greenfield projects are obviously more challenging, and decisions must be made relating to: (i) the *level of automation*, which can range anywhere from manual score sheets or spreadsheets to fully automated systems; (ii) *responsibility* for the system, which is usually the IT department, if the scoring system resides on a mainframe, but can be elsewhere if a parameterised system is being used, and in either case it can be outsourced; and (iii) the necessary *staff and customer education*, to explain the system and allay any fears that may arise. Customer service can be aided if decline reasons and a means of contesting the system decision are provided. These tasks are much easier for brownfield developments.

With the physical implementation, lenders must take into consideration factors relating to: (i) *data*, especially where characteristics are missing, or there has been operational or population drift; (ii) *resources*, ensuring the appropriate staff is available when needed; and (iii) *migration*, dealing with work in progress. Testing is also required to ensure that everything is working according to design, and should be done in a manner that will not affect the current process. The primary things to guard against are: (i) *implementation errors*, where any of the loaded scorecard or strategy parameters are wrong; (ii) *operational drift*, changes to any aspect of the calculations or upstream processes that affect the results. Once the final system has been implemented, there should also be some means of handling individual queries, as staff members will often recognise cases where the decision being returned is abnormal.

24 Overrides, referrals, and controls

Credit scores are powerful tools, but they are not a panacea. Checks and balances are needed to ensure that the system is being used as intended, and its integrity is protected. The section is treated under the headings of:

(i) **Policy rules**—Reasons for using rules instead of scores.
(ii) **Overrides**—Instances where the score and system decisions may be contested.
(iii) **Referral reasons**—Instances where the decision may be questioned; extra input can be called for, based upon a combination of score and policy.
(iv) **Controls**—Mechanisms that can be used to protect against staff errors, corporate espionage, organisational Alzheimer's, etc.

The first three sections relate to the major stages in the decision process: (i) *pre-score decision*, policy rules, which determine whether the cases will be considered at all; (ii) *score decision*, based purely on the score and associated strategies; (iii) *system decision*, a combination of score and policy, as determined by the system; and (iv) *final decision*, the end result, based upon score, policy, and any manual intervention. For the latter three, the decision can be changed in between each. Both policy and judgment can be used to override score decisions (score overrides), whereas only judgment can be used to override system decisions (system overrides).

24.1 Policy rules

Prior to the advent of credit scoring, decisions were either made subjectively, or were based upon policy rules. These include: (i) *product rules*, that eliminate applicants who do not meet particular qualifying criteria, such as age, income, or minimum loan amount; (ii) *credit rules*, factors known to be associated with high credit risk, such as extreme levels of delinquencies; and (iii) *fraud prevention rules*, that insist upon verification of applicant details, etc.

Assuming that cases are within the scorecard's scope, in an ideal world the scores would be the sole basis for decisions, and policy rules would become redundant. That is unlikely, and lenders instead strive to integrate as much data into the scores as possible, and minimise the number of policy rules. Whether a specific attribute is treated through score or policy will depend upon its frequency and severity, as illustrated in Figure 24.1.[1] High-frequency events are modelled, but may not be included if their severity is low. Low-frequency/low-severity attributes have little or no effect, and can be ignored. And finally, rare but severe events are impossible to model, but can be recognised using policy rules that either force the lowest

[1] This policy/score matrix is based upon one presented by Evren Üçok of Mercer Oliver Wyman during late 2005, and is being used with permission.

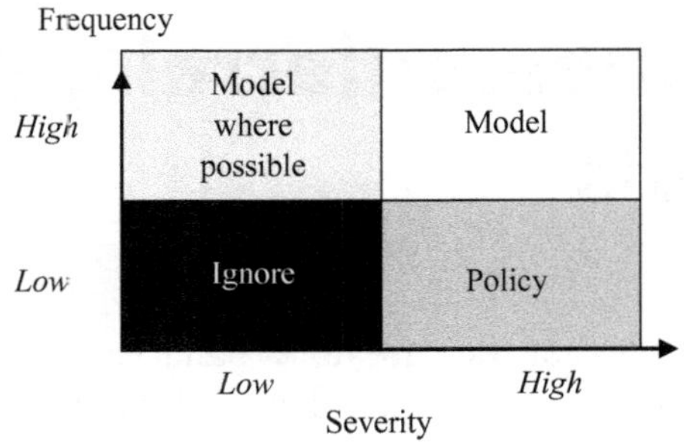

Figure 24.1. Credit policy/score matrix.

possible grade, or worst possible decision. When credit scoring is first implemented, there is a distrust of the new tool, and most of the credit policies will be those that guided underwriters in the past. Over time, the policies should be limited to focus on cases where scorecards are known to be inadequate. The process of getting to this point can, however, be a long one, with regular revisions as new data becomes available.

Besides rare but severe events, other factors that would still demand policy rules are known scorecard weaknesses and new data fields. Scorecard weaknesses may have been highlighted by the underwriters or other analysis, and be the result of operational drift or problems with assumptions made in the scorecard development. New data fields arise mostly when there are new links within the organisation, or to the credit bureau, such as when shared information becomes available and lenders can use 90-days-past-due with other lenders for the first time. In both of these latter cases, the policies are only temporary fixes, until the models can be redeveloped (McNab and Wynn 2000). In contrast, the situation regarding rare events may be addressed in a subsequent development if—and only if—the cases become less rare. This can result from improved efficiencies in gathering information, more aggressive marketing strategies, or changing economic conditions.

24.2 Overrides

Overrides are not done by cowboys . . . uhhh, well, maybe they are at times. In credit scoring, the overrides refer to decisions, not horses, but sometimes the mentality is the same. They occur when decisions are modified, and according to Lewis (1992) can be done by: (i) *policy rules*, set by the lender for specific subgroups; or (ii) *people*, who make subjective calls based upon information, or intuition. Refers are similar, except no specific decision is made; instead, further input or actions are required, that may or may not result in an override.

While scoring's primary purpose is to guide decisions that affect the customer directly, it can also be used for risk-based processing, to override information, verification, or documentation requirements. The most obvious example is super scores, which are very high or very low pre-bureau scores that are so far removed from the cut-off that the bureau data is unlikely to change the decision. If that is the case, then why incur the extra expense?

The motivation for pre-bureau strategies is greatest where bureau data is expensive, difficult to process, or of questionable value. As its cost and reliability has improved, this motivation has reduced. Today, its omission may introduce unacceptable risks, especially where the bureaux are providing fraud checks and other verification services.

Subjective overrides

Subjective overrides are synonymous with system overrides, and occur any time a staff member, or members, overturns the system decision. They may be: (i) *information-based*, motivated by data not already included in the score (such as proof of an inheritance or lottery win, or a personal visit from the customer with detailed and/or audited financial statements); or (ii) *intuition-based*, based neither on policy nor information, but instead on a gut feeling, or perceived insight into the applicant that the system has missed. The former is desirable, while the latter is to be discouraged.

Where the scorecards are known to be reliable, which includes most first-world high-volume consumer credit environments, controls are required to limit subjective overrides. They cannot always be banned outright, as they may be necessary to enable customer contests, and are an effective means of obtaining feedback from front-line staff. It is instead better to limit them, to say 3 per cent of system decisions, and monitor them to ensure that the thresholds are not exceeded (whether at region, branch, underwriter, or some other level). Most score overrides will be clustered around the cut-off, and although the analysis will be hampered by small numbers, the data can be reviewed to identify possible shortcomings in, and abuse of, the system.

In contrast, where cases are scored for guidance, subjective overrides are part and parcel of the process. This occurs mostly in: (i) high-value/low-volume lending; (ii) instances where scorecards developed in one situation are being used elsewhere; (iii) greenfield scoring implementations, where scoring is a new concept and underwriters still have significant latitude; and (iv) any other instances where the scores are known to be questionable. Even then, the underwriter may still be limited on the extent of the override, and have to motivate it.

Wherever feasible, there should be well-defined override reasons, and an override procedure with appropriate levels of authority. Some of the possible high- and low-score override reasons are provided below:

High-score overrides

Poor history—Review of past dealings indicates poor performance, making the customer *persona non grata* with the lender.

Adverse bureau—There are judgments or other factors on bureau, that the lender does not believe are represented adequately in the scorecards, usually because the numbers were too small for their risk to be adequately represented.

Affordability/overextended—Although the lender is creditworthy, the required loan repayments would push him/her over the edge. Responsible lending requires that lenders take affordability into consideration, and most will not lend if the required repayment is greater than some limit, such as 30 per cent of gross income (see Chapter 35, Fair Lending).

Collateral—The lender believes the asset's value will be difficult to realise if the customer defaults, say because of the age of an automobile, or the condition of a property.

Low-score overrides

Good existing customer—Companies with multiple products may not be capable of incorporating all of the information into a single score, so allowances may be made to take into consideration good current and past dealings with that customer.

V.I.P.—Rejection might jeopardise the relationship with a valued customer, whether the applicant, a relative, or associate; for example, an application from the daughter of the owner of a large commercial customer. Exceptions are often made!

Staff—Loans to staff members, whether of the lender, or of other companies with special arrangements. This can cause special problems when the person changes jobs.

Security—Credit scoring plays a huge role in both secured and unsecured lending, and any security mitigates the risk. It could be the asset being purchased, a third-party guarantee, cession of other assets, or a sizeable deposit.

Interested third-party pressure—A relationship with another entity may be jeopardised if the loan is rejected, such as brokers that source deals (home loans), and traders arranging finance for their customers (motor-vehicle finance). These are to be guarded against.

Unrecognised assets—Strategies are often based on income, and although a customer may not have sufficient to justify the requested loan, he/she may be financially wealthy but illiquid, or be expecting a windfall that can be confirmed (e.g. house sale).

Staff members' apparently lower risk can usually be attributed to their loan repayments coinciding with the salary date, and taking precedence over other obligations. Amounts extended may be greater than allowed to the general public, and when problems occur they will be more severe.

Dealers/brokers can become very familiar with lenders' scoring systems, to the extent that they can identify, with some accuracy, which cases will be accepted and which not—especially when they have their own access to the credit bureaux. This can cause some 'push back' when implementing new scorecards (Wiklund 2004), and unscrupulous dealers may massage application details to gain acceptance.

24.3 Referrals

The primary difference between overrides and referrals is one of semantics: with overrides, the decision is overturned; with referrals, it is just being questioned and may or may not be overturned. According to McNab and Wynn (2003), referrals are usually a form of policy, used for the purposes of validation, manual review, or changing account conditions. Customer contests would also fall into this camp.

24.3.1 Validation

Lenders cannot always rely upon application details being the gospel truth, and will try to ensure that certain key details are correct, especially those that: (i) are most subject to manipulation; and (ii) can cause the greatest damage if they are wrong. Telephone and/or bureau calls are made to verify details like income and phone numbers, and to protect against identity theft. While most of this was once required purely as part of good business practice, today it is increasingly demanded by 'Know Your Customer' legislation.

Documentation/security procedures

The primary mechanisms used to check application details are requests for the applicant to supply documentation, and telephone calls to key parties:

Identity documents—Birth certificate, passport, driver's licence, or other personal identification document.
Bank statements and payslips—Used to verify income.
Municipal accounts—Used to verify whether the applicant lives at the given address.
Security calls—Phone calls made to numbers provided on the application form, to ensure that the applicant can be contacted. Difficulties arise where the applicant does not have access to the work telephone, for example factory workers, teachers, or salesmen.
Employer calls—Phone calls made to the employer, to verify employment and income. There may be problems with getting co-operation, especially where the number of employees is large, and the human resources department is small and ill-equipped to deal with large numbers of enquiries. This is NOT their business.
Trade references—Applicants may be requested to provide details of other companies with whom they have credit relationships, who will be contacted to confirm.

Demands for physical documentation can be onerous in environments where: (i) customer contact is limited, such as for Internet-based transactions; or (ii) the documentation requirements are inappropriate, such as in micro-lending environments, where applicants may be self- or unemployed, do not have phone numbers or fixed formal addresses, and suffer from poor literacy.

Suspicion triggers

Validation could be done on every applicant, but this can be very expensive. There are often suspicion triggers, to indicate that further investigation is necessary.

Demographic inconsistencies—Possible embellishment of the application form, like where the income is inconsistent with the applicant's age, education, place of residence, etc.
Contact inconsistencies—Unconfirmed address or phone number, which can be checked using databases provided by phone companies, credit bureaux, or other vendors.
Duplicate applications—Other applications from the same customer may already be on the system, perhaps as the result of mistaken double capture, but it could also indicate fraud.
Fraud indicators—One or more of the customer's details—personal identification number, phone number, address, etc.—appear on a fraud database (see Chapter 31, Fraud).

On the latter point, even if it was a confirmed fraud, fraudsters tend to be transient. Legitimate individuals will eventually take over their addresses and phone numbers, and these databases must be updated accordingly. Fraud syndicates are also very sophisticated, and can have the capability of forging identity documents, municipal accounts, and payslips; and phone operators, to impersonate legitimate employers. They will adapt quickly to any changes in validation procedures, especially if staff members are involved.

24.3.2 Account conditions

Just because the above hurdles have been overcome does not mean the job is done. The final hurdle is whether or not: (i) the applicant passes certain minimum criteria; and (ii) what the applicant wants is consistent with what the lender is willing to provide. The following is far from exhaustive:

First, when making certain offers, lenders will have product rules. Ideally, applicants falling outside the target group should not be applying in the first place, but some may slip through and possibly be accepted. This applies especially to sponsored schemes, where the loan is being subsidised, and special campaigns, meant for one particular target market. These qualifying criteria should be made very clear, whether posted in the shop-front window, in pamphlets available in the branch, a poster on the wall, or on the application form. The issues governing the criteria may relate to: (i) *profitability*, sub-economic loan size; (ii) *affordability*, insufficient income; (iii) *residence*, applicant lives outside target area; or (iv) *target group*, applicant does not qualify for a specific scheme. Product rules may change over time, and companies may target new markets that were previously excluded.

Second, where assets are being financed, there are often rules regarding the security/collateral, due to concerns about how much will be recovered if it is repossessed and sold: (i) *value*, limits on the amount that can be financed, say the purchase price of the asset, and perhaps a small margin for transaction costs; (ii) *age*, with second-hand goods, especially motor vehicles, age of the asset will be an issue, because of reduced marketability; (iii) *condition*, for home loans, a property assessor may recommend that the loan be declined if the property is suspect. If the customer is offering other collateral, there may be similar concerns about its liquidity and price volatility.

And finally, terms and conditions . . . For amounts, terms, and/or conditions that the lender is uncomfortable with, the customer may instead be offered lower-risk options (a 'down-sell'), including a reduced loan amount/term, greater security, or alternative product, rather than turning down the loan.

24.4 Controls

A sophisticated infrastructure has now been developed for use in managing a business. Is this not something worth protecting? Given that the greatest strengths of financial services companies lie in their ability to process information, the answer is a definite 'Yes'. This section on controls is split into two subsections: (i) *the playing field*, the risks that may arise, and tools that can be used to protect against them, at any stage in the decision process; and (ii) *scorecard controls*, actions that may be taken to protect the scorecards, whether in terms of documentation, implementation, or other.

24.4.1 The playing field

This area can be viewed as a combination of several dimensions, something like a SWOT analysis, but more like a twelfth-century mediaeval tale: *Damsels in Distress*, what is being protected, or rescued; *White Knights*, the lenders, who protect their interests on a variety of fronts (which is quite straightforward, and is not covered further); *Black Knights*, the villains, be they mortal or not; *Their Weapons*, the tools used during their onslaught; and *Our Shields*, the tools available to protect the fair damsels.

Damsels in Distress

Controls are meant to ensure the firm's viability as an ongoing concern. This may be somewhat over-dramatic, but less so than the analogy to protecting the life and virtue of a damsel in distress, and there are parallels. The first question that needs to be asked is, 'What is being protected?' Protective measures are required for: (i) *processes*, the physical implementation of systems, scorecards, and strategies; (ii) *documentation*, of the scorecard development, final point allocations, strategies (cut-off, pricing, etc.), scoring infrastructure, and so on; and (iii) *knowledge*, individuals' understanding of the organisation's workings, its strengths, and its weaknesses.

Black Knights

The next question is, 'Protect the damsels from what?' Companies face several threats, not just relative to scoring, but in almost any area of the business, including: (i) *Fraud*—Financial services companies are the biggest targets of fraudsters, who need knowledge of business processes, often provided by insiders, to crook the system. (ii) *Corporate espionage*—Having first player advantage is worth something, but comes at a price. In many industries, the costs of R&D—especially the mistakes made through the learning curve—are great. Many companies will leave pioneering to others, and some use underhanded means to gain access to their experience. (iii) *Staff errors*—Even the most well-intentioned people make mistakes, especially in complex environments. Testing can assist to identify where such errors are more likely to occur, so that adjustments can be made. (iv) *Malcontents*—Individuals with malicious intent, usually disgruntled employees, who can use their knowledge of the system to cause damage to the company. (v) *Organisational Alzheimer's*—Details are fresh in people's minds when systems, scorecards, and strategies are first implemented. As time goes by, employees either forget, or find better pastures, and documentation is either insufficient or grows legs and walks. When problems arise and must be fixed, both mechanic and manual are gone. (vi) *Loose Lips*—After first implementation, employees are ever vigilant. Over time however, it becomes more difficult to distinguish between what is common knowledge and what is not, and they are more likely to share details of what were once sensitive details.

Their Weapons

The risks can arise from a number of sources, mostly relating to knowledge of the system. Fraudsters and competitors can get information from a variety of sources: (i) *Staff movements*—Staff mobility can be high, and employees take knowledge with them, ranging from simple knowledge about verification checks to technical details about the scorecard development process, and how shadow limits can be manipulated. (ii) *Contractors and consultants*—People employed on a temporary basis get privileged access to information. Prime cases are scorecard developers, credit bureaux, external auditors, and consultants, who have access to a number of different companies. (iii) *Business intelligence*—Likened to military intelligence, this is industrial espionage, where information is obtained via nefarious means from naïve or disgruntled employees or internal documentation. This applies not only to knowledge of processes, but also details of employees, who may be headhunted. (iv) *Public sources*—Information available to the general public, either published by the company, the media, or other sources, that may not be referring to a specific company. This information is not considered sensitive, yet may provide an impetus for lagging competitors.

Our Shields

White knights are not totally defenceless. There are a number of different tools that they can use for protection against the threats: (i) *Authority levels*—Determine when permissions are needed, and who is to provide it for systems changes, communications, contractual agreements, etc. (ii) *Access control*—Set hurdles that restrict access to the things being guarded, including the use of physical security, safes, passwords, etc. (iii) *Confidentiality agreements*—Legal agreements, including restraint of trade, that prevent staff and others from divulging proprietary information to outsiders. (iv) *Documentation*—Prevent organisational Alzheimer's, by ensuring that critical aspects are well documented. (v) *Change control*—Put in place a procedure to be followed whenever changes are made to processes and/or systems. (vi) *Audit*—Analysis based upon existing documentation and staff knowledge, which is done by internal or external auditors to ensure that everything corresponds to the documentation, approved procedures, and/or best practice.

24.4.2 Scorecard and strategy controls

The controls mentioned earlier are general, and most could apply elsewhere in the organisation. Some should be considered in more detail though, as they relate directly to scorecard development and implementation, and the staff/consultants that are involved. First, confidentiality agreements should be in place to prevent people from divulging any proprietary knowledge gained as a result of the relationship with the company. This is difficult to enforce for staff members that change jobs, and while restraint of trade agreements are a possibility, they are unlikely tools, as they are usually too severe for the type of work being done.

Second, scorecard development documentation is needed, covering any and all assumptions made during the process and the information they were based on. This includes details of the sampling, good/bad definition, characteristic analyses, segmentation, reject inference, and the models themselves. For in-house developments, this documentation is highly sensitive, and even auditors may not have access to it. Separate sets of documentation may be created for use inside and outside the area. Third, authority levels should be set for signing off each stage, without which the development may not continue. Some of the major milestones are: sample selection, good/bad definition, segmentation, final scorecards, associated strategies, testing, and implementation.

Fourth, there should be password protection for both the testing and implementation phases, to ensure that only individuals empowered to make the changes have access, not only to scorecards, but also to policies and strategies. Penalties for implementing unauthorised changes should be steep. Scorecard changes should be rare, while occasional changes to strategies and policies will be more common.

Finally, the controls for strategies will be much like those for scorecard developments, except the extent of secrecy may be a bit less. The documentation should justify the strategies, or changes that are made, including any empirical analysis. The champion/challenger process is the key tool, used to provide proof that proposed changes have benefits over the dominant strategies.

24.4.3 Override controls

When railroads first rolled westwards, cowboys used to revel in racing trains, but as technology progressed, the horse became a sure loser. In like fashion, with sufficient effort, some underwriters can out-perform a credit scoring system, but it is costly, and over time the system will evolve to a point where it is futile. To protect against this, lenders must produce reports to monitor overrides, and highlight areas where the system's logic may be incorrect.

Overrides can be limited by penalising underwriters who exceed a specified maximum. It can be set high initially, and then reduced as the system improves and gains acceptance. Lack of acceptance may not be limited to underwriters but extend to management, who continue to rely upon a myriad of legacy policy rules. This dilutes the scorecards' effectiveness, and efforts must be made to identify and remove policy rules that are not adding any value, while simultaneously invoking some very few new rules, more appropriate for the scoring environment.

Changes to processes can also cause problems with dealers, brokers, and other external agents, who are channels for new business. They are often significant constituencies and business partners, and become familiar with lender policies, to the extent that they know whether or not an applicant will be accepted prior to submitting the application. For them, a decline can be quite personal, because it means lost margins or commissions. Because they do not assume any risk, their view of an applicant will be totally different from that of a lender. They may cause significant push back, motivating for low-score overrides, so extra care is needed.

24.5 Summary

With credit scoring, lenders need to put in checks and balances, and mechanisms for exception handling. Lending used to be based on credit policy rules and intuition, but even with scores, policies must be retained to deal with *rare but severe events* that cannot be effectively modelled, known *scorecard weaknesses*, or *new data* yet to be incorporated.

The *pre-score, score, system,* and *final decisions* are the major stages in the decision process. The pre-score decision relies on policy rules to pre-screen cases that should not be processed further. Score and system decisions are preliminaries that can be overridden. The system decision is a combination of score and policy, and the final decision brings in subjective input, and an assessment of information not embodied in the scores. *Low-side overrides* are rejects that are accepted, *high-side overrides* are accepts that are rejected, and *referrals* are cases that demand extra scrutiny. In some instances, applicants are *scored for guidance*, and underwriters have greater latitude, but lenders may still impose restrictions. Overrides are a necessary evil, especially to allow customers to contest the decisions, and to recognise external information not captured by the scores. Nonetheless, controls must be put in place to guard against abuse.

Referrals are overrides' poor cousins, which demand a greater emotional and physical investment. They are usually done to: (i) prompt further investigation of *borderline cases*; or (ii) protect against *fraud or embellishment*, by validating details, or requesting supporting documentation. While there may be a temptation to validate every case, *suspicion triggers* can also be used to highlight those warranting greater attention. Thereafter, there may still be issues that block the application, like product rules, insufficient security/collateral, or unacceptable loan terms and conditions.

Controls are required both for the broader business, and the scorecard development. The game comes complete with villains, heroes and heroines, weapons and shields. The primary things to be protected are the processes, documentation, and knowledge of the system, which can be threatened by all sorts of *human foibles*, including errors, deceit, loose tongues, poor memories, wild exuberance, and wanderlust. Protection can be provided by scribes, gatekeepers, and overseers, who ensure the ongoing integrity of the system. Effective monitoring must also be put in place to track overrides, and ensure that the system is working according to plan. There will always be a human element, but it should be kept within limits. Ultimately, the goal is to have scores driving the decisions, with a minimum of policy rules and human input.

25 Monitoring

> If something cannot be measured, it cannot be managed.
>
> *Lawlor and Haynes* (2003)

Now comes mention of that dirty word, monitoring! Unfortunately, it has to be done! Its purpose is to determine how well the models are working, and what is happening within the processes in which they are being used. The following discussion relates mostly to bespoke application scorecards, but other situations must be considered.

Table 25.1 provides a list of most of the reports used to monitor credit risk, whether specifically related to credit scoring or not. The three right-most columns are generalised labels for the types of processes being monitored: *selection*, which accept or reject possible new entrants (e.g. application scoring); *portfolio*, used to manage existing accounts (e.g. behavioural scoring); and *entry*, instances where all through-the-door cases are ranked (e.g. collections).

Service delivery monitoring is not covered. Key aspects of this are tracking the time it takes to: (i) return a decision; and (ii) advise the customer. These, and other operational issues, should form part of the service-level agreement with the area responsible for providing transaction-level credit decisions.

Other texts on credit scoring usually split the reports into two broad groups. Thomas et al. (2002) distinguish between monitoring and tracking reports, whereas Mays and Nuetzel (2004) use the American OCC front- and back-end distinction (Table 25.2):

Front-end reports—Focus upon process monitoring and population stability, with no reference to the performance of booked accounts. These reports can be generated as soon as scorecards are implemented.

Back-end reports—Focus upon the performance of booked accounts, and backtesting of scorecards and the selection process. Accounts must have some time to mature, before any reporting can be done.

These terms may cause some confusion, as the front-end and back-end labels are also used to distinguish between the new-business and collections functions respectively.

With both of these categories, a distinction can be made between *snapshot reports*, which reflect the situation at one point in time, and *drift reports*, which cover two or more successive periods. Other dimensions include: (i) *row contents*—score, characteristic, process stage; (ii) *column*

Table 25.1. Report types and applications

Report type	Report name	Selection	Portfolio	Entry
Portfolio analysis	Delinquency distribution		✓	
	Transition matrix		✓	
Performance monitoring	Scorecard performance	✓	✓	✓
	Vintage analysis	✓		✓
Drift	Population stability	✓	✓	✓
	Score shift	✓	✓	✓
Decision process	Decision process	✓		
	Decision by score	✓		
Adherence	Policy rules	✓		
	Override reasons	✓		

Table 25.2. Front- and back-end reporting

	Snapshot	Drift
Front-end	Process monitoring, score distribution, override reasons	Population stability, and any multiple snapshots
Back-end	Portfolio analysis, performance tracking	Vintage analysis

contents—count or monetary value, number or percentage, stand-alone or cumulative; and (iii) *function*—scorecard, override, process monitoring.

Reporting functions

Documenting the monitoring of credit scoring processes is made difficult by the nature of the beast: (i) performance data is used to develop the models, but some time is needed before enough has accrued for backtesting; (ii) initial validation is done to ensure drift has not been so great as to invalidate the scorecards or strategies; (iii) business's primary interest in the process is the inputs and outputs, not its inner workings; and (iv) a major concern is organisational adherence to the decisions provided by the system. This provides the basis for the following reports:

Front-end reports

Score drift—Tracks how population and operational drift are affecting the scoring system's results, both at characteristic and final score level.

Selection process—Covers inputs and outputs of a selection process, in particular, the volumes being processed and taken to book.

Override reasons—Covers changes in the decision as cases move through the selection process, to ensure maximum benefit is being obtained from the scores.

Back-end reports

Portfolio analysis—Details the spread of accounts across delinquency statuses, and movements between statuses over time.

Performance tracking—Used to check the power, accuracy, and stability of the scorecards, and requires sufficient time for accounts to mature.

Override performance—Checks how well cases perform if the score or system decisions have been overturned.

Portfolio chronology log—A blow-by-blow account of events that may affect the reported results, whether related to marketing, infrastructure, strategy, or the broader economy.

The reports presented here are not fixed and final, but can be reformatted and combined as required. Acceptance will be much wider, and quicker, if they are: (i) *easy to interpret*, whether tables or graphs; (ii) *consistent*, from month to month; and (iii) *easy to generate*, as often as required. The primary directive is the basic, 'keep it simple, stupid!' Analysts may have to drill deeper, so reporting systems should have sufficient flexibility for them to create different views of the data, whether by changing the rows, columns, time period, or subpopulation. This requires a parameterised system, where these elements are under their control.

25.1 Portfolio analysis

This textbook focuses primarily on credit scoring and the associated systems, but at this point it helps to take a step back. The jumping off point is reports that would be produced in an environment without credit scoring, especially the portfolio analysis reports used to track the composition of the existing book. They could be generated for any characteristic—like value, market segment, geographical region, channel, time on books, maturity, or any combination thereof. Portfolio risk is the primary focus here, and there are two main reports:

Delinquency distribution—Shows the spread of accounts according to some measure that has traditionally been considered highly indicative of risk, such as days-past-due.

Transition matrix—Shows the movement between different buckets over a specified time period, where the buckets are defined by delinquency, value, and other attributes.

25.1.1 Delinquency distribution

The primary portfolio analysis report is the Delinquency Distribution Report, which is the bedrock of any back-end reporting. The example provided in Table 25.3 gives a good indication of a portfolio's risk profile. If it looks familiar, it is because the rows also formed the basis of the roll-rate analysis, done to set the good/bad definition in Chapter 15. The table is split into three sections: (i) the *number and value* of accounts in each of the main buckets; (ii) the *percentages* that fall into each; and (iii) the *cumulative percentage* of 'non-dormant' accounts, that are at a given delinquency status, or worse. The last section is important for giving a quick indication of portfolio risk.

Table 25.3. Delinquency distribution report

Current deln	Accounts		Column %		Cum Col%	
	Number	Value '000s	Number	Value	Number	Value
Current	418,492	248,412	75.3	76.4	100.0	100.0
Overdue	30,567	18,809	5.5	5.8	20.3	23.4
30 days	12,227	7,783	2.2	2.4	14.5	17.6
60 days	12,783	8,107	2.3	2.5	12.2	15.2
90 days	13,894	8,756	2.5	2.7	9.7	12.7
120+	15,561	12,323	2.8	3.8	7.1	10.0
Legal	21,675	20,106	3.9	6.2	4.1	6.2
Dormant	40,255	649	5.5	0.2		
Totals	555,766	324,297	100	100		

Table 25.4. Transition matrix

Current state	Future state					
	Current (%)	Late (%)	Default (%)	Dormant (%)	Closed (%)	Total (%)
Current	83.6	7.0	2.2	3.1	4.1	100.0
Late	66.2	14.3	8.5	4.0	7.0	100.0
Default	12.9	8.5	23.5	37.6	17.5	100.0
Dormant	48.5	4.0	2.0	12.0	33.5	100.0
Closed	0.0	0.0	0.0	0.0	100.0	100.0

The delinquency distribution report provides nothing by itself, other than a quick overview. Further value can be obtained by comparing delinquency distributions, either over time (drift), or across different segments of the book. In each case a point of reference is required, whether the most recent distribution, or a portfolio average. The report usually provides the basis for loss-provision calculations, assuming that the business has already determined the provision rates to be associated with each delinquency status. A behavioural risk score could serve the same purpose, but the business may be resistant to using it for provisioning.

25.1.2 Transition matrix

If changes in the delinquency distribution are tracked over time, then lenders will have some idea of shifts within the book, but not know the exact mechanics. Transition matrices can provide greater insight, by detailing movements between the different states within a given time period, whether in terms of number of accounts, or monetary value (see Section 9.2.1).

Table 25.4 illustrates a summary set of account states: (i) *current*, open and has been used in the recent past; (ii) *late*, payment is slightly overdue, perhaps 30 or 60 days; (iii) *default*, payment is very late, perhaps 90 days or more; (iv) *dormant*, account is open, and if there is any outstanding amount, it is immaterial; and (v) *closed*, an exit state, where the customer has been lost.

The other exit state is write-off, which in the example has been combined with default. If possible, lenders should distinguish between 'closed good' and 'closed bad', as the latter are those where the monies were recovered, but with unwarranted difficulty or expense.

There could also be a separate class for non-performing loans that have been passed on to recoveries or legal, and for other statuses that are considered important. In a collections/recoveries environment, the same approach could be used with different classifications: (i) *reactivated*, normalised, and no longer in the system; (ii) *outsourced*, sent to an external agency; (iii) *written-off*, all cost-effective avenues have been exhausted, and the loss has been taken to book; and (iv) *in progress*, still in the system.

The example in Table 25.5 puts greater focus on delinquency states, and the percentage of accounts moving between 'No Borrowing' (NB), 'Closed', 0 to 3 'Months in arrears', and 'Write-off'. Closed and write-off are exit states (or absorption states), black holes that accounts enter, never to return. In order to simplify the example, it has been assumed that arrears on 1-, 2-, and 3-periods past-due accounts either move into the next status, or are rectified.

Table 25.6 shows the Markov process that the book will go through, if this transition matrix is correct. The details for month 0 show the portfolios' current distribution, with no write-offs

Table 25.5. Past due—transition matrix

Start	Transactions (%)						
Closed	NB (%)	Closed (%)	0 (%)	1(%)	2(%)	3(%)	W/off (%)
NB	79.0	2.5	18.5	0.0	0.0	0.0	0.0
Closed	0	100.0	0.0	0.0	0.0	0.0	0.0
0	8.0	2.0	80.0	10.0	0.0	0.0	0.0
1	8.0	1.0	71.0	0.0	20.0	0.0	0.0
2	8.0	1.0	61.0	0.0	0.0	30.0	0.0
3	8.0	1.0	51.0	0.0	0.0	0.0	40.0
W/off	0	0.0	0.0	0.0	0.0	0.0	100.0

Table 25.6. Past due—Markov process

Month	Account distribution (%)						
	NB	Closed	0	1	2	3	W/off
0	10.0	0.0	78.3	7.8	2.6	1.3	0.0
1	15.1	2.1	71.6	7.8	1.6	1.0	0.8
2	18.5	4.1	66.7	7.2	1.6	0.6	1.4
3	20.7	6.1	62.7	6.7	1.4	0.6	1.8
6	23.2	11.8	54.8	5.7	1.2	0.5	2.8
12	21.8	22.1	45.6	4.7	1.0	0.4	4.4
24	16.4	38.7	33.6	3.4	0.7	0.3	6.9
60	6.7	66.8	13.6	1.4	0.3	0.1	11.1
120	1.5	81.8	3.0	0.3	0.1	0.0	13.3
240	0.1	85.8	0.2	0.0	0.0	0.0	13.9
360	0.0	86.0	0.0	0.0	0.0	0.0	14.0

or closures, and each of the following rows is for subsequent periods. Note that it takes some time before the defaults are realised. After a year, only 4.4 per cent of accounts have been written-off, but this increases to 14.0 per cent with the balance of accounts closing. Granted, this is after 30 years, and any expected profits should have been made before then. In the meantime, new customers will join each month, who could possibly be reflected with some adjustments.

Segmentation

The above analysis was done using past-due statuses, but the same could be done using behavioural scores. More likely however, is that the analysis will be done using some combination of characteristics, including but not limited to:

Past due—The number of periods since the last payment was received.
Status codes—Serious status codes, such as legal, provision raised, written-off, closure, or other categories.
Account age—Younger accounts are usually riskier and older accounts more stable.
Application score—Adds value with younger accounts, where little history is available.
Behavioural score—Adds value with established accounts, where internal account performance data is already available.
Months since last activity—May indicate dormancy, where chance of reactivation is small and probability of closure is high.
Outstanding balance—Movements may differ, depending upon how much is at risk, if only because lender strategies often use this as a criterion.
Product holding—The dynamics may differ, depending upon the types of other accounts held, and behaviour on those accounts. This is especially true for cheque accounts.
Economic conditions—The model may be affected by interest rates, unemployment rates, or other factors in the broader economy.
Time—Separate matrices may have to be used to model seasonality, in particular holiday periods, and especially Christmas.
Mover/stayer—This involves identifying near-exit states, where a large portion of accounts will stay indefinitely (stayers), which improves lenders' ability to model those cases that change states (movers).

According to Thomas et al. (2001), the mover/stayer concept was first used in labour mobility studies, and later for consumer purchasing behaviour. In the consumer credit environment, stayers include accounts that are: (i) *low risk*, with an established history (like full-payers); (ii) in *recoveries*, where rectification is unlikely; and (iii) *dormant*, that are unlikely to close or be reactivated.

25.2 Performance tracking

Credit scoring models are developed using both observation and outcome data, so it makes sense to use the same for monitoring. The delinquency distribution report provides a very effective overview of a portfolio's risk, but no indication of what happens subsequently. Performance tracking reports serve this purpose, and are covered here under three headings:

Scorecard performance—Analyses the distribution of delinquencies, by score.
Vintage analysis—Analyses changes in the portfolio performance over time.
Score misalignment—A tool to measure discrepancies, at characteristic level.

It may be some time before these reports can be produced. For application processing, 12 to 18 months may be needed before sufficient performance can be obtained to make a direct comparison with the development sample (vintage analysis can be used in the meantime using shorter outcomes—see Section 25.2.2). In contrast, timeframes may be relatively short in collections and recoveries processes, perhaps as little as three months.

25.2.1 Scorecard performance

After scorecards have been implemented, lenders' primary interest is in whether they are performing as expected. This is often described in terms of predictive: (i) *power*, to rank cases according to risk; and (ii) *accuracy*, to provide reliable estimates of bad rates, or good/bad odds. Figure 25.1 provides an illustration of the difference between the two. The natural log odds is plotted against the score, for both the expected (original) and actual (recent) performance. When a scorecard loses its ranking ability, the slope of the line flattens. In contrast, if the predictive accuracy changes the line will shift up or down, but may remain parallel to the original. More often than not, a combination of the two forces is at play.

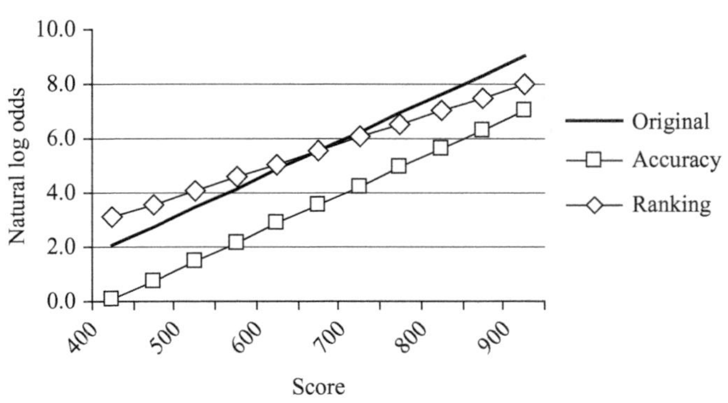

Figure 25.1. Scorecard performance drift.

The same framework could be used to test scorecard performance across key market segments and/or delivery channels, with the average population performance as the baseline. It may highlight certain sub-segments where the scorecards are less effective.

Deterioration in ranking ability is more serious, and if serious enough, the scorecard may have to be redeveloped. If the predictive accuracy has changed, the task is to recognise it, and update the strategies as soon as possible.

The basis for this analysis is the scorecard performance report, as shown in Table 25.7, which comprises: (i) score ranges; (ii) the number and percentage of booked accounts that fall within each range; and (iii) the percentage of those within the range that fall into each performance category. The example provides the percentages falling in each score range, but many lenders prefer the cumulative distributions used for the Gini and KS calculations. 'Not-taken-ups' and/or 'dormancies' may be presented as separate columns, if they are also an issue.

Performance comparisons

Such snapshots provide little! More can be gained from determining drift relative to a benchmark, such as the development sample, or one or more prior periods. It must be done on a like-for-like basis, and when monitoring application scores, the choice of benchmark and outcome period must be considered. In the example, the scorecard is providing some value, but a Gini

Table 25.7. Scorecard performance report

Score range		Booked		Row percentages		
Low	High	#	Col (%)	Good (%)	Ind (%)	Bad (%)
0	780	64	0.2	36.8	13.2	50.0
781	825	77	0.2	42.3	12.0	45.7
826	855	137	0.4	46.8	10.6	42.6
856	880	259	0.7	50.2	9.1	40.7
881	900	438	1.2	53.8	7.7	38.5
901	925	809	2.1	58.0	6.4	35.6
926	945	2,045	5.4	61.5	5.5	33.0
946	965	2,623	6.9	65.5	4.5	30.0
966	985	3,699	9.8	69.4	3.7	26.9
986	1,005	4,407	11.6	72.6	3.0	24.4
1,006	1,025	4,880	12.9	75.8	2.5	21.7
1,026	1,050	6,369	16.8	79.1	2.1	18.8
1,051	1,090	7,785	20.5	82.0	1.7	16.3
1,091	High	4,294	11.3	83.3	1.5	15.2
Totals		37,886	100	75.0	2.9	22.1
Gini coefficient		20.2%	KSstatistic		15.2%	

coefficient of 20.2 per cent is not much, if 35 per cent is considered the (thumb-suck) minimum acceptable for an application-scoring development. This comparison is not on a like-for-like basis though, as the benchmark is for performance inclusive of reject inference, and here there are no rejects. The choices are to either apply reject inference, or to strip rejects from the benchmark. Both of these are impractical, and even if the latter were attempted, it assumes that similar strategies have been employed.

Likewise, a common outcome period should be used. This can hamper comparisons against the development sample for application scoring developments. It has become easier though, as today most application processing systems allow review of account performance at specific time periods after acceptance, say every six months (see 'Vintage Analysis', in next section). For a proper comparison, the same performance characteristics should be extracted for each sampled case.

Other issues to be considered are changes within the business, and its operations. In application scoring, the greatest attention is focused around the cut-off. Below the cut-off, the number of cases may be small or non-existent, especially if the score decision is being strictly enforced. If not, policy rules and manual overrides may be either complementing (intelligent overrides) or subverting the scores, thus influencing the performance of marginal accepts. For behavioural scoring, and application scores used in risk-based pricing or limit setting, there will be an interest in the full score spectrum.

Performance definitions

It takes some time before sufficient performance is available for proper comparison using the development good/bad definition, and in the interim, different performance definitions can be used as the accounts mature. For example, for early performance monitoring during the first six months after a deal is booked, the percentage of accounts that are 30+ and 60+ days-past-due may be tracked. The 30-days category will capture *first-payment defaulters*, where special attention may have to be paid to identify potential fraud. It can also provide an early indication that the score cut-off needs to be reassessed, or that a change elsewhere (policy, marketing, infrastructure) is affecting risk. After six months to a year, attention will be shifted onto the 60+ and 90+ categories.

> Early delinquencies can be influenced by a number of factors, including problems with the postal system—especially strikes. Thomas et al. (2002:160) mention a 1980s UK postal strike, where many cardholders believed that no payment was required if the statement was not received. The 30- and 60-days-past-due categories increased significantly, from 10 to 30 and 4 to 7 per cent respectively, but these normalised soon after the strike was over.

Even once cases have matured, lenders may still wish to track using other definitions. The most likely alternatives are bad/not bad definitions, using either: (i) the bad definition, as per the good/bad definition; (ii) a hard bad definition; (iii) a definition specified by an external authority, like the Basel II default definition.

25.2.2 Vintage analysis

When monitoring new entrants, lenders want some inkling of their performance, as soon as possible after they are through the gates. This applies to both new-business acquisition (bad, attrition) and collections and recoveries (reactivation, hard action). New entrants are binned into groups—by entry month, quarter, or some other period—which are assessed at fixed 'time since entry' (TSE) intervals thereafter (such as 'time on books', 'account age', or 'time in collections').

The choice of TSE period will be determined by the volume of new entrants, as the cell values are difficult to interpret if the numbers are small. The reporting system's design should allow aggregation over different periods, and some longer-term trends may only become apparent when aggregated by year.

Entrants that fall within the same group are either called cohorts, or are said to be of the same vintage. An example of a cohort/vintage analysis report is provided in Table 25.8. It can be used to monitor the outputs of any process with a defined entry point, including both selection and recovery processes. In the example, no reference is made to scores, but it is possible to create separate reports for each score range. When focused specifically upon delinquencies, these reports are sometimes called 'dynamic delinquency' or 'dynamic trend' reports. The same concepts can also be used to track other performance aspects, such as balances and closure.

Each Roman legion was comprised of 10 cohorts, each of which had between 300 and 600 men. In scientific studies, a cohort is a group of individuals with a common characteristic, which in credit and insurance is often limited to people born within the same year. The term vintage is usually used with respect to the year, or season, when grapes are harvested, and the age of wines. It is also often applied to other product batches of a common age, and is entirely appropriate in account origination; just like fine wines, accounts also mature.

Table 25.8. Cohort/vintage analysis—by bad rate

Accept date	Time on books							
	@03	@06	@09	@12	@15	@18	@21	@24
Q1/CCY1	2.8	4.8	7.2	8.6	9.4	9.6	9.7	10.0
Q2/CCY1	3.1	5.8	7.9	9.4	10.2	10.6	11.0	
Q3/CCY1	2.9	4.9	7.1	8.8	9.6	10.3		
Q4/CCY1	2.8	4.7	7.3	8.6	10.0			
Q1/CCY2	2.9	4.9	7.1	9.2				
Q2/CCY2	2.6	4.8	7.9					
Q3/CCY2	2.8	5.3						
Q4/CCY2	3.4							

Such reports always have the entry date and TSE on the two axes, but the cell definitions can vary (if the cell is blank, it means that there is of yet no performance data for that cell). Bad rates and good/bad odds are the norm, but the cells could display any other percentage, ratio, value, or count. Some possibilities are total value, average account balance, and limit utilisation. Any analysis should take into consideration changes over the period, whether to strategies, infrastructure, or the environment.

According to Thomas et al. (2003:158) the cohort (date) definition should be consistent with the product, and consistently applied. Possible date characteristics include the application, approval, booking (account opening, draw down, or activation), or expected first-payment date. Examples are the 'activation' date for credit cards, and 'drawdown' for loan products.

According to McNab and Wynn (2003), a vintage analysis can be analysed along three dimensions: (i) the *rows*, or life-cycle effect; (ii) the *columns*, or new-account effect; and (iii) the *diagonals*, or portfolio effect. The life-cycle effect refers to changes as new entrants mature, as illustrated in Figure 25.2, which shows the bad rate curves for different cohorts. The portfolio effect (reflected along the diagonal of Table 25.8) is similar, but looks at the current portfolio's composition going backward from the most recent month.

The life cycle and portfolio effects are both time effects, where the bad rate increases with the TSE. Each of them has a different point of departure, and approaches the problem from different directions—forwards and backwards in time respectively.

Finally—and most importantly, because this is where the real value is obtained from this analysis—Figure 25.3 shows the new-account effect, which tracks entrants at specific TSEs (3, 6, 9, etc.). Any shifts will likely be the result of changes to the process, strategies applied, the economy, or other factors. This applies no matter what is being measured (bad rate, attrition rate, limit utilisation, etc.), and adverse changes can highlight a need for corrective action.

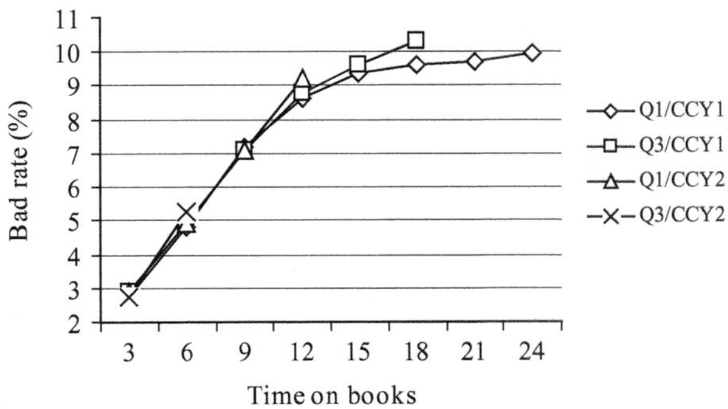

Figure 25.2. Life-cycle effect.

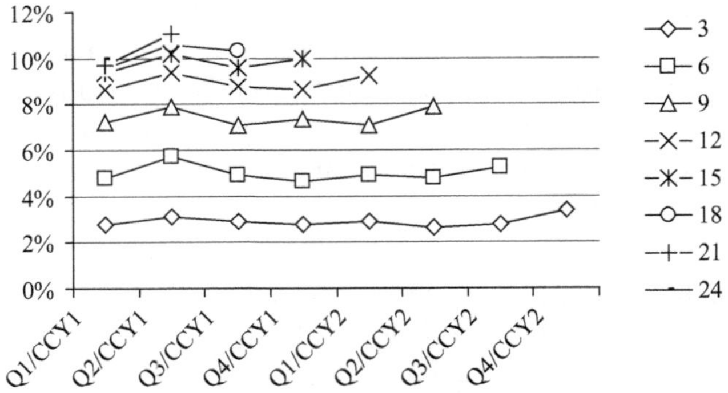

Figure 25.3. New-account effect.

Gerard Scallan of ScorePlus, a UK consulting agency, calls these 'triangular matrices', and suggests an innovative 'fill-in-the-blanks' approach for cells below the diagonal. It works by developing a regression model on non-blank cells, to estimate the log odds of each empty cell as a function of the row, column, and diagonal numbers. The key is a piece-wise approach, which recognises the non-linear nature of the problem. For example, the life-cycle effect may be flat for very new and very mature accounts, but linear in between. Economic variables like interest and unemployment rates may also be incorporated.

Growth and contraction

Care must be taken here, as even if the risk distribution of new accounts does not change, the portfolio's risk can be affected by significant changes in new-business volumes. Where new account volumes are high, the loss rates as a percentage of the total will improve initially, but deteriorate as the recently acquired accounts mature. Likewise, if a portfolio is shrinking, the loss rates will increase, because the higher proportion of mature accounts will be generating bad debts.

By extension, if a lender is trying to grow its portfolio by targeting riskier markets, it should ensure that it sets provisions accordingly. There is a tendency to raise provisions only when potential losses become obvious, which may only occur after two or more years. For new-business risk to be provided for effectively, it is much more prudent to start raising provisions as part of account origination.

25.2.3 Score misalignment

'When should the scorecards be redeveloped?' The easy answer for this oft-posed question is, 'When the business really starts feeling uncomfortable with the scores, and the problem cannot be fixed by adjusting cut-offs!' This answer may not go down so well, but as stated by Thomas et al. (2002:161), 'There is no simple answer or simple statistical or business test that can be performed to decide when corrective action is required and what it should be'.

While the knee-jerk reaction is to suggest redeveloping the scorecards at the first sign of problems, lenders can take steps to determine their extent, and whether a simple realignment is possible. The primary tools used are the score misalignment report and graph, examples of which are presented in Table 25.9 and Figure 25.4. If a scorecard is working correctly, then the risk of all cases falling within a score range should be the same; there should be no characteristics with attributes of significantly different risk. If so, the points should be adjusted or the characteristic should be included within the scorecard if it is not already there.

The illustration for 'Home Phone Given (Y/N)' assumes the scores have a 200-point baseline, and 20-point odds doubling. It uses the natural log odds as the basis of comparison, and the two misalignment columns show what score adjustment would be required, to bring them into line. The average log odds for the 'Yes' and 'No' groups are slightly different, and it would

Table 25.9. Score misalignment

Score	Log odds			Misalignment	
	Portfolio	Home phone	No home phone	Home phone	No home phone
Points		0	−18	3.0	−5.0
180	3.47	3.58	3.29	3.2	−5.2
190	3.81	3.89	3.66	2.3	−4.3
200	4.16	4.29	3.97	3.7	−5.4
210	4.51	4.58	4.34	2.2	−4.7
220	4.85	4.97	4.66	3.5	−5.5
230	5.20	5.29	5.01	2.7	−5.6
240	5.55	5.70	5.38	4.3	−4.6
250	5.89	5.98	5.71	2.5	−5.3
260	6.24	6.33	6.09	2.7	−4.3
270	6.58	6.67	6.40	2.5	−5.3
280	6.93	7.04	6.78	3.1	−4.4

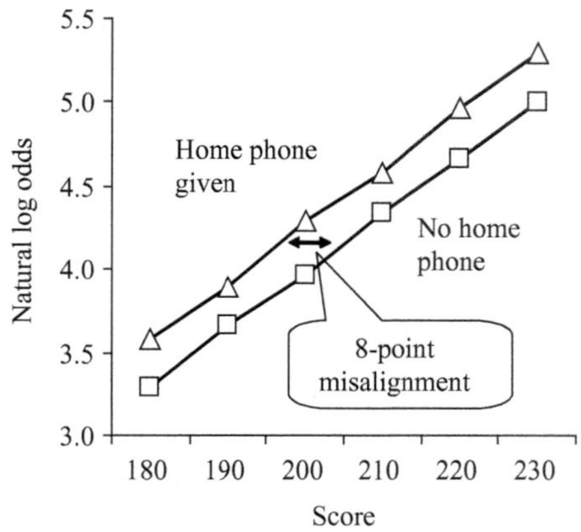

Figure 25.4. Score misalignment.

require an adjustment of about 8 points to fix. Rather than trying to answer, 'How did this occur?' lenders will instead focus on, 'What can be done about it?' Redeveloping the scorecard is always a possibility, but: (i) it is expensive, both financially and emotionally; and (ii) it may not be feasible, because of a lack of data.

If the problem is limited to one or two characteristics, an alternative is to extend the life of the existing scorecard by doing a realignment, but this should be done with care. Scoring models are developed using statistical techniques, which take into consideration correlations between all of the characteristics, and it can be extremely dangerous to start modifying individual point allocations. Steps must be taken to ensure that the modified model is truly an improvement!

> Thomas et al. (2002:162) suggest that where there is some concern about the changes, they can be implemented on a champion/challenger basis initially, and in full only once the business is more comfortable with the adjustments.

The manner in which the realignment is done can vary, depending upon whether the lender has a concern with one or both groups. In the example, the current point assignment for 'No Home Phone' is currently −18. If the lender is comfortable with the score for 'Home Phone = Y', then the simple solution is to adjust that score 8 points down, to −26. If, on the other hand, it wants to keep all scores in line with the original, it would either: (i) reduce 'No Home Phone' to −26, and increase the score constant by 3; or (ii) reduce 'No Home Phone' to −23, and assign 3 points to 'Home Phone Given'.

Scorecard degradation is a slow process in stable environments. Many scorecards have been used, and have performed effectively, for seven years or longer, and were only updated because the business thought that they were too old. Indeed, Lewis (1992:116) suggested that lenders should consider how much the scorecards have to degrade before they will consider replacing them, and try to predict when this replacement is necessary. There are, however, times where the business will be forced to redevelop the scorecards, either as a reactive move to address infrastructure, market, and process changes, or as a proactive move to take advantage of new data sources.

> Lewis (1992:116) also suggested that lenders 'can plan the assembly of the sample data that will be needed for the development of the new system'. This is no longer such an issue in environments with sophisticated application processing systems, that can store and monitor applications. It may, however, still be a concern for lenders that have not made the same investment in systems.

25.3 Drift reporting

Ultimately, the scorecards will be used within a process, and there will be a lot of questions relating to fit within the process, and drift over time. Changes can occur because of strategic, operational, marketing, or economic drift, and if the changes are great enough, there may be sufficient motivation to redevelop the scorecard. The primary tools for tracking the extent of drift and its source(s) are:

Population stability reports—Track changes in characteristic distributions, especially the score, and measure the variance between observed and expected frequencies. The greater the similarity, the more likely the scores are still working as expected.

Score shift report—Identifies the source of changes in the score distribution, which can be done at score or attribute level; the smaller the shift, the better.

25.3.1 Population stability report

Something oft restated within this text is the fact that credit scoring is backward-looking, and environmental shifts can be detrimental. Shifts can be monitored though, using population-stability reports like that shown in Table 25.10. It presents the empirical cumulative distribution functions (ECDF) for the development sample, and one or more recent periods. The ECDFs are also used to calculate a stability index and/or KS statistic of the development sample,

Table 25.10. Population stability report

Score range Low	Score range High	This month (%)	This quarter (%)	Last quarter (%)	Year ago (%)	Development (%)
0	780	4.6	7.8	7.3	5.6	4.9
781	825	7.2	13.0	12.6	8.3	10.3
826	855	16.5	20.0	16.9	18.1	19.7
856	880	23.3	28.2	25.8	26.2	32.2
881	900	32.1	40.1	34.9	35.2	42.8
901	925	46.4	51.0	46.1	51.9	57.8
926	945	59.3	62.9	59.6	61.9	65.9
946	965	79.0	74.4	74.2	81.0	74.9
966	985	84.0	85.3	85.6	84.8	81.0
986	1,005	93.2	91.7	93.0	94.1	93.6
1,006	1,025	99.6	96.1	96.4	99.4	96.9
1,026	1,050	99.8	98.3	98.5	99.8	98.6
1,051	1,500	100.0	100.0	100.0	100.0	100.0
Below cut-off		23.3	28.2	25.8	26.2	32.2
Above cut-off		76.7	71.8	74.2	73.8	67.8
Total processed		19,157	16,432	16,800	15,731	28,074
Average score		920	916	920	915	915
Stability index		0.11	0.09	0.07	0.05	
KS statistic (%)		11.4	6.8	11.7	7.6	

relative to each of the other periods. The example also provides several other elements, to aid the interpretation of the results:

Above and below cut-off—Provides an indication of what the reject rates would be, if the score cut-off (881 in the example) were strictly enforced.
Average score—Used to monitor the overall risk of through-the-door applicants. It may be simplistic, but can nonetheless indicate important trends not already highlighted by the above and below cut-off figures.
Stability index—The distribution is benchmarked against the development sample: values less than 0.10 indicate minimal change; up to 0.25, moderate change; and over 0.25, change that may give rise to concern (see Section 8.2.3).
KS statistic—Likewise, except it provides the maximum percentage difference between the recent and development ECDFs.

The example focuses upon the total applications for each of the periods, but it is also possible to drill down into the decisions made. Mays and Nuetzel (2004) provide an example that has separate blocks for total applications, not taken up, approved, and booked, across the different periods. The latter three are presented as a percentage of total through-the-door applications over the period. Further analysis can highlight changes that might have their origins in: (i) customers accepting *competitors' offers*; (ii) *process problems*, which cause customers to go elsewhere; and (iii) *policy rules and judgmental overrides*, which overturn the score decision.

25.3.2 Score shift report

The population-stability index can indicate that the score distribution is changing, but provides no insight into the causes. Analysts can drill down further by calculating score shifts, which for attributes are calculated as:

Equation 25.1. Attribute score shift $S_i = \left(\frac{O_i}{\sum O} - \frac{E_i}{\sum E}\right) \times \beta_i$

where i is an index for an attribute, β_i is the point allocation, and O and E are the recent and development sample frequencies respectively. Thereafter, the score shifts for the characteristic and scorecard are the sum of the prior level values:

$$S_c = \sum_{i=1}^{k} S_i \quad \text{and} \quad S = \sum_{c=1}^{l} S_c$$

where k is the number of attributes within the characteristic, and l is the number of characteristics within the scorecard. The calculation for a single characteristic is illustrated in Table 25.11.

Table 25.11. Score shift calculation

Accom status	β_i	Development	Recent	Score shift
Own	20	5,000	6,000	1.47
Rent	0	4,000	3,500	0.00
LWP	5	1,500	1,200	−0.14
Other	0	1,000	1,100	0.00
Totals	5.8	11,500	11,800	1.33

Unfortunately, although the characteristic-level analysis is easier, it can be misleading. Point allocations (β_i) can have opposite signs, such that attribute-level shifts offset each other, which forces a thorough review of all attributes This can be addressed by: (i) using absolute values for the score-shift calculation; or (ii) restating all of the points as positive values for the final scorecard. For the latter, if the value for 'live with parents' were −5, the points could be restated as Own = 25, Rent = 5, LWP = 0, and Other = 5 (and also deduct 5 points from the constant). The former is the more common approach.

This analysis provides little, other than an understanding of what is causing changes in the score distribution. If the shifts are significant, there are typically only three possible actions: (i) *do nothing*; (ii) do a *score realignment*, which can be dangerous; (iii) *redevelop* the scorecards using more recent data. There are only rare instances where the shift is the result of operational issues that can be corrected.

25.3.3 Characteristic analysis—booking rates

The characteristic analysis report was presented in Section 9.4 (Basic Scorecard Development Reports) and Section 16.2.1 (Characteristic Analysis Report), as a tool used to: (i) assess characteristics' predictive power; and (ii) for coarse classing. There, the focus was on performance statuses, but the same format can also be used for selection statuses. In particular, booking rates once the scorecards are implemented and operating. Significant shifts can indicate problems that were not identified during initial testing, or arose thereafter.

Thomas et al. (2002:153) mention the case where data capture operators classify 'Occupation' into a number of predefined categories. If the distribution changes significantly, a possibility is that new operators are not being properly trained and/or motivated, or the checking mechanisms have become lax. A key indicator of data quality issues is changes in the 'Other' category, which may vary over time, or across different offices within the same organisation.

The characteristics and attributes used should correspond closely to those used in the scorecard, but there may be differences. Table 25.12 shows a basic snapshot for 'Judgments on Bureau', where the booking rate for applicants with judgments is, understandably, much lower than average. This provides little value by itself, but instead needs to be benchmarked against the development sample, or prior periods. Mays and Nuetzl (2004) stress the importance of

Table 25.12. Characteristic analysis—booking rates

Attribute	Judgments on bureau					
	Through the door	Col %	Booked	Col %	Booking rate	WoE
Total	85,268	100.0	60,450	100.0	70.9	
No match	4,381	5.1	2,988	4.9	68.2	–0.127
No judgment	76,738	90.0	56,254	93.1	73.3	0.120
Judgments	4,150	4.9	1,207	2.0	29.1	–1.781
Information value × 100			18.90			

monitoring both through-the-door and booking rates, as it can highlight both inconsistencies and opportunities:

Example 1—*Customer age.* If the proportion of applicants under 25 increases significantly, but the proportion booked for that group remains constant, it may indicate that a credit policy rule is blocking a marketing campaign targeted at the youth market. Effective communications are crucial for these conflicts to be avoided.

Example 2—*Fixed or variable interest rate.* If the deals booked show a shift from fixed to variable interest rates, while the proportion of customers applying for each remains unchanged, then it could indicate that the product preference of lower-risk applicants has shifted, whether because of differential pricing, changed perceptions of future interest moves, or changing risk-tolerance levels.

25.4 Selection process

The origins of credit scoring were in new-business processing, which is still the cornerstone of retail credit risk management. Makuch (2001:8) estimates that 'as much as 80 per cent of the "controllable and measurable" risk is observable at the point of underwriting.' Stated differently, greater gains are to be had from stopping bad business at the gate than from taking evasive action once it is inside. As a result, lenders invest heavily in their application processes, and monitoring the inputs, outputs, and what happens in between. The exact form of the reports will vary from organisation to organisation, whether in terms of the level of detail, number of reports used, and whether they are snapshots or drift reports.

Report types

In Sections 25.4.1 through 25.4.4 the selection-process monitoring-reports are described according to the information provided as rows in each:

Decision process—Shows the flow of applications from point of entry through to final decision, inclusive of override monitoring.

Score distribution—Presents the score by system or final decision, to show how it has influenced the decision, and the extent of score overrides.
Policy rules—An overview of individual policies, and their impact upon accept/reject rates.
Manual overrides—An analysis of override reason codes and their influence upon accept and reject rates.

The business's greatest interest is in the inputs and outputs of the process, and any interest in the mechanics is to ensure that the desired results are being obtained. Thus, it will focus mostly on the decision-process monitoring reports, which make little or no reference to the scorecards. At the top is the entire through-the-door population; and at the bottom, booked deals. Along the way, there are several different points where cases are either lost, or re-channelled. This process is covered in more detail in Module F: Risk Management Cycle, but is covered briefly here in Section 25.4.1. The other reports go further, to provide some insight into what influenced the decision during the process.

Required characteristics

Setting up these monitoring reports is easier if certain pieces of key information are available, for each case being decisioned:

Undecisioned reason—Code to indicate why a case has either not been scored, or has been removed from the process prior to a final decision being made.
Scorecard—If there is more than one scorecard, it is important to know which one was used to derive the score.
Score—The total points generated by the scorecard, possibly including the raw score, calibrated score, and/or risk grade.
Score decision—The accept/reject/refer decision, based upon the score and the cut-off strategy that was in place when the application was processed.
Policy rule—Code, indicating which rule was invoked to modify the score decision.
System decision—The accept/reject/refer decision, based upon the combination of score and policy (automated rules), prior to any human intervention.
Override reason—Code indicating the reason given to justify a manual override.
Final decision—The accept/reject decision advised to the customer, subsequent to both policy and human intervention. For many applications, the final decision will not have been made at the time of reporting.
Take up—Customer decision, as reflected by whether or not a deal was booked and/or funds were drawn. This requires some means of matching records on the application processing and account management systems.

Even though account performance is not referred to in front-end reports, there is still a time element. If the lead-time for lenders and/or customers to make up their minds is significant, the undecisioned and unbooked deals for recent months will be misstated. It may take weeks or months for booking rates to stabilise, especially for high-ticket lending, like home loans.

In an ideal world, there will be separate fields for each of these, and perhaps others, but lenders usually have to make do with what they have. Where applicable, the product type, interest rate, repayment period, and other features offered to the customer should be recorded, as these will influence the take-up rates. Also, key dates or times associated with the decision process—like application, capture completion, system decision, final decision, and booking—are needed to help identify what could be costly bottlenecks in the process; the longer it takes to provide a decision, the more likely that business will be lost.

A key hurdle is obtaining data from the various sources, especially credit bureaux or other agencies. With automation, this is less of an issue, and the focus is instead on referrals, and communications with the customer. There are, however, instances that can be difficult to automate, such as where there are several principals for a small-business loan.

Decision tracking

A useful tool for monitoring the decision overrides is a transition matrix, such as that provided in Table 25.13. This shows the extent of system overrides, which are summarised by the override rates provided in the last row: (i) *low-score* or reject overrides (200/1700 = 11.8%); *high-score* or accept overrides (200/8300 = 2.4%); and *total* overrides as a percentage of decisions made (400/1000 = 4.0%). The same format can be used to check the system decision and/or final decision against the score decision. If the override rates are thought to be unreasonably high, the lender may wish to drill deeper into the policy rules and override reason codes. For the latter, lenders may want to go even further, to find the greatest culprits; whether at market segment, branch, or individual underwriter level.

Some confusion may arise when using the term 'override'. For the purposes of this textbook, score and system overrides refer to what is being overridden, while policy and manual overrides refer to how it is being done. Score overrides may be done through automated policies or manually, while system overrides can only be done manually. Also, the terms low-score, reject, and

Table 25.13. Override monitoring

Final decision	System decision		
	Accept	Reject	Total
Accept	7,800	200	8,000 80%
Reject	200	1,500	2,000 20%
Total	8,300 83%	1,700 17%	10,000
Override rates (%)	2.4	11.8	4.0

upward overrides are often used synonymously, as are high-score, accept, and downward overrides. This is not an issue in environments where scores prevail, but if policies play a large role, it can cause confusion.

25.4.1 Decision process

The starting point when monitoring any selection process is the total through-the-door population, or at least those that have been recorded. The illustrations presented here focus upon a simple snapshot of the numbers and percentages for a single period, and are treated as several small reports, which could possibly fit onto a single page (Table 25.14). In practice, these components would be combined in some fashion to track drift over time, and every lender will develop its own format. The process is covered under several cascading headings:

Through-the-door—Number and percentage of cases processed through the system, which may be presented as a raw number, column per cent, and/or cumulative percentage.

Not decisioned—Cases that did not make it through the system, for various reasons (see below). If their numbers are large, failure to report on them can hide operational inefficiencies.

Decisioned—Cases for which a decision was provided. Of primary concern are the final accept/reject and take-up rates, as well as override monitoring, to track adherence to the system.

Not decisioned

Undecisioned cases should form a small proportion of the total through-the-door cases, but if their numbers are significant, an analysis may be able to identify inefficiencies that can be rectified, or business that is being unnecessarily lost due to bottlenecks in the system. Decisions may be lacking because the applications are:

Table 25.14. Through-the-door and not decisioned

Description	#	%
Through-the-door	5,810	100.0
Not decisioned	610	10.5
Decisioned	5,200	89.5
Not decisioned	610	100.0
Out of scope	240	39.3
Withdrawn	70	11.5
Incomplete	300	49.2
Work in progress	60	9.8

Out-of-scope—Cannot be handled by the process, because: (i) the system is not designed for it; and/or (ii) the business unit is not responsible for cases of that nature, and they are rechannelled to the appropriate area.
Incomplete—Required information or documentation is missing, such as application form details (statutory fields, not signed) and supporting documentation (identification, financial statements, legal authorities).
Withdrawn—At customer request, whether because the documentation cannot be obtained, the need for the product no longer exists, or funds have been obtained elsewhere.
Work-in-progress—Still in the system, and awaiting a final decision.

Decisioned

The primary focus of process monitoring is decisioned cases, in particular:

(a) The score decision, based upon scores and associated cut-offs.[1]
(b) The system decision, as provided by the process, based upon both score and policy.
(c) Manual overrides, differences between system and final decisions that result from human intervention.
(d) The final decision, advised to the customer.
(e) The customer decision, to determine whether acceptance was mutual, and resulted in a booked deal.

In some dynamic multiproduct environments, the reporting is extended to the product requested, up-sell or down-sell offers, and product approved. Products will usually have different terms and conditions, like revolving versus fixed-term, and unsecured versus secured. Appropriate fields must be provided, to report adequately on them.

Rather than detailing different reports here, various dimensions are described around which reports can be developed. They can be set out as rows or columns, and be combined with each other, or other dimensions, such as predictive characteristics (characteristic analysis), score (override monitoring), or time (drift analysis).

Score decision (Table 25.15)—Accept/reject/refer, which is a combination of score and the lender's cut-off strategy. The counts and percentages may be provided not only for the totals, but also the final rejects and booked deals.

[1] In some instances a pre-score decision may be included to handle out-of-scope, statutory policy declines, and other cases.

System decision (Table 25.16)—Accept/reject/refer, but this time it is the score decision, overlaid with policy rules. Further detail may be provided, regarding whether the decision is based upon: (i) *statute*, where the deal is not allowed by law; (ii) *policy*, where one or more preset rules is the deciding factor; or (iii) *score*, based purely upon the cut-off score.

Overrides and refers (Table 25.17)—All decisions can, more or less, be split into six categories, based upon the system and final decision. Greater focus may be put on cases where the two do not agree: (i) *accept overrides*, system accepts that were declined; (ii) *reject overrides*, system declines that were accepted; and (iii) *refers*, that may be accepted or rejected. Further value can be had from including extra columns for the counts and percentages of booked deals.

Table 25.15. Score decision

Description	#	%
Total	5,200	100.0
Accepts	3,575	68.8
Refers	675	13.0
Rejects	950	18.3

Table 25.16. System decision

Description	#	%
System decisions	5,200	100.0
Accepts	3,400	65.4
Refers	800	15.4
Rejects	1,000	19.2
Statute	50	5.0
Policy	200	20.0
Score	750	75.0

Table 25.17. Overrides and refers

Description	#	%
System accepts	3,400	100.0
Accept/Accept	3,000	88.2
Accept override	400	11.8
System rejects	1,000	100.0
Reject/Reject	800	80.0
Reject override	200	20.0
System refers	800	100.0
Refer/Accept	600	75.0
Refer/Reject	200	25.0

Table 25.18. Final decision and take-ups (booking)

Description	#	%
Final decision	4,400	100.0
Rejects	600	13.6
Accepts	3,800	86.4
Booked	3,300	86.8
Not taken up	500	13.2

Final decision and take-ups (Table 25.18)—The final decision is either reject or accept. An 'accept' does not automatically mean that a deal has been concluded, so they are further split into: (i) *taken-up* (booked), where an account has been opened, and utilised to a minimum acceptable extent; and (ii) *not-taken-up* (NTU, or not booked), where the customer either does not respond to the deal, declines the terms offered, or does not use the facility once opened.

A special case is dormancies, where the facility is never used, or is used briefly. The latter occurs where people avail themselves of special offers, or are trying to establish a credit record. It may only be possible to analyse these in back-end performance reporting, but where possible they should be treated as part of selection-process monitoring.

25.4.2 Decisions by score

Where credit scoring is used as part of selection processes, lenders want to know the extent of the scores' influence, which can be determined by analysing the reject, NTU, and booked deal rates, by score. An example is provided in Table 25.19, which highlights a distinct difference above and below the cut-off of 925. The graph in Figure 25.5 uses a different dataset, but is a graphical representation of the same type of report. It shows the number of accepts and rejects by score; and the associated low- and high-score override rates, assuming a score cut-off of 500. As would be expected, overrides are clustered around the cut-off.

Once again, much of the benefit of this analysis comes not from looking at a snapshot, but from analysing drift over time. Lenders' greatest interest is in changes to override and booking rates. If the changes occur suddenly, they are most likely the result of a new policy, a process change, or a significant change in the profile of the through-the-door population. Gradual changes will result from modest changes in the level of adherence; either the underwriters are buying into the new tool, or are losing confidence in it; or the controls are being tightened, or loosened. A decrease in booking rates around the cut-off can also be indicative of problems with the time taken to provide a decision.

25.4.3 Policy rules

As stated earlier, the system decision is affected by statute, company policy, and score. For all intents and purposes though, both statute and company policy can be treated under the heading of 'policy rules'. These are usually applied after the score decision has been determined,

Table 25.19. Final decision and take-ups by score

Score range		Thru-the-door		Selection status (%)		
Low	High	#	Col %	Reject	NTU	Booked
0	780	1,321	2.8	93.4	1.8	4.8
781	825	973	2.1	87.4	4.7	7.9
826	855	1,038	2.2	80.6	6.2	13.2
856	880	1,206	2.6	72.8	5.7	21.5
881	900	1,351	2.9	64.2	3.4	32.4
901	925	2,086	4.5	57.1	4.1	38.8
926	945	2,282	4.9	3.6	6.7	89.6
946	965	2,964	6.4	2.9	8.6	88.5
966	985	3,974	8.6	2.1	4.8	93.1
986	1005	4,842	10.4	1.9	7.0	91.0
1,006	1,025	5,088	11.0	1.4	2.7	95.9
1,026	1,050	6,693	14.4	1.1	3.7	95.2
1,051	1,090	8,197	17.6	1.0	4.0	95.0
1,091	High	4,448	9.6	0.7	2.8	96.5
Totals		46,463	100.0	13.9	4.5	81.5

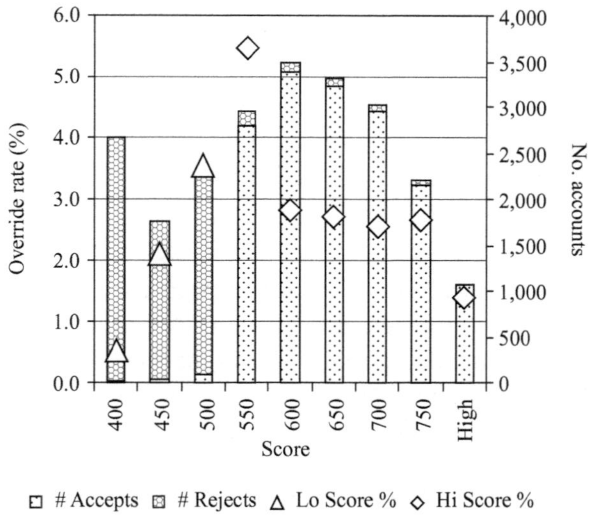

Figure 25.5. Final decision and score overrides by score.

and can be viewed as automated score overrides (see Sections 24.1 and 24.2). Any reporting is meant to determine: (i) the extent of adherence to policy rules; (ii) whether they are providing any value; and (iii) if it is possible to change policies, to optimise the benefits obtained from the scores.

Table 25.20. Policy rules—accept/reject

June CCY3–June CCY4	Total Apps	Rejects		
		No.	Col %	Row %
Total	50,025	13,711	100.0	27.4
Policy accepts	30,581	1,470	10.7	4.8
Policy declines	12,097	10,544	76.9	87.2
Policy refers	7,347	1,697	12.4	23.1
Policy declines	12,097	10,544	76.9	87.2
Unrehabilitated insolvent	25	24	0.2	96.0
Under age	45	29	0.2	64.4
Loan to asset value >110%	792	769	5.6	97.1
Repayment/income >20%	1,422	1,297	9.5	91.2
Adverse bureau >3	1,674	1,527	11.1	91.2
Poor asset condition	516	446	3.3	86.4
Poor past experience	445	341	2.5	76.6
Positive ID not provided	380	339	2.5	89.2
Failed score	6,798	5,772	42.1	84.9
Rehabilitated insolvent last 5 years	101	71	0.5	70.3
Fail score—V.I.P. indicator	862	366	2.7	42.5
Loan to value >80% and marginal accept	202	93	0.7	46.0
Possible fraud	157	48	0.4	30.6
Problems on existing account	2,558	605	4.4	23.7
Adverse on bureau >1	422	106	0.8	25.1
Repayment/income >15%	2,256	304	2.2	13.5
Age >60—check payment term	789	104	0.8	13.2

Table 25.20 illustrates an application monitoring report, used to check underwriters' adherence to policy. Slavish adherence is only expected for statutory rules; otherwise, there will always be times where overrides are justified, especially if customers contest the decisions. Very low levels of adherence can indicate that a policy rule is redundant, which can be confirmed if sufficient performance data is available. If the default rates are low, there may be sufficient justification to delete a policy-reject rule, or alternatively change it to a policy refer. Any instance where both bad rates and reject rates are high must be retained as policy rejects, and new policies must be implemented for any high-risk pockets that are identified.

25.4.4 Manual overrides

When credit scoring is first implemented, underwriters usually have significant latitude to do judgmental overrides. It is usually an interim measure to ensure the system is working, and over time there will be stricter enforcement of the system decisions. Even so, such overrides can still provide valuable feedback, especially if groups are identified where low-score overrides perform significantly better than accepts marginally above cut-off.

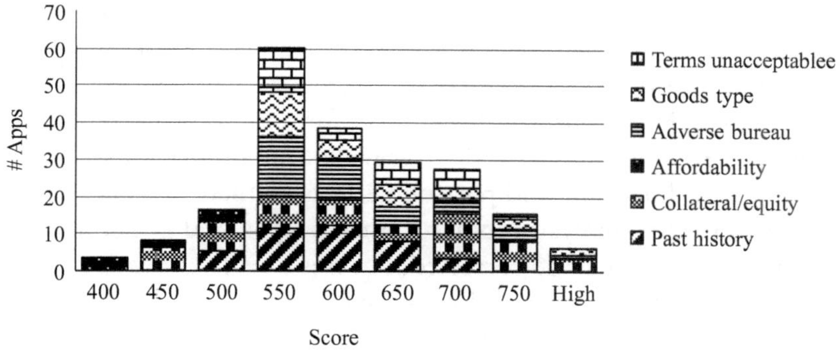

Figure 25.6. Override reason codes by score.
Note: Example applies to secured asset finance.

Care must be taken when interpreting the results, as overrides are a special group. The better performance may not be credited specifically to those characteristics, but to a combination of the underwriters' abilities, and applicants' persistence, in light of better knowledge of their own financial circumstances. Unfortunately, it is difficult to do the same analysis for high-score overrides, where no performance data is available.

In either case, reason codes can aid the tracking of manual overrides, but the possible options should be kept to a minimum. If there are many, staff members may pick one at the top of the list, or a default 'other' category. The number of overrides may also be small, making meaningful analysis difficult. A common way of reporting on override reasons is by score ranges, as presented in Figure 25.6. Other possibilities are to analyse across branches, market segments, lending officers, or other dimensions, especially if there are problems, and the lender wants to know where they are arising.

25.5 Summary

Credit scoring is a powerful tool, but one that must be watched. The reports used for monitoring form part of a broader suite that is used to monitor various aspects of credit risk. There are two broad groupings: (i) *front-end reports*, that focus upon process monitoring, population stability, and adherence; and (ii) *back-end reports*, that allow backtesting of the scorecards, analysis of the performance of booked accounts, and portfolio analysis to review the risk of the entire book. The reports may be presented as snapshots showing the situation at one point in time, or as drift reports covering two or more periods.

Some of the most commonly used reports originated prior to credit scoring, and still serve a valuable role, in particular those used for portfolio analysis. *Delinquency distribution reports* provide overviews of portfolios' risk profiles, by focusing on delinquency and other statuses. In contrast, a *transition matrix* shows movements between the different categories over a given time period, and is used to derive a Markov chain used in forecasting. While the delinquency statuses are the usual and most powerful suspects in this analysis, behavioural scores, application scores, status codes, balance, account age, and other characteristics can also play a role.

Performance reports are powerful tools, which rely upon links between observation and outcome. Comparisons should be done on a like-for-like basis, in terms of both the groups being compared, and definitions used. The *scorecard performance* report provides proof that it is working as intended, but requires sufficient time for booked deals to mature. In the meantime, *vintage analysis* reports can be used to provide a preliminary indication, and allow a view of the new account, life cycle, and portfolio effects. If there are sufficient cases, it is also possible to use characteristic-analysis reports to drill deeper, and possibly to check for *score misalignment*. Misalignments may be corrected, but this a risky endeavour.

In the absence of performance, it is still possible to monitor the through-the-door population. Of greatest interest is drift relative to the development sample, in terms of the: (i) *score distribution*, tracked using a population-stability report, and the associated index; and (ii) *score contributions*, at attribute, characteristic, and/or scorecard level. The greatest risk is operational drift that changes the meaning of the characteristics used, whether due to changes in their calculation or upstream processes.

Much monitoring is done of the decision process, and what affected the decision along the way. Possible decisions are: (i) *accept*, offer to provide the product on certain terms; (ii) *reject*, refuse the product; and (iii) *refer*, get more information before making a decision. After a pre-score decision, the process involves a score, system, and final decision, which may all be the same. A score is used to provide a *score decision*; which may be overturned by policy rules, to provide a *system decision;* which in turn, may be overturned by manual overrides, to provide the *final decision*; and the customer may upset everything, by not taking up the offer, and walking away without a *booked deal*. Monitoring will also assess relationships with different characteristics, including score range, market segment, and geographical region. *Policy rules* are monitored to ensure they are serving the desired purpose and *manual overrides* are tracked to ensure the system has the required level of acceptance. Overrides should be kept to a minimum, but are necessary to handle customer queries, and can provide valuable feedback on possible faults and staff acceptance of the system.

26 Finance

Ultimately, lenders' main goal is to make a profit, whether by increasing revenue, decreasing expenses, or both. Thus far, very little has been mentioned about these key dynamics. This module's final chapter takes a side step, to look at key subject areas directly related to finance:

Loss provisioning—An overview of provision calculations.
Direct estimation—How to estimate provisions based on historical loss values.
Component approaches—Ditto, but uses loss probability and severity estimates.
Scoring for profit—Use of credit scoring to make decisions based on profit estimates.
Risk-based pricing—Benefits and pitfalls of pricing for risk.

26.1 Loss provisioning

In everyday English, the term 'provisions' relates to advance preparations made for a specified, or unknown, eventuality. The most common usage refers to food and supplies, assembled for situations where they may not be readily available, especially for military campaigns, camping expeditions, disaster contingencies, or reality television's survivor programmes. In the financial world, provisions are an accounting concept, referring to monies set aside for probable future losses.

In the credit environment, provisions are raised for expected loan losses. They are a key part of prudent credit risk management; even the earliest moneylenders grappled with the problem mentally, but did not have the proper tools until after modern accounting was developed in the fifteenth century. Although frowned upon by investors, provisions can also be adjusted during an economic cycle; lenders' profits are understated in good years, to create a 'war chest' that can be drawn upon during less favourable times. Loss provisions fall into two categories:

(i) **General provisions**, that are not associated with specific cases, but are instead made against broad classes of accounts.
(ii) **Specific provisions**, made for probable losses where the lender knows problems exist, including classes like known and suspected fraud, legal, and recoveries.

The tax treatment of provisions varies from country to country. According to McNab and Wynn (2003), the UK tax authorities allow the deduction of specific provisions only.

Table 26.1. Provision calculation

	Delinquency	# Accts	Average balance	Balance (000s)	Provision rate (%)	Provision raised
General provision	Current	20,000	1,000	20,000	0.2	40.00
	30 days	2,000	1,500	3,000	1.5	45.00
	60 days	1,000	1,700	1,700	4.3	73.10
	90 days	500	1,800	900	16.5	148.50
	120 days	250	1,000	250	41.5	103.75
	150 days	125	750	94	65.4	61.31
	180 days	50	500	25	95.0	23.75
	Totals	23,925		25,969	1.9	495.41
Specific	Recoveries	275	2,100	578	45.4	262.19
	Legal	350	2,300	805	75.3	606.17
	Totals	625	4,400	1,383	62.8	868.35

In each case, a percentage of the asset value (meaning the outstanding loan balance) will be set aside for the possibility that some or all of its value will not be realised, and be written off.

Provision calculation

Provisioning has traditionally been very simplistic. Historical loss rates are determined for various buckets of accounts (usually defined by delinquency status and other key risk indicators, such as account age), which are then used to provide estimates. An example of a provision calculation is provided in Table 26.1. The rows are the delinquency statuses, which here are split into: (i) *general provision*, for days-past-due; and (ii) *specific provision*, for accounts with serious statuses. The columns provide the details of: (i) the number and average balance of accounts; (ii) their total balance; (iii) the provision rate, however it was determined; and (iv) the provision held.

According to McNab and Wynn (2003), upon whose example Table 26.1 is based, in practice, the provision-calculation report would have more rows for: (i) further days-past-due categories, perhaps out to one year; and (ii) splits by risk scores, using the application score for new accounts, behavioural score for established accounts, and perhaps a combination of the two for points in between. Frauds should be treated as an operational loss, not a credit loss, and be reported on separately. Known frauds are usually written off the moment they are identified, because of the extremely low recovery rates. The fraud charge for the year may be stated as a percentage of sales, of outstanding balances, or both, and the treatment will vary from company to company (McNab and Wynn 2003:148).

Table 26.2. Bad debt charge

A Current year provision		868.35
B Prior year provision		838.48
C Marginal change	B−A	29.87
D Write-offs		323.23
E Annual charge	C+D	353.10
F Average book value		12,978
G Provision rate	E/F	2.7%

Provisions are calculated at regular intervals, and combined with actual write-offs to determine the bad-debt charge for each period, as illustrated in Table 26.2. It is calculated as the difference between the provisions for the current and prior years (C), plus any write-offs that occurred during the year (E). The bad-debt rate (G) is then the bad-debt charge as a percentage of the portfolio's average book value for the period. Thereafter, there are two types of approaches that can be used to determine the loss-provision rates, the choice of which varies depending upon the environment and available tools.

(a) **Direct approaches**—Estimate losses directly, using data that should be readily available from the accounts department.
(b) **Component approaches**—Derive estimates for loss drivers, often using statistical techniques.

Both groups rely heavily on an analysis of historical data, and usually try to get maximum benefit out of available behavioural and application scores. The resulting 'expected loss' values are then used not only for provisioning, but also for risk-based pricing, portfolio valuation, and as part of scoring-for-profit (see Section 26.4).

26.2 Direct loss estimation

Traditional methods for deriving provisions are fairly simple, and use data that should be readily available from the accounts department. Historical loss percentages are calculated for each risk bucket using money values, without referring to the number of accounts, which are then used as the basis for future estimates. There are two basic direct loss-estimation methods:

(i) **Net-flow method**—Assumes that accounts either get worse, or are paid off in full.
(ii) **Transition matrix/Markov chain**—Allows balances to move between various states.

26.2.1 Net flow approach

Thus far, means have been presented for deriving provision values for each period (see Table 26.1), but not the provision rates to be used in those calculations. Net-flow models are the traditional approach, which rely upon straightforward analysis of historical figures. They try to determine what will happen to the accounts over their lifetime, and assume two possible outcomes: either the accounts will be paid off in full, or will default and be written off.

An example is provided in Table 26.3. The calculation is based on the distribution of account balances (not number of accounts) by past-due status for two consecutive months. It is then used to determine: (i) the *net roll rates* from one bucket to the next; and (ii) *provision rates*, calculated as the products of the net roll rates.

Net roll rates can be determined using one or more pairs of consecutive periods. In the latter case, the values for each pair are combined under t and $t + 1$, prior to doing the calculations.

The rows and columns used are:

Current—Total value of accounts that are not delinquent, for both months.

New spend—Account balances that are current in the second month that were not on the books in the first, especially new business.

Current (i)—Amount that is current in the second month, net of new spend. Separate provisions are made for month 5 by itself, and for subsequent months to infinity.

Delinquency statuses—The days-past-due buckets, running through to write-off. In practice, the amount of detail would be much greater, and include other characteristics. Each bucket has a write-off provision (write-off is assumed at 150-days-past-due).

Roll rate—The percentage of the balance that moves into the next delinquency bucket, which for the 30-day bucket is 160/800, or 20 per cent.

Table 26.3. Net-flow model

Delinquency	Time (t)	Time ($t + 1$)	Roll rate (%)	Provision rate (%)	Forecast time (t)	Time (t) write-off
Current	16,000	16,300				
New spend		4,300				
				0.675	to infinity	108
Current (i)	16,000	12,000	75.00	0.225	5	36
30 days	800	800	5.00	4.50	4	36
60 days	240	160	20.00	22.50	3	54
90 days	160	120	50.00	45.00	2	72
120 days	120	96	60.00	75.00	1	90
Write-off		90	75.00	100.00		
Totals	17,320	13,266		2.29		396

Provision rate—The percentage of that bucket's balance that is expected to be written off; calculated as the cumulative product of the roll rate for that bucket and all worse statuses. For the 60-day bucket, it is 100%×75%×60%×50%, or 22.5%.
Forecast time—The number of periods into the future when write-off is expected.
Time (*t*) write-off—The monetary amount expected to be written off at time 't'.

McNab and Wynn (2003) describe two roll-rate methods that are very similar: (i) use calculations like those above; or (ii) drop the cumulative product calculations, and instead use a drift report format providing values for time $(t + 2)$ and further periods out towards infinity. If the same roll rates are used, the end result should be almost the same. Note, however, that many lenders will use the first approach without the 'to infinity' calculation, but with more periods, perhaps out to one year.

The difference between one year and to infinity is small. For Table 26.3, the provision rates are 0.585 per cent versus 0.675 per cent for the final leg and 2.20 per cent versus 2.29 per cent overall. The 'to infinity' calculation is like 'perpetuity' discounting (a 'perpetuity' is a bond with perpetual coupon payments, whose value is calculated by dividing the coupon amount by the discount rate). In this instance, the infinite time series is collapsed by multiplying the provision rate for the 'current' bucket by $R_0/(1\text{-}R_0)$, where R_0 is the proportion of accounts that are up-to-date and remain current. For the example, the provision rate for all periods after time $(t + 5)$ is $0.675\% = 0.225\% \times 75\%/(1 - 75\%)$, or \$108 of the original \$16,000 in the current bucket at time (t). When included with the other provisions, the end result is a total provision of \$396 for the full book of \$17,320, or a provision rate of 2.29 per cent.

While this approach is relatively easy, the assumptions made along the way make it extremely naïve. Mays (2004:132–3) and McNab and Wynn (2003:154–155) list some of the problems with the inherent assumptions:

Assumption	Reality
(a) Accounts in each bucket either get worse, or are repaid in full!	Many accountholders pay sufficient to make their accounts current, while others make part payments, such that the accounts improve only marginally, or stay in the same delinquency bucket for many months.
(b) Spending occurs only on accounts that are current!	Spending also occurs on delinquent accounts. Many will still be operating within their limits, or transacting under the floor limits. Roll rates over 100 per cent are possible.
(c) All accounts in a given bucket have the same roll-rate probability.	Segments always exist where rates differ. In particular, loss rates are highest for accounts mid-way through their life cycles, and lower for new and mature accounts.
(d) Loss probability and loss severity are the same!	There may be segments of the book with high probabilities but low severity, or vice versa.
(e) The future will be like the immediate past!	Lenders will experience changing credit quality over time, whether due to changes in the economy, marketing, credit strategy, legal, accounting, or other forces.

Table 26.4. Transition matrix with money values

States	Time ($t+1$)						Total
Time (t)	Repaid	0	1	2	3	W/off	
Current	1,660	73,040	8,300				83,000
1	200	7,800	500	1,500			10,000
2	100	2,950	150	300	1,500		5,000
3	40	460	20	80	200	1,200	2,000
Total	2,000	84,250	8,970	1,880	1,700	1,200	100,000

Mays (2004:132) highlights that credit quality can be heavily affected by rapid expansion or contraction of a portfolio, especially when there are shocks, like the bulk purchase or sale of loans. This will not be reflected in net flow models. For new accounts, it takes time for bad debts to materialise; and for mature accounts, most have already occurred. 'The likelihood of loss is much higher in the middle of loan life than at the beginning or end.'

26.2.2 Transition matrix/Markov chain

The use of Markov chains to derive loss values can address several of the problem assumptions, in particular: (i) movements between the different states can be recognised and incorporated; (ii) spending is recognised in the state where it occurs; (iii) seasonal or cyclical changes in the risk profile can be incorporated; and (iv) extra states can be defined, to ensure that the transition matrix has the Markov property.

The first step is to derive a transition matrix showing movements between states, expressed in money terms, over the period of interest. The simple example in Table 26.4 is limited to three past-due buckets, with extra roll-on states for 'repaid' and 'write-off'. It may look easy to derive, but special effort must be taken to:

(a) Ensure that sufficient buckets have been provided. Once again, extra splits may be included for other risk determinants, such as application scores, behavioural scores, and account age. Separate matrices may also be used to account for seasonality.
(b) Avoid the curse of small numbers, by using data from multiple periods, perhaps two or more years. The greater the number of transition states, the greater this need becomes.

The data in Table 26.4 can then be used as the basis for calculating roll rates from one period to the next, as illustrated in Table 26.5, where: (i) separate rows have been provided for the two absorption states, 'repaid' and 'write-off', to ensure that the matrix is square; and (ii) the total for each of the rows is 100 per cent.

Table 26.6 then shows a summary of the resulting Markov chain. Month 0 is the current distribution of the book, as per the row totals in Table 26.4, which does not include any accounts

Table 26.5. Transition matrix with roll rates

State	Time ($t+1$)					
Time (t)	Repaid (%)	0 (%)	1 (%)	2 (%)	3 (%)	W/off (%)
Repaid	100					
0	2	88	10			
1	2	78	5	15		
2	2	59	3	6	30	
3	2	23	1	4	10	60
W/off						100

Table 26.6. Markov chain for past due

Month	State					
	Repaid (%)	0 (%)	1 (%)	2 (%)	3 (%)	W/off (%)
0	0.0	83.0	10.0	5.0	2.0	0.0
1	2.0	84.3	9.0	1.9	1.7	1.2
2	3.9	82.6	8.9	1.5	0.7	2.2
3	5.8	80.8	8.8	1.5	0.5	2.7
6	11.2	75.3	8.2	1.4	0.5	3.6
12	20.8	65.4	7.1	1.2	0.4	5.1
24	36.5	49.3	5.3	0.9	0.3	7.7
60	63.9	21.1	2.3	0.4	0.1	12.2
120	79.5	5.1	0.6	0.1	0.0	14.7
240	84.2	0.3	0.0	0.0	0.0	15.5
360	84.4	0.0	0.0	0.0	0.0	15.5

repaid or written off. Each subsequent row shows the distribution of accounts in future periods. As can be seen, it takes some time before the defaults are realised. After a year, only 5.1 per cent of accounts have been written off, but this increases to 15.5 per cent after many years, with the rest of the balances being repaid in full. Granted, this is after 30 years for this portfolio, and it can only be hoped that the desired profits have been made before then. In the meantime, there will be new customers joining each month who are not reflected here, but could be with some adjustments.

26.3 Loss component estimation

All of the above approaches focused upon deriving loss rates directly. Loss forecasting can also be done by deriving estimates for the various loss components. Several approaches are available, which vary by:

(i) The hazard definition used for the target variable, like default, loss, or write-off.
(ii) Whether it is a current-status or worst-ever definition (the loss remains the same, but the probability and severity components adjust).
(iii) How the components are split out.

The most comprehensive definition for retail portfolios, as already described in Chapter 3, treats probability-of-default (PD), exposure-at-default (EAD), loss-given-default (LGD), and maturity ($f(M)$), as separate elements. Mays (2004) presents this as a combination of loss probability and loss severity. The two approaches should provide almost the same result, but differ in that Mays' approach:

(i) Deals directly with loss, and does not consider default.
(ii) Treats EAD, LGD, and $f(M)$ under the single heading of 'severity'.
(iii) May be sufficient for provisioning, but does not meet the strict standards required by Basel II for capital allocation purposes.

In general, credit scoring's use for capital allocation purposes has lagged in its use for loss provisioning and forecasting. Basel II's 'use test' is, however, putting pressure upon banks to make better use of such tools. While this has little impact upon other types of credit providers, they will likely benefit from the new methodologies and skills being developed.

Unfortunately, it is not possible to give this topic a totally comprehensive treatment, as a number of different tools can be used, and the number of permutations and combinations is huge. An overview of some of the 'probability' and 'severity' modelling options are provided instead.

26.3.1 Loss probability modelling

The first stop is probability modelling, if only because it has the greatest degree of consensus. Under this heading fall:

Loss timing—Use of existing scores, combined with a further analysis of losses, to derive lifetime-loss probabilities.
Loss scoring—Development of a bespoke regression model.
Loss extrapolation—Based upon existing application and/or behavioural scores, or other bespoke scores derived using a definition that is more lenient than loss.

The choice of approach will depend upon the data, scores, and skills available. Other approaches can also be used, such as Markov chains or survival analysis, but these have been touched on elsewhere, and are not covered again here.

Loss timing

Mays (2004:135) presents an illustration of a loss-timing curve that would be applicable for application scoring of prime first-lien mortgages, which typically do not mature until between three and five years after being booked. This is quite a long period, but the concepts are the same for almost any retail credit portfolio. The timing is greatest for mortgages, due to the nature of the asset being financed (it is the most valuable asset that many people will ever own), and the time frames being considered (maturities of up to 30 years).

Prime first-lien mortgages are loans to good quality customers that are buying a new home, as opposed to subprime loans for risky customers, readvances and second mortgages, and other secured or unsecured lending.

An example of a loss-timing curve is provided in Figure 26.1, which shows: (i) the period when the loss occurred (x-axis); (ii) the percentage of total loss-making accounts in each period (y-axis); and (iii) details for the total book and each of the risk groups (curves within the graph).

The proportion of losses in the early periods is small, increases until it peaks several periods into the deal, and then declines thereafter. What is less obvious is that the shape of the curve varies according to the risk; as risk increases, the timing is more likely to be sooner than later. The problem is how to determine these loss probabilities, given that it takes some time before the shape of these curves becomes evident.

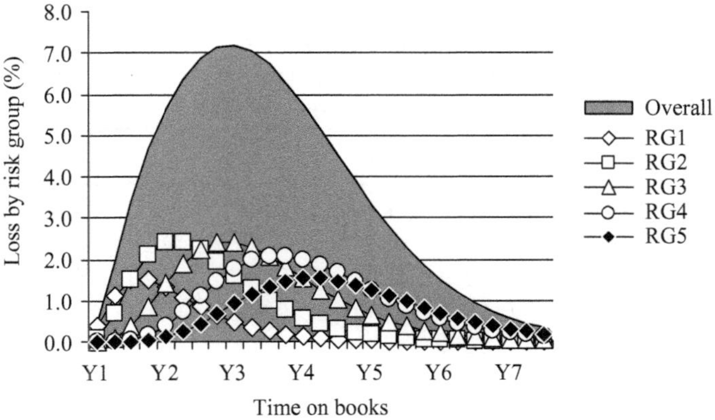

Figure 26.1. Loss-timing curves.

This loss timing can then be used to adjust the PD for each risk grade, and will be referred to here as $f(M)$. As time goes by, losses are realised, and the chance of something going wrong over the remaining period reduces. This is better stated as:

Equation 26.1. PD% maturity adjustment $$f(M) = \sum_{t=M+1}^{T} D_t \Big/ \sum D$$

where D is the number of defaulted or loss cases, M the current age of the deal, T the total term of the deal from start to finish (which may exceed the contractual term), and t is an index for each time period. The resulting value will always be 100 per cent at the start of a deal, and reduce to zero at the end. It is based upon historical values, and some smoothing may be required to get a generalised function that can be used for forecasting.

Vintage loss analysis

The loss-timing illustration in Figure 26.1 is a hypothetical snapshot, which is a composite of several years' data. Mays (2004:139) describes a vintage-loss analysis that shows drift in loss timing over time, and is used as part of the lifetime-loss-probability calculation. Table 26.7 provides a similar example, that shows the:

Accept year—Year when the deals were taken to book.
Years on book—Number of years until loss occurs.
Cell values—Number of loss-making accounts, as a percentage of total accounts accepted during a year.
Average—Simple average loss probability for that number of years on book.
Cumulative average— The cumulative percentage of accounts that result in losses, the final of which is used as the basis for the lifetime-loss probability (1.77 per cent in the example).

Table 26.7. Vintage loss analysis

Risk indicator 3 (of 5)							
Accept year	Years on book						
	1	2	3	4	5	6	7
CCY1	0.13	0.22	0.51	0.27	0.27	0.27	0.15
CCY2	0.12	0.24	0.48	0.27	0.16	0.11	
CCY3	0.21	0.20	0.36	0.24	0.20		
CCY4	0.28	0.68	0.52	0.31			
CCY5	0.08	0.22	0.39				
CCY6	0.18	0.29					
CCY7	0.11						
Average	0.16	0.31	0.45	0.27	0.21	0.19	0.15

Mays suggests that vintages with hazard rates significantly different from the average (as in CCY4 in the example) should be given extra attention, as they may arise from abnormal economic, process, or marketing shocks. If it is thought to be an isolated event, that vintage could be ignored, or the probability could be built into future forecasts, by adjusting the probabilities for each year by the likelihood of that event recurring in that year.

While this type of analysis can be applied to the entire taken-up book, it is most effective when applied to sub-segments of the population. In the example, the calculations have been done for one of five predefined application-score ranges. The same analysis would be repeated for the other four, and the resulting values would then be used to derive a lifetime loss probability.

Lifetime loss scoring

Another approach is to develop a bespoke lifetime-loss model, which is similar to a credit-scoring model, except: (i) the target variable is 'loss/no loss' instead of 'good/indeterminate/bad' or 'default/not default'; and (ii) the time frame being considered is much longer. Rather than one to two years, the window must be at least as long as the average loan life. The final result will be very similar to a behavioural or application model, and for the most part the same characteristics will dominate—especially those relating to gearing, affordability, and credit history. It presupposes that the lender has access to predictive modelling resources—staff, data, computing power, and so on—to do the development.

Mays (2004:136) describes a bespoke loss model, developed for the mortgage portfolio mentioned on page 2. In that environment, the most predictive variables are: (i) the applicant's *bureau score* at the time of application (creditworthiness); (ii) *repayment to income* (ability to repay); and (iii) the *loan-to-value ratio* (motivation to repay). The minimum time frame required would be in the region of 7 to 8 years, and ideally over 12 years. Once the model is developed, the results can then be grossed up to provide the loss probability for the full period, perhaps up to 20 or 30 years. For example, if a model is developed using 9 years of data and 80 per cent of losses are thought to occur within this period, then the resulting probabilities are multiplied by 125 per cent (being 100/80). The only concern is whether sufficient losses were used in the model development.

> Mays (2004) shows how models developed using logistic regression can be adjusted to assess loss probabilities at a fixed point after opening, irrespective of account age today. If accounts aged up to eight years are modelled, and account age is included as an explanatory characteristic, then the probability of loss within eight years can be derived from the score calculated, using, *ceteris paribus*, the point allocation for accounts aged eight years. This is similar to using a control characteristic except, rather than ignoring the characteristic, the worst possible points are used.

Lifetime-loss models are more difficult to develop than most credit scoring models, because losses are rarer than 'bads' or 'defaults', and more time is needed before there are enough of them to develop a model. In many cases there will never be enough, and estimation must be done using something akin to risk-score extrapolation, covered next. Problems can also arise because of operational drift over the period, but otherwise, the resulting model should be quite stable. Other approaches may be simpler, but according to Mays (2004), bespoke models have advantages in that:

(i) The resulting estimates will be *more accurate*, and based upon characteristics that are specific to loss probability.
(ii) *Variables* can be used that may not otherwise be allowed or feasible, including certain restricted demographic details and characteristics not available at the time of application.

Banks may, however, have to make choices here, as Basel II's 'use test' requires that any models used to derive PD estimates for capital allocation be actively used for decision-making with the business. In that instance, lifetime-loss models cannot be used, but the loss risk can be extrapolated based on normal credit scores.

Note here that Basel II is quite new and interpretations differ. It may be possible to include allowed characteristics in the main model, and extraneous and banned characteristics can then be included as a final stage, used only for loss probabilities.

Risk extrapolation

There may be instances where: (i) lenders want to get the maximum mileage out of risk scores that are already available; and/or (ii) there are insufficient loss cases to develop a bespoke loss model, and a more lenient definition has to be used. In both cases, the risk score is used as the basis for estimating what proportion of 'bads' or 'defaults' will ultimately result in losses.

The risk extrapolation example in Table 26.8 shows the mapping of a behavioural risk score onto loss probabilities. To do it, both bad and loss counts had to be determined over prior years. Lenders will not always be so lucky though, and may have to use very sketchy data, and/or subjective expert input, especially if the time taken before losses are realised is long.

This approach can be used to map between probabilities for any number of different wayward credit behaviour definitions, whether good/bad, default, loss, write-off, or others.

26.3.2 Loss severity modelling

The next step is to determine what happens once default has occurred. In Chapter 3, the expected-loss calculation was summarised as a function of exposure-at-default, probability-of-default,

Table 26.8. Risk extrapolation

Risk group	Counts			Rates (%)		
	Total	Bad	Loss	Bad/tot	Loss/bad	Loss/tot
0	1,355	209	178	15.4	85.2	13.1
1	7,517	592	504	7.9	85.1	6.7
2	12,004	535	433	4.5	80.9	3.6
3	17,672	479	328	2.7	68.5	1.9
4	25,244	410	269	1.6	65.5	1.1
5	47,610	402	263	0.8	65.4	0.6
6	53,872	260	156	0.5	59.8	0.3
7	67,223	166	112	0.2	67.2	0.2
8	59,993	70	49	0.1	70.0	0.1
9	37,224	19	12	0.1	63.2	0.0
Totals	329,714	3,142	2,302	1.0	73.3	0.7

loss-given-default, and maturity:

Equation 26.2. Expected loss $\text{EL} = \text{EAD} \times \text{PD}\% \times f(M) \times \text{LGD}\%.$

The PD% and *f(M)* portions of this have already been covered, and now the EAD and LGD elements need greater attention.

The EAD dynamics vary by the type of product. For fixed-term lending, it is a function of current exposure, interest rate, repayments, and time-to-default, while for transaction products, it is a function of the current exposure and limit granted. Predictive modelling approaches at account level are often unreliable, and lenders will instead do the calculations at portfolio level. For example, an EAD percentage can be calculated as total exposures at default, as a percentage of the total one year prior to default.

The calculation given in Equation 26.2 can, for the most part, be restated as:

Equation 26.3. $\text{Exp. Loss} = \text{Exposure} + \text{Interest} - (\text{Recovered} - \text{Costs} + \text{Mitigation})$

Exposure (E)—Loan balance outstanding on date of default, which is the cumulative total of principal, interest, and charges (including penalties).

Interest (I)—Funding cost that accrues after the default event. Any interest accruing on the outstanding balance is treated as interest in suspense. For loss forecasting, either the contract or cost-of-funds rate may be used, usually via a present-value calculation.

Recovered (R)—Amounts recuperated directly from the client, whether upon disposal of the repossessed asset, or from other sources.

Costs (C)—Any expenses incurred during the recovery process, including legal and admin fees, and the cost of restoring or maintaining the assets prior to sale.
Mitigation (M)—If credit insurance has been taken out, there will be a recovery from the insurance company, or an area responsible for insurance.

Retail lenders may opt to self-insure, and only use external insurers for high-value deals, or to insure a portion of the portfolio. The insurance will usually cover a specified percentage of the EAD, interest costs, and recovery costs, less any amounts recovered. According to Mays (2004), insurance is usually only used for mortgages where the loan-to-value is greater than 80 per cent.

Using these elements, the formula for the LGD rate can be restated very simplistically as:

Equation 26.4. LGD rate $$\text{LGD\%} = \frac{(E + I - (R - C + M))}{E}$$

In most unsecured lending environments, the distribution of accounts by LGD will surprise; the bulk of accounts will either be close to no loss, or full loss, as illustrated in Figure 26.2. The question that then has to be answered is, 'How can the estimates be derived?'

For bond defaults, LGD can be determined in two ways: (i) market valuation, which uses the market price of the security on, or shortly after, the date of default; (ii) workout, which involves discounting post-default cash flows to date of default. The latter is the obvious choice for retail portfolios, as the individual deals are not traded, and even portfolios are illiquid.

Thereafter, LGD percentages are calculated for buckets with markedly different profiles. Some of the most powerful characteristics used to define the buckets would be: (i) *default reason*, like days-past-due or insolvency; (ii) *score at time of default*, whether a behavioural, collections, or bureau score; (iii) *security*, in terms of type of asset, or guarantee; and (iv) *borrower's balance sheet*, which applies more to middle-market lending to businesses (loan-to-value, assets versus liabilities, investments). Another possibility would be to develop a model

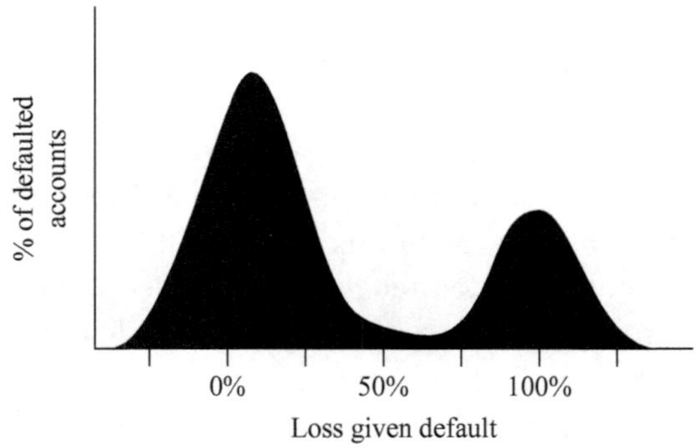

Figure 26.2. LGD distribution.

using the cash flows (E/I/R/C/M) associated with different outcome types (cure, restructure, liquidation); which for the results to be meaningful requires sufficient cases for analysis, and a very good understanding of the LGD dynamics. Certain comments can also be made about the various cash flows that will impact upon the LGDs, as follows:

Recoveries

Recovery rates differ significantly between unsecured and secured lending. For the former, lenders are totally reliant upon obtaining repayment directly from the borrower, even if legal action is taken. In contrast, for secured lending, like motor-vehicle finance or home loans, lenders can recover some of the outstanding balance from disposal of the asset, and will be affected by changes in asset values. If disposal proceeds are greater than the outstanding balance, inclusive of legal costs, penalty fees, and interest, the balance should be refunded to the borrower.

While it is possible to model the recovery rates, purely as a function of deal and borrower characteristics, other approaches may also be used to estimate asset prices. According to Mays (2004), estimation can be done by extrapolating current trends using published regional forecasts, or by using some process like Monte Carlo simulation to generate a variety of different estimates.

Asset price estimation brings with it new complications: (i) changes in asset values affect default probabilities; and (ii) asset values are usually driven by economic variables, whose relationships change over time, especially as new factors come into play. One example is the extreme fall in house prices that occurred in the American mid-west during the mid- to late 1980s, which was driven largely by low oil prices. In England, the bubble during the late 1980s was driven by Thatcherite legislation and tax breaks promoting home ownership. In South Africa, the property bubble caused by the gold price spike during the late 1970s only burst in 1984, as a result of increasing interest rates and political instability.

Recovery rates will also vary depending upon the economy, especially economic downturns. The market can become flooded with distressed assets, like houses and second-hand cars, especially if the downturn was preceded by some irrational exuberance, and an asset price bubble. There are, however, also cases where recovery rates are negatively correlated with a buoyant economy, especially when price reductions for new assets affect the second-hand market. This is especially true: (i) with advances in technology driving down costs; and (ii) an improvement in exchange rates, affecting the price of imports.

Recovery costs

In many cases, the costs associated with maintaining and disposing of assets can be significant, and seriously impact upon the amounts finally recovered: *motor vehicles*—maintenance and storage costs pending sale; *home loans*—all normal costs associated with running a property, which, besides repair and insurance costs, also include regular charges by municipalities (rates,

Table 26.9. Expected loss summary—money values

RG	Orig-asset value	Original loan	Current balance	Expos @ Default	Interest cost	Recovery	Costs	Mitigation	Loss given default	Expected loss (000s)
1	104.0	87.3	70.6	68.1	4.2	33.4	14.1	19.3	33.7	304.28
2	226.5	190.4	156.1	145.9	9.1	83.1	31.2	36.8	66.4	515.26
3	367.4	306.2	257.5	230.0	14.0	151.9	49.9	44.0	98.0	612.62
4	496.2	414.9	358.5	306.3	19.6	231.8	70.7	41.4	123.4	645.81
5	611.1	515.6	456.8	375.2	23.8	322.2	87.2	30.9	133.1	520.60
Totals	1805.3	1514.3	1299.5	1125.5	70.9	822.4	253.1	172.4	454.6	2598.55

All values stated as millions, except Expected Loss stated as thousands.

utilities) and housing complexes (levies). According to Mays (2004), historical costs are the most likely basis, but a predictive model could be used if related factors can be identified.

Risk mitigation

Elsewhere in this text, credit insurance was mentioned as a revenue source. Here, it is viewed as a security blanket, for when things go wrong. It only provides value if it is true risk mitigation though. In subprime lending, credit insurance is often just another revenue source to offset the potential for high losses; it is compulsory, and there is no reinsurance. In contrast, for high-ticket home loans an insurance company may underwrite the policy, and the lender will keep a percentage of the insurance premium. Not all home loans will have credit insurance though; borrowers may select it as an option when they take out the loan, or it may be compulsory for certain high-risk categories, such as loan-to-value ratios greater than 80 per cent.

26.3.3 Forecast analysis

Unfortunately, it is not possible to provide a detailed means of building the loss forecast, because every lender/product will vary in terms of the available data. There will, however, be many commonalities in the forecasts' outputs. The following gives a good idea of what one might look like.

Table 26.9 focuses on the money values generated by the forecast, under the headings of: exposure, interest, recovery, costs, and mitigation. The numbers have been manufactured for a 15-year period, based upon a variety of assumptions, some of which were discussed earlier. The end goal was the expected-loss figure. A review of the averages, presented in Table 26.10 below, can provide greater insight, and certain things are immediately evident:

Column per cent—The proportion of accounts being taken on each period was consistently 10, 15, 20, 25, and 30 per cent (not shown) in risk grades 1 to 5 respectively, yet the current portfolio distribution reflects a shift from the higher- to lower-risk categories. This is consistent with expectations; riskier accounts are more loyal, while low-risk accounts are more likely to settle early.

Table 26.10. Expected loss summary—averages

RG	Count	Col (%)	Loss prob	Avg Age	Time Rem (%)	Original loan	Current balance	Paid down	Exp. @ default (%)	Loss given default (%)	Exp. loss rate
1	1,357	13.6	6.5	71	17.8	64,310	52,007	80.9	96.5	49.6	0.43
2	1,847	18.5	3.7	66	25.3	103,066	84,537	82.0	93.5	45.5	0.33
3	2,114	21.1	1.9	58	39.1	144,825	121,829	84.1	89.3	42.6	0.24
4	2,277	22.8	1.1	49	56.4	182,215	157,426	86.4	85.4	40.3	0.18
5	2,405	24.1	0.6	41	72.4	214,387	189,927	88.6	82.1	35.5	0.11
Totals	10,000	100.0	2.3	55	45.6	151,430	129,950	85.8	86.6	40.4	0.20

Loss probability—Was purely a function of the loss rates associated with each of the risk grades, so there are no surprises.

Average age—As the risk decreases, so does the average account age. This is again consistent with lower-risk accounts being more likely to settle early.

Time remaining—This is the average $f(M)$, used to adjust the PD rates. The lower the account age, the greater the probability that something may still go wrong.

Original loan and current balance—These are simple averages.

Paid down—The relationship between the current and original loan balances increases with risk, purely because riskier accounts are slightly older on average.

EAD—The EAD as a per cent of the original loan amount reduces, because of the longer time-to-default associated with lower risk.

LGD—Ratio of the loss forecast to EAD, which was mostly affected by an assumption of increased recovery rates at lower risk levels.

Expected-loss rate—The expected loss stated as a percentage of current balance, which decreases as risk decreases.

It also helps to check the account distributions by some key measures. Figure 26.3 illustrates the loss-given-default distribution, where several patterns are noted:

(i) **Negative values**—Some customers will repay the loans inclusive of penalties and costs, resulting in a net profit.

(ii) **Multi-modal distribution**—There is a third minor peak around 72 per cent, because losses for accounts with no credit insurance are more severe.

(iii) **Tails**—There are a lot of cases where losses run very high, due to high maintenance costs and a long lead-time prior to disposal. Some of the negative values were also quite large, and have been lumped together under negative 12 per cent.

The time taken before write-offs are made against provisions is also of interest. According to Figure 26.4, which is based upon the assumptions used for those accounts not currently in

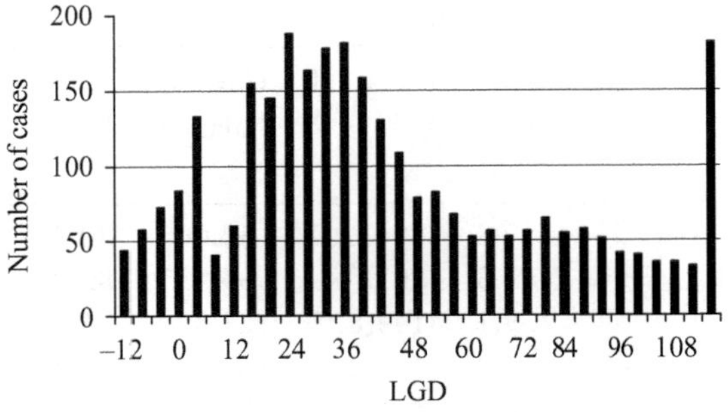

Figure 26.3. Example LGD distribution.

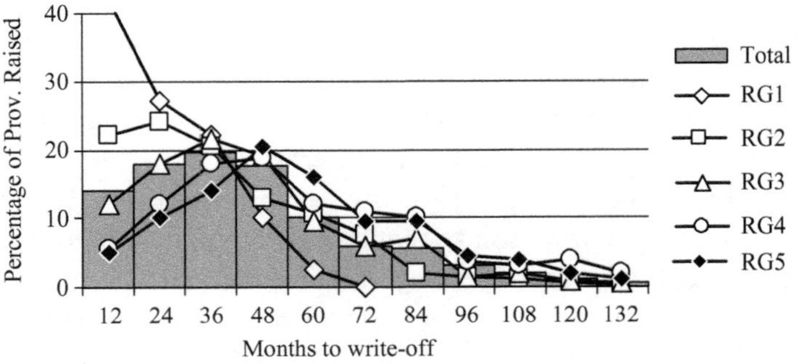

Figure 26.4. Time to write-off.

default, full write-offs will peak in about 3 years, and within 10 years everything will have worked its way through the system. This holds true only for the existing book, and ignores any new business being taken on.

Something that should also always be kept in mind, even if it is not readily apparent from the numbers, is that losses are typically high in the middle, and low at both ends of the loan term. By extension, without adequate loss provisions, profit will be overstated when new business is taken on, especially for a rapidly growing portfolio. For this example, the loss provision required for a portfolio being maintained at the same size was 0.35 per cent of current balances, whereas for a portfolio growing at 50 per cent per annum, the value would be 0.41 per cent, even without a change in the through-the-door risk profile.

26.4 Scoring for profit

When credit scoring was first implemented for new-business processing, the primary benefits were better and more consistent decisions, at lower cost. The focus was upon minimising risk, but since the mid-1990s the focus has shifted onto businesses' true interest of improving profits,

Table 26.11. Profit drivers

Element	Income	Expense
Risk	Recoveries	Write-offs
Late payment	Penalties	Collection
Balance	Interest income	Cost of capital
Activity	Transaction fees	Transaction processing
Insurance	Credit insurance	Underwriting
Marketing	Cross-sell	Acquisition

at both product and customer level. This is not always an easy task though, as: (i) the profit drivers are many; (ii) they can change significantly over time; and (iii) they are heavily influenced by lender strategies. The following section looks at profit under the headings of:

(i) **Profit drivers**—The key influences, such as risk, balance, activity, late payments, insurance, and acquisition.
(ii) **Profit-based cut-offs**—A relatively simple way of using profit considerations to drive a selection process.
(iii) **Profit modelling approaches**—A brief overview of more sophisticated approaches, used to consider profit in decision-making.

26.4.1 Profit drivers

Risk may influence profit, but it is only one of several profit drivers. The full set is covered here, under the headings of risk, balance, activity, late payment, credit insurance, and marketing (Table 26.11):

Risk—The most obvious loss element, around which most of this textbook revolves. Risk is evidenced by delinquencies, and offset by recoveries.

Late payment—Penalties, whether interest rates or fees, are often set at punitive levels to discourage late payments, and can be a significant source of revenue. There are, however, also collections costs, associated with contacting customers about the payment. This cost is proportional to the time spent in collections, and tends to be spread across those customers in the collections queue, not the full customer base.

Balance—Account balances are the primary driver behind net interest income, being the difference between interest earned and the cost of capital. The major dangers are low utilisation, and early settlement.

Activity—The number of times that the customer transacts, or uses the facility. This applies primarily to credit card and cheque accounts, where it generates both transaction income (merchant and service fees) and processing costs.

Insurance—Many lenders will offer insurance, either to protect the asset (home, motor, etc.), or the repayment stream from the individual (life, sickness, job loss). It may be done entirely by the lender, or be fully or partially underwritten by an insurance company.

Acquisition costs—The costs of acquiring new business are not insignificant. This includes marketing and application processing costs, and may include special offers, such as tokens, prizes, or low introductory rates. These costs should be offset by income over the life of the deal, as well as profit from cross-sales after the customer has been accepted.

Many cardholders are transactors who are not viewed as profitable. Although merchant fees are significant, they are heavily offset by funding costs due to the interest-free period, say 60 days. Banks view credit cards as part of a broader product offering, and these low-risk customers are subsidised by other areas. Unfortunately though, transactors often have a low credit demand, and are resistant to cross-sells except perhaps for savings products.

If any of these seem familiar, it is because many of the same concepts were already covered under loss forecasting. In this instance though, lenders' focus is on marginal income and expenses. Fixed costs associated with customer service and the ongoing management of the business must be ignored, unless there is some marginal element associated with new accounts.

The ideal customer!

While the above gives an indication of the profit drivers, the relationship between profit and risk is less evident. Some generalisations can be made though:

Low-risk customers: (i) have a lower demand for credit, and are more likely to settle early; (ii) tend to be price sensitive; (iii) are sought after by other lenders; and (iv) are likely to take advantage of special introductory offers and move on.

High-risk customers: (i) have a propensity for credit, and borrow for longer; (ii) are less price sensitive; (iii) are more loyal; (iv) generate other income, including penalty fees and credit-insurance income.

The ideal customer could then be described as someone who has a high ongoing balance, misses the odd payment but does not default, takes out credit insurance, and probably has a low bureau score. Indeed, they are often the messiest, and closest to the cliff's edge. In contrast, loss-making customers can lie at two ends of the spectrum: (i) very high-risk customers that result in charge-offs; and (ii) low-risk customers that have low balances or pay early, readily shift their accounts between lenders, do not take out credit insurance, and when transacting, take maximum advantage of interest holidays (Table 26.12).

Table 26.12. Risk by profit drivers

Risk profile	High	Low
Credit propensity	High	Low
Price-sensitive	Low	High
Loyalty	High	Low
Other income	High	Low

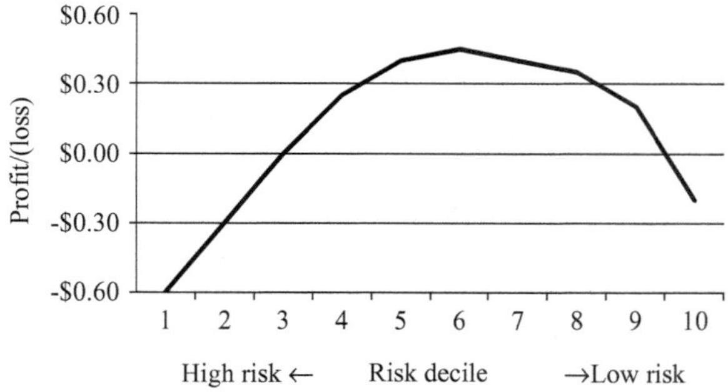

Figure 26.5. Risk versus profit per account.

Thus, it is a fine line to tread when the most profitable customers and the biggest loss-makers have similar profiles. This is illustrated in Figure 26.5; profit increases initially as risk improves, but reaches a peak and then decreases, eventually becoming a loss. In some environments, the profit structure may surprise; a credit card portfolio might be 70 per cent loss makers, 25 per cent to offset the loss, and 5 per cent to generate any profit.

McNab and Wynn (2003) provide a similar Risk versus Profit graph with respect to profit-based marketing campaigns, but similar relationships exist throughout the risk management cycle. In most situations the graph turns down at lower-risk levels, but does not go negative.

26.4.2 Profit-based cut-offs

One of the simplest yet most effective ways of incorporating profit is to identify the cut-off score where the expected profit goes south of the border. To do so, values are derived for the various profit drivers by risk score. Inputs into the process may already be known, have to be researched, or have to be assumed. Lenders may have a good idea of the numbers, but the number of assumptions can be worrying, and must be documented.

An example of a profit-based score cut-off analysis is provided in Table 26.13, where the score cut-off would be about 274. Each of the dollar values is stated on a per account basis, with the final column being the marginal contribution from an extra account in that score range. It may not cover all of the elements mentioned in Section 26.4.1, but it considers those

Table 26.13. Profit-based cut-off

Score	Number of applicants	Loan amount requested ($)	Insurance income ($) ($)	Interest income (%)	Bad rate (%)	Provision rate	Provision ($)	Operating cost ($)	Contribution ($)
232	7,481	3,110	237	530	40.8	45.0	1,399	238	−870
247	6,209	2,713	200	445	30.5	33.5	910	195	−460
254	5,319	2,772	204	452	26.2	28.7	797	179	−320
259	4,695	2,725	203	438	23.2	25.3	690	167	−216
264	5,604	2,648	187	420	20.8	22.6	599	158	−150
268	6,256	2,721	198	435	18.7	20.3	551	151	−69
273	6,928	2,686	187	423	16.7	18.0	483	143	−16
277	8,157	2,771	194	439	14.9	15.9	442	137	54
282	9,200	2,794	192	441	13.2	14.0	392	131	110
286	9,505	2,868	202	456	11.7	12.4	356	126	176
291	10,508	2,888	197	463	10.4	10.9	315	122	223
295	11,180	2,891	191	452	9.2	9.5	276	117	250
300	12,607	2,870	187	447	8.1	8.3	239	113	282
304	13,512	2,969	194	460	7.1	7.3	217	110	327
309	14,160	2,861	182	445	6.3	6.4	182	108	337
314	15,195	2,840	169	431	5.5	5.5	157	105	338
319	22,708	2,765	157	414	4.7	4.7	129	101	341
326	23,471	2,781	154	417	3.8	3.8	105	99	367
333	27,166	2,705	146	403	3.1	3.0	81	96	372
342	31,695	2,663	136	387	2.4	2.3	60	94	369
352	25,834	2,552	119	364	1.8	1.6	42	91	350
364	24,325	2,490	110	349	1.3	1.1	28	90	341
382	20,131	2,288	82	294	0.7	0.6	14	86	276
Total	321,847	2,709	156	407	6.2	5.6	155	110	298

most pertinent to the situation at hand, which in this instance is a hypothetical personal-loan portfolio for a bank:

Loan amount requested—The amount of the loan that the customer applies for.
Bad rate—Number of cases expected to be bad, based upon scorecard development data, inclusive of reject inference.
Provision rate—Expected loss as a percentage of the loan amount requested, which also recognises post-default recoveries, interest costs, and collections costs.
Interest income—The revenue that is expected over the life of the loan, which should take cognizance of potential early settlements.
Funding costs—The cost of capital required to fund the deal. The rate to be used should be available from the finance department, and cover both debt and equity funding.

Insurance income—Amounts paid by customers for credit insurance. If a third party underwrites the credit risk, then the calculations must reflect this.
Operating cost—Most operating costs are fixed and would not be considered, but marginal costs associated with account management and mailing would be included.
Contribution—The marginal net profit/(loss) expected if that applicant is accepted, which is the sum of the other items. All cases that provide a net profit should be accepted.

Please note, that the focus is on marginal income and expense. Sunk costs are ignored! Exactly what is treated as sunk costs will vary. Mailing costs for a marketing campaign would be included if it is still in the planning stages, but treated as sunk if they have been incurred, and cannot be recovered.

26.4.3 Profit modelling approaches

One might ask, 'So why not just calculate the profit over a period, and develop a model to predict it?' This has been tried, but do not underestimate the investment required! In an ideal world, lenders should base their decisions upon the lifetime value of the customer, which is the net-present-value of all profits expected from a customer, across all products. This is difficult to derive though, because of the complexity and time horizons involved.

Banks' customers may have four to eight products on average, and while the most profitable may be personal loans, the bank will still have to provide a cheque account or credit card that may run at a loss. Most of the bad debts are seen early in the customer life cycle, whereas the income takes longer to accrue, and customers' future lending requirements are less certain.

Instead, lenders tend to (perhaps unwisely) focus on shorter-term profits from individual products, but even this can be problematic. Thomas (2000) and Liu (2001) highlighted several issues, which are expanded upon here:

Profit definition—What defines profit? The delineation between profitable and unprofitable is often unclear. Contributors are often poorly understood, and data may not be readily available or accessible, especially if proper analysis requires transaction and accounting data. Issues arise with the time horizons to be assessed, in areas where account relationships last for many years. Also, the required result is an estimate, which is more difficult to derive than a risk ranking.
Variables—A large number of factors influence profit. Default propensity may be a function of acceptance, credit limit, and collections decisions, but for profit this list grows to include marketing, pricing, and others.

Data warehousing—Sophisticated cost-accounting and allocation processes are required. In order to do it properly, the system must: (i) be designed to contain all of the profit elements; (ii) record revenues and costs at the transaction level; (iii) be integrated throughout the entirety of the risk management cycle; and (iv) be capable of amalgamating the information at the customer level, or perhaps even family level.

Outcome period—When considering historical data, only data on profit to date is available. This may provide an incomplete picture, as customer relationships do not fit neatly within the one-year time frame, and much longer windows are often required. If the window is too short, the censoring may cause any strategies based on the model to maximise short-term profit, but lose the customer relationship.

Outcome drift—Because profit is influenced by so many decisions, the results of a profit model will be highly sensitive to changes in the underlying assumptions, and it is likely to have a much shorter life span than pure default models. Profit is also a function of economic conditions, and other factors that are difficult to incorporate.

Profit scoring may be an elusive ideal, but there are other advanced methods for dong profitability analysis. Thomas (2000) includes it as one of four types of profit modelling approaches used in practice:

(i) **Profit score**—Relies upon developing a regression or other model that predicts profit using available data, like that available at time of application.

(ii) **Markov chains**—While these are being used effectively for modelling defaults at the account level, they suffer from the 'curse of dimensionality' when applied to more complex real world situations—especially because the numbers in individual cells at the lowest level make the model unstable. Also see Thomas et al. (2001), who proposed an approach that combined estimates of the outstanding balance, past-due status, and other information.

(iii) **Survival analysis**—Uses approaches borrowed from the insurance industry and elsewhere, to estimate how many units will survive from one period to the next. The approaches include proportional hazard and accelerated life models, that use information from the first few months of loans to estimate longer-term profit.

(iv) **Matrix approach**—Separate measures/models are used to rank accounts according to the major profit elements, like risk (defaults), revenue (usage), response (cross-sales), and retention (will he stay or will he go now?). Different groups within this mix are then identified, and the lender estimates the profitability for each. If it were used to prescreen a mailing campaign, only those cells that are profitable would be mailed.

According to Thomas et al. (2001), at that time almost all commercial attempts at modelling profitability used a matrix approach to combine risk and return surrogates, like behavioural

Table 26.14. Matrix approach to risk versus revenue limit setting

Risk	Low	Revenue ($) medium	Hi ($)
Low	50	100	200
Medium	25	50	100
Hi	None	25	50

scores and outstanding balances (Table 26.14). Different strategies are set for each cell in the matrix, perhaps without even doing an analysis to determine expected profit.

Thus, most of the approaches mentioned are a long way off, if not the stuff of legend, for most lenders in the consumer-credit environment. Even so, as scoring methodologies develop further, they may become more readily accessible, especially if they are provided as part of vendor software.

26.5 Risk-based pricing

Sometimes it's like climbing the Matterhorn. The higher you get, the more you get out of it, but you are closer to a sheer edge, and risk losing everything with a few wrong steps.

David Edelman

Directly related to 'scoring for profit' is risk-based pricing (RBP), where interest rates and fees charged on individual loans are varied, to compensate for risks specific to those deals. The concept was borrowed from the insurance industry, where premiums are a function of actuarial calculations. Wholesale lending has used RBP for many years, but it has only taken hold in the consumer and small-business markets since the mid-1990s. Many retail-credit lenders are striving to implement RBP, but there are pitfalls along the way. Very little literature exists on the topic with regard to credit scoring, and what follows is based on a very small number of articles.

26.5.1 Mechanics and implementation

Scroggins et al. (2004) did a survey of 156 American banks to determine what factors they take into consideration when segmenting customers, both for credit risk assessment and pricing. Approximately 60 per cent of the respondents indicated that the segmentation was used for loan pricing, and 55 per cent for setting loan terms. The primary factors that emerged were, in order of importance: (i) credit risk; (ii) collateral; (iii) loan purpose; (iv) relationship with bank; (v) potential profitability; (vi) loan size; (vii) loan maturity; and (viii) competition.

Credit risk was also top of the list in a 1993 study by the same authors, but is now of even greater importance. Even so, the only articles that could be found on practical issues associated with RBP implementation were by David Edelman, who indicates that lenders tend to 'graduate' from flat-rate into RBP, passing through several stages along the way:

(i) Increased rates for higher-risk customers, near cut-off only.
(ii) Lowered standard rate, plus lower rates for low-risk customers.
(iii) Decrease in cut-off score, and acceptance of previously declined cases.

By the time a lender has reached stage (iii), there is a marked change in the nature of the business, including substantial impacts upon collections, legal, and other areas, as demands upon them increase. For account origination, traditional flat-rate lending follows a process something like:

(i) A lender places a product advertisement, stating a standard or typical rate.
(ii) A borrower applies, stating what his/her requirements are.
(iii) Information is obtained from the bureau, and other sources.
(iv) An accept/reject decision is made.
(v) The lender advises terms, usually with a standard or typical rate.
(vi) If the terms are mutually acceptable, then the deal proceeds.

At first, credit scoring's only impact was to automate the accept/reject decision. If the lender accepted an application, there was a good chance that the borrower would take up the loan. The moment pricing is varied by risk: (i) the risk has to be assessed before the price can be set; (ii) borrowers' propensity to accept the offered terms changes; and (iii) the definition of a 'typical rate' changes. On the latter point, there will be issues relating to fair marketing, which may be the subject of legislation.

In the United Kingdom, the Office of Fair Trading requires that the advertised typical rate, or better, be offered to at least 66 per cent of accepted applicants, which was increased from 50 per cent in October 2004.

The most obvious way of doing RBP is cost-recovery pricing, which attempts to allocate costs to each and every deal and charge accordingly. This applies not only to expected-loss recovery, but also to operating costs and capital charges. Indeed, capital charges will likely even include a premium, to ensure that lenders' ongoing capital growth keeps pace with the economy. An alternative is to exclude the capital charges, and use a pure loss-recovery approach to set minimum prices. Lenders may also adjust prices to attain certain strategic objectives, such as using higher rates to maximise short-term profits, and lower rates to grow market share and improve longer-term profits.

While all of this sounds logical and is appropriate for securitisers, it can give rise to problems for lenders who book deals for their own account, and have to deal with ongoing customer relationships. Certain factors must be considered. For example, there are some groups that are relatively high risk, but become loyal and highly profitable customers, such as students, immigrants, and previously excluded groups that do not have many options. It would be best to use the expected lifetime value of a customer, but this is a holy grail that is extremely difficult to determine. Also, if a strict cost-recovery approach is used, the required interest rate often exceeds usury for customers previously just above cut-off. The lender has the choice of excluding those customers (higher cut-off), or lowering the rates charged.

26.5.2 Behavioural changes

And finally, the use of risk-based pricing affects humans, whose behaviour may change once the terms are varied. This applies not only to customers, but also front-line staff, underwriters, and collectors.

Customers

The most important potential behavioural changes lie with the customers. The greatest factor is information asymmetries, which can influence the situation in two ways. First, adverse selection results when applicants who know they are higher-risk take up the offer, in spite of the higher price. One possible solution would be to move the prices even higher, but Edelman (2003b) suggests that the extent of adverse selection is not that great. If it is occurring, it should be possible to detect it, by analysing the dynamic delinquency reports for each risk band. Also, if payment-protection insurance is offered, applicants that perceive themselves as higher-risk are more likely to take it up, which provides a mitigating income stream.

Second, when there are rate queries, people who contest are often better than average risks, either because they are financially aware enough to question, or they have a better understanding of what they can afford. They are also likely to limit the query to why the rate is different, and not the extent of the difference.

Customers can also be price-insensitive, and accept higher rates when they could do better, whether as a result of: (i) customer loyalty; (ii) low loan values, where it is not worth the effort of shopping around; and/or (iii) the customer being financially unaware, and not realising that the rate is inappropriate. For the purposes of responsible lending, the onus is put upon the lender to make the best possible offer. While this may not seem right, regulators take a dim view of *caveat emptor* excuses.

And finally, the risk of early repayment increases with the interest rate. Given that account origination is expensive, it follows that lenders will have an interest in maintaining the relationship for as long as possible, and this may mean adjustments to RBP to recognise it. This is especially true in an environment of falling fixed interest rates, where there is significant motivation for customers to refinance.

Front-line staff

Another major group to be considered is front-line staff. Most people responding to an advertised rate will logically think that they will get that rate if they apply. It the rate offered is higher, or perhaps even lower, pressure will be put on staff members to explain the difference, and communicate company policy. Problems will be even greater for existing customers if their interest rates change. Edelman (2003b) suggests three measures to assist front-line staff:

Staff education—Inform staff about why RBP has been implemented, what possible scenarios can arise, and how to handle them in terms of customer communication and dispute mediation.

Referral process—Implement a process where problematic or borderline cases can be referred to a specialist underwriting team for a decision, and ensure that staff members are familiar with the process.

Feedback—Ensure that both front-line staff and underwriters are kept informed about how the arrangement is operating. Management must also ensure communication, to determine whether policies need to tightened, relaxed, or changed in any way.

In general, the goal is to avoid front-line staff becoming bogged down with RBP problems, so that they can instead focus on their key goals—whether customer service, relationship management, or intelligence gathering. At the same time, staff will often provide feedback quicker than the quickest analytics, as customer complaints are received.

Underwriters

Lenders' final backstop in the risk assessment process is the underwriters, who have to scrutinise the different deals. In general, in an automated environment, the underwriter's function becomes one of *verifying both data and policy rules*—especially where decisions are contested. If one or the other is suspect, then the underwriter may (motivate to) override the system decision—especially where a single policy rule is tripped.

Once RBP is introduced, there is another variable in the equation. If underwriters' core function is to assess risk, can they adequately take price into consideration in their evaluations? Should the interest rate be ignored? Does a higher rate mean increased risk? Is it sufficient to mitigate the risk? In general, there is only a loose correlation between price and risk on existing deals, because: (i) non-risk issues are also taken into consideration when setting prices; (ii) different objectives may be used on different deals; and (iii) both the objectives and the premiums/discounts charged may vary over time.

Lenders usually set interest rates according to the risk at a point in time. Upon later review, circumstances may have changed dramatically, and even if not, an underwriter may have no clue as to why a particular rate was chosen in the first place. Rather than relying solely upon the interest rate, lenders are well advised also to provide reason codes, or risk bands, to aid underwriters' interpretation.

Collectors

Another affected area is collections, and decisions have to be made about whether the price should be taken into consideration when choosing collections actions. In general, price has little bearing upon collectability, and collectors usually focus on higher-balance accounts at a given level of risk. Irrespective, some collectors may associate higher interest rates with higher risk, and focus their efforts on high interest accounts, even though they provide greater compensation at the same level of risk. Their decisions may also be influenced by the interest rate when a customer requests an extension.

Edelman (2003b) also makes the following suggestions for both the underwriting and collections areas:

Communication—There should be regular discussions with risk, finance, and marketing, to provide feedback on the goals and results of RBP. This is easier said than done, as the level of understanding is often low.

Objectives—Set clear objectives, whether in terms of profitability, bad debt reduction, operating cost management, or others. This requires a good understanding of key business drivers, especially profitability, and requires a shift away from traditional goals of pure loss avoidance.

Tactics—Wherever possible, use an agreed set of tactics that are applied consistently, and only overridden using an agreed procedure.

Monitoring—Effective means of monitoring both the strategies and the underwriters/collectors are needed, including turnover times and other operational monitoring. Underwriters should also be tracked by accept and bad rates, and collectors by promise-to-pay and recovery rates.

26.5.3 Strategic issues

As indicated earlier, RBP is usually associated with charging higher-risk customers more, and lower-risk customers less. The latter is expected to increase volumes, especially where the rates are loss leaders, intended to encourage take up by better customers. There are some strategic issues though:

- Different lenders will usually try to target the same small group of people with lower interest rates, and success will be a function of both the brand and the prices.
- The targeting of low-risk customers may result in them benefiting from a virtuous cycle, as lenders will continuously be trying to attract them with lower rates.
- There will be a group that will not take up the offer at any price, purely because they do not have the need.

Rather than adjusting price to match risk, risk can instead be adjusted to match price, whether by adjusting the amount, term, fees, or collateral/guarantees. Another possibility is to change the process, either raising or lowering the hurdles for things like proof of income and/or address. With risk-based processing, these requirements could be waived (if allowed) if the risk is low, or be more strictly enforced if it is high. For example, for low-risk customers, mortgage lenders can rely on a drive-by valuation instead of a full valuation, to speed up the approval process.

In general, champion/challenger experimentation is the best way of arriving at strategies for any given product or process. This applies whether for account origination, account management, collections, or elsewhere. Lenders can try out different combinations of price, qualifying criteria, documentation and collateral requirements, collections strategies, and so on. The key is to have a set of metrics, which can be used to compare the results and support one strategy over another. This puts a heavy emphasis upon profitability modelling, and having a firm understanding of interest and insurance income, bad debts, operating costs, and the likely effects of a change in price.

Lenders have greater flexibility with repeat business, but this brings with it greater responsibility when setting rates. The fact that the lender already has a history with a borrower usually implies that the risk is lower, but customers may request variations in loan amounts or terms. There may have been changes to applicant quality, base rates, or lenders' policies and procedures, which demand higher rates.

Most RBP is done at time of application or account review, but some lenders will adjust rates on existing accounts. Care must however be taken due to potential customer dissatisfaction, especially from the uncertainty created. The most common is for a default event to trigger increased rates. The logical extension is to have ongoing RBP, changing interest rates monthly or quarterly according to the latest risk assessment. This would either provide a target for borrowers to work towards, or welcome surprises for those customers who exhibit the desired behaviour. It may seem like the realm of science fiction with even greater challenges, but many lenders already use a variation of it to renew overdraft facilities; if certain prescribed criteria are met, the facility is automatically renewed at the stated rate, otherwise an increased rate may be charged, but with the possibility of having it reduced when the criteria are again met.

Finally, the use of RBP has implications for scorecard developments, as regards sampling and acknowledging RBP's impact upon customers' risk profiles. The primary directive is to provide a risk ranking, and the score's reliability may vary depending upon the extent of sampling in different risk regions. According to Edelman (2003a), in the absence of risk-based pricing, there is a need for greater sampling where the greatest benefit is expected to be achieved: (i) around the cut-off for selection processes (application scoring); and (ii) in the medium- to high-risk regions for steady-state processes (behavioural scoring). For the former, providing greater accuracy for clear-cut accepts and rejects provides little value, as it is highly unlikely that the scores will affect the decisions. An old score or bureau score can help to identify where the focus should be, and then be used for stratified random sampling, to adjust the sampling without increasing the total size of the sample.

Once risk-based pricing is introduced, scores' reliability must extend across the risk spectrum (possibly excluding clear-cut rejects). The scores will hopefully also provide better predictive accuracy, especially if strict expected-loss recovery pricing is to be done. One could argue that sampling should be greater around all of the cut-offs, but credit scoring is fuzzy; as risk reduces, the points become more difficult to identify! A further factor to consider is the effect of lenders' strategies upon borrowers' risk profiles, like higher risks resulting from higher interest rates. When developing new scorecards, lenders should try to neutralise the effects of their own strategies by using control variables, or some other means.

26.5.4 Effects on consumers

All of the above has focused upon the firm and the customer, without covering the effect on consumers as a whole. According to Edelberg (2003), RBP started becoming widely adopted from the mid-1990s onwards. She analysed data from Surveys of Consumer Finances (USA) for the period 1983 to 1998, which included details on home loans, motor vehicle finance, credit cards, educational loans, and general consumer loans.

Interestingly, Edelberg (2003) noted that interest rates for educational and collateralised loans were similar, which may result because: (i) they are subsidised by government or others, because the loan is towards a productive purpose; or (ii) educational loans are not discharged in the event of bankruptcy, which means there is greater creditor protection.

Several changes were noted from 1995 onwards, which was when mortgage securitisers Fannie Mae and Freddy Mac first adopted credit scoring and, shortly thereafter, risk-based pricing. Their research showed that:

- There were increased premiums per unit of risk over the period, where the risk unit was a 1-per-cent increase in the probability of bankruptcy.
- Fewer high-risk customers were denied credit.
- Debt levels increased more for low-risk customers than high-risk customers.

Over the period, the range of interest rates charged doubled, as higher-risk borrowers were charged higher rates. The increase was particularly evident for home loans, where for each risk unit interest rates increased more than two-fold for first mortgages, and five-fold for second mortgages. Rates also increased more than two-fold for motor-vehicle loans, and by factors of 0.48 and 0.30 for credit cards and education loans respectively, with no change exhibited for general consumer loans.

Borrowing increased most for low-risk households, whose interest rates dropped, but even though access to credit increased for higher-risk households—especially for secured lending—their total borrowing increased less, or fell. In combination, RBP seemed to force higher-risk lenders onto secured products, where interest rates are lower. Also, it seems to have been adopted quickest for products that are easily securitised, and where secondary markets exist for trading the debt.

26.6 Summary

Business's main function is to provide stakeholder value, and if done well, the company should be making a profit. There are times however, when the measurement of profit is problematic—especially at the transaction level. The area entrusted with ensuring the long-term financial health of the organisation is finance, which will be looking at ways of increasing revenue, reducing costs, or both, whether at the individual account, product, market, or organisational level.

In the credit arena, there are a number of different aspects of immediate interest to the finance department. First, finance needs to have some indication of what loss provisions to raise. When loans are taken onto the books, lenders know that there will be a certain level of losses associated with them, but the extent and timing is less clear. In general, there are two broad categories of provisions: *specific provisions* made against individual loans; and *general provisions* that provide reserves for all other credit losses.

General provisions can be determined using either direct or component approaches. Direct approaches include *net-flow models*, *transition matrices*, and *Markov chains*. The net-flow approach is simplest, but assumes that accounts either get worse, or are paid off in full. In contrast, transition matrices recognise the movements between states in both directions, and can be used in Markov chains to model the state of the portfolio at different points in time into the future.

Component approaches try to estimate losses in terms of probability and severity. *Loss probability* modelling may focus upon: (i) loss timing; (ii) a bespoke scoring model for losses; and/or (iii) loss extrapolation, using existing risk measures. *Loss severity* then focuses on other aspects of the expected-loss calculation (EAD, LGD, and maturity), and takes into consideration exposures, post-default funding costs, recoveries and associated costs, and proceeds from any risk mitigation in place.

If credit scores are powerful risk-assessment tools, and risk is a major profit element, then it makes sense for lenders to score for profit. This requires an understanding of the primary *profit drivers*, including risk, balance, late payment, activity, insurance, and marketing. Care must be taken as the most profitable customers are also of higher than average risk. If a lender wishes to maximise profit, the cut-off may be moved into areas previously considered unacceptable on a risk basis.

Profit modelling is done, but with varying degrees of success. Factors that affect its use include: (i) the *profit definition*; (ii) having sufficient *data-warehousing* space; (iii) determining an appropriate *outcome period*; and (iv) *drift*, as each element is unstable in its own right, and the aggregate is even more unstable. In general, the four main approaches used for modelling profit are profit scores, Markov chains, survival analysis, and matrix approaches that combine risk, revenue, and other metrics. The matrix approach is the most commonly used, and requires the fewest assumptions.

Finally, RBP is used by lenders to vary prices according to risk. Its acceptance has been greatest where loans are securitised and traded, but it is being accepted more broadly over time. Cost-recovery pricing might seem like a valid option, but the final prices charged cannot be the result of a purely analytical exercise—they must take into consideration market forces, and the broader customer relationship. It is not only the customers' behaviour that may be influenced by RBP, but also that of front-line staff, underwriters, and collectors. Even so, RBP empowers lenders to enter new markets, enabling greater access to credit for riskier customers, albeit either at higher rates or with greater security.

Module G

Credit risk management cycle

27 Marketing

> Despite many years of recognizing the mutual advantages of communication between credit and marketing strategies, it still happens too infrequently and to too little effect.
>
> *Thomas* et al. 2002:154

Marketing is the starting point of this module's little adventure. It includes functions like research and development (R&D), advertising, and sales. It is usually a vibrant area, often with a cowboy mentality, that is driven by sales volumes, asset growth, or some other income statement or balance sheet measure. This is also an area that consumes considerable analytical tools, and data mining resources.

Usually, any improvement in new business volumes will be positive for the business, but can have a negative effect where it results in unexpected costs, especially credit losses. Ultimately, the goal is to have a prospective customer take up an offer, which may be pre-approved, or be the result of an application. If the latter, the first step is to bring the customer and the application form together, either by making the form widely available, sending it out, or bringing the customer in. Like so many parts of managing a business, there are costs involved, and some means are more effective than others at reaching the intended audience.

So, what is scoring's relevance in this function? First, because marketings' bang-per-buck can be increased. Second, because marketing actions will have a direct downstream impact on areas where scoring plays a major role.

27.1 Advertising media

Products cannot be sold unless customers realise they have a need, and think that the product can fill the void. All the usual advertising media can be used, which can be divided into several broad classifications, based on the type of media and their reach, as set out in Table 27.1. The term 'reach' refers to how many people will see, hear, or read it, and can be treated under two headings:

Broad-based—Media that reach a large number of people, such as radio, TV, newspapers, and magazines. Tailoring can only be done if it has the desired target audience. These media are usually very expensive, and often fail to reach the right people.

Personal—Directed at a specific individual, usually by post, and today increasingly via the Internet. Focus can be put on a specific target market, as long as the marketer has effective screening tools. Costs can be contained, but it may be difficult to identify the individuals that should be targeted.

Table 27.1. Advertising media

	Broad	Personal
Print	Newspaper, magazine	Direct mail
Tele	Television, radio	Telephone
Cyber	Internet	E-mail, ATM
PtoP	Network marketing	Dealer

In contrast, 'media type' refers to the mechanism used to communicate with potential customers, which is treated under four headings:

Print—Put to paper, whether newspaper, magazine, or snail mail.
Tele—Old technology, done from a distance over TV, radio, or phone.
Cyber—New technology, done using computers—Internet, e-mail, or ATMs.
Person to person (PtoP)—Door-to-door and walk-in contacts. These might involve third parties, like dealer networks, brokers, agents, and perhaps even network marketing.

These marketing channels do not always provide the desired results, which will vary, depending upon a number of different factors:

Medium—Type of advertising used, and level of use.
Appeal—Effectiveness of the message.
Need—Does the product offered fill a gap for the target market?
Response—Number of applications received.
Acceptance—Number of applications accepted.
Risk—Potential bad debt losses.
Value—Potential return from open and active accounts.

More often than not, marketing campaigns will result in a trickle of new business. They may, however, result in a flood, where the volumes can create strains on downstream processes, such as application processing.

27.2 Two tribes go to war—quantity versus quality

War does not determine who is right—only who is left.

Bertrand Russell (1872–1970)

Culture is something normally associated with ethnic groups, like Zulus, Hindis, or Blackfoot, usually relating to their traditions, dress, songs, stories, and so on. Its true meaning is much broader though, and could be defined as a set of assumptions that are common to a group, and are passed on to new entrants into that group. Such assumptions will have developed over a

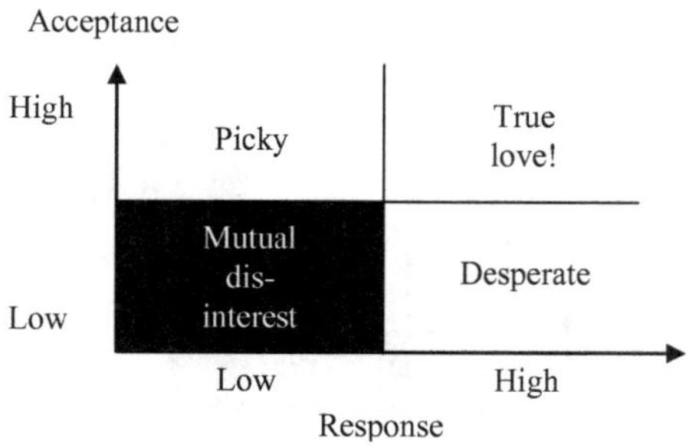

Figure 27.1. Response versus acceptance.

period of time, and will often have played a role in the ongoing survival of the group, or in enhancing its well-being. The behaviour, dress, artefacts, and institutions of the group are evidence of the culture, but do not define it.

The term is also often applied to companies, or even to departments within companies. The two areas of lending organisations that usually have substantially different cultures are marketing and credit. Marketing is the liberal; the young Turk looking to change the world; the tout that gets business to come to the door, and is measured by how many join the queue and are let through the door. The goal is to attract the greatest number of creditworthy customers at least cost, and the outcome is measured by both the quantity and quality of the applicants (see Figure 27.1).

In contrast, credit is the conservative; the wizened hand holding the reins, saying, 'Slow down!'; the bouncer cum gatekeeper, who ensures that only those deserving of credit are allowed through the door. It is measured by how well those who enter behave once inside, either in terms of credit losses, or the associated profitability. A very simplistic representation of the relationship is:

Equation 27.1. Expected profit $= P(\text{Good}) \times R - (1 - P(\text{Good})) \times B$
where R is the expected revenue, and B the amount borrowed

The key is managing the level of $P(\text{Good})$. Quality control is very important, but being too strict may turn away good business. The balance between risk and possible return has to be balanced (see Figure 27.2). This is where the conflict arises. In the traditional world, marketing's task was to attract the business, while credit controlled the risk—two goals that are at odds with each other. Nowadays, the relationship is more co-operative: marketing's goal is to attract applicants that have a high likelihood of being accepted; credit's is to maximise asset and revenue growth, while still controlling the risk. Both are interested in the overall profitability of the organisation, and the risk/return trade-offs. Some obvious areas where conflicts between credit and marketing require a meeting of the minds are:

Campaigns—Where advertising is broad based, there may be many applications that are either not creditworthy, or will not be profitable. Capacity constraints can arise where extremely high volumes are generated. Disclaimer clauses are required in advertising to indicate that applicants must meet minimum qualifying criteria.

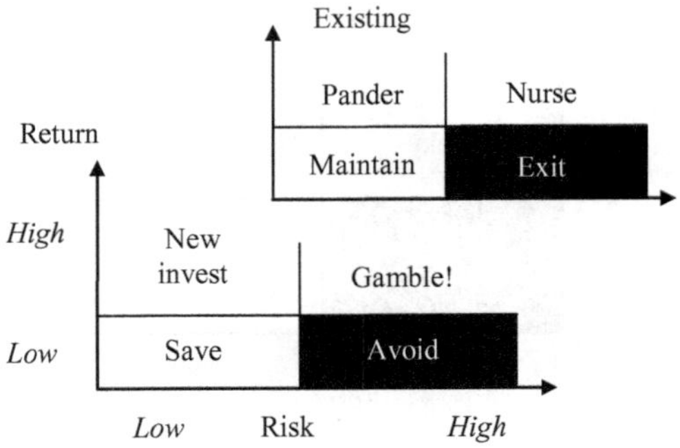

Figure 27.2. Risk versus return.

Application forms—Marketing often controls what is requested in the application forms, and may change the forms without considering the impact upon the decision processes. Indeed, many scorecard developers have experienced the pain of delivering a final model that relies upon a key field, recently dropped from the application form.

New markets—There may be little past experience in that market, whether defined by income, geographic area, or some other factor. It may be possible to apply existing risk assessment tools, but this must be done with great care.

New products—There may be no comparable experience for that product, making applicants impossible to assess using existing decision methodologies. Any credit evaluation would have to be done using a set of policy rules.

Credit also has to control the impact of marketing actions on the application-processing function, and both volumes and acceptance rates may vary greatly depending upon the campaign. Ideally, marketing strategies should also take into consideration the cost of application processing. Best practice is to score all applications, including pre-approvals, in order to maintain a richness of data.

Conflicts such as these are best dealt with by ensuring significant two-way communication between credit and marketing. Acceptance rates can be significantly improved if marketing has knowledge of credit processes, and designs well-targeted campaigns using appropriate media. Likewise, if credit is aware of upcoming campaigns, it can ensure that the application processing area is properly resourced, and may be able to tailor policies for the occasion.

27.3 Pre-screening

Luck is what happens when preparation meets opportunity.

Chorafas (1990)

One of the most widely used advertising media for financial services is direct mail. Companies obtain mailing lists from different sources, either at a price from third parties, or gratis from the company's own systems. The lists are then scrubbed, to optimise both the response and acceptance rates. According to McNab and Wynn (2003), initial list scrubbing would include:

Duplicate names—Required where lists from different sources are combined, to remove names that appear more than once.
Existing customers—Already have the product being offered.
Non-target—Fall outside the target group, possibly defined by income, age, geographic, or other parameters.
Bad on bureau—High risk, based on judgments or other bureau info.
Poor past or current performance—On other products with the lender.

The effectiveness of direct-mail campaigns will vary greatly. In cases where there is no existing relationship and the market is saturated, a simple 1 per cent response rate may seem a miracle. In contrast, where it is an existing customer and the campaign is well targeted, it can go to 30 per cent plus.

A number of different scorecards can be applied to enhance the effectiveness of the marketing campaign. Twaits (2003) splits these into three broad categories: risk, response, and value (Figure 27.3):

Risk—Do a preliminary risk assessment using the available characteristics. The effectiveness will depend on how much information is available, and how appropriate the model is for the target population.
Response—Based upon the results from prior campaigns, determine whether a person is likely to respond. This may also include churn scoring to determine whether or not an opened account will stay open.
Value—Determine whether an accepted applicant will provide value for the lender. The goal is profits, but the lender may instead use proxies, related to the expected borrowing patterns and/or revenue generation.

This framework was presented specifically for marketing, whereas the broader framework is the 4 Rs (risk, response, revenue, and retention), covered in first chapter. In each of these

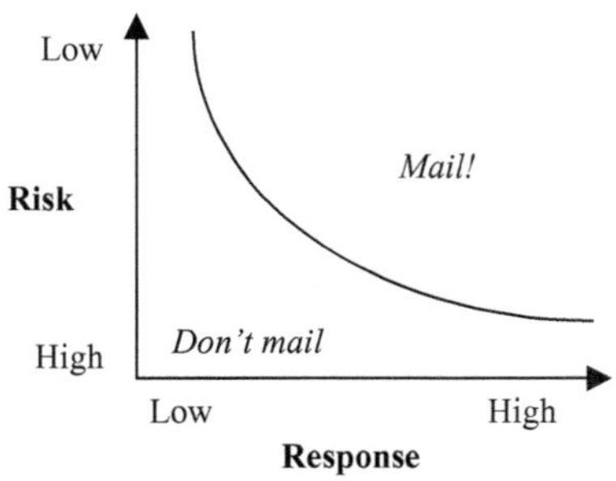

Figure 27.3. Risk versus response.

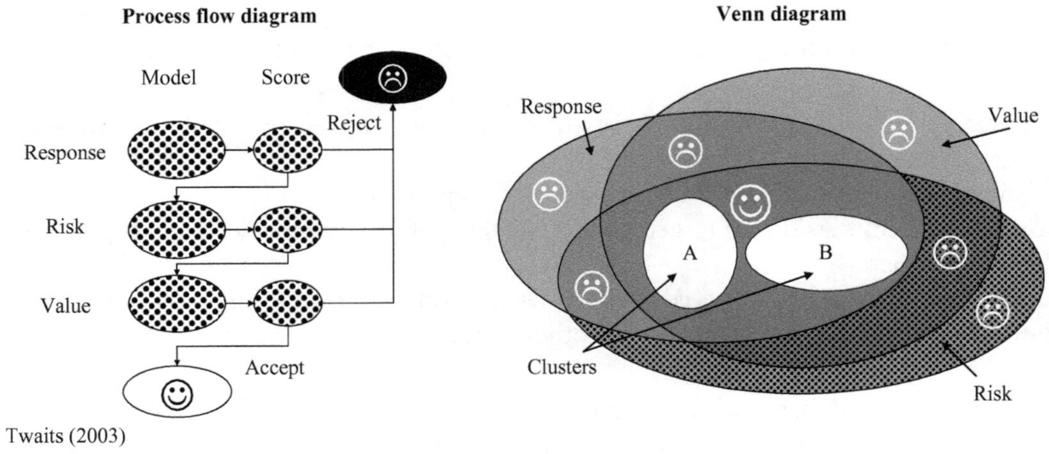

Figure 27.4. Risk, response, value scoring.

cases, there should be a correlation, hopefully strong, between the score and the target variable being measured. The scores will then be used to decide whether or not to stuff an envelope or make a call. Where applicable, especially where campaigns relate to existing customers, it is important to keep a hold-out sample of say 10 per cent, which can act as a benchmark to gauge the campaign's effectiveness.

The use of these types of scoring is illustrated in Figure 27.4, both as process-flow and Venn diagrams. The process-flow diagram is a simple one, where a single hurdle rate for each scorecard must be met for an account to be accepted. This not the norm however, as the scores are usually combined in conditional score matrices, where each cell represents a statement of the form 'If ($S_{Risk} > A$ and $S_{Risk} <= B$) and ($S_{Response} > Q$ and $S_{Response} <= R$) and ($S_{Value} > X$ and $S_{Value} <= Y$) then do J'. The Venn diagram illustrates both this and the identification of clusters where different approaches may be necessary, such as the use of trendier themes for younger prospects, and a focus upon security and flexibility for older applicants.

27.4 Data

Marketing presents specific challenges with respect to data, because the nature of the beast is different. There may be tons of possible data, yet not much to work with. According to Twaits (2003), problems exist with:

Location—Information sits in a variety of places and formats (Oracle, SAS, DB2, SQL Server, etc.). A great deal of time and effort needs to be expended just to bring this information together.
Quality—How good the information is, in terms of accuracy, age, and applicability.
Understanding—Whether the information's meaning is understood, as well as its applicability to the current problem.

The data sources that would be used for pre-screening are:

Application processing system—Application data, from all sources.
Account management system—Behavioural data, for both performance data and other or past product holdings.
Credit bureau data—Performance on accounts held with other companies.
Aggregated data—Information held at postcode, or some other level.
Fulfilment data—Marketing database, used to track outbound campaigns and their fruits.

Different data will be used for the different models, as illustrated in Table 27.2. For example, application processing, account management, and credit bureau information would be compiled for a risk scorecard, while value scorecards may also bring in demographic data at postal code level.

Marketing scorecards are most effective when developed ad hoc for specific campaigns, yet data compilation can take days or weeks, especially if it resides in different locations. With modern technology, it makes more sense to automate the compilation process, so that data can quickly and easily be stored in a data mart available for all functions, as illustrated in Figure 27.5. This also has the advantage that reporting functions can be developed to identify trends over time.

Table 27.2. Data extraction

Data source	Score type		
	☠	☎	$ £ ¥ €
Application processing	✓		✓
Account management	✓	✓	✓
Credit bureaux	✓		✓
Aggregated		✓	✓
Fulfilment		✓	

☠ = Risk, ☎ = Response, $ £ ¥ = Value (Twaits 2003)

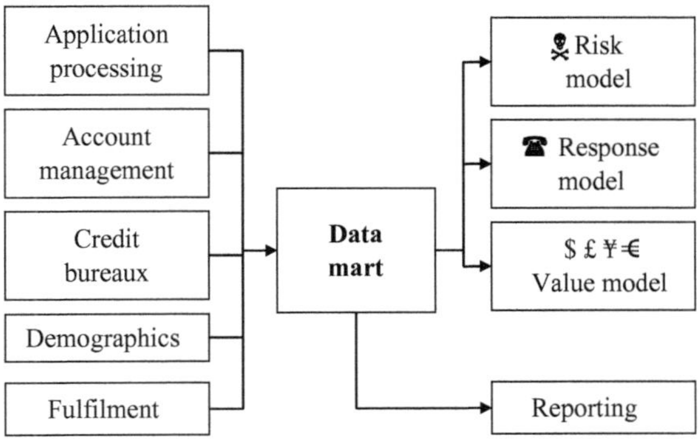

Figure 27.5. Data mart.

While the framework provided by Twaits (2003) is quite comprehensive, it does not recognise the provision of leads from outside sources. Even so, many of the lists are so commonly available, and widely used, that they do not offer a competitive advantage. Any company that is seriously in the game of customer relationship management should invest in processes that allow it to leverage off of every customer contact, whether outbound or inbound.

27.5 Summary

Marketing is the first stage in the credit risk management cycle, and is responsible for bringing potential new customers through the door. Messages are composed and directed to reach a target market. The media used may be broad-based or personal; and in either case may involve print, tele-, cyber-, or person-to-person communications.

Each option has a cost, and lenders try to ensure maximum bang-per-buck. Ingenious ways have been found, which have engendered a cowboy mentality and *quantity/quality issues*, that often cause a marketing/credit conflict in the areas of campaigns, application forms, new markets, and new products. During recent years, marketing has worked hard to harness the power of data, but is restricted because most of the relevant data lies outside of the organisation, and there are access restrictions. Even so, marketers will try to milk what data is available. This is especially true for direct mailing and phone contacts, where costs vary with each individual approach. These costs can be reduced using data available for each potential customer. Lenders wish to target those most likely to: (i) take up the offer; (ii) repay the debt; and (iii) result in a net profit.

Lenders will thus assess response, risk, and return. The first step is *pre-screening*, to rid the list of: duplicate names; existing customers; recently targeted and non-target customers; bad on bureau; and poor past performance. *Statistical techniques* are then used to derive risk, response, and return estimates, and the models can be integrated as hurdles or matrices, further to eliminate unlikely or unprofitable candidates. *Data sources* include application-processing and account-management systems, credit bureau data, aggregated data, and fulfilment data. Problems can, however, arise because of the location, ownership, quality, and understanding of the data. If possible, data marts should be established, which can be updated and accessed for future campaigns. Many lenders will use lists provided by outside vendors, but at times these are so widely used that they provide little value.

28 Application processing

> Money, it turned out, was exactly like sex, you thought of nothing else if you didn't have it, and thought of other things if you did.
>
> *James Baldwin* (1924–1987)
> US Writer, in 'Nobody Knows my Name'.

Throughout many of the previous sections, the processing of applications has been discussed, but at no point has the processing of applications really been discussed. It has instead been brought up in bits and pieces, even though it is the most critical point in the risk management cycle. According to Makuch (1998:3), 'It is estimated that as much as 80 per cent of the "measurable and controllable risk" is decided upon at the time of underwriting'.

Application processing relates primarily to new-business origination. It is one of the first—and often only—contacts that customers will have with the company. It is much like a first date, which governs first impressions and, in like fashion, customers may not just relish the memories but also tell their friends. In its most primitive form, subjective decisions must be made based on information provided on a paper form, and there may only be a few applications per day. In today's modern environment however, application forms are transmitted by fax, Internet, courier, satellite, or fibre-optic cable to a central area that may deal with thousands daily. Volumes can also vary greatly depending upon changes in the economy, or the number and effectiveness of recent marketing campaigns, so some flexibility and planning is required. The process is affected by many factors, which can be classified as inward- and outward-looking:

INWARD-LOOKING—The process's effectiveness for originating new business, including:

- **Data accuracy**—Whether the information provided by the customer was correct, and also captured correctly.
- **Turnover times**—Time required to make and deliver decisions, which may range from nanoseconds to days, or even weeks.
- **Override rates**—Percentage of score and system decisions that are being overridden, both low- and high-score overrides, and efficiency of the referral process.
- **Take-up rates**—How many accepted applicants become active customers, which is often a function of the application process design.
- **Fulfilment efficiency**—Time taken from the accept decision to when the customer receives the product, whether evidenced by account opening or drawdown.

OUTWARD-LOOKING—Less measurable factors, relating to customer interaction, including:
Flexibility—Ability to handle non-standard customer requests.
Sensitivity—Able to communicate decisions, positive and negative.
Transparency—Openness about processes used, and what affected a specific decision.

Long turnover times have a huge impact upon take-up rates. When many people apply for finance they want it now, and often give their business to the first lender that says 'Yes'. Hence the popularity of street-corner lenders, with names like FastCash and Ea$y.

These are overriding considerations that relate to the entire application process. In this section, it is traced from beginning to end, in a few pages. Whether done in-house or outsourced, most aspects of the process remain the same. There are basically six parts to the process, which are treated here as three groups of two:

Gather—Getting relevant information from interested customers.
Acquire—Completion and submission of an application form.
Prepare—Putting information into usable form/data capture and sanitation.
Sort—Obtain any other information required, rank each case, and make a decision.
Enquire—Get other relevant information from internal and external sources.
Decide—Whether to accept the application, or what to offer.
Action—Advise the customer, and deliver the goods.
Advise—Communicate the decision, and perhaps up-, down-, or cross-sell, if appropriate.
Fulfil—Deliver the promised product—cash, card, or chequebook.

The level of detail that follows may seem excessive for a text that focuses on credit risk, but is justified by the process's importance in the credit risk management cycle (CRMC). An understanding of the operational aspects can assist anybody that deals with it. The descriptions are in very simplistic terms, because of similarities with the processing of certain types of foodstuffs and raw materials.

28.1 Gather—interested customer details

Simply stated, the goal of the first part of the process—gathering—is to obtain information from interested customers, and do as much pre-processing as possible, to ensure that the next stage goes smoothly (Figure 28.1). It could be compared to harvesting apples . . . Only those apples with an expected value at market are passed on for sorting and grading, and any that are obviously rotten or damaged should be discarded (or be put to other uses).

28.1.1 Acquire applicant details

Assuming that the marketing department has done its bit to produce a population of interested customers, all that is needed are channels through which they can apply for the product. These can be categorised according to: (i) whether or not the customer received assistance to complete the form; and (ii) the medium used to submit it.

Assistance

Was assistance provided to the customer when completing the application form? There are three possibilities:

Customer direct—No intermediaries. Customers submit the application directly to, or deal directly with, the lender. The only assistance might be from relatives or acquaintances, with no other interest in the transaction.

Staff assisted—Customers are aided by bank staff, either because staff members are capturing details directly onto the computer, or customers have problems interpreting questions on the form.

Interested third party—Somewhere in the process, there is a dealer, broker, agent, etc., who has an interest in the customer getting the finance. This is most common for financing asset purchases, like motor vehicles and home loans.

Medium

The application forms can come through different channels, the two broad classifications being paper-based and electronic, with the former requiring the extra data-capture stage to put information into a usable form.

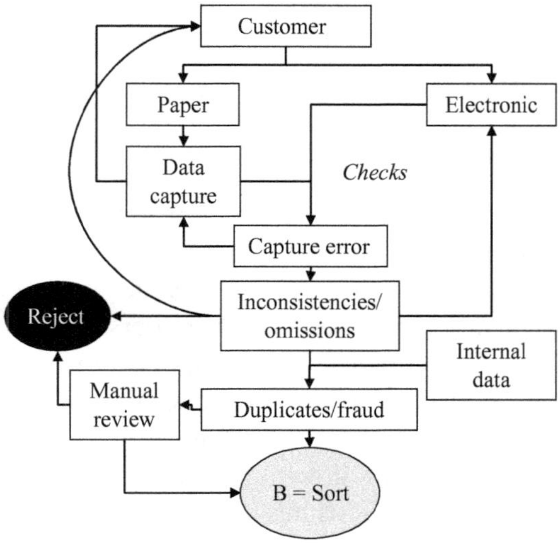

Figure 28.1. Gather.

Paper-based—An application form is received, and details are transcribed into electronic form by a data capture area. The primary paper-based channels are mail (snail, internal, courier, etc.) and fax. It has the distinct disadvantage that errors can occur during the capture process.

Electronic—Application details are put directly into an electronic form, either by the customers or someone assisting them, thus reducing the chance of errors. The number of electronic channels has grown as the cost of communications has reduced and infrastructure has improved—especially for Internet applications. Although the application may be accepted without putting pen to paper, many products will require a signature at some stage later in the process.

Once received, it is possible to check all applications for any undetected errors and omissions, and possible misrepresentation/fraud. The next few paragraphs cover paper-based processing in a bit more detail, because it poses a crucial extra stage with its own complexities.

28.1.2 Paper-based capture

Where the medium is paper-based, one person fills in a form and another has to transcribe it into an electronic format. This 'data capture' is a tedious operational process that manages the pieces of paper, which are collected, captured, and filed. It may sound fairly simple, but there are organisations that receive hundreds, or even thousands, of applications per day, and need efficient operations to manage the process. And while one of the prime directives is to ensure data quality, this process adds a step that, by its very nature, can introduce inaccuracies.

There are significant benefits to be gained through division of labour along a production line, and one of the production line positions is the data-capture operator. Their primary purpose is to capture data onto the computer as accurately as possible. Their lives can be made much easier by breaking the task into two parts, interpret and transcribe.

Interpret—pre-capture screening

Paper-based applications can have a variety of different forms, depending upon the company and its marketing campaigns. The main dimensions are: *distribution*—branches, mail, handouts at commuter stations, or magazine inserts; *size*—A4, leaflet; and *length*—single- or multiple-sheet. No matter what the form, customers may not understand questions, have handwriting that cannot be interpreted, or fail to complete it fully. Applications may also vary by condition—torn, crumpled, soiled, soggy, poor fax, or just plain poor handwriting.

The goal of the pre-capture screening process is to make data capture as smooth as possible. This can involve simple interpretation of difficult to read details, or extend to phoning the customer where this fails. Each of these actions has a cost, and it may be necessary to trash some of the applications, especially where mandatory fields have not been completed, or the form has not been signed.

Transcribe—physical capture

At some point in the future, lenders will move into the realm of optical character recognition, where paper forms can be fed into machines that automatically and accurately transcribe the information into electronic form—as text, not as images as is done currently. Until then, however, data has to be captured manually. It may be a simple process, but it requires distinct skills—in particular, fast and accurate typing skills, and a mentality that has a focus upon detail.

The data capture operator's job is tedious, and a great deal of effort must be invested in motivating and monitoring them. The design of the capture system—both computer and workflow—can facilitate the process. Capture screens should at least request the fields in the same order as they appear on the form, and possibly have a screen layout that is exactly like the form. One key measure is capture operators' speed, but as speed increases, so too does the possibility of error. This can be monitored either by: (i) *manual review*, another person checks the form against captured data; or (ii) *dual capture*, capturing the same application twice, and comparing the results. Monitoring adds extra expense, but can be controlled by sampling applications captured by each operator. The sampling rates can vary according to the operators' experience.

28.1.3 Pre-scoring screening and sanitation

It may be an irritation, but information verification and application sanitation are keys to flotation, to ensure no degradation or self-deprivation, but instead self-preservation without devastation. Apply imagination when checking the notation, with information rotation and a few confirmations. No default eradication or loss elimination, just part of the drive to profit elevation—and job salvation.

Before doing a credit evaluation, credit providers must ensure: (i) that the quality of the data is good; and (ii) that money will not be wasted on bureau calls unnecessarily. Paper-based applications may have been screened once already prior to capture, but will have to be screened again during the capture process.

Fine-filter—field checks

Data quality is a key issue in credit scoring, whether for scorecard developments or day-to-day transaction processing. Wherever possible, checks should be carried out to ensure that details provided are consistent with what the system expects, and any inconsistencies should be corrected. Field checks can include, amongst others:

Numeric fields—Some can be set so that text will not be accepted at any time.
Postal codes—Check format for that country, for example 11AA 1AA in UK, A1A 1A1 in Canada, five digit zip in USA, four digit in South Africa, etc.
Date fields—Check for valid date, and that it is reasonable.
'Time at' characteristics—Ensure correct format, for example YY, MMM, or YYMM.

These checks can occur: (i) *mid-capture*, as the details are entered; or (ii) *post-capture*, prior to saving/submitting. When done mid-capture, the terminal will usually beep to alert the capturer to a problem (touch typists do it without looking), along with a message to provide some guidance, so that the error can be corrected immediately. When done post-capture, any violation would cause faulty details to be highlighted on the screen. If the problem cannot be rectified immediately, the system design should allow the capture operator to save what has been captured, and carry on with something else until the problem has been solved.

Coarse filter—application checking

The next step is to check for obvious deal breakers. If a problem is identified, the application will either be declined outright, or will not go further until the problem is rectified. These rules can include:

DECLINES—decline outright and advise customer.

Prohibited by statute—By law, the applicant does not have the capacity to enter into a contract, for example minors, unrehabilitated insolvents, mentally insane, etc. It may, however, not be possible to ascertain some of these from an application form (insane?).

Lender policy—Applicant does not meet the criteria set by the lender for that product, perhaps based on income, age, address, or employment status.

Application unsigned—If not signed, the form may not be valid in the eyes of the court, as evidence of a contract.

Permission not granted—Where applicable, the applicant could deny permission to obtain data from, or share data through, the credit bureau(x). Bureau data is crucial for most decisions, but the number of denied cases is usually small. If considered at all, a different set of rules would apply, and the chances of acceptance would be lower.

REFERS—customer may be contacted to correct the details, but if no success, then decline.

Mandatory field checks—If not already pre-screened. These are fields that are absolutely essential for the account to be opened, like name and address.

Scored field checks—If key fields are blank, or a predefined number of scored characteristics are missing.

Cross-field check—Ensure that certain fields (like age, income, time at employment, etc.) are consistent, as a check against embellishment/fraud.

Cross-filter—internal databases

The application is now available in electronic form, and to the best of anybody's knowledge the process can continue. But what if there is readily available information that raises suspicion about possible fraud, an error in the process, or troubled past dealings? Further checks of internal data sources are needed:

Suspected fraud—Search fraud databases for possible match on name, address, or contact details.
Application—Search for duplicate applications. These may be genuine, especially for home loans and motor vehicle finance where the customer shops around, and applications are submitted via dealers, brokers, or agents.
Past history/performance—The applicant may already have serious delinquencies on other accounts held with the lender.

28.2 Sort—into strategy buckets

We now have an application that has been cleaned and scrubbed, and is ready to be presented to the next stage—sorting all of the cases into buckets. Any and all cases falling into a given bucket (scenario) should then receive the same treatment (strategy). The definition of the scenarios and strategies is something that is done upfront by the business, and may change over time.

This sorting process involves several stages, as illustrated in Figure 28.2:

Enquire—Obtain information from other sources, primarily the credit bureaux, but also other databases.
Measure—Segment and score each case, to provide the risk and other measures required to make a decision.

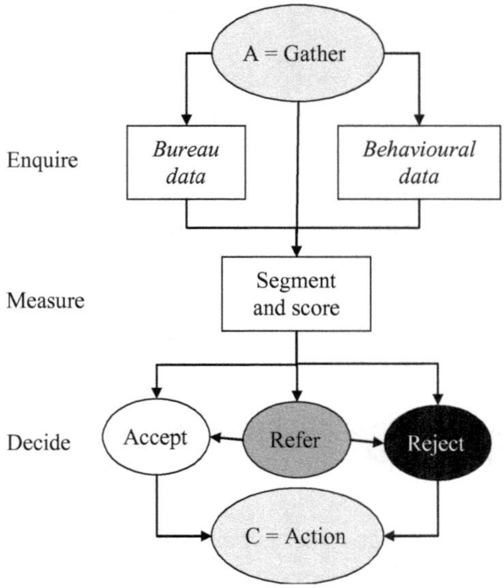

Figure 28.2. Sort.

Decide—Use scores and policy to assign each case to a strategy bucket. The buckets may be limited to reject, refer, or accept, but can extend to the maximum credit limit, interest rates, loan terms, cross-sell, or other options.

Enquire—internal

Lenders often have multiproduct relationships with their customers, or large numbers of repeat customers, who are much lower risk than off-the-street business. These deals will probably be approved even if there are some problems, which is not surprising, given that new-business acquisition costs can be 10 times those of extending loans to existing customers. Internal performance data can help to identify the really bad apples, and to define terms and conditions to be offered for additional or repeat business.

Bad apples are the greatest concern, assuming that they were not already weeded out during the earlier sanitation process. Some customers will apply for loans in spite of serious arguments, delinquencies, or even write-offs in the past, either because they are taking a chance, or they are desperate. They may also be *persona non grata* because of strong suspicions of illegal or fraudulent dealings. Scoring is insufficient for these cases, so policies are needed to override score decisions.

Enquire—external

According to McNab and Wynn (2003), there are a variety of reasons for getting information from outside sources:

Performance elsewhere—Details of past and current financial dealings are key inputs for assessing creditworthiness. Credit bureaux compile consumers' credit histories, based on court records, existing account performance, and enquiry records (see Section 12.3, on Credit Bureau Data).

Existing commitments—A check of applicants' financial commitments elsewhere, which is done as part of responsible lending, to protect against customers over-committing themselves.

Identity verification—Check identity details against another source, as protection against both fraud and money laundering. This can include ensuring that the personal identification number is valid, and that the name, address, and contact details are correct—or at least consistent with those provided elsewhere.

Fraud prevention—Ensure details do not exist on external fraud database.

There is also an issue of exactly when credit bureau information should be included as part of the process, as there is a cost involved. The two possibilities are:

Enquire on all—Obtain bureau data for every applicant. It increases the total cost of bureau calls, but may be offset by improved efficiencies as an extra step in the process is avoided.

Selective enquiries—Do pre-bureau screening, to weed out those applications where bureau information would not change the decision. Lenders usually do this where decline rates are very high, or very low, relative to the average.

Pre-bureau scoring has the disadvantage that it adds another stage into the process, and further complication. The key factor to consider is how this extra step will impact upon the bottom line, including:

Cost per call—This will vary according to the bargaining power of the lender, competitive pressures between bureaux, and improvements in technology. Internal staff costs, like the time spent on the telephone obtaining the information, should also be considered.

Decline volumes—If the number of declines is relatively low, then the extra complication is not warranted.

Number of metrics—The greater the number of measures used in the decision process, the more complicated pre-bureau screening becomes.

Value at risk—The greater the amount, the greater the potential loss and need for extra information. A policy rule should be in place to demand a bureau call for any application above a certain amount.

Strategy dependence—To what extent does the extra value provided by the bureau information influence strategies, like maximum loan amounts?

Customer service—Is customer service impacted upon in any way, perhaps by increasing the time required to make decisions? This will be a consideration if the process is manual, or the probability of bureau downtime is high.

Measure and decide

We now have sufficient information from the application form, and internal and external databases, to use scores, segmentation, strategy, and policy to make a decision. This could be compared to game playing: *segment*, decide which game to play; *scores and cut-offs*, setting the rules; *strategy*, actual game play; and *policies*, the referee, guiding the course of game play:

Segment—Necessary where there are substantial differences in the type of customers, especially where different infrastructure is employed (marketing, account management, collections). Separate models may be needed for different subgroups within a single channel, because the relevance of certain characteristics changes.

Score—The primary scores of interest are credit risk scores, especially application scores and bureau scores. These can, however, be supplemented with churn, profitability, revenue, usage, or other scores, to weed out unprofitable cases or adjust terms. Each score is split into bands that can be used to apply strategies.

Strategy—What to do, when! In its simplest form, this is a single scorecard cut-off to say 'Yay' or 'Nay'. In more complex forms, it involves multiple cut-offs used to vary terms

Table 28.1. Strategy tables

Accept/reject						Terms of business					
Single		Multiple				Single		Multiple			
S	D	S	1	2	3	S	D	S	1	2	3
1	R	1	R	R	A	1	0	1	0	0	2
2	A	2	R	A	A	2	1	2	0	1	3
3	A	3	A	A	A	3	2	3	1	2	3
4	A	4	A	A	A	4	3	4	2	3	4

of business; not just the maximum loan amount, but also the interest rate, repayment period, collateral required, or other terms of business. Table 28.1 provides simplistic representations of strategy tables that might be used for single and multiple score cut-offs, whether to provide an accept(A)/decline(R) decision, or a risk indicator (e.g. decline(0), or accept (1–4)).

When setting strategies, cut-offs can be adjusted for different subpopulations, to achieve business objectives. A lower cut-off may be set for younger applicants, to make inroads into that market; or a higher cut-off for a new market that has not yet been tested.

28.3 Action—accept or reject

After gathering and sorting, the decisions must be actioned (see Figure 28.3). This has two primary parts: *advise*, communicating the decision to the customer; and *fulfil*, delivering the goods, or not, as the case may be. Both parties will of course be hoping for an accept, in which case there may be further steps prior to fulfilment—like documentation and delivery. The lender may also wish to up- or cross-sell the applicant. The hard part is dealing with declines, and issues around decline reasons, down-sells, and the appeals process.

28.3.1 Declines

Lenders once operated as black boxes; people would apply for a loan, but had no idea of what influenced the lender's decision. It is not an issue for those who get what they want, but very perplexing for the poor person who is turned away or down-sold. Lenders themselves did not see the necessity of providing decline reasons. The costs can be high, and many believed the extra opacity acted in their own best interests, to prevent against fraud or reckless borrowing. What is increasingly accepted in all walks of life however, is that transparency is good, whether for governments, companies, churches, or lenders.

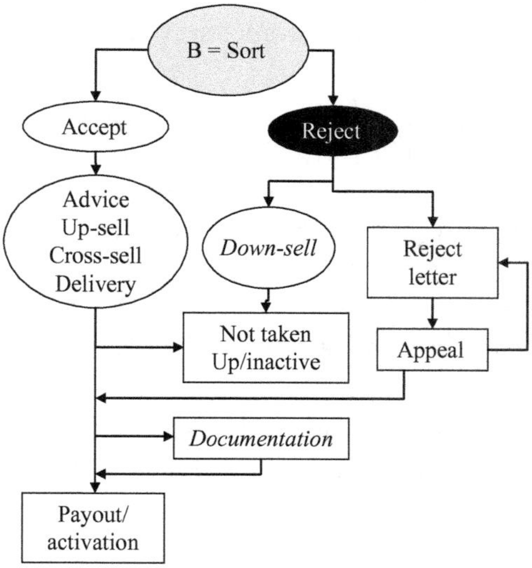

Figure 28.3. Action.

When underwriters make credit decisions, the subjectivity makes it difficult to give specific decline reasons. In contrast, credit scoring enables lenders to provide objective reasons, and societal demands for them to do so have been increasing, including requirements by law. Declined borrowers will have two basic questions, and the lender has to decide whether, and how, they are to be answered:

How was the decision made? Explain whether the decision was based on human judgment, scores, or both. The fact that applications are scored is often stated up-front, prior to application, but may be restated in the communication giving the decision.

What affected it the most? Indicate why the application was declined. It may be possible to get away with saying 'declined on score', 'declined on policy', or 'declined on statute'. Regulations may, however, demand greater detail on factors that negatively influenced the decision, like 'bad on bureau', or 'poor past dealings'.

Down-sells

Many customers will request a product, an amount, or terms of business that the lender is not comfortable with. Rather than rejecting the customer outright, the lender may make a counter-offer. While this will often work, it should be done with great care. The borrower may be offended, to the extent that the down-sell is more damaging to an existing relationship than an outright decline.

Appeals

Being declined for a loan can sometimes have the same devastating effect upon a person's hopes as being sentenced to jail, and in both cases the person can appeal. The extra resources—people and processes—demanded by an effective appeals process can add significant overheads, but be offset by a significant improvement in customer-service levels. Customers may contest decisions on the basis of:

Bureau details—May be contested directly with the bureau, or with the lender. The details may be incorrect, or not properly represent the individual's circumstances.
Further information—This may include financial statements, bank statements, or other information.
Security—Where the application is for a bank loan, the applicant may offer collateral, guarantees, or have somebody stand surety for the loan.

Whether or not the appeal causes the decision to be overridden depends on the lender's policies, and possibly the individuals (customer, underwriter) involved.

28.3.2 Accepts

The only time money is provided immediately upon acceptance is when the funds applied for are to be paid directly into a bank account, and all other formalities have been completed. Otherwise, there are further hurdles to cross.

Documentation

It is quite possible that this maze has been navigated without anybody ever putting pen, photocopier, or laser-jet to paper, but now is the time. The types of documentation that may be required by the lender are:

Identity documents—Copies of one or more of a birth certificate, driver's licence, passport, or other personal identification document.
Contract—A signed piece of paper that confirms that there really is an obligation to pay. This may or may not be required, as the application form, combined with other documentation, will often suffice.
Proof of ownership—Documentation showing that the borrower is the legal owner of the asset, especially if it is to be security for a loan.
Proof of purchase—The invoice and/or receipt, if the customer has already paid for an asset and reimbursement is required.
Insurance—Against unforeseen events, either against the asset (home, motor, etc.) or the repayment stream from the individual (life, sickness, job loss).

Documentation is meant not only to protect against risk, but also to meet legal 'Know Your Customer' requirements that protect against money laundering. It comes at a cost though, as it makes distance lending more difficult, and increases the cost of account origination generally.

Fulfilment

The primary concern now is that the product be delivered to the right person, and in good time. Just how it is delivered depends upon the product. There are two broad product classifications, based upon who initiates the draw down, and whether or not there is a transaction medium:

Initiation

Customer initiated—An account is opened with a credit limit, but the customer decides on the timing and amount of any transfer out of that account. This includes most transaction products, and many revolving credit facilities.

Lender initiated—The lender pays out the funds, either directly to the individual, or to a third party (for asset purchases). These are almost always high-value loans, at least as far as the customer is concerned, where problems in delivery can cause high anxiety.

Medium

Transaction products—Require some medium in order to transact on the account, *paper* (cheque), *plastic* (card), or both. Potential for fraud arises if the medium falls into the wrong hands. Lenders should take extra steps to ensure that it is received by the correct person, either by using secure delivery channels (registered mail, courier, branch collection), or pre-activation identity checks (phone calls to activate).

Non-transaction products—Fixed-term and revolving facilities, that do not accommodate third-party transactions: *personal loans*, are provided directly to the applicant, either cash, cheque, or into a bank account; and *asset loans*, funds provided either directly to the applicant or to the seller, but only after formalities are completed.

Not-taken-up (NTU)/inactive

There are many instances where people will be approved for a product, but then nothing happens. Either the applicant never completes the final steps before the loan is paid out (personal or asset loan), or never uses the facility that is granted (card, revolving credit, cheque overdraft). This may occur because:

No longer required—The application was related to a specific purpose, and the applicant has changed his/her mind.

Uncompetitive offer—Customer received a better offer from another lender.

Lost communication—The customer never received the acceptance notice.

Blocked communication—An agent, or other person involved in the process neglected to pass on the advice, instead directing the business elsewhere.

Documentation lacking—The applicant is unable to provide the required documentation, for example proof of identity and/or affordability.

Lenders must monitor what is happening to NTUs, as it may highlight missed opportunities—in particular, where there are competitive or communications issues. The last category, 'Documentation Lacking', becomes increasingly problematic with legislation requiring lenders to 'Know Your Customer', and can be exclusionary. Subprime, low-income, and emerging market customers are those least likely to have this documentation readily available.

Up-sells

Just as lenders may be willing to provide a lesser amount, a lower-status product, or more demanding terms, the opposite is also possible. Up-sells can occur where: (i) communications with the customer are easy, or necessary, prior to finalising the deal; or (ii) the terms can be easily changed after the transaction is approved. The communications part is not a problem where the applicant comes into a branch for an answer, or is still waiting on the other side of a computer, or ATM, screen. It is more difficult where communications are by mail, especially if all the customer is waiting for is a piece of plastic with the requested limit.

Cross-sells

Now that the applicant has been accepted for one product, the lender may wish to offer more. Cross-sell opportunities depend upon the credit provider's total product offering, the borrower's profile, and his/her existing product holdings. Pre-approved cross-sales are best done with some type of caveat, such as being valid only if accepted within a given time frame. If it is not fully pre-approved, then some type of pre-screening should be done to ensure a high probability of acceptance. The customer relationship may be damaged if the sales pitch is done and the customer is then declined.

Do not underestimate the power of cross-sells! It is an exceptional tool for lenders to grow market share, and if done properly can be done with acceptable risk. Potential borrowers should be wary of the barrage of offers they may receive though, as they will quickly lead to financial problems if all of them are accepted.

Approval in principle

For unsecured lending, the possibility of a future upgrade is usually possible, no matter what the means of communication. However, where a customer is shopping for high-ticket items, such as motor vehicles and home loans, it usually helps to know how much credit is available upfront. This knowledge can either expand their choices when shopping, or save the time and frustration of being turned down after a choice is made. Lenders hoping to finance high-ticket purchases need to have, and advertise, the capability of providing an 'approval in principle', and their ability to provide greater assurance when customers are shopping for major assets.

Credit insurance

A major tool used to mitigate credit losses is credit insurance, which can provide a substantial revenue contribution. It is similar to homeowner's or motor vehicle insurance, except the repayments are forgiven in the event of death, illness, job loss, or other predefined events. The insurance premium may be paid up-front or charged monthly, and tends to add between 2 and 5 per cent to the borrower's effective annual interest charge. Of that, perhaps one-half to two-thirds might go to lenders' bottom line. The customers most likely to take out credit insurance are those who are at greatest risk, are the most price insensitive, and tend to be in a weak bargaining position. For that matter, in some high-risk environments like subprime lending, credit insurance is compulsory, which sadly indicates the lop-sided balance of power between borrower and lender in that market.

28.4 Summary

The most critical point in the risk management cycle is application processing, where lenders govern how much risk they take on. Fifty years ago it was a manual process, but credit scoring has allowed its automation. It is often the only contact that customers have with the company, so it is important to create good impressions. When managing the process, lenders must consider both: (i) *inward-looking factors*, like data accuracy, turnover times, override rates, fulfilment efficiency, and take-up rates; and (ii) *outward-looking factors*, like process flexibility, sensitivity, and transparency.

The process is comprised of three parts: (i) *gather*—acquire and prepare; (ii) *sort*—enquire and decide; and (iii) *action*—advise and fulfil. Gathering can be done directly from the customers, perhaps with staff assistance, or via agents, dealers, or brokers. It may be paper-based or electronic, where the former poses extra challenges due to extra data-capture and data-quality issues. Both involve screening, to ensure appropriate information is available for a risk assessment, including bureau data and information on past dealings. Sorting involves measuring each case in order to decide upon a course of action: accept, reject, or refer. The decision will also be a function of the terms of business, such as the interest rate and loan term.

Actioning involves both communication and delivery. The difficult part is advising declines, including issues relating to decline reasons and the appeals process. For accepts, the goal is to ensure that the customer takes up the offer. Mechanisms differ, depending on: (i) whether there is a transaction medium; and (ii) whether take-up is initiated by the customer or the lender. 'Not-taken-ups' are a possibility, perhaps because the product is no longer required, the offer was uncompetitive, communication was lost or blocked, or the customer cannot obtain documentation. Lenders may also *up-sell* and *cross-sell* to accepts, and *down-sell* to rejects. *Approval in principle* may be given to customers that are shopping for high-ticket items, and need to know how much credit they qualify for. *Credit insurance* can also be used to mitigate the risk, and provide lenders with another income stream.

29 Account management

Once customers have taken up the product, lenders enter the realm of 'account management', which takes on different meanings depending upon the context in which it is used. In its truest sense, it covers all front- and back-office functions used to manage existing account relationships, including billing, payment processing, limit management, renewals, collections, recoveries, tracing, etc. Within the credit risk management cycle (CRMC) however, it refers more specifically to the management of non-delinquent accounts—collections and fraud excluded. The goals are to manage individuals' appetites for credit, and to try to keep them coming back for more.

Behavioural scores are key tools here. A major difference between application and behavioural scoring is that: (i) the former gathers information from as *many sources* as possible—application form, credit bureau, past and existing dealings, and so on; while (ii) the latter uses various aspects of *account performance* as predictors, and relies less on demographics and bureau details. The one exception is customer scoring, a form of behavioural scoring that combines the performance of all products into a single score. This data distinction is not cast in stone though. There are increasing demands to increase the number of data sources used to assess existing accounts, including bureau, other-product, customer-supplied, and other data. Customer scores can be used to assess the overall customer relationship, and provide the basis for estimates where no dedicated score is available.

This section starts with a brief look at different limit types (agreed, shadow, and target), and then borrower types and associated lender functions (set out in Table 29.1). Borrowers are split into two groups, based upon: (i) how they get the funds (*limit availment*) and: (ii) what happens subsequently (*account repayment*). The former applies to transaction products (cheque/credit card) where there are takers, askers, and givers. In contrast, the latter (repeater, repayer, keeper, stealer) applies more broadly across other types of lending products. Each requires different mechanisms to manage them, the most obvious examples being:

Over-limit management (takers)—Set maximums for over-limit excesses.
Limit-increase requests (askers)—Set maximums for limits that may be agreed based purely on readily available behavioural information.
Limit-increase campaigns (givers)—Determine what will be offered, as part of a marketing drive.
Limit reviews (repeaters)—Do periodic reviews to determine whether or not the facility will continue, and on what terms.

The asker/taker/giver framework is used informally by some organisations, but was not found documented anywhere. The author learnt of it from an association with Experian UK.

Table 29.1. Borrower types

Type	Definition	Lender action
Based on Limit availment		
Taker	exceeds the limit without permission	Authorisations/referrals
Asker	requests increase in the limit	Limit management
Giver	offered limit without asking	Marketing
Based on account repayment		
Repayer	pays back the funds in full	Cross-sales
Repeater	renews or extends the facility	Renewals
Keeper	is negligent in repaying	Collections
Stealer	has no intention of repaying	Fraud

29.1 Types of limits

Just as there are different ways in which customers will request or abuse limits, so too, there are tools to manage them. The lender's goal is to grow the limits and balances while managing the risk and customer satisfaction, all of which can be passive, reactive, or proactive. The labels used for the types of limits will vary from company to company, but can be classified as:

Agreed limit—That agreed with the customer for normal operation of the account, but which may fall away if any of the terms and conditions of the agreement are broken. It may also be called an arranged, declared, or known limit.

Shadow limit—Operates in the background, as the upper bound for over-limit excesses. This is used where no permission has been sought, whether because the excess is an oversight, or the customer is loath to go through the formalities of applying.

Target limit—Maximum limit that will be granted on customer request without excessive formalities. For good customers, this is usually higher than both the agreed and shadow limits, and may cover multiple products. Some lenders use a shadow limit for both.

The shadow and target limits may also be referred to as confidential limits, as they often operate in the background and are not readily known to customers. Both of them are set according to limit strategies that the lender wishes to apply. These will be based upon the behavioural risk indicator, combined with the current limit and/or some income or turnover figure (credit turnover, disposable income). Examples are 'Current Limit' + X, 'Current Limit' × 1 plus Y per cent, or 'Average Monthly Deposits' × Z.

The chart in Figure 29.1 is a simplistic illustration, where a percentage is added to the agreed limit, which for riskier accounts tries to reduce the amount at risk. The shadow and target limits are higher than the agreed limit, at least for accounts that are not delinquent. Serious delinquency implies that the agreement has been breached, and the agreed limit no longer holds. The lender

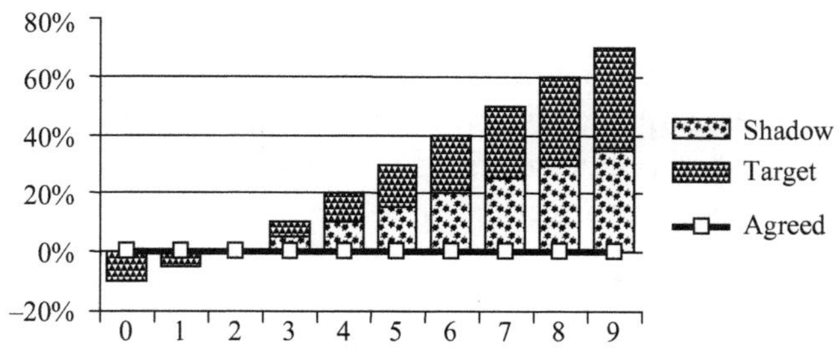

Figure 29.1. Limit strategies.

is then within rights to refuse all future transactions until the account is brought back into order. This also presents an opportunity to reduce the limit, or adjust other terms.

Use of other scores

A general disadvantage of the agreed/shadow/target limit framework is that limits are upwardly mobile, but downwardly sticky—once a limit is granted, it is difficult to take away. The alternative is for the lender proactively to change a declared limit upwards and downwards based upon a combination of risk, usage, attrition, and other considerations, but this would confuse customers, and not be well received when the move is downwards.

Behavioural risk scores can also be combined with other factors to determine these limits. A usage score can be used to push up limits for high-usage customers, and down for low usage, at the same level of risk. Note that behavioural scores are based upon current utilisation and circumstances, and do not provide an indication of what will happen when customers' circumstances change. Conservatism is especially wise when setting strategies for customers that appear to have little or no need for those limits already granted.

The lender can also use a customer score to set strategies for multiple products. These would put an absolute maximum upon the combined limits and/or required repayments, and possibly also restrictions at product level. For example, a lender may wish to limit total exposure to 10,000, of which at most 50 per cent can be provided on transactional products. This can cause some conflict within the organisation though, as each product area will seek to influence the strategies to their own benefit.

Just how these strategies are set is not straightforward, as the relationship between the loan amount, term, and repayment amount complicates matters. Can fixed-term and credit card limits for the same amount be treated equally? Home loan and revolving credit repayments? Each of these products will have a different value to the customer, and a different risk profile. Some time should be dedicated to devising customer-level strategies that are appropriate for the organisation and its combination of products.

Triage

Lenders need not wait until problems arise, but may also use scores and cash-flow data to identify customers that require debt counselling. Those customers are then contacted, to determine whether there are financial difficulties, and, where necessary, are offered advice on cash-flow triage. This refers to the allocation of cash flows to achieve the greatest benefit, like paying down loans with the highest interest rate first—which may be done by individuals and companies alike. Very little literature is available on the topic, but early indications are that it is providing customer-service and public-relations wins for lenders offering such advice (also see Section 29.2.3, Informed-customer effect).

The concept of triage (from French *trier*, to sort) has been borrowed from the field of medicine, where it refers to the management of disaster and war-time scenarios. Casualties are sorted into deceased, immediate, delayed, and minor categories in order best to allocate scarce medical resources. It was originally developed by Dominique Jean Larrey to aid Napoleon's armies. Today it refers to any process where sorting is done to achieve the greatest benefit with limited resources.

29.2 Over-limit management (takers)

There will always be people who take without asking, whether by design or otherwise. Some will treat the borrowed article with care and respect, and put it back in the same state as it was found. Others will use it with abandon, with little regard for object or owner. Many transactional lending products allow customers to exceed the agreed limit to some extent, either because it is not possible to control over-limit excesses completely, or because it is a service for which charges can be levied. This section covers over-limit management, and is restricted to transaction products (cheque and plastic).

29.2.1 Cheque accounts—pay/no pay

The late 1800s saw the widespread use not only of cheque accounts, but also of overdraft facilities in England. The nature of the account is such that people can write cheques that push the balance either into an unarranged overdraft, or over the arranged overdraft limit. Initially, banks would return cheques unpaid when there were *insufficient funds* in the account (NSF), but they soon learnt that this could cause problems with customers and others, especially where the accountholders were influential members of the community and/or good customers of the bank.

Banks thus developed a referral process to make pay/no pay decisions—the cheque would be referred to the branch manager, or an underwriter, to assess whether the bank was willing to accept that risk. The only difference between then and now is that cheques are being replaced by electronic transactions, and the human element has largely been eliminated from the decision process.

Most people will seldom, if ever, issue an NSF cheque, but for others it can be a way of life. In the United States, during 2002, there were over one billion NSF cheques issued—more than

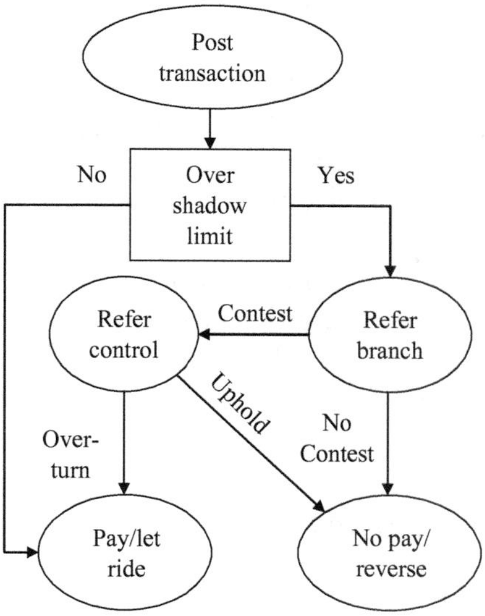

Figure 29.2. Pay/no pay.

three per transmission account. These can generate significant fees, which combined with excess fees might average $150 to $200 per troubled account (Sheshunoff 2002).

The process used by different banks may differ, but usually boils down to: (i) *post* the transaction to the account; (ii) if over shadow limit, then *refer* to the branch, or responsible person, who may wish to contest; (iii) *contests* are taken up with a central control area, which may agree to overturn the system decision; and (iv) if there is no contest, or the control area upholds the system decision, then the posted item is *returned NSF* (Figure 29.2).

The key tool here is the shadow limit, up to which the bank will honour transactions. The higher the limit, the fewer the referrals and associated cost, but there is a trade-off because the higher limits increase the value at risk. Lenders can also work on maximising penalty revenues, but this requires sophisticated tools to score and monitor high-risk/high-reward customers. It also presents ethical issues, especially where customers are ill-informed.

Uncleared effects

Because of the way in which inter-bank cheque clearing systems work, there is a delay between the time when a cheque is deposited into a bank account, and when the bank actually receives the funds. Most banks will reflect the funds in the account up-front and pay interest on it, but there are extra risks if they allow those funds to be withdrawn. The cheque may be returned, either because it is NSF, fraudulent, or the account is closed. Fraudsters regularly play upon these delays, using knowledge about the banking system, and how long it takes a cheque from one bank to clear through another.

Banks protect themselves against this possibility by putting a freeze on any withdrawals against that deposit for several days, a balance category called uncleared effects, and the maximum balance available at any one time will be the cleared balance. This can cause a great deal of unnecessary inconvenience for customers, especially where the bank allows a flat 10 days for a salary cheque to clear, even though the other bank is across town. Fortunately, most employers have now arranged automated salary deposits, but many people still receive money that they rely upon by cheque.

In order to improve customer service, banks may allow withdrawals against uncleared effects, mostly for customers with established track records. This can be very dangerous though, as it creates opportunities for fraud—especially where there are deposits for abnormally large amounts. Banks need to ensure that withdrawals against uncleared effects are reasonable, given the circumstances. This could be done using an absolute maximum value, a shadow limit, a multiple of average debit turnover, or some other measure.

29.2.2 Credit cards—authorisations

An advantage of recently developed products is that they do not have the baggage of inflexible legacy infrastructure, and are able to adopt whatever is newest and sexiest at the time. So it was with credit cards, which have authorisations processes that are real-time, online, all the time (or so it is hoped). Many merchants will not accept cheques due to the extra risks, or the costs associated with ensuring that there are sufficient funds. With credit cards, it is as simple as a swipe through a machine.

It has not always been this way, as credit card transactions also used to involve a lot of paper and telephone calls. The procedure would be:

- Customer *presents card* for purchase.
- A *transaction slip* with card and purchase amount details is completed if, (i) amount is less than the *floor limit*, or (ii) merchant gets *authorisation*, which requires a phone call to the card issuer for an authorisation number, that is noted on the transaction slip.
- Merchant *submits* transaction slip to card issuer for payment.
- Customer is *billed* through card issuer's account systems.
- Merchant *receives funds* after a predetermined period.

The decision processes used for card authorisations and cheque pay/no pay decisions are basically the same. The main difference is that the merchant is assuming the role otherwise played by an employee, to carry out any further checks.

Automation has made the authorisations process more efficient than the pay/no pay process though. This is not only because speedpoints have eliminated most of the paper shuffling, but also because there is no longer a need for a manual check on those millions of transactions that are well within accounts' agreed limits, and many more that are within tolerances that the card issuer is comfortable with (see Figure 29.3). Person-to-person contact is now used only where voice authorisations are required, primarily as tools against fraud prevention.

The use of floor limits is also being increasingly questioned. Over-limit and delinquent accounts can still purchase up to this limit, irrespective of account status, thus increasing the

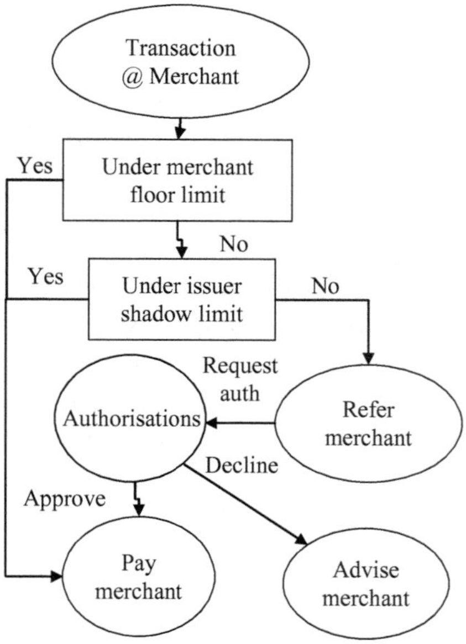

Figure 29.3. Card authorisations.

amount-at-risk outside of lender control. Floor limits are a legacy of how the transactions were originally processed, and have been retained to handle those circumstances where telecommunications links are slow, or unavailable. Some card products are now being offered where there is no floor limit (compulsory authorisation), and chip & PIN cards will also make floor limits redundant, because balance records will be stored directly on the card.

Cash advances

A facility that requires special mention is cash advances, as they are often treated separately. Credit cards are usually used as a payment mechanism when buying goods and services, and interest will only accrue after 30 or 60 days, depending upon the agreement. Customers who want cash loans will get them from other sources, as interest on cash advances starts accruing immediately, and the interest rate charged is almost always much higher than for equivalent bank loans. Where cash advances are made, especially for the first time, it can be an indicator that the customer has exhausted other sources. There is an implied higher risk that might have come about so quickly that it is not represented in any of the risk measures calculated for that account. Most card issuers will have different rules to govern cash advances—including different shadow limits.

29.2.3 Informed customer effect

According to a paper by Alex Sheshunoff (2002), something that lenders should take into consideration with pay/no pay decisions is the 'informed-customer effect'—people who have to choose between equally bad choices will pick the one that is best understood. Thus, if there is a choice between paying penalties on the bank, telephone, utility, medical, education, or other account, the customer may not choose that with the lowest penalty, but that which is most transparent in its policies. Non-bank entities benefit from this significantly, as their policies are usually very straightforward.

In contrast, customers often have a poor understanding of banks' NSF policies, because: (i) shadow limits are cloaked in secrecy; (ii) fees are poorly advertised, and seen to be punitive; and/or (iii) bank staff are not geared to handle queries from customers. In general, banks often fear that public knowledge will lead to abuse and possible fraud, and will only see the downsides of extra transparency, instead of the potential benefits. They instead focus on controlling risk, with little regard for optimising the revenue, customer-service, and compliance elements.

By using both behavioural risk and revenue scores, combined with customer education, it is possible to achieve several goals simultaneously. There are two elements to the customer education process:

(1) Advise customers that over-limit and late-payment situations are wrong, and that these will have a negative impact upon their credit standing. The impact of this message will be greatest where penalty charges are high and credit bureaux are well developed, or the environment is changing in that direction.
(2) Advise the customers of company policy with respect to excess and late-payment situations, especially the fees. These need to be fair and consistently applied, otherwise further uncertainty and dissatisfaction will result. Customers can then make informed decisions about where trade-offs should be made. A strong case can be made here for making shadow limits known.

This approach, combined with appropriate segmentation and strategies, allows for ethical maximisation of penalty fee income. Quoted studies indicated that properly informed customers might shift as much as $45 in penalty fees from non-bank entities to banks. Old National Bankcorp and United Bankshares in the United States claimed to have increased theirs by as much as 50 per cent.

29.3 More limit and other functions

29.3.1 Limit-increase requests (askers)

Customers' needs change over time, and accepted customers will likely be back for more. This is entirely natural, as initial limit strategies tend to be conservative, due to the usual difficulties of predicting the future. Once the customer has been around for some months however, the account performance will provide a much clearer picture of what the future holds, and the lender will slowly become more receptive to providing higher limits.

In this realm, the target limit comes into play. This is the maximum limit that the lender is willing to entertain for a given customer, which hopefully optimises revenue without inordinately increasing the risk. There are two possible scenarios:

Permanent limit increase—Becomes the new agreed limit for the account.
Temporary limit increase—Put in place to accommodate a customer's short-term requirements, and is reset to the agreed limit after an agreed period.

The decision may be based solely upon information available on that account, but could also bring in information from elsewhere, whether the original application, or performance data from other accounts. At the extreme, it might involve a customer application and bureau calls, which—although inconvenient and costly—could indicate lower risk, and accommodate even higher limits.

The goal today however, is to offer customers increased limits with as little inconvenience as possible, which means avoiding the use of application forms. In an ideal world, these extra processes should only be invoked if: (i) a customer requests a limit higher than the target limit; (ii) there is a good chance that the higher limit will be accepted and used, if offered; and (iii) the extra costs and complexity are sufficiently offset, by improved revenues and customer perceptions.

29.3.2 Limit-increase campaigns (givers)

Borrowing money is something avoided by many people, who associate today's pleasure with tomorrow's pain because of the claim against future income. This aversion is especially acute amongst groups that, at some point, either had little access to affordable credit or had borrowed and were stung by a change in economic or personal circumstances. It also arises from people's fear of being rejected. It used to be that the bank manager was a respected figure in the community—along with the judge, sheriff, mayor, and local factory/mine owner—and this still applies in many small communities. People wanting to borrow money would enter the hallowed banking halls, and present themselves to let their case be heard. The situation itself can be embarrassing, and rejection even more so. Where a person in dire straits has already borrowed and is coming back for more, the Dickensian scene of Oliver Twist and 'Please Sir, can I have some more?' comes to mind.

The point is that people are more likely to ask for a loan, or loan increase, if: (i) they are confident about it being approved; or (ii) there is less embarrassment and/or disappointment associated with rejection. The first implication is that lenders should focus marketing on customers with an appetite for credit, and a high probability of acceptance. The second is that process design should take into consideration the effect of rejection on applicants' emotions.

For the former, it is quite possible for lenders to come up with a simple rule-set that can be applied to the existing customer base, such as 'balance exceeded 80 per cent of limit in the last three months and no delinquencies in the last year'. The decision can, however, be made much

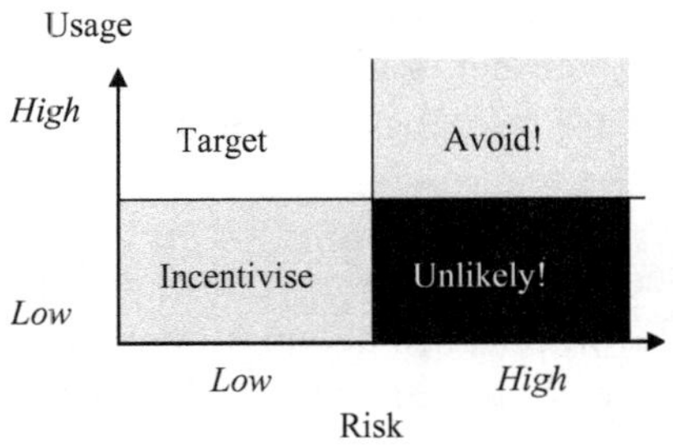

Figure 29.4. Risk versus usage.

more scientific by the combined use of risk and usage scorecards (see Figure 29.4). There are three ways in which these can be applied:

Pre-approve—Approve the limit upfront, but only put it in place once the offer is accepted. This provides a better response rate, but requires greater lender confidence and stricter rules.

Pro-active—Grant the limit, and then advise the customer that the limit has been increased. This is very cheap and effective, although some customers may not wish the higher limits. Responsible lending issues also present a concern, and lenders may restrict proactive limit increases to one per year.

Pre-screen—Invite selected customers to apply, and assess them further using details available at time of application. This is more expensive, but allows higher limits to be granted, based upon more complete information.

In each of these cases, one of the difficulties will be managing how identified customers will be treated in subsequent campaigns, both for those that take up the offer, and those that do not. All of them should be excluded from future campaigns for some minimum period, say six months, while those that take up the offer will only be re-included after they have again met the qualifying criteria.

Strategies can also be defined for accounts with high attrition probabilities. These customers may not respond to a limit-increase offer, but may reactivate the accounts if they know that the lender is aware of them, and that the funds will be available in need. Offers can again be determined using simple rule-sets, or a combination of risk and attrition scores. Many lenders are known for being very good at sales, but poor at after-sales service. Any tools that can highlight problems on individual accounts can also assist in account retention. Information from sources other than just the account management system should be considered, including customer-scoring and customer-contact databases (see Chapter 12, Internal Systems).

29.3.3 Limit reviews (repeaters)

For fixed-term products, the limit is granted with the view that the amount will be repaid by a given future date. For others no future date is given, but personal loans are not perpetuities.

Lenders usually exercise some caution, and review the facility at regular intervals to see if there have been any changes in the customer's financial position. Just how this is done, and how it impacts upon terms going forward, will vary depending upon the product. With *card products* there is an *expiry date*, at which point the issuer decides whether or not to reissue the card, and what limit to provide going forward. This is a good time to increase limits for accounts that have been performing well, and have indicated an appetite for more credit. The decision will be almost exclusively based upon the past performance of the account, perhaps with a call to the credit bureau. For *overdrafts* and *revolving credit*, the task may be more onerous, including an *annual review*, with calls for financial statements and copies of the most recent payslip. These actions are justified, as the more stringent risk management allows lenders to offer larger amounts at lower rates.

29.3.4 Cross-sales (repayers/repeaters/leavers)

A lender may not only try to maximise customers' use of one product, but also try to initiate use of another. The number of possible product combinations increases with the breadth of the product offering, especially for banks that offer cheque accounts, personal loans, credit cards, home and motor vehicle loans, savings and investment products, and others. This is really the realm of marketing, except the target market is the existing, or recently approved, customer base, and the goal is to focus efforts where they will provide the best results. Indiscriminate campaigns can be expensive, unrewarding, and even damaging. The combination of information on existing product holdings, utilisation, demographics, and the like provides a powerful tool for deriving what the customer will want next—not only for credit products, by those with a credit appetite, but also for savings and investments products, by those without. For credit products, the selection tools would be some combination of risk, response, revenue and retention models, like those used to target totally new customers. Most of this can be assessed relatively cheaply using internal data, but bureau data could add considerable lift.

29.3.5 Win-back (leavers)

In its most extreme form, win-back campaigns may be required to handle public relations disasters, like strike action, computer glitches, or natural disasters. This includes the use of any number of possible, and often imaginative, media to reach the customer. Most cases that lenders deal with are not so dramatic. Their primary interest is attrition, resulting from: (i) the need disappearing; (ii) service dissatisfaction; (iii) competitive offers; and/or (iv) being forced out. With the latter, it is a case of good riddance. For the others, the lender should make some effort to keep the customer; the effort can be highly rewarding, especially considering the comparatively higher cost of attracting new customers.

Leavers can exit via three avenues: *early settlement*, *dormancy*, and *account closure*. The most obvious, and costly, is early settlement, which can have a significant impact on deal profitability, especially for asset finance. Simply stated, a home or motor vehicle is sold, and the loan repaid, possibly in a fraction of the contractual period. With luck, the lender will finance its replacement, but this is not guaranteed. The lender has three ways of proactively trying to address early-settlement risk:

(i) be more generous with pricing, and other terms for new business, where early settlement risk is high (which are usually low credit risk);
(ii) try to identify existing accounts that may settle early, and make them offers prior to the event; and/or
(iii) ensure that the new business process can identify existing customers, so that if they apply for a new loan, the best possible offer will be made to them.

The final point is particularly important, especially where applications are submitted by agents/dealers, who may steer the deal to a competitor. Steps could be taken to contact the customer by phone, to advise that the loan has been approved.

Stern (2002) splits out several factors that drive prepayments for high-ticket home loans: loan age, discount/premium, interest rate (motivates refinancing, as fixed rates reduce), burnout (exposure to refinancing incentives), loan size (better quality customers, hence greater refinancing alternatives), loan quality (spread, original LTV, and asset price appreciation), geography, and seasonality. The dominant factor is loan age, which is characterised by an initial ramping up of prepayments during the first few months, followed by a gradual decline over the remaining life of the loans.

With transaction products, accounts may lie dormant for extended periods. Where there are no charges associated with holding the product, there is little motivation to close. The customer may forget about the account, or just keep it open for the off chance that it may come in useful at some point in the future. For lenders, these accounts present computer, billing, and other costs.

While win-back strategies can be developed around simple markers, scoring has the advantage of being much more scientific, and once developed allows lenders to develop strategies that are much easier to understand and apply. The tactics used are limited only by people's imaginations, and may include, amongst others, special offers, reduced prices, and prizes. This is often done at great expense, and one case is known of overseas trips being offered . . . to people who took their business elsewhere anyways. Why should a customer pass up a free meal?

29.4 Summary

While the new business process is used to control the front-door, the account-management process is used to control what is inside. It is critical, not only to ensure that customers behave, but also customer satisfaction. This is the realm of behavioural scores, a subcategory of which is customer scores. These are used for *limit setting*, *pay/no pay decisions* (cheque) and *authorisations* (credit cards). Different strategies may be employed depending upon limit availment (taker, asker, giver), and account repayment patterns (repayer, repeater, keeper, stealer).

Different limit types are used: *agreed limits*, known to the customers; *shadow limits*, used in the background to control excesses; and *target limits*, for marketing. These can be set as functions of the current limit, or some turnover or income measure. Usage and customer scores may also influence the process. There are also special cases, like the treatment of uncleared effects, and assisting customers with cash-flow triage. The limit-management functions must cover: over-limit excesses, limit-increase requests, limit-increase campaigns, limit reviews, and cross-sales.

Shadow limits are tools used to manage *over-limit excesses* (takers) for transaction products. For *cheque accounts*, they drive pay/no pay decisions, which determine whether cheques will be returned NSF. Banks will also allow time for deposits to clear, and may allow withdrawals against uncleared effects, once a track record has been established. For *credit cards*, such limits are used to drive authorisations. Smaller transactions are governed by floor limits, and may be allowed regardless. Speedpoints have aided the efficiency of this process by eliminating the paper. Cash advances are a special case, because interest is charged immediately, and at higher rates than elsewhere, which implies higher risk. For all lending products, lenders should also be cognizant of the *informed-customer effect*; when borrowers have equally bad choices, they will choose the one that is best understood. Given that penalty fees can be a major source of income, lenders' late-payment policies should be as transparent and as fair as possible.

Target limits define maxima that lenders will entertain, without asking for extra information. *Limit-increase requests* (askers) may be processed automatically, but if the amount requested exceeds the threshold, a formal application and further information will be required. For *limit-increase campaigns* (givers), the limit might be used for: pre-approvals, pro-active increases, or mail-shot pre-screening. For *limit reviews* (repeaters), the limit aids the renewal decision. Card products have an expiry date, while overdraft and revolving credit have review dates. Some lenders strive for evergreen limits, which are only reviewed if there are problems. *Cross-sales* (repayers, repeaters, and leavers) can also be facilitated, by setting offered limits for a product, based on behaviour on one or more others. Finally, *win-back strategies* (leavers) are used to prevent early settlement, dormancy, and/or closure.

30 Collections and recoveries

All progress is based upon a universal innate desire on the part of every organism to live beyond its income
Samuel Butler (1835–1902), author of *The Way of All Flesh*, in *Notebooks* (1912)

According to psychological studies, death, public speaking, and asking for money are people's greatest fears. It is no wonder then, that so many businesses have such a hard time collecting what is due. They are afraid to ask! Fortunately, in the credit industry this only applies to late payers, who enter the realm of collections and recoveries (C&R). Lenders' challenge is to decide upon the appropriate treatment, as some self-cure, some just need a nudge (or counselling), and only a few require drastic action. Their numbers may be low, but costs can be high, and credit providers have to develop extremely thick skins.

30.1 Overview

The primary distinguishing feature between C&R and account management is URGENCY! The three functions can be briefly summarised as:

(1) **Account management**—Deals with accounts that are up to date in their payments, or are being maintained in a satisfactory manner. The focus is to ensure customer satisfaction, and *grow* the relationship.
(2) **Collections**—Deals with early delinquencies and first-time offenders. The focus is to rectify any problems, and *maintain* the relationship.
(3) **Recoveries**—Deals with hard-core delinquencies and repeat offenders. The focus is to get the money back, and often to *sever* the relationship.

Delinquent accounts will be passed from one stage to the next, based upon rules decided upon by the lender. Downstream moves imply a greater degree of risk; the more time that passes, the less likely the amount will be collected. Indeed, there is an ongoing race between creditors . . . When problems arise, the first to knock on the customer's door is typically the first to be repaid, leaving little for latecomers. The C&R functions may be outsourced, by those who do not: (i) wish to invest management time in playing the 'black hat' that wants his money back; and/or (ii) have the necessary volumes to make it pay.

Delinquency reasons

There are a variety of reasons for an account being delinquent, and just because a person misses a payment, does not mean that he/she is bad:

Payment oversights—Where no debit order is set up, and the accountholder forgets to make the payment, perhaps because of being away on holiday.

Technical arrears/excesses—Arise because of delays in the payments system, like where the required payment date is the 20th and the customer makes a transfer on the 19th but it is not credited to the account until the 25th.

Incorrect details—Errors made when setting up debit orders, either by the accountholder or lender. These become evident as first-payment defaults, and should not recur once fixed.

Poor financial planning—Customer commits beyond his/her income. It may be the result of a holiday or special purchase, or totally irresponsible borrowing with little hope for a quick settlement, which is often aggravated by irresponsible lending.

Personal distress—Job loss, marital strife, personal or family illness, loss of home to flood or fire, or any other event that may cause both personal and financial trauma. This is the most difficult case, and often requires special arrangements to be made.

Disputes—Either with the lender, or vendor. Lender disputes often relate to perceived over-charging, or errors with respect to fees, interest rates, payment processing, etc. In contrast, vendor disputes are where the customer has a problem with goods delivery or quality, and refuses to pay.

Skips/gone aways—Customer changes jobs/addresses, without providing new contact details. This may be an innocent oversight, but often indicates serious problems. Some customers make repeated moves, and can be very difficult to track.

Sinatra doctrine—Customers do it their way, by denying responsibility for purchase, or paying when they feel like it.

Insolvency—The worst possible scenario for all concerned—voluntary or forced bankruptcy.

Process

Successful collections rely not only upon the customer's willingness and ability to repay, but also the collector's ability and resolve to collect. The challenge is to be able to contact the debtor, and be first at the door on payday. A very simplified overview of the internal processes for both collections and recoveries is provided in Figure 30.1.

Upon entering each area, the account is assessed to see if there is any marker indicating that special treatment is required. If not, it is passed through an automated decision process, to determine the action to be taken. The action's end result can be: (i) a *payment*, that either regularises or settles the account; (ii) a *promise-to-pay*, that requires further monitoring; or (iii) *no response*, making it necessary to pass the account to the next stage (C&R and legal). The tracing and fraud areas complement C&R: tracing, to find a gone-away customer; and fraud, to determine whether or not the case is fraudulent.

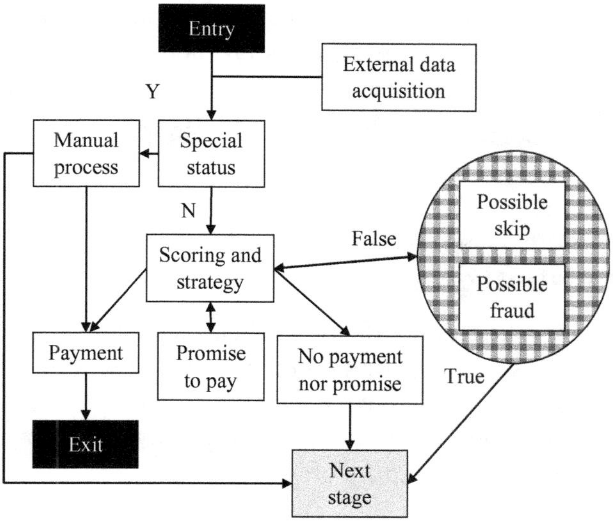

Figure 30.1. C&R—flowchart.

Core systems requirements

The success of both areas, whether in-sourced or outsourced, is heavily dependent upon the technology being used, and requires efficient and cost-effective access to:

Communications infrastructure—Telephone links to the regions being serviced, and automatic queuing systems to manage outbound calls.
Customer contact and tracing information—Phone directory, property register, etc. In the absence of correct contact details, resources for finding skips are crucial.
Payment history—Own and other accounts, including credit bureaux, etc. Agencies with significant mass can also create their own historical information, for example use information from contacts on one company's accounts to assess another's.
Scoring—Systems to risk rank and prioritise accounts.

Agencies

One or both of C&R can be outsourced, especially when there are large numbers of small value accounts. Agencies will have their own tracing and fraud areas, and significant infrastructure to back them up. Most lenders will try to manage the collections process themselves and outsource (or sell) hard recoveries, but both functions may be outsourced where lending is secondary to the core business. This arrangement does cause some confusion, because in the outsourced world, 'collections' agencies may do both C&R.

The outsourced environment can be very demanding, especially where there is significant inter-agency competition—which may even be agencies on the other side of the planet. Significant competitive advantages come from being able to leverage off of information provided by several lenders, so agencies will also try to maximise market share, to achieve critical mass.

30.2 Triggers and strategies

Courts will often go easy on first-time offenders when they break the law, and sentence them to 'corrections', a minimum security prison where the goal is rehabilitation. It is similar with lenders, except borrowers are sentenced to 'collections'. In both cases, offenders have to break some rule before being considered for entry, and typically they do not want to be there. Accounts will end up in collections because of:

Over-limit/excesses—Spending has exceeded the agreed credit limit.
Missed payments—Repayments are less than expected, or are not received at all.
Returns/dishonours—Transactions have been declined, perhaps because of poor financial planning, or late receipt of the paycheque that month.
Special statuses—Extraordinary circumstances, for example deceased, dispute, legal, insolvent, etc.

These can occur in different combinations, and the level of risk and treatment will differ for each, for example over limit only, missed payment only, both, both and statused, first-payment defaulter, etc.

A distinction is made between first-payment defaulters that are debit orders and others, as account details may be incorrect and easily fixed. Otherwise, there is a strong probability of fraud. Many lenders have specialised teams working only on this subgroup, especially where it involves high-ticket items that can be easily moved across borders.

If Collections is corrections, then Recoveries is death row . . . or at least life imprisonment, or banishment to a desert isle; excepting that the goal is still to get back as much money as possible at least expense. This, of course, implies that the task has become more difficult, which is to be expected when trying to break a relationship. It is much like a bad divorce with a lot of acrimony, and disputes over who keeps what.

Strategy

The strategies used for collections and recoveries can vary on a number of fronts, each of which relates to how communications with delinquents are structured:

Content—What is being suggested, like full payment, partial settlement, or legal action.
Tone—Hard or soft, friendly or formal.
Delivery—Statement message, special letter, phone, email, SMS, or any other means. Each has a cost, which must be offset against its effectiveness.
Timing—Wait times between actions, scheduling of campaigns, movements of accounts into and out of the stage.
Extent—Degree of effort expended, whether in terms of number of collections actions, attempts at contact, or skip tracing efforts.

The choice of strategy can be influenced by a variety of factors. Besides delinquency and/or score, other considerations could be the age of account (new/established), prior history (first-time versus repeat offender), or balance outstanding (high/low).

Table 30.1 provides an example of a strategy table that could be used in a collections environment. It does not cater for all possible cases though, as lenders may also wish to vary choices by other factors, for example high value, debtor ceased communications, VIP indicator, special disputes, etc.

Practical considerations

In many instances, credit providers rely upon contacting people at home, especially where credit users either do not have a work phone, or access is poor (teachers, factory labourers, salesmen). If so, there is a short window of about three hours, between 6 and 9 pm (or Saturdays), where the customer can be contacted and be politely interrupted in the process of eating supper, watching television, or putting the kids to bed. When time is short, and countless operators are being employed to make the calls, lenders want to focus upon those calls most likely to have a positive result—starting with having somebody pick up the phone, and then hopefully having a 'right party contact'. When on the phone, operators must have up-to-date information regarding contact details and past communications. When was the last contact was made? Was a promise-to-pay received? Was it honoured?

The strategies used will have an impact upon the resources required—collectors, skip tracers, legal, phone lines, dialler, etc. In tele-collections environments, any auto-dialler system must

Table 30.1. Collections strategy table

Value ➜	Low			High		
Risk ➜	Low	Mid	Hi	Low	Mid	Hi
1 down	S1	S1	S2	S1	S2	R1
2 down	S2	S2	R1	S2	R1	R2
3 down	R1	P2	N1	P1	N1	N2
4 down	N1	N1	L	N2	L	L
5 down	L	L	L	L	L	L

S = statement message, R = reminder letter, P = phone call, N = default notice, L = Legal.

not only be able to prioritise the calls, but also ensure that they are channelled to the right operators, as the skills and mentalities required for the different strategies can be totally different—softly-softly with late payments (soft collections), or hard-line with hardened delinquents (hard collections). It is like the difference between the police and the army—one is trained to resolve problems, the other is trained to kill.

30.3 Scoring

Collections and recoveries can function without any form of scoring. In primitive environments, a shotgun approach is used to contact as many people as possible using available resources, with adjustments based purely on current delinquency. As sophistication grows, other factors are used to drive strategies, such as past delinquencies, type of debt, and balance outstanding. When scoring is added to this pot, it provides a further and crucial dimension, which enables further process efficiencies. The goal has changed somewhat from behavioural scoring though, as the choice is not just between good and bad, but good, bad, and worse—and how much is recovered, if any. Resources are directed to maximise recovery amounts, in particular by driving a 'predictive dialler' that governs the prioritisation of outgoing calls.

Even with collections scores, other factors will still play a role in the choice of strategies, in particular the amount at risk and cost of each action. The collections action must be identified, which provides the greatest value for:

Equation 30.1. $\textbf{Net return} = \textbf{value} \times \textbf{prob. of recovery} - \textbf{cost of action}$

This formula is very simplistic. Modifications are required to take into consideration partial recoveries and/or amounts that can be realised by selling the debt.

Thus, where it is almost certain that the account will self-cure, no action should be taken. All it will do is incur a cost! If recovery is probable, but uncertain, and the amount is low, then statement messages or letters may be used. In contrast, if recovery is unlikely and the amount is high, then more expensive actions may be considered, including phone contact and legal action. The same applies to skip tracing. Why try to find somebody if they are unlikely to pay anyway? A single legal notice may be the most that can be justified!

> *Self-cures* are delinquent accounts that would be repaid without any collection action. They often arise because of technical arrears. *Worked cures* are those that require collections actions. Collections scorecard monitoring should track both.

Scoring can also be used for portfolio valuation—both by seller and buyer. Some companies will not just outsource their recoveries, but will sell their seriously delinquent accounts. A due diligence process is necessary to determine how much they are worth; the company that incurred the bad debt wants to get back as much as possible, while the company purchasing the portfolio wants to make a profit. Purchasers typically base the assessment on bureau data, as there is little other information of use. Some will develop bespoke bureau scores, based upon their past experiences with purchased portfolios.

C&R versus behavioural

Although both collections and behavioural scores measure the customer's propensity to repay, there are major differences. C&R scores have greater urgency, are more dynamic, and have fewer restrictions. 'Urgency' implies that there will be a greater focus upon these accounts, because the probable loss has increased identifiably and substantially. 'Dynamic' implies that the time horizons are shorter. C&R are not willing to wait for a full year for an outcome, and instead want results within one to three months (Table 30.2). Even sooner would be better, but that may be expecting too much.

There are also fewer restrictions in C&R scoring, in particular on the characteristics that may be modelled. In the behavioural scoring environment, the focus is on characteristics that reflect customer behaviour, not lenders' actions (which are instead addressed through strategy). In the collections area, this barrier falls away—collections actions can also be included in the scores.

Each time a collections action is undertaken, there is a *customer response*, and both can be included when the score is recalculated. Thus, getting a promise-to-pay will have a positive influence on the score—at least to the extent that customers, in general, honour their promises. Collections agencies can also use *lender details*, like annual turnover, number of employees, age/control of company, type of industry, etc., whether assessing a single account or a portfolio.

Collections scorecard classifications

There are two possible ways to differentiate between scorecards and their use in the collections area: prior-stage versus stage-bespoke (Table 30.3), and entry versus sequential. Which are used will depend upon: (i) the relationship between account management, collections, and recoveries; (ii) the technological sophistication of each area; and (iii) a decision on the amount of time and effort necessary for tailoring scorecards to specific circumstances.

Table 30.2. Collection scoring summary

	Collections	Recoveries
Population	1 down	Handed over
Entry def'n	G = less than 90 days	G = Paid X%
	B = 90 days plus	I = Paid 100 − X%
		B = No payment
Sequential	G = same/prior level	G = same/prior
	B = next level	I = promise
		B = no pay/promise
Actions/strategies	Statement messages	
	Mail contact	Mail contact
	Phone contact or pass to Recoveries	Phone contact
		Legal action

Prior-stage scorecards are those developed for an earlier stage in the risk management cycle, a prime example being behavioural risk scorecards used for collections (the value for recoveries would be limited, as these very high-risk accounts would not be well represented in a behavioural model). Lenders will use prior stage scores where it is not economical to develop stage bespoke scorecards, especially where the technology is not in place. While not as powerful, a prior-stage score has the advantage of reducing the amount of scorecard development that is required, while still achieving most of the benefits of decision automation. It can also be incorporated into a stage-bespoke score, either via a matrix or as a predictive characteristic.

Stage-bespoke scorecards are those developed specifically for that stage—collections or recoveries. Greater infrastructure is required, but these have the definite advantage of being tailored to the task. Two types of stage-bespoke scores can be used, entry and sequential, as illustrated in Figure 30.2. The distinction is similar to that between application and behavioural scores, except the customer is a very reluctant participant. In both cases, delinquency status will be the primary criterion in the good/bad definition, but elements about contact history, promises- to-pay, and broken promises will also play a role.

An entry score is used to set initial strategies on entry. The levels chosen to define good and bad may well depend upon the area's subjective view of success or failure (McNab and Wynn 2003). For collections, bad may be 90-days delinquent, while good is fully recovered. In contrast, for recoveries, both good and bad could be based upon the percentage recovered, or the focus shifted to predicting that percentage.

The other possibility is a sequential score, which is used for ongoing management within the area. It is recalculated regularly—perhaps every month, week, or every time a collections action is taken—with the goal of preventing past-due accounts from getting even worse. The good/bad definition will be as simple as 'does the account become even more delinquent over

Table 30.3. Bespoke versus prior stage

Scorecard	Stage applied		
	Account management	Collections	Recoveries
Behavioural	Bespoke	Prior	
Collections		Bespoke	Prior
Recoveries			Bespoke

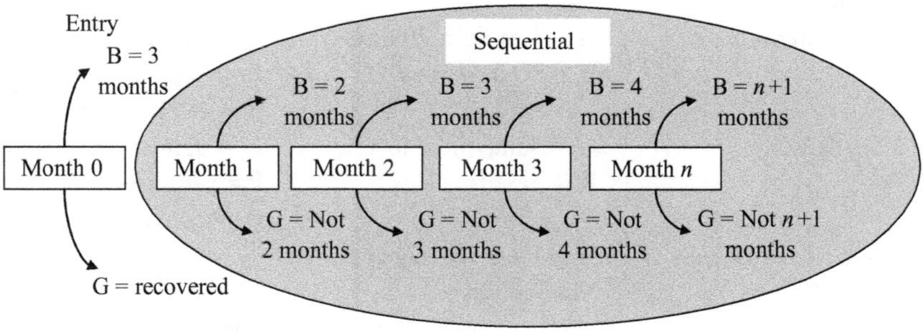

Figure 30.2. Entry versus sequential definitions.

the next month, or not?', and there may be a separate scorecard for each level of delinquency. Scores would then have to be transformed onto a common scale in order to compare them.

Champion/challenger

Collections and recoveries were the first areas where champion/challenger methodologies were used in the 1990s. Different strategies can be tested and implemented relatively quickly, purely because their effects are quite transparent in collections. It is, however, wise to put some limitation on how often these changes can be effected. A strategy can be tested as a challenger, but once implemented as champion, at least 60 days should be allowed to see what impact it has on the entire book, before making any further changes. Failure to do so may cause instabilities, which make monitoring difficult.

A further difficulty here is the strategy effect—if a strategy dictates that resources are directed at a specific group of accounts then their risk is reduced, and any groups being ignored, or receiving insufficient attention, will have higher risk. What then? In dynamic areas like collections, it is necessary to recognise the effects of one's own actions. One way is to use a statistical method like neural networks (NNs), which can self-train. Another is to do rapid redevelopment, which involves making regular updates using new data, but with minimal changes to assumptions. These create feedback loops that are essential for scoring to be used effectively. Finally, a further possibility is to have separate models for each possible strategy, and choose the strategy that yields the highest score.

Reporting

The C&R's ultimate goal is to recoup more money in less time, with less effort. As with any thing, reporting should focus on key results measurables (KRMs), and how they are affected by their primary determinants. The KRMs here are cure rates (self- and worked), recovery rates, time to recovery, and cost/recovery efficiencies, which are primarily determined by the score range, account balance, amount delinquent, region/collector, and other factors.

30.4 Summary

While application scoring guards the front door, and account management is used inside, C&R guard the back. The trigger for entry into C&R processes will be one or more of: (i) over limit/excesses; (ii) missed payments; (iii) returns or dishonours; and (iv) special circumstances, like insolvent or deceased. First-payment defaulters are a special case, as there is a greater potential for fraud.

The primary distinguishing feature of C&R is how quickly actions have to be taken, much like managing a nightclub. *Collections* is responsible for controlling guests who get a bit out of hand, but are still valuable paying guests. In contrast, *recoveries* plays the tough, who has to ask undesirables to leave and guard against breakage of furniture, fittings, and other patrons. Collections is used primarily to handle payment oversights, technical arrears and excesses, incorrect details, poor financial planning, personal upsets, and possibly disputes; whereas recoveries will deal with insolvencies, skips/gone-aways, and customers that deny responsibility for the debt.

The two areas are very similar, in terms of the theoretical process; for both, the possible outcomes are *payment*, *promise-to-pay*, or *no response*. The difference is the severity of the actions that have to be taken, with recoveries more likely to involve legal and other costs (and possibly even broken bones, in less savoury environments). Process requirements include communications links (automatic diallers), customer contact details, payment histories (internal and bureau), and scoring systems, to risk rank and prioritise accounts. In many instances, lenders will rely on agencies to perform one or both of the functions.

The C&R strategies relate primarily to customer communications, in particular, the content, tone, delivery, and timing, as well as the expenses incurred for communications and legal costs. Scores can be used as drivers, and decision trees are also a possibility. While scoring can be and is used, it is not the primary force. Both the value at risk and cost of each action play a role, and lenders' primary goal is to ensure that resources are allocated to maximise recoveries. This is also an area where delinquent loans are bought and sold, and scores can be used to aid portfolio valuations.

The C&R differs significantly from account management, in that actions are required more urgently, and the time horizons are shorter. As a result, there are several marked differences between C&R and behavioural scoring. In particular, the observation and outcome windows are shorter, and the strategy effect is greater (making the scorecards less stable). There are, however, fewer restrictions on what data can be used within the models. Lender strategies can be incorporated as predictors, as well as customers' responses to past collections actions. Lenders can use prior-stage scores, like behavioural scores, but bespoke collections scores are better suited: (i) an entry score, when the account first enters; and (ii) sequential scores, that are segmented based upon time in the area. Champion/challenger strategies can be employed to good effect, as the results become quickly evident, but the shift in resources affects the validity of the scores.

31 Fraud

Credit evaluation is based on dealings with Joe Average, who even if financially challenged, is at least honest. Unfortunately, credit people and systems are ill-prepared for Joe Fraudster, who relies upon deception and trickery to part people from their loot. It has thus become best practice for lenders to have dedicated fraud teams within the business. Both credit and fraud are responsible for prevention and cure in their respective areas, but fraud has the added task of investigating suspects. The two areas could be likened to bobby-on-the-beat policing and detective work. Credit deals with losses resulting from bad decisions, domestic problems, unforeseen circumstances, or perhaps even negligence—but the cases are relatively straightforward, and at least lenders stand a chance of getting the money back. In contrast, fraud deals with cases involving criminal intent that are difficult to identify, and where the funds are extremely difficult to recover—to the extent that it is best to write off fraud losses, once they are identified.

Lenders are often reluctant to prosecute 'known' fraudsters, because this is difficult and expensive to prove, and/or it is a source of potential embarrassment if reported in the press. Instead, greater effort is put into prevention, by: (i) identifying fraudsters' *modi operandi*; (ii) developing rules to block them; and (iii) checking databases containing details of known or suspected frauds. There is also a significant trade-off between the cost of the fraud, and that of preventive measures. It is unlikely that businesses will ever be able to eradicate fraud totally, as the cost would be too high to bear, in terms of required resources and lost business.

Credit cards and cheque accounts are not the only victims. Fraud can occur with any type of consumer credit product—home loans, personal loans, retail finance, online merchandising, etc. Further, it is not always the obvious type, where individuals take the money and run. Fraud syndicates invest much time and effort researching lenders' operations, including the use of spies to learn their policies and procedures. They are thus are able to adapt very quickly to protective measures taken against them.

Fraud significantly increases the cost of doing business, to the extent that lenders can be bankrupted. Beyond the obvious financial losses, there is also the cost of developing and maintaining a fraud-prevention infrastructure, and the effect of fraud checks upon customer service. In an area where the results are so uncertain, there will always be: (i) *false positives*, where the investigation can cause unnecessary delays and upset; and (ii) *false negatives,* which slip through nonetheless. All of this has an unreasonable impact on John and Jane Public, just because of the dishonesty of a few. As a result, fraud prevention is viewed as a non-competitive issue, requiring co-operation across the industry. When it comes to sharing data, the fraud area has greater latitude than credit, whether from the authorities, the credit bureaux, or other lenders.

Within the business, the challenge is to separate fraud from credit losses, as the former are considered operational in nature. How can one loss be distinguished from the other? At what stage does embellishment become fraud, if at all? As a result, the responsibility for fraud often falls in or near credit. Further, if there were problems coming up with sufficient bads for a credit scorecard, it is even worse for fraud. This is in spite of the huge fraud losses occurring internationally, and the opportunities for fraud seem to be growing, especially in the online world.

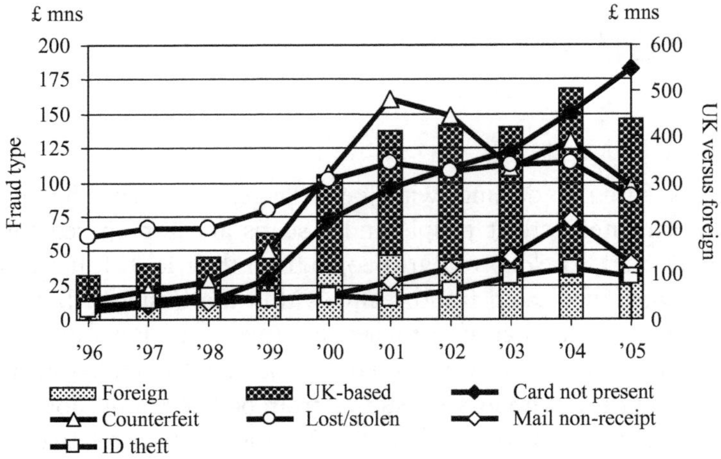

Figure 31.1. UK plastic fraud losses.

Fraud trends

Fraud follows new products and technologies, like a pride of lions eager to sense weakness in a herd of zebra. Over time, business learns to counter it, but it never goes away entirely. Credit card fraud losses in the United Kingdom are a case in point. They were £165, £83, and £135 mn in 1991, 1995, and 1998 respectively; the dip in 1995 resulted from co-operation that started with the formation of the Plastic Fraud Prevention Forum (McNab and Wynn 2003). The figures increased further to £188, £317, and £411 mn for 1999 through 2001 respectively. Figures for the United States are not as readily available, but according to Fair Isaac, identity fraud has been on the rise, and cost US lenders more than $1 billion in 2002 (no indication is provided of product types included in this figure). Stevens (1998) gives a figure of 700,000 cases of identity fraud in a single year.

The make-up of card fraud losses has also changed, as fraudsters have adapted to technology and payment methods. APACS (2006) provided UK fraud statistics, shown in Figure 31.1. For many years the largest category was *lost or stolen cards*, but in 2000 *counterfeit cards* took first place, as it became easier to copy cards using new technology to skim details. At about the same time, growth in Internet eBusiness opened up opportunities for card-not-present fraud, which took over first place in 2003. Of this, a significant amount is transacted in other countries, but this peaked at 33.6 per cent of the total in 2001, and by 2005, reduced to 18.8 per cent. The success is attributed to improved fraud detection, and the formation of a specialist banking-industry sponsored police unit.

The 2002 APACS report had 'Application' and 'Other' categories, but these were both small, and application fraud was decreasing rapidly. It is assumed that application fraud is under control and no longer considered a threat in the card environment.

According to NeuralT (2002), about £250 per card is lost on stolen cards before the card issuer is notified of a problem.

When viewing the numbers, the significant growth in card usage must be recognised; losses as a percentage of turnover reduced from a peak of 0.330 per cent in 1991, to 0.183, 0.141, and 0.112 per cent in 2001, 2004, and 2005 respectively. The 21 per cent fall in 2005 is particularly significant. Chip and PIN technologies are credited with bringing counterfeit, lost/stolen, and mail-not-received fraud under control since 1999, which helps to explain the huge shift to card-not-present fraud.

31.1 Types of fraud

The types of fraud are as many and varied as the products and technologies that are targeted. This section attempts to provide an overview of fraudsters' *modi operandi*, and covers several dimensions, including but not limited to:

- The *product* being targeted—cheque, card, etc.
- The fraudster's *relationship* to the account—first-, second-, or third-party.
- The business *process* that is affected—application or transaction processing.
- *Timing*—short- versus long-term, offered or manipulated limit.
- Types of *identity misrepresentation*—embellishment, theft, and fabrication.
- How the fraudster *acquires* the article—lost or stolen, not received, at hand, skimming.
- How the article or details are *used*—counterfeit, not present, or altered.
- *Technologies* involved—ATM fraud, Internet.

Many of the fraud types referred to in this section were covered by McNab and Wynn (2003), but they have been broadened, renamed, and regrouped to make them applicable to a broader range of products than just credit cards and personal loans.

Product being targeted

Transaction products (cheque, credit card, and debit card) are the most susceptible to fraud, but others are also at risk, including personal loans and asset finance. Accountholders can also be conned, perhaps into the purchase of bogus or non-existent goods, and in the process borrow funds, or provide a cheque or credit card as payment. These cases of *caveat emptor* fall outside of the scope of this book though, which instead focuses on fraud types where the lender: (i) can suffer a direct loss; (ii) can be held liable; or (iii) has a responsibility to prevent it.

Relationship to account

The next distinction that can be made is between first-, second-, and third-party fraud, the difference being the relationship to the accountholder:

First party—The legitimate accountholder. This covers identity embellishment when applying for new facilities, and first-payment defaults where the accountholder had no intention of repaying.
Second party—Another party legitimately and willingly involved in the transaction, such as the payee or merchant. For example, cheque or card transaction details may be altered to a higher amount after the fact.
Third party—An unrelated party, who has no legitimate role in the transaction or process. This category presents the greatest losses to lenders, including lost, stolen, never received, not present, impersonation, and other categories.

Business process affected

A lot of fraud goes undetected, or is only detected when it is too late. There are, however, two primary points in the business process where fraud-prevention measures can be taken—application processing and transaction processing. These are also used as labels for broad types of fraud:

Application—Any fraudulent activities that involve manipulation of the account-origination process, typically relying on identity misrepresentation. It can involve actions at any point in the process, and complicity on the part of employees.
Transaction—Involves the plastic or paper ('the article') used to transact on the account, or the account details. This includes lost and stolen, not received, counterfeit, and not present. Fraud-prevention measures are taken as part of transaction processing, and the authorisations/referrals process.

Manner and timing

With application fraud, there will be variations in how the fraudulent transactions are effected. Timing may be: *short-term*, where the fraudster absconds immediately, or as soon as possible; or *long-term*, where they manage the relationship until lenders' controls are relaxed, and/or limits are increased as high as possible. The manner will vary with timing. For short-term, it may be reflected in a first-payment default (revolving credit, fixed-term, and credit card), a withdrawal against uncleared effects, if and when it is allowed (cheque and credit card); or kiting, to create fictitious balances with banks that are withdrawn (cheque and others). For long-term, a fraudster may run an account legitimately before absconding, or a syndicate may engage in kite flying—often involving multiple accounts with different lenders—to trick them into increasing limits on seemingly profitable accounts.

Treatment of security

With asset finance, the purchased item usually acts as security for the loan. This presents other risks, related to whether or not the lender will be able to repossess the asset and recover all or some of the outstanding balance from its disposal:

Goods irrecoverable—Borrower absconds with the goods, or sells them. This is both possible and profitable with movable items like motor vehicles, which can end up in other countries, and even on other continents.
Security misrepresented—An asset's value is misstated. For new purchases, the fraudster pockets the difference between the loan amount and purchase price. In any event, only a portion of the outstanding balance is recovered, if it is repossessed.

Identity misrepresentation

The risk of fraud exists wherever people are able to misrepresent themselves, pretending to be somebody or something they are not, in order to gain an advantage. Application fraud is the most common, but fraudsters may also target existing accounts. The types of identity misrepresentation are:

Identity embellishment/massaging—First-party fraud, where the identity is genuine, but customers misstate their details to get what they want.
Identity theft/impersonation—Third-party fraud, where the fraudster masquerades as a real individual (the 'mark'). This is the proverbial 'wolf in sheep's clothing'.
Identity fabrication/empty house—Third-party fraud that involves the creation of an individual, or entity, which does not exist.

A product that is available for detecting identity fraud is Raven™, which was developed by Fair Isaac and is available through TransUnion in Canada only. This combines bureau, HAWK Fraud Detection System, and application data.

Embellishment/massaging

While fraud is normally associated with criminal intent, embellishment can occur where there is a genuine intention to repay. This is where applicants' details are massaged to improve the

Chorafas (1990) tells of an infamous instance during the 1980s. William Stoecker was a 32-year old former welder and college dropout turned speculator. Banks lent him and his Grabill Corporation almost $500 mn for leveraged buyouts. When some banks became suspicious and called in their loans, others were still knocking at his door, eager for the potential business. The company failed in early 1989, after non-payment of loans.

chances of acceptance. It may be difficult to distinguish between credit and fraud losses; 'little lies' would fall into the credit arena, and 'big lies' into fraud. Just where to draw the line is uncertain!

The greatest opportunities for embellishment arise where customers, dealer/brokers, or even staff, can change inputs. Manipulation can be minimised by obtaining data from automated and reliable sources, such as internal systems, and the credit bureaux. Even so, Wiklund (2004) highlights that even bureau scores can be manipulated. It takes a month or so before a new credit line is reflected within a customer's profile, and if the newly acquired funds are used to reduce other lines, the score will be bumped up temporarily.

Impersonation/identity theft

Identity theft has become the new-age scourge, and victims are often not aware that they have been targeted, until legal action for non-payment is threatened. The theft requires a fairly intimate knowledge of personal details, and can involve forging identification documents, and stealing account statements out of the post, usually to apply for a new account. It also includes practices like changing mailing and contact details on existing accounts.

Unfortunately, both lender and mark usually only become aware of the identity theft after default occurs, and the lender tries to take legal action. This may be months, or even years, after account opening, as fraudsters will sometimes maintain seemingly normal accounts for extended periods. According to Experian UK, identity fraud goes undetected for an average of 16 months, and the longest time recorded was 4½ years (1618 days).[1] Consumers are able to protect themselves though. According to McNab and Wynn (2003), the UK CIFAS system includes a fraud category called 'protective registration', which allows individuals 'who have been subjected to attempted, or perceived, threat of identity theft' to protect themselves against third-party use of their identity data.

Part of the problem with fighting identity theft has been its treatment within society. According to Stevens (1998), identity theft was only made a federal offence in the United States in 1998. Even then, victims that report identity theft to the police often have difficulty getting help. Criminal laws do not recognise them as victims, and they often have little knowledge of how their details were obtained, and no proof that they did not borrow in their own names. Although federal law does not hold victims liable for bills incurred by imposters, people struggle to prove their innocence, and may spend years trying to fix their credit records. Stevens puts most of the blame for this situation on lenders' laxity when doing screening, and their overwhelming emphasis upon the Social Security Number, without further identity confirmation.

Fabrication/empty house

Identity fabrication refers to creating an individual that does not exist, which in its simplest form is the use of a false name at a valid (or invalid) address. In some cases the fraudster(s) may even occupy the address temporarily, or use details of deceased individuals and create histories for them. Fraudsters need greater skills to do identity fabrication, but the fraud is more difficult for lenders to detect.

---

[1] *Credit Risk International*, March 2004, p. 8.

Handling of transaction media

With transaction accounts, extra risks are presented by the medium used to operate the account. It can be lost, stolen, intercepted, redirected, or skimmed, and the fraudster may use it as is, alter the details, counterfeit it, or transact without it. These are treated here as two broad categories: acquisition—how the fraudster acquires what is required; and utilisation—how it is subsequently used.

Acquisition

Lost or stolen—The genuine article (card or cheque) is lost by, or stolen from, the accountholder. Most fraudulent transactions will occur within a very short period after the loss, usually hours, if not minutes. Fraudsters want to transact before the article is stopped—not just to get the funds, but also because the risk of detection increases once the loss is reported. With credit cards, accounts are closed, and the card details loaded to a hot card file. With cheques, individual cheques, or series thereof, will be stopped.

Not received—The genuine article is never received, either because it was redirected (address change) or intercepted (stolen in transit). *Redirection* relies upon identity theft, to change the address and other contact details, and in the lending industry applies primarily to credit cards. *Interception* relates primarily to mail theft, usually of credit cards en route to accountholders, and cheque payments en route to payees.

Skimmed—Card or cheque details are obtained when they are provided for a genuine transaction, either directly from the medium, or from documentation relating to the transaction. It can be done by anybody that can gain access to the information—waiters, bank employees, Internet hackers, or people rummaging through refuse for account statements and transaction slips. With credit cards, a separate card reader may be used to record details from the magnetic strip that are then forwarded (sold) to fraudsters, who use them to create counterfeit articles, or commit card-not-present fraud.

At hand—The accountholder still has possession of the article, either the card or the chequebook. In this case, the fraud is second- or third-party . . . either the transaction details were altered by the payee/merchant, or the account details were skimmed.

Utilisation

Counterfeit—Facsimiles are created that are presented by somebody pretending to be the true accountholder. People with the required details and skills can forge cheques and other commercial paper. Likewise, credit cards can be cloned, using skimmed details.

Not present—No article is presented. Instead, account details are provided without any physical proof that the account exists. With cards, this refers to any transactions where the account details are provided over the phone, or via the Internet (card-not-present). With cheque accounts, this will only happen if there are transfers occurring between multiple accounts in different names (cross firing/kite flying).

Altered—Where the article, or the details of the transaction, are changed to benefit the fraudster. This normally applies to cheques, where either the amount or payee is changed. It would also apply to merchants that change the amounts on credit card transactions.

Unaltered—The real article is used as is. This covers most of the lost, stolen, and not-received fraud mentioned earlier.

Technologies Involved

Just as there are different products that can be targeted by fraudsters, there are also different technologies that can be either targeted or employed.

ATM fraud—Is a special case of lost or stolen card fraud, where the fraudster obtains the card as part of a normal ATM transaction. The PIN is obtained by 'shoulder surfing', and the card through: card swap, exchange of cards through sleight of hand; card trap, device inserted into the ATM to capture and hold the card; or pick-pocketing, obtaining the card by stealing the wallet. Another possibility involves violence, where the card is stolen, and the PIN coerced from the cardholder.

Internet—Any form of fraud that somehow involves the Internet. This can apply to: (i) websites that are selling bogus or non-existent goods; (ii) hackers using spyware to obtain account and password details; and (iii) phishing, use of emails and bogus websites to con individuals into divulging details, including personal identifiers, account numbers, and security codes.

31.2 Fraud detection tools

The number of fraud types makes this look like a one-sided game, but it is not. Lenders can, and do, take significant actions to prevent it, which is especially challenging given fraudsters' mercurial nature. Crooks (2005) makes several high-level suggestions, about how lenders can improve their fraud responses:

Cross-product—Fraud detection should work across all products maintained by a business, and not focus on individual areas.

Data sharing—Working together with other lenders provides better results than lender specific initiatives.

Flexible—The systems should be modular, and capable of being adapted to add new capabilities, and data sources.

Analytics—Make best use of any tools that may be available for analysing flows of money into and out of accounts, and to profile customers and merchants.
Syndicates—Rather than focussing on the actions of individual fraudsters, instead try to tie them to the organised fraud network to which they belong.
Law-enforcement—To the maximum extent possible, try to manage fraud such that cases can be readily handed over to law enforcement agencies for prosecution. If anything, this aspect is crucial for fighting organised crime.

These philosophies have given rise to a number of specific fraud-prevention tools and approaches, which are growing and evolving as fraud trends change:

Internal negative files—Databases containing details of known, and possibly suspected, fraudsters. It also includes hot card files, containing details of lost and stolen cards, which are distributed to merchants.
Shared databases—Pooled data, meant to aid fraud prevention, including negative files. Contributing lenders can share details on all new credit applications, so that new applicants' details can be checked for possible manipulation. One example of this is Detect®, a service offered by Experian in the United Kingdom.
Rule-based verification—An expert-system approach, which relies upon a customised set of policy rules, derived based on past experience with fraud. Different rule-sets are developed for account origination and account management, and if a rule is breached, the case is tagged for further investigation. This can include checks against the applicant's income, amount requested, applicant's address, voter's roll, phone numbers, etc.
Scoring—The use of statistically-derived models to aid fraud prevention. These are covered in Section 31.4.
Pattern detection—Comparisons are made across applications or transactions, in an attempt to identify patterns that indicate fraud. These patterns may be known from past experience, or unknown.

Fraud avoidance schemes, such as CIFAS in the United Kingdom and SAFAS in South Africa, are only designed to find links with past frauds. They are not effective for the detection of new fraudsters, or those who can avoid being linked to past frauds.

The types of pattern detection can be split further into three types:

Application cross-checking—Checks information across multiple applications. This will look for applications with the same applicant, address, phone number, or other details which may indicate that a syndicate is operating.

Transaction cross-checking—Checks information across transactions to the same, and possibly multiple, accounts. The goal is to identify accounts where: (i) known patterns occur—such as deposits and withdrawals of the same amounts; or (ii) the patterns fall outside of the normal bounds, for an account of that type.

Merchant reviews—A merchant accepting credit card payments may check for: (i) multiple and/or unusually large orders of the same product; (ii) use of the same credit card number under different names and dates; and (iii) the appearance of account numbers in sequence, in an attempt to find a valid number.

31.3 Fraud prevention strategies

Just as there are different tools, there are also different strategies that can be used. The number of possibilities will vary according to the product. Once again, many of these methods were mentioned in McNab and Wynn (2003), but have been renamed or regrouped.

Account origination—security checks

The first, and most obvious, place to do checks is during the front-end origination process, to protect against identity misrepresentation and mail theft:

Security calls—Checks to ensure that critical applicant information is correct, including employer, address, income, and contact phone numbers.

Proof of receipt—Where plastic is sent through the mail, registered mail is often used, so that receipt can be confirmed if there are problems.

Account activation—The plastic has been sent in the mail, but before it can be used the cardholder must confirm receipt and personal details by phone, or by completing and returning a slip included with the card.

Some addresses may present very high interception risks, especially residences where there is easy access to a communal area where mail is left. These require extra security, and may be dropped entirely from unsolicited card campaigns.

Payment medium—security measures

Whenever plastic or paper is used to transact on an account, extra risks are created. Lenders are, however, able to take precautions, to protect against lost- and stolen-article fraud, counterfeiting, and alterations.

Personal identification numbers (PIN)—Used with any card that can be used to withdraw cash directly from an automated teller machine.
Anti-counterfeiting—Include watermarks, holograms, special engraving, card verification numbers, and other features applied to plastic or paper to inhibit counterfeiting. Silicon chips will also become common in the near future.
Treatment of alterations—Refers more specifically to cheque accounts, where the tolerance relative to cheque alterations can be modified.

Account management—authorisations/referrals

Beyond trying to increase the security of the physical transaction media, lenders can also include fraud checks with each transaction. Human intervention is used in an otherwise automated process, if there is any suspicion of possible fraud. For cheque accounts, the *clearing period* allows fraud checks to be done only after the cheque has been presented, and in suspicious circumstances the accountholder is contacted to ensure that the cheque is valid. For credit cards, a *telephonic identity check* can be done when the card is presented at a merchant. The cardholder is asked one or more questions, whose answers would only be known to the genuine accountholder. They may be based on information provided at time of application (e.g. mother's maiden name), identity and current contact details, or recent transactions.

These checks are expensive, and are only done where past experience has indicated fraudulent activity is a possibility. A set of rules, which hopefully include fraud scores, is employed to invoke a telephone call, based on transaction types, values, geographical location, type of merchant, past transactions, and/or other characteristics. The types of instances where authorisations may be required are:

Over floor limit—All transactions above a merchant's predefined threshold require specific authorisation. Where a specific merchant is suspected of fraudulent activity, this limit can be reduced, or removed.
Over credit limit—Although primarily used to manage credit risk, this limit also acts as a cap on fraudulent activity with lost and stolen cards, for transactions over floor limit.
First-time use—When presenting the card for the first time, the customer will be asked to confirm his/her identity. This is used primarily where cards are sent through the post, and protects against false 'never received' claims.
Suspicious transactions—If the transaction falls outside the normal pattern of spending for that account.
No article—Card-not-present cases require greater diligence, and every transaction may be verified. Cases also occur where cheques are presented in unusual ways, for example written out on a sock or a piece of scrap paper, in which case special arrangements must be made.
Account closed—Used where plastic is lost or stolen. Any subsequent transaction needs to be confirmed by the accountholder

31.4 Fraud scoring

While scoring is used primarily to control credit risk, it can also provide an effective tool for detecting possible frauds. Unfortunately, very little information is readily available on fraud scoring per se. The following sections are based primarily upon what was provided by McNab and Wynn (2003) and NeuralT (2002), and some personal experience of the author.

There are two main types of fraud scoring. Application fraud scoring is used at time of application (including Internet fraud), and transaction fraud scoring during transaction processing. The former applies to any type of product, whereas the latter applies primarily to cheque accounts and credit cards. Practically everything that has been said in this textbook about credit scoring can also be applied here, but with several differences:

Small numbers of confirmed frauds can make it difficult or impossible to develop a scorecard.

Data quality may be even more suspect than in other areas. Unlike in the credit risk area, where accounts more than 90-days delinquent are automatically flagged bad, fraud scoring is reliant upon the manual setting of fraud indicators against confirmed frauds. Some lenders will be less than diligent in this area, which further reduces numbers. Once a *modus operandi* has been detected and blocked, fraudsters are very quick to adapt. Whatever infrastructure is used, there must be sufficient flexibility for the lender to parry fraudsters' changing strategies.

Reaction times must be very quick. It is no good to identify possible fraud, and then wait for two weeks to do anything. This places significant demands upon infrastructure and staff.

False positives can either cause customer dissatisfaction or marketing opportunities, depending on the circumstances.

Reasons for any delays cannot be given. Customers cannot be told that they are being investigated as potential fraudsters! All that can be done is to confirm that he/she is not, and possibly correct any databases in the background.

The company most associated with fraud scoring is HNC Software, which became a subsidiary of Fair Isaac after its purchase in 2002. Its flagship product is Falcon™, a transaction fraud scoring system that runs on companies' in-house systems. Another more recent product is Gemini™, an application fraud scoring service hosted by Equifax that was introduced in Canada in 1998 and the United States in 1999. Both use neural networks (NNs) to track changing fraud patterns.

Application fraud scoring

Application fraud is not as common as transaction fraud, and the number of confirmed frauds available to develop a model can be very small, unless the organisation is large, or data is being pooled. Otherwise, this is much the same as application risk scoring, with some differences.

First, there is *no reject inference*, as the number that will be rejected purely because of the fear of potential fraud is extremely small, and fraudsters usually have sufficient knowledge of the system to ensure that their applications will not be rejected. Second, the *type of fraud* needs to be noted, as this may affect the scoring and pattern-detection models. Third, care must be taken to ensure that the models are *opaque* to outsiders, meaning that they do not focus upon one or two obvious characteristics, but instead have a spread.

And finally, many—sometimes most—of the identified cases will be false positives, possible frauds that are genuine transactions, where the customer is inconvenienced by the extra checks and delays. This can cause great frustration, especially where the application is urgent. It is difficult to turn such situations into a positive customer experience, as the reason for the delay cannot be explained.

The biggest problem with fraud scoring is making the distinction between fraud and credit losses. While first-payment default is quite a blatant indication of possible fraud, the situation is less clear where payments have been made for several months. Some frauds may involve the planning and resources of a covert military operation, and take months to build up a relationship to the point where the lender relaxes its policies. These may relate to withdrawals against uncleared deposits, or allowing shadow limits beyond a certain level. While it may be possible to identify some of these accounts at time of application, it is still necessary to take precautions as part of transaction processing.

Transaction fraud scoring

While credit risk may be assessed behaviourally on a regular monthly basis, this can be insufficient to prevent fraud. Most of the relevant patterns only become evident in an analysis of detailed transactions, not monthly totals. As a result, most fraud scoring is done as part of transaction processing. This involves scoring and reviewing countless transactions, possibly while a customer is waiting in-store. There are several key features required for an effective system:

- **detection methods**, including scorecards, pattern detection, and policy rules;
- **strategy** design and testing abilities;
- **systems** that can apply these and prioritise cases by value and risk;
- **investigators** who can resolve cases quickly.

For each of these, there must be sufficient flexibility to handle changing circumstances. Once again, the problem with numbers can be problematic, both because of the low number of frauds, and difficulties in confirming them. It may, however, be possible to pool information across lenders, especially where products have the same basic features and transaction types. This will provide a much more robust solution than a small lender operating in isolation. Different types of fraud may also require different responses, for example lost and stolen card, card-not-present, cheque fraud, etc. Multiple sets of scorecards and policy rules can be applied, depending on the circumstances.

Unlike traditional credit scoring, which usually relies upon statistical techniques like linear or logistic regression, transaction fraud scoring often relies upon NNs, which have the advantage that they: (i) can handle large volumes of data; (ii) are highly predictive; (iii) deal with interactions; and (iv) can (supposedly) be 'trained' to adapt to tricks being used by the fraudsters, as they adapt to companies' fraud prevention efforts. Care must be taken to ensure that interactions are properly modelled, and that the final model performs well out-of-sample (NeuralT 2002).

Credit card environment

In the card environment, there is a distinction made between real-time and post-authorisation scores, which are calculated immediately prior to and after transactions respectively. These are used to detect fraud where card transactions are being posted in rapid succession—perhaps within minutes or hours of each other. For example, after two transactions in 10 minutes, the post-authorisation score may flag the account for real-time authorisation, prior to the third transaction. If this score then indicates a high probability of fraud, the merchant is required to obtain a telephonic authorisation, and the cardholder is asked to confirm identity. After four or five transactions, the authorisation may be declined regardless.

The number of characteristics used to identify possible transaction fraud is limited. These include number of transactions over the past 24 hours, the transaction amounts, merchant type, card present (Y/N), and some others. It may be wise to have separate scorecards for the card-not-present and lost-and-stolen card categories. Another issue is managing the volume of transactions being investigated. There are periods, like Christmas, when people's shopping patterns change, and transaction volumes increase considerably. Lenders are hard pressed to maintain the same vigilance, and some reliance has to be put upon a prioritisation and queuing system.

Cheque account environment

The treatment of cheque accounts differs from credit cards. Card authorisations have the luxury of being able to ask the merchant to contact them telephonically. The customer's identify is then confirmed while still in the shop, prior to approval. In contrast, cheque referrals rely more on the grace period provided while cheques clear (perhaps 10 days), albeit this is often waived on established accounts. Cheque accounts have also been around for much longer, and tried and tested procedures have been developed to deal with many of the fraudulent practices that have occurred in the past. That said, it still occurs, whether via a stolen chequebook, identity theft, use of an account with fraudulent intent, or altered details.

The time windows for cheque account fraud can also be very long, as fraudsters may use sophisticated kite-flying techniques to make accounts look genuine. An effective fraud detection system must thus be able to analyse information over weeks, or months, irrespective of whether scoring or pattern detection is being used. This is one situation that can be turned to the bank's benefit, as investigation may identify marketing opportunities—like where the customer has received a large inheritance, and is looking for an appropriate investment.

31.5 Summary

Fraud is considered an operational risk, not credit risk, with the primary difference being the use of *deception*, and the continual search for soft targets. It is difficult to identify though, and as a result, credit and fraud risk must be considered simultaneously. The primary targets are transaction products, especially cheque and credit card, but no product is immune. In the credit card arena, there was a shift first from 'lost and stolen' to 'counterfeit' fraud, and more recently to 'card-not-present' fraud. The changes have resulted primarily from the success of some fraud prevention measures, which have been offset by the fraud opportunities provided by new technologies (card skimmers, Internet).

Fraud types are many and varied, and are split on a number of different dimensions: *product* (cheque, card, asset finance); *relationship* (first-, second- and third-party); *business process* (application, transaction); *timing and manner* (immediate and initial limit only, or long-term to increase limit offered); *security* (misrepresented or on-sold); *identity misrepresentation* (embellishment, impersonation, and fabrication); *handling of transaction media*, which varies by acquisition (lost or stolen, not received, at hand, and skimmed) and utilisation (counterfeit, not present, altered, or unaltered); and *technologies* involved (ATM (card swap, card trap) and Internet). Fraud syndicates may also penetrate lender operations better to understand their processes, and staff members are not immune to temptation.

Fraud is a lucrative area for the criminally inclined, due to the difficulties that lenders have in successful identification and prosecution. As a result, lenders spend much on prevention, and are usually very willing and eager to share information. The tools employed include *internal negative files* (including hot card files), *shared databases*, *rule-based verification*, *pattern recognition* (application and transaction cross-checking, and merchant reviews), and *scoring*. Given fraudsters' mercurial nature, the approaches must be flexible, provide information for ready analysis, and facilitate the provision of information to law enforcement agencies. Fraud prevention strategies can be employed: (i) at *account origination* (security calls, proof of receipt, account activation); (ii) to the *transaction medium* (PIN numbers, anti-counterfeiting measures, and treatment of alterations); and/or (iii) via the *account-management process* (over floor limit, over agreed limit, first-time use, suspicious transactions, no article, and account closed).

Fraud scoring is possible, but often suffers because of small numbers (further confounded by problems with identification and confirmation), data quality issues, fraudsters' adaptability, short reaction times when applying strategies, large numbers of false positives, and problems explaining delays to customers. The most well accepted technique is NNs, primarily because of its ability to adapt to changing fraud patterns. Application fraud scoring is rare, largely because there are usually insufficient numbers to develop a model. Lenders instead rely on fraud databases, inconsistency checking, and policy rules to guide their investigations. If the applicant is found to be genuine, then any databases should be updated to reflect it (address, phone number).

Most fraud scoring is done at transaction level, which is dependent upon detection methods (scorecards, pattern detection, policy rules), strategies, systems, and investigators. In the *credit card* environment, both real-time and post-authorisation scores may be used, depending upon what has happened before; the need for speed is greater, due to the nature of the product. *Cheque accounts* usually have a clearing period before funds can be drawn, but a risk is posed for established accounts where it is waived. Fraudsters can be very sophisticated in their use of kite flying to inflate the limits offered, which may take several years. Models and rules will yield a lot of false positives, and wherever possible, the information should be used to identify marketing opportunities.

Module H

Regulatory environment

32 Regulatory concepts

> Corporate Governance is concerned with holding the balance between economic and social goals and between individual and communal goals. The corporate governance framework is there to encourage the efficient use of resources and equally to require accountability for the stewardship of those resources. The aim is to align as nearly as possible the interests of individuals, corporations and society.
>
> *Sir Adrian Cadbury in 'Global Corporate Governance Forum'*, World Bank, 2000.

There are many concepts used in business for which there seem to be no precise definitions. Several that gained prominence since 1970 were 'best practice', 'good governance', 'business ethics', and 'social responsibility'. They do not fall neatly under legal, but are related, because many of society's demands of business fall under these headings. For that matter, they do not relate directly to credit scoring, but provide motivations for, or restrictions against, its use within the business.

32.1 Best practice

Very little literature exists on best practice as a concept, but it is fairly easy to define. Simply stated, it refers to practices (processes, techniques, methodologies, and the use of technology, equipment, and resources) that have a proven record of success at providing a desired result. Organisations spend years gaining experience, but often do not know whether the correct lessons have been learnt. It may seem right, but be suboptimal. Best practice can originate in-house, but it is usually borrowed experience used to pre-empt the learning curve, whether to keep in step with leaders, or satisfy the demands of a regulatory authority or industry body.

By doing so, companies can improve process efficiencies, correct errors that have been made in the past, improve relationships with stakeholders, and lessen the cost of the learning curve when entering new terrain. It has thus become a growth industry for experts and researchers, to document and disseminate best practice in practically every known field of endeavour. They may be commissioned by companies trying to gain a competitive edge or by industry bodies and regulators trying to establish an industry 'code of best practice'. Care must however be taken, as its proponents: (i) may have their own agenda that the purported best practice supports, and/or (ii) hide behind it without having a proper understanding of its workings or the ability to recognise circumstances where it may not be appropriate.

32.2 Good governance

In contrast to best practice, good governance is a concept for which there is a fair amount of literature, but as of yet, no single model. Even the definition of 'governance' is elusive. Most available definitions will split it along two dimensions, and perhaps the best definition is:

> The process of decision-making and the process by which decisions are implemented (or not implemented).
>
> *Unattributed*

The definition was probably first provided in the 1980s as part of the 'Post-Washington Consensus', which promoted open-market policy reform on behalf of the World Bank and the UK Department for International Development. It is now widely quoted, including in an article called, 'What is Good Governance?' on the UNESCAP and other websites.

This definition can be used in many contexts, including corporations, local government, national government, non-governmental organisations, international agencies, and so on. It is typically viewed as requiring participation, the rule of law, transparency, responsiveness, consensus, equity and inclusiveness, effectiveness and efficiency, and accountability.

The difference between governance and good governance is accountability! The concept was initially used with respect to governments,[1] but over time it has been extended to corporations, to ensure that executives: (i) respect the power they have been given over organisational resources and: (ii) strive to utilise them to the long-term benefit of shareholders. Milton Friedman was the first to attempt a definition of corporate good governance (CGG) in 1970:

> He [corporate executive officer] has direct responsibility to his employers. That responsibility is to conduct business in accordance with their desires, which generally will be to make as much money as possible while conforming to their basic rules of society, both those embodied in law and those embodied in ethical custom.[2]

Friedman's definition reflects the near robber-baron mentality that dominated his economic philosophy, which presumed that anything legal, and done without deception or fraud, was ethical. While well-suited to companies focused upon short-term profits, it can be contrary to the long-term best interests of all concerned. Problems are most acute where there is a misalignment of interests between management and other stakeholders, in particular where incentives reward short-term performance.[3] The emphasis has thus shifted to control over the corporate executive, as Roy C. Smith's 2003 description indicates:

> [CGG is] the practice of exercising self-restraint to reduce the risks within performance-oriented corporations and, when necessary, to rein in the considerable powers granted by the board to the CEO.

---

[1] Much of it stemmed post the Watergate scandal in the United States.

[2] 'The Social Responsibility of Business is to Increase its Profits'. *New York Times Magazine*, 13 September 1970. Cited in Lutz (2002).

[3] Adam Smith had already noted in 1776 that managers are the agents of owners (an 'agency problem'), and that although the arrangement usually works, checks and balances are required.

This way of thinking did not evolve overnight; it has evolved as many people applied their minds to the problem, especially after the spectacular booms and busts of many companies since Friedman's heyday—first after a spate of corporate failures in the United Kingdom in the period 1990 to 1992 (Maxwell Communications, British and Commonwealth (BCCI), and Polly Peck),[4] and then in the early- to mid-2000s in the United States (Enron and WorldCom) and Europe (Parmalat, in Italy). CGG has thus become a concern for shareholders, banks, the general public, and government, especially as problems often arose from corruption or mismanagement.

In 1992, Sir Adrian Cadbury[5] presented a report (now known as the Cadbury Code of Best Practice[6]) on corporate governance for the United Kingdom, which became the cornerstone for thinking on the topic worldwide. His report defined governance as the way organisations are: (i) *directed*—encompasses leadership, strategic intent, operating principles, and values, and provides an overarching framework for the manner in which the organisation's leaders operate; and (ii) *controlled*—covers systems, processes, guidelines, policies, and procedures, and other aspects relating to day-to-day functioning, where there is less judgmental freedom. It recommended, amongst others: (i) the separation of the chief executive and chairman posts; (ii) the appointment of a significant number of independent non-executive directors, as well as audit and remuneration committees; (iii) that institutional shareholders exercise their voting rights; and (iv) that it be the board's duty to ensure reporting on the company's position is balanced and understandable (Sparkes 2003).

According to GlobalChange.com, the purpose of corporate good governance is to build 'TRUST' amongst all stakeholders: Transparent—totally open; Responsible—acting in the broader and longer-term interests of all; Uncompromising—commitment to highest moral positions; Successful—achieves great results by combining excellence with values; and Temperate—taking care to avoid major risks, wild decisions and extravagance. 'The Future of Corporate Governance', http://www.globalchange.com/corporategovernance.htm.

In 1999, the OECD presented its 'Principles of Corporate Governance', which built upon the Cadbury Code. It provided a more comprehensive structure setting out specific responsibilities for different stakeholders, including the executive, managers, shareholders and others, as well as rules and procedures for making corporate decisions. In doing so, it provided the tools for both setting corporate goals, and monitoring performance.

American legislation came late, and was more draconian after the many corporate disasters in the early 2000s. The Sarbanes-Oxley Act, which came into force on 30th July 2002, went further, by providing legislation instead of a code of practice. It has specific clauses to protect the objectivity of securities analysts and their research, and to strengthen penalties for securities

[4] Common features of these corporate failures were: lack of control, possible fraud, concentration of power, weak boards of directors, and over-optimistic financial reporting.

[5] Sir Adrian was the chairman of Cadbury Schweppes from 1965 to 1989, and was a director of the Bank of England from 1970 to 1994.

[6] This was followed by codes presented by Greenbury (1995), Hampel (1998), Stock Exchange Code, Turnbull (1999), and Higgs (2003).

law violations. In particular, it requires the chief executive and financial officers to certify that their companies' financial statements are correct; anybody certifying non-compliant or false reports will face criminal charges. The new hard-line stance was evidenced by the 2005 jailing of WorldCom's former CEO, and has had some unintended consequences regarding individuals' willingness to accept top positions within American corporations, and international companies' willingness to operate on American soil, or list on their exchanges.

Whereas company executives used to be laws unto themselves, there are now requirements for fairness, accountability and due diligence, transparency and continuous disclosure, and adherence to standards. All of this does, of course, have benefits: shareholders are believed to be willing to pay a 30 per cent premium for well-run companies.[7]

32.3 Business ethics and social responsibility

The exclusively economic definition of the purpose of the corporation is a deadly oversimplification, allowing overemphasis on self-interest at the expense of consideration of others.

Kenneth Andrews (ed.), Business Ethics: Managing the Moral Corporation, Harvard Business School Publishing, 1989.

The call for good governance has gone beyond accountability to owners and shareholders, and expanded to other stakeholders, including employees, suppliers, customers, community, and government.

'Stakeholder Theory' was popularised during the 1980s by R. Edward Freeman, who defines a stakeholder as 'any group or individual who can affect or is affected by the achievement of the organisation's objectives'. It can be seen as increasing the number of parties to the social contract (Lutz 2002).

This leads us to two related concepts that can be confused with good governance, and pose conflicts with traditional views on the role of business in society: business ethics and corporate social responsibility.

Business ethics—The determination of what is morally right or wrong in business situations and acting accordingly.

In 1987 Michael Cook stated that there are two opposing views of ethics: either anything legal is ethical; or anytime one's conscience gnaws, it is unethical. The true position is likely somewhere in between. Most early twentieth century literature on business ethics was either a critical attack by socialists against 'the amorality of business thinking' (Clark 2006), or part of a call for social responsibility by academics schooled in philosophy or theology (McGrath 2003).

Corporate social responsibility (CSR)—The company's respect for and conduct towards the wants, needs, and concerns of other stakeholders.

[7] This was reported in a study by the McKinsey Quarterly with respect to companies in emerging economies.

The CSR requires that their responsibilities broaden from a strict focus on company profitability, to broader economic, sociological, and environmental concerns—both at home and abroad. There are two types: ethical conduct and philanthropy. While many companies are happy to limit CSR to the former, others are happily engaging in the latter—especially where the recipients are close to home. Besides the 'feel good' factor and potential publicity gain, there is also the possibility that social investments will have indirect long-term benefits for the business.

Milton Friedman condemned any attempts by business to meddle in social affairs, and thought charitable functions were better left to government and NGOs.

According to Sparkes (2003), there are three quite crucial factors that are motivating companies to become socially responsible: (i) possible negative influences upon the brand and corporate image, as evidenced by Nike for its employment practices overseas; (ii) the growth of socially responsible investment funds; and (iii) growing political consensus to encourage corporate social responsibility, which is evidenced by pension and investment funds increasing use of their voting rights to press on social responsibility issues.

These concepts reflect radical shifts in the public's view of the role of the corporation, and calls for change have taken on near religious proportions. The end result is a significant broadening of what was once a narrow focus: companies are now responsible for more and accountable to more. Businesses are no longer expected to be just profit-driven organisations responsible only to shareholders, but also social institutions accountable to a broader range of stakeholders.

Gardner (2001) refers to the 'Principles of Global Corporate Responsibility', whose purpose is to 'promote positive corporate responsibility consistent with the responsibility to sustain the human community and all creation'. The principles were developed jointly by the Taskforce on the Churches and Corporate Responsibility (Canada), the Ecumenical Council for Corporate Responsibility (UK), and the Interfaith Centre of Corporate Responsibility (USA). The latter is very active, and filed 135 of the 261 shareholder resolutions in the USA during 2001 (Sparkes 2003).

In the United Kingdom the cries for accountability occurred after events like the Herald of Free Enterprise ferry sinking at Zeebrugge, the King's Cross Underground station fire in London, and the Piper Alpha oilrig explosion. A common feature was 'gross deficiencies in general management practices' (ALARM 2001).

Over the last 50 years, companies have been facing increasing pressure to live up to public expectations, whether expressed through lobby groups, industry codes of practice, or specific legislation. While it is still accepted that maximising long-term shareholder value is businesses' primary goal, business ethics and social responsibility are not precluded. Indeed, public opinion has forced management to become more broadly accountable to civil society, at least to the extent of operating within societal norms in all aspects of business. The challenge then becomes one of engendering a culture of accountability amongst their rank and file, which is particularly problematic when most measures of performance are monetary.

32.4 Compliance hierarchy

Who goeth a borrowing goeth a sorrowing.

Benjamin Franklin

Human interactions are defined by relationships. Where the relationships are close, problems can usually be sorted out directly with the individuals concerned—whether with little brother, the neighbour, the schoolteacher, or grumpy grandpa. If the parties are much more distant however, the task becomes trickier. The modern world relies upon rules, regulations, laws, and statutes to govern people's actions and interactions, with the goal of ensuring that society continues to operate smoothly and with minimal disruptions. This extends to the relationship between institutions and civil society, which can be done using a number of different mechanisms:

Law set by statute—Specific laws written to govern a situation, sometimes being a summation of the legal precedent.
Legal precedent in civil law—Accepted treatment that has been determined by cases that have been decided by the courts in the past (also known as 'common law').
Code of practice—Standards set by a representative body for an industry or discipline, usually developed to address common problems.
Policies and procedures—Standards applied by a company that embody lessons learnt in the past, and which may also borrow from industry best practice.
Unwritten code—Practice developed over time, which may or may not be documented, but nonetheless serves the purpose and has become a part of the culture.

If a written code-of-practice is not adhered to, the penalty is often the denial or removal of group membership and the benefits of associated services. Although limited, this may be significant, especially if continued membership implies a stamp of approval, certifying adherence to legal requirements. These bodies are also often tasked with communicating with the outside world—government, the public, and others.

This framework applies to a large number of human endeavours, including professional associations (doctors, lawyers, accountants, estate agents) and industries (oil and gas, banking, transportation). It also applies to credit scoring and lenders that use it. Lenders have a responsibility to ensure adherence to regulations, which can be aided by taking certain steps: (i) appoint a compliance officer, whose sole responsibility is compliance; (ii) perform regular compliance audits; (iii) make sure that managers and staff understand their responsibilities; and (iv) consult with unions and other stakeholder bodies, that may be affected by changed practices and procedures.

32.5 Summary

At the highest level, credit scoring is governed by many of the same concepts governing business as a whole, and more often than not, is a tool that assists lenders towards those ends. This section provided an overview of two pairs of high-level concepts, which will aid understanding of upcoming topics. The first pair is: 'best practice', which are practices with a proven record of success at meeting a certain end, which includes credit scoring as it is used for retail credit risk assessment; and 'good governance', which refers to management's accountability to shareholders and other stakeholders, with respect to how they direct and control the enterprise, and ensure that their interests are aligned.

The other pair is 'business ethics' and 'social responsibility', both of which represent a significant shift in what is expected of businesses' role in society, compared to the robber-baron philosophies of old. These have no bearing on whether credit scoring is used, but do affect how it is used. 'Business ethics' relates to the conduct of business in a manner that is morally right, while 'social responsibility' refers to enterprises' conduct towards the needs, wants, and concerns of broader society, which may be limited to ethical conduct, or extend to philanthropy.

Finally, lenders have to comply with rules and regulations, which come in a variety of forms. *Legal statutes* and *precedents* are first in the compliance hierarchy, as failure to abide by them can result in lawsuits, fines, and adverse publicity. It is best when industries can self-regulate though, and many will implement *codes of practice* to pre-empt legislation. Individual enterprises may also implement *policies and procedures* based upon best practice, and in some instances there may even be an *unwritten code* that governs their actions.

Each of the next five sections covers a different aspect of the regulatory environment, each of which has evolved over time. The preferred state of affairs is where companies are self-regulating, and a minimum of external intervention is required, but there are bad apples in every barrel. Industries will typically set up governing bodies, with their own codes of practice, while legal precedents will define the playing field. More often than not though, legislators will implement statutes, which can vary greatly in effectiveness, and there is a fear that lenders may suffer from excessive regulation.

33 Data privacy and protection

George Orwell's book '1984', written in 1948, presents a world governed by an all-knowing and all-seeing 'Big Brother'. This was not the entertaining and sometimes humorous TV reality program, but a sinister prediction of how technology would assist a totalitarian state to control people's lives. While the real model for the story was the then Soviet Union, the title became a metaphor for increasing public concern—if not paranoia—about western governments' use of technology to compile cradle-to-grave dossiers on their citizens, and its potential impacts on civil liberties.

This section considers data privacy and protection. First, it considers some of the background including: (i) *information as property*, and the differential treatment afforded personal and commercial data; and (ii) *legal background* (Tournier Case, Fair Credit Reporting Act, OECD Guidelines, etc.). Thereafter, it covers the *data privacy principles* that underlie much of the legislation, covering the data's manner of collection, reasonableness, quality, use, disclosure, security, and subjects' rights.

33.1 Background

From the 1950s to 1970s, much thinking—especially in the United States—was devoted to data privacy as it relates to the public sector. The focus only shifted to the private sector as data was increasingly used to drive selection in application processing (accept/decline) and direct marketing (mail/no mail), whether for financial, health, insurance, or other services. Owens and Lyons (1998) highlighted two aspects of this thinking:

Information as property—Control over personal information is crucial for negotiating relationships with society at large, and most modern thought views personal information as being property of the individual. The concept of 'property rights' cannot apply though, because by its very nature information is something that is shared, and it is impossible to exclude other parties. Indeed, a pure property-rights view assumes that possession is 9/10ths of the law, and individuals would be powerless without help from legal and regulatory frameworks to limit how data pertaining to them is used by others.

Personal versus commercial—Data privacy applies to both individuals and corporations, but for corporations its sole purpose is to protect commercial interests, whereas for individuals it is an important moral value, and necessary for the protection of human dignity. Each bit of information is seen as a view into the person's soul; an invasion of privacy. The term 'data privacy' is thus usually short-form for 'personal data privacy rights,' and the focus is upon maintaining them in relationships with the private sector.

Something that is peculiar to banking relationships is the power imbalance between borrower and lender. Potential borrowers are asked for all sorts of personal information, some of which may be sensitive. Many people do not like these little snippets being known, and may do everything in their power—including not borrowing—to avoid being put into a potentially compromising position. Banks have thus gone out of their way to ensure that this information is kept between them and their customers, and most developed countries provide a framework to protect against shady operators. This fiduciary relationship with respect to data is also known as a 'duty of secrecy', which is an implied contractual obligation. It is the one thing that counters the power imbalance, is accepted in law in most countries to varying degrees, and applies most strongly to banks. Most companies will honour it, but the growth of technology, the profit motive, and overzealous employees have often caused transgressions. It used to be governed primarily through legal precedents in civil law, but today many countries have specific statutes in place.

33.1.1 Historical overview

The concept of privacy as a personal right is a recent one. While some of the early thought evolved in the late 1800s, the first legal precedent was the 1924 Tournier case in England (see Section 33.1.2), which defined acceptable practices regarding data privacy for banks and certain other financial institutions. It became accepted throughout the United Kingdom and Commonwealth, and was cited in case law in many other countries, including the United States.

The situation in Europe was much different. According to Arnold (1996), German civil law upholds the doctrine of 'Treu und Glauben', and their duty of banking secrecy was strengthened by the 1986 removal of the obligation for banks to provide information to the tax authorities. In Switzerland (1934) and Austria (1979) there is specific legislation making it a criminal offence to breach a customer's right to secrecy. In some of these cases, this legislation was so strict that their banking systems became havens for drug runners and despots trying to hide their ill-gotten gains. These restrictions are, however, being relaxed, as issues regarding crime prevention take precedence. Governments are putting the onus upon financial services, and other companies, to 'Know Your Customer' and report any suspicious transactions. Most developed countries now have laws with significant penalties for banks and others that—unwittingly or not—are being used for money laundering.

According to the OECD's website, in 2004, half of the OECD member countries either already had data privacy laws (Austria, Canada, Denmark, France, Germany, Luxembourg, Norway, Sweden, and the United States) or had prepared draft bills (Belgium, Iceland, the Netherlands, Spain, and Switzerland).

The first statutory legislation specifically directed at credit was the USA's Fair Credit Reporting Act (1970), which focused specifically upon consumer reporting agencies.[1] Besides confirming

[1] Given that the credit bureaux are the primary conduits for information sharing between lenders, the legislation practically addressed data privacy in the credit industry as a whole.

the value of the credit bureaux' services, one of the American Congress's findings stated in the Act was that, 'There is a need to insure that consumer reporting agencies exercise their grave responsibilities with fairness, impartiality, and a respect for the consumer's right to privacy'. The Act thus protects against practices that were common when underwriting was still judgmental, and sources of information included newspaper clippings and barbershop gossip.

The 1970s was defined by increasing automation, and legislation appropriate for this new and rapidly changing environment was required. Rather than following the Americans' sectoral example though, most other countries developed more general privacy legislation, often covering both public and private sectors. It soon became apparent that the differing treatments would hinder transborder data flows, at a time when they were growing rapidly in the finance and insurance sectors.

In 1980, the Organisation for Economic Co-operation and Development (OECD) presented a set of guidelines that were influential in shaping and standardising the national legislation in the different countries, as well as countless industry codes of practice. This was followed by the Council of Europe's Convention on Data Privacy in 1985 and the European Union Data Protection Directive in 1995. Over the period, many countries either structured or adapted their own legislation to accommodate the recommendations. Examples of legislation used in various countries are provided in Table 33.1.

33.1.2 Tournier case (1924)

Legal textbooks often cite decades-old cases from foreign jurisdictions. The concepts are often so basic that they are globally applicable, and in some instances different countries have practically the same treatment in law, even though it came about through different means. The great grandpappy of all privacy legislation in the United Kingdom and Commonwealth, and which has also been cited in the United States and elsewhere, is what is commonly referred to as the 'Tournier case'—legal precedent set by 'Tournier v. National Provincial and Union Bank of England [1924] 1 KB 461', Judge LJ Bankes presiding.

According to Owens and Lyons (1998), the Tournier case set a precedent for contract law, effectively stating that there was an implied 'duty of confidentiality' in the contract (written or

Table 33.1. Data privacy legislation

Country	Name
USA	Fair Credit Reporting Act (1970)
	Personal Information Privacy Act (1997)
Canada	Privacy Act (1985)
UK	Data Protection Act (1985/88)
Hong Kong	Personal Data (Privacy) Ordinance (1995)
Australia	Privacy Act (1988), Privacy Amendment (Private Sector) Act (2000)
South Africa	Promotion of Access to Information (2000)
	National Credit Act (2006)

unwritten) between banker and customer, unless 'modified by express contractual terms'. In the intervening years, it has also been applied to other types of financial services companies. When reading the details, it must be remembered that this was a time when bankers knew customers personally, and were often privy to confidential information, even if there was no lending relationship.

Tournier had a bank overdraft of just under £10, which he contracted to repay in weekly instalments of £1 each. He had no fixed address, as he was employed on a probationary three-month contract as a 'traveller' (salesman), and used the address of his employer. Default occurred during an away trip, and the employer was contacted to ascertain Tournier's whereabouts. In conversation, the branch manager disclosed that Tournier made frequent overdrafts, and ventured the opinion that Tournier was betting heavily, as a cheque had been endorsed to a bookmaker. Tournier was not hired, successfully sued the bank, and was able to regain his credit standing.

Judge Bankes held that the banker's duty of secrecy was not only a moral duty, but also a legal duty arising out of contract that does not terminate when the customer relationship is terminated, and applies to all information collected before and during the relationship, but not after it ends.

> In my opinion it is necessary . . . to direct the jury to what are the limits and what are the qualifications of the contractual duty of secrecy implied in the relation of banker and customer [no precedent currently exists] . . . On principle I think that the qualifications can be classified under four heads: (a) where disclosure is under compulsion by law; (b) where there is a duty to the public to disclose; (c) where the interests of the bank require disclosure; (d) where the disclosure is made by the express or implied consent of the customer.

It is these exceptions that subsequently governed what information the banks could provide:

Compulsion by law—Where 'a proper authority derived from statute or an order of the court' was exercised. No obligation can be placed upon the bank to contest the application, probe supporting evidence, or in any way hinder the police investigation.

Duty to the public—Where the account belongs to a revolutionary body, the client is suspected of treason, or the account is being used in connection with trade with an enemy of the state.

Interests of the bank—To collect or sue for an overdraft, whether through cession or legal action, or to defend management's actions when explicitly questioned, but NOT to be used for marketing purposes by any third parties, including companies in the same group.

Customer consent—If the customer provides consent for the information to be divulged. An assumption used to be made that any request for bank facilities implied customer consent to obtain bank references from other banks, but specific consents are usually required.

For many years, this was the only case governing data privacy as it relates to financial institutions, and strict interpretation of the exceptions provided a major obstacle to sharing credit performance data between banks. These principles were, however, encompassed within the UK Data Protection Act (1985/8), and in the 1990s some banks started seeking customer consent to provide credit performance to the credit bureaux. The benefits were soon realised.

33.1.3 OECD data privacy guidelines

During the 1970s, the rapid growth in automated data processing caused several countries to implement or consider data privacy legislation, but it was soon recognised that the disparate treatments had the potential of limiting transborder data flows. In 1978 the OECD commissioned the drafting of guidelines that would act as minimum standards and aid the free flow of information between countries.

The 'Guidelines on the Protection of Privacy and Transborder Flows of Personal Data' ('OECD Guidelines') were prepared by a group of experts under the chairmanship of The Hon. Justice M.D. Kirby, Chairman of the Australian Law Reform Commission, and were presented on 23rd September, 1980. The eight principles provided a consensus of views, based upon law reform efforts in most OECD states at that time (Owens and Lyons 1998). These were intended to be adopted by member countries, whether within existing or new legislation, to govern data privacy in both the public and private sectors. It should come as no surprise that the principles both encompass and broaden the Tournier principles. The following is quoted almost verbatim from the OECD Guidelines, and if it seems familiar it is, because it has been used as the basis for data privacy legislation in several countries, including the United Kingdom:

Collection limitation principle—There should be limits to the collection of personal data, and any such data should be obtained by lawful and fair means, and, where appropriate, with the knowledge or consent of the data subject.

Data quality principle—Personal data should be relevant to the purposes for which it is to be used, and to the extent necessary for those purposes should be accurate, complete, and kept up-to-date.

Purpose specification principle—The purposes for which personal data are collected should be specified, not later than at the time of data collection; the subsequent use should be limited to the fulfilment of those purposes, or other not-incompatible purposes, and as are specified on each occasion of change of purpose.

Use limitation principle—Personal data should not be disclosed, made available or otherwise used for purposes other than those specified in accordance with [the purpose specification principle], except: a) with the consent of the data subject; or b) by the authority of law.

Security safeguards principle—Personal data should be protected by reasonable security safeguards against such risks as loss or unauthorised access, destruction, use, modification or disclosure of data.

Openness principle—There should be a general policy of openness about developments, practices, and policies with respect to personal data. Means should be readily available for establishing the existence and nature of personal data, and the main purposes of their use, as well as the identity and usual residence of the data controller.

Individual participation principle—An individual should have the right: a) to obtain from a data controller, or otherwise, confirmation of whether or not the data controller has data relating to him; b) to have communicated to him data relating to him within a reasonable time; at a charge, if any, that is not excessive; in a reasonable manner; and in a form that is readily intelligible to him; c) to be given reasons, if a request made under subparagraphs (a) and (b) is denied, and to be able to challenge such denial; and d) to challenge data relating to him, and if the challenge is successful, to have the data erased, rectified, completed or amended.

Accountability principle—A data controller should be accountable for complying with measures that give effect to the principles stated above.

Although the OECD Guidelines make no reference to any regulatory body or requirements for registration, they do define 'data controller' as 'a party who, according to domestic law, is competent to decide about the contents and use of personal data regardless of whether or not such data are collected, stored, processed or disseminated by that party or by an agent on its behalf'.

Besides recommending that member countries adopt legislation that encompasses these principles, the Guidelines also suggested that: (i) member countries encourage industry self-regulation; (ii) provide the means for individuals to protect their rights; (iii) provide sanctions and remedies for any organisations not adhering to the principles; and (iv) and protect against unfair discrimination.

33.1.4 Council of Europe convention

While the OECD Guidelines were just that, guidelines, in 1985 the Council of Europe (CoE) went further and published its 'Convention for the Protection of Individuals with Regard to Automatic Processing of Personal Data' ('CoE Convention'), which called on signatory countries to implement domestic legislation. The primary goal however, was to secure personal data privacy rights, not foster transborder data flows, with respect to any data that is automatically processed or stored. Within this can be seen several of the basic concepts that today underlie the national legislation of several countries:

Article 5 on 'Data Quality' states that, 'Personal data undergoing automatic processing shall be: (i) obtained and processed fairly and lawfully; (ii) stored for specified and legitimate purposes and not used in a way incompatible with those purposes; (iii) adequate, relevant and not excessive in relation to the purposes for which they are stored; (iv) accurate and, where necessary, kept up-to-date; (v) preserved in a form which permits identification of the data subjects for no longer than is required for the purpose for which those data are stored'.

Article 6 on 'Special categories of data' states that, 'Personal data revealing racial origin, political opinions or religious or other beliefs, as well as personal data concerning health or sexual life, may not be processed automatically unless domestic law provides appropriate safeguards. The same shall apply to personal data relating to criminal convictions'.

Articles 7, 8, 10, and 12 cover issues of data security, rights of personal access and contest, sanctions and remedies, and transborder data flows respectively. The latter is noteworthy!

Although it states that countries may not prohibit transborder data flows, it requires them to ensure equivalent levels of data protection in the other country.

33.1.5 EU data protection directive 95/46/EC

The original OECD Guidelines were done in cooperation with both the Congress of Europe and the European Community, which strive for consistency between member states. It was, however, only in 1995 that the OECD Guidelines and CoE Convention were effectively brought together, under what has become known as the EU Data Protection Directive ('EU Directive').[2] Its stated purpose is to protect personal privacy, but not at the expense of inhibiting transborder data flows.

Article 6 of the EU Directive practically repeats the CoE Convention's Article 5 verbatim, while Article 7 provides criteria that must be met for data processing to be legitimate: (i) unambiguous personal consent; (ii) in performance of a contract; (iii) result of a legal obligation; (iv) necessary to protect vital personal interests; (v) in the public interest; or (vi) for a legitimate purpose where personal rights are not infringed. Article 25 covers transborder data flows specifically, and requires an assessment of personal privacy protection in the receiving country.

The EU Directive was not received with open arms by all. According to Owens and Lyons (1998), many countries wished to retain national legislation with which they were familiar. When the draft directive was first made available in 1991, the United Kingdom thought it went too far beyond its own national legislation, while Germany thought it did not go far enough. Most of these issues have either been addressed, or are being addressed on a case-by-case basis, often with legal co-operation between countries on specific transborder inter-company arrangements.

In a 1994 case quoted by Owens and Lyons (1998) Citibank was processing the German Railwaycard, used to obtain rail-fare discounts in Germany, at offices in the United States. There were public concerns about the use of personal information to market credit products. The problem was solved by a contractual arrangement that ensured German law would be applied, including the ability to enforce in German courts.

33.1.6 Special circumstances

The principles may be relaxed in certain circumstances. In general, access to information may be aided or restricted where:

Aided—(i) the data is provided to the individual to whom it pertains, to somebody to whom consent has been granted, or to a representative of either; (ii) it is reasonable relative to the original purpose, and sensitivity, of the information provided; (iii) the records do not contain any personal identifiers, and are being used for statistical purposes only; or (iv) it is necessary to carry out a contractual obligation.

[2] This is more formally referred to as 'Directive 95/46/EC on the Protection of Individuals With Regard to the Processing of Personal Data and on the Free Movement of Such Data, OJ L 281/31 of Nov. 23, 1995'.

Restricted—(i) the requests are frivolous; (ii) it seriously breaches the privacy of others; or (iii) it would reveal intentions, and prejudice commercial negotiations, or the determination of liability in legal proceedings.

Furthermore, any and all limitations may be lifted where:

Legal—The data is required by statute or court order.
Public interest—It relates to people considered to be financial threats to the public, whether because of dishonest or fraudulent activities or past bankruptcies; or there is a justifiable belief that there is a significant threat to life or health of any party.
National interest—It is prejudicial to the actions of the police, courts, or internal revenue, the security and maintenance of order in prisons and other places of detention, or the national security or international relations of the state.

An attempt has been made not to repeat these special circumstances in the sections that follow, as they tend to apply generally. If more specific detail is required, please refer directly to the legislation for the country of interest.

Some definitions

Several definitions should also be provided here. This is not an exhaustive list, as many of the other definitions used in the legislation are quite straightforward:

'Personal data'—Any data that can be associated with a specific and identifiable individual. It excludes data with no personal identifiers, held for statistical analysis only.
'To process'—To perform any action upon personal data, such as collection, storage, modification, retrieval, disclosure, and so on. It applies primarily to automatic processes, but usually also extends to manual processes.
'Data subject'—Individual to whom the personal information pertains.
'Data controller'—Person within the organisation who is responsible for what personal data is processed, and with whom the general public and the regulators will communicate.
'Data protection commissioner'—In countries where applicable, the regulator responsible for ensuring the data privacy principles are upheld.

33.2 Data privacy principles

The rest of this section sets out a data privacy framework that can be applied across countries. It is based largely on the OECD principles covered in Section 33.1.3, but was also influenced by a reading of the legislation for the United Kingdom, Australia, Hong Kong, Canada, and the United States.

Manner of collection—Obtained lawfully, fairly, without deception, and with the knowledge of the individual.
Reasonableness—Relevant, not excessive, and justifiable.
Quality—Adequate, accurate, up-to-date, and not held for longer than necessary.
Use—State the purpose of the data at time of collection, and use it only for that or a reasonable or directly related purpose.
Disclosure—Not disclosed without consent, unless certain conditions are met.
Subjects' rights—To general knowledge of what data organisations hold and why, as well as to query, access, and contest, their own personal data.
Security—During both processing and disposal of personal data, and ensuring that trans-border data flows are afforded equivalent privacy rights.

33.2.1 Manner of collection

Information about individuals may be obtained directly from them, via the credit bureaux, or from other sources. The concerns are then: 'How it is collected?' 'Does the individual know?' and 'How is it associated with the individual?'

A peculiar aspect of the Australian Privacy Amendment Act (2000) is that the eighth principle requires anonymity, meaning that 'wherever lawful and practicable' individuals should be able to conduct business with an organisation without identifying themselves. It is presumed that this applies primarily to cash transactions or transactions where there is no future obligation or contractual relationship.

Means of collection

This covers collection from the individual and other sources, where it must be ensured that the means of collection is lawful, fair, and free of deception. Individuals should be aware that the data is being collected, and informed of the correct purpose for which it will be used. Data obtained using trickery or spying is out of bounds. The information may be relevant, but is it fair?

When a person is providing information, there are certain things they should know, and some effort may be required to educate them. Who is requesting the information? How will it be used? Will they be able to view their own information? Whom do they contact if they have questions, or issues? Will other parties have access to the information? Are there any laws requiring the information to be collected, or consequences if any of it is not provided? Any efforts to address these questions will go a long way in terms of customer relations.

Notice and consent

The primary concern with data protection is not just ensuring privacy, but also making sure that individuals know that somebody may try, or is trying, to process information about them. Legislation will typically cover the 'Consent required!' and 'Notice required!' approaches, albeit often using other labels. 'Consent required!' refers to instances that demand the *individual's permission* before any processing can occur, either in whole or in part. Requests for consent may be presented in three forms:

Compulsory—Default answer is 'Yes' with no option of saying 'No'. Used where consent is a precondition of any further processing, such as doing information searches on credit or health information, for finance and insurance respectively.

Opt out—Default answer is 'Yes', individual must specifically say 'No'. Used where a very high proportion of people would choose the option, which usually implies that it is very reasonable relative to the purpose.

Opt in—The default answer is 'No', individual must specifically say 'Yes'. Used where many, if not most, people would NOT choose the option, especially where there is extra sensitivity, or extra cost.

In contrast, 'Notice required!' refers to instances where individuals need only be *informed of their rights*. This approach applies especially to medical or law enforcement agencies, which assume a certain compulsion to obtaining the data. It may also be used by lenders in certain instances, where it is infeasible to request consent, and the action to be performed is reasonable in the circumstances.

Consent

Most privacy barriers to data collection can be circumvented by obtaining the consent of the data subject. Application forms will have clauses advising that: (i) a bureau enquiry will be made when assessing the application; and (ii) subsequent performance on the account will be advised to the credit bureau(x). These are usually provided as compulsory-consent clauses, but some companies may provide them as opt-out clauses. There are, however, cases where there is greater compulsion to collecting the data, and the applicant need only be given notice of:

- If provided personally by the individual, is provision compulsory or voluntary, and if compulsory, what are the consequences if refused?
- Why is it being collected, and who will use it?
- Whether the data obtained can be contested, and how?

This approach applies primarily to medical or law enforcement agencies, but may also apply to lenders that are gathering extra information, because they have been forced to take legal action to recover funds.

Two cases that demand higher standards before data can be obtained are: (i) sensitive information, and (ii) investigative reports. Sensitive information includes racial or ethnic origin, political opinions, religious or other beliefs, trade union membership, physical or mental health, sexual orientation, and criminal record or any record of court proceedings. This is meant to protect against potential unfair discrimination.

Investigative reporting can also be especially intrusive, and the quality of data suspect. Information can be obtained from gossips, spies, barkeeps, and newspaper clippings on political party allegiances, marital infidelities, and failure to pay parking tickets, amongst others. It can be defined more formally by paraphrasing the American FCRA, where 'investigative' is associated with 'information on a person's character, general reputation, personal characteristics, or mode of living obtained through personal interviews with neighbours, friends, associates, or others with such knowledge'. Investigative reporting was curtailed by the FCRA, which provided a major boost for credit scoring and automated decision-making in the United States.

Personal identifiers

Countries that have a personal identifier usually allow it to be used by all government, financial, medical, and other service organisations. This by itself can present special risks with respect to identity theft, and in the United States the Personal Information Privacy Act (1997) had the primary purpose of protecting against misuse of the Social Security Number. Several other countries do not have a personal identifier, including the United Kingdom and Australia. There is some indication that the United Kingdom may adopt their National Insurance number, but this is yet to be seen (Wilkinson 2003). In Australia, there are further restrictions built into their Privacy Act (1988), which specifically prohibits the use of personal identifiers provided by external companies.

33.2.2 Reasonable data

> Institutions are required to have, and follow, internal credit policies to discipline the credit investigation and granting process. Such policies would not be served by considering extraneous information of uncertain relevance taken in another context.
>
> *Owens and Lyons* (1998)

Legislation also demands that the personal data being processed be reasonable, which may be answered by asking the following three questions: (i) Is the data really needed to meet the business objective? (ii) Is it what is required, and no more? And (iii) Can it be justified to the individual concerned? This will of course vary, depending upon the industry and the purposes for which it is used. For credit providers, this includes marketing, credit, and ongoing account management.

The first two questions are often put in terms of 'relevant', and 'not excessive', while the latter could be called 'justified'. To expand further, the data is relevant if it provides value in the process, and is logical. Shoe size may be relevant to a cobbler, but not to a lender. Many pieces of information may have real or spurious correlations with credit risk, attrition, or response, but not be appropriate.

It is not excessive if the amount being collected is no more than might be reasonably expected for the specified purpose. Lenders do not wish to put together extensive files of personal information that would rival the old East German Stasi or Soviet KGB. Information held about other individuals, whether close relatives, neighbours, or business associates would be considered excessive for most commercial transactions (spouse and guarantors excluded).

And finally, it is justified if lenders can hold their heads up high and tell customers why it is required. This is most relevant for sensitive personal information, where stricter rules apply. Even then however, asking for consent can usually circumvent any restrictions. This applies especially to the life insurance industry, where information on health status is highly relevant. It is less relevant for credit scoring, and according to Thomas et al. (2002), a lot of information that may be predictive, such as health status or driving convictions, is not used, because it may be a political hot potato.

33.2.3 Data quality

Any organisation that relies upon data has an obligation not only to itself, but also to the people that it serves, to ensure the quality of information being used. Data should simultaneously be: (i) adequate—sufficient for the purposes, not incomplete; (ii) accurate—correct and not misleading; and (iii) recent—represents the current situation as closely as possible, is kept up-to-date, and is not held for longer than necessary. These concepts were already covered in Chapter 11 (Data Consideration and Design), but are treated again here, in order to touch on the legal requirements.

Adequate

'Adequacy' is a difficult concept to come to terms with, because the legislation provides little indication of what it means. It is effectively the counterpoint to 'not excessive'. If data is used to make a decision that affects an individual, then the amount and type of data should be sufficient to make an informed decision. Otherwise, the interests of neither individual nor decision-maker are being served.

Accurate

Data should provide a proper representation of the borrower's situation. Most people are familiar with the horror stories associated with blatant data inaccuracies, especially in an age where computers have multifaceted impacts on people's day-to-day lives and society in general. The most horrifying are when computer glitches cause petty bureaucrats to tell poor unfortunates that they are dead.[3] In the credit world, people have been known to be recorded as dead, but worse yet, they may have judgments or other transgressions falsely recorded against their names.

---

[3] This might itself cause a heart attack or job-loss, because the boss objects to having dead people around the office, especially if he already has enough problems with the living.

The bureaux rely upon matching routines, using names, addresses, and other personal identifiers to match account performance, judgments, and other information to specific individuals. The process does not always work though, and at times the match is with records of relatives, or unknown people with a similar name. Another possibility is that the match is correct, but the information is incorrectly recorded. An example is delinquent accounts that have been brought up-to-date, but the adverse status has not been lifted. It is the responsibility of both lender and bureau to ensure that these statuses are correct.

Some credit bureaux have invested heavily in data enhancement capabilities, which have generated surprising improvements in their 'correct' match rates, and hence their risk assessment capabilities. According to Taylor (2004), when lenders upgraded to Equifax's compliant processing system, they experienced an average improvement of 15 per cent over recently developed non-compliant systems. It also significantly reduced the number of complaints to Equifax, as previously, two-thirds of them related to third-party data being included in individuals' credit records.

Recent

Data has a shelf life, much like farm produce, making it necessary to keep it fresh. There are usually no guidelines governing refresh rates, other than some broad statement that data should be recent. In contrast, limits on data-retention periods are often legislated, especially for negative and adverse credit information.[4] This will usually be 5 or 10 years, but for some industries or countries the periods may be much shorter.

The retention period may be shortened: (i) for public-relations purposes; (ii) to reduce the storage cost; or (iii) because the information is perceived to provide little value after the shorter period. In South Africa, the legislated limit was five years, but rapid economic and political change and public pressure after 1994 caused the bureaux to shorten the holding period to three years, and in 2006 the National Credit Act reduced the limit to one year for judgmentally set adverse statuses, like 'slow payer'.

33.2.4 Use of data

The depth and breadth of data available to organisations is growing exponentially, and with it its possible uses, especially where it is available across different but related industries. Both marketers and customers benefit from data-driven target marketing, as it allows for both more appropriate offers, and reduced costs. The problem is that people are becoming frustrated with the deluge of spam clogging their phone, voicemail, email, snail mail, SMS, and other personal

[4] Some unscrupulous credit providers have been known to reload an old judgment under a new date so that the consumer is penalised for longer periods.

communication lines. Many customers would rather not to be contacted for anything unrelated to the product being requested.

A contentious issue is the use of credit information for underwriting automotive and household insurance. The February 2003 issue of 'In Brief', published by the Texas Senate Research Center in Houston, highlighted increasing concerns and public complaints regarding the inappropriate use of credit information for insurance scoring. Most insurers use generic FICO bureau scores, and only a few insurers had developed bespoke models.

As a result, data usage is 'limited to the purposes for which it was obtained, or a directly related purpose'.[5] According to Owens and Lyons (1998), the most sensitive personal data is that relating to health and finance, in particular the use of: (i) health information in credit decisions; and (ii) any personal information for direct marketing.[6] The purposes for which personal data is to be used should be stated at or before time of collection. If companies wish to use it for unrelated purposes, the individual must be informed, whether by obtaining consent or providing notice. Notice may be sufficient where it is not practical to obtain consent, but the requirements differ, depending upon whether the data will be used by the company for its own marketing, or disclosed to a third party:

Own behalf—Allow individuals opportunities to opt out of future campaigns with every marketing contact, and advise of the procedure and contact details each time.[7]
Third party—Advise the individual prior to disclosure or processing for the first time, and provide the opportunity to contest at no charge.

33.2.5 Disclosure of information

Privacy legislation often treats 'use' and 'disclosure' under a single heading, but the two are really quite different. 'Use' is by the entity collecting the data and 'disclosure' is to third parties. Disclosure should only occur either with the consent of the individual, or at the request of some legal authority. The former does, however, have an impact on lenders' ability to function. Information sharing has become crucial for companies trying to do credit risk, fraud, and other assessments, but legal restrictions may impose unreasonable barriers.

Information sharing

Application forms usually have consent clauses, relating to payment profile searches and marketing contacts. In countries like the United States, Canada, England, and South Africa, it

[5] It was this aspect of the UK Data Protection Act that caused problems with the use of voters' roll data for credit decisions (Wilkinson 2003, see also Chapter 12).

[6] The research quoted by Owens and Lyons indicates that data privacy complaints, and general concerns by the Canadian public are relatively few, with the exception of where it affects them directly, whether at home or in the workplace.

[7] The implementation and management of this opt-out facility may pose a problem (as can be attested by anybody who has tried to get himself removed from countless email lists).

is possible for organisations to share account performance information (Chapter 14, Information Sharing), but only if the customer gives consent. There may thus be consents for: (i) searches to be done at the credit bureaux; and (ii) the provision of performance details. The most common treatment is for lenders to request compulsory consents; if customer refuses, then the application proceeds no further. If, on the other hand, the law were to demand that individuals be able to opt out, it then becomes a challenge to: (i) do costly manual risk assessments on a small number of accounts; and (ii) to modify the infrastructure, so that payment profiles for these customers are not shared.

33.2.6 Subjects' rights

As stated at the start of Section 35.4, there has been much public concern about what information is held by organisations, first public, and then private. Although, in law, information usually belongs to the entities that collect it, it is not theirs to do with as they please. There are certain responsibilities that go along with maintaining these databases:

Openness—Transparency with the general public regarding the type and purpose of personal information processed by the organisation.
Query, access, and contest—Individuals' rights to determine whether personal information is held about them, what information is held, and have it corrected in need.

Legislation usually presents these as two separate principles, yet they are directly related—openness is a prerequisite for the rest.

Openness

There is an automatic tendency to associate the expression 'freedom of information' with allowing individuals access to data held about them, but this is not the case. It instead refers to the *general public's ability to enquire about the activities of public and private organisations*, especially as they may impact on them as individuals, society, and the environment. In the context of data privacy, the general public has a right to know about any practices and policies relating to personal data held by organisations: 'Is any personal information held?' 'What type of information?' 'How and where was it obtained?' 'What will is it used for?' 'To whom may it be disclosed?' 'Who and where is the data controller?' Any entity that processes or holds personal data has a responsibility to make this information readily available—meaning accessible, without unreasonable effort or cost, in an understandable form. Some countries also demand that they register with the local data privacy commissioner. A prime example is the United Kingdom, which requires separate registration for each of the purposes for which data will be used.

Thomas et al. (2002) highlight a key cultural difference between the mentalities in the United States and the United Kingdom. In the United States, information will be made available unless there is good reason to restrict it. In the United Kingdom, and other countries, the information will be protected unless there is good reason to make it available.

While this comment is valid, it is based solely upon the names of the legislation in the two countries: the United Kingdom's Data Protection Act (DPA of 1988) that applies to both public and private institutions, and the USA's Freedom of Information Act (FOIA of 1967) that applies to government agencies. The mistaken assumption is that 'freedom of information' applies to personal information, as the primary purpose of the FOIA is to demand public access to information regarding the operations of government agencies, not the personal information they hold. It was the later Privacy Act (1974) that allowed individuals access to information pertaining to them held by American government agencies.

Query, access, and contest

Legislation also usually provides individuals with the rights of query, access, and contest: (i) query whether personal data about them is being held; (ii) have access to it; and (iii) contest it if they believe it is incorrect. These facilities must either be provided free, or at a reasonable charge, and the details must be provided in an intelligible form. This right is not unfettered though; restrictions may apply, and if so the organisation should provide specific reasons to the applicant in writing, and still allow as much access as is reasonable in that instance.

When individuals contest their records, data controllers have a responsibility to correct, delete, or ensure that the individuals' requests are attached. The latter can be difficult, as these occurrences are rare and the reasons vary greatly. As a result, lenders and credit bureaux often make no allowance for it. If the data is successfully contested, the individual should be notified, along with any individuals to whom the data was disclosed over the preceding months.

Many companies, in particular credit bureaux, will charge the public to obtain access to their own information, or to query the data held. The charges may be justified in terms of the infrastructure and manpower required, but there is still an obligation to ensure the service is provided as cheaply as possible.[8] The regulatory authority may set limits, such as one free enquiry per year, or a refund of charges if data is successfully contested.

33.2.7 Data security

Any entity that collects and holds information is holding it in trust. This implies that there are not only responsibilities in terms of proper use and disclosure, but also in terms of ensuring that individuals' rights are protected.

Controlled access

One of the greatest fears is unauthorised access to personal information. Third parties may use illegal means to access data, which may involve the complicity of employees within a credit

[8] Keeping this charge as low as possible is a matter of ethics based upon the presumption that personal information belongs to the individual, even though no property rights apply.

bureau, bank, credit card issuer, finance house, or other credit provider. The data may be used: (i) as a form of industrial espionage; (ii) to defraud the business and/or customers; or (iii) to gain intelligence to be used for purposes other than credit. There are two levels of security that are affected: (i) security around the holding of data, that involves 'lock and key' and password controls; and (ii) security around its disposal, that may require that paper documents containing certain types of personal information be shredded.

Transborder data flows

The OECD's 1980 guidelines were meant to aid the free flow of information across national borders, whether to related or unrelated companies, but it was not their intention that these flows would be unfettered. Companies sending information must ensure that individuals' privacy rights are protected, which implies that the same, or higher, standards will be applied in the other country. This can be difficult where practices vary greatly, and as such, there is a need to standardise the legislation.

33.3 Summary

Modern economies are being driven by the availability of information. In credit this relates to information that can be used in credit assessments, but which may also be used for other purposes. This falls under the heading of data privacy, where certain distinctions should be made. First, of most concern is personal information (natural persons), especially that relating to health or finances, as opposed to information relating to enterprises (juristic persons). Second, in credit there is a power imbalance, because lenders require individuals' information to make loan decisions. In truth, personal information belongs to the individual to whom it pertains, but possession is 9/10ths of the law. As a result, data controllers have fiduciary duty to use it responsibly, and protect it.

Banks are held to even higher standards. For many years, the 1924 Tournier case provided legal precedent, stating that banks have a 'duty of secrecy', and that personal data could only be divulged where there was: (i) compulsion by law; (ii) a duty to the public; (iii) in the interests of the bank; or (iv) with customer consent. In 1970, the USA implemented its *Fair Credit Reporting Act*, which is sectoral legislation governing the credit bureaux. In contrast, Europe adopted a broad-based approach in the OECD's 1980 *Data Privacy Guidelines*, the Council of Europe's 1985 *Data Privacy Convention*, and the European Union's 1995 *Data Protection Directive*, all of which have been assisting standardisation of legislation across various countries, and transborder data flows.

Regulations governing data privacy and protection typically cover data's: *manner of collection*—lawful, fair, and free of deception; *reasonableness*—relevant and not excessive; *quality*—accurate, adequate, and recent; *use*—limited to the purposes for which it was obtained; *disclosure*—maintenance of a duty of secrecy; *individuals' right to know*—openness about the activities of organisations, especially regarding what data is held and why, and individuals' right to query, access, and contest their own personal data; and *security*—both in terms of physical security, and ensuring similar standards are upheld if data is transmitted across borders. Much of this was presented in the OECD guidelines, which set out a number of principles covering data collection, data quality, purpose specification, use limitation, security safeguards, openness, individual participation, and accountability.

In all instances, there are cases where exceptions are made. In particular, lenders have greater latitude where the data is being provided with the consent of the individual concerned, or it is reasonable relative to the purpose for which the data was originally collected. For credit, consent is usually compulsory, or opt-out, and for marketing, opt-in. Different standards are applied to marketing, depending on whether the data is for own use or aiding third parties. Lenders are compelled to disclose personal data where it is required by law, or is in the public or national interest. At the same time, although lenders are required to be open with the public, they are entitled to refuse frivolous requests. Standards for data collection are higher where sensitive information is involved, or it is obtained via investigative reporting.

34 Anti-discrimination

People are warned against the use of generalisations, yet this is one of the most fundamental tools used by humans, to compensate for a lack of information in new situations. Some generalisations are pre-programmed as basic instincts, and may cause either repulsion or attraction in a given situation. Others are learnt as a part of people's cultural upbringing; something is believed to be true just because family, friends, or government have said so. And finally, generalisations based upon people's own limited past experiences may be used in certain situations.

The problem comes when people believe that their generalisations are more valid than any new information. Bigotry is the extreme case, but even normal people operating in everyday circumstances have biases that cause them to discriminate unfairly against others. It might have been borderline acceptable in 1950, but not in the early twenty-first century, after one of the greatest global cultural changes was the realisation that people have more similarities than differences. Today there is a huge focus upon equality in the workplace, the schoolyard, the sports field . . . and the provision of credit. As a result, many countries have implemented anti-discrimination legislation, also referred to as 'fair access' or 'equal opportunity' legislation. The credit industry may be covered by specific legislation, or by general legislation, presented under one of three banners:

(prohibition of) Unfair discrimination—To prejudice, based upon knowledge of a group membership, defined by colour, culture, gender, religion, nationality, or some other personal characteristic.

(promotion of) Equal opportunity—To make available, or offer to all, without discrimination.

(protection of) Human rights—To ensure basic freedoms, facilities, and protection.

The key phrase here is 'unfair discrimination', but the definition is unclear. All human decisions are based upon discrimination, whether subjective or objective, but poorly founded subjective decisions can result in unfair discrimination.

34.1 Discrimination—what does it mean?

Before carrying on further, the meaning of 'discriminate' as it is used in legislation should first be clarified:

USA: Equal Credit Opportunity Act (1974)—'It shall be unlawful for any creditor to discriminate against any applicant, with respect to any aspect of a credit transaction—on the basis of race, color, religion, national origin, sex or marital status, or age (provided the applicant has the capacity to contract).'

UK: Human Rights Act (1998)—'The enjoyment of the rights and freedoms . . . shall be secured without discrimination on any ground such as sex, race, colour, language, religion, political or other opinion, national or social origin, association with a national minority, property, birth or other status.'

Does 'discriminate' refer to 'unfair treatment' or 'ability to differentiate'? The legislation's intention is to protect against the former, while credit scoring provides the latter. The question arises, 'Can an objective statistically derived model result in unfair discrimination?' There are several possible views.

View 1—Decisions based on credit scoring cannot be considered valid unless a causal link can be shown.

While this view is often expressed, it tends to be one used by the uninformed, and is not reflected anywhere in law. In general, the use of statistical methods that rely upon identifying correlations has become broadly accepted. The people most discriminated against are those with characteristics most highly correlated with poor payment performance.

View 2—Credit scoring discriminates unfairly, because many of the factors that lead to poor credit performance, like job insecurity, poor health, and unstable family lives, are more prevalent in minority communities. Where these are the result of historical imbalances, it only serves to perpetuate those imbalances.

Contrary to this belief, credit scoring's widespread adoption has increased the availability of credit to those previously excluded, as lenders exploited the confidence provided by the new tools to grow not just their portfolios, but the total credit market pie. Already in the 1970s, at about the time the US's Community Reinvestment Act was implemented, even minority community leaders had come to accept that: (i) the decisions are objective, and based upon past experiences of the credit provider that can be proven; and (ii) it allowed the credit providers to lend in areas where they might otherwise not lend at all. Even so, there were still concerns about unintentional consequences, and during the 1990s the attention of American regulators shifted from banks' disparate treatment of minorities, to the unintentional disparate impact of credit scoring on them (Barefoot 1997). This is something that is extremely difficult to prove, and by the early 2000s the focus had shifted away from the statistical models, to the personal prejudices that influence the override process.[1]

[1] Refer to the 1999 settlement between the Department of Justice and Deposit Guaranty Bank in the United States.

View 3—Any characteristic may be used as long as there is a business need, and it can be shown that its contribution is statistically significant, and there are no other characteristics to replace it.

This view is sometimes applied in practice, but care must be taken. Gender is a prime example of where an overriding business need may be evident. For insurance, it is generally accepted that women have a longer life expectancy, and may have lower motor vehicle accident rates, and thus qualify for lower life and automotive insurance premiums respectively. For credit however, there is little ready acceptance of gender differences in payment performance. In some subprime and third-world environments the distinction can be pronounced; women are more stable and better payers than men, and there is often little other credit related information to make up for it. The problem comes when decline reasons have to be given, 'Sorry sir, but one of the reasons you were turned town was because you are male'. The lender has to be prepared to substantiate the inclusion of this characteristic to the individual concerned, and the general public.

View 4—Credit scoring is an acceptable tool, as long as sensitive characteristics are not included in the model, such as race, religion, gender, and others. Individuals have little or no control over these traits, many of them defined at birth, and as such should not be penalised for them.

This is the view adopted by most legislation, or the interpretation thereof. Although most of the predictive power lies in consumers' personal credit histories, demographic characteristics could still indicate sub-groups where character and personal distress risks are greater, which could tip the scale where other negative factors are evident. Some of these characteristics are forbidden by statute, but there are often areas where their use is disputed. When in doubt, a general rule is that: (i) they should be avoided, and instead be replaced with others that capture consumers' personal behaviour; and/or (ii) they may be used, as long as the weightings are small within a broader risk assessment.[2]

> For some key groups, there will be pressure to increase credit availability. Ideally, if a group gets negative points, the answer is not to remove the characteristic(s), but instead to adjust the strategies for that group. Possibilities include lower score cut-offs, increased marketing, improved customer education, and so on. For many groups however, this is not legally possible, as the defining characteristic may not be used.

View 4 is the one most commonly held in first-world environments, where there is a wealth of credit-related information. The combination of legislation and public pressure have put the onus on lenders to find other characteristics that better represent the risk, or at least represent it using more acceptable characteristics. Demographic details have become less crucial as scoring

[2] One possible way of lessening the impact of sensitive characteristics is to stage them in, after all other possible characteristics have been considered.

Table 34.1. Unfair discrimination legislation

Country	Name
USA	Equal Credit Opportunity Act (1974/6)
Canada	Human Rights Act (1985)
United Kingdom	Sex Discrimination Act (1975)
	Race Relations Act (1976)
	Human Rights Act (1998)
South Africa	Protection of Equality and Prevention of Unfair Discrimination Act (2000)
Australia	Covered at Provincial Level

models have benefited from improved access to external data, and advances in information sharing. In contrast, view 3 still applies in some emerging environments, where borrowers' creditworthiness is more opaque. They do not have the same depth of credit histories—whether because the market is financially or technologically unsophisticated—and lenders will be severely prejudiced if anti-discrimination legislation is implemented and interpreted in the same strict fashion.

Anti-discrimination legislation provided a major boost to credit scoring (see Table 34.1). The first was the USA's Equal Credit Opportunity Act (1974), which is peculiar in that it is sectoral legislation specific to the credit reporting industry (it demanded that lenders be able to show what factors influenced a decision, which can be easily done with traditional models). In contrast, the UK initially implemented two general acts, covering discrimination on the basis of sex and race, albeit primarily relating to employment practices.[3] Their Human Rights Act (1998) later provided broader legislation, covering society as a whole. Further specifics for the UK and other countries are provided in Chapter 38 (National Differences).

34.2 Problematic characteristics

The primary characteristics that are considered *verboten* in credit decisions are race, religion, national origin, and sexual orientation. This does not mean that they cannot be asked—only that they cannot be used as part of the decision. Some are still required to keep track of customer demographics, and to report on business being done with disadvantaged groups. Even if the law did not specifically prohibit these characteristics, credit providers would still be loath to include them, because of the potential outcry if Joe and Jane Public were to find out. The treatment of many other characteristics will vary, depending upon the jurisdiction, and the type of scoring being done. These include age, gender, marital status, and others. The following are some brief notes on some commonly requested characteristics, whose use has been prohibited or restricted:

[3] Thomas et al. (2002) do, however, mention these in the context of credit decisions.

Gender—While many concerns regarding sexual discrimination relate to the treatment of women, in credit scoring, the legislation may have a negative impact on those it is meant to protect. In many environments, women are more responsible payers.

Marital status—May be prohibited, but it is more likely to be restricted and/or qualified. If included as a scored characteristic, the categories may be limited to married, unmarried, and separated, with no allowance for divorced or widowed. For joint accounts and 'community of property' marriages, bureau searches are required on the spouse.

Age—If not already stated in law, the credit provider should ensure that applicants in the 60+ category get the highest points possible; otherwise, it may be accused of discriminating against the elderly. The repayment term of the home loan may be adjusted (say from 30 to 20 years) to compensate, which may affect affordability.

Telephone numbers—There are almost no limitations on how contact details may be used in predictive modelling, with one exception. Legislation may prohibit lenders from prejudicing applicants whose phone numbers are unlisted, which is possible if lenders check for matches with telephone listings.

Private or government assistance—The value of extra income may not be reduced because of its source, including alimony, child support, welfare, or government grant (USA).

In some jurisdictions where there is no direct legislation, contentious characteristics may be used, with the proviso that they: (i) form part of a comprehensive credit risk assessment; (ii) be reasonable in the context of that assessment; and (iii) are justifiable based upon past dealings with similar applicants. At the same time, there is an expectation that lenders will access as much other information as possible, in order to reduce the importance of sensitive demographic characteristics normally associated with groups that are perceived to be disadvantaged.

Note here that there is a small conflict between anti-discrimination and fair-lending legislation. Ill health, marital strife, and job loss are the primary drivers of delinquency, and are more prevalent in some sectors of society than others. Banning the use of demographic characteristics may limit the lender's ability properly to assess borrowers' ability to weather life's upsets, and the resulting financial storms, which should be part and parcel of any comprehensive 'affordability' assessment.

34.3 Summary

At one time, lenders' decisions were entirely subjective, based upon personal knowledge of the customer. Decision makers can be the victims of their own biases though, and it is the borrowers—usually those that can least afford it—that suffer because of lenders' prejudices. As a result, most countries have implemented anti-discrimination legislation, under the heading of 'prohibition of *unfair discrimination*', 'promotion of *equal opportunity*', or 'protection of *human rights*'. These acts may be directed specifically at credit, or, more broadly, at employment and other practices.

Credit scoring allows lenders to discriminate between good and bad credit risks, which some observers see as problematic, because the causal factors cannot be shown. Even so, the public and legislators have accepted the use of predictive models that rely on correlations and not causation. Indeed, there is an acknowledgment that credit scoring facilitates objective decisions, and allows lenders to operate in areas where they might otherwise not lend at all. Some of the characteristics are still contentious, but may still be used if a business need can be shown. In other instances, they may be forbidden totally.

The primary characteristics that are forbidden are *race*, *religion*, *national origin*, and *sexual orientation*, while *gender*, *marital status*, and *age* are high on the list. In general, the consensus is that demographic characteristics, over which consumers have no control, should be replaced by those that they can change, in particular behaviours and statuses specific to each person, that do not include group memberships. There may, however, be instances where contentious characteristics can still be used, such as where: (i) there is a lack of other credit-related information; (ii) a sound business reason can be proved; (iii) the data is just one part of a broader assessment.

In general, such restrictions have compromised lenders' ability to assess credit risk, but have forced them to search for other data, relating to the behaviour of each individual, to replace it. The resulting models are better, but not as good as they could be if all data were used. There will always be pockets of society that are less able to handle life's upsets, and cannot be identified using the data currently available. Indeed, this situation creates a conflict between equal opportunity and fair-lending (affordability) legislation. Improvements are, however, still possible, especially if access to information on individuals' financial status (assets/liabilities, income/expense) is improved.

35 Fair lending

> It is only the poor who pay cash, and that not from virtue, but because they are refused credit.
> *Anatole France* (1844–1924) French Writer, in 'A Cynic's Breviary'.

The first laws governing lending related to the charging of interest, and over time a distinction was made between interest and usury. Indeed, many countries either had, or have, a Usury Act that sets the maximum rate of interest that may be charged and guidelines for its disclosure. Nowadays, there is a bevy of fair-lending legislation that goes much further. These are often referred to as Consumer Affairs, Unfair Business Practices,[1] or Consumer Credit Acts, which may cover marketing, account origination, account management, and collections practices. Such legislation is meant not only to protect consumers from borrowers, but also from themselves. In this domain, a distinction is made between three types of lending practices, something like 'illegal', 'barely legal', and 'legal', but not quite:

Predatory lending—Victimisation of borrowers, through deceptive practices highly-prejudicial loan terms, and lack of regard for their ability to repay. Usually associated with nefarious practices used by loan sharks and some subprime lenders.

Irresponsible lending—Negligent practices that mislead borrowers, or attract those prone to over-indebtedness. Usually associated with overzealous and poorly thought out strategies.

Responsible lending—Acceptable practices that ensure borrowers can afford the repayments and know the consequences, and still try to accommodate as many people as possible.

Predatory lending has been the focus of most legislation in the past. This is changing though, as modern technology has allowed lenders to be ever more aggressive and prone to irresponsible lending. Legislation has adapted by becoming stricter: Australia has required affordability checks for some years; the United States has some legislation at state level; the United Kingdom is motivating it at national level; and South Africa implemented sectoral legislation covering credit provision in 2006/2007.

[1] Debt collections practices often have a separate act to govern the types of methods used, including the times at which the debtor can be contacted, provisions for the debtor to break contract, etc. Outsourced collection may receive different treatment.

35.1 Predatory lending

For the love of money is the root of all evil.
2 Timothy 6:10

The rich ruleth over the poor, and the borrower is servant to the lender.
Proverbs 22:7

The most negative image of a lender is the loan shark, who cares nothing about anything but exacting exorbitant returns from a poor public. It exists alongside a number of quasi-acceptable lending practices that are lumped under the generic heading of predatory lending, an appropriate name given that sharks are the most efficient predators. It is usually associated with subprime lending—but some subprime lending may be quite legitimate.

In some environments, seemingly exorbitant interest rates are acceptable, especially for loans of small amounts with short repayment terms. This applies to subprime/emerging markets with much higher than average risks, where people cannot gain access to credit under normal terms. These lenders have also developed practices peculiar to their environments, which at times may be/seem unethical, for securing repayments.

For the purposes of this text, predatory lending practices are those that victimise borrowers for the personal gain of the lender. It includes fraudulent, deceptive, discriminatory, or highly unfavourable lending practices that are either illegal or not in borrowers' best interests. The key is whether or not the lending is appropriate for the environment, and acceptable within that society. There are several signs of predatory lending, which include:

(i) Interest rates far in excess of normal usury limits, usually triple-digit when annualised.
(ii) No consideration of borrowers' ability to repay.
(iii) Upfront loading of credit insurance, to cover death, disability, or job loss.
(iv) Administration and other fees that are excessive relative to the loan amount.
(v) Prepayment penalties that inhibit early repayment.
(vi) Steering borrowers into loans of higher cost than necessary.
(vii) Mandatory arbitration clauses that inhibit legal recourse against the lender.
(viii) Compulsory refinancing, at regular intervals with further fees.

These practices are deemed undesirable, but may be part and parcel of doing business in some sectors. The choice for legislators is between banning the practices outright, and setting guidelines for their use (the latter being preferred).

Some subprime lenders take advantage of customer ignorance by not offering the best possible rate, in spite of a favourable risk assessment. Many people, especially minorities and the financially unsophisticated, are price insensitive. They often believe that their credit ratings are worse than they actually are, need the money, and will not question the rates offered. Lenders may also neglect to report positive payment performance to the credit bureaux, in spite of this facility being available, in order to prevent good payers from building up a credit history that will allow them to escape the rip-off market.

35.2 Irresponsible lending

Once, bankers could be compared to doctors in the way they attempted to provide finances to a customer in his or her best interest. Now, they are more like bartenders, knowingly serving alcohol to people that are already drunk.

Antony Elliot, executive of the Centre for Financial Innovation, England.[2]

Credit scoring is backward-looking; it bases its decisions upon the performance of past applicants, and for the most part, prejudices future applicants that already show signs of stress at time of application. But what about those where the cracks have not yet appeared? What about people who are applying for a loan without fully considering the effect it will have upon their disposable income and lifestyle? Many may repay their loans, but the extra commitments can cause significant social stresses, with severe consequences for some—anxiety, relationship problems, mental health problems, etc.

The same effects could result from cash purchases, but cash retailers are not held accountable if customers' money could be better spent elsewhere. Credit providers are held to a higher standard, because they are making claims against future income, not current or past income, where the money is at hand.

Lenders' failure to consider the possible effects of their actions upon individuals is called 'irresponsible lending'. It is also recognised that many individuals are prone to 'irresponsible borrowing', meaning to indebt oneself with little consideration for one's own ability to repay. In the early 2000s, the UK Department of Trade and Industry published reports on over-indebtedness, its causes and remedies (DTIUK 2001/3/3a, Kempson 2002). The conclusions are based upon an attitudinal survey done in 2002, and the results are summarised in Tables 35.1 and 35.2. Some of the irresponsible lending practices blamed for fuelling over-indebtedness were:

[2] Sean Poulter, 'High Street Loan Sharks', p. 4, Daily Mail (England), 9 May, 2005.

Table 35.1 UK survey results.

Definition	% of households
Credit facility	75
Outstanding commitments	47
Perceived financial difficulties[3]	20
. . . from over-commitment[4]	10
Arrears on at least 1 account[5]	13
Heavy credit users	7+

- Loan agreements, which are unclear regarding terms and conditions.
- Lack of credit checks and affordability assessments, prior to increasing limits on credit cards, overdrafts, and personal loans.
- Offering pre-approved loans.
- Motivating transfer of debt, by offering higher limits and lower rates.
- Reducing minimum payments.
- Unsolicited issuing of cheques, which can draw on credit card accounts, with poor or no explanation of how interest and fees accrue.

The survey was done by Market & Opinion Research International (MORI) between March and May 2002 and included 1,674 respondents from across Britain. A further 189 people aged 18–24 were surveyed separately. The report makes no mention of how the 'householders' were selected for the survey.

The DTIUK 2003 report concluded that there were three indicators of heavy credit use, which were likely to lead to financial difficulties:

- Four or more current credit commitments, excluding mortgages and unutilised credit facilities (at greatest risk are cases with six or more commitments of over £500).
- Spending 25 per cent or more of gross income on consumer credit commitments.
- Spending 50 per cent or more on consumer credit and mortgage commitments.

Such indebtedness levels may seem acceptable, but as can be seen in Table 35.2,[6] they were associated with very high levels of arrears. There is thus an obligation upon lenders to take extra care when making a loan, to ensure that the borrower: (i) fully understands the consequences of entering into the commitment; and (ii) can afford the repayments.

[3] Kempson (2002:39).

[4] Ibid, p. 51. The DTIUK (2003) report indicates however, that the principal causes of financial difficulties are job loss and relationship breakdown.

[5] Many respondents were reluctant to admit to financial difficulties, even though they were in arrears on three or more commitments.

[6] DTIUK (2003:12).

Table 35.2. UK over-indebtedness

Definition	% of households	% in arrears
4 or more credit commitments	7	50
25% consumer credit	5	50
50% consumer credit+mortgage	6	40

Many lenders dispute the practicality and benefit of affordability assessments, but consumer credit legislation will probably demand them, where it does not already do so. There may be broad guidelines, which force lenders to apply their minds to the problem, or specific legislation that requires lenders to change their processes and reporting. Its effectiveness will depend upon effective information sharing between lenders, and customers' willingness and ability to provide reliable information about their financial situation. Indeed, for the latter, applicants must certify that, to the best of their knowledge, the information provided is accurate, to ensure that lenders avoid accusations of negligence.

35.3 Responsible lending

. . . while we clearly have to *address the real abuses* that exist, . . . we also need to preserve and encourage to the greatest extent possible, consumer access to credit, meaningful consumer choice, and competition amongst responsible lenders in the provision of financial services *to low and moderate income families.*

John D. Hawke Jr.,
US Comptroller of Currency, at US House
Committee on Banking and Financial Service, 24 May, 2000.

While predatory and irresponsible lending are undesirable, there is little guidance about what should be done—or so says the European Parliamentary Financial Services Forum (EPFSF 2003). According to them, the lender has no crystal ball of what the future holds for the borrower, and the available information may provide an incomplete picture—especially for irresponsible borrowers, who do not divulge full details. The best most lenders can do is lend in good faith. The EPFSF report argues that there are several criteria that drive responsible lending:

Due diligence—Consider creditworthiness and ability to repay.
Financial inclusion—Try to accommodate those that are less creditworthy.
Transparency—Ensure that details are clear, and not misleading.
Customer education—Ensure that the customer fully understands.

These criteria will apply at every step of the risk management cycle. Best practice recommendations for different stages in the risk management cycle include:

Solicitation—Make sure advertising/targeting is appropriate, and that marketing materials are transparent, without small print and jargon.
Acquisition—Apply the best decision tools available to internal and external data, to assess the borrower's ability to repay and guard against possible over-indebtedness, and if declined, ensure the applicant understands why.[7]
Management—Assist borrowers through customer education throughout the life of the credit agreement, while also trying to identify over-indebtedness at the earliest opportunity, using the credit bureaux as necessary.
Collections—Treat borrowers fairly and sympathetically, and advise them of collection methods that may be used, and options if they experience financial difficulties; while at the same time supporting and promoting free money advice services that can assist them.

Most of these are also in the DTIUK (2003) report, with the further recommendation that lenders make the greatest possible use of payment-profile (shared performance) data to assess potential over-indebtedness. This goes beyond the strict use of application and bureau scores, to consider further the required repayments relative to the applicant's income, and how much will be left over to feed the family.

Harbin (2004) states that Experian UK developed a Consumer Indebtedness Index (CII), using existing bureau data (limit utilisation, credit activity, debt profile, and a post-code aggregate). While effective, it has been criticised because it does not include income data specific to the individual. Work is currently underway to develop an estimated disposable-income figure and an affordability index. Harbin S. (2004). *Credit Risk International*, March 2004, pp. 21–23.

35.4 Summary

For years, the primary legislation governing lenders was usury legislation that set maximum interest rates, but legislation has grown over time to cover lending practices. A distinction is made between predatory lending (illegal), irresponsible lending (barely legal), and responsible lending (legal). Predatory lending is the most undesirable, and is the primary target of legislation.The greatest problem area is subprime lending, where loan sharks and others have a reputation of taking advantage of the less fortunate.

In contrast, irresponsible lending may not be illegal, but is unethical, in that it takes little consideration of the customer, whether in terms of affordability or personal circumstances.

[7] The robber-baron approach to pricing is that businesses should 'charge what the market will bear'. For responsible lending, the view is instead to ensure that the charge is sufficient to provide a fair risk-adjusted return.

Even reputable banks are subject to the temptation of reckless-lending practices, which include: poor credit checks and affordability assessments (especially for limit increases); offers of pre-approved loans; offers of higher limits and lower rates to attract business; reducing minimum payments; and having credit agreements with vague conditions.

The ideal is for lenders to engage in responsible lending, and legislation has been implemented in some countries to promote it. The criteria include ensuring: (i) creditworthiness and ability to repay; (ii) attempts to accommodate those that are less creditworthy; (iii) that information presented to the client is clear, and not misleading; and (iv) that the customer understands the commitment. Requirements for responsible practices extend throughout the credit risk management cycle (CRMC), including solicitation, acquisition, management, and collections, each of which has its own particular features, where lenders are expected to operate in an ethical manner.

Affordability assessments are a key component of responsible lending, but can be difficult. Lenders are reliant upon information obtained from their customers, credit bureaux, and own systems—which may not provide a proper representation: (i) borrowers may not understand, or misrepresent their own financial situation; and (ii) the record of credit obligations may be incomplete, either because of matching problems, or because the data is unavailable. Most lenders have found that affordability plays much less of a role in their assessments than past payment performance, but that is a purely empirical view, that ignores the potential social stresses that can be created by over-indebtedness, especially amongst those least able to cope with life's upsets.

36 Capital adequacy

> Bungee jumpers, skydivers and Indy 500 drivers are in the business of taking risk. Bankers are in the business of managing risk.
>
> *John D. Hawke Jr.*, US Comptroller of the Currency, 1999.

Anybody with a knowledge of economic history will be familiar with runs on the bank, which have caused many unnecessary bank failures during depressions and times of economic crisis. A rumour of financial problems at a bank can cause its depositors to storm its doors, waving their savings books and demanding their money. What may have been a minor problem becomes major, because cash in the vault is limited, and many other assets illiquid. In a world where a sound banking system is a major mainstay of both economy and society, such runs are unhealthy.

Capital adequacy is a measure of banks' financial strength and ability to absorb such shocks. It is usually stated as the ratio of equity to assets,[1] which lenders hope to keep as low as reasonably possible, because equity is more expensive than debt and higher capital requirements demand a higher return on assets. Regulators' demands for minimum requirements have a history of their own.

Capital adequacy—a historical overview

Runs on the bank occur sporadically and are difficult to predict, but are usually associated with depressions and recessions:[2] Worldwide—1893–1899 and 1929–1937; USA—1839–1843, in the fledgling American state banks and some local governments; 1873–1879 after a crisis of confidence following cronyism during the railway boom; and 1980s, with the savings and loan crisis; Latin America—1820s, 1870s, and 1980s; Southeast Asia—1997/1998. Europe's banking systems have not suffered to the same extent, as they were dominated by banks that were either state-run, or very large and well capitalised.

The amount of capital used to fund banks has declined over time. According to Berger et al. (1995), US banks had a 50/50 debt to equity mix in 1840, but this was already declining steadily when the National Banking Act was implemented in 1863, and continued southward as various factors combined to reduce risk, including 'improved geographical diversification, development of regional and national money markets; and introduction of clearing houses and other mutual guarantee associations'. The Act required a capital ratio of 10 per cent, whereas banks at the time had an average capital ratio of about 40 per cent. According to Burhouse et al. (2003), some reference was also made to the number of individuals in a bank's service area.

---

[1] The capital-adequacy ratio is a gearing ratio restated: Assets = Debt + Equity, Gearing = Debt/Assets, Adequacy = Equity/Assets, Gearing + Adequacy = 100 per cent. For financial institutions, 10 per cent capital adequacy makes more sense than 90 per cent gearing.

[2] Sylla (2001), and various Internet sources.

The founding of the Federal Reserve system in 1914 also assisted, by allowing banks to discount assets with the Reserve to meet liquidity needs, rather then selling them at distressed prices. By 1940, capital ratios had settled in the 6 to 8 per cent range, and remained there into the 1980s. According to Ong (1999:5), similar patterns occurred in Europe and elsewhere. In the United States, runs on the bank stopped with the creation of the Federal Deposit Insurance Corporation (FDIC) in 1933. Other countries developed explicit or implied government guarantees at different times.[3]

These safety nets brought with them their own risks, in particular the moral hazard that without the scrutiny of depositors and creditors, banks would invest in riskier assets. The responsibility for controlling bank behaviour then shifted to regulatory agencies, which imposed capital requirements. The requirements not only protected banks against unexpected shocks, but also caused them to indulge in less risky behaviour (Allen and Gale 2003). Regulation has been problematic, due to problems defining what is adequate. During the 1930s and 1940s, some simple capital ratios were considered as possible measures, but up until the 1970s, these were deemed insufficient. Instead, supervisors set the required capital for each bank judgmentally.

Historically, the absolute minimum that has been considered adequate is about 4 per cent, but this is for large or sovereign first-world banks. Burhouse et al. (2003) states that in the United States, for the years from the Second World War to the 1970s, the ratios ranged from 5 to 8 per cent, but this was an era of stability. The onset of stagflation, and the failures of the Franklin National Bank in 1974 and First Pennsylvania Bank in 1980, indicated that even large banks were not invulnerable. This has been exacerbated by continuing trends for banks to take on riskier loans. In 1981, the federal banking agencies implemented regulatory minimum capital requirements of between 5 and 6 per cent. American regulators also looked to France, United Kingdom, and West Germany, who had implemented risk-adjusted capital standards in 1979, 1980, and 1985 respectively. The Federal Reserve Board followed by issuing guidelines for the calculation of risk-adjusted capital ratios.[4] Even so, there are still concerns, as consolidation since the 1980s means more public money controlled by fewer banking hands.

Savings and Loans (S&L) are community-based organisations, similar to building societies and credit unions, whose goal is to amass local savings to finance home loans for members of the community. According to Chorafas (1990), during the 1970s and 1980s, S&Ls were ill prepared for the structural changes to the mortgage-finance industry. The advent of securitisation, and the availability of funds from pension funds and institutional investors, caused their returns to drop, while deregulation caused increased competition

[3] A 1996 Basel Committee survey stated that of 70 countries without deposit insurance, all but one were developing economies, that instead tend to have partial insurance for retail depositors. Goldstein (1997:fn28)

[4] The guidelines were similar to later Basel I requirements: cash and equivalents, 0 per cent; money market instruments, 30 per cent; mortgages, 60 per cent; other bank assets, 100 per cent.

for depositors' funds. The crisis was obvious in 1984, and was exacerbated further, during the mid- to late 1980s, by significant weakness in the oil patch, and in the agricultural sector in large parts of the United States. Significant retail loan losses were experienced, in particular for mobile homes in the southwest. When combined with mismanagement, corruption, and poor regulation, this led to massive S&L insolvencies. By 1988, of the 3,000 S&Ls, 1,000 were losing money, and half of these were insolvent.

The question should then immediately arise, 'What does this have to do with credit scoring?' Data privacy and unfair discrimination legislation were the first to force credit scoring upon lenders, but for banks, capital adequacy is taking it even further. The carrot is that, if banks can show that their internal ratings provide a meaningful differentiation of risk, they can be used to calculate what is hoped to be a substantially lower capital requirement—depending upon the risk of each bank's portfolio of course. The regulation being referred to is the Basel Accord, whose goal is to level the playing field between internationally active banks, by ensuring that they follow best practice with respect to capital adequacy. The following is split into three sections: (i) the original Basel Accord (Basel I) of 1988; (ii) the New Basel Accord (Basel II) of 2004; and (iii) the risk-weighted asset calculation.

36.1 Basel capital accord 1988 (Basel I)

The Basel Accord is the brainchild of the Bank for International Settlements, and was written by the specially appointed Basel Committee on Banking Supervision. Basel I was finalised in 1988 and banks were given until 1992 to comply. Over the next five years it was implemented in 12 countries, which were the G10 countries plus Luxembourg and Switzerland. The purpose was to level the playing field between, and improve the financial stability of, internationally active banks in the G10, by standardising and increasing reserves held for credit risk. Goldstein (1997) noted the practical effects that it had on increasing banks' capital reserves, and putting a focus on the riskiness of banks' assets, while Clementi (2000) commented that it went a long way towards addressing financial stability, by providing: (i) a framework for determining the riskiness of assets; (ii) a definition of capital-weighted assets; and (iii) a minimum capital-adequacy ratio. Clementi (2000) also noted that between 1990 and 1992, 'the percentage of U.S. banks that were well-capitalized increased from 86 per cent to 96 per cent, despite an economic recession and weak banking conditions'.

Basel I was very simplistic in its approach. It assigned risk-weights to four possible asset classes: 0 per cent—sovereign debt of OECD countries, or their central banks (S); 20 per cent—other banks and public sector institutions in OECD countries (B); 50 per cent—any loans secured by residential property (R); and 100 per cent—all other loans (O). Once applied to the total lending in each category, the sum provides the total risk-weighted assets (RWA):

Equation 36.1. Basel I $$\text{RWA} = (0\% \times S) + (20\% \times B) + (50\% \times R) + (100\% \times O)$$

The required capital ratio was set at a minimum of 8 per cent, with at least 4 per cent being tier-one capital ('core capital', meaning shareholders' equity and disclosed reserves), and the balance tier-two capital (undisclosed reserves and subordinated debt).

Equation 36.2. Minimum capital reserve requirement $$8\% < \frac{T_1 + T_2}{\mathrm{RWA}_{\mathrm{Credit}}}$$

In total, Basel I was implemented in over 100 countries. Already during the drafting of the 1988 accord, banks outside the G10 were making plans to comply, without any coercion from their regulators, as compliance was expected to bring with it: (i) higher credit ratings; and (ii) lower funding and other transaction costs when dealing with international banks. Regulators in most countries eventually required compliance however.

Some emerging market countries set higher requirements, either initially, or through a later increase. These include: 9 per cent—Columbia; 10 per cent—South Africa and UAE; 12 per cent—Singapore and Argentina. These higher percentages were to compensate for higher credit and market risks and lower reporting standards in developing countries. The level of 8 per cent was intended for the G10 industrialised countries only (Goldstein 1997). There is a concern that the higher requirements make emerging-market banks less competitive in international markets.

36.2 New Basel capital accord 2004 (Basel II)

As a first attempt, Basel I was not perfect, but it was a step in the right direction. While it did have the desired effect of increasing total capital reserves, the evidence as regards improving financial stability was not always clear-cut (van Roy 2003). Its two main faults were that it did not provide any further risk differentiation within the four broad asset categories, and it gave only limited recognition to the risk mitigating effects of security.

The Basel II process started with the publication of the first consultative paper (CP1) in June 1999, with CP2 following in January 2001, CP3 in April 2003, and final publication of Basel II in June 2004. Banks are expected to comply by end 2006 for the standardised and foundation IRB approaches, and by end 2007 for the advanced IRB approach, and the Advanced Measurement Approach for operational risk. The United States will probably be adopting the advanced approach one year later, but only for major banks, due to the Senate's concerns of how it will impact the competitiveness of its smaller banks.

As Figure 36.1 illustrates, there was a huge increase in scope between the 1988 and 2003 accords. Basel I focused on credit risk, while Basel II went further to cover credit, operational, and market risk, and what it calls the 'pillars' of (i) minimum capital requirements,

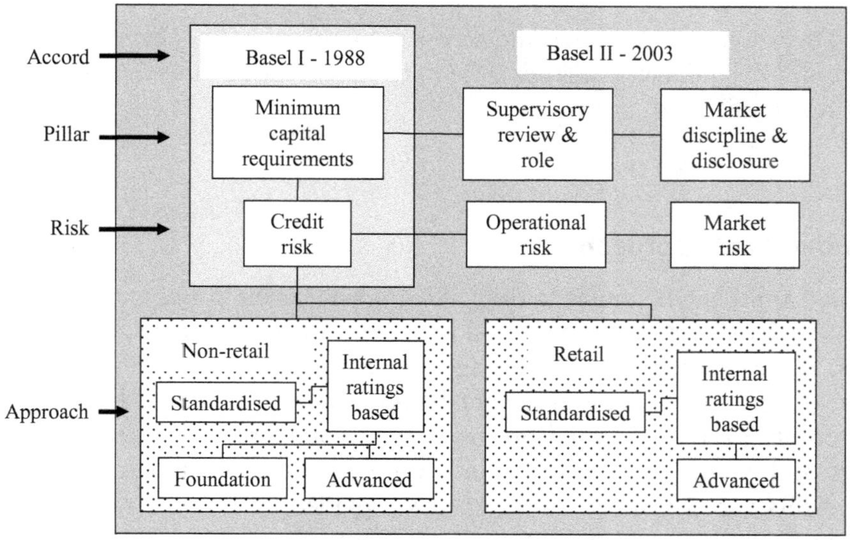

Figure 36.1. Basel I versus Basel II.

(ii) supervisory review and the role of bank supervisors, and (iii) market discipline through enhanced disclosure.

Equation 36.3. Basel II minimum requirement $$8\% < \frac{T_1 + T_2}{\mathrm{RWA}_{\text{Credit + Operations+Marketing}}}$$

> Operational risk is defined in Basel II as 'the risk of loss resulting from inadequate or failed internal processes, people, and systems, or from external events'. This would also include fraud. Market risk refers to the risk of adverse movement in the prices of traded securities, and had already been included in Basel I in a 1996 amendment.

According to Clementi (2000), Basel II has focused on modernising the risk-weighted assets (RWA) denominator, but with no attention paid to the capital numerator (see Equation 36.3); the task of addressing both would have been too great, and even the minimum percentages have been left unchanged. As it stands, Basel II poses significant challenges, and requires substantial investments in information technology and risk-assessment capabilities, both for initial compliance and ongoing improvements thereafter. In addition, banks that have traditionally operated product silos are under pressure to take customer and portfolio views.

The focus here is still credit risk, and the RWA calculation. A major change is that banks can now also use their own credit risk models, but the Accord recognises that not every bank will have the same level of sophistication by the proposed G10 implementation at end 2006. It was thus necessary to provide for both simple and more complex methodologies, which marks the difference between both the standardised and internal-ratings based (IRB) approaches, and the foundation and advanced approaches.

According to Jackson (2002), when the second Quantitative Impact Study was done in 2001, 22 out of 138 banks were already able to complete the advanced IRB approach, and this proportion was likely to grow.

36.2.1 Standardised approach

The standardised approach is similar to the 1988 accord, except it uses external ratings for 11 risk categories, with risk weights of up to 150 per cent for regular loans, and even higher for past due loans. The categories are: (i) sovereign and central banks; (ii) non-government public sector entities; (iii) multilateral development banks; (iv) private-sector banks; (v) securities firms; (vi) corporates; (vii) residential property; (viii) commercial property; (ix) higher risk; (x) other assets; and (xi) off balance sheet. For sovereign, interbank, and corporate lending, the regular risk weights vary according to the grades provided by external ratings agencies, and may be reduced to recognise the risk mitigating effects of collateral, guarantees, and credit derivatives. Unrated corporate exposures get a 100 per cent rating, which also applies to all BBB+ to BB− loans. Depending upon jurisdiction, some banks may be allowed to apply 100 per cent to all corporate loans.

For retail portfolios, the risk weight is 75 per cent, if it can be shown that the portfolio is well diversified; and this reduces to 35 per cent if it is secured by residential property that is worth substantially more than the loan. The threshold will vary by jurisdiction, but usually applies to loan-to-value ratios of 80 per cent or less. Also, under Basel I, the 100 per cent weighting was often applied only to the exposure above the threshold; whereas for Basel II, this is usually being interpreted as applying to the full amount of the loan.

36.2.2 IRB approach

The IRB approach is a highly mathematical 'value at risk' (VaR) approach, where the weights are derived based on the banks' own risk measures. There are two possible IRB approaches (see Table 36.1): *foundation*—banks calculate their own PDs, and the other VaR elements are provided by the national regulator, based upon data pooled from different banks; and *advanced*—all of the VaR elements must be based upon internal estimates. The foundation approach will

Table 36.1. Basel II—IRB approach

	Wholesale		Retail advanced
	Foundation	Advanced	
PD	Internal	Internal	Internal
LGD	Supervisor	Internal	Internal
EAD	Supervisor	Internal	Internal
M	Supervisor	Internal	N/A

be favoured by smaller banks that feel they are not able to provide their own estimates, due to lack of data, whether because of low volumes, or lack of infrastructure.

According to Cespedes (2002), the Basel II approach is based on Merton's model, but assumes a portfolio of infinite counterparties. Jackson (2002) states that it is derived from the CreditMetrics model, which has the same origin. Merton's 1974 single-factor model is based solely upon the probability of default, and assumes that default occurs when debt exceeds the asset value of the firm, and that future asset values are normally distributed.

36.2.3 Exposure classes

For retail exposures, it is assumed that banks will have the necessary data; whereas for all other exposures, there is a choice. There are several exposure classes, which group assets that are subject to the same types of risks. There are five main exposure classes, and the treatment may vary for each: retail, corporate, interbank, sovereign, and equity.

Retail exposures are defined as those where there are a large number of exposures, which are *managed on a pooled basis*, including SME lending. Retail has three further subclasses: (i) residential mortgages, (ii) revolving credit to €100,000, and (iii) 'Other', which includes most fixed-term lending, including motor vehicle finance. The probability-of-default (PD), exposure-at-default (EAD), and loss-given-default (LGD) estimates are calculated using banks' own credit scores (behavioural and application) and product characteristics, and maturity does not form a part of the calculation.

The Basel Committee defines SMEs as enterprises with annual sales of less than €50 million, which may be modified by national regulators. In many instances it will be impossible to determine sales, and banks will instead use a total exposure figure. SME loans may only be classified as Retail if no individual exposure within the portfolio is greater than €1,000,000.

According to Allen et al. (2003:3), SMEs are receiving particular attention in countries where they 'comprise a significant component of the industrial sector (e.g. Germany)'. Capital requirements are up to 20 per cent lower than for the majors, if only because of the greater diversification.

In contrast, the corporate, interbank, and sovereign classes relate to wholesale lending where the *number of deals is small, values are high, and each deal receives much greater attention.* Corporate is split into specialised lending (project finance, object finance, commodities finance, and commercial real estate) and other corporate. The last category is split further into income-producing and high-volatility real estate. For wholesale classes, customer and transaction risk should be rated separately, with the latter reflecting the LGD. Loan maturity and size of business also form part of the calculation.

36.2.4 Default definition

In Chapter 3, brief mention was made of the distinction between a 'good/bad definition' used for scorecard developments, and a 'not default/default definition' used for finance, regulatory, and other calculations. Good/bad definitions are usually tailored to specific portfolios, whereas default definitions are standardised, to facilitate comparison across portfolios, or over time. The Basel II definition of default can be summarised as: (i) 90 days past due, or in excess of the agreed limit, for any material obligation; (ii) if it is known, that there is a high probability of loss; or (iii) the debt was written-off, or sold at a material loss. It is a worst-ever definition applied over a one-year period, that is used to facilitate comparison across banks, and to standardise the calculation of regulatory capital.

> Internal ratings and default and loss estimates must play an essential role in the credit approval, risk management, internal capital allocation, and corporate governance functions of banks using the IRB approach. *Rating systems and estimates designed and implemented exclusively for the purpose of qualifying for the IRB approach and used only to provide IRB estimates are not acceptable.* It is recognised that banks will not necessarily be using exactly the same estimates for both IRB and internal purposes. For example, pricing models are likely to use PDs and LGDs relevant to the life of the asset. Where there are such differences, a bank must document them and demonstrate their reasonableness to their supervisor.
>
> The 'use test'—Basel II Framework paragraph 444

The 'use test' requires that, in order to qualify for the IRB approach, its component inputs must be based upon measures (e.g. scores or grades) used within the business, for credit risk measurement and management (limit setting, pricing, collections, etc.), strategy and planning, and reporting. It is generally accepted, and expected, that the initial estimates should be derived in a manner that *best suits business needs*, and then be adjusted as necessary. According to the BCBS (2006b:fn1), 'Use test compliance generally concerns internal use of these estimates prior to their transformation or adjustment for regulatory capital purposes'. Thus, for credit scoring, scorecards should be developed using a good/bad definition appropriate for managing a group of interest, and then calibrated, or mapped, to provide PD estimates for the appropriate Basel II default class (usually Retail). In like fashion, the EAD and LGD estimates will be a function of the score, as well as other information (such as agreed limit for EAD, and collateral for LGD).

36.2.5 Ratings implications

Credit scores fall under the heading of 'internal ratings', and as such, Basel II has provided credit scoring with a huge boost. It does, however, bring with it new pressures, in particular with regards to how the scoring models will be used in this process, and ensuring that standards are maintained:

- Credit scoring was initially used to provide risk rankings (power), and lenders are now also expected to provide reasonable default rate estimates (accuracy).
- For each of PD, LGD, and EAD, processes and controls must be in place for the quantification stages: (i) *data*—assembly required for model development; (ii) *estimation*—determining the linkage between the predictors and PD/LGD/EAD; (iii) *mapping*—of the model onto the bank's data and obligors, for IRB purposes; (iv) *application*—to the portfolio, and ensuring results make sense. Each of these must be fully documented with regular updates. The approaches used for estimation will usually be empirical, and the extent to which judgment is used to modify ratings must also be documented.
- Banks must prove that their models distinguish between different levels of risk. There must be at least seven risk grades, and two default grades, and a meaningful distribution, where no one risk category contains more than 30 per cent of exposures. Many banks are modifying their processes to accommodate up to 25 risk grades (almost in line with rating agency grades) that will cover all portfolios.
- Most banks will be able to provide PD estimates, but defaults are rare events, and many—especially smaller banks—may struggle with EAD and LGD. Some banks are pooling data to come up with estimates and provide data for analysis going forward.
- Banks are expected to strive continually to improve their risk assessment processes. This includes incorporating all possible available data to come up with their risk estimates, such as financial statements, industry reviews, account behaviour, bureau data, and subjective inputs, where they are feasible.
- The executive and senior management are expected to understand how the models work, and the effect of any changes. Where banks use models or grades provided by outside consultancies, the same requirements hold. If the required information is not forthcoming, banks will be more likely to develop their own bespoke models.

It should be noted that for estimation, the IRB approach is not prescriptive about how the credit scoring models are developed and used. The primary requirement is that the risk grades should have substantially the same meaning over time, and that it should be possible to determine their volatility over time.

36.2.6 Implementation issues

Basel II's entire approach is one of conservatism, and it is typically considered prudent to adjust estimates downwards, to approximate what might happen during an economic shock (such as for the downturn LGD). Also, banks are required to do back-testing, which implies keeping track of model results and final projections, in order to compare them with subsequent performance. This has severe implications, as many banks were only keeping historical data for two to three years, whereas Basel II prefers banks to use data for an economic cycle (7 to 10 years) where possible, but not less than 5 years, or 7 years for wholesale LGD and EAD.

As a result, banks have been making major infrastructure changes driven purely by this regulatory requirement. While Basel II is generally viewed in a positive light for many smaller lenders, it is yet to be seen whether it will generate sufficient benefits to cover the implementation and ongoing maintenance costs (not an issue in the United States, where it will only be implemented for larger banks). According to Fernandes (2005:42), the IRB Foundation approach would only provide capital relief of about one-twelfth of the Basel I requirements, which may or may not suffice. If nothing else, Basel II has significantly increased the barriers-to-entry into the banking industry, at least until there is greater agreement on, and readily-available software to do, the required calculations. Only then will the general public be able to reap the combined benefits of increased financial stability, and reduced borrowing costs.

36.3 RWA calculation

Initially, it was hoped that this text could be written without providing the exact details of the capital requirement calculation, as at the time of initial writing (CP3) it was in a state of flux, and further changes were possible. Since the final publication though, there have been no substantial changes, other than the introduction of some new concepts. This section is restricted to covering the requirements for the core corporate class and the three retail classes. The calculations are summarised in Table 36.2. There are two types of inputs: (i) those derived from internal ratings, such as credit scores; and (ii) statutory inputs provided by Basel II. These are then used in a two-stage calculation; first the correlation, and then the capital requirement. Note that the one-year PD features throughout the calculations.

Correlation (ρ)

Investment across a large number of assets has an advantage, as diversification reduces risk, except where the asset values are highly correlated. Although correlations amongst well-diversified banking portfolios are not high, they still exist. The Basel formula tries to accommodate these, by providing standard correlation values for high- and low-quality portfolios of each asset class, and the PD is used to determine which value in the range will be used:

Equation 36.4. Correlation $$\rho = \left(\rho_{\mathrm{Lq}} \times \left(\frac{1 - e^{-Q \times \mathrm{PD}\%}}{1 - e^{-Q}}\right)\right) + \left(\rho_{\mathrm{Hq}} \times \left(1 - \frac{1 - e^{-Q \times \mathrm{PD}\%}}{1 - e^{-Q}}\right)\right)$$

The Basel input values for ρ_{Hq} and ρ_{Lq} acknowledge that the correlations within high quality portfolios are higher than within low quality portfolios. Thus, as the PD ranges from 0 to 100 per cent, the correlation (ρ) moves within the range provided, which is anywhere from 3 to 24 per cent.[5] The exceptions are qualifying 'retail revolving credit' and 'home loans', where constants of 4 and 15 per cent are used respectively. For non-retail assets, there is also an

[5] High-value commercial real estate is higher, ranging from 12 to 30 per cent.

Table 36.2. Basel risk-weighted asset calculation

Block	Variable	Description	Corporate	Retail		
				Revolving	Mortgage	Other
Inputs						
Internal ratings	PD	Probability of default	2.3%	3.0%	1.0%	2.5%
	LGD	Loss given default	40%	80%	75%	60%
	EAD	Exposure at default	100	100	100	100
	M	Maturity	1.0			
	T	Size (in millions of Euros)	50			
Basel inputs	Q	Exponent factor	−50	−50	−50	−35
	r1	Low quality	12%	4%	15%	3%
	r2	High quality	24%	4%	15%	16%
Formulae						
Correlation	SA	= 4% × (1 − (T − 5)/45)	0.0%			
	A	= (1 − EXP(Q × PD))/(1 − EXP(Q))	0.675			0.583
	R	= r1 × A − r2 × (1 − A) − SA	15.9%	4.0%	15.0%	8.4%
Capital requirement	B	= (0.08451 − 0.05898 × LOG(PD))^2	0.0330			
	MA	= (1 + (M − 2.5) × b)/(1 − 1.5 × b)	1.0000	1.0000	1.0000	1.0000
	u1	= NORMSINV(PD)	−2.0047	−1.8808	−2.3263	−1.9600
	u2	= SQRT(R) × NORMSINV(0.999)	1.2321	0.6187	1.1968	0.8967
	u3	= SQRT(1 − R)	0.9171	0.9798	0.9220	0.9570
	U	= NORMSDIST((u1 + u2)/u3) × MA	19.98%	9.88%	11.03%	13.33%
	K	= U × LGD	7.99%	7.91%	8.27%	8.00%
Risk assets	RWA	= K × 12.5 × EAD	99.88	98.84	103.37	99.95

R = Correlation, SA = Size adjustment, K = Capital requirement, MA = Maturity adjustment

adjustment for companies with turnover between €5 and €50 million, to recognise lower correlations amongst smaller firms. This reduces the capital requirements by an average of 10 per cent, and 20 per cent for the smallest companies (Jackson 2002).

Equation 36.5. Size adjustment $\rho = \rho - 0.04 \times \left(1 - \frac{S-5}{45}\right)$

According to Allen et al. (2003:fn3), 'the assumption of an inverse relationship between PD and correlation is quite controversial'. Most academic studies have found that low-quality firms are more subject to systemic risks arising from market shocks, and hence have higher correlations than high-quality firms.

Capital requirement

To calculate the percentage of capital required, the primary inputs are the PD and correlation, which are then applied to the LGD.[6] Something that is introduced here is the 'standard -

[6] Cespedes (2002) notes one of the model's weaknesses is that LGD is not given the same level of attention as the PD. It is a simple ratio, which does not take into consideration correlations within the underlying markets.

normal cumulative distribution', and its inverse. These are represented in Equation 36.6 as Φ and Φ^{-1} respectively, and can be obtained in MSExcel using the =NORMSDIST and =NORMSINV functions:

Equation 36.6. Capital requirement $K\% = \text{LGD}\% \times \Phi\left(\dfrac{\Phi^{-1}(\text{PD}\%) + \sqrt{\rho} \times \Phi^{-1}(99.9\%)}{\sqrt{1-\rho}}\right)$

The term '$\Phi^{-1}(99.9\%)$' is meant to indicate that the required value should have a confidence level of 99.9 per cent. In other words, the resulting capital requirement should be high enough to cover almost any disaster. For sovereigns, banks, and corporates, there is also a maturity adjustment (M), to increase the capital requirement for terms of one year or more.[7] The maximum value for M is 5.

Equation 36.7. Maturity adjustment $K\% = K\% \times \dfrac{(1 + (M - 2.5) \times b)}{(1 - 1.5 \times b)}$

where $b = (0.08451 - 0.05898 \times \log(\text{PD}))^2$

A future margin-income adjustment is made for some retail revolving credit exposures, to account for the risk mitigating effect of the high levels of future marginal income. This applies in particular to credit card portfolios, where the income levels are sufficient to reduce the level of risk.

Equation 36.8. Future margin-income adjustment. $K\% = K\% - 0.75 \times \text{PD}\% \times \text{LGD}\%$

In July 2005, this was further adjusted to recognise the lower risk associated with guarantors. Initially, the lower of the obligor and guarantor PDs was to be used; a treatment, which fails to recognise that both parties must default before there is a loss. Thus, there is a double-default adjustment for corporate, retail SME, and public sector bodies:

Equation 36.9. Double-default adjustment. $K\% = K\% \times (0.15 + 160 \times \text{PD}\%_{\text{Guarantor}})$

At the end of this, the resulting capital requirement percentage will average somewhere between 7 to 9 per cent, and is then multiplied by 12.5 (inverse of 8%), before being applied to the EAD, to come up with the risk-weighted assets.

Equation 36.10. Risk-weighted assets $\text{RWA} = K\% \times 12.5 \times \text{EAD}$

Finally, in June 2006, another adjustment was made to the risk-weighted assets value, increasing it by a multiple of 1.06 for the IRB approaches (Basel Committee for Banking Supervision 2006a:fn11). Its purpose was to keep the overall level of capital held by the banking system constant, as it was realised that Basel II was a big leap. This adjustment will probably be modified in future.

As concluding remarks, please note that operational and market risk have not been covered here. Also, while the intention is that Basel requirements should not push up capital requirements

[7] In July 2005, par. 321–324 of the Accord were adjusted to allow an exemption for shorter maturities under certain conditions, such as where the instrument is for less than one year, and there is no intention of reissuing the loan thereafter.

for internationally active banks, there will be higher requirements for those with higher-risk portfolios. Banks adopting the IRB approach should be able to reduce these requirements, and the greatest savings are likely to be achieved in the retail arena, where the quantification of risk is most advanced.

36.4 Summary

The focus of this textbook has been on credit scoring for the retail credit industry in general, where banks play a major role. Economies are volatile creatures though, and subject to the vagaries of cycles, that affect banks' liquidity, and public confidence in them. The extreme case is runs on the bank, which have characterised every depression since banking began. This is unhealthy for financial institutions, which are pillars of modern economies, responsible for safeguarding savings, and channelling them into productive pursuits.

As a result, regulators work to ensure the integrity of the financial system by requiring banks to keep sufficient capital to weather downturns. Capital-adequacy ratios were initially set as flat percentages of banks' assets, usually somewhere between 4 and 10 per cent, but this took no cognizance of the riskiness of banks' assets, which varied depending upon the nature of their portfolios. The first attempt to address asset-value volatility was the 1988 Basel Accord, which prescribed adequacy ratios for loans to sovereigns (0%), banks (20%), residential property (50%), and other (100%). This was highly unsophisticated though, as it recognised neither the possible risk profiles within each of those groups, nor the risk-mitigating effects of security. At the same time, banks' risk-grading capabilities were improving, and some means was needed to recognise and reward them.

The Basel II Accord was spawned of this need, which uses a 'value at risk' approach to calculate an expected loss, based upon a PD, EAD, LGD, and maturity (M (wholesale only)). Not all banks have the same capabilities though, and different approaches are provided for: (i) *standardised*, for use by less sophisticated banks, which like the Basel I Accord has fixed percentages for different asset classes; and (ii) *internal-ratings based*, which relies upon banks' own risk grading systems.

The latter provides two further options: (i) *foundation*, internal ratings are used to provide PD, but regulators' estimates for EAD, LGD, and M; and (ii) *advanced*, internal ratings are used throughout. Either approach can be used for wholesale portfolios, which are split into project finance, object finance, commodities finance, commercial real estate (income-producing and high-volatility), and other corporate. The carrot for using the advanced approach is supposedly lower capital-reserving requirements, but it is likely that banks will gain greater benefits from improved risk-management and pricing capabilities. Retail portfolios are limited to the advanced approach, and are split into residential mortgages, revolving credit to €100,000, and other.

Credit scoring's first significant boosts came from the USA's Fair Credit Reporting and Equal Credit Opportunity Acts during the 1970s, and is now gaining a further boost from Basel II, as it provides the basis for internal ratings for retail banking. Basel II has its own default definition, which even if not used directly to develop models, is highly correlated with their bad definitions.

The 'use test' requires that the models be used in day-to-day decision making, and not be used for IRB purposes only. Compliance comes with new challenges though, because: (i) lenders' models are expected to provide not only predictive power, but also accurate estimates; (ii) processes, controls, and documentation must be in place for data, estimation, mapping, and application; (iii) continual improvements in risk-grading processes are expected; (iv) processes are being modified to provide greater granularity, in many instances allowing up to 25 risk grades; (v) some banks may struggle with EAD and LGD estimates; and (vi) management is expected to understand models' workings.

Unlike prior approaches, where capital requirements were calculated as a flat percentage of asset values, with Basel II they are calculated as a function of asset correlations and the PD and LGD estimates, and then adjusted for the EAD and *M* values. Correlation is important, because it recognises diversification within a portfolio, which is less for corporate and home loans, and greater for small-business loans (Basel II does not recognise diversification across portfolios or borders).

At the time of writing (2006), financial regulators in most developed and developing countries have agreed to adopt Basel II, and banks are striving to comply by the 2007 deadline. The primary holdouts are the United States, which is questioning its appropriateness for smaller banks; and many underdeveloped or emerging markets that do not have the technical abilities. At the same time, many banks that initially opted for the standardised or foundation approaches are considering, or have adopted, more demanding approaches; not because of the potential *lower capital requirements*, but in the hope that compliance will provide an implicit stamp of approval that facilitates *lower borrowing costs* in international financial markets. There are significant costs being incurred, and only time will tell whether they are justified. The largest banks will probably benefit most, as they have the mass to warrant the infrastructure investment. Financial stability will come at the cost of increased barriers to entry into the financial-services industry, and tougher trading conditions for smaller banks.

37 Know your customer (KYC)

In 1957, Eliot Ness published his memoirs, which were presented both as a TV series (1959 to 1963, starring Robert Stack; and 1993/1994, starring Tom Amandes), and a movie (1987, starring Kevin Costner). This was The Untouchables, set during the 1930s prohibition, which told the story of a small band of Chicago policemen, whose primary nemesis was Al Capone—one of the most infamous gangsters of that era. Capone was finally indicted on tax evasion charges in 1931, and sentenced to 11 years in prison. The case was a precedent, because a hardened mobster, who was involved in practically every type of racketeering and organised crime possible during that era, was jailed for a crime that many consider acceptable.

Racketeering—Any act, threat, attempt, or conspiracy to commit a crime for financial gain. These usually involve a pattern of criminal activities that may involve one or more of murder, kidnapping, gambling, arson, robbery, bribery, and extortion. It also includes dealing in pornography, prostitution, slavery, narcotics, firearms, stolen goods, or other prohibited items, depending upon jurisdiction.

Organised crime—complex of highly centralized enterprises set up for the purpose of engaging in illegal activities. (*The New Encyclopaedia Britannica* 1986, Vol. 8, 994). It is usually characterised by:

(1) the commission of serious crimes, that: (i) have a motive of profit and power; (ii) with little regard for the interests of the community; (iii) often under the guise of legitimate activities; and (iv) executed over a long period of time; and
(2) involve a social structure: (i) with three or more people; (ii) that mimics commercial activities; (iii) involves planning and tactics; and (iv) influence is exerted through force, intimidation, bribery, or corruption, against or in the police, judiciary, politics, media, and business.

Crimes with the goal of financial gain thus have an inherent weakness, because *if the law can't attack the racket, it can go for the dough*. As a result, the criminally minded have developed countless schemes to turn dirty money into clean:

Money laundering—'Term used to describe investment or other transfer of money flowing from racketeering, drug transactions and other illegal sources into legitimate channels so that its original source cannot be traced'

Black's Law Dictionary, 6th edition

Most industrialised countries have implemented anti-money laundering legislation, and have established supervisory bodies to which any suspicious transactions must be reported.[1] Banks and others have effectively been drawn in as tools in crime prevention. In most instances, there are severe penalties for failure to report an offence, which may extend to criminal prosecution. In more recent years, many countries have also implemented 'Know Your Customer' (KYC) legislation, which puts an onus of due diligence onto banks, and others that process large amounts of money on behalf of individuals and institutions. In the early 2000s, this demand increased even further, to combat the funding of terrorist organisations.

In some ways, the principles of data protection and KYC conflict: the former limits data collection, the latter increases it; the former demands secrecy, the latter transparency. In general though, KYC requirements are catered for by data-protection clauses, which require entities holding personal information to provide details to authorities, where demanded by law.

Why is this being discussed in a book on credit scoring? There are three reasons. First, the primary point of contact with the customer is at time of *account opening*, when it is possible to demand identification details and supporting documentation. It thus impacts upon automated account origination processes, by putting in extra hurdles. Second, banks must have mechanisms in place during the *account-management* process, to identify suspicious transactions. Third, it is possible to use *mathematical and statistical techniques* to identify abnormal transactions. The most common tools used are pattern-identification methodologies though, which are beyond the scope of this book.

37.1 Due diligence requirements

While KYC requirements involve extra costs, compliance with the legislation has had the benefit of protecting banks against direct and indirect losses suffered due to poor controls, in particular those resulting from fraud. In October 2001, the Basel Committee for Banking Supervision issued 'Customer Due Diligence for Banks' ('Basel KYC'), which provides guidelines for national legislation. These guidelines effectively provide best practice that 'embrace[s] routines for proper management oversight, systems and controls, segregation of duties, training and other related policies'. They highlight that KYC practices would assist in protecting against:

Operational risk—of losses arising from failed processes, people, or systems;
Reputational risk—of adverse publicity affecting confidence in the bank, especially amongst depositors of small banks, who may withdraw funds, or demand higher interest rates;
Legal risk—of lawsuits and unenforceable contracts, arising from failure to adhere to mandatory KYC legislation;
Concentration risk—arising from loans being made to a small number of related parties.

[1] In the United States it has long been a requirement that all transactions above a certain limit, currently US$10,000, has to be reported.

Banks should have policies and procedures to protect themselves against being used by criminal elements. The required due diligence falls under four headings:

(i) **Customer-acceptance policy**—Should be clear, and include descriptions of higher-risk customers by background, country of origin, public or high profile position, linked accounts, business activities, etc.
(ii) **Customer identification**—The beneficiary of the account must be adequately identified, failing which the account should not be opened. Regular reviews should also be done, to ensure that records are kept up-to-date.
(iii) **Ongoing monitoring of high-risk accounts**—This applies especially to 'publicly exposed persons', who may be prone to corruption. Banks should be alert to external information, and senior management should approve significant transactions.
(iv) **Risk management**—Day-to-day monitoring to identify abnormal transactions. This requires investigating any transactions whose value exceeds thresholds set by the bank, for a class of accounts or transaction types.

The requirements will not be covered in full detail, but some comments on noteworthy items will be made:

- Copies of customer identification documents should be retained for five years after the account is closed and details of transactions for five years after the transaction is effected.
- Due diligence requirements are greater for high net-worth individuals with private banking relationships. It is noted that some countries are also implementing laws prohibiting bribes to foreign officials.
- Record reviews can be triggered by a request for a significant increase in facilities, change in documentation requirements, or substantial change in account behaviour.
- For introduced business, banks must not open the account unless they can satisfy themselves about the identity of the beneficiary. This applies to intermediaries, lawyers, and foreign institutions unless it can be shown that the introducer's KYC standards are as high, or higher. Otherwise, the introducer must provide the required documentation, so that the banks' own policies and procedures can be employed.
- Risks are higher for non face-to-face business, yet banks should still try to maintain the same KYC standards, possibly by requesting extra documentation, demanding that documents be certified, or requiring independent contact with a third party.

37.2 Customer identification requirements

The Basel KYC guidelines are not prescriptive with regards to the identification requirements, other than to state that the ID presented should be the one least capable of being forged. Instead, in February 2003, the Working Group on Cross-Border Banking provided its 'General

Guide to Account Opening and Customer Identification' (Working Group Guide) as an attachment, which provides much more detail for both individuals and institutions.

For individuals, banks should request: legal names, residential address, contact details, date and place of birth, nationality, occupation and employer name, personal identification number, account type, and signature. The information should also be confirmed by insisting upon: original identification documents (ID number, date of birth, etc.); account and utility statements (address); follow-up contacts, by phone or mail; and confirmation of document authenticity, through certification by embassies or notary publics.

There is, however, some latitude provided here for smaller transactions: (i) requirements can be limited to name and address, for once-off or occasional transactions below an established minimum value; and (ii) policies should not be so 'restrictive that they result in a denial of access to the general public to banking services, especially for people who are financially or socially disadvantaged'. Unfortunately, however, the implementation and interpretation of these requirements tends to be quite strict. This can cause problems in some environments, especially where people are highly transient, or have little proof of address or employment, like South Africa and Brazil. Financially excluded individuals may pass credit risk scorecards, only to be turned down because they cannot present the required documentation.

For institutions, the requirements are similar, with the exception that banks are expected to: (i) identify the beneficiaries behind the legal entities; (ii) obtain copies of financial statements, incorporation documents, the memorandum of association, and others; (iii) obtain a bank reference; and/or (iv) confirm details through a known firm of lawyers or accountants, or a business-information or other external service. These requirements have little or no impact upon automated decision-making, so they are not discussed further.

38 National differences

The following provides an indication of some of the national differences between various English-speaking countries, which are summarised in Table 38.1. The first line provides the name used in that country, to refer to shared performance data. The acronyms in the second line indicate what personal identifier is used: social security number (SSN), social insurance number (SIN), identification number (ID). The items for 'bureaux since' and 'debt % of net national product' are based on a survey done by Japelli (1999).

38.1 United States of America

The United States was the trendsetter for most of the early legislation that had a direct impact upon credit scoring, especially its Fair Credit Reporting Act and Equal Credit Opportunity Act. It is the only country where information sharing is so ubiquitous that they do not even have a name for it, beyond 'shared information'. The three major bureaux operating in the United States are Experian, Equifax, and Trans Union, who all use the Social Security Number as a personal identifier.

Consumer Credit Protection Act (CCPA)

The CCPA is a broad heading for all of the acts implemented specifically for consumer protection. The act is divided into sections including:

Table 38.1. National differences

	AU	CA	UK	US	ZA
Environment					
Personal identifier	SSN	SIN	None	SSN	ID
Payment profile name	N/A	Positive info	White data/ CAIS	Shared information	Payment profile
Debt % of NNP (1980)	7.7	14.4	5.7	16.1	?
Credit bureaux					
Private bureaux since	1930s	1919	1960s	1890s	1901
Trans union	Other bureaux	Y	N	Y	Y
Equifax		Y	Y	Y	N
Experian		N	Y	Y	Y

Subchapter I	**Consumer credit cost disclosure** a.k.a. Truth in Lending Act	
§ 1601	Part A	General provisions
§ 1631	Part B	Credit transactions
§ 1661	Part C	Credit advertising
§ 1666	Part D	Credit billing
§ 1667	Part E	Consumer leases
§ 1667	Subchapter II	Restrictions on garnishment
§ 1679	Subchapter IIA	Credit repair organisations
§ 1681	Subchapter III	Credit reporting agencies a.k.a. Fair Credit Reporting Act
§ 1691	Subchapter IV	Equal Credit Opportunity (Act)
§ 1692	Subchapter V	Debt Collection Practices a.k.a. Fair Debt Collection Practices Act
§ 1693	Subchapter VI	Electronic fund transfers

Of these, Subchapters III and IV are most relevant to credit scoring, as they both promoted its use.

Fair Credit Reporting Act (FCRA) 15 USC § 1681

The USA implemented the FCRA in 1970, to set rules for their credit bureaux. These rules ensured data privacy and accuracy, and limited the scope of the information that could be used by the credit bureaux to credit-related information, including positive information.

According to Staten and Cate (2004), the regulatory authorities in the United States recognise the value of a voluntary credit information-sharing arrangement, and are sensitive to the potential overheads that can be created by unnecessary regulation. As a result, Congress has been loath to impose new requirements, unless there is a clear indication of a problem.

Equal Credit Opportunity Act (ECOA) 15 USC § 1691

The ECOA was first implemented in 1974, and modified in 1976. For consumer credit, it was the first bespoke anti-discrimination legislation ever, and prohibited unfair discrimination on the basis of *race*, *colour*, *religion*, *national origin*, *sex*, *marital status*, *age*, or *because an applicant receives income from a public assistance program*, or *the good faith exercise of any right under the Consumer Protection Act*. Regulation B uses the expression, 'empirically derived, demonstrably and statistically sound' for statistical models, and considers all others as judgmental. It allows the use of age, as long as older applicants are not prejudiced (they must receive the highest possible points). Lenders must, however, protect against disparate impact (which could occur with the use of postal code or home ownership), but may still use such factors if it can be shown that they: (i) serve a legitimate business need; and (ii) cannot be replaced by other factors. Section (a)(2)(I) of the act also requires that declined applicants be provided with specific reasons for a decline, and that broad categories such as 'Policy' and 'Score' are insufficient.

Data privacy

American data privacy legislation focuses more on the public sector than the private sector. The two most well known Acts apply to government agencies only:

- Freedom of Information Act (FOIA) 5 USC § 552 (1967, 1975, 1996);
- Privacy Act 5 USC § 552a (1974).

For the private sector, there is neither comprehensive data privacy legislation, nor a monitoring agency. The Personal Information Privacy Act of 1997 does apply, but is rather limited in its scope, as its primary purpose is to protect against the misuse of the Social Security Number, and other personal details, for commercial purposes.

Patriot Act (2001)

The Patriot Act was implemented on 26 October 2001, in the immediate aftermath of 9/11. It was a rushed piece of legislation, which effectively gave significant powers to law and intelligence services. Section 326, also referred to as 'Verification of Identification', or 'International Money Laundering Abatement and Anti-Terrorist Financing Act', requires financial institutions to implement certain KYC requirements, including to: *(1) verify the identity of any person opening an account; (2) maintain records of the information used to verify the person's identity; and (3) determine whether the person appears on any list of known or suspected terrorists or terrorist organizations.* Compliance was required by 1 October 2003. An industry initiative was the 'Customer Identification Program', which required the creation and maintenance of a database that is checked against a government watch-list. There are public concerns regarding data privacy, and what might happen to the poor unfortunates who share names with those on the list.

38.2 Canada

The credit bureaux operating in Canada are Equifax and TransUnion. Experian does not have a Canadian presence. Canada uses a Social Insurance Number as a personal identifier, and the sharing of positive information is allowed. According to Stevens (1998), the selling of lists and 'header information' for marketing purposes has never been allowed in Canada.

Data privacy

Canada has two national acts and commissioners responsible for data privacy. First, the Information Commissioner and 'Access to Information Act' ensure rights of access to information held by government institutions. Second, the Privacy Commissioner and 'Privacy Act'[1] govern broader aspects of data privacy, in both the public and private sectors. In general though, Canada has sectoral legislation for the private sector. Provincial acts and offices for information and data privacy are in place, mostly for the public sector. No registration is required for entities holding personal information.

---

[1] Privacy Act, (R.S. 1985, c. P-21).

There are 'constitutional limitations on federal power to effect privacy legislation' (Owens and Lyons 1998). However, banks are federal institutions, and can be regulated at a national level. In 1992, the Bank Act was modified, effectively to put a federal ban on banks' and trust companies' use of personal information to sell insurance. This was not so much to protect the individual, but instead to protect the insurance industry from unfair competition.

In 1994, Quebec implemented detailed and far reaching regulations in its 'Act respecting the protection of personal information in the private sector', or Bill 68.[2] The Canadian Standards Association (CSA) also issued a set of privacy principles in 1996 under the heading of '*Model Code for the Protection of Personal Information*'.[3] These can be paraphrased as:

Accountability—Responsible data controller(s) shall be appointed.
Identifying purposes—Purpose of use must be given at time of collection.
Consent—Individual consent is required for collection, use, or disclosure.
Limiting collection—Fair, lawful, and limited to that necessary.
Limiting use, disclosure and retention—Used only for the stated purpose, and not to be held longer than necessary.
Accuracy—Accurate, complete and up-to-date.
Safeguards—Protected by appropriate security safeguards.
Openness—Transparency about information management policies and practices.
Individual access—Right of individual access and contest.
Challenging compliance—Right to challenge compliance with the data controller.

Unfair discrimination/equal opportunity

Human Rights Commissions exist at national level, and in each of the ten provinces. No specific reference is made to credit, except as one of the many facilities offered to the public.

Canadian Human Rights Act (R.S. 1985, c. H-6)

Prohibits discrimination on the basis of *race, national or ethnic origin, colour, religion, age, sex, sexual orientation, marital status, family status, disability, and conviction for which a pardon has been granted.* This applies to the provision of goods, services, facilities, or accommodation, customarily available to the general public.

38.3 United Kingdom

The major credit bureaux in the United Kingdom are Experian and Equifax. Information sharing is allowed, but banks only started sharing with the rest of the consumer-credit industry

[2] Quebec's privacy legislation was the first in North America that applied to the private sector generally. It takes the European approach, but without the requirement of mandatory registration or licensing (Owens and Lyons 1998).

[3] The Canadian Bankers' Association (CBA) issued a similar, but more detailed, set of principles in the same year.

from the mid- to late-1990s. The terms 'white data' and 'positive information' are often used to describe shared performance data. The most well known scheme is CAIS, Customer Account Information Sharing.

The UK environment is much different from the United States, Canada, and South Africa, as there is no national personal identifier. Their National Health Number has been considered, but thus far has not been accepted for broader use. Reliance is put upon the customer name and address, and computer routines are used to do the matching, which is aided by some data enhancement. Also, they use the term 'county court judgment' (CCJ), and in Scotland a 'court decree'.

Consumer Credit Act 1974

This act requires that the Director General of Fair Trading license most businesses that offer goods or services on credit, or lend money to consumers, and provides certain protections to consumers. It requires, amongst others, that: (i) borrowers be provided a detailed written quotation of the true interest rate; (ii) be allowed a cooling-off period; and (iii) that all agreements be in writing.

Data Protection Act 1984, updated 1998

The United Kingdom has a Data Protection Commissioner, who is a specific individual with the responsibility of ensuring that this act is enforced. The Act is divided into six parts: (I) *preliminary*, definitions and a general framework; (II) *rights of data subjects and others*, rights of access to data by individuals, and the required procedures; (III) *notifications by data controllers*, details for registration of data controllers, that is all entities holding personal information must be registered; (IV) *exemptions*, including crime prevention, taxation, required by law, etc.; (V) *enforcement*, possible remedies, and how to invoke them; and (VI) *miscellaneous and general*. The principles set out in the act are that any data held by an organisation should be:

(i) Fairly and lawfully obtained.
(ii) Used for limited purposes.
(iii) Adequate, relevant and not excessive.
(iv) Accurate, and where necessary kept up-to-date.
(v) Not kept longer than necessary.
(vi) Processed in accordance with the data subject's rights: (i) to know what is being held; (ii) to whom it may be disclosed; and (iii) in some cases, the data source.
(vii) Secure.
(viii) Not transferred to countries without adequate protection.

The legislation also requires that (i) every organisation holding personal information register with the Data Protection Commissioner; (ii) state the purpose for which data is being held; and (iii) a data controller be appointed. Separate registrations are required for each purpose.

Unfair discrimination

There are three unfair discrimination acts that can be considered relevant to credit scoring:

- Sex Discrimination Act, 1975
- Race Relations Act, 1976
- Human Rights Act, 1998

The latter is the most far reaching, and states that 'The enjoyment of the rights and freedoms . . . shall be secured without discrimination on any ground such as *sex*, *race*, *colour*, *language*, *religion*, *political or other opinion*, *national or social origin*, *association with a national minority*, *property*, *birth* or *other status*'.

38.4 Australia

Australia has a Federal Privacy Commissioner, as well as privacy commissioners in several states. Data privacy in most states is governed by the 1988 Commonwealth Privacy Act (CPA), which was updated in 2000 to include ten National Privacy Principals. The largest credit bureau in Australia is the Credit Reference Association of Australia (CRAA), followed by Credit Bureau Australia (CBA), and Dun & Bradstreet (D&B). Unlike most other English-speaking countries, the Australian CPA does not allow information sharing, albeit this has been under review since 1997. Currently, only derogatory and enquiry details can be used in credit scores, or be provided to bureaux subscribers.

Privacy Act, 1988, and Privacy Amendment (Private Sector) Act (2000)

The Australian data privacy principles are covered under the headings:

(i) **Collection**—Fairly and lawfully obtained, relevant, and obtained with knowledge or consent of individual;
(ii) **Use and disclosure**—Used for limited purposes, or done with consent of the individual;
(iii) **Data quality**—Accurate, complete, up-to-date.
(iv) **Data security**—Protect existing data, and delete if no longer required.
(v) **Openness**—Demands transparency, with respect to data held and management policies.
(vi) **Access and correction**—Allows individuals to view and contest personal data.
(vii) **Identifiers**—Prohibits the use of personal identifiers provided by external agencies.
(viii) **Anonymity**—Where feasible, individuals should be able to conduct business without identifying themselves.
(ix) **Transborder data flows**—Limits transfer of data to foreign countries.
(x) **Sensitive information**—Regulates collection of sensitive personal data.

Anti-discrimination and equal opportunity

There are several acts and commissions in Australia covering disabilities, gender, and race, and some states have their own. These are all broad-based, where credit is only one of many aspects covered.

38.5 Republic of South Africa (RSA)

There are two major consumer credit bureaux operating in South Africa: Experian and TransUnion (see Credit Bureau Association, end of section). Each resident has an Identification (ID) Number, which is broadly used as a personal identifier. Information sharing is allowed, but banks only started sharing with the rest of the consumer-credit industry from the early 2000s. The South African industry has been heavily influenced by developments in the United Kingdom, and has adopted many of the same terms, including 'payment profile' to describe shared performance data.

Legislation

South Africa differs from the other countries covered here, as it is the only developing economy within the group. Prior to 1999, consumer credit legislation in South Africa was limited, and greater emphasis was put on industry codes of practice. This has been changing rapidly though, as there have been increasing pressures to enact specific legislation, adapted for the post-apartheid era.

Consumer Affairs (Harmful Business Practices Act), 1988

'to provide for the prohibition or control of certain business practices; and for matters connected therewith'.

First enacted in 1988, this is very broad legislation covering many aspects of business operations, and will likely be modified in future to cover aspects of data privacy and unfair discrimination. It may ultimately provide much of the same protection afforded by the American 'Consumer Credit Protection Act'.

Promotion of Access to Information (PAIA), Act 2 of 2000

The PAIA was implemented on 21 January, 2000, to replace 1997's draft Open Democracy Bill, whose purpose was to make government more transparent. The new Act is broader, and sets out principles governing access to, and use of, personal information held by any government or private body.

This Act only covers certain aspects of data privacy, in particular access to information held by the government and other organisations. Its purpose is to empower individuals to exercise and protect their rights, and it sets out the procedures through which they can request information about themselves. Affected organisations are obliged to provide the information upon request, unless otherwise stated within the Act. The limitations stated in Section 9 deal with privacy, commercial confidentiality, and good governance.

Prevention of Organised Crime (POCA), Act 121 of 1998

This was the first RSA legislation introduced to combat organised crime, money laundering, racketeering, and criminal gang activities. It criminalized certain activities, and provided for the confiscation of the proceeds of unlawful activities.

Financial Intelligence Centre (FICA), Act 38 of 2001

The POCA had some shortfalls when it came to addressing money laundering. It was supplemented by FICA, which came into full operation on 30 June 2003. It affects all companies, requiring them to retain customer details for five years, and report any suspicious transactions to a Financial Intelligence Centre.

National Credit Act (NCA) of 2006

This piece of far-reaching legislation was implemented to replace the Usury Act of 1926 (amended 1968), and the Credit Agreements Act governing 'hire purchase' agreements (Act 75 of 1980). There was an exemption to the Usury Act in 1992, which allowed loans of small amounts at higher interest rates—called 'micro-lending'. Its purpose was to foster lending for development purposes, but most was used for consumption, and unfair practices abounded. In 1999, the Microfinance Regulatory Council (MFRC) was formed to govern the industry, and the two major credit bureaux were co-opted into hosting the National Loans Register (NLR), starting in 2002. The initiative was successful, and emboldened the MFRC to broaden its scope to become the National Credit Regulator. With the new Act, the Usury Act exemption falls away, and the entire credit industry will be governed by the same piece of legislation.

The NCA's primary focus is 'fair credit', but it also covers many elements of 'data protection'. It created a mad rush by lenders to get new customers and load limits prior to its implementation, because: (i) marketing practices are restricted; and (ii) the new affordability assessments will make qualifying for credit more difficult. The Act also specifies the types of fees lenders can levy, and penalty fees are prohibited by omission. This puts significant pressure on lenders to improve their front-end risk assessment capabilities, and implement risk-based pricing.

The credit bureaux have also been impacted significantly, in particular as regards: (i) *disclosure requirements*, as they must allow the public to query, and contest, their own records; (ii) *data retention periods*, which have been shortened; and (iii) *matching requirements*, as judgments may only be matched if there is an ID number. The Act calls for the creation of a National Credit Register, which unlike the NLR, will be used primarily for regulatory purposes. Lenders are required to use this data for an affordability assessment, failing which, a defaulting borrower may be excused the debt. Although not definite, it is likely that the NLR will be integrated with other bureau data in future.

Industry bodies

Banking Council of South Africa

Governing body for the South African banking industry, which, like most, has its own Code of Conduct. Section 4.1 of the Code of Banking Practice covers confidentiality, and reads almost exactly like the Tournier exceptions. Section 4.2 relates more specifically to providing data to the credit bureaux.

Consumer Credit Association (CCA)

The CCA was originally comprised of retailers lending to the consumer market, but has expanded to include banks, telecoms companies, and micro-lenders. Retailers were sharing data for many years, but it was only in the early 2000s that any of the banks started sharing.

Credit Bureau Association (CBA)

There are 10 or more credit bureaux operating in South Africa, with ITC and Experian dominating the market for consumer credit information. The major player for company data is KreditInform, while Compuscan is active in the micro-lending market. All of the credit bureaux are represented by the CBA, which has a Code of Conduct endorsed by the Business Practices Committee. The code covers compliance procedures, disclosure of information to customers, procedures in the event of disputed accuracy, data retention periods, and disclosure to other parties. Even so, there is a public perception that the credit bureaux are perpetuating historical inequalities. This has been worsened, because the information provided has on occasion been inaccurate, and because bureau data is sometimes used in employment screening. It is hoped that many of these perceptions will improve once the bureaux comply with the new NCA requirements.

Glossary

This glossary is intended to inform the reader, as to how words and expressions have been used within this book. Some are commonly understood, but may have various other connotations. In some instances, new terms have been derived for concepts saldom or never covered elsewhere. Several Internet-based dictionaries, and the Collins English Dictionary (21st Century edition), have been used as reference materials during its compilation. All of the mentioned items are nouns unless otherwise stated. Otherwise, the following abbreviations are used:

abbr.	Abbreviation	pref.	Prefix
acr.	Acronym	pl.	Plural
adj.	Adjective	syn.	Synonym
ant.	Antonym, opposite	rel.	Related
inf.	Informal, slang	v.i.	Intransitive Verb
n.	Noun	v.t.	Transitive Verb
~	Repeat immediately prior text	~~	Ditto, but twice

4 Rs the four basic elements that drive lenders' profits, and can be measured: risk, response, revenue, and retention.

5 Cs the five elements underlying traditional credit risk assessment: capacity, capital, conditions, character, and collateral.

abscond (*v.i.*) to depart secretly, with the hope of not being found, whether to: (i) avoid penalties or prosecution; (ii) renege on responsibilities (family, financial, work); or (iii) enjoy the spoils of misappropriated assets.

absorption process of taking in; ~ ***state***, state defined within a transition matrix that receives, but does not dispatch (syn. *exit state*).

accept (*v.t.*) to respond in the affirmative; (*n.*) case that is accepted (ant. *reject*); ***~ance rate***, percentage of applicants that are accepted; ~ ***override***, a system accept, that is turned down by an underwriter or policy; ***~/reject***, (*adj.*) related to the immediate result of a selection process (rel. *all good/bad, known good/bad, reject inference*); ~~ ***model***, scorecard developed specifically to differentiate between accepts and rejects, and used as part of the reject inference process.

account a record of financial transactions; ***~ing risk***, risk that arises from the possibility that an entity's financial statements are not an accurate representation of its financial situation, whether the result of errors or misrepresentation; ~ ***management***, processes used to manage the account relationship, for example limit setting and authorisations/referrals; ~ ***origination***, processes used to acquire new business, such as R&D, marketing, application processing, account opening.

accuracy (i) correctness, precision; (ii) extent to which a model's estimates agree with actual values; ~ ***rate/ratio***, see 'Gini coefficient'; ~ ***test***, any test used to measure how close a model's estimates are to the actual values.

adaptive (*adj.*) gradual process of adapting to new conditions; ~ ***control***, adjusting parameters or structure in response to external disturbances or changes in the process, esp. for industrial processes (rel. *feedback loop*).

advanced approach one of Basel II's IRB approaches, which requires use of internal ratings to derive estimates for PD, EAD, and LGD (rel. *foundation approach*).

adverse (*adj.*) contrary to own interests; ~ ***selection***, choices contrary to one's interests that result from asymmetric information, esp. where consciously exploited by an opponent; ~ ***on bureau***, existence of bureau data element that indicates past delinquencies/defaults.

affordability ability to do something without causing financial distress, or other undesirable consequences; ~ ***assessment***, evaluation of a borrower's ability to repay.

agency company to whom a specific type of function is outsourced, in particular collections and recoveries, and credit rating agencies; ~ ***rating grade***, risk grade provided by a credit rating agency, either for an obligor, or a specific bond issue (rel. *internal rating*).

aggregate (*v.t.*) to gather together into a body or whole; ~***d data***, **1** totals for different time periods or classes (e.g. region or industry) that aid analysis; **2** summary statistics for a class that can be used to aid the assessment of an individual within that class.

agreed limit maximum amount that can be borrowed, as agreed between both borrower and lender.

algorithm a logical procedure, involving rules and/or mathematical formulae, that is used for problem solving; ~***ic***, (*adj.*) based upon a logical or structured process (ant. *heuristic*).

all good/bad (*adj.*) related to the combination of both known and inferred performance (rel. *known good/bad, accept/reject, reject inference*); ~ ***model***, scorecard that represents the entire population, inclusive of reject inference.

Altman, Edward financial economist and professor; ~***'s z-score model***, a model developed in the 1960s to measure bankruptcy risk, based upon financial statement information.

annualise (*v.t.*) to convert an amount or percentage, relating to an accrual or accumulation over time, into its yearly equivalent.

anti-discrimination (*adj.*) related to a class of legislation that guards against unfair discrimination.

appl~*y* (*v.t.*) to request provision (credit, insurance, services, assistance), or admission (employment, education, membership), esp. formally and in writing; ~***icant***, person or company that applies; ~***ication***, a formal written request, whether paper-based or electronic; ~~ ***form***, document used to provide identification and contact details, background, motivations, and other information that will aid the assessment; ~~ ***processing system***, computer system used to collect information and make decisions; ~~ ***scoring***, use of scoring technologies in the account origination process.

approv-*e* (*v.t.*) accept, authorise, sanction; ~***al in principle***, to accept, within certain limits, in order to aid marketing or to assist a customer in making a major asset purchase.

archive collection of preserved information; ~ ***method***, 'as is' storage of records, to be accessed by retrospective searches (rel. *split-back method*).

arrears **1** unpaid or overdue debt; **2** at the end of the period (e.g. 'interest paid in arrears'); ~ ***status***, duration and/or legal status of arrears.

artificial intelligence computer programs or statistical processes that mimic the functioning of the human brain.

asymmetric information differences in available information, that affect game players competitive advantage.

ATM automated telling machine; ~ ***fraud***, any fraud related to the use of ATMs.

attribute **1** trait, property, or feature held by an individual case; **2** one of several possible categories for a particular characteristic, such as 'age <30' or 'home phone = 'Y'' (syn. *bin*, *group*).

attrition gradual loss of numbers (members, units, accounts, customers) over time (rel. *churn*); ~ ***scoring***, used to rank accounts according to probability of account closure or dormancy.

augmentation form of reject inference that adjusts the accepts' weights, so that they represent both accepts and rejects (rel. *reweighting*).

AUROC (*acr.*) Area Under the Receiver Operating Characteristic; a measure of test Validity, which is the area under the ROC, inclusive of the area below the diagonal.

authorisation approval to go ahead with a transaction; ~***s***, area responsible for dealing with merchants, when credit card transactions fall foul of lenders' predefined credit and fraud rules; ~ ***number***, reference provided to merchant as confirmation; ~ ***score***, a score that is calculated before, or after, an authorisation is approved.

autocorrelation correlation between the error terms of cases occurring in a series, esp. time series.

automated telling machine machine used to dispense cash, which usually requires the use of a plastic card and PIN (abbr. *ATM*).

back **1** behind; **2** less often seen or used; ~***-end processes***, collections and recoveries, responsible for ensuring repayment when problems are encountered; ~***-end reports***, OCC term, used for reports that focus on performance monitoring; ~***-office functions***, parts of the business that are essential for effective handling of the account, but with which the customer seldom has contact; ~***test***, comparison of actual to expected results, after actual results become available; ~***ward-looking***, (adj.) related to an empirical assessment of historical data, with no human input on the current situation, whether subjective or via current market prices (ant. *forward-looking*); ~***ward elimination***, regression procedure that starts with all variables and incrementally removes those that add the least value (rel. *stepwise*).

bad **1** (*adj.*) related to an undesirable state; **2** *(n.)* observation that does not have the desired outcome; ~ ***debt***, loan not repaid, write-off; ~ ***debt provision***, income statement charge, made in anticipation of future losses; ~ ***rate***, percentage of observations that are bad (rel. *default rate*).

balance **1** position relative to the equilibrium; **2** current value owing, or due, on an account; ***~d sample***, sample that uses equal numbers of goods and bads; ***~ sheet***, a report summarising an economic entity's financial condition at a point in time, including assets, liabilities, and net worth (rel. *income statement, financial statements*).

bankrupt financially ruined, insolvent; ***~cy scoring***, use of statistical models to measure risk of insolvency.

Basel (Basle), major city in Switzerland; ***~ Accord***, sets out uniform capital requirements for banks in different countries, to ensure that they compete on a level playing field, which relies upon a calculation of risk weighted assets; ***~ Committee on Banking Supervision***, group established to standardise banking regulations across jurisdictions (abbr. *BCBS*); ***~ I***, first Basel Accord, from 1988, that has specified risk weights for different asset categories; ***~ II***, second Basel Accord, formulation of which started in 1999, to allow use of internal ratings (rel. *standardised*, *internal ratings based*).

batch group of cases that is treated together; ***~ enquiry***, bureau searches done simultaneously for a large group, for either the current date, or a retrospective date.

Bayes, Thomas (1702–1761). British mathematician and Presbyterian minister; ***~' theorem***, a proof that the probability of A given B can be determined using the probabilities of A, B, and B given A; ***~ian***, (*adj.*) related to Bayes' theorem, and typically associated with the mathematics involving conditional probabilities.

BCBS (*acr.*) Basel Committee for Banking Supervision.

behavioural pertaining to the conduct of an individual or entity; ***~ scoring***, use of data on internal account conduct to provide a credit risk assessment for limit setting, authorisations, collections; ***~ risk indicator***, a derivative of the behavioural score, where the scores are split out into risk bands.

benchmark **1** point of reference; **2** standard against which results are compared.

Bernoulli, Jakob (1654–1705) Swiss mathematician and scientist; ***~'s law***, a.k.a. law of large numbers, theorem that the properties of a large number of random observations will approach the averages for the population; ***~ trial***, an experiment where independent observations are made of a phenomenon that only has two possible outcomes, typically referred to as success or failure.

bespoke custom, tailored, made to order; ***~ scorecard***, empirical risk ranking tool developed specifically for a customer, product, and/or process (ant. *generic scorecard*, syn. Am. *custom scorecard*).

best practice practices (processes, techniques, methodologies, and the use of technology, equipment and resources), with a proven record of success at providing a desired result.

bias (irrational) tendency, inclination; ***~ed***, leaning to one side, prejudiced (ant. *unbiased*); ***~~ sample***, not representative of the relationships that it is intended to represent (ant. *unbiased* or *representative sample*).

bin receptacle containing like items (syn. *class, group*).

binary (*adj.*) consisting of two (units, components, elements, terms) or based on two (syn. *dichotomous*); ***~ outcome***, result with two possible values: good or bad, zero or one, etc.

binomial (*adj.*) consisting of two possible outcomes, success or failure; ~ ***distribution***, the distribution showing the probability of a given number of successes in a two-outcome experiment; ~ ***test***, used for hypothesis testing of expected and observed success rates in a single group.

bivariate (*adj.*) involving two variables; ~ ***statistics***, any numbers representing the relationship between two variables.

black ~ ***box***, process where the inner workings are not known, or easily interpretable (rel. *opaque*, ant. *white box*); ~ ***data***, negative or adverse data, usually referring to judgments recorded by the credit bureaux; ~ ***hats***, bad guys, or people with evil intent, a reference to old cowboy movies; ~***list***, **1** a register of people who are out of favour; **2** negative payment performance data on the credit bureaux.

bond certificate of debt that states the repayment terms, security, and other conditions; ~ ***issue***, bonds sold at one point in time under the same terms; ~~***r***, government or company that raises funds through a bond sale; ~~~ ***rating***, credit rating given to an issuer, usually as an aggregation of all of its issue ratings.

book **1** (*v.t.*) to create or note an asset in a register; **2** (*n.*) the total portfolio of all assets.

Boolean (*adj.*) relating to George Boole (1815–1864); ~ ***algebra***, logic system that relies solely upon Boolean values to represent all problems; ~ ***logic***, logic using Boolean algebra that drives most computers; ~ ***variable***, any number that can only have values of 1 or 0, to indicate true or false.

bootstrap sampling (bootstrapping) sampling with replacement, such that an observation can appear two or more times in the same sample, usually done where the available sample is small, and/or a large number of different samples have to be generated.

break discontinuity; ~***point***, **1** level at which a different strategy is employed; **2** upper or lower bound of a range; **3** point in a process where there is a pause or decision.

brownfield a redevelopment, or development done off of an existing base (ant. *greenfield*).

bureau agency that compiles and distributes information (rel. *registry*); ~ ***data***, information, obtained from a bureau, relating to individuals or enterprises; ~ ***manager system***, computer system used to obtain, store, and retrieve bureau information, in order to avoid repeated bureau calls.

business a productive enterprise, and the people involved; ~ ***ethics***, the determination of what is morally right or wrong in business situations, and acting accordingly; ~ ***report***, report compiled regarding the operations and financial dealings of a company, usually as a precursor to investment, or entering into an agreement; ~~ ***score***, credit score derived using information provided in a business report; ~ ***risk***, possible failure to achieve business targets as a result of misreading the economic or competitive environment, or having inappropriate strategies/resources to produce/sell a product or service.

buy (*v.t*) acquire at some expense; ~ ***data***, **1** purchase of data at cost; **2** acceptance of high risk cases, in order to see how they perform (rel. *random supplementation*).

C&R *Collections and Recoveries*, the two business functions that deal with delinquent accounts.

CAIS (*acr.*) Credit Account Information Sharing (UK).

calibrate (*v.t.*) **1** to mark a scale, to indicate measurement units; **2** to ensure that scores have the same meaning, according to some measure (rel. *transform*); **3** to transform scores into risk estimates, whether directly, or through mapping.

Calinski–Harabasz statistic ratio of between-group variance to within-group variance, which if maximised provides the optimal clustering.

campaign series of actions with a given objective, usually with constraints relating to targets, time, and resources.

capital equity, wealth; ~ ***adequacy***, a measure of banks' financial strength and ability to absorb shocks, usually stated as the ratio of equity to assets; ~~ ***requirement***, proportion of equity and subordinated debt that a bank must use when making loans, as required by the banking regulator; ~ ***structure***, mix of assets, liabilities, and equity in their various forms.

card **1** credit card; **2** 'plastic' used as a transaction medium; ***~-not-present***, (*adj.*) credit card transactions where no physical card has been presented; ~ ***swap***, ATM fraud, involving an exchange of cards; ~ ***trap***, ATM fraud, involving a machine blockage to trap the card.

cardinal (*adj.*) related to an essential component; ~ ***number***, discrete numeric value, denoting a count of the number of members in a set.

CART (*acr.*) Classification and Regression Trees, a mathematical procedure used to derive decision trees.

categor~y, class or group, with a common attribute; ***~ical***, (adj.) relating to categories (rel. *binary, nominal, ordinal*); ~~ ***data***, data that classifies cases into distinct, mutually exclusive groups based upon common qualitative attributes.

cash money in highly liquid form, such as notes, coins, and funds readily available from financial institutions; ~ ***advance***, use of a credit card to draw cash, and not purchase goods or services; ~ ***flow***, **1** movement of liquid funds; **2** cash payments and receipts; ~~ ***statement***, financial statement detailing funds' movements; ~~ ***triage***, allocation of income to liabilities that will provide greatest relief.

caveat emptor, [Latin] buyer beware.

censor to exclude, ban, or cut portions, whether by design or not; ***~ed data***, information that is required but is not available, like reject performance, or outcomes outside the observation window (rel. *truncated data*).

ceteris paribus [Latin] all else being equal.

CHAID (*abbr.*) Chi-squared Automatic Interaction Detection, a mathematical procedure used to derive decision trees.

challenger contender for peak position (ant. *champion*); ~ ***strategy*** proposed new strategy, that is tested against an existing dominant strategy.

champion **1** entity or idea in peak position earned through competition (ant. *challenger*); **2** defender or promoter, esp. member of the company executive who takes a direct interest in a project, and ensures that it gets adequate resources; ~ ***strategy***, dominant strategy, currently employed by a lender; ***~/challenger***, (*adj.*) related to a means of experimentation, involving controlled use of proposed strategies on a small portion of the population, and results comparison against a control group.

change to make (*v.t.*) or become (*v.i.*) different in some way; ~ ***control***, rules put in place to govern changes to systems; ~ ***management***, a process of ensuring that changes are implemented in an orderly and transparent fashion, such that implementation errors and stakeholders' fears are minimised.

channel **1** a conduit through which something can move from point A to point B, which may occur in the natural environment, or in man-made processes; **2** different mechanisms used in marketing or decision-making, to move an applicant from one stage of a process to the next.

character risk chance of loss or damage arising from individuals' dependability, including possibility of irresponsible borrowing, frivolous disputes, skip, moral hazard, and fraud (rel. *personal distress risk*).

characteristic **1** distinguishing trait; **2** a data element that describes an observation (rel. *variable*); ~ ***analysis***, a table used to analyse a binned characteristic (rows), as it relates to the sample/population, and one or more measures (columns, for example, counts, rates, weights of evidence, point allocations, average scores); ~ ***selection***, process of choosing which characteristics will be considered for inclusion in a model.

charge-off (*v.t.*) to write off a debt, or a portion thereof; (*n.*) non-performing loan.

cheque written order for a bank to pay a specified amount of money out of an account.

child node in a decision tree, any node that has a parent.

chi-square test calculation used to evaluate a theory or hypothesis, through a comparison of actual and expected results (rel. *goodness of fit*).

CH statistic (*acr.*) Calinski–Harabasz statistic.

churn customer attrition, usually associated with: (i) loss to a competitor; or (ii) the opening of accounts solely to take advantage of special offers.

CIFAS (*acr.*) Credit Industry Fraud Avoidance System (UK).

class **1** (*v.t.*) to group into categories; **2** (*n.*) a group that is defined by one or more qualities (rel. *attribute*); ***~ification tree***, **1** series of if/then/else statements used to group; **2** graphic data visualisation tool, that branches from a root node, through child nodes, to terminal nodes in order to describe the relationship between characteristics in a dataset and a target variable (rel. *decision tree*).

clear (*v.t.*) **1** to free of obstruction; **2** to process a transaction and receive funds; ***~ed balance***, amount available to be withdrawn, excluding any 'uncleared effects'; ***~ing system***, infrastructure used to process transactions between parties, for example cheques between banks, card transactions between merchants and card issuers, etc.

closed user-group facility or service whose rules prohibit access by non-members.

cluster to be (*v.i.*), or keep (*v.t.*), close together; ~ ***analysis***, a statistical technique used to identify groups of cases that have similar attributes.

coarse class fine classes that have been grouped and will be treated as one, in order to reduce the number of attributes when developing a model (rel. *fine class*; syn. *bin*, *group*).

cod~*e*, **1** symbol(s) used to denote words, ideas, or concepts; **2** letter or number used to denote a discrete attribute; **3** set of procedures, designed to instruct a computer; ***~ing***, transcription of paper-based applications into an electronic form, so named because many of the application details are captured as codes.

coefficient (*maths*) any numeric value that appears as a constant or multiplier in a formula, for example 5 and 7 in y = 7+5x; ***~ of determination***, a measure that indicates what percentage of the error term is explained by a model (syn. *R-squared*).

cohort **1** unit within a Roman legion; **2** people with a common cause; **3** group with similar characteristics, esp. age; ***~ analysis***, see *vintage analysis*; ***~ performance***, outcome performance data from other lenders/products, used in the reject inference process.

collateral asset(s) pledged as security for a loan.

collections management of delinquencies to recover funds and retain the customer relationship; ***~ agency***, company to whom the recoveries (and possibly also collections) function is outsourced.

collinearity condition where independent variables are highly correlated, which with statistical regression techniques can bias the results, often resulting in a 'wrong sign' problem.

complian~*t*, done, or having a form that is, in accordance with requirements, including legal statute and precedent, industry codes of practice, accepted standards, rules, and company policies and procedures; ***~ce***, 1 acting as required; **2** area within an organisation, responsible for ensuring that requirements are adhered to; ***~~ hierarchy***, combination of legal statutes, legal precedents, codes of practice, policies and procedures, and unwritten codes that govern business's actions.

compulsory consent a permission that must be provided before an action can be carried out, such as for inclusion in mailing lists provided to third parties.

concentration risk risk of having large exposures to assets that face similar risks, possibly to a single group of companies, or assets in the same region or industry.

confidence level of trust; ***~ interval***, the range of values that can hold true for a population at a given ~ level, based on sample results; ***~ level***, level of certainty, usually 95 or 99 percent, that a test result is accurate or representative (rel. *significance level*); ***~ limit***, the upper or lower bound of a ~ interval.

confidential (*adj.*) secret, private; ***~ity***, level to which something is kept secret from outside parties; ***~~ agreement***, legal document, obliging one party not to divulge certain information; ***~ limit***, limit not disclosed to the customer, that is used to govern over-limit excesses and limit increases (rel. *shadow* and *target limits*).

consent (*adj.*) related to giving permission; ***~ clause***, statement in a legal agreement or other document that gives permission to perform a specified action; ***~ required***, instances where regulations insist that consent is obtained (rel. *notice required*).

consolidated score a score that has one or more other predictive scores as inputs, esp. where the input scores predict the same aspect of behaviour, for example, credit risk.

constrained judgment subjective decisions that are limited by policy rules, or aided by statistically derived models.

consumer person that uses goods or services; ~ ***credit***, monies provided to individuals for the purchase of consumer goods and services, usually on an unsecured basis.

contact details address, phone, and email details provided to ensure that a company or individual can be contacted in need.

context circumstances relevant to a fact or event; ~ ***sensitivity***, ability to adapt appropriately to specific circumstances.

contingency table statistical table containing observed frequencies for two variables.

continuous closely joined, occurring in an unbroken series (rel. *ordinal, discrete, nominal*); ~ ***variable***, numeric variable that can have an infinite number of possible values, and where the differences between values have meaning.

contribut~*e*, (*v.t.*) to play a part in providing a result; ***~ion***, **1** marginal change resulting from a given factor; **2** amount of information added by a single attribute (rel. *information value*); **3** marginal increase in net profit provided by an individual account; ***~or***, a lender, or other business, that provides information to a credit bureau, and usually has access to that information.

control variable any variable included in a scorecard development to recognise some effect, but which is not used in the final model, effectively neutralising it.

converted characteristic derived from another characteristic, in order to put it into a form that makes sense in a credit scoring model.

corporate (*adj.*) **1** related to a group of people acting as one, who are treated as an individual in law; **2** a class of enterprises that are distinguished by their large size; **3** under Basel II, an enterprise class, including specialised lending (project finance, object finance, commodities finance, and commercial real estate) and other corporate; ~ ***good governance***, risk reduction via the exercise of corporate self-restraint, including limiting the CEO's power within the organisation; ~ ***lending***, provision of large loans to a small number of enterprises, each of which receives individual treatment; ~ ***social responsibility***, companies' respect for and conduct towards the wants, needs, and concerns of other stakeholders.

correlation extent to which the values for two characteristics vary in tandem with one another; ~ ***coefficient***, a measure of the linear relationship between two variables, ranging from -1 to $+1$, where 0 denotes no relationship.

cosmetic (*adj.*) intended to improve appearances; ~ ***scorecard changes***, exclusion of characteristics, or changes in coarse classing, intended to make point allocations easier to understand and explain.

Council of Europe association of states established in 1949 to promote co-operation, human rights, and economic and social progress; ~ ***Convention (1985)***, a call for signatory countries to implement data protection legislation.

counterparty other party involved in a contract or transaction; ~ ***risk***, risk of non-performance by the other party.

credit **1** promise to pay in the future, in order to buy or borrow in the present; **2** right to defer payment of debt; ~ ***active***, making use of credit facilities from one or more lenders; ~ ***analyst***, employee that analyses borrowers' creditworthiness, and provides inputs on whether or not to extend credit, and on what terms; ~ ***appetite***, need for credit by individual borrowers; ~ ***bureau***, agency, usually privately owned and operated, that pools and distributes data from various sources, including publicly available data, subscriber data, and own data (syn. *credit reference agency, credit reporting agency*; rel. *public credit registry*); ~ ***card***, plastic card used to pay for goods and services, which is associated with an account where credit is available; ~ ***cycle***, expansion and contraction of credit within the economy, as part of an economic cycle; ~ ***history***, record of debt and payment habits, used to assess creditworthiness; ~ ***insurance***, arrangement where in return for one or more payments, a credit obligation will be forgiven in the event of death, illness, job loss, or similar; ~ ***intelligence***, information about (potential) borrowers, used to aid credit decisions; ~ ***provider***, moneylender, or goods/service provider who allows purchases on account; ~ ***rating agency***, service that assigns risk grades to the debt of public and/or private firms; ~ ***registry***, see 'public credit registry'; ~ ***report***, document that details an individual's or juristic's credit history, and current credit status; ~~***ing agency***, see 'credit bureau'; ~ ***risk***, any risk arising because of a real or perceived change in a counterparty's ability to meet its credit commitments when due (rel. *default risk*, *counterparty risk*); ~~ ***cycle***, changes in overall credit quality within the economy over time; ~~ ***management cycle***, the sequence of business functions that deal with credit risk, which for retail credit include marketing, application processing, account management, collections, and recoveries (abbr. CRMC); ~ ***scoring***, use of statistical techniques to assess credit risk, at any stage in the CRMC, but esp. for new business; ~ ***spread***, interest margin that compensates the lender for credit risk, usually calculated as the loan rate, less the risk-free rate of return; ~ ***user***, borrower or consumer requiring credit for asset purchase; ~ ***underwriter***, see 'credit analyst'.

creditor entity to whom money is owed (ant. *obligor*); ~***s' rights***, legal recourse available to creditors to collect on defaulted loans.

CRM (*acr.*) Customer Relationship Management.

CRMC (*acr.*) Credit Risk Management Cycle.

cross-fire **1** shots at the same target(s) from different positions; **2** transact on multiple accounts as part of a swindle.

cross-sell to offer additional products to accepted applicants or existing customers.

cross-sectional data macroeconomic data for different geographical regions.

c-statistic measure of separation or divergence, which for binary outcomes is the same as the Area Under the Receiver Operating Characteristic curve (AUROC).

cumulative (*adj.*) increasing in gradual steps; ~ ***distribution function***, cumulative total, as a percentage of the overall total, for each point in an ordered distribution; ~ ***total***, sum of all values to a point in an ordered distribution.

cure (*v.t.*) to restore to health or good condition.

current (*adj.*) **1** recent; **2** up to date; ~ ***account***, cheque account.

custom (*adj.*) made to order, tailored, bespoke (hence ~*ise*, ~*isation*).

customer purchaser of goods or services; ~-***direct***, (*adj.*) involving direct customer communications, with no intermediaries or assistance; ~ ***number***, a unique number assigned to each customer, so that all accounts held, irrespective of product type, can be identified; ~ ***relationship management***, process or function focused upon managing customer interactions; ~ ***scoring***, use of statistical models to derive a single risk measure for customers, that covers all products, inclusive of any savings or investments; ~***-supplied***, (*adj.*) obtained from the customer, including application forms and supporting documentation.

cut-off threshold, defined using some measure, used to determine whether or not an action is performed, or which action; ~ ***score***, score below which cases are rejected or referred.

data a collection of facts and figures relating to individual cases, which is used to draw conclusions, especially in formalised processes involving computers; ~ ***acquisition***, process of extracting and transporting data for some purpose; ~ ***analysis***, manual evaluation done to describe, summarise or compare'; ~***base***, a large store of information that can be readily accessed and used; ~ ***capture***, transcription of data into electronic format; ~ ***collection***, process of acquiring and capturing primary data from one or more sources; ~ ***controller***, person who is lawfully responsible for determining the contents and use of data; ~ ***decay***, degradation in data quality, purely as a result of age; ~ ***design***, definition of data required for a system or model development; ~***-driven***, (*adj.*) performance of actions that vary depending upon the data presented for each case; ~ ***enhancement***, credit bureaux' use of supplementary information to better match individuals with their credit histories; ~ ***hungry***, requires more data to achieve similar results; ~ ***mining***, interrogation of large amounts of data to: (i) find relationships or patterns; or (ii) test hypotheses; ~ ***preparation***, process of designing and creating the sample used for a scorecard development; ~ ***privacy***, expectation that data relating to specific individuals will not be unreasonably disseminated; ~~ ***principles***, guidelines covering manner of collection, reasonableness, quality, use, disclosure, subjects' rights, and security; ~ ***protection***, defence against improper use of or access to data, and assurance of data quality, esp. as regards personal data; ~ ***quality***, ability of data to serve the purpose for which it was obtained, which requires it to be relevant, accurate, complete, current, and consistent; ~ ***quantity***, the depth and breadth of data, which can be a function of the data's accessibility and the homogeneity of the group being assessed; ~ ***reduction***, process of reducing the number of cases and/or variables in a dataset, to aid understanding, subsequent analysis, and processing requirements; ~ ***retention period***, amount of time before data is removed from a system; ~ ***scrubbing***, the process of removing or modifying records in a dataset to address data quality issues (e.g. duplicate records, incomplete records, invalid field entries); ~***set***, a collection of data for immediate use, usually sampled from a larger database, or collected with respect to a larger population; ~ ***subject***, entity to whom data pertains; ~ ***type***, characteristic of a data field, whether defined in statistical (continuous, discrete, cardinal, ordinal), practical (currency, count, score), or computer (character, floating point, integer) terms; ~ ***visualisation***, use of graphical tools for data analysis.

decision a position, opinion, judgment, or course of action to be taken that is derived after due consideration; ~ ***automation***, the use of computers to make decisions, so that company strategies are applied quickly and consistently; ~ ***engine***, system used to make decisions, based upon available information and company strategies; ~ ***matrix***, set of rules governing the course of action to be taken based on two or more pieces of information, esp. scores (syn. *strategy matrix*); ~ ***science***, use of scientific principles and tools to make decisions; ~ ***support***, business function that assists the decision-making process, including people and systems; ~ ***tree***, logical or graphical representation of a decision process (rel. *classification tree*).

declared limit borrowing limit on an account, that is known to the customer (rel. *agreed limit*).

decline see *reject*.

deconstruct (*v.t.*) to break something into parts to achieve greater understanding of how it works, and/or to develop a means of simulating a process.

default 1 failure to honour financial commitments; **2** severe form of delinquency, where there is a high probability that the credit provider will be forced to take legal action and/or write off the debt; ~ ***correlation risk***, possibility that a group of borrowers will default together, which increases the overall risk of a portfolio; ~ ***data***, data on historical defaults, whether in the wholesale or retail markets; ~ ***event***, any event that causes an obligation to be classed as a default; ~ ***rate***, proportion of loans in a portfolio that default within a specified time period, whether historical or forecast.

degrees-of-freedom 1 minimum number of variables required to describe a statistic (abbr. *d.f.*); **2** the number of attributes for a characteristic, less one, which is used to determine the threshold chi-square statistic for confidence level tests as d.f. increases the threshold also increases.

delinquent see '*arrears*'.

delivery provision of final goods and/or services to their intended recipients; ~ ***platform/ system***, computer hardware and software used to host companies' processes, and deliver results to end users, or other systems.

demographic related to the characteristics that describe a population; ~ ***data***, **1** information relating to people, for example age, income, language, obtained from applications or questionnaires, that is used to draw broad generalisations about groups; **2** data obtained from national census and marketing research.

denominator bottom part (divisor) of a fraction, *the denominator of 5/8 is 8* (ant. *numerator*).

dependent influenced by other forces (ant. *independent*); ~ ***staging***, treatment of variables in blocks, where each stage is integrated into the next, which results in greater emphasis being given to variables in earlier stages; ~ ***variable***, factor (to be) explained by changes in one or more independent variables (syn. *response/target variable*).

derive (*v.t.*) **1** to come from; **2** to deduce through the use of logic; ***~d characteristic***, created from other readily available characteristics.

derogatory (*adj.*) prejudicial; related to low opinion; ~ ***data***, information that can only be prejudicial, esp. when obtained from external sources (syn. *negative* or *black data*).

descriptive statistical technique finds patterns that describe the data, usually by finding logical groupings of records or characteristics (rel. *factor analysis, cluster analysis*, *predictive statistical technique*).

deterministic (*adj.*) related to a process where the outcome can be exactly determined, given knowledge of some or all inputs (ant. *stochastic*).

develop (*v.t.*) to create or improve gradually; ***~ment***, project that is worked on incrementally; ***~~al evidence***, any materials supporting the development results.

dichotomous (*adj.*) **1** sharply distinguished or opposed; **2** divided into two mutually exclusive classes (rel. *binary*).

discharge (v.t.) **1** release of contents; **2** relieve of burden or cargo; **3** satisfaction or dismissal of obligations, esp. contractual.

disclosure **1** divulgence or making known; **2** provision of data to third parties.

discontinuity a break in something that is otherwise continuous or linear.

discrete (*adj.*) separate or distinct (rel. *continuous*); ~ ***characteristic***, any data element provided as a whole number, or count (rel. *nominal variable*).

discrimin~*ate*, (*v.t.*) **1** to separate, distinguish, tell apart, esp. in the sense of a model's ability to ~ between goods and bads; **2** to treat unfairly due to personal prejudices (rel. *bias*); ***~ant analysis***, statistical technique(s) that allow the separation of observations into distinct known groups, such as good and bad.

dishonour (*v.t.*) to refuse to honour a cheque, debit order, or other transaction, presented by a third party against a customer account.

disparate (*adj.*) unlike, different; ~ ***impact***, prejudicial effect of seemingly neutral policy or practice on certain sub-segments, esp. protected groups; ~ ***treatment***, **1** unlike treatment of people by a process, esp. when based on gender, race, national origin, religion, age, or other prohibited bases; **2** discrimination inherent in what should be a fair and neutral test, either because of underlying correlations, or human influences.

distance space between two points, whether physical or temporal; ~ ***lending***, use of technology to provide credit to individuals in different geographical regions; ~ ***to default***, measure of risk that is a function of 'value' as a ratio of 'volatility'.

distressed (*adj.*) under severe financial stress; ~ ***debt***, **1** loans owed by borrowers experiencing financial difficulties; **2** junk or non-investment grade bonds; ~ ***restructuring***, renegotiation of loan terms, usually to the detriment of the lender.

distribution the spread of cases across possible values, or classes, for a characteristic.

divergence separation, deviance, difference; ~ ***measure***, any summary statistic that measures the difference between two distributions, such as the information value, chi-square, or Gini coefficient; ~ ***statistic***, a measure of separation for continuous variables, calculated as the squared difference between the means, divided by the average variance for the two groups.

documentation **1** supporting facts and figures recorded in a physical or electronic format; **2** evidence or proof, esp. as it relates to scorecard and systems developments, that records details of information used, assumptions made, and the end product implemented.

domain expert person with knowledge about a given subject area, who is called upon to help specify input, process, output, and control requirements, esp. when developing knowledge-based computer systems (syn. *subject-matter expert*).

down being or moving lower (ant. *up*); ***~-sell***, offer of other products to declined applicants, with less advantageous terms (e.g. higher interest rates, less payment flexibility, lower amounts); ***~stream***, subsequent stages in a process; ***~turn***, weakening of economic activity, for example two or more consecutive quarters of negative real-GDP growth; ~~ ***LGD***, loss estimate provided for downturn scenario, esp. for Basel II purposes.

drift (*v.i.*) **1** to move, as if carried by air or water; **2** (*n.*) gradual, but sometimes sudden, change from original or expected state; **3** (*n.*) changes in the market, economy, operations, data sources, or outcomes, that impact upon the appropriateness of the process or its parameters; ~ ***analysis***, review and comparison of summary data at different points in time; ~ ***report***, presentation of data to aid comparisons over time.

drop-down box field on a computer screen, where various possible options are provided when the user clicks on it.

dual (*adj.*) relating to, or denoting, two (2); ~***ity***, existence of two opposing forces, or concepts (syn. *dichotomy*, *polarity*); ~ ***processing***, use of two systems to process the same inputs, usually as part of testing, to ensure quality control.

dummy variable a binary variable used in regression modelling, to represent a single attribute of a binned characteristic.

duplication **1** making of an exact copy; **2** process of creating a twin record, and apportioning the original weight between the two (rel. *fuzzy*).

duty of confidentiality/secrecy implied contractual obligation that personal details should not be provided to third parties, which arises from a fiduciary relationship between bank and borrower.

dynamic delinquency report vintage analysis used to track delinquencies, as part of new business monitoring (rel. *vintage analysis*, *cohort analysis*).

EAD (*acr.*) Exposure at Default.

early (*adj.*) **1** near the temporal beginning; **2** ahead of schedule; ~ ***performance monitoring***, tracking of payment behaviour during the first months after account opening; ~ ***settlement***, repayment of a loan in full, prior to the end of its contractual term.

ECDF (*acr.*) Empirical Cumulative Distribution Function, the cumulative total for each score as the score increases, stated as a percentage of the overall total.

EDW (*acr.*) Electronic Data Warehouse.

efficiency curve see 'Lorenz Curve'.

EFTPOS (*acr.*) Electronic Funds Transfer at Point of Sale, the system used for clearing credit card and debit card transactions.

eigenvalue the amount of variance in a set of variables, that is explained by a given factor or component, as determined from the correlation or covariance matrix.

electronic data warehouse data storage facility, that brings data together from across business units, and allows general access within the business (abbr. EDW).

embellishment **1** enhancement, usually superfluous; **2** minor first-party fraud, where applicants provide incorrect details to improve chance of acceptance (syn. *massaging*).

emerging market developing country, or other market that is undergoing a growth phase, and may come onto a par with developed markets.

empirical (*adj.*) **1** derived from observation and measurement; **2** based upon past experience (ant. *judgmental*); ~ ***analysis***, use of numerical techniques to analyse historical data.

enquiry or **inquiry** a request to an external bureau, for information about a prospective or existing customer (syn. *search*); ~ ***count***, number of enquiries recorded against each customer by the bureau.

enterprise business venture, undertaking with a purpose; ~ ***lending***, provision of credit to business undertakings; ~ ***market***, small business, middle market, and large companies.

entry scorecard scorecard applied when an account first enters a stage of the risk management cycle, and the resultant score is retained.

equal identical or equivalent; ~ ***Credit Opportunity Act (1974)***, American anti-discrimination legislation specific to credit; ~ ***opportunity legislation***, laws against unfair discrimination, esp. on the basis of colour, religion, language, gender, national origin, etc.

error **1** mistake or inaccuracy; **2** difference between expected and observed values, esp. model predictions and actual outcomes (syn. *residual*); ~ ***of commission***, inaccuracies arising from duplication, or incorrect capture, calculation, or matching; ~ ***of omission***, inaccuracies arising from missing data or records; ~ ***term***, element in a formula that forces equality between predicted and actual, usually represented by the letter '*e*'.

EU (*acr.*) European Union; ~ ***Data Protection Directive 95/46/EC (1995)***, integration of OECD Data Privacy Guidelines and Council of Europe Convention, with the purpose of protecting data privacy, but not at the expense of transborder data flows.

eugenics selective breeding used to improve the genetic stock of a species, esp. humans, a now discredited concept initially proposed by Sir Francis Galton in the 1880s, which became popular amongst European intelligentsia into the early twentieth century.

event warning advice given immediately, whether from internal or external sources, if an event indicative of high risk occurs.

evergreen (*adj.*) plant that renews its foliage irrespective of season; ~ ***limit***, credit limit that is automatically renewed if it meets certain criteria.

ex ante, [Latin] **1** beforehand; **2** based upon expectations of the future; ***ex post***, [Latin] **1** after the fact; **2** based upon past history or actual events.

excess breach of the limit agreed on an account.

exit state the black hole of a transition matrix, that cases never exit once entered, for example, account closure or write-off (syn. *absorption state*).

exogenous (*adj.*) originating outside of a system (ant. *endogenous*); ~ ***factor***, anything that occurs outside of a system, that may influence its functioning and/or output.

experiment an act conducted to test an idea; ***~al design***, plan specifying how a test will be performed, in order to ensure that the results are reliable, interpretable, and usable; ***~ation***, a process of changing a design or parameters to test hypotheses, where either knowledge or performance improvements are being sought.

expert system a system developed to capture and exploit the knowledge of experts in a field, for example, using inputs from different doctors to develop a means of identifying illnesses based upon the symptoms.

exposure **1** vulnerability to damage, loss, or hardship, arising from external sources; **2** potential loss, including the current investment, and any amounts that can be demanded in future; **3** the higher of the current debit balance, and the arranged limit; ~ ***class***, 1 a group of assets, subject to the same risks; 2 under Basel II, the groupings of retail, corporate, interbank, sovereign, and equity; ~ ***at default***, real or estimated exposure on the date of default (abbr. EAD).

extension account management practice of forgiving late payments, and adding them on to the end of the loan period.

external (*adj.*) related to being outside of certain boundaries; ~ ***data***, data obtained from sources outside of a system or organisation; ~ ***rating***, any rating provided by an external agency, esp. those provided by credit rating agencies and bureaux, or regulatory authorities.

extrapolation 1 to estimate beyond the range already known; 2 a reject inference technique that uses known good/bad performance.

facility 1 something available to serve a particular function; 2 loan provided by a bank.

factor analysis a multivariate statistical technique used for data reduction, which simplifies available data by creating factors to summarise correlated characteristics.

fair done in a manner considered acceptable by each party to a transaction, and to the regulatory authorities; ~ ***credit***, provision of credit using accepted practices; ~ ***Credit Reporting Act (1970)***, American legislation that governs credit bureaux' operations; ~ ***lending***, provision of funds on a basis that is considered non-discriminatory, and upon terms considered reasonable in those circumstances; (rel. *responsible lending* and *equal opportunity*).

false 1 untrue; 2 observation classified incorrectly; ~ ***negative***, positive (undesirable result, or bad) incorrectly identified as negative (syn. *type I error*, *false alarm*); ~ ***positive***, negative (desired result, or good) incorrectly identified as positive (syn. *type II error*).

feasib-*le*, (*adj.*) capable of being achieved; ***~ility study***, investigation done in advance to determine whether a project can meet objectives within certain parameters.

feedback 1 results from an experiment; 2 information provided by a process, used to make decisions regarding process changes; ~ ***loop***, mechanism that drives the return of information, and adjusts inputs to maintain stable output (rel. *adaptive control*).

FICO (*acr.*) Fair Isaac Company; ~ ***score***, generic credit bureau score, derived using a model developed by FICO.

fiduciary (*adj.*) involving trust; ~ ***duty***, level of trust expected because of the relationship, contractual or otherwise, between two parties.

field an area on an application form, or in a database, that is meant to contain data, often used in the same sense as 'characteristic', and/or 'variable'.

final (*adj.*) 1 occurring at the end; 2 related to last revision; ~ ***model***, last and hopefully best representation, to be used for an intended purpose; ~ ***scorecard***, find model, to be implemented in a production process.

financial (*adj.*) related to dealings with money; ~ ***exclusion***, lack of access to formal credit markets; ~ ***inclusion***, availability and promotion of access to formal credit markets, especially for underserved sectors of the population; ~ ***intelligence***, information about individuals' financial transactions, used to guard against criminal and terrorist activities; ~ ***ratio***, financial statement numbers stated as proportions of each other, used to facilitate evaluation/comparison of enterprises' financial health; ~~ ***scoring***, use of financial ratios in predictive models, to measure credit risk; ~ ***sophistication***, extent to which money matters are managed in a complex and refined manner; ~ ***spreading***, capture of financial statements into a common format, to aid comparison; ~ ***statements***, reports that summarise data and provide an indication of financial status, including balance sheet, income statement, cash flow statement, statement of retained earnings, supplementary notes, and management commentary.

fine class **1** a very narrow range within a characteristic, for example Age = 21 (syn. *initial enumeration*, rel. *coarse class*); **2** (*v.t.*) to create a detailed frequency distribution, used to do subsequent grouping.

first (*adj.*) before anything else; ~ ***party***, main person or entity with whom a contract is agreed; ~~ ***fraud***, fraud committed by known accountholder; ~***-payment default***, instances where the initial payment on a new loan are missed, which may be technical arrears, but can also indicate possible fraud.

Fisher, Sir Ronald Aylmer (1890–1962). English statistician, renowned for his work in multivariate analysis; **~'s linear discriminant analysis,** statistical technique used to determine group membership.

fixed term time period that is known and agreed; ~ ***loan***, made for a known term, and usually repaid in regular instalments.

flat maximum optimal result that will never be exceeded, but which may be approached by any number of suboptimal solutions.

floor limit the maximum amount that can be transacted without requiring authorisation, esp. with respect to credit cards.

footprint geographic or physical area covered (electronic signal, postal code, sonic boom, etc.); ***leave a ~***, to leave a mark when performing some action, for example an increase in the search count that results from a bureau enquiry.

forecast (*v.t.*) to suggest a possible value at a future date (rel. *predict*).

foreign key data item in a database, that can be linked to a primary or matching key in another database, in order to cross-reference and/or supplement data.

forgiveness period amount of time before defaults, judgments, dishonours, and other transgressions are excused, which may be set by law or company policy.

form **1** structure or shape; **2** document containing spaces into which information is entered; ~ ***design***, derivation of information required and its layout, in order to facilitate ease of completion, and maximise the value of information provided.

forward to the front, whether in space or time (ant. *backward*); ~***-looking***, (*adj.*) **1** taking a view on the future; **2** incorporates subjective human input relating to the future, either directly or via the market value of traded securities; ~ ***selection***, regression procedure, that starts from scratch and incrementally includes variables that add the most value to the model.

foundation approach one of Basel II's allowed IRB approaches for calculating risk weighted assets, which uses internal ratings to derive PD estimates, and values specified by the national regulator for other elements (rel. *advanced approach*).

fraud transaction involving deception or trickery, that gives the perpetrator an unfair advantage, usually with criminal intent and the goal of personal gain; ~ ***detection***, process of identifying potential fraud; ~***ster***, swindler, person that commits fraud; ~ ***syndicate***, criminal organisation that specialises in fraud; ~ ***warning***, cautionary advice indicating potential for intentional deception by a customer.

free-form field unstructured field, where the applicant may write any possible answer.

front (*adj.*) most often seen or used; ~***-end reports***, OCC term, used for reports that focus on through-the-door process monitoring, with no attempt to track performance; ~***-end processes***, marketing and application processing functions, responsible for attracting new business; ~***-office functions***, customer interface, parts of the business dealing directly with customers.

F-statistic **1** statistic used to assess the difference between the means of two groups; **2** label sometimes applied to the information value.

fulfilment final stage of account origination process, where the customer is provided with the product applied for; ~ ***data***, details relating to telemarketing and direct mail contacts, and their outcome.

fuzzy (*adj.*) indistinct; ~ ***logic***, reasoning that can use partial truths (probabilities) as opposed to definitive true/false values (rel. *Boolean logic*); ~ ***parcelling***, performance manipulation technique, where rejects are split into good and bad portions, esp. for use with reject inference (syn. *duplication* and *partial reclassification*).

G10 group of 10 industrialised countries that are members of the International Monetary Fund (IMF), including Belgium, Canada, France, Germany, Italy, Japan, Netherlands, Sweden, United Kingdom, and the United States (Switzerland joined as the 11th member in 1984, but is not part of the IMF).

GAIN (*acr.*) Gone Away Information Network (UK).

Galton, Sir Francis (1822–1911). English polymath, who introduced the statistical concepts of regression and correlation.

GBIX (*acr.*) good, bad, indeterminate, and exclude; ~ ***characteristic***, variable derived by applying the good/bad definition to a dataset; ~ ***definition***, see 'good/bad definition'.

generated characteristic characteristic derived from two or more other characteristics, in an attempt to address interactions.

generic **1** one-size-fits-all; **2** applied generally (ant. *bespoke*); ~ ***scorecard***, developed using data from many sources, and used for many products and companies.

genetic algorithm mathematical procedure based upon evolutionary principles, such as mutation, deletion, and selection.

geocode see 'lifestyle indicator'.

Gini, Corrado (1884–1965). Italian economist; ~ ***coefficient***, a measure of separation, usually used to assess income disparties, but used in credit scoring to assess models' predictive power. (rel. AUROC).

gone-away applicant cannot be found using contact details given (address, phone, email), either due to a move without a change of address notification, or possible fraud (syn. *skip*).

good observation that has the desired outcome (ant. *bad*); ***~/bad definition***, a set of rules that defines good, bad, indeterminate, and excluded cases.

good governance the process of decision-making and the process by which decisions are implemented or not implemented (rel. *corporate ~*).

goodness of fit the extent to which a distribution has an expected pattern, which is measured using a chi-square statistic.

grade position in a scale, according to some quality (rel. *score*).

granular (*adj.*) composed of small parts, like grains in a bushel of wheat; ***~ity***, amount of detail with which information, such as a score, is presented.

greenfield project that is a totally new development (ant. *brownfield*).

group **1** set with one or more similar attributes, that is treated or acts as a unit; **2** aggregation of various categories for a categorical characteristic, or a range within a numeric characteristic; ***~ lending***, loans made to several individuals, which is a feature in the micro-finance market.

hard (*adj.*) related to extreme firmness, or great effort; ***~ code***, (*v.t.*) to include parameters within a program's source code; ***~ collections***, relating to actions used to deal with late delinquencies, where the chance of recovery is low, the probability of legal action/loss is high, and recovering the money takes precedence over maintaining the customer relationship (syn. *recoveries*, rel. *soft collections*).

hazard undesirable situation or event (rel. *survival*); ***~ definition***, set of rules that define the binary variable used in risk modelling (rel. *good/bad definition*); ***~ rate***, percentage of cases within a group that encounter that hazard within a given time frame.

hetero (*pref.*) different (ant. *homo*); ***~geneous***, (*adj.*) of different kinds; ***~~ population***, group that is characterised by more differences than similarities; ***~scedastic***, having residuals/ errors that vary across the range of possible predicted values.

heuristic (*adj.*) **1** based upon discovery and invention; **2** learning process based on trial and error (ant. *algorithmic*); **3** rules of thumb, intuition, expert opinion, or common sense, in particular where exact relationships are unknown, and especially when determining a knowledge base for artificial intelligence.

high (*adj.*) towards the upper end of some scale; ***~ net worth***, (*adj.*) related to wealthy individuals, with say more than US$1mn in liquid assets; ***~-score override***, a score accept that is declined either by a policy rule, or manual override (syn. *accept override*); ***~-ticket***, expensive, with a high price-tag, often luxury goods—such as motor vehicles, homes, yachts, airplanes.

hire purchase payment of goods on an instalment basis, with ownership being transferred only once paid in full.

historical (*adj.*) related to what has happened in the past; ***~ method***, with rating grades, an analysis of default and loss rates using historical data; ***~ sample***, data for a selection of past cases, which is used for data analysis and/or model development.

H–L statistic see 'Hosmer–Lemeshow statistic'.

hold-out sample group of in-time cases that are used to validate model results (rel. *out-of-time sample*).

home domicilium, residence; ~ ***collected credit***, form of subprime lending, where instalments are collected from clients at their homes (rel. *subprime lending*); ~ ***loan***, finance provided to purchase a house, apartment, or other personal residence.

homo (*pref.*) same (ant. *hetero*); ~***genous***, (*adj.*) **1** same or similar; **2** of the same kind (ant. *heterogeneous*); ~~ ***data***, collection of data objects that can be treated and interpreted in the same fashion; ~~ ***population***, group that is characterised by more similarities than differences; ~***scedastic***, having residuals/errors that exhibit a constant pattern across the range of possible predicted values (ant. *heteroscedastic*).

Hosmer–Lemeshow statistic measure of goodness of fit between observed and expected probabilities, calculated as the sum of the squared z-statistics over a series of risk groups, where the z-statistic is the normal approximation to a binomial distribution.

host **1** (*v.t.*) to provide resources and/or facilities for a function or service; **2** (n.) computer system that provides or facilitates services to end-users or other computer systems; ~***ed solution***, bespoke scorecard or system operating on an external computer.

hot card lost or stolen credit card; ~ ***file***, record of lost or stolen cards, that is distributed to card merchants.

householding **1** consolidation of data at address level; **2** use of related-party information, whether for individuals or companies.

human rights obligations and duties of society to individuals, such as freedom, justice, security, etc.

hurdle obstacle to be overcome, prior to moving on to a next stage; ~ ***rate***, percentage that must be exceeded before inclusion in the next stage is considered.

hybrid (*adj.*) of mixed origin; ~ ***model***, model developed by combining models of different types, or an existing model with more data or views, esp. expert input.

hypothesis suggested explanation for an observed phenomenon; ~ ***test***, use of a statistical technique to prove a null hypothesis, as opposed to an alternative hypothesis.

identifier number or code used by an entity uniquely to identify itself to outside parties, such as personal identifiers and company registration numbers.

identity (*adj.*) related to characteristics specific to an individual or entity; ~ ***fabrication***, creation of a fictitious identity; ~ ***fraud***, identity misrepresentation, with the purpose of committing fraud; ~ ***misrepresentation***, misstatement of one's identity, intended to mislead; ~ ***theft***, the use of other people's personal information without their knowledge, esp. to commit fraud or other illegal acts; ~ ***verification***, process of ensuring that identity details are correct for an entity.

idiosyncratic (*adj.*) related to an individual case; ~ ***factor***, characteristic that is peculiar to individual cases; ~ ***risk***, a risk that arises from the unique circumstances of a particular case, and can be mitigated through diversification (ant. *systemic risk*).

implementation process of installing hardware, software, or models, to achieve a given end; ~ ***error***, mistake made during installation that has adverse consequences.

in-sample (*adj.*) pertaining to data used to determine the coefficients for a model (ant. *out-of-sample*, rel. *training sample*).

inbound (*adj.*) approaches by (prospective) customers to the business, esp. for call centres and customer service queries (ant. *outbound*).

income earnings received from economic activities and investments; ~ ***statement***, report summarising income and expenses for a given period (rel. *balance sheet, financial statements*).

independent not influenced by other forces (ant. *dependent*); ~ ***staging***, separate treatment of blocks of variables, in no particular order, the results of which are later integrated; ~ ***variable***, factor that is tested to determine whether it is predictive of an observed outcome (ant. *dependent variable*, rel. *predictor, observation*).

indeterminate **1** uncertain, ambiguous; **2** case whose performance cannot be definitively classified as good or bad.

index **1** unique integer that identifies a record or group within a series; **2** ratio of attribute odds to population odds, a statistic often provided in characteristic analyses (rel. *weight of evidence*).

industry risk uncertainty arising from risks common to an industry, that are likely to affect all enterprises operating in that sector similarly.

infer (*v.t.*), deduce, conclude based upon available evidence; ***~ence***, act or process of inferring (see '*reject inference*'); ***~red performance***, assignment or adjustment of rejects' outcome performance; ***~red policy reject***, case where both reject and bad probabilities are extremely high, for which no reject inference is performed.

information data (summarised), communications, or instructions that inform; ~ ***asymmetry***, differences in the quality and quantity of available information, that affect decision-makers' competitive positions; ~ ***goods***, information as a traded commodity; ~ ***rent***, extra utility achieved from having information not available to other players; ~ ***services***, business activities relating to the provision of data, analysis, analytical software, or other data-related services; ~ ***sharing***, the pooling and joint use of account performance data by credit providers (rel. *payment profile, shared information*); ~ ***value***, Kullback divergence measure, as used to assess characteristics' predictive power.

informed customer effect customers' tendency to choose the option that is best understood, all else being equal.

initial enumeration a first pass at creating a set of counts in a frequency distribution, which has as much detail as possible (syn. *fine classing*).

insurance agreement to reimburse in case of loss, in exchange for an upfront or regular stream of payments; ~ ***scoring***, use of credit data, scores, and techniques to determine insurance claim and policy lapse probabilities.

integral cumulative area under the curve above the x-axis, calculated at different values of x as x increases ($F(x) = \Pr[X \leq x]$), which can also be considered as the derivative's inverse [$F(x)$ is the integral of $f(x)$ if $dF/dx = f(x)$]. It may be done for both continuous $\left(F(x) = \int_{-\infty}^{\infty} f(\mu)d\mu\right)$ and discrete characterstics $\left(F(x) = \sum_{i=0}^{x} f(X)\right)$.

integrate (*v.t.*) **1** to combine or merge; **2** to combine different pieces of data or information, into a single measure.

interaction **1** influence of variables upon each another; **2** where a predictor's effect upon the response variable varies, depending upon the value of one or more other predictors; **3** where different predictive patterns exist, for different subgroups within the population.

interest charge for borrowing money; ~ ***in suspense***, accrued interest on non-performing accounts, that cannot be treated as income; ~ ***margin***, difference between the interest rate earned on a loan, and the cost of capital; ~ ***rate***, cost of borrowing, stated as a percentage of the outstanding balance, usually over a period of one year.

intermediate model a model that is developed and influences a scorecard development (*known good/bad* and *accept/reject models*), but is not part of the final deliverable (*all good/bad model*).

internal (*adj.*) within some defined limits, esp. within a company; ~ ***data***, data generated by a company's operations, obtained neither from the customer nor another outside party; ~ ***ratings***, scores or grades derived by lenders to represent obligors' credit risk (rel. *agency rating*); ~~ ***based***, (*adj.*) related to being derived from risk measures, used internally by a company as a part of their: (i) business processes, (ii) strategy and planning, and (iii) reporting (*acr.* IRB).

Internet means of accessing the World Wide Web; ~ ***fraud***, any fraudulent activity that makes use of the Internet, whether to gain confidential information, or transact.

irresponsible (*adj.*) negligent, not accountable, barely legal; ~ ***borrowing***, indebting oneself, without consideration of one's own ability to repay; ~ ***lending***, practices that have the result of misleading and/or over-indebting borrowers.

investigative report any report containing 'information on a person's character, general reputation, personal characteristics, or mode of living, obtained through personal interviews with neighbours, friends, associates, or others with such knowledge'—Fair Credit Reporting Act.

investment grade rating agency grade, considered good enough for investors who are restricted in their investments, usually BBB or better (ant. *speculative grade*).

IRB (*acr.*) Internal Ratings Based; ~ ***approach***, a means allowed by Basel II to derive banks' risk weighted assets, which uses banks' own internally derived estimates, and includes the foundation and advanced approaches (rel. *standardised approach*); ~ ***component***, element used in the IRB approach, including probability-of-default (PD), exposure-at-default (EAD), loss-given-default (LGD), and maturity (*M*).

ISIC (*acr.*) International Standard Industrial Classification; a numeric code used to classify the industries within which companies operate.

issue see 'bond issue'.

jackknifing removal of a single case from the development sample, done repeatedly and at random, to validate a model developed on a small sample (rel. *bootstrapping*).

judgment **1** opinion, decision; **2** determination by a court; **3** court order demanding repayment of debt; ~***al***, (*adj.*) based upon human judgment (ant. *empirical*, syn. *subjective*); ~~ ***overlay***, the use of judgment, to adjust either the risk assessment, or the decision provided by a rule-based decision system; ~~ ***scoring***, use of human judgment, to perform a risk assessment using available information (syn. *manual underwriting*).

juristic related to law; ~ ***individual/person***, legal entity, that is treated as an individual in law, esp. registered companies (ant. *natural person*).

kite **1** series of linked loans, investments, and/or other financial transactions, made by one or more entities, that pose inordinate and often poorly understood risk to financial institution(s); **2** negotiable paper with nothing to back it, used as part of a swindle, usually to facilitate access to credit; ~***ing***, making of fictitious deposits; ~ ***flying***, a longer-term pattern of fictitious deposits and withdrawals, intended to maximise the available credit limit (syn. *cross firing, paper hanging*).

K-nearest neighbours machine learning technique that determines group membership by finding cases that are most similar to an 'unseen' case, for which membership is unknown.

Know Your Customer class of legislation requiring credit providers to ensure proper identification of customers, meant to protect against criminal and terrorist activities (acr. KYC, rel. *financial intelligence and control, anti-money laundering*).

known (*adj.*) **1** specified and identified; **2** existing and readily quantifiable; ~ ***fraud***, financial loss proven to be the result of intentional deception; ~ ***good/bad***, (*adj.*) related to performance of accepted accounts, where performance is known (rel. *all good/bad, accept/reject, reject inference*); ~~ ***model***, scorecard developed using known performance only, used as part of the reject inference process; ~ ***performance***, an observed outcome exists, esp. for cases both accepted and taken-up in a selection process, for which no reject inference is necessary (ant. *no performance*); ~ ***to inferred odds ratio***, ratio used to assess the appropriateness of the inferred good/bad odds.

Kolmogorov, Andrei Nikolaevich (1903–1987). Soviet mathematician; ~-***Smirnov statistic***, measure of separation, calculated as the maximum absolute difference between the cumulative percentages of bads and goods, across the entire score range.

Kullback, Solomon (1903–1994). American cryptanalyst and mathematician; ~ ***divergence measure***, measure of separation used to compare two frequency distributions, which makes no assumption about rank ordering, and which in credit scoring is used to rate: (i) the predictive power of individual characteristics (information value); or (ii) the extent to which the through-the-door population has changed over time (stability index).

KYC (*acr.*) Know Your Customer.

legacy **1** something handed down, or passed on; **2** something that has survived, often with an implication of known weaknesses; ~ ***systems***, old systems and software that are still being used.

legal (*adj.*) related to the law; ~ ***origin***, **1** origin as defined by law; **2** type of legal system classified according to origin, for example English, Germanic, Napoleonic, Nordic, and Soviet.

lending practice customary way that one or more lenders operate, which may be responsible, irresponsible, or predatory.

LGD (*acr.*) Loss-Given-Default.

lie factor **1** difference between what is presented and the truth, or what is said and done (rel. *misrepresentation*); **2** difference between the size of an effect shown in a graphic, and in the underlying data, usually expressed as a ratio.

lien legal claim over an asset that has been pledged as security, for the repayment of debt.

life state, manner, and/or duration of existence; ***~-cycle effect***, increasing bad rate, that is associated with increasing time since acceptance (rel. *time effect*); ***~style indicator***, a code assigned to an address, that is indicative of the type of neighbourhood, for example, old-money, happy families, workers' quarters (syn. *geo-* or *mosaic code*); ***~time customer value***, the net present value of profits, expected to be made from a customer relationship, across all products.

lift improvement in predictive power, provided by changes to the scorecard development process, data used, or other variation.

likelihood ratio for any diagnostic test, the ratio of the probability of a positive result for a true positive, relative to the same probability for a true negative.

limit maximum amount that may be borrowed on an account (rel. *agreed ~, declared ~*); ***~ review***, call for updated information, in order to reassess the agreed limit.

linear (*adj.*) lying in a straight line; ***~ probability modelling***, use of linear regression to model a binary outcome; ***~ programming***, an operations research technique, initially used for military logistics, that seeks to maximise or minimise a value, while not violating given constraints; ***~ regression***, formula explaining a linear association between characteristics (e.g. $y = a + bx$), or statistical technique used to derive it.

liquid (*adj.*) **1** readily exchangeable into another form; **2** funds are readily available to meet commitments; ***~ate***, to convert assets into cash; ***~ity risk***, **1** risk that an asset will not be readily exchangeable, due to a lack of market demand, hence affecting the value realised; **2** risk that a credit obligation cannot be met, due to short-term lack of available funds.

list array of names, words, or other objects; ~ **scrubbing** to remove cases from a list that clearly fall outside of the target population (e.g. duplicate records, already customers, bad on bureau).

loading physical process of implementing a scorecard on a host system.

loan shark person that lends money at exorbitant rates, without any consideration of borrowers' ability to repay, often involving intimidation or violence to collect.

local knowledge information peculiar to certain cases or environments, that cannot be captured in automated processes; ***~ system***, infrastructure used to collect, store, and interrogate customer intelligence, obtained from branch staff, collections, over-limit management, customer relationship management, and elsewhere.

logistic pertaining to a logarithmic function; ***~ regression***, statistical method, used to develop models to calculate the probability that one of two possible outcomes will occur.

logit **1** logistic unit, or natural log odds, $\text{logit}(p) = \log(p/(1-p))$; **2** logistic regression that provides logit estimates.

Lorenz, Max Otto (1876–1959). American mathematician; ~ **curve** a graph used in economics to illustrate income inequalities, that has been adopted by credit scoring to show the ability of a model to discriminate between good and bad accounts (rel. *Gini coefficient*, syn. *efficiency curve*, *trade-off curve*, *power curve*).

loss 1 amount by which expenditure exceeds revenue; 2 amount written-off; ~ ***given default***, estimate of amount written off, assuming that default occurs (acr. LGD, rel. *probability of default*, *exposure at default*); ~ ***probability***, proportion of cases that are expected to result in losses; ~ ***provisions***, funds set aside to cover potential bad debts, or a reduction in asset values; ~ ***severity***, extent of the loss, assuming a loss occurs; ~ ***timing***, distribution of losses over time, from a given reference point.

low-score override a score reject, that is accepted either by a policy rule, or manual override (rel. *reject override*).

mail order type of retail business where consumer goods are ordered and delivered by mail.

Malthus, Thomas Robert (1766–1834), Church of England minister, famed for his hypothesis that population growth would exceed production growth; ~***ian***, (*adj.*) related to Malthus's hypothesis.

map set of rules for transforming values from one scale or set to another; ~***ping table***, a table specifying the relationship between the two formats, for example $X = C$, $Q = W$.

MAPA (*acr.*) monotone adjacent (violators) pooling algorithm.

margin 1 edge, or rim and area immediately adjacent; 2 increment over minimum or prior value; ~***al***, 1 related to small increments; 2 at, or near, a limit or cut-off; ~~ ***accept/reject***, selected and non-selected applicants respectively, at or near the cut-off; ~~ ***risk***, change in credit quality (as measured by the bad rate or good/bad odds) implied by a small change in score, esp. near the cut-off.

mark 1 sign, symbol, or distinguishing feature; 2 (*inf.*) individual or entity being defrauded.

market collection of entities that buy and sell goods/services; ~***ing***, collection of business functions/actions associated with new business acquisition, including segmentation and solicitation; ~~ ***mix***, blend of product, price, package, promotion, and place, used for a given target market; ~ ***risk***, uncertainty arising from fluctuations in market prices.

Markov, Andrei Andreyevich (1856–1922). Russian mathematician; ~ ***chain***, mathematical representation of a Markov process, that consists of: (i) the current states; (ii) a transition matrix, and (iii) the number of stages in the sequence; ~ ***process***, stochastic process, where changes in states have the Markov property; ~ ***property***, memoryless.

mass (*adj.*) related to a large number of people; ~ ***customisation***, tailoring of products to meet individual needs on a large scale.

master/niche see 'mother/child'.

match (*v.t.*) to link database records for applications, accounts, and/or customers, using a common piece of information; ~***ing key***, data item used to link records, such as a personal identification number or customer number.

matrix rectangular representation of information, using rows and columns (rel. *contingency table*); ~ ***approach***, use of a tabular format for data analysis, or to assign strategies.

maximum likelihood estimation process used when doing logistic regression.

mean average; ~ ***reversion***, tendency of default rates associated with different credit ratings to gravitate towards the mean, when they are analysed over successive periods, from a given observation point.

measure of ~ ***(in)dependence***, any statistic used to indicate correlation between two variables, that will always show the extent (random ↔ perfect, independent ↔ dependent), and often the direction (positive/negative); ~ ***divergence***, statistic used to measure the difference between expected and observed values; ~ ***separation***, measure of dependence used to assess the predictive power of a test, characteristic, grade, or score.

medium (pl. *media*), something used to transmit, transform, or cause some desired effect.

memoryless (*adj.*) the Markov property, where a transition matrix contains all of the information required to provide a reasonable estimate of the future, using information only about the present, and not the past.

merchant **1** trader, retailer; **2** business entity that accepts credit cards as a form of payment.

Merton, Robert C., American economist and mathematician; ***~'s model***, calculation used to value a firm as a European put option on its assets, with a strike price equal to its liabilities.

micro (*pref.*) very small; ***~finance***, **1** provision of government, or donor-subsidised, finance to poor communities in third-world and emerging market environments; **2** responsible lending for small amounts; ***~lending***, term used for subprime lending in South Africa.

middle market business customers who are neither large nor small, and fall somewhere between retail and wholesale (rel. *SME, corporate*); ~ ***lending***, provision of finance to medium-sized companies, which Dun & Bradstreet defines for the United States as the approx. 113,000 companies with sales between $10 mn and $500 mn.

misclassification inclusion in an incorrect category; ~ ***costs***, financial implication of incorrect classification; ~ ***matrix***, contingency table containing the number and percentages of true positives, true negatives, false positives (type I errors), and false negatives (type II errors), given certain assumptions (rel. *percent correctly classified*).

misrepresentation false or misleading statement of the true nature of an asset, person, enterprise, scenario, or circumstances.

missing not present, cannot be found; ~ ***data***, **1** errors of omission, whether blank predictors or missing records; **2** rejects and not-taken-ups with no performance, which may be missing: (i) completely at random, (ii) at random, or (iii) not at random; ***~ness***, conditions surrounding the state of being missing.

mitigate (*v.t.*) lessening of severity, either naturally, or through conscious action.

mixture model **1** model based upon proportions of a total, instead of raw values; **2** modelling of a probability density function, using the distributions of its constituent subpopulations; **3** model developed using cases whose true classification is not known.

model **1** a representation of a person, object, entity, or process, including financial and statistical models (rel. *scorecard*); **2** (*v.t.*) process of creating a representation; ~ ***risk***, potential for errors and costs resulting from use of an incorrect or inappropriate representation.

modus operandi [Latin] mode of operation.

money primary medium of exchange within an economy, that also acts as a store of value and unit of account; ~ ***laundering***, movement of money earned from illegal activities through legitimate channels, in order to make its origin untraceable.

monoton~*e*, (*adj.*) unvarying, or monotonous; ~~ ***adjacent (violators) pooling algorithm***, tool used to create groups, that enforce a monotonic relationship (abbr. MAPA); ***~ic***, (*adj.*) **1** consistently increasing and never decreasing, or vice versa; **2** relationship between two characteristics, where an increment or decrement in one, is always associated with a unidirectional change in the other (ant. *non-monotonic*).

moral hazard the risk that the behaviour of one or more parties to a contract will change once it has been finalised, such as insurance causing increased recklessness.

mortgage pledge of a house, or other property, as security for a loan (a.k.a. *mortgage bond*, or *bond*); ~ ***insurance***, credit insurance, used specifically to ensure repayment of a mortgage under specified adverse circumstances, such as death or illness; ~ ***securitiser***, company that creates and sells securities backed by home loans, for example, Fannie Mae and Freddie Mac.

mother/child means of developing scorecards for different subpopulations, where a mother scorecard is developed for the full sample, which is used in the child scorecards for each subpopulation.

MSExcel Microsoft Excel, a computerised spreadsheeting package.

multi (*pref.*) many; ***~collinearity***, the existence of a relationship between two or more predictors, which makes it difficult to gauge their influence on the target, and affects the robustness of the final model; ***~variate***, (*adj.*) involving three or more variables; ~~ ***analysis***, use of statistical, mathematical, or graphical techniques, to analyse the relationship between three or more variables simultaneously; ~~ ***regression***, production of an equation or algorithm, to explain the relationship between two or more predictors and a response variable.

naïve (*adj.*) simple, primitive, inexperienced, uninstructed; ~ ***model***, simple representation that provides a constant result, or extrapolates a trend line, to provide a benchmark.

natural (*adj.*) product of nature, not artificial or imitation; ~ ***individual/person***, in law, a human being, susceptible to physical forces, for example consumers, sole proprietors, and partnerships (ant. *juristic person*).

negative case that does NOT have the specified attribute(s), normally associated with rare, and difficult to identify maladies (rel. *good, not default*, ant. *positive*); ~ ***correlation***, association where two values move in opposite directions; ~ ***data***, see 'default data'.

net flow the value, number, or percentage of accounts that move to a worse status; ~ ***model***, model based upon net values, that flow from one delinquency bucket to the next.

neural network data mining technique, that attempts to mimic human cognitive processes.

NGO (*acr.*) non-governmental organisation; a not-for-profit organisation, usually involved in socially-responsible projects.

no performance lack of an observed outcome, esp. for rejects and not-taken-ups in a selection process, where reject inference is necessary (ant. *known performance*).

nominal (*adj.*) related to giving a name; ~ ***variable***, data element, whose possible values are represented by labels (names) or codes (letters/numbers) that provide no indication of relative rank (rel. *categorical*).

non (*pref.*) not; ~***-linear***, (*adj.*) related to something that is not linear; ~~ ***regression***, statistical models used to represent relationships that are not linear, and the associated development techniques; ~***-monotonic***, (*adj.*) related to a sequence of values that both increase and decrease; ~***-parametric***, (*adj.*) related to statistical methods that make no assumptions about the underlying data; ~***-performing loan***, (abbr. NPL) **1** a bad account, where interest is no longer being treated as income; **2** an account written off, or in recoveries.

normal usual, standard; ~ ***approximation***, alternative and less accurate approach for assessing a binomial distribution with large numbers, which assumes a normal distribution; ~ ***distribution***, bell-shaped distribution of a random continuous variable, that can be defined by its mean and standard deviation (a.k.a. *bell curve* or *Gaussian distribution*); ~***ise***, (*v.t.*) **1** cause to conform with a norm; **2** to transform a variable, such that the result has a normal distribution; ~~***d scores***, scores that are adjusted, such that values have meaning with respect to a standard measure (e.g. *P*(Good), log odds).

not sufficient funds (*adj.*) of, or relating to, any instance where the amount of money available in an account is insufficient to meet claims upon it (acr. NSF); ~ ***cheque***, any cheque that takes an account over its agreed limit, or the amount the bank is willing to extend, esp. where the cheque is dishonoured.

not taken up accepted applicant, who does not open or use the offered product, perhaps because a better deal was obtained elsewhere (abbr. NTU).

notice a communication that does not require a response; ~ ***required***, instances where the other party need only be informed of their rights, or the action to be performed.

NPL (*acr.*) non-performing loan.

NSF (*acr.*) not sufficient funds.

NTU (*acr.*) not taken up.

null zero, of no value; ~ ***class***, group for which no dummy variable is created; ~ ***hypothesis***, tentative explanation that is tested in an experiment (ant. *alternative hypothesis*).

numerator top part (dividend) of a fraction, *the numerator of 5/8 is 5* (ant. *denominator*).

numeric (*adj.*) related to numbers; ~ ***variable***, variable comprised only of numbers, esp. measures or counts.

oblig~***ation***, **1** moral or legal requirement; **2** loan that must be repaid, or commitment that must be upheld; ~***e***, (*v.t.*) compel morally or legally; ~***or*** **1** person or entity bound to fulfil an obligation; **2** debtor, borrower. (rel. *counterparty*)

observation **1** set of details recorded at a point in time, for a given case; **2** (*adj.*) pertaining to records containing predictive information, used to develop a statistical model; ~ ***point/window***, period(s) over which data is gathered.

OCC (*acr.*) Office of the Comptroller of the Currency, the American banking regulator.

OECD (*acr.*) Organisation for Economic Co-operation and Development, focuses on growth and development of member countries, which include the G10 and about 20 other high-income countries; ~ ***Data Privacy Guidelines (1980)***, set of principles meant to foster transborder data flows, including collection limitation, data quality, purpose specification, use limitation, security safeguards, openness, and individual participation.

one-tailed test a statistical significance test, used to determine whether an observed value occurred by chance, but testing in one direction only, greater than or less than (rel. *significance test*, *two-tailed test*).

online (*adj.*) related to a direct connection to a computer; ~ ***enquiry***, bureau searches that involve a direct connection to the bureau, whether manual or automated; ~ ***fraud***, any fraud involving the Internet.

opacity extent to which something is opaque (syn. *opaqueness*, ant. *transparency*); ***opaque*** not easily understood, or seen through (ant. *transparent*); ***openness 1*** transparency; **2** general public's ability to enquire about the activities of public and private organisations.

operation a process, or series of actions, used to produce goods or services; ***~al drift***, changes in processes, calculations, and systems, that influence outputs; ***~al risk***, uncertainties arising from operational problems within a process, including those related to people, technology, infrastructure, and fraud, often associated with mistakes and unforeseen circumstances; ***~alise***, to implement a plan or system within a company's operations; ***~s***, generic term referring to back-office functions that ensure the smooth running of an organisation, for example billing, communications, systems, etc.; ~~ ***research***, analysis of business or industrial problems, using techniques such as linear programming and critical path analysis.

opt (*v.t.*) to choose, or select; ~ ***in***, to choose to be involved in, or partake; ~ ***out***, to choose not to partake.

order logical arrangement of elements; ***~ed regression***, a form of regression appropriate for ordinal target variables, which may be logit or probit.

ordinal (*adj.*) related to order (rel. *continuous, nominal*); ~ ***variable***, descriptor that denotes relative position, but not distance, from those occurring before or after.

ordinary least squares process used for linear regression, that finds coefficients for independent variables that minimise the sum of the observations' squared error terms.

organised crime illegal activities that make use of complex business-like structures.

origination process that creates, or starts; ~ ***process***, new business process, used to receive and process applications, open accounts, and/or disburse funds.

out-of- (*pref.*) not a member, or part; ***~sample***, pertaining to observations that were not part of an analysis, esp. where used for testing (ant. *in-sample*, rel. *holdout sample*); ***~time***, pertaining to observations drawn from a different time period than the training sample.

outbound (*adj.*) approaches by the business to customers, esp. by call centres for marketing and collections (ant. *inbound*).

outcome **1** result, what transpired; **2** (*adj.*) pertaining to performance data used in predictive modelling; ~ ***point***, the date at which the final outcome is observed; ~ ***window***, months between observation and outcome; ~ ***variable***, see 'target variable'.

outsource (*v.t.*) to contract outside parties to perform functions, that would otherwise be done internally; ~ ***agent***, external entity to which certain support and/or service functions are contracted.

over (*pref.*) above a specified boundary; ~***draw***, (*v.t.*) to withdraw an amount of funds, greater than what is available in an account; ~***draft***, facility where borrowers can overdraw (usually on a cheque account) to an agreed limit, and only pay interest on the balance outstanding; ~-***fit***, (*v.t.*) to create a model that works well on the training sample, but not on the hold-out sample and/or the general through-the-door population once implemented; ~-***fitted model***, resultant model; ~***indebted***, owing money, to the extent that repayments cannot be met, or it causes severe personal stress; ~***ride***, (*v.t.*) **1** to change the decision that would normally result, by any means; **2** any case where this has been done; ~***write***, (*v.t.*) to update a data field with a new, and usually more recent, value.

P(?) notation used to represent a probability; **P(Good)**, probability that an equivalent observation will be good; **P(Good|X)**—ditto, but for some group of accounts with certain characteristics, X; **P(Accept)**, probability that an applicant will be accepted.

paramet-*er*, **1** measurement or value, that something else depends upon; **2** constant, limiting, or governing independent variable; **3** any value used to describe a statistical population, like the mean or variance; ~***ric***, (*adj.*) **1** related to parameters; **2** related to statistical methods, that make assumptions about the underlying data (ant. *non-parametric*).

parcel (*v.t.*) **1** divide into parts; **2** assign cases to different categories; ~***ling***, a performance manipulation technique, which may be polarised, random, or fuzzy, that assigns no performance cases to good, bad, and possibly other performance categories.

Pareto, Vilfredo (1848–1923). Italian engineer, economist, and sociologist; ~ ***principle***, also called the 80/20 principle, it refers to the phenomenon that the bulk of an effect is often generated by a small number of cases.

parsimony **1** frugality, stinginess; **2** simplicity, brevity; **3.** philosophical principle that the simplest answer is usually the best one (syn. *Ockham's Razor*).

partial observability unavailability of outcome performance data for a portion of the population, especially for cases rejected by selection processes (rel. *reject inference*).

past due payment not yet received, usually stated as days or months (rel. *delinquent*).

pawn (*v.t.*) to provide personal property as collateral for a loan; ~ ***shop***, business establishment where people can pawn goods.

pay/no pay a situation, where a lender has to decide whether or not to pay funds on behalf of a client, esp. for cheques that put an account over the agreed limit.

payday loan sub-prime short-term loan, to be repaid on the next salary date.

payment **1** amount of money paid; **2** (*v.t.*) transaction involving paying money; ~ ***history***, obligor's record of honouring obligations (rel. *payment profile*, *credit history*); ~ ***medium***, cheques or plastic used to transact; ~ ***profile***, **1** series of numbers and/or letters indicating an account's past due status over preceding months; **2** individual's credit history with a bureau's contributors.

PD (*acr.*) probability of default.

Pearson, Karl (1857–1936). English mathematician; ~*'s* ***product moment correlation coefficient***, measures the strength of a linear relationship between two continuous variables.

percent correctly classified true positives and true negatives as a percentage of the total sample, measured at a given cut-off score (rel. *misclassification matrix*).

perfect (*adj.*) **1** without blemish; **2** accurate or exact; ~ ***equality***, instance where the Lorenz curve lies along the diagonal, esp. where all incomes are equal; ~ ***inequality***, instance where the Lorenz curve covers the entire area above or below the diagonal.

performance **1** outcome, result, response, behaviour, target; **2** extent to which decisions or efforts provide the desired results; ~ ***manipulation***, deliberate changes made to outcome performance, esp. as part of the reject inference process; ~ ***status***, label used to indicate whether results were good, bad, or indeterminate.

perpetuity **1** forever, eternity; **2** a loan with interest payments, but the capital amount is never repaid, the primary example being those issued by the French government to finance the Napoleonic wars.

persona non grata, [Latin] unwelcome individual.

personal (*adj.*) pertaining to an individual; ~ ***character***, individual morality and ethics; ~ ***data***, any data that can be associated with a specific and identifiable individual; ~ ***distress***, illness/death, domestic dispute, job loss, and personal disaster; ~ ***disaster***, accident, fire, flood, or other natural disaster, esp. where it causes loss of assets or income; ~ ***identifier***, a number or code known to a person, used to identify and link data specific to that individual, esp. at national level, such as the Social Security Number (USA), Social Insurance Number (Canada), or Identification Number (South Africa); ~ ***identification number***, number used by a cardholder to access funds, intended as a security measure against fraudulent use (abbr. PIN).

phishing fraudsters' use of emails and bogus websites, to masquerade as a bank or other institution, in order to deceive individuals into divulging personal details, like personal identifiers, account numbers, and security codes.

PIN (*acr.*) personal identification number.

plastic (*inf.*) any type of credit card, irrespective of who issues it.

platform **1** level or raised area used for a specific purpose, such as a waiting, work, or launch area; **2** computer hardware or software used to fulfil a specific function.

poach (*v.t.*) **1** to hunt illegally; **2** to target other businesses' customers.

point-in-time (*adj.*) related to the immediate future, with no reference to the economic cycle (ant. *through-the-cycle*); ~ ***estimate***, an approximation calculated using data for a time period much shorter than an economic cycle.

policy rule(s) that guide decisions and actions, by individuals, enterprises, and governments; ~ ***accept/reject***, cases where the decision either is, or is assumed to have been, overridden by a policy rule; ~ ***rule***, definition of a scenario, and the action to be taken in that instance.

political risk risk that arises from real or potential changes to a country's political framework, esp. where they may effect the economy, and/or local business environment.

pool (*v.t.*) to combine; (*n.*) combined resources or stake; ***~ed basis***, reliant upon grouping cases, and treating each group as one; ***~ed data***, data provided by different parties, in order to develop a model that will be used by all; ***~ing algorithm***, process used to group cases together, esp. as part of characteristic transformation, using bivariate statistics.

population all cases in a group of interest, from which samples can be drawn; ***~ drift***, any changes in score distribution resulting from market or infrastructure changes, which can affect model stability; ***~ flow***, graphical or logical representation of the population's distribution across different performance categories (good, bad, reject, etc.); ***~ shift***, see 'population drift'.

portfolio collection of financial assets; ***~ analysis***, review of a portfolio's composition, value, and returns, esp. to determine whether changes are required to enhance performance.

positive **1** (*adj.*) advantageous, good, tending in same direction (ant. *negative*); **2** (*n.*) case that has the specified attribute(s), normally associated with rare, and difficult to identify maladies (rel. *bad, default*); ***~ correlation***, association where two variables tend to move in the same direction; ***~ data***, see '*payment profile*'.

power **1** ability or capacity to achieve specified ends; **2** scorecard's ranking ability; ***~ curve***, see 'Lorenz curve'.

pre- (*pref.*) in advance of; ***~approve***, (*v.t.*) to accept prior to submission through normal process, either as part of marketing, or to assist customers with major asset purchases; ***~bureau***, (*adj.*) related to actions done prior to a bureau call; ***~capture screening***, manual process of vetting applications, and removing those with a low chance of being selected, either because there are faults with the application, or it is a poor quality applicant; ***~screen***, (*v.t.*) process of vetting prospective clients, before making them an offer.

predatory lending victimisation of borrowers through deceptive practices, highly-prejudicial loan terms, and lack of regard for their ability to repay.

predict (*v.t.*) to suggest a result, usually based upon available information and past experience (rel. *describe*); ***~ive accuracy***, ability of a model to provide an accurate probability estimate (rel. *ranking ability*); ***~ive models***, statistically derived models used to rank risk and/or provide estimates; ***~ive power***, extent to which a characteristic or model explains a target variable; ***~ive statistical technique***, method of harnessing independent (predictor) variables, to provide estimates of an independent (target) variable; ***~ion error***, extent to which actual results differ from those expected; ***~or***, predictive variable (syn. *independent variable, observation characteristic*).

prepayment repayment prior to that contractually agreed (rel. *attrition, early settlement*); ***~ risk***, probability that accountholders will settle obligations early (syn. *early-settlement risk*).

primary key data item that has a unique value for every record in a database, and can be used uniquely to identify each record.

prime best, superior; ***~ interest rate***, offered to best customers, a benchmark against which other lending rates are set (usually for variable rates).

principal component uncorrelated variable derived to simplify a dataset, and aid understanding; ~ ***analysis***, mathematical technique used for variable reduction.

private (*adj.*) restricted to a smaller group within the community (ant. *public*); ~ ***firm***, company whose equity is not publicly traded; ~***-firm model***, any model used to assess the credit risk of private companies.

probability likelihood of a future event, ranging from impossible (0) to certain (1); ~ ***distribution***, set of probabilities associated with all possible values of a given variable; ~ ***of default***, (acr. PD) likelihood of future default status arising (rel. *exposure-at-default*, *loss–given-default*).

probit probability unit, based upon the inverse cumulative distribution function (rel. *probit*); ~ ***model***, regression that provides estimates, that assume a Gaussian distribution.

process 1 series of actions to achieve an end; **2** any action performed upon data, such as collection, storage, modification, retrieval, disclosure, and so on; ~ ***components*** factors that drive and control a process, including data, systems, models, strategies, analytics, and reporting.

product good or service offered to customers; ~***ion system***, infrastructure used to produce a product to be delivered; ~ ***moment***, see 'Pearson's ~'; ~ ***rules***, company policy that determines who the product is offered to, and under what terms, for example min. applicant income, min. loan amount, max. term, geographical area covered, etc.

profit the difference between revenue and expenditure; ~ ***drivers***, key elements that influence profits, like risk, balance, activity, late payments, insurance, and acquisition; ~ ***modelling***, representation of profit expectations, given certain assumptions; ~ ***scoring***, use of scoring models, to assess whether individual transactions will be profitable.

project task requiring considerable effort, and/or resources; ~ ***steering committee***, individuals who determine a project's necessity, and ensure that sufficient resources are available for its completion, including the sponsor, champion, and representatives of different business areas; ~ ***team***, individuals directly involved in the project, including the project manager, scorecard developer, internal analysts, functional experts, and technical resources.

promise-to-pay promise by the customer to pay all or part of an overdue amount.

propensity general tendency; ~ ***scoring***, use of a scoring model, to rank the probability of people acting in a certain way, esp. response scorecards, for example taking up a product, responding to a mailing, ordering from a catalogue, etc.

provision preparation or allowance made in advance of an expected or unexpected future event (rel. *loss provisions*); ~ ***rate***, percentage of asset value, set aside for potential future losses.

prox~***ied risk***, a risk that is represented by items that are known, and represented within the system; ~***y***, something used to represent something else.

public (*adj.*) open to the community as a whole (ant. *private*); ~ ***credit registry***, government agency that is a repository for credit related information, either to assist credit provision, or to aid in monitoring the financial system (acr. PCR, rel. *credit bureau*); ~ ***firm***, company whose equity is publicly traded; ~~ ***model***, any model used to assess the credit risk of publicly traded companies; ~***ly traded***, exchangeable on the open market.

pure judgment subjective assessment, with no model or template.

push back contest of a decision, whether by internal staff, an external agent, or the customer.

P-**value** measure of statistical significance, which states the confidence with which a hypothesis can be said to be true.

quantification act of determining a quantity, or estimate thereof; ~ ***process***, mapping of risk ratings into IRB parameters; ~ ***stages***, steps required to derive empirical estimates, including data, estimation, mapping, and application.

R&D see '*research and development*'.

racket illegal enterprise, undertaken with the goal of financial gain; ***~eer***, person involved in such an enterprise.

random (*adj.*) without any pattern; ~ ***number generator***, function within a computer program that returns a different unrelated number, usually between 0 and 1, each time it is invoked; ~ ***parcelling***, a form of performance manipulation, where cases are assigned at random into other categories, usually done on a stratified basis using some score; ~ ***sample***, group of accounts chosen to represent a population, without regard to individual attributes; ~ ***supplementation***, reject inference technique, where cases below cut-off are accepted at random, to determine how they perform.

rank position within a sequence; ***~ing ability***, extent to which a predictive model can effectively rank observations by some measure (rel. *predictive power*); ~ ***order***, (*v.t.*) to sort into a sequence.

rare event infrequent events, that hamper: (i) the development of a predictive statistical model, if it is the target variable; (ii) whether the risk of that event is adequately represented in the final model, if it is a predictive variable.

rapid redevelopment regular scorecard updates using new data, but with minimal changes to assumptions.

ratings risk assessments, esp. internal and credit rating grades; ~ ***delay***, time lag, before new credit related information is reflected in ratings; ~ ***drift***, small changes in obligors' aggregated ratings over time, which may be upward or downward; ~ ***migration***, movements between grades within a transition matrix, within a given timeframe; ~ ***momentum***, tendency of rating grades to move in the same direction as the last change.

re-age (*v.t.*) to reset the delinquency counter, for accounts that have been in the same delinquency bucket for several months, or where arrangements have been made with the customer.

reasonable (*adj.*) **1** capable of reason; **2** not unfair, unbiased; ~ ***data***, data that is relevant, justified, and not excessive, for the purposes that it is being used; ***~-model test***, a check of model results against those of another, real or presumed, produced in similar circumstances.

recent sample a selection of accounts from the most recent three or so months prior to starting the development, used to test predictor and model stability.

receiver operating characteristic (abbr. ROC) tool used to measure the reliability of the prediction of a binary outcome, which provides the probability that a 'Good' and 'Bad', chosen at random, will have the correct rank orderings relative to one another; ~ ***curve***, a graph of true positive versus false positive rates, across the range of scores or grades (rel. *Lorenz curve*, *area under curve*).

reclassification a performance manipulation technique, which may be either rule- or score-based, that assigns certain cases to bad, and either ignores the rest, or leaves them to be treated using another technique (rel. *parcelling*, *reweighting*).

reciprocity agreement governs the sharing of bureau data, usually limiting the availability of payment profile data to contributing subscribers.

record **1** physical or electronic details of an individual case; **2** an element within a database, containing characteristics of an account, individual, or some other item.

recover (*v.t.*) to obtain repayment from a borrower, after a default event; ***~ies***, business function that manages serious delinquencies to recover funds, usually associated with a break in the customer relationship (rel. *collections*); ***~y rate***, proportion of defaulted balances recovered, whether actual historical or expected future values.

recursive partitioning algorithm a statistical technique used to derive classification trees.

redline to refuse a loan(s) because of presumed and possibly unsubstantiated risks associated with a characteristic(s), esp. individuals/properties in low income suburbs.

reduced form model model that relies upon the value of a firm's traded debt securities, and assumes that the modeller only has information available to the market as a whole.

refer (*v.t.*) neither accept nor reject, usually requiring more information, a separate process, or input from a higher authority, to make the final decision; ***~ral***, case being referred; ***~~s***, area or process, that deals with over-limit management for cheque accounts.

refresh rate frequency of updates, esp. as regards data.

registry official written record of names, events, transactions, or assets (see '*public credit registry*').

regression **1** tendency of progeny to be more like the population than their parents; **2** production of a formula or algorithm, that explains a response variable as a function of one or more predictors (rel. *linear regression* and *logistic regression*); ***~ formula***, an equation that explains the relationship between a dependent variable, and any number of independent variables; ***simple ~***, uses a single predictor; ***traditional ~***, model where points are assigned to individual attributes within each characteristic.

reject case refused by a selection process, or discarded by a production process; ***~ inference***, process used to deduce what the performance of a refused applicant would have been, had it been accepted; ***~ override***, any accepted application, that would otherwise have been declined, had normal rules been applied; ***~ rate***, percentage of cases rejected.

relationship lending provision of loans based upon personal knowledge of the customer and his/her needs, which implies judgmental evaluations (ant. *transactional lending*).

reputational risk risk of potential damage from a deterioration of a person's standing in the eyes of the community, or a firm's standing in the eyes of its stakeholders.

research and development (*abbr.* R&D) function responsible for developing and testing new products, often done as part of marketing for financial services.

residual 1 amount that has not been explained by a model; 2 the difference between the predicted and actual values for a case (syn. *error*); ~ ***mass***, amount remaining, after all other obligations have been settled.

response result, reaction, answer; ~ ***function***, the term on the left-hand side of a regression equation, which is either the target variable, or a function thereof; ~ ***scoring***, use of statistical models, to assess whether individuals will respond to marketing and other approaches; ~ ***variable***, dependent, outcome, or target variable (ant. *predictor*).

responsible lending lending done in a fair and acceptable manner, which considers individuals' needs and ability to repay.

restructure (*v.t.*) to renegotiate terms of a loan agreement, such as the repayment amount, remaining period, interest rate, collateral, outstanding capital, or other aspect.

retail (*adj.*) related to a large number of individual or small lot sales to a mass market (ant. *wholesale*); ~ ***credit***, lending to individuals or small businesses, where common strategies are applied to all members within a portfolio (rel. *consumer credit*).

retention power to keep or hold in place; ~ ***scoring***, use of statistical models to assess whether accounts will stay open and active.

retrospective review of past events or statuses; ~ ***enquiry***, obtaining details for an individual or account at some historical date, whether from the credit bureau or own systems.

returned items payments into an account, whether by cheque or debit order, that are not honoured by the bank and have to be reversed (rel. *dishonour*, *not sufficient funds*).

revenue monies received from sales of goods and services; ~ ***scoring***, use of statistical models to assess whether accounts will generate sufficient revenue to make them worthwhile.

review to reassess an account, in particular where there is a credit limit; ~ ***date***, date when the continued operation of an account is to be reassessed.

revolv~e, (*v.t.*) to turn around; ***~er***, cardholder who uses the account as a borrowing facility (ant. *transactor*); ***~ing credit***, lending product that allows amounts to be repaid and redrawn, within an agreed limit, as the customer requires.

reweight (*v.t.*) to change the weight of observations within a dataset, in order to achieve some end; ***~ing***, performance manipulation technique, esp. in reject inference process (rel. *parcelling, reclassification*).

risk 1 uncertainty of future outcomes; 2 possibility of loss, unexpected or undesirable results, desired benefits not being achieved, or opportunities being missed; ~ ***band***, range of scores or grades treated on a like basis; ***~-based pricing***, use of risk measures, as a basis for varying loan prices or terms; ***~-based processing***, changes to the process based upon a preliminary risk assessment; ***~-free rate***, baseline interest rate, that can presumably be earned free of risk, usually the yield on government bonds; ~ ***indicator***, a letter, number, or symbol used to identify a risk band or grade; ~ ***mitigation***, any factor that reduces loss probability or severity, such as credit insurance; ~ ***ranking***, any sorting of cases or scenarios, according to some perception or measurement of risk; ***~-weighted assets***, a restatement of banks' assets that recognises the underlying risk, for the purposes of determining minimum capital reserves.

robust (*adj.*) 1 sturdily built; 2 able to withstand a changing environment; 3 straightforward and commonsensical.

ROC (*acr.*) Receiver Operating Characteristic.

roll (*v.i.*) to turn over and over; ~ ***rate***, percentage of cases that move from one state to another, over a given time period; ~~ ***report***, documentation of roll rates by category (rel. *transition matrix*).

root node in a decision tree, the top or first node that has children but no parents.

R-squared statistic indicating how much of the error, relative to the mean, is explained by a model (syn. *coefficient of determination*).

S&P (*abbr.*) Standard and Poor's.

SAFAS (*acr.*) South African Fraud Avoidance System (RSA).

safety net **1** strong net used to catch workers/performers in the event of accidental or intentional falls; **2** laws, regulations, institutions, and other frameworks, implemented to ensure the soundness of a national banking system, including deposit insurance, payment guarantees, and asset discounting.

sample a part that has been selected to be representative of the whole (a 'population'), for the purposes of a study, test, or analysis; ~ ***design***, a blueprint for the construction of a sample, including quantities, stratification, and time periods; ~ ***selection bias***, model inaccuracies arising from unrepresentative samples, because one or more groups were over-, under-, or not represented; ~ ***size***, number of cases used for a study; the greater the number, the more reliable the results; ~ ***window***, period of time over which observations are collected.

sampling process of creating a sample; ~ ***rate***, percentage of accounts chosen to represent a segment (rel. *weight*).

scal~*e*, **1** established measure or standard; **2** (*v.t.*) to bring a measure into line with a standard; ***~ing***, standardisation of scorecard results, esp. to provide final scores with meaning (rel. *normalisation, alignment, calibration*).

scenario possible sequence of events, situation, or set of attributes, which may require specific actions.

score **1** a number that represents either quality, or performance; **2** the total of points accumulated on a scorecard; **3** (*v.t.*) to allocate and total points; ~ ***accept/reject***, applications with scores above/below the cut-off; ~ ***break***, score value that defines the split between two risk bands; ***~d decision***, strategy applied to a specific case, chosen purely based on score(s); ***~d for guidance***, processed through a decision system, but the results are not strictly enforced, and the underwriter has the final call; ~ ***distribution***, counts of cases falling into each score, or score range; ~ ***drift***, changes in score, resulting from changes in the through-the-door population, the economy, internal processes, or other factors; ~ ***matrix***, contingency table, whose rows and columns are defined by scores, esp. where cells represent some outcome measure; ~ ***reject***, any case with a score below the accept cut-off; ~ ***sheet***, piece of paper used to mark and record scores.

scorecard **1** piece of paper used to record points earned during a contest; **2** table indicating points to be allocated to different attributes; **3** regression model used as part of a rating process; ~ ***alignment***, adjustment of point allocations, and/or the final score, so that the latter can be directly compared with those provided by other scorecards (syn. *normalisation*, *scaling*); ~ ***development***, process of producing a scorecard; ~ ***vendor***, outsource agency that develops predictive models for use by business, esp. bespoke models.

search (*v.t.*) **1** to seek; **2** (*n.*) enquiry made against an external database, esp. credit bureau (syn. *enquiry*).

second after first; ~ ***party***, someone not party to a contract, but who has a defined role, for example, subcontractor, service provider, or intermediary; ~~ ***fraud***, fraud committed by somebody with a defined role in the process, such as a merchant.

secur~e, (*adj.*) safe, protected from danger or risk; ***~ed loan***, amount lent against security, esp. assets like fixed property or motor vehicles; ***~ity***, **1** a person or thing that provides comfort, such as a guarantee or collateral; **2** certificate or contract, indicating rights to a stream of interest or dividends; ***~itise***, (*v.t.*) to convert loans or other assets into tradable securities, esp. where similar assets are combined; ***~itisation***, process or act of securitising.

segment **1** one of several parts of a whole; **2** group with similar attributes that requires separate treatment; ***~ation***, act or result of segmenting, which is done to improve assessment/ planning, when the whole is made up of substantially different parts; ~~ ***drivers***, factors that influence the scorecard segmentation, including marketing, customer, data, process, and model fit factors.

select (*v.t.*) to choose, esp. in some order of preference; ***~ion criteria***, basis for defining preferences; ***~ion process***, procedure designed for choosing, such as new business processing [application scoring, accept or not accept], and marketing [response scoring, contact or not contact].

self-cure a delinquent account that is, or will probably be, repaid without any collections action.

self-fulfilling prophecy any prediction, that by being made, makes it come true or significantly increases the probability thereof.

sensitivity proportion of true positives, to true positives plus false negatives (rel. *specificity*).

sequential occurring in a series; ~ ***definition***, good/bad definition, that varies by delinquency status, or time since entry, usually requiring different scorecards (collections, recoveries); ~ ***scorecards***, series of scorecards, that are applied at different levels of delinquency (rel. *entry scorecard*).

shadow limit limit that operates in the background, to govern over-limit abuses.

shared (*adj.*) apportioned, or used jointly; ~ ***information***, see 'shared performance data'; ***~-performance data***, details of borrowers' repayment patterns, positive and negative, shared between lenders (rel. *credit history*, *payment profile*, *information sharing*).

significance in statistics, extent to which evidence indicates something did not occur by chance alone; ~ ***level***, probability of wrongly rejecting a hypothesis, often denoted by the Greek character alpha (α) in statistical formulae (rel. *confidence level*); ~ ***test***, a statistical procedure used to determine whether observed values could occur by chance.

skim (*v.t.*) **1** to pass quickly over, often as a means of gathering something lying at or near the top; **2** to obtain details from a credit card or other transaction medium, esp. for fraudulent activities.

skip see '*gone-away*'.

small business companies limited in size or scope, whether by sales turnover, number of employees, asset size, or geographical reach; **~~ *lending***, provision of finance to small companies, which Dun & Bradstreet defines for the United States as the approx. six million companies with sales turnover between $100k and $10mn.

SME (*acr.*) Small and Medium Enterprise; independent business, limited in terms of employees, assets, and revenue, usually with limited geographical reach; thresholds vary, but include: EU—less than 250 employees, €50mn revenue, and €43mn assets (Recommendation 2003/361/EC); **USA** varies by industry, but max. 500 employees and $28.5mn revenue (refer to Small Business Administration website); **Basel II** less than €50mn revenue, or in the absence of revenue data, less than €1mn exposure.

soft collections relating to actions used to deal with early delinquencies, where the chance of recovery is high, and the focus is on maintaining the customer relationship (ant. *hard collections*).

solicit (*v.t.*) to try to influence or persuade people towards some end, usually as an invitation to do business; **~*ation***, process of inviting potential customers to do business.

Somer's D see 'Gini coefficient'.

sovereign (*adj.*) supreme authority, usually referring to a monarch or national government; **~ *debt***, money borrowed by a national government, whether local or foreign; **~ *risk***, credit risk associated with lending to any national government.

Spearman, Charles (1863–1945). British behavioural psychologist, known for both the rank order correlation coefficient, and factor analysis; **~*'s rank order correlation coefficient***, measure of agreement between two sets of relative rankings for the same set of cases, with values ranging from –1 to +1.

specificity proportion of true negatives, to true negatives plus false positives (rel. *sensitivity*).

speculative grade credit rating grade, usually BBB- or worse, considered too poor for investors who are restricted in their investments (rel. *junk bond*, *investment grade*).

speedpoint machine that enables instantaneous processing and authorisation of merchants' credit card transactions, via a telephone link.

spread (*v.t.*) **1** distribute, disperse, scatter; **2** (*v.t.*) to lay out data on a screen; **3** (*n.*) difference between bid and offer prices; **~*sheet***, piece of paper that displays data as rows and columns, or a computer program that uses a similar format.

split-back method storing of events, transactions, account load dates, and other details, so that a record can be recreated at a future date, and returned during a retrospective search (rel. *archive method*).

stability index Kullback divergence measure, when used to assess the change in a probability distribution over time, such as the holdout and recent samples' score distributions.

stag~*e*, **1** position in a process; **2** group of characteristics considered for potential inclusion in a model; **3** (*v.t.*) to develop a model for a subgroup of characteristics, esp. for dependent staging; **~*ing***, process of treating scorecard characteristics in groups, which may be dependent or independent.

standard reference point, basis for comparison; ~ ***error***, a measure of sampling error, which indicates how well an estimate matches the value for the total population; ***~ise***, bring into line with a standard; ***approach***, simplest approach allowed under Basel II, which applies specified risk weights to different asset classes (rel. *internal ratings based*).

statistic data that can be represented as an objective value; ***~s***, a science dealing with the collection and analysis of facts, esp. for individuals and economies; ***~al method***, any tool used in the field of statistics to analyse data; ***~al model***, a model derived using statistical methods, to explain a real world situation.

steady state **1** a stable condition, which is either unchanging, or changes at a constant rate; **2** with Markov chains, a point where the frequency distribution becomes a constant.

step **1** one of a series of moves towards a goal; **2** selection of a single variable for inclusion in a model, according to some optimisation formula; ***~ping***, process of considering variables for inclusion; ***~wise***, (*adj.*) proceeding in steps; ***~~ regression***, forward selection or backward elimination, with an evaluation of moves in the opposite direction.

stochastic (*adj.*) **1** probabilistic; **2** related to a probability distribution (ant. *deterministic*); ~ ***process***, process whose outputs are not independent, yet whose relationship cannot be represented by a deterministic formula; ~ ***variable***, random data element, whose possible values can be represented as a probability distribution.

stopping rule condition used to specify when an algorithm should consider itself satisfied, and either stop, or move on to the next stage.

store place of business, dedicated to the retail sale of goods or services; ~ ***card***, a plastic card used for transacting at various branches of a retailer, or group of retailers; ~ ***credit***, retailers' provision of goods on a 'buy now, pay later' basis.

strategy **1** plan intended to aid the achievement of an objective; **2** action to be taken in a given scenario; ~ ***curve***, graphical equivalent of the strategy table; ~ ***effect***, the impact of strategies upon a population's dynamics, especially where it affects the risk being measured; ~ ***inference***, analysis of repeated games, in order to determine players' strategies; ~ ***manager***, software used to manage strategies applied by a lender, or investor; ~ ***matrix***, see 'decision matrix'; ~ ***setting***, process of determining what actions will be performed in different scenarios; ~ ***table***, report showing trade-offs between accept and bad rates at different cut-off scores.

strat*~um*, (pl. *strata*) group that possesses similar qualities; ***~ified random sampling***, sampling process, where different strata are sampled separately at different rates, to ensure that each is adequately represented.

structur*~al model*, **1** any theoretical model, that is intended to describe the structure of the environment; **2** credit risk model, which assumes that the firm's structure is embodied in its financial statements; ***~e***, arrangement of parts (ant. *unstructured*); ***~ed data***, information that is presented in a highly structured form, such as that sourced from application forms, account management systems, credit bureaux, and financial statements.

subjective (*adj.*) **1** proceeding from, or taking place, in a person's mind; **2** based on, or influenced by, personal opinions and biases; **3** difficult to prove empirically (ant. *objective*);

~ ***decision***, a decision that is not based upon objective factors; ~ ***scoring***, see '*judgmental scoring*.'

subprime (*adj.*) less than best, or of unacceptable risk; ~ ***lending***, provision of finance to high-risk groups at above average rates, usually for small amounts with no security, with extra effort put into securing repayment (rel. *payday loan*, *home collected credit*); ~ ***rates***, interest rates that exceed those normally accepted within a country, esp. when outside of Usury Act or other legislated limits.

subscriber somebody that receives information from a credit bureau, and usually also contributes.

super known good/bad model model of known performance, that uses both observation and cohort performance data as predictors, and which is used as part of the reject inference process.

survival failure to succumb to hazards (rel. *hazard*); ~ ***analysis***, review of mortality rates over time; ~ ***function***, series of survival rates at different time horizons; ~ ***rate***, proportion of cases that survived a given length of time (rel. *default rate*, *hazard rate*).

swap set observations that are assigned different group memberships depending upon the circumstances, such as accept/reject statuses that are affected by process changes, or differences between actual and expected good/bad statuses.

swindle deception or trickery, done for financial gain.

system collection of interoperating elements that exist, survive, or produce; ~ ***accept***, case for which the system decision is 'Accept'; ~ ***decision***, decision made by a system, esp. policy- and score-based selection processes; ~ ***reject***, case declined by the system.

systemic (*adj.*) related to a whole system or body, as opposed to a part; ~ ***risk***, risk of an event having severe and unexpected consequences in other areas of a financial market or system, and which cannot be addressed by diversification, for example, failure of one market participant causing others to fail, or of natural or man-made events leading to widespread panic.

target point to aim for, goal; ~ ***definition***, specification of response variable, usually a good/bad indicator; ~ ***drift***, change in relationship between predictor and response variables over time; ~ ***limit***, maximum amount that a lender will grant as a limit, if the customer requests it; ~ ***variable***, outcome status to be predicted (syn. *dependent variable*).

technical arrears arrears that are inconsequential, or quickly rectified, usually resulting from problems with the payment system, or details that are incorrectly loaded.

terminal node in a decision tree, any node that has no children.

terms and conditions statement of how the relationship will operate, and what will happen if any of the terms are breached.

terms of business combinations of loan amount, interest rate, repayment period, collateral, and other factors, that have been agreed between lender and borrower.

testing process undertaken to discover errors, faults, and weaknesses, whether in design or implementation.

text mining identification and extraction of meaningful information, from large amounts of text data, such as email messages, collectors' notes, and web pages.

thin file cases for which little information exists, esp. as regards credit bureau records for youth, new immigrant, and underserved markets.

through-the-cycle (adj.) related to an economic cycle, which is usually seven or more years; ~ ***estimate***, value or probability derived for an economic cycle, usually stated as an annualised figure.

through-the-door (*adj.*) inbound approach of prospective business; ~ ***population***, all customers that apply for a product.

third after second; ~ ***party***, somebody other than the two parties to a contract/transaction, who is not legitimately or willingly involved; ~~ ***fraud***, fraud committed by a third party; ~~ ***insurance***, insurance that protects against claims by third parties.

tick-box a field on a form, with a limited number of possible answers, where the applicant marks a response.

time dimension related to age or duration; ~ ***effect***, increasing bad rate associated with age of account; ~ ***to default***, estimate of time remaining before default occurs, assuming default is a certainty.

Tournier case English legal case from 1924, a precedent that defined the circumstances under which banks were allowed to divulge customers' details.

trac~e, (*v.t.*) to find individuals who have absconded; ***~ing***, function responsible for tracking down delinquent customers.

trade (*v.t.*) **1** exchange of goods and/or services; **2** craft or employment; ~ ***creditor***, supplier of goods or services, to whom money is owed; ***~d debt***, loans for which there is a secondary market, which includes securities issued by governments and their agencies, public utilities, companies, and asset-backed securities; ***~-off curve***, see 'Lorenz curve'; ***~line***, line of credit, usually associated with companies' trade credit, but also used with respect to individuals.

traffic light signalling device used to control traffic and pedestrian flow, green = go, red = stop, yellow = caution; ~ ***approach***, a framework that defines three risk levels (green = low risk, yellow = medium risk, red = high risk), esp. for PD validation.

training ensuring that model results make sense, and are not overfitted to the sample; ~ ***data/sample***, set of data containing predictor and response variables, that is used to develop a predictive model (rel. *holdout sample*).

tranche **1** portion or instalment; **2** one of a series of financial instruments, that differs only by its issue or expiry date.

transact to conduct business; ***~or***, credit cardholder that repays balance in full, every month.

transaction **1** exchange of money and/or goods; **2** disbursement or receipt; ~ ***medium***, article used to effect transactions on an account, usually plastic or cheque; ~ ***scoring***, use of scoring at transaction level, such as for credit card purchases.

transactional (*adj.*) related to transactions; ~ ***data***, data at transaction level; ~ ***lending***, loan provision, where decisions are based upon an automated assessment of borrowers' payment histories (ant. *relationship lending*); ~ ***product***, accounts used to conduct everyday financial transactions, esp. cheque, transactional savings, and credit/debit cards.

transborder across borders, esp. between countries; ~ ***data flows***, movements of data between countries.

transcribe (*v.t.*) to make a written copy, esp. with a change of form, such as speech to writing, paper-based to electronic, music to paper, or between languages or scripts.

transform (*v.t.*) **1** to convert to another form; **2** to convert the original characteristics into variables, that are used in the statistical modelling process; **3** modify scores to give them new meaning, esp. when mapping onto a different scale (rel. *calibrate*); ***~ation***, process of transforming.

transition change from one state to another; ~ ***matrix***, a contingency table, containing the probabilities of accounts moving between states (rel. *roll rate*, *Markov chain*); ~ ***probability***, likelihood of a case moving from one state to another.

transparent (*adj.*) **1** easily seen through or understood (ant. *opaque*); **2** related to ready availability of data appropriate and sufficient for an assessment.

triage **1** process used to sort the injured, in order to obtain best benefit from scarce medical resources; **2** any process used to allocate scarce resources based on need and benefit, whether for oneself or for others.

true correct, genuine, consistent with what is known, esp. as regards statements, assertions, or test results; ~ ***negative***, negative correctly identified as negative (syn. *hit*); ~ ***positive***, positive correctly identified as positive.

truncate (*v.t.*) **1** shorten; **2** cut-off; **3** remove data from a dataset, whether by design or by accident, from either end of the risk spectrum; ***~d data***, data or attributes that are excluded, or not properly treated, because there is insufficient experience (rel. *censored data*).

two-tailed test a statistical significance test, used to determine whether two values can be treated as equal, and if the observed value falls too far above, or below, the expected value, the null hypothesis is rejected. The two critical values are determined using $\alpha/2$ and $1- \alpha/2$ (rel. *significance test*, *one-tailed test*).

type I error prediction of positive result that is incorrect (syn. *false positive* or *false alarm*); **type II error** prediction of negative result that is incorrect (syn. *false negative*).

uncleared effects recently deposited cheques and other items, where insufficient time has elapsed to be sure that the transaction has been successfully processed by the other institution.

under (*pref.*) below; ***~served market***, a portion of the population with little access to formal financial markets, esp. poor and minority groups; ***~write***, (*v.t.*) to assess risk, and make a decision on whether or not to extend, guarantee, or purchase a credit facility, and if so, then under what conditions.

univariate (*adj.*) related to a single variable; ~ ***statistics***, any numbers meant to describe a variable, such as mean, median, mode, and standard deviation.

unstructured data information that cannot be readily assessed, because the key factors cannot be deconstructed into analysable data, for example notes to financial statements, information on management competence and future company prospects, local knowledge about the customer and regional circumstances, etc. (ant. *structured data*).

up being or moving higher (ant. *down*); ~***-sell***, offer of better, more advantageous products to qualifying applicants, usually with higher limits, lower interest rates, and more flexible repayment terms (rel. *down-sell*, *cross-sell*); ~***stream***, prior stages in a process.

use test a stipulation within Basel II, that the risk measures used to determine capital requirements also be used to drive banks' decision-making.

usury exorbitant or unlawful rate of interest; ~ ***Act***, regulations governing the maximum rates of interest that may be charged.

utilisation extent to which something is used; ~ ***scoring***, measures existing or prospective customers' propensity to use product offered.

validation **1** process undertaken to ensure that a model is valid for out-of-sample and/or out-of–time groups, whether immediately after model development, upon implementation, or at any point thereafter; **2** review of a development project that covers both qualitative (conceptual soundness), and quantitative (predictive power, explanatory accuracy, and stability) factors.

value at risk (*abbr.* VaR), **1** measure used to assess the potential reduction in one or more asset values over a stated time period, given certain assumptions; **2** (*adj.*) related to the calculation of future asset values at given dates and confidence intervals, using mathematical approaches that recognise asset price volatility.

variable **1** (*adj.*) having different possible values; **2** (*n.*) data element used in statistical modelling process; ~ ***reduction***, process of selecting variables for (possible) inclusion in a model, esp. associated with factor analysis.

vendor **1** seller; **2** person or entity that exchanges goods or services for money, such as credit bureaux and scorecard development services.

Verhulst, Pierre-François (**1804–1849**). Belgian mathematician and doctor in number theory, famed for identifying the logistic function; ~ ***curve***, the S-shaped logistic curve.

verification **1** proof of correctness, obtained through investigation or analysis; **2** check done to ensure that actual is in line with expected.

vintage of a common age, esp. wine; ~ ***analysis***, report that treats accounts of a similar age, for example accounts six-months old, on a like basis (syn. *cohort analysis*).

voice authorisation done over the phone at the merchant's place of business, if the card issuer wishes to confirm the identity of the cardholder.

voters' roll register of people eligible to vote.

wayward **1** badly behaved; **2** not conforming to expectations of, agreements with, or wishes of others; ~ ***definition***, definition of bad, default, or loss used in risk modelling.

weight **1** measure of a mass's downward force; **2** number indicating the relative importance of a record within a sample; ~***ed average***, mean that has been adjusted for values' relative importance, esp. frequencies or weights; ~***ed sample***, sample representing the full population, where each record's weight indicates how many cases it represents; ~ ***of evidence***, indication of predictive power, calculated using 'percentage of X in group to total X values for both X = positive and X = negative (rel. *information value*).

white (*adj.*) lacking colour; ~ ***box***, process where the inner workings are known, transparent, easily interpretable (ant. *black box*); ~ ***data***, see 'payment profile'.

wholesale (*adj.*) **1** related to bulk sales to a small number of customers (ant. *retail*); **2** corporate, interbank, and sovereign lending; ~ ***credit***, lending of large amounts to countries, companies, and projects, that requires individual attention to each deal.

window **1** transparent opening; **2** period of opportunity; **3** period over which observations are made, or that is allowed before outcomes are measured.

withdrawn rating instance where a rated company no longer receives a rating, which happens most often for smaller borrowers who either no longer need debt, or decide it is too expensive.

workout **1** resolution of conflict between two parties; **2** repayment or renegotiation of distressed debt, to avoid foreclosure or bankruptcy; ~ ***LGD***, determination of LGD based upon discounted post-default cash flows, even if no recovery is made.

worst-ever (*adj.*) related to the most undesirable status over a given time period.

write-off **1** an asset that is irrecoverable (syn. *bad debt*); **2** (*v.t.*) (*write off*) the act of placing an asset into this irrecoverable class.

wrong sign (*adj.*) related to regression coefficients that have +/- signs contrary to those indicated by the data; ~ ***problem***, output of regression calculation, where multicollinearity causes distortions in the coefficients, with one or more having wrong signs, and others having coefficients with correct signs, but exaggerated values.

Yates's correction adjustment to chi-squared calculation, which is done if expected frequencies are low.

yield **1** (*v.t.*) to furnish a return on an investment; **2** (*n.*) discount rate required, or interest, coupon or dividend rate returned on an investment.

z-score standardised score, with a mean of zero and standard deviation of one, which effectively expresses the score as the number of standard deviations from the mean (syn. *standard score*).

z-statistic measure of the number of standard deviations from the mean.

Bibliography

Aguais S.D. (2005). 'Point-in-Time vs. Through-the-Cycle Ratings'. Barclays Capital: London. Presentation to Risk Management 2005 Conference: Geneva.

Aguais S.D., Forest L.R. Jr., Wong E.Y.L., and Diaz-Ledezma D. (2003). 'Point-in-Time versus Through-the-Cycle Ratings'. (Draft) In: Ong, Michael K. (ed.) *The Basel Handbook—A Guide for Financial Practitioners*. Risk Books: London.

Akerlof G.A. (1970). 'The Market for "Lemons": Quality Uncertainty and the Market Mechanism'. *Quarterly Journal of Economics* 84(3), August, 488–500.

ALARM (2001). 'Corporate Governance in the Public Sector: Role of Risk Management'. Association of Local Authority Risk Managers: Sidmouth, Devon.

Albright S.C., Winston W.L., and Zappe C. (2003). *Data Analysis and Decision Making with Microsoft Excel*. Thomson Books: Pacific Grove, CA.

Allen F. and Gale D. (2003). *Capital Adequacy Regulation: In Search of a Rationale*. Wharton School, University of Pennsylvania and Department of Economics, New York University.

Allen L. (2001). *Credit Risk Modelling of Middle Markets*. Zicklin School of Business, Baruch College: New York.

Allen L., DeLong G., and Saunders A. (2003). 'Issues in the Credit Risk Modeling of Retail Markets'. Working Paper No. FIN-03–007. NYU Stern School of Business: New York. <http://ssrn.com/abstract=412520>.

Alpert M. and Raiffa H. (1969). 'A Progress Report on the Training of Probability Assessors'. Unpublished manuscript, Harvard University. In: Kahneman, D., Slovic, P., and Tversky, A. (1982, eds), *Judgment under Uncertainty: Heuristics and Biases*, 294–305. Cambridge University Press: New York.

Altman E. I. (1968). 'Financial Ratios, Discriminant Analysis, and the Prediction of Corporate Bankruptcy'. *Journal of Finance* 22, September, 589–610.

Altman E.I., Haldeman R.G., and Narayanan P. (1977). 'ZETA Analysis—A New Model to Identify Bankruptcy Risks of Corporations'. *Journal of Banking and Finance* 1, Summer, 29–54.

Anderson R.A. (2003a). 'Use of Bureau Data'. *Credit Risk International* 12, 9.

——**(2003b).** 'Micro-Lenders: Subprime Lending in South Africa, Implications for Credit Bureaux and Credit Scoring'. Conference Paper for Credit Scoring and Control XIII.

——**(2005).** 'MAPA Mapping: Scorecard Calibration using a Monotone Adjacent Pooling Algorithm'. Presentation at Credit Scoring and Control IX, University of Edinburgh.

Anderson T.W. and Goodman L.A. (1957). 'Statistical Inference about Markov Chains'. *Annals of Mathematical Statistics* 28, 89–109.

APACS. Association for Payment Clearing Services: London.

——**(2002).** 'Fraud in Focus'.

——**(2006).** 'Plastic Card Fraud'.

Arnold L. (1996). 'The Banker's Duty of Secrecy', in partial fulfilment for LLM degree at Rand Afrikaans University: Johannesburg, South Africa.

Arsham H. (2002). *Applied Management Science: Making Good Strategic Decisions*. Merrick School of Business, University of Baltimore. <http://home.ubalt.edu/ntsbarsh/opre640/opre640.htm>.

Asch L. (2000). 'Credit Scoring: A Tool for More Efficient SME Lending'. *SME Issues* 1(2), November.

Ash D. and Meester S. (2002). 'Best Practices in Reject Inferencing'. Conference on Credit Risk Modeling and Decisioning: Philadelphia, PA.

Åström K.J. and Wittenmark B. (1995). *Adaptive Control*, 2nd Ed. Addison-Wesley: Reading, MA.

Bailey M. (2003). 'Rapid Scorecards and Practical Shortcuts'. *Credit Risk International* June, 37–39.

Bala A. (2001). *Scientific Method and Dialogue Among Civilizations*. University Scholars Program and Department of Philosophy, University of Singapore.

Banasik J., Crook J.N., and Thomas L.C. (2001a). 'Sample Selection Bias in Credit Scoring Models'. Working Paper 01/5, Credit Research Centre, University of Edinburgh.

——**(2001b).** 'Recalibrating Scorecards'. *Journal of the Operational Research Society* 52(9), 981–988.

Barber B. and Odean T. (1999). 'The Courage of Misguided Convictions: The Trading Behaviour of Individual Investors'. *Financial Analysts Journal* 55(6), November/December, 41–55.

Barefoot J.S. (1997). 'Shooting the Rapids of Credit Scoring and Fair Lending'. (Compliance Clinic) *ABA Banking Journal* 89(9), 32.

Barron J.M. and Staten M.E. (2003). 'The Value of Comprehensive Credit Reports: Lessons from the U.S. Experience'. In: Miller, M. (ed.) *Credit Reporting Systems and the International Economy*, 283–287. MIT Press: Cambridge, MA.

Basel Committee for Banking Supervision (BCBS), Bank for International Settlements.

——(2001). 'Customer Due Diligence for Banks'. October 2001.

——(2003a). 'Consultative Document: The New Basel Capital Accord'. April 2003.

——(2003b). 'Draft Supervisory Guidance on Internal Ratings-Based Systems for Corporate Credit'. July 2003.

——(2004).'International Convergence of Capital Measurement and Capital Standards: A Revised Framework'. June 2004.

——(2005). 'Studies on the Validation of Internal Ratings Systems'. Working Paper No. 14, February 2005.

——(2006a). 'International Convergence of Capital Measurement and Capital Standards: A Revised Framework—Comprehensive Revision'. June 2006.

——(2006b). 'The IRB Use Test: Background and Implementation'. *Basel Committee Newsletter* 9, September.

Basharin G.P., Langville A.N., and Naumov V.A. (1989). *The Life and Work of A.A. Markov*. Decision and Control Laboratory, University of Illinois: Urbana. <http://decision.csl.uiuc.edu/~meyn/pages/Markov-Work-and-life.pdf>.

Bayes T. (1763). 'An Essay towards solving a Problem in the Doctrine of Chances. By the late Rev. Mr. Bayes, communicated by Mr. Price, in a letter to John Canton, M.A. and F.R.S.'. *Philosophical Transactions of the Royal Society of London* 53, 370–418.

Berger A.N. and Udell G.F. (2001). 'Small Business Credit Availability and Relationship Lending: The Importance of Bank Organizational Structure'. Finance and Economics Discussion Series, 2001–36. Federal Reserve Board: Washington, DC.

Berger A.N., Herring R.J, and Szegö G.P. (1995). 'The Role of Capital in Financial Institutions'. Working Paper, University of Pennsylvania: Wharton, PA.

Bernanke B.S. (2006), 'Basel II: Its Promise and Its Challenges'. Remarks made at the 42nd annual conference on Bank Structure, 18 May 2006 in Chicago. Federal Reserve Board: Washington, DC. <http://www.federalreserve.gov/boardDocs/speeches/2006/20060518/default.htm>.

Bernstein S.N. (1926). '*Sur l'extension du théoremé limite du cacul des probabilies*', etc. *Math Annalen*, Bd. 97, 1–59.

Birch D. (2002). *Tallys and Technologies, Monetary Innovation Goes Back a Long Way*. THOCP 27 April 2006: Guildford, Surrey, UK. <http://www.thocp.net/timeline/papers/tallies.pdf>.

Bisgaard S., Hoerl R., Neagu R., and Snee R. (2002). *The Theory and Practice of Applying Six Sigma to Business Processes.* KW Tunnell Consulting: King of Prussia, PA.

Black F. and Scholes M. (1973). 'The Pricing of Options and Corporate Liabilities'. *Journal of Political Economy* 81(3), 637–654.

Blöchinger A. and Leippold M. (2006). *Testing Probability Calibrations for Credit Scoring Models.* Credit Suisse and University of Zürich.

Blume M.E., Lim F., and MacKinlay A.C. (1998). 'The Declining Credit Quality of US Corporate Debt: Myth or Reality?' *The Journal of Finance* 53(4), August, 1389–1413. Reprinted in Shimko (1999).

Bounsaythip C. and Rinta-Runsala E. (2001). 'Overview of Data Mining for Customer Behavior Modeling'. Research Report TTE1-2001–18. VTT Information Technology: Espoo, Finland.

Boyce, G.H. (1995). Information, Mediation, and Institutional Development: The Risk of Large-Scale Enterprise in British Shipping, 1870–1919. *Manchester University Press.*

Boyle M., Crook J.N., Hamilton R., and Thomas L.C. (1992). 'Methods for credit scoring applied to slow payers'. In: Thomas, L.C., Crook, J.N., and Edelman, D.B. (eds) *Credit Scoring and Credit Control*, 75–90. Clarendon Press: Oxford.

Breiman L. Friedman J.H. Olshen R.A. and Stone C.J. (1984). *Classification and Regression Trees.* Wadsworth Publishing: Belmont, CA.

Bridges S. and Disney R. (2001). *Modelling Consumer Credit and Default: The Research Agenda.* Experian Centre for Economic Modelling, University of Nottingham.

Bugera V., Konno H., and Uryasev S. (2002). *Credit Cards Scoring with Quadratic Utility Function.* Department of Industrial and System Engineering, University of Florida: Gainesville, FL.

Bunn D. and Wright G. (1991). 'Interaction of Judgmental and Statistical Forecasting Methods: Issues and Analysis'. *Management Science* 5(37), 501–518.

Burhouse S., Feid J, French G., and Ligon K. (2003). *Basel and the Evolution of Capital Regulation: Moving Forward, Looking Back.* Federal Deposit Insurance Corporation: USA.

Burns P. and Ody C. (2004). 'Forum on Validation of Consumer Credit Risk Models'. Summary of forum held on 19 November 2004, hosted by the Payment Cards Center and Wharton's Financial Institutions Center: Philadelphia, PA.

Calinski T. and Harabasz J. (1974). 'A Dendrite Method for Cluster Analysis'. *Communications in Statistics* 3, 1–27.

Carruthers B.G. and Cohen B. (2002, 2006). *The Mechanisation of Trust: Credit Rating in 19th Century America.* Department of Sociology, Northwestern Univeristy: Evanston, IL.

Carty L.V. (1997). *Moody's Rating Migration and Credit Quality Correlation, 1920–1996.* Moody's Special Report, #25097. MIS Global Service Research: New York.

Carty L.V. and Fons J.S. (1994). 'Measuring Changes in Credit Quality'. *Journal of Fixed Income* 4, June, 27–41.

Cate F.H., Litan R.E., Staten M., and Wallison P. (2003). *Financial Privacy, Consumer Prosperity, and the Public Good.* AEI-Brookings Joint Centre for Regulatory Studies: Washington, DC.

Cespedes J.C.G. (2002). 'Credit Risk Modelling and Basel II'. *Algo Research Quarterly* 5(1), Spring, 57–66.

Chen G.G. and Åstebro T. (2003). *Bound and Collapse Bayesian Reject Inference when Data are Missing Not at Random.* University of Waterloo and University of Toronto.

Chen K.H. and Shimerda T.A. (1981). 'An Empirical Analysis of Useful Financial Ratios'. *Financial Management* 10(1), Spring, 51–60.

Chorafas D.N. (1990). *Risk Management in Financial Institutions.* Butterworth & Co.: London.

Clark J. (2006). *Business Ethics.* SME Financial Systems: Dublin. <http://www.sme.ie/articles>.

Clementi D. (2000). 'Crisis Prevention and Resolution—Two Aspects of Financial Stability'. Deputy Reserve Bank Governor's inaugural lecture at the South Bank University Centre for Monetary and Financial Economics, 6 September. Bank of England: London. <http://www.bankofengland.co.uk/publications/speeches/2000/speech97.htm>.

CoFC (1999). 'Retail Credit Scoring and Override Measurement'. Council on Financial Competition, Corporate Executive Board: Washington, DC.

Coffman H.T. (1986). 'The Proper Role of Tree Analysis in Forecasting the Risk Behavior of Borrowers'. MDS Reports 3,4,7. Management Decision Systems: Atlanta, GA.

Collins English Dictionary, 21st Century Edition, 5th Ed. (2000). Harper Collin Publishers: Glasgow.

Consumer Affairs Committee (CAC 2003). 'Report no. 1249 of 2003, on the Role of the Credit Bureau Association and Adherence to its Code'. *Staatskoerant*, 11 April 2003, No. 24739: Pretoria, South Africa.

Consumer Federation of America (CFA 2002). 'Credit Score Accuracy and Implications for Consumers'. Report done in conjunction with the National Credit Reporting Association: Bloomingdale, IL.

Cramer, J.S. (1987) 'Mean and Variance of R^2 in Small and Moderate Samples'. *Journal of Econometrics* 35, 253–266.

——**(2002).** 'The Origins of Logistic Regression'. Discussion Paper 02–119/4, Tinbergen Institute: Rotterdam. <http://ideas.repec.org/p/dgr/uvatin/20020119.html>.

Crook J.N. and Banasik J. (2002). 'Does Reject Inference Really Improve the Performance of Application Scoring Models?' Working Paper Series No. 02/3, Credit Research Centre, The School of Management, University of Edinburgh.

Crook J.N., Hamilton R., and Thomas L.C. (1992). 'The Degradation of the Scorecard over the Business Cycle'. *IMA Journal of Mathematics Applied in Business and Industry* 4, 111–123.

Crooks T. (2005). '6 Steps to Staying Ahead of the Enemy'. *ViewPoints*, October, 3–5.

Cyert R.M., Davidson H.J., and Thompson G.L. (1962). 'Estimation of Allowance for Doubtful Accounts by Markov Chains'. *Management Science* 8, 287–303.

Dahiya S., Puri M., and Saunders A. (2003). 'Bank Borrowers and Loan Sales: New Evidence on the Uniqueness of Bank Loans'. *Journal of Business* 76, 563–582.

Dahlin M. (2000). *Technology Trends Data.* University of Texas: Austin, TX. <http://www.cs.utexas.edu/users/dahlin/techTrends>.

Dawes R.M. and Corrigan B. (1974). 'Linear Models in Decision-Making'. *Psychological Bulletin* 81, 95–106.

Dawson J., Kern K., and Leung J. (2000). 'Genetic Algorithms and Evolution Strategies'. Presentation at University of Calgary. <http://members.shaw.ca/julie.leung/cpsc/533/main.htm>.

Delianedis G. and Geske R. (1999). *Credit Risk and Risk Neutral Default Probabilities: Information about Ratings Migrations and Defaults.* The Anderson School at UCLA.

Desai V.S., Convay D.G., Crook J.N., and Overstreet G.A. (1997). 'Credit Scoring Models in the Credit Union Environment Using Neural Networks and Genetic Algorithms'. *IMA Journal of Mathematics Applied in Business and Industry* 8, 323–346.

Desai V.S., Crook J.N., and Overstreet G.A. (1996). 'A Comparison of Neural Networks and Linear Scoring Models in the Credit Environment'. *European Journal of Operational Research* 95, 24–37.

Djankov S., McLiesh C., and Shleifer A. (2004). *Private Credit in 129 Countries.* World Bank: Washington DC and Harvard University: Cambridge, MA.

DTIUK. Consumer Affairs Directorate, Department of Trade and Industry: London.

——**(2001).** 'Report by the Task Force on Tackling Overindebtedness'.

——**(2003).** 'Task Force on Tackling Overindebtedness, 2nd Report'. Consumer Affairs Directorate, Department of Trade and Industry: London.

——**(2003a).** 'Ministerial Response to 2nd Report'. 190 Ev 184.

Duckenfield M. (2003). *Globalisation or Globaloney? Just How New is Globalisation?* School of Public Policy, University College London.

Durand D. (1941). *Risk Elements in Consumer Instalment Financing.* National Bureau of Economic Research: New York.

Dwyer D.W., Kocagil A.E., and Stein R.M (2004). *The Moody's KMV EDF RiskCalc v3.1 Model: Next Generation Technology for Predicting Private Firm Risk.* Moody's KMV Company: USA.

Edelberg W. (2003). 'Risk Based Pricing of Interest Rates in Household Loan Markets'. FEDS Working Paper No. 2003–62. Federal Reserve Board: Washington, DC.

Edelman (2003a). 'Risk Based Pricing for Personal Loans: How it Works and How to Make it Work'. Presentation at *Credit Scoring and Control VIII*, University of Edinburgh.

——**(2003b).** 'Working with Some Behavioural Aspects of Risk-Based Pricing'. *Credit Risk International*, November.

Engelman B., Hayen E., and Tasche D. (2003). 'Measuring the Discriminative Power of Rating Systems'. Discussion Paper Series 2: Banking and Financial Supervision, No 01/2003. Deutsche Bundesbank: Frankfurt am Main.

Engle-Warnick J. (2001). *Inferring Strategies from Observed Actions: A Non-Parametric Binary Tree Classification Approach.* Nuffield College, Oxford University.

EPFSF (2003). *Responsible Lending and Financial Inclusion.* European Parliamentary Financial Services Forum: Brussels.

Ericsson J. and Olivier R. (2001). 'Liquidity and Credit Risk'. Working Paper, Financial Market Group, London School of Economics.

Falkenstein E. (2002). 'Credit Scoring for Corporate Debt'. In: Ong, Michael K. (ed.) *Credit Ratings: Methodologies, Rationale, and Default Risk*, 169–188. Risk Books: London.

Falkenstein E., Boral A., and Carty V. (2000). *RiskCalc™ for Private Companies: Moody's Default Model Rating Methodology.* Moody's Investors Service, Global Credit Research.

Feelders A.J. (2000). 'Credit Scoring and Reject Inference with Mixture Models'. *International Journal of Intelligent Systems in Accounting, Finance and Management* 9, 1–8.

Feldman R. (1997). 'Credit Scoring and Small Business Loans'. *Community Dividend*, Spring. Federal Reserve Bank of Minneapolis.

Fernandes J.E. (2005). 'Corporate Credit Risk Modelling: Quantitative Rating System and Probability of Default Estimation'. Banco BPI: Lisbon.

Fisher, R.A. (1922). 'The Goodness of Fit of Regression Formulae, and the Distribution of Regression Coefficients'. *Journal of the Royal Statistical Society* 85, 597–612.

——**(1925).** *Statistical Methods for Research Workers.* Oliver and Boyd: Edinburgh, Scotland.

——**(1936).** 'The Use of Multiple Measurements in Taxonomic Problems'. *Annals of Eugenics* 7, Part II, 179–188.

Fitzpatrick P.J. (1932). 'A Comparison of Ratios of Successful Industrial Enterprises with those of Failed Firms'. *Certified Public Accountant* 12, 598–605, 656–662, 727–731.

Fox J. (2005). *Introduction to Non-Parametric Regression.* Lecture Notes, Department of Sociology, McMaster University: Hamilton, ON.

Fractal Analytics (2003). 'Comparative Analysis of Classification Techniques: A Fractal Whitepaper'. Fractal Analytics: Mumbai, India. <http://www.fractalanalytics.com>.

Frascati L. (2006). *Fundamentals of Mortgage Law, and Equity of Redemption.* Ezine Articles: Burnaby, BC.

Furletti M. (2002). 'An Overview and History of Credit Reporting'. Discussion Paper 02–07, Issued by the Payment Cards Center of the FRB of Philadelphia.

Galil K. (2003). 'The Term Structure of Default'. Part of Ph.D. Dissertation, Tel Aviv University.

Gardner C. (2001). *Balance Scorecard Ethics.* NSW: Australia.

Garson D.G. (2005). 'Discriminant Function Analysis'. Part of PA 765 (Quantitative Research in Public Administration) course notes. North Carolina State University: Raleigh, NC. <http://www2.chass.ncsu.edu/garson/pa765/discrim.htm>.

Gehrig T. and Stenbacka R. (2002). *Information Sharing in Banking: An Anti-Competitive Device?* Universität Freiburg: Germany; and Swedish School of Economics: Helsinki, Finland.

Gini C. (1910). *'Indici di Concentrazione e di dipendenza'. Atti della III Riunione della Societ`a Italiana per il Progresso delle Scienze*. Reprinted in his 1955 *'Memorie di metodologia statistica, I, Variabilitá e Concentrazione'*, 3–120. Libreria Eredi Virgilio Veschi: Rome.

Glößner P. (2003). 'Calculating Basel II Risk Parameters for a Portfolio of Retail Loans'. Thesis submitted in partial fulfilment of the requirements for an MSc. in Mathematical Finance. Kellogg College, University of Oxford.

Goldstein M. (1997). *The Case for an International Banking Standard: Policy Analyses in International Economics*. Institute for International Economics: Washington, DC.

Good I.J. (1950). *Probability and the Weighting of Evidence*. Grin: London.

Goody J. (1977). 'The Domestication of the Savage Mind'. Part of 'Themes in Social Sciences' series. Cambridge University Press.

Gratzer K. (2006). *Insolvent, thus a swindler? The Insolvency Law and Imprisonment for Debit in Sweden*. XIV International Economic History Congress: Helsinki.

Gray A., Mars M., Hinder J., and Klug H. (2003). 'Internal Ratings Validation Study'. Prepared for the British Bankers, Risk Management, and International Swaps & Derivatives Associations. PricewaterhouseCoopers: London.

Groebner D.F. and Shannon P.W. (1989). *Business Statistics: A Decision-Making Approach*. Merill Publishing Company: Columbus, OH.

Gross D.B. and Souleles N.S. (2002). 'An Empirical Analysis of Personal Bankruptcy and Delinquency'. *Review of Financial Studies* XC, 319–347.

Hand D.J. (1998). 'Reject Inference in Credit Operations'. In: Mays, E. (ed.) *Credit Risk Modelling: Design and Application*, 181–190. AMACOM.

Hand D.J. and Henley W.E. (1993/4). 'Can Reject Inference Ever Work?' *IMA Journal of Mathematics Applied in Business and Industry* 5, 45–55.

Hayslett H.T. (1968). *Statistics*, 3rd Ed. 1974. Made Simple Books: Oxford, UK.

Heckman J.J. (1979). 'Sample Selection Bias as a Specification Error'. *Econometrica* 47(1), 153–162.

Henley W.E. (1995). 'Statistical Aspects of Credit Scoring'. Ph.D. Thesis, Open University: Milton Keynes, UK.

Hidy R.W. (1939). 'Credit Rating before Dun & Bradstreet'. *Bulletin of the Business Historical Society* 13, December, 81–88.

Holland J.M. (1974). 'Genetic Algorithms and the Optimal Allocation of Trials'. *SIAM Journal of Computing* 2, 88–105.

——(1975). *Adaptation in Natural and Artificial Systems*. University of Michigan Press: Ann Arbor, MI.

Hosmer D.W. and Lemeshow S. (1989). *Applied Logistic Regression*. John Wiley & Sons: New York.

Hoyland C. (1995). *Data-Driven Decisions for Consumer Lending: Credit Scoring Techniques for Risk Management*. Lafferty Publications: Dublin.

Hsia D.C. (1978), 'Credit Scoring and the Equal Credit Opportunity Act'. *The Hastings Law Journal* 30, November, 371–448.

International Finance Corporation (IFC 2001). 'Making Small Business Lending Profitable'. Proceedings from the Global Conference on Credit Scoring, 2–3 April 2001: Washington, DC.

Jackson P. (2002). 'Bank Capital: Basel II Developments'. *Financial Stability Review* 13, December, 103–109. Bank of England: London.

Japelli T. and Pagano M. (1999). 'Information Sharing in Credit Markets: International Evidence'. Working Document R-371. Red de Centros de Investigacion, Banco Interamericano Desarollo (BID): Washington, DC.

——(2005). 'Role and Effects of Credit Information Sharing'. Centre for Studies in Economics and Finance, Working Paper 136. University of Salerno: Italy.

Jarrow R.A. and Protter P. (2004). 'Structural versus Reduced Form Models: A New Information Based Perspective'. *Journal of Investment Management* 2(2), 1–10.

Jarrow R.A. and Turnbull S. (1992). 'Credit Risk: Drawing the Analogy'. *Risk Magazine* 5(9), October, 63–70.

——(1995). 'Pricing Derivatives on Financial Securities Subject to Credit Risk'. *Journal of Finance* 50(1), 53–86.

Jentzsch N. (2005). 'Integrating Microfinance into Formal Financial Markets'. German Federal Ministry for Economic Cooperation and Development: Bonn. Workshop notes for African Microfinance Conference, 1 September 2005: Cape Town, South Africa.

Joanes D.N. (1993/4). 'Reject Inference Applied to Logistic Regression for Credit Scoring'. *IMA Journal of Mathematics Applied in Business and Industry* 5, 35–43.

Johnson E.L. and Nemhauser G.L. (1992). 'Recent Developments and Future Directions in Mathematical Programming—Technical'. *IBM Systems Journal* 31(1), March, 79–93.

Johnson R.W. (2004). 'Legal, Social, and Economic Issues in Implementing Scoring in the United States'. In: Thomas, Edelman, and Crook (eds) *Readings in Credit Scoring: Recent Developments, Advances, and Aims*, 5–15. Oxford University Press.

Kamesam S.V. (2002). 'Co-Operative Banks in India: Strengthening through Corporate Governance'. Speech given by the Deputy Governor of the Reserve Bank of India at the National Convention of Urban Co-operative Banks, 5 July 2002: Hyderabad.

Keller P. (2005). *Six Sigma DeMYSTiFieD: A Self-Teaching Guide.* McGraw Hill: New York.

Kempson E. (2002). 'Overindebtedness in Britain: A Report to the Department of Trade and Industry'. Personal Finance Research Centre: UK.

Kitchenman W.F. (1999). *U.S. Credit Reporting: Perceived Benefits Outweigh Privacy Concerns.* Tower Group: Needham, MA.

Kolmogorov A.N. (1933a). *Grundbegriffe der Wahrscheinlichkeitsrechnung.* Springer: Berlin.

——(1933b). '*Sulla determinazione empirica di una legge di distribuzione*'. *Giornale dell' Istituto Italiano degli Attuari*, 483–491.

Kronberg D., Tjur T., and Vincents B. (1998). *Credit Scoring: Discussion of Methods and a Case Study.* Copenhagen Business School: Denmark.

Kullback S. (1958). *Information Theory and Statistics.* John Wiley & Sons: New York.

Lacour-Little M. and Fortowsky E. (2004). 'Credit Scoring and the Fair Lending Issue of Disparate Impact'. In: Mays, E. (ed.) *Credit Scoring for Risk Managers: The Handbook for Lenders*, 157–167. South-Western Publishing: Mason, OH.

Lavis G. (1993). *Prudential Control over the Banking System: Failure or Success.* Institute of Technology: Sydney.

Lawlor H. and Haynes A. (2003). *The Company You Keep: The South African Credit Executive's Manual on Corporate Governance.* Credit Blende: Edenglen, South Africa.

Lewis E.M. (1992). *An Introduction to Credit Scoring*, 2nd Ed. Athena Press: San Rafael, CA.

Libby R. (1975). 'Accounting Ratios and the Prediction of Failure: Some Behavioral Evidence'. *Journal of Accounting Research*, 150–161.

Little R.J.A. and Rubin D.B. (1987). *Statistical Analysis with Missing Data.* John Wiley & Sons: New York.

Liu Y. (2001). *New Issues in Credit Scoring Application.* Institut für Wirtschaftsinformatik, Georg-August-Universität Göttingen: Germany.

Lorenz, M. O. (1905). 'Methods of Measuring the Concentration of Wealth'. *Publications of the American Statistical Association* 9, 209–219.

Love I. and Mylenko N. (2003). 'Credit Reporting and Financing Constraints'. World Bank Policy Research Working Paper 3142, October.

Lutz D. (2002). 'Beyond Business Ethics'. Presented at the 2nd Annual Conference of the Business Ethics Network of Africa, 28–29 November, 2002, Pan Africa University: Lagos, Nigeria.

Makuch W.M. (2001). 'Scoring Applications'. In: Mays E. (ed.) *Handbook of Credit Scoring*, 3–21. Glenlake Publishing: Chicago, IL.

Malthus T.R. (1798). *Essay on the Principle of Population: As it Affects the Future Improvement of Society, with Remarks on the Speculations of Mr. Godwin, M. Condorcet, and Other Writers*. Self-Published: London.

Markov A.A. (1913). '*Ischislenie veroyatnostej*'. St. Petersburg, 1900, 1908, 1913. Moscow, 1924. 2nd ed. translated in 1912 as '*Wahrscheinlichkeits-Rechnung*'. Teubner: Leipzig, Berlin.

di Martino P. (2002), *Approaching Disaster: A Comparison between Personal Bankruptcy Legislation in Italy and England (1880–1930)*. University of Bristol: UK.

Massey, F.J. Jr (1951). 'The Kolmogorov-Smirnov Test of Goodness of Fit'. *Journal of the American Statistical Association* 46, 68–78.

Maurer W.M. (2006). *Pious Property: Islamic Mortgages in the United States*. The Russell Sage Foundation: New York.

Mays E. (1998), ed. *Credit Risk Modelling: Design and Application*. Glenlake Publishing: Chicago, IL.

——(2001), ed. *Handbook of Credit Scoring*. Glenlake Publishing: Chicago.

——(2004), ed. *Credit Scoring for Risk Managers: The Handbook for Lenders*. South-Western Publishing: Mason, OH.

Mays E. and Nuetzel P. (2004). 'Scorecard Monitoring Reports'. In: Mays, E. (ed.) *Credit Scoring for Risk Managers: The Handbook for Lenders*, 201–217. South-Western Publishing: Mason, OH.

McCahill L.J. (1998). 'Organizational Issues in Building and Maintaining Credit Risk Models'. In: Mays, E. (ed.) *Credit Risk Modelling: Design and Application*, 14–22. Glenlake Publishing: Chicago, IL.

McGrath R.V. (2003). 'A Survey of Perspectives on Business Ethics and Business Education'. Ph.D. program in Educational Studies, Lesley University: Boston.

McNab H. and Wynn A. (2000, 2nd Ed. 2003). *Principles and Practice of Consumer Credit Risk Management*. Financial World Publishing: Canterbury, Kent.

Meehl P.E. (1954). *Clinical versus Statistical Prediction: A Theoretical Analysis and a Review of the Evidence*. University of Minnesota Press: Minneapolis, MN.

Merton R.C. (1973). 'Theory of Rational Option Pricing'. *Bell Journal of Economic and Management Science* 4(1), 141–183.

Miller M. (2001). 'Credit Reporting Systems around the Globe'. Presentation at a Policy Seminar for the Inter-American Development Bank, World Bank: Washington, DC.

——(2003), ed. *Credit Reporting Systems and the International Economy*. MIT Press: Cambridge, MA.

Miu P. and Ozdemir P. (2005). 'Practical and Theoretical Challenges in Validating Basel Paramenters: Key Learnings from the Experience of a Canadian Bank'. McMaster University and Bank of Montreal: Toronto.

Myers J.H. and Forgy E.W. (1963). 'The Development of Numerical Credit Evaluation Systems'. *Journal of American Statistical Association* 58(303), September, 779–806.

Neural T. (2002). 'Scoring Technologies for Fighting Fraud'. Neural Technologies: Petersfield, Hampshire. <http://www.neuralt.com/nt3/pressoffice/articles/scoring_technologies_for_fighting>.

Neves E. (2003). *Credit Risk International*, June, 17–19.

Nisbet R., Krantz D., Jepson C., and Fong G. (1982). 'Improving Inductive Inference'. In: Kahneman D, Slovic P., and Tversky A. (eds) *Judgment Under Uncertainty: Heuristics and Biases*, 445–462. Cambridge University Press: Cambridge.

OECD (2003). 'Guidelines on the Protection of Privacy and Transborder Flows of Personal Data'. First adopted 23 September, 1980. Organisation for Economic Co-operation and Development: Paris.

Olegario R. (2002). *Credit Reporting Agencies: Their Historical Roots, Current Status, and Role in Market Development*. University of Michigan Business School: Ann Arbor, MI.

Olsson C. (2002). *Risk Management in Emerging Markets*. Pearson Education: London.

Ong M.K. (1999). *Internal Credit Risk Models: Capital Allocation and Performance Measurement*. Risk Books: London.

Ong M.K. (2002), ed. *Credit Ratings: Methodologies, Rationale, and Default Risk*. Risk Books: London.

Owens R. and Lyons S. (1998). 'Privacy and Financial Services in Canada'. Research Report prepared for the Task Force on the Future of the Canadian Financial Services Sector: Ottawa.

Overstreet G.A.J., Bradley E.L., and Kemp R.S. (1992). 'The Flat-Maximum Effect and Generic Linear Scoring Models: A Test'. *IMA Journal of Mathematics Applied in Business and Industry* 4(1), 97–109.

Pareto V. (1896). '*Cours d'économie politique*'. Lecture notes from the Université de Lausanne, 3 volumes, 1896–1897.

Pearson K. (1894). 'On the Dissection of Asymmetrical Frequency Curves'. *Philosophical Transactions of the Royal Society of London A* 185, 71–110.

Pesaran H.M., Schuermann T., Treutler B., and Weiner S.M. (2004). 'Macroeconomic Dynamics and Credit Risk: A Global Perspective'. Cambridge Working Papers in Economics 0330, Faculty of Economics, University of Cambridge.

Pinches G., Mingo K., and Caruthers J. (1973). 'The Stability of Financial Patterns in Industrial Organisations'. *Journal of Finance* 28(2), May, 389–396.

Pine B.J., Peppers D., and Rogers M. (1995). 'Do You Want to Keep Your Customers Forever?' *Harvard Business Review*, March/April, 103–114.

Platts G. and Howe I. (1997). 'A Single European Scorecard'. Proceedings of Credit Scoring and Credit Control V, Credit Research Centre: Edinburgh.

Poirier D.J. (1980). 'Partial Observability in Bivariate Probit Model'. *Journal of Econometrics* 12, 209–217.

Princetich M.T. and Tobin K.M. (1998). 'Using Consumer Credit Information'. In: Mays, E. (ed.) *Credit Risk Modelling: Design and Application*, 14–22. Glenlake Publishing: Chicago, IL.

Public Record Office (2000). *Data Protection Act 1998: A Guide for Records Managers and Archivists*. Public Record Office: Richmond, Surrey.

Rabin M. (2000). *Inference by Believers in the Law of Small Numbers*. Economics Department, University of California: Berkeley, CA.

Rechenberg I. (1973). *Evolutionsstrategie Optimisierung technischer Systeme nach Principien der biologischen Evolution*. Frommann-Holzboog: Stuttgart.

Rhyne E. (2001). *Mainstreaming Microfinance: How Lending to the Poor Began, Grew, and Came of Age in Bolivia*. Kumarian Press: Bloomfield, CT.

Rubin D. (1976). 'Inference and Missing Data'. *Biometrika* 63, 581–592.

Saisana M. (2004). *Construction of Composite Indicators: Step 5, Normalisation*. OECD European Commission Directorate General Joint Research Centre: Ivira, Italy. <http://farmweb.jrc.cec.eu.int/ci/S5_standardization.htm>.

SAS e-Intelligence (2000), 'Building Consumer Credit Scorecards with Enterprise Miner'. SAS Institute: Cary, NC.

Schönbucher P. (2003). *Credit derivatives pricing models: Models, Pricing, and Implementation.* John Wiley & Sons: West Sussex.

Schreiner M. (2002a). *Scoring: The Next Breakthrough in Micro-Credit.* Microfinance Risk Management, Centre for Social Development, Washington University: St. Louis, MO.

——(2002b). *Benefits and Pitfalls of Statistical Credit Scoring for Microfinance.* Microfinance Risk Management, Centre for Social Development, Washington University: St. Louis, MO.

Schuermann T. (2004). *What Do We Know about Loss Given Default.* Wharton Financial Institutions Centre, University of Pennsylvania.

Schuermann T. and Jafry Y. (2003a). *Measurement and Estimation of Credit Migration Matrices.* Wharton Financial Institutions Centre, University of Pennsylvania.

——(2003b). *Metrics for Comparing Credit Migration Matrices.* Wharton Financial Institutions Centre, University of Pennsylvania.

Scroggins W.A., Fielding W.T., and Clark L.J. (2004). *Borrower Segmentation and Credit Scoring in Bank Consumer Lending.* Jacksonville State University: Jacksonville, NC.

Shafer G. (1996). *The Significance of Jacob Bernoulli's Ars Conjectandi for the Philosophy of Probability Today.* Rutgers University: Newark and New Brunswick, NJ.

Sheshunoff A. (2002). *New Overdraft Scoring and Collections Strategy Creates Big Payoff.* Alex Sheshunoff Management: Austin, TX.

Shimko D. (1999). *Credit Risk: Models and Management.* Risk Publications: Haymarket, London.

Siddiqi N. (2006). *Credit Risk Scorecards: Developing and Implementing Intelligent Credit Scoring.* John Wiley & Sons: Hoboken, NJ.

Simon H. (1971). 'Designing Organizations for an Information-Rich World'. In: Greenberger M. (ed.) *Computers, Communications, and the Public Interest*, 40–41. The Johns Hopkins Press: Baltimore, MD.

Smith A. and Elkan C. (2004). *A Bayesian Network Framework for Reject Inference.* Department of Computer Science, University of California: San Diego La Jolla.

Smith R.C. (2003). *Is Corporate Governance Worth the Cost of Improving It?* Stern School of Business: New York.

Smirnov N. (1939). 'On the Estimation of the Discrepancy between Empirical Curves of Distribution for Two Independent Samples'. *Bulletin, Moscow University* 2(2), 3–16.

——(1948). 'Table for Estimating the Goodness of Fit of Empirical Distributions'. *Annals of Mathematical Statistics* 19, 279–281.

Somer R.H. (1962). 'A New Asymmetric Measure of Association for Ordinal Variables'. *American Sociological Review* 27, 799–811.

Sparkes R. (2003). 'From Corporate Governance to Corporate Responsibility: The Changing Boardroom Agenda'. *Ivey Business Journal*, March/April. University of Western Ontario. <http://www.iveybusinessjournal.com/view_article.asp?intArticle_ID=406>.

Spearman C. (1904a). 'The Proof and Measurement of Association between Two Things'. *American Journal of Psychology* 5, 72–101.

——(1904b). ' "General Intelligence", Objectively Determined and Measured'. *American Journal of Psychology* 15, 201–293.

Srinivisan V. and Chakrin L.M. (1987). 'Credit Granting: A Comparative Analysis of Classification Procedures'. *Journal of Finance* 42, 665–683.

Stanton T.H. (1999). *Credit and Loan Scoring: Tools for Improved Management of Federal Credit Programs.* Center for the Study of American Government, John Hopkins University: Baltimore, MD.

Staten M.E. and Cate F.H. (2004). *Does the Fair Credit Reporting Act Promote Accurate Credit Reporting?* Joint Center for Housing Studies, Harvard University: Cambridge, MA.

Statsoft (2003). *Electronic Statistics Textbook*. Tulsa, OK. <http://www.statsoft.com>.

Stern H. (2002). 'Residential Jumbo Loan Level Prepayment Model'. *Quantitative Perspectives*, April. Andrew Davidson & Co., Inc.: New York.

Stevens J.T. Jr (1998). *How Identity Theft can Ruin Your Good Name*. Crimes of Persuasion: Schemes, Scams, Frauds. Les Henderson: Azilda, ON, Canada. <http://www.crimes-of-persuasion.com/index.htm>.

Sylla R. (2001). *A Historical Primer on the Business of Credit Ratings*. Stern School of Business: New York. Prepared for Conference on 'The Role of Credit Reporting Systems in the International Economy', 1–2 March, 2001, World Bank Headquarters: Washington, DC.

Tamilia R.D. (2002). 'The Wonderful World of the Department Store in Historical Perspective—A Comprehensive Bibliography Partially Annotated'. Centre de Recherche en Gestion, Document 08–2002. University of Quebec in Montreal. <http://www.unites.uqam.ca/esgcrg/papers/2002/08–2002.pdf>.

Tanigawa T. and Zhao Q. (1999). *A Study of Efficient Generation of Decision Trees using Genetic Programming*. The University of Aizu: Aizu-Wakamatsu, Japan.

Taylor S. (2004). 'The Data Protection Act 1998: a Blessing in Disguise for the UK?' *Credit Risk International* March, 24–27.

TBYA (2006), 'A Brief History of Bankruptcy in the U.S'. *The Bankruptcy Yearbook and Almanac*. New Generation Research: Boston, MA.

Thomas L.C. (2000). 'A Survey of Credit and Behavioural Scoring: Forecasting Financial Risk of Lending to Consumers'. *International Journal of Forecasting* 16, 149–172.

Thomas L.C., Edelman D.B., and Crook J.N. (2002). *Credit Scoring and its Applications*. Society for Industrial and Applied Mathematics. SIAM Publishing: Philadelphia, PA.

——**(2004), ed.** *Readings in Credit Scoring: Recent Developments, Advances, and Aims*. Oxford University Press.

Thomas L.C., Ho J., and Scherer W.T. (2001). 'Time Will Tell: Behavioural Scoring and the Dynamics of Consumer Credit Assessment'. *IMA Journal of Mathematics* 12, 89–103.

Thomson W. (1926). *History of Banking*. New Era Publishing: London.

Twaits A. (2003). *Understanding Customer Behaviour through the Extraction, Storage, Modeling and Reporting of Customer Data*. Amadeus Software Limited: Oxfordshire.

Van Roy P. (2003). *The Impact of the 1988 Basle Accord on Banks' Capital Ratios and Credit-Risk Taking: An International Study*. European Centre for Advanced Research in Economics and Statistics (ECARES), Université Libre de Bruxelles: Belgium.

Varian H.R. (1998). *Markets for Information Goods*. University of California: Berkeley, CA.

Vaughan S. and Berry K.E. (2005). 'Using Monte Carlo Techniques to Demonstrate the Meaning and Implications of Multicollinearity'. *Journal of Statistics Education* 13(1). <http://www.amstat.org/publications/jse/v13n1/vaughan.html>.

Verhulst P.F. (1845). '*Recherches mathematiques sur la loi d'accroissement de la population*' [Mathematical Researches on the law of Population Growth Increase]. *Nouveaux memoirs de l'Academie royale des sciences et belles-lettres de Bruxelles* 18: Brussels, Belgium.

Verstraeten G. and van den Poel D. (2004). 'The Impact of Sample Bias on Consumer Credit Scoring Performance and Profitability'. Working Paper 2004/232, Faculteit Economie en Bedrijfskunde, Universiteit Gent: Belgium.

Vitanyi P.M.B. (1988). 'Andrei Nikolaevich Kolmogorov'. *CWI Quarterly* 1, 3–18. CWI & University of Amsterdam: Holland.

Wan S.L (2001). *The New Basel Capital Accord and its Supervisory Inspirations*. Monetary Authority of Macau.

Weldon G. (1999). 'Inferring Behaviour on Rejected Credit Applicants'. Sigma Analytics and Consulting: Atlanta, GA. Prepared for 'Statistics, Data Analysis, and Modelling', SUGI 24, SAS Conference.

Wells H.G. (1922). *A Short History of the World.* MacMillan and Co.: New York.

Wiginton J.C. (1980). 'A Note on the Comparison of Logit and Discriminant Models of Consumer Credit Behaviour'. *Journal of Financial and Quantitative Analysis* 15, 757–770.

Wiklund D. (2004). 'How to Successfully Manage a Credit Score Development Project'. In: Mays E. (ed.) *Credit Scoring for Risk Managers: The Handbook for Lenders*, South-Western Publishing: Mason, OH.

Wilcox J.W. (1971). 'A Simple Theory of Financial Ratios as Predictors of Failure'. *Journal of Accounting Research* 9(2), 389–395.

Wilkinson G. (2003). 'Retaining Voters' Roll for Credit Decisions'. Conference Notes for Credit Scoring and Credit Control VIII, University of Edinburgh.

Wilks S.S. (1938). 'Weighting Systems for Linear Functions of Correlated Variables when there is no Dependent Variable'. *Psychmetrickia* 8, 23–40.

Williams R.H., Zimmerman D.W., Zumbo B.D., and Ross D. (2003). 'Charles Spearman: British Behavioural Scientist'. *Human Nature Review* 3, 114–118.

Witt B.R. (1999). *Evaluating Credit Scoring Systems.* Farleigh Wada & Witt: Portland, OR.

Wolfs F. (2002). *Introduction to the Scientific Method.* University of Rochester: New York.

Yamauchi E. (2003). 'An Empirical Analysis of Risk Premium in Credit Spreads of Corporate Bond'. Research Report for M.Sc. Finance, Graduate School of System and Information Engineering, University of Tsukuba: Japan.

Yobas M.B., Crook J.N., and Ross P. (1997). 'Credit Scoring Using Neural and Evolutionary Techniques'. Working Paper 97/2, Credit Research Centre, University of Edinburgh.

Young R. (2005). 'Credit Bureaus in Latin America: Expanding Financial and Other Services to the Base of the Pyramid'. *MicroFinance Matters*, May. United Nations Capital Development Fund: New York.

Zandi M. (1998). 'Incorporating Economic Information into Credit Risk underwriting'. In: Mays, E. (ed.) *Credit Risk Modeling: Design and Application*, 165–168. Glenlake Publishing: Chicago, IL.

Appendices

Appendix A: Chi-square table

This table provides the critical χ^2 values for various combinations of confidence level and degrees of freedom. These are used to determine whether or not a hypothesis should be accepted with any degree of reliability.

Example 1: There are five categories and the χ^2 value is 2.55. An observer wants 90 per cent certainty that the observed and expected distributions are substantially the same and, any variations are random. If d.f. = 4 and $p = 0.90$ then $\chi^2_{critical}$ is 1.06. The hypothesis is rejected, but might be accepted if we only required 50 per cent certainty.

Example 2: There are 11 categories and the χ^2 value is 55.20. An observer wants 99 per cent certainty that the distributions are substantially different, and variations are not random. If d.f. = 10 and $p = 0.01 = (1-0.99)$ then $\chi^2_{critical}$ is 23.21. The hypothesis is accepted.

Chi-square (χ^2) table

d.f.	0.99	0.975	0.95	0.9	0.7	0.5	0.3	0.1	0.05	0.025	0.01	0.001
1	0.00	0.00	0.00	0.02	0.15	0.45	1.07	2.71	3.84	5.02	6.63	10.83
2	0.02	0.05	0.10	0.21	0.71	1.39	2.41	4.61	5.99	7.38	9.21	13.82
3	0.11	0.22	0.35	0.58	1.42	2.37	3.66	6.25	7.81	9.35	11.34	16.27
4	0.30	0.48	0.71	1.06	2.19	3.36	4.88	7.78	9.49	11.14	13.28	18.47
5	0.55	0.83	1.15	1.61	3.00	4.35	6.06	9.24	11.07	12.83	15.09	20.52
6	0.87	1.24	1.64	2.20	3.83	5.35	7.23	10.64	12.59	14.45	16.81	22.46
7	1.24	1.69	2.17	2.83	4.67	6.35	8.38	12.02	14.07	16.01	18.48	24.32
8	1.65	2.18	2.73	3.49	5.53	7.34	9.52	13.36	15.51	17.53	20.09	26.12
9	2.09	2.70	3.33	4.17	6.39	8.34	10.66	14.68	16.92	19.02	21.67	27.88
10	2.56	3.25	3.94	4.87	7.27	9.34	11.78	15.99	18.31	20.48	23.21	29.59
11	3.05	3.82	4.57	5.58	8.15	10.34	12.90	17.28	19.68	21.92	24.72	31.26
12	3.57	4.40	5.23	6.30	9.03	11.34	14.01	18.55	21.03	23.34	26.22	32.91
13	4.11	5.01	5.89	7.04	9.93	12.34	15.12	19.81	22.36	24.74	27.69	34.53
14	4.66	5.63	6.57	7.79	10.82	13.34	16.22	21.06	23.68	26.12	29.14	36.12
15	5.23	6.26	7.26	8.55	11.72	14.34	17.32	22.31	25.00	27.49	30.58	37.70
16	5.81	6.91	7.96	9.31	12.62	15.34	18.42	23.54	26.30	28.85	32.00	39.25
17	6.41	7.56	8.67	10.09	13.53	16.34	19.51	24.77	27.59	30.19	33.41	40.79
18	7.01	8.23	9.39	10.86	14.44	17.34	20.60	25.99	28.87	31.53	34.81	42.31
19	7.63	8.91	10.12	11.65	15.35	18.34	21.69	27.20	30.14	32.85	36.19	43.82
20	8.26	9.59	10.85	12.44	16.27	19.34	22.77	28.41	31.41	34.17	37.57	45.31

d.f.	0.99	0.975	0.95	0.9	0.7	0.5	0.3	0.1	0.05	0.025	0.01	0.001
21	8.90	10.28	11.59	13.24	17.18	20.34	23.86	29.62	32.67	35.48	38.93	46.80
22	9.54	10.98	12.34	14.04	18.10	21.34	24.94	30.81	33.92	36.78	40.29	48.27
23	10.20	11.69	13.09	14.85	19.02	22.34	26.02	32.01	35.17	38.08	41.64	49.73
24	10.86	12.40	13.85	15.66	19.94	23.34	27.10	33.20	36.42	39.36	42.98	51.18
25	11.52	13.12	14.61	16.47	20.87	24.34	28.17	34.38	37.65	40.65	44.31	52.62
26	12.20	13.84	15.38	17.29	21.79	25.34	29.25	35.56	38.89	41.92	45.64	54.05
27	12.88	14.57	16.15	18.11	22.72	26.34	30.32	36.74	40.11	43.19	46.96	55.48
28	13.56	15.31	16.93	18.94	23.65	27.34	31.39	37.92	41.34	44.46	48.28	56.89
29	14.26	16.05	17.71	19.77	24.58	28.34	32.46	39.09	42.56	45.72	49.59	58.30
30	14.95	16.79	18.49	20.60	25.51	29.34	33.53	40.26	43.77	46.98	50.89	59.70

The $\chi^2_{critical}$ values were calculated using an MS Excel spreadsheet.

Appendix B: Student *t*-test table

This table provides the critical *t*-test values for various combinations of confidence level and degrees of freedom. These are used to determine whether or not a hypothesis should be accepted with any degree of reliability.

The numbers provided here are for a one-tailed test, meaning that they are appropriate for testing the extremes, like $\Pr[X < x]$ and $\Pr[X > x]$. There are, however, a lot of cases where we wish to test whether a value falls somewhere in the middle, $\Pr[-x < X < x]$, in which case the *p*-value will be half of the confidence level.

Student *t*-test table

	p-values											
d.f.	0.99	0.975	0.95	0.9	0.7	0.5	0.3	0.1	0.05	0.025	0.01	0.001
1	0.02	0.04	0.08	0.16	0.51	1.00	1.96	6.31	12.71	25.45	63.66	636.62
2	0.01	0.04	0.07	0.14	0.44	0.82	1.39	2.92	4.30	6.21	9.92	31.60
3	0.01	0.03	0.07	0.14	0.42	0.76	1.25	2.35	3.18	4.18	5.84	12.92
4	0.01	0.03	0.07	0.13	0.41	0.74	1.19	2.13	2.78	3.50	4.60	8.61
5	0.01	0.03	0.07	0.13	0.41	0.73	1.16	2.02	2.57	3.16	4.03	6.87
7	0.01	0.03	0.06	0.13	0.40	0.71	1.12	1.89	2.36	2.84	3.50	5.41
10	0.01	0.03	0.06	0.13	0.40	0.70	1.09	1.81	2.23	2.63	3.17	4.59
15	0.01	0.03	0.06	0.13	0.39	0.69	1.07	1.75	2.13	2.49	2.95	4.07
20	0.01	0.03	0.06	0.13	0.39	0.69	1.06	1.72	2.09	2.42	2.85	3.85
25	0.01	0.03	0.06	0.13	0.39	0.68	1.06	1.71	2.06	2.38	2.79	3.73
30	0.01	0.03	0.06	0.13	0.39	0.68	1.05	1.70	2.04	2.36	2.75	3.65
40	0.01	0.03	0.06	0.13	0.39	0.68	1.05	1.68	2.02	2.33	2.70	3.55
50	0.01	0.03	0.06	0.13	0.39	0.68	1.05	1.68	2.01	2.31	2.68	3.50
75	0.01	0.03	0.06	0.13	0.39	0.68	1.04	1.67	1.99	2.29	2.64	3.43
100	0.01	0.03	0.06	0.13	0.39	0.68	1.04	1.66	1.98	2.28	2.63	3.39
150	0.01	0.03	0.06	0.13	0.39	0.68	1.04	1.66	1.98	2.26	2.61	3.36
200	0.01	0.03	0.06	0.13	0.39	0.68	1.04	1.65	1.97	2.26	2.60	3.34
1000	0.01	0.03	0.06	0.13	0.39	0.67	1.04	1.65	1.96	2.24	2.58	3.30

The values presented here were calculated using an MS Excel spreadsheet.

Index